The Melanocortin Receptors

The Receptors

The Melanocortin Receptors, edited by *Roger D. Cone,* 2000
The GABA Receptors, Second Edition, edited by *S. J. Enna and Norman G. Bowery,* 1997
The Ionotropic Glutamate Receptors, edited by *Daniel T. Monaghan and Robert Wenthold,* 1997
The Dopamine Receptors, editedby *Kim A. Neve and Rachael L. Neve,* 1997
The Metabotropic Glutamate Receptors, edited by *P. Jeffrey Conn and Jitendra Patel,* 1994
The Tachykinin Receptors, edited by *Stephen H. Buck,* 1994
The Beta-Adrenergic Receptors, edited by *John P. Perkins,* 1991
Adenosine and Adenosine Receptors, edited by *Michael Williams,* 1990
The Muscarinic Receptors, edited by *Joan Heller Brown,* 1989
The Serotonin Receptors, edited by *Elaine Sanders-Bush,* 1988
The Alpha-2 Adrenergic Receptors, edited by *Lee Limbird,* 1988
The Opiate Receptors, edited by *Gavril W. Pasternak,* 1988
The Alpha-1 Adrenergic Receptors, edited by *Robert R. Ruffolo, Jr.,* 1987
The GABA Receptors, edited by *S. J. Enna,* 1983

The Melanocortin Receptors

Edited by

Roger D. Cone

Vollum Institute, Oregon Health Sciences University
Portland, OR

Humana Press
Totowa, New Jersey

999 Riverview Drive, Suite 208
Totowa, New Jersey 07512

This publication is printed on acid-free paper. ∞
ANSI Z39.48-1984 (American National Standards Institute) Permanence of Paper for Printed Library Materials.

For additional copies, pricing for bulk purchases, and/or information about other Humana titles, contact Humana at the above address or at any of the following numbers: Tel.: 973-256-1699; Fax: 973-256-8341; E-mail: humana@humanapr.com

Cover design by Roger D. Cone and Linda Cordilia.

Printed in the United States of America. 10 9 8 7 6 5 4 3 2 1

Library of Congress Cataloging in Publication Data

Main entry under title:
The melanocortin receptors / edited by Roger D. Cone.
p. cm. -- (The Receptors)
Includes bibliographical references and index.
ISBN 0-89603-579-4 (alk. paper)
1. MSH (Hormone)--Receptors. 2. Proopiomelanocortin--Receptors.
I. Cone, Roger D. II. Series.
[DNLM: 1. Receptors, Corticotropin--physiology. WK 515 M5165 2000]
QP572.M75M44 2000
573.4'54--dc21
DNLM/DLC 99-38475
for Library of Congress CIP

Preface

The melanocortins have a fascinating history, first as pituitary peptide hormones, and more recently as neuropeptides. The study of the melanocortin peptides and their receptors has contributed many "firsts" to biomedical research. Based on the frog skin pigmentation assay, melanotropic activity was first identified in pituitary extracts early in this century; in many ways these experiments heralded the beginning of modern pituitary endocrinology. The melanocortin peptides were also among the first biologically active peptides to be purified and sequenced in the 1950s by Bell, Lerner, Li, Harris, and Geschwind. Cloning of the complete proopiomelanocortin precursor gene by Nakanisha and Numa in 1979 provided one of the first examples of a prohormone precursor encoding a variety of different neuropeptides and peptide hormones.

More recently, work in the field has largely been focused on the receptors for the melanocortin peptides. My own interest in receptors for the melanocortin peptides derived from a structural question rather than any knowledge of, or interest in, the biology of these peptides. In 1989, the structure of the luteinizing hormone receptor was published, and this made it clear that the large glycoprotein hormones were binding to a large extracellular domain attached to the canonical hydrophobic seven-membrane spanning domain known at the time to be the conserved structure for such G-protein coupled receptors as rhodopsin, the β-adrenergic receptor, and the substance K receptor. Though the substance K receptor clearly was capable of binding the hydrophilic substance P peptide without a large extracellular domain, I remember many discussions among scientists at the time, particularly John Potts and Henry Kronenberg at the MGH, that perhaps the extracellular motif of the glycoprotein hormone receptors was a conserved domain that could be involved in the binding of many large peptide hormones such as PTH and ACTH. Ultimately, Kathleen Mountjoy in my laboratory, with important reagents from Jeff Tatro and Seymour Reichlin, Vijay Chhajlani in Sweden, and Ira Gantz at the University of Michigan were able to disprove this hypothesis with the cloning of a family of five different receptors for the melanocortin peptides.

Around the time of the cloning of the melanocortin receptors, there was skepticism about whether many interesting biological findings would result from continued studies of the melanocortin receptors. The mechanism of action of the MSH-R and ACTH-R in pigmentation and adrenal steroidogenesis, for example, seemed to be fairly well understood. Happily, over the last eight years, around every corner a remarkable new finding has arisen regarding the melanocortin receptors, their mode of action, and their physiological roles. Continuing with the listing of "firsts," the MC1-R was the first example of a G-protein-coupled hormone receptor to be constitutively activated by naturally occurring mutations, the agouti and agouti-related proteins are the first examples of endogenous antagonists of the GPCRs, and the MC4-R is the first GPCR to be demonstrated to be involved in the central control of energy homeostasis. If the annual number of publications in the melanocortin field is any indicator, the tenfold increase over the last five years suggests a tremendous newfound interest in the remarkable complexity of action and function of the melanocortins.

Organizing *The Melanocortin Receptors* has given much pleasure owing to the many fine colleagues I have had the privilege of working with, quite a number of whom have provided chapters for this volume. I would like to express my sincere thanks to the authors, and to the editors at Humana Press for making this book happen. Finally, I thank my wife Midge and children Miriam, Anna, and David for their continued encouragement and patience, and for genuinely sharing in the excitement of scientific discovery.

Roger D. Cone

Contents

Part IV. Receptor Function

Part V. Receptor Regulation

Part VI. Future Vistas

Contributors

ZALFA ABDEL-MALEK • *Department of Dermatology, University of Cincinatti Medical Center, Cincinatti, OH*

ROGER A. H. ADAN • *Rudolf Magnus Institute for Neuroscience, Utrecht University, Utrecht, Netherlands*

MARTINE BÉGEOT • *Hospital Debrousse, Lyon, France*

MARK BIRCH-MACHIN • *Department of Dermatology, University of Newcastle, Newcastle, UK*

BRUCE A. BOSTON • *Department of Pediatrics, Oregon Health Sciences University, Portland, Oregon*

WENBIAO CHEN • *Massachusetts Institute of Technology Center for Cancer Research, Cambridge, MA*

ROGER D. CONE • *Vollum Institute, Oregon Health Sciences University, Portland, Oregon*

ADRIAN J. L. CLARK • *Department of Chemical Endocrinology, St. Bartholomews Hospital, London, UK*

ALEX N. EBERLE • *Department of Research, University Hospital, Basel, Switzerland*

SILVIE FROIDEVAUX • *Department of Research, University Hospital, Basel, Switzerland*

CARRIE HASKELL-LUEVANO • *Department of Medicinal Chemistry, University of Florida, Gainesville, FL*

EUGENE HEALY • *Department of Dermatology and Dermatopharmacology, Southampton General Hospital, Southampton, UK*

GUOXIA HAN • *Department of Chemistry, University of Arizona, Tucson, AZ*

VICTOR J. HRUBY • *Department of Chemistry, University of Arizona, Tucson, AZ*

ROBERT A. KESTERSON • *Department of Molecular Physiology and Biophysics, Vanderbilt University School of Medicine, Nashville, TN*

AARON B. LERNER • *Department of Dermatology, Yale University School of Medicine, New Haven, CT*

DONGSI LU • *Department of Pathology, Washington University School of Medicine, St Louis, MO*

KATHLEEN G. MOUNTJOY • *Research Center for Developmental Medicine and Biology, University of Auckland, Auckland, New Zealand*

JONATHAN L. REES • *Department of Medical and Radiological Sciences, University of Edinburgh, Edinburgh, UK*

José M. Saez • *Hospital Debrousse, Lyon, France*
Walter Siegrist • *Department of Research, University Hospital, Basel, Switzerland*
Jeffrey B. Tatro • *Endocrine Division, New England Medical Center Hospital, and Tufts University School of Medicine, Boston, MA*
Dag Inge Vage • *Department of Animal Science, Agricultural University of Norway, Norway*
William O. Wilkison • *Zen-Bio Inc., Research Triangle Park, NC*

Part I

Historical Perspectives

CHAPTER 1

Proopiomelanocortin and the Melanocortin Peptides

Alex N. Eberle

1. Introduction

The "melanophore stimulants" were discovered about 80 yr ago when, with surgical ablation experiments, the pituitary gland was shown to be involved in the control of skin color of amphibia. The pars intermedia was soon recognized as the origin of the biological principle, then also named "intermedin," which induced darkening of amphibian skin (for a short historical review see ref. *1*). In the 1950s, the development of an isolated frog skin bioassay by Shizume et al. *(2)* paved the way for the isolation *(3)*, molecular characterization, and sequence determination of the melanocyte-stimulating hormones (MSHs; melanotropins) from pig by Lee and Lerner *(4)*, Geschwind, et al. *(5)*, Harris and Lerner *(6)* and Harris and Roos *(7)*. In subsequent years, α-and β-melanocyte-stimulating hormones were isolated from bovine, equine, sheep, macaque, camel, dogfish, and salmon pituitary glands and their sequences determined (reviewed ref. *8*). The advent of molecular cloning and sequencing techniques of the gene(s) of the melanotropin precursors made it possible to determine or confirm many more MSH sequences.

The isolation and sequence determination of adenocorticotropic hormone (ACTH; corticotropin) *(9,10)* and of sheep β-lipotropic hormone (β-LPH; β-lipotropin) *(11,12)* as well as of γ-lipotropin (γ-LPH) *(13)* demonstrated that the sequence of α-MSH was part of the ACTH sequence, whereas the sequence of β-MSH was comprised within that of β-/γ-LPH. These findings led to the hypothesis that the longer peptides may serve as precusors for the shorter forms. In the 1970s, the C-terminal *18–39* portion of ACTH, named

The Melanocortin Receptors Ed.: R. D. Cone

corticotropin-like intermediate lobe peptide (CLIP) *(14)*, was shown to be a biologically active hormone produced and secreted in the pars intermedia of the pituitary gland, and soon afterward the C-terminal *61–91* part of β-LPH, β-endorphin *(15–17)*, was also isolated from the pars intermedia and demonstrated to display potent opioid activity. This was additional evidence that ACTH and β-LPH are processed to smaller, biologically active peptides in the pars intermedia, whereas in the pars distalis of the pituitary gland, they are formed, stored, and secreted in their intact longer forms.

In 1971, the observation by Yalow and Berson *(18,19)* that ACTH-producing thymoma released high molecular weight forms of immunoreactive ACTH (''big ACTH''), which possessed low ACTH bioactivity and could be transformed enzymatically into ACTH, was the starting point for the subsequent discovery of a common precursor molecule for ACTH/α-MSH and β-LPH/β-MSH/β-endorphin *(20–22)*. In 1979, Nakanishi et al. *(23)* reported the nucleotide sequence of the cDNA coding for bovine proopiomelanocortin (POMC) and demonstrated that POMC contains a hitherto unknown MSH-like peptide sequence, named γ-MSH, as well as N-terminal peptides. The most recent addition to the three different MSH sequences is that of δ-MSH found in the POMC precursor of the dogfish *(24)*. The structural feature characterizing all MSH sequences and that of ACTH is the core tetrapeptide His-Phe-Arγ-Trp, which is crucial for the interaction with the receptors of these peptides and hence for their biologic activity. While the term *opiomelanocortins* is generally used for any or all of the POMC peptides, the term *melanocortins* only relates to ACTH/MSH-derived peptides (further details on the nomenclature are found in ref. *8*).

In the last twenty years, POMC molecules from many different species were analyzed and their cDNA sequenced. POMC or its mRNA was also detected in many different tissues in the mammalian body, indicating that the bioactive peptides do not only function as circulating hormones released from the pituitary gland or as neuropeptide regulators in the brain but that they may also be formed in the periphery, although in very tiny amounts, where they exert para- or autocrine effects.

Receptors for MSH on melanoma cells were first characterized biochemically by photocrosslinking *(25–27)* but the true breakthrough came in 1992 with the cloning of the first two melanocortin (MSH, ACTH) receptors from mouse and man by Cone and collaborators *(28)* and of the human MSH receptor by Chhajlani and Wikberg *(29)*. Subsequent cloning experiments added to these receptors, now named MC1 and MC2 receptors, three more subt ypes, namely, MC3 *(30,31)*, MC4 *(32)*, and MC5 *(33–36)* receptors. The cloning of these receptors has opened a new era of research into MSH and

ACTH peptides, from focused structure–activity studies of ligands to the distribution of their site of action. Receptor subtype-specific agonists and antagonists became available, and receptor-deficient animals were generated, leading to the discovery of novel functions for MC receptors, for example, the tonic inhibition of feeding behavior, the prevention of the orexigenic effect of MCH, and the mediation in the CNS, at least partially, of the leptin signal *(37–40)*.

This chapter focuses on the melanocortin peptides as ligands, their structures and chemical characteristics, their use as tools for biologic and biochemical studies in vitro and in vivo, and their biosynthesis and regulation through their common precursor, proopiomelanocortin. The literature cited in this text mainly considers publications about MSH peptides of the last ten years; additional citations of the earlier literature are found in a review published in 1988 *(8)*.

2. Structure and Chemistry of the Melanocortin Peptides

Melanocortin peptides or their POMC precursor have been identified in the pituitary gland, the brain and various peripheral tissues of all classes of vertebrates either by bioassay, radioimmunoassay, immunocytochemistry, *in situ* hybridization or PCR. However, the quantitative isolation and structural analysis has been confined to the pituitary gland of a few species. The protocols for isolation of MSH and ACTH peptides were reviewed in *(8)*. An elegant three-step isolation procedure was developed by Bennett *(41)*, who homogenized tissue extracts at low pH (<1) and immediately adsorbed the peptides on ODS-silica cartridges, thus avoiding proteolysis of the peptides. Elution from ODS was followed by separation on CM-and QMA-ion exchange cartridges, and the final purification was achieved by reversed-phase high performance liquid chromatography (HPLC). This made it possible to separate, for example, all different forms of POMC-derived intermediate lobe peptides in a single run *(42)*. The addition of mass spectrometry to HPLC simplified the determination of the structure of novel melanocortin peptides and further increased the sensitivity so that POMC peptides could be determined from a single neuroendocrine cell *(43)*.

2.1. α-, β-, γ-, and δ-MSH Peptides

The structures of the naturally occurring melanotropins, α-MSH, β-MSH, γ-MSH, and δ-MSH, elucidated by peptide sequence determination or inferred from the POMC cDNA are listed in Tables 1–3. As indicated above, the structural relation between the peptide pairs ACTH/α-MSH, β-LPH/β-MSH, and *N*-POMC/γ-MSH originates from the biogenetic processing of the POMC precursor molecule via the larger corticolipotropic peptides *N*-POMC → γ_3-MSH → γ-MSH, ACTH → ACTH[1–17] → α-MSH, and β-LPH → γ-LPH → β-MSH, as illustrated by Fig. 1 (*see also* 6.3). In the following, a short description is given for the naturally occurring melanocortin peptides.

2.1.1. α-MSH

The different forms of α-MSH are listed in Table 1. Normally, its sequence corresponds to the first 13 N-terminal amino acid residues of ACTH and is identical in all mammals from which it has been isolated (e.g., pig, ox, sheep, horse, monkey, camel, and rat) or determined from POMC cDNA sequences (e.g., mouse and guinea pig). The only α-MSH molecules that differ in structure are that of *Xenopus laevis* whose Ser^1 residue is replaced by Ala, that of the dogfish whose Val^{13} is replaced by Met, and that encoded by the second POMC gene of teleost fishes (e.g., salmon B or trout B) where the C-terminal valine is replaced by the tripeptide structure-Ile-Gly-His-OH and the N-terminal acetyl may be a His residue (Table 1). In the sea lamprey, an α-MSH peptide derived from the ACTH sequence and encoded for by the POC (proopiocortin) precursor *(65)* could not be detected in the pituitary gland; instead two longer forms of melanotropin are generated and secreted, MSH-A (19 residues) and MSH-B (20 residues) that are encoded for by a different precursor, POM (proopiomelanotropin) *(66)*. MSH-B is the more potent peptide in melanophore assays and is therefore grouped within the list of α-MSH peptides although it differs in structure, in particular because of the long N-terminal extension. MSH-A resembles more closely the β-MSH molecules.

The N-terminal serine residue of α-MSH is *N*-acetylated in many species and the C-terminal valine almost always contains a carboxamide group. These modifications render stability to the α-MSH molecule against exopeptidases and in many bioassays they increase the potency of the peptide *(8)*. A certain percentage of pituitary α-MSH is processed to the diacetylated form [*N,O*-bisacetyl-Ser^1]-α-MSH (or simply diacetyl-α-MSH). The physiologic function of the second acetyl group is still unclear since it may result from a spontaneous N→O shift of the acetyl group at the N-terminal Ser, a process also observed by the synthetic chemist under certain conditions. In most bioassays, there is no potency difference between α-MSH and its diacetylated form. However, recruitment of lactotrope cells secreting prolactin in a primary culture of rat anterior pituitary cells by α-MSH, diacetyl-α-MSH, or *N*-acetylated β-endorphin was markedly increased by the latter two peptides and to a lesser extent by α-MSH *(68)*. No lactotrope recruitment was observed with desacetyl-α-MSH or β-endorphin. This is one example of an assay where

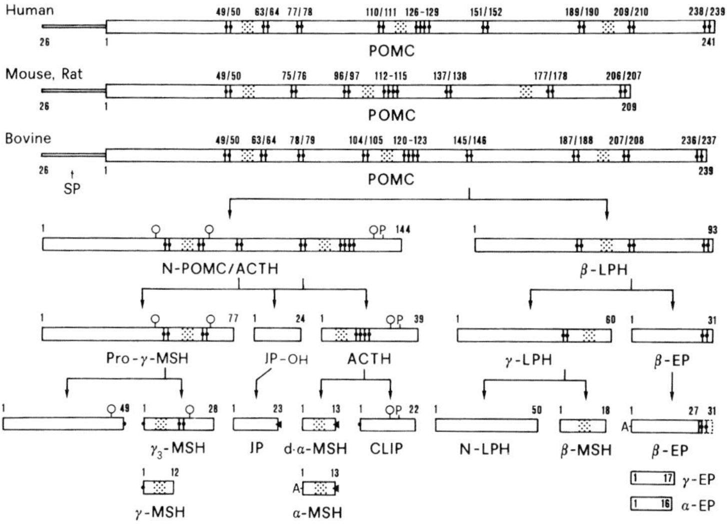

Fig. 1. Sites of cleavage in POMC of man, mouse/rat and ox and pattern of processing of POMC in melanotrophs of the bovine pars intermedia (note the different processing of γ_3-MSH and γ-LPH in the mouse and rat). Stippled area = MSH core sequence; (●), basic residue; (◀), amide; (○), site of glycosylation; A, acetyl; d, desacetyl; P, phosphate; SP, signal peptide. Glycosylation and phosphorylation of *N*-POMC/ACTH and acetylation of β-EP are only partial.

diacetyl-α-MSH has a considerably higher activity than α-MSH. In amphibian and reptile pituitaries, in fetal pituitaries of mammals and in the brain of many species desacetyl-α-MSH is the predominant form *(8)*. The released peptide from amphibian pituitaries, however, is acetylated *(67,68)*. Whether ACTH[1–14] found in frog brain besides α-MSH *(70)* serves a physiologic function on its own or is just an intermediate product in the maturation process of POMC fragments is not yet clear.

Mammalian α-MSH is a basic peptide with a p*I* of 10.5–11.0 and with acetylated N-terminus and a C-terminal amide group has a molecular weight of 1664.93 (without counterions). It is relatively heat stable and can be warmed to 80°C in physiologic solution for a short time period without loss of biologic or immunologic activity *(8)*. On the other hand, the Met^4 residue is prone to oxidation (*S*-oxide), which dramatically decreases the biologic activity of α-MSH. The single oxidation to the *S*-oxide (sulfoxide) is reversible by incubation with thioglycolic acid *(8)*, whereas the double oxidation to the *S*-dioxide (sulfone) cannot be reversed without harming the molecule.

Table 1
Structure of α-MSH Peptides from Different Species

Species[a]	Structure[b]	References
α-MSH	1 2 3 4 5 6 7 8 9 10 11 12 13	
	Aq	
Mammals	Ac-Ser-*Tyr*-Ser-Met-Glu-*His*-*Phe*-*Arg*-*Trp*-Gly-Lys-*Pro*-Val-NH_2	44
Mammals[c]	Ac-Ser-*Tyr*-Ser-Met-Glu-*His*-*Phe*-*Arg*-*Trp*-Gly-Lys-*Pro*-Val-NH_2	6
Mammals[d]	H-Ser-*Tyr*-Ser-Met-Glu-*His*-*Phe*-*Arg*-*Trp*-Gly-Lys-*Pro*-Val-NH_2	49
Rana ridibunda/catesbeiana[e]	Ac/H-Ser-*Tyr*-Ser-Met-Glu-*His*-*Phe*-*Arg*-*Trp*-Gly-Lys-*Pro*-Val-NH_2	53
Xenopus laevis[e]	Ac/H-Ala-*Tyr*-Ser-Met-Glu-*His*-*Phe*-*Arg*-*Trp*-Gly-Lys-*Pro*-Val-NH_2	54
Salmon A, trout A, sturgeon[e,f]	H-Ser-*Tyr*-Ser-Met-Glu-*His*-*Phe*-*Arg*-*Trp*-Gly-Lys-*Pro*-Val-NH_2	56-60
Salmon B, trout B[g]	H-His-Ser-*Tyr*-Ser-Met-Glu-*His*-*Phe*-*Arg*-*Trp*-Gly-Lys-*Pro*-Ile-Gly-His-OH	57, 58
Dogfish[h]	H-Ser-*Tyr*-Ser-Met-Glu-*His*-*Phe*-*Arg*-*Trp*-Gly-Lys-*Pro*-Met-NH_2	63
MSH-B	1 2 3 4 5 6 7 8 9 10 11 12 13 14 15 16 17 18 19 20	
Lamprey	H-Val-Gln-Glu-Ser-Ala-Asp-Gly-*Tyr*-Arg-Met-Gln-*His*-*Phe*-*Arg*-*Trp*-Gly-Gln-*Pro*-Leu-Pro-NH_2	64, 66

[a]Latin names of the different species are shown in Table 8.

[b]Common residues in *italics*.

[c]Originally isolated from pig. An identical structure was found in the ox *(45)*, horse *(46)*, monkey *(47)*, rat *(48)*, camel *(49)*, sheep *(50)*, and, as inferred from the POMC cDNA sequence, in mouse *(51)* and guinea pig *(52)*.

[d]Originally detected in camel pituitaries. The desacetylated form also exists in the fetal pituitary and the brain of adult mammals.

[e]Desacetyl-α-MSH is the predominant storage form in the neurointermediate lobe of amphibia and fishes. The released form is either α-MSH or desacetyl-α-MSH but may also be diacetyl-α-MSH *(55,67,68)*.

[f]Identical sequence for other teleost fishes, such as gar *(60)*, carp *(61)*, and sockeye salmon *(62)*.

[g]Originally isolated as Ac-Ser-α-MSH variant from salmon pituitaries, without N-terminal His *(56)*.

[h]Also exists in the form of free C-terminal acid.

2.1.2. β-*MSH*

The structure of the β-MSH peptides of the different vertebrates is more variable than that of α-MSH (Table 2). In mammals and elasmobranch fishes, the β-MSHs are octadecapeptides, whereas in amphibia and teleost fishes they are shortened at the N-terminus by one residue. The p*I* of mammalian β-MSHs ranges from 5.2 to 5.8 and hence differs considerably from the basic p*I* of α-MSH. All β-MSHs share six constant residues: Tyr^5, His^9-Phe^{10}-Arg^{11}-$Trp^{12,}$ and Pro^{15}. Certain species such as *Xenopus*, salmon, and trout express two different POMC-mRNAs (*see* Table 8) which explains the existence of two different forms of β-MSH, for example, in teleost fishes (Table 2); in Xenopus, the β-MSHs from both POMC precursors are identical. It has been shown earlier that the human pituitary does not produce β-MSH octadecapeptide but secretes γ-LPH into the circulation *(80)*. On the other hand, β-MSH octadecapeptide was demonstrated in the brain by microsequencing of the peptide isolated from hypothalami *(71)*, and it also occurs in a variety of POMC-producing tumors from where it may be secreted together with ACTH, CLIP, or unprocessed POMC precursor molecule *(81)*.

The rat and mouse POMC sequences do not contain a pair of basic residues at their β-LPH positions, and hence no β-MSH is formed in the neurointermediate lobe of these species *(42)*. It cannot be excluded that in other tissues, β-LPH is processed by different enzymes and that β-MSH-like molecules are formed, for example, in POMC-producing tumors *(8)*. Guinea pig POMC has two dibasic residue pairs in the C-terminal region of β-MSH: a Lys-Arg, which is the primary processing site, and an Arγ-Lys, which is part of the β-MSH molecule. The same Arγ-Lys sequence, which corresponds with the N-terminal processing site for the γ-MSH molecules (*see below*), is also part of the N-terminal region of β-MSH of the frog *Rana esculenta*. It cannot be excluded that in both the guinea pig and the frog shorter variants of β-MSH may be formed.

2.1.3. γ-*MSH*

The γ-MSH sequence does not occur in the POMC precursor of all vertrebrate species, as opposed to the α-and β-MSH sequences. γ-MSH is notably absent in the salmon *(82)*, the trout, and the gar; a remnant form is present in the sturgeon and a form similar to mammalian γ-MSH occurs in the dogfish. The sea lamprey, however, has no equivalent peptide to γ-MSH.

The γ-MSH sequence exists as dodecapeptide, named [Lys]-γ_1-MSH or simply γ-MSH, and as a longer form with 22 to 31 amino acid residues, named γ_3-MSH (Table 3). The latter is found in all mammals, whereas the former does not exist in the mouse, rat, and guinea pig because the corresponding dibasic residue pair for processing of γ_3-MSH to γ-MSH is missing in these species.

Table 2
Structure of β-MSH Peptides From Different Species

Species[a]	Structure[b]	References
β-MSH	1 2 3 4 5 6 7 8 9 10 11 12 13 14 15 16 17 18	
Human[c]	H-Asp-Glu-Gly-Pro-*Tyr*-Arg-Met-Glu-*His-Phe-Arg-Trp*-Gly-Ser-*Pro*-Pro-Lys-Asp-OH	71
Macacus	H-Asp-Glu-Gly-Pro-*Tyr*-Arg-Met-Glu-*His-Phe-Arg-Trp*-Gly-Ser-*Pro*-Pro-Lys-Asp-OH	47
Porcine	A-Asp-Glu-Gly-Pro-*Tyr*-Lys-Met-Glu-*His-Phe-Arg-Trp*-Gly-Ser-*Pro*-Pro-Lys-Asp-OH	5, 72
Equine	H-Asp-Glu-Gly-Pro-*Tyr*-Lys-Met-Glu-*His-Phe-Arg-Trp*-Gly-Ser-*Pro*-Arg-Lys-Asp-OH	73
Bovine	H-Asp-Ser-Gly-Pro-*Tyr*-Lys-Met-Glu-*His-Phe-Arg-Trp*-Gly-Ser-*Pro*-Pro-Lys-Asp-OH	74
Ovine	H-Asp-Ser-Gly-Pro-*Tyr*-Lys-Met-Glu-*His-Phe-Arg-Trp*-Gly-Ser-*Pro*-Pro-Lys-Asp-OH	50
Camel	H-Asp-Gly-Gly-Pro-*Tyr*-Lys-Met-Glu-*His-Phe-Arg-Trp*-Gly-Ser-*Pro*-Pro-Lys-Asp-OH	49
Guinea pig[d]	H-Asp-Asp-Gly-Ser-*Tyr*-Arg-Met-Glu-*His-Phe-Arg-Trp*-Gly-Thr-*Pro*-Arg-Lys-Gly-OH	52
Rana ridibunda	H-Asp-Arg-Lys-*Tyr*-Lys-Met-His-*His-Phe-Arg-Trp*-Glu-Gly-*Pro*-Pro-Lys-Asp-OH	76
Rana catesbeiana	H-Gly-Arg-Lys-*Tyr*-Lys-Met-His-*His-Phe-Arg-Trp*-Glu-Gly-*Pro*-Pro-Lys-Asp-OH	77
Xenopus laevis	H-Asn-Gly-Asn-*Tyr*-Arg-Met-His-*His-Phe-Arg-Trp*-Gly-Ser-*Pro*-Pro-Lys-Asp-OH	54, 84
Salmon A	H-Asp-Gly-Ser-*Tyr*-Lys-Met-Asn-*His-Phe-Arg-Trp*-Ser-Gly-*Pro*-Pro-Ala-Ser-OH	56
Salmon B	H-Asp-Gly-Ser-*Tyr*-Arg-Met-Gly-*His-Phe-Arg-Trp*-Gly-Ser-*Pro*-Thr-Ala-Ile-OH	78
Trout A	H-Asp-Gly-Ser-*Tyr*-Lys-Met-Asn-*His-Phe-Arg-Trp*-Ser-Gly-*Pro*-Pro-Ala-Ser-OH	58
Trout B	H-Asp-Gly-Ser-*Tyr*-Arg-Met-Gly-*His-Phe-Arg-Trp*-Gly-Ser-*Pro*-Thr-Ala-Ile-OH	58
Sturgeon	H-Asp-Gly-Ser-*Tyr*-Lys-Met-Asn-*His-Phe-Arg-Trp*-Ser-Gly-*Pro*-Pro-Lys-Asp-OH	59
Gar	H-Asp-Gly-Ser-*Tyr*-Lys-Met-His-*His-Phe-Arg-Trp*-Ser-Arg-*Pro*-Pro-Lys-Asp-OH	60
Dogfish	H-Asp-Gly-Asp-Asp-*Tyr*-Lys-Phe-Gly-*His-Phe-Arg-Trp*-Ser-Val-*Pro*-Leu-Lys-Asp-OH	24, 63
Spotted dogfish	H-Asp-Gly-Ile-Asp-*Tyr*-Lys-Met-Gly-*His-Phe-Arg-Trp*-Gly-Ala-*Pro*-Met-Asp-Lys-OH	79
MSH-A		
Lamprey	H-Asn-Pro-Glu-Leu-*Tyr*-Gln-Met-Asn-*His-Phe-Arg-Trp*-Gly-Gln-*Pro*-Pro-Thr-His-Phe-OH	64, 66

[a]Latin names of the different species are shown in Table 8.
[b]Common residues in *italics*.
[c]From nonpituitary tissue.
[d]Mouse *(51)* and rat *(75)* POMC do not have a dibasic residue pair for processing of γ-LPH to β-MSH.

Table 3
Structure of γ-MSH and δ-MSH Peptides from Different Species

Species[a]	Structure[b]	References
<u>γ-MSH</u>[c]	1 2 3 4 5 6 7 8 9 10 11 12	
Mammals[d]	H-Lys-*Tyr*-Val-Met-Gly-*His*-*Phe*-*Arg*-*Trp*-Asp-Arg-*Phe*-NH_2	83
Frog (Rana)[e]	H-Lys-*Tyr*-Val-Met-Ser-*His*-*Phe*-*Arg*-*Trp*-Asn-Lys-*Phe*-NH_2	85
Xenopus laevis	H-Lys-*Tyr*-Val-Met-Thr-*His*-*Phe*-*Arg*-*Trp*-Asp-Lys-*Phe*-NH_2	86
Dogfish	H-Asn-*Tyr*-Val-Met-Thr-*His*-*Phe*-*Arg*-*Trp*-Asp-Lys-*Phe*-NH_2	24
Leech	H-*Tyr*-Val-Met-Gly-*His*-*Phe*-*Arg*-*Trp*-Asn-Lys-*Phe*-NH_2	87
<u>γ_3-MSH</u>[f]	1 2 3 4 5 6 7 8 9 10 11 12 13 14 15 16 17 18	
Human	H-Lys-*Tyr*-Val-Met-Gly-*His*-*Phe*-*Arg*-*Trp*-Asp-Arg-*Phe*-Gly-Arg-Arg-Asn-Ser-Ser-	88
	-Ser-Ser-Gly-Ser-Ser-Gly-Ala-Gly-Gln-OH (CHO on Asn)	
	19 20 21 22 23 24 25 26 27 28 29 30 31	
Macacus	H-Lys-*Tyr*-Val-Met-Gly-*His*-*Phe*-*Arg*-*Trp*-Asp-Arg-*Phe*-Gly-Arg-Arg-Asn-Ser-Ser-	89
	-Ser-Gly-Ser-Ala-His-Gln-OH (CHO on Asn)	
Bovine	H-Lys-*Tyr*-Val-Met-Gly-*His*-*Phe*-*Arg*-*Trp*-Asp-Arg-*Phe*-Gly-Arg-Arg-Asn-Gly-Ser-	83, 90
	-Ser-Ser-Ser-Gly-Val-Gly-Gly-Ala-Ala-Gln-OH (CHO on Asn)	
Porcine	H-Lys-*Tyr*-Val-Met-Gly-*His*-*Phe*-*Arg*-*Trp*-Asp-Arg-*Phe*-Gly-Arg-Arg-Asn-Gly-Ser-	91
	-Ser-Ser-Gly-Gly-Gly-Gly-Gly-Gly-Gly-Gly-Ala-Gly-Gln-OH (CHO on Asn)	
Rat	H-Lys-*Tyr*-Val-Met-Gly-*His*-*Phe*-*Arg*-*Trp*-Asp-Arg-*Phe*-Gly-Pro-Arg-Asn-Ser-Ser-	92
	-Ser-Ala-Gly-Gly-Ser-Ala-Gln-OH (CHO on Asn)	
Mouse	H-Lys-*Tyr*-Val-Met-Gly-*His*-*Phe*-*Arg*-*Trp*-Asp-Arg-*Phe*-Gly-Pro-Arg-Asn-Ser-Ser-	51, 93
	-Ser-Ala-Gly-Ser-Ala-Ala-Gln-OH (CHO on Asn)	
Guinea pig	H-Lys-*Tyr*-Val-Thr-Gly-*His*-*Phe*-*Arg*-*Trp*-Gly-Arg-*Phe*-Gly-Arg-Gly-Asn-Ser-Ser-	52
	-Gly-Ala-Ser-Gln-OH (CHO on Asn)	
Xenopus laevis B	H-Lys-Tyr-Val-Met-Thr-His-Phe-Arg-Trp-Asn-Lys-Phe-Gly-Arg-Arg-Asn-Ser-Thr-	86
	-Gly-Asn-Asp-Gly-Ser-Asn-Thr-Gly-Tyr-OH (CHO on Asn)	
Sturgeon	H-Asn-*Tyr*-Val-Met-Ser-*His*-*Phe*-<u>His</u>-*Trp*-Asn-Thr-*Phe*-Gly-Gln-Arg-Met-Asn-Gly-	59
	-Thr-Pro-Gly-Gly-Ser-OH (CHO on Asn)	
<u>δ-MSH</u>	1 2 3 4 5 6 7 8 9 10 11 12	
Dogfish	H-Asp-Gly-Lys-Ile-*Tyr*-Lys-Met-Thr-*His*-*Phe*-*Arg*-*Trp*-NH_2	24, 94

[a] Latin names of the different species are shown in Table 8.

[b] Common residues in *italics*.

[c] In this review, γ-MSH is equivalent to [Lys^0]-γ_1-MSH. Synthetic γ_1-MSH is equivalent to des-[Lys^0]-γ-MSH; it occurs naturally in the leech. Synthetic γ_2-MSH corresponds to γ_1-MSH extended by a C-terminal Gly-OH residue and has not been reported as natural peptide.

[d] Isolated from bovine neurointermediate lobes.

[e] Isolated from neurointermediate lobes of *Rana esculenta (72)* and from the brain of *Rana ridibunda (73)*.

The C-terminal half of γ_3-MSH displays considerable sequence variations between different species whereas the N-terminus shows a much higher conservation of structure, with Tyr^2, His^6-Phe^7-Arg^8-Trp^9, and Phe^{12} as invariant residues in all species. γ_3-MSH is glycosylated at its Asn^{16} residue (in the sturgeon at Asn^{17}). In *Xenopus laevis*, the molecular weight of the glyco group was determined to approx 1800 *(86)*, yielding a total mass of both γ_3-MSH-A and γ_3-MSH-B of about 5000 kDa.

The γ-MSH dodecapeptide, which is processed from γ_3-MSH by cleavage at the dibasic residue pair, contains a carboxamide group and an N-terminal Lys. The carboxamide group is most likely generated in the same way as described for α-MSH, by hydroxylation and subsequent cleavage of the C-terminal Gly of the γ-MSH[1–13] intermediate product. The N-terminal Lys at position 1, which is the result of the different type of cleavage of the dibasic residue pair Arg-Lys as compared to Arg-Arg-, Lys-Lys, or Lys-Arg found at other processing sites of POMC. In the dogfish *(24)* and the leech *(87)*, γ-MSH appears to be processed to a peptide without N-terminal lysine. All mammals that produce γ-MSH share the same sequence. In amphibia, the Gly^5 residue of mammalian γ-MSH is replaced by either Ser or Thr and the Arg^{11} residue by Lys.

2.1.4. δ-*MSH*

The δ-MSH structure has hitherto only been found in the POMC of dogfish *(24)*. It corresponds with the molecule originally isolated in 1981 by McLean and Lowry *(94)* who named it γ-MSH. However, sequencing of dogfish POMC demonstrated that its γ-MSH structure is almost identical with mammalian γ-MSH and that the dodecapeptide containing the tetrapeptide His-Phe-Arg-Trp-NH_2 as C-terminal carboxyamide represents a novel type of melanotropin not found in POMC of other species analyzed to date *(24)*. The chemistry and physiological function of δ-MSH has not yet been elucidated.

2.2. *Adrenocorticotropic Hormone*

The adenocorticotropic hormone (ACTH; corticotropin) also forms part of the melanocortin family of peptides because it contains the melanotropic His-Phe-Arg-Trp structural element, which is essential for interaction of ACTH with its receptor, the MC2-receptor. The amino acid sequence of ACTH from selected species are listed in Table 4. All mammalian ACTH[1–24] sequences are identical, except for the guinea pig whose ACTH contains an Ala in position 24 instead of Pro. Nonmammalian vertebrates contain a few modifications in the 1–24 region, but many more alterations are found in the 25–39 region, also among mammalian species. The 25–39 part of the ACTH

Table 4
Structure of ACTH Peptides From Different Species

Species[a]	Structure[b]	1	2	3	4	5	6	7	8	9	10	11	12	13	14	15	16	17	18	19	20	References[c]
Human	H-	Ser-	*Tyr*-	*Ser*-	*Met*-	*Glu*-	*His*-	*Phe*-	*Arg*-	*Trp*-	*Gly*-	*Lys*-	*Pro*-	Val-	*Gly*-	Lys-	*Lys*-	*Arg*-	*Arg*-	*Pro*-	Val-	10, 95, 99
Macacus																						89
Porcine																						10, 95, 98
Bovine, ovine																						10, 95
Rat, mouse																						51, 75
Guinea pig																						52
Rana ridibunda																					-Ile-	76
Rana catesbeiana																					-Ile-	77
Xenopus laevis A		-Ala-																			-Ile-	96
Salmon A	H-His-													-Ile-		-His-					-Ile-	97
Dogfish														-Met-		-Arg-						24

Species[a]	21	22	23	24	25	26	27	28	29	30	31	32	33	34	35	36	37	38	39			
Human	-*Lys*-	*Val*-	Tyr-	Pro-	Asn-	Gly-	Ala-	*Glu*-	Asp-	Glu-	Ser-	Ala-	Glu-	Ala-	Phe-	Pro-	Leu-	Glu-	Phe-OH			
Porcine											-Leu-											
Bovine, ovine													-Gln-									
Rat, mouse						-Val-			-Asn-													
Guinea pig				-Ala-					-Glu-													
Rana ridibunda			-Phe-		-Thr-	Asp-			-Glu-			-Ser-		-Ile-				-Leu-				
Rana catesbeiana			-Phe-		-Thr-	Asp-			-Glu-			-Ser-		-Ser-			-Ile-	-Leu-				
Xenopus laevis A							-Val-		-Glu-					-Ser-			-Met-	-Leu-				
Salmon A				-Ala-	Ser-	Ser-	Leu-		-Gly-	Gly-	Asp-	Ser-	Ser-	Glu-	Gly-	Thr-	Phe-	Pro-	Leu-	Gln-	Ala-	OH
Dogfish						-Ser-	Phe-					-Val-		-Asn-	Met-	Gly-	Pro-	-Leu-				

[a]Latin names of the different species are shown in Table 8.
[b]Common residues in *italics*.
[c]The earlier literature on the ACTH sequence determination was reviewed in refs. *10* and *95*, together with the corrected sequences of some mammalian species.

molecule appears to play mainly a role in vivo in that it protects the 1–24 portion from degradation. This explains why ACTH[1–24] has a higher in vitro potency than ACTH[1–39] but a somewhat lower in vivo activity *(95)*. Since ACTH[1–39] also serves as precursor for the corticotropin-like intermediate lobe peptide (CLIP) or ACTH[22–39], which has a different physiologic function and bioactivity profile, the amino acid changes at the C-terminus of ACTH should rather be assessed in regard to CLIP function in the different species. However, the receptor for CLIP has not yet been characterized.

The chemistry of ACTH resembles that of α-MSH; oxidation of Met^4 dramatically reduces the binding activity of ACTH and hence almost completely abolishes the bioactivity of the molecule. Alterations within the His-Phe-Arg-Trp tetrapeptide core also considerably affect the bioactivity profile of ACTH (e.g., replacement of Trp by Trp(NPS), Phe or Ile), whereas shortening of the ACTH[1–24] sequence to modified 1–18 sequences such as [D-Ser^1,Lys^{17},Lys^{18}]-ACTH[1–18]-amide*(101)*or [β-Ala^1,Lys^{17}]-ACTH[1–17]-NH-$(CH_2)_4$-NH_2 *(102)* increases the in vivo activity of the molecule. A comparison of the bioactivity profiles of different ACTH fragments and analogs are found in the review by Schwyzer *(95)*; few truly novel structure–activity data have been accumulated since then, but the recent cloning of MC2-R and the need for subtype-specific ligands will soon yield new classes of compounds with corticotropic and/or lipolytic activity.

3. Physiology and Assays of the Melanocortin Peptides

3.1. Physiology of the Melanocortins

The occurrence, content, and distribution of melanocortin peptides in the pituitary gland, the brain, peripheral tissues, and in the circulation of different species has been reviewed *(8)*, along with the hypothalamic network regulating biosynthesis and secretion of the different melanocortins. It has been demonstrated that melanocortins are pleiotropic peptides affecting a number of different cells in the nervous system and many peripheral organs by interaction with several subtypes of melanocortin receptors. There is a vast clinical literature about the involvement of the hypothalamopituitary-adrenal axis and its regulatory factors corticotropin releasing factor (CRF), ACTH, and cortisol, as well as their receptors in various pathophysiologic conditions, for which, however, specialized reviews should be consulted (*see* Chapter 3). Table 5 lists only the most prominent physiologic effects of the melanocortins and the text below concentrates on some of the recent findings.

Table 5

Summary of the Different Pysiological Effects Induced by Melanocortin Peptides

Melanocortins	Ligand; receptor subtype; comment
Pigment migration Melanogenesis	α-MSH, β-MSH, ACTH, agouti; MC1-R; color change by circulating melanocortins, melanogenesis by circulating and paracrine melanocortins.
Melanoma and other tumors	α-MSH, β-MSH/γ-LPH, ACTH, CLIP; MC1-R; tumor growth regulation, induction of melanogenesis, biosynthesis of ligands by tumor.
Skin cell differentiation Immunosuppression	α-MSH, ACTH; MC1-R, MC2-R (?); keratinocytes differentiation, regulation of cutaneous inflammation and hyper-proliferative skin diseases.
Trophic action on CNS and PNS; regeneration	α-MSH, ACTH; MC3-R, MC5-R (?); effects on brain and body growth and development, regneration of peripheral and central nervous system.
Brain: memory, behavior, transmission	α-MSH, γ-MSH, ACTH; MC3-R, MC4-R; MC5-R; avoidance, grooming, social and sexual behavior, stretching-yawning, synaptic transmission.
Mediation of food intake regulation	α-MSH, agouti; MC4-R; tonic inhibition of feeding behavior, antagonism of orexigenic signal of MCH, mediation of leptin signal.
Fat tissue: lipolysis Metabolic effects	α-MSH, ACTH, γ-LPH: MC2-R, MC5-R (?); lipolysis in adipocytes (rabbit: MSH; rat: ACTH), effect on sugar and calcium plasma levels.
Adrenal cortex: trophic effect, corticosteroids	ACTH, α-MSH, pro-γ-MSH; MC2-R, MC5-R; trophic effect on fetal and regenerating adrenal, stimulation of corticosteroid secretion.
Immune system: anti-inflammatory activity	α-MSH, MSH(11-13), ACTH; MC1-R, MC5-R (?); antiinflammatory and antipyretic activity, partly mediated via unknown receptor (?).
Regulation of exocrine gland function	α-MSH, ACTH; MC5-R; Stimulation of sebaceous, preputial, Harderian and lacrimal gland function.
Cardiovascular system: heart, natriuresis	γ-MSH, α-MSH, ACTH; MC3-R, MC5-R (?); increase in natriuresis, increase in heart rate and arterial pressure.
Gonads: development Eye: permeability; oth.	α-MSH, ACTH; MC3-R (?), MC5-R; development of gonads, androgen secretion, stimulation of retina, increase in BAB permeability.
Pituitary gland: co-regulation of anterior lobe	α-MSH, ACTH; MC3-R, MC5-R (?); Modulation of activity of hypothalamic releasing factor of anterior lobe peptides (PRL, LH, GH).

3.1.1. Pigment Migration

In lower vertebrates such as *Xenopus* or *Rana*, adaptation of skin color to the light intensity of the environment is induced by release of α-MSH from melanotrope cells of the neurointermediate lobe of the pituitary gland. The release is under inhibitory control of dopamine, γ-aminobutyric acid (GABA), and NPY secreted from synaptic terminals of suprachiasmatic nuclei and mediated by cAMP *(103)*. α-MSH release is stimulated by magnocellular nuclei releasing thyrotropin releasing hormone (TRH) and corticotropin releasing factor (CRF) that also act via cAMP. Acetylcholine,which induces α-MSH secretion, is produced by the melanotrope cell and acts in an auto-excitatory feedback on melanotrope M1 muscarinic receptors *(103)*. POMC-peptide release is driven by $[Ca^{2+}]_i$ oscillations regulated by hypothalamic neurotransmitters and acetylcholine via receptor-mediated stimulation of Ca^{2+} influx through N-type calcium channels *(103)*. Frog neurotensin induces a concentration-dependent increase in α-MSH release from perifused frog pars intermedia cells *(104)*,whereas melanostatin *(105)*, NPY and SPYY inhibit melanotropin release from neurointermediate lobes *(106)*.

3.1.2. Melanogenesis

In the mammal, the melanogenic effect of systemic melanocortin peptides is well documented *(8)*. More recently, it has been demonstrated that melanocortin peptides are generated in the skin; for example, α-MSH was localized to keratinocytes, melanocytes, and possibly Langerhans cells *(107)*. ACTH was also present and showed strongest staining at differentiated keratinocytes. It seemed that the convertases PC1 and PC2 are expressed in the skin *(107)*, a prerequisite for POMC processing. Using HPLC, fragments of ACTH were found such as ACTH[1–10], acetylated 1–10, and ACTH[1–17], which increased dendricity in cultured human melanocytes *(107)*. ACTH peptides were reported to occur in greater amounts in the skin than α-MSH *(108)*. Another important finding relates to the regulatory function of agouti protein (AP; in the human: agouti signal protein, ASP) in melanogenesis by mammalian melanocytes where AP reduces total melanin production and elicits the synthesis of pheomelanin rather than eumelanin *(109)*. Expression of α-MSH-induced tyrosinase, TRP-1 and TRP-2, are inhibited by AP. Furthermore, whereas differentiation of murine melanoblasts in hair follicles is stimulated by α-MSH or forskolin, AP inhibits this process by inhibiting the expression of the melanogenic transcription factor, microphthalmia, and its binding to an M box regulatory element *(110)*. The balanced expression of melanocortin peptides and AP or ASP *(111)*, and the occurrence of MC1 receptor variants in the different species, for example, in humans *(112)* or in the pig *(113)*, are the basis for the variability of mammalian pigmentation. In addition, the regulation of MC1 receptor expression is regulated by ultraviolet (UV)-light and proinflammatory cytokines *(114)*.

3.1.3. Role in Melanoma

The role of melanocortin peptides in melanoma is reviewed in Chapter *17*. In humans, immunoreactive α-MSH was found in melanoma, and the extent of its production and release could be correlated with the expression of MC1 receptors *(115)*. Stimulation of MC1-R by α-MSH in human melanoma inhibited the tumor necrosis factor alpha (TNF-α)-stimulated expression of the intercellular adhesion molecule-1 (ICAM-1), with 90% inhibition at 10 n*M* and 50% at 1 n*M* hormone concentration *(116)*. On the other hand, treatment of human melanocytes by [Nle^4, D-Phe^7]-α-MSH led to increased adhesion to fibronectin by reorganization of the actin stress fiber cytoskeleton *(117)*. The chemical nature of melanomα-derived α-MSH has not unequivocally been proven: in cultivated cells (but not in the culture medium) IR-α-MSH was determined to concentrations of 0.4 to 2.3 n*M*, depending on the melanogenic status of the tumor cells *(118)*. In melanoma tumors, IR-α-MSH existed in the form of bioactive 16-kDa and 5- to 9-kDa molecules. In our own laboratory, similar high molecular weight MSH molecules were found that exerted a melanogenic effect but did not displace bioactive MSH radioligand in the receptor binding assay *(W. Siegrist, unpublished observations.)*

3.1.4. Immunosuppression in the Skin

α-MSH and ACTH are produced by human keratinocytes and the biosynthesis is upregulated by interleukin-1 (IL-1), UV-light, or phorbol ester *(119)*. This means that skin-derived melanocortin peptides predominantly originate from keratinocytes. MC1 receptor expression was reported in various cell types of the skin, immunocompetent and inflammatory cells, keratinocytes, melanocytes, and dermal microvascular endothelial cells *(120)*. The latter produce increased levels of IL-8 upon stimulation with α-MSH *(121)*. In addition, α-MSH seems to modulate keratinocyte proliferation and differentiation and downregulates the production of IL-1 and IL-6 in monocytes and macrophages. These and other findings suggest that α-MSH is part of the mediator network that regulates inflammation and hyperproliferative skin diseases *(120)*.

3.1.5. Trophic Action on the Nervous System

The effect of melanocortin peptides on brain development, the developing motor system as well as on the regeneration of the peripheral nervous system (PNS) or the central nervous system (CNS) is well documented *(8,122)*. Besides effects on neurons, melanocortin peptides, in particular ACTH[1–17], appear to activate astrocytes by increasing cAMP levels *(123)*; the concomitant mitogenic effect was, however, mediated by a different signaling mechanism. Recent studies focused on the trophic effect of melanocortins on the outgrowth of neurites from PNS and CNS *(124)*. It was demonstrated that maximal axonal and dendrite outgrowth occurred at $10^{-8}M$ α-MSH, whereas cholera toxin B-labeled neurons only required $10^{-10}M$ α-MSH for maximal axonal outgrowth, possibly because of different MC receptor expression. Lichtensteiger et al. *(125)* demonstrated that MC receptor expression shows marked region- and stage-specific, often transient, ontogenetic parameters in the developing rat brain. The early presence of MC4-R in the CNS and PNS and the transient regional peaks of mRNA expression, often concomitant with periods of neural network formation, suggest a role of MC4-R in early ontogeny although MC3-R may be involved in analogous processes during postnatal development *(126)*.

3.1.6. Memory and Behavior

The role of MSH and ACTH peptides on active and passive avoidance behavior of experimental animals and on human psychopathology has been reviewed *(8,127)*. The melanocortins exert a positive effect on short-term memory, activate sexual behavior, stimulate aggression and social behavior, grooming, stretching and yawning, and improve locomotor activity in animals. The MSH/ACTH[4–9] sequence was shown to be crucial for most of these effects. Modified 4–9 analogs, for example, Org 2766, are much more potent stimulants of several forms of behavior and they improve retrieval of information in rats even after brain damage *(128)*. Modulation of excessive grooming behavior, whose structural requirements for the ligand differ from those of avoidance behavior, appears to be mediated by MC4 receptors, as concluded from a recent study comparing the effect of MC3/4 agonists and antagonist *(129)*. The ACTH[4–9] sequence also affects human behavior insofar as performance is increased, which is accompanied by elevated electrophysiological activity *(8)*. Furthermore, the peptide interferes with human sleep by disturbing it *(130)*. [It should be noted, however, that sleep regulation is thought to be mainly controlled by the balance between gonadotropin releasing factor (GRF) and CRF rather than the melanocortins.] Several other psychopathologic disorders such as anxiety and depression are linked to the hypothalamo-pituitary-adrenal axis and partly to melanocortin peptides, but CRF was shown to have direct actions in the brain in these disorders and therefore CRF antagonists are favored for pharmacotherapeutic intervention as an alternative to antidepressants *(131)*. The development of melanocortin-derived sequences for the treatment of mild forms of dementia, actively pursued for many years, appears to have been discontinued, despite the findings that peripherally and centrally applied melanocortin analogs and fragments positively influence cholinergic, adrenergic, and dopaminergic neurotransmission and several neurochemical parameters *(8)*. For example, α-MSH and α-MSH[11–13] had long been shown to induce a marked increase in the firing rate of dopaminergic neurons *(132)*. It cannot be excluded that novel subtype-specific agonists will be found displaying a better pharmacotherapeutic potential for improving memory.

3.1.7. Modulation of Food Intake Regulation

The demonstration that melanocortins exert a tonic inhibition on food intake by targeted disruption of the MC4 receptor *(38)* and by pharmacologic studies with MC4 receptor agonists and antagonists *(37)* has assigned an important role to this peptide-receptor system in the development of obesity, insulin resistance, and type II diabetes. Whereas experimental animals lacking MC4-R develop obesity *(38)*, mice that cannot produce the MCH peptide are hypophagic and lean *(133)*. This functional antagonism between α-MSH and MCH, originally discovered in the melanophores of teleost fishes *(134)*, has become an important element within the peptidergic neuroregulatory network of food intake behavior. Although it is not yet clear whether in humans the role of these peptides in food intake regulation corresponds with that found in the mouse, the first observations of genetic defects in the POMC prohormone (abolished POMC translation or lack of POMC processing to functional ACTH/α-MSH) demonstrated that the lack of melanocortin peptides leads to severe early-onset obesity, adrenal insufficiency and red hair pigmentation in humans *(135)*.

3.1.8. Fat Tissue and Adrenal Cortex

The lipolytic activity of ACTH and LPH was studied in the 1960s by Rodbell and later by Ramachandran and by Ng (reviewed in ref. *8*). The data of these authors showed that ACTH elicits high lipolytic acitivity in adipocytes of different species (e.g., rat, mouse, hamster, guinea pig, rabbit), but α-MSH only elicited a high response in the rabbit and the guinea pig *(136)*. This differential response to melanocortin peptides may reside in the different structural requirements of MC receptor subtypes expressed in these species, or it is possible that adipocytes of some species express exclusively one subtype of MC receptor (e.g. MC2-R), whereas others in addition (or exclusively) express for example, MC5-R. Through which receptor subtypes other direct metabolic processes of melanocortins (*see* ref. *8*) are mediated, is not yet clear.

The role of melanocortins in adrenal development and function is presented in chapter 3 of this volume and will not be detailed here. Structure-activity studies with various ACTH/MSH fragments and analogs demonstrated that MSH peptides exert only weak steroidogenic activity in fasciculata/reticularis cells of the adrenal cortex *(8)*. α-MSH is relatively more potent in stimulating aldosterone secretion from glomerulosa cells. It is of particular interest that salt-loading of rats diminishes the response of adrenocortical cells to α-MSH, whereas in salt-deprived animals α-MSH becomes almost as potent as ACTH *(8)*. These findings indicate a differential expression of MC receptor subtypes in the different zones of the adrenal cortex as well as altered expression in the different dietary states of the animal. It is now known that bovine glomerulosa cells express both MC2 and MC5 receptors, whereas the latter are not found in fasciculata/reticularis cells *(137)*.

3.1.9. Immune System

The antipyretic activity of MSH peptides in the rabbit studied by Lipton and coworkers *(138)* demonstrated that α-MSH and its C-terminal 11–13 fragment antagonize the fever-inducing effect of interleukin-1β (IL-1β) when administered centrally or, at higher doses, also peripherally and hence exert central antiinflammatory activity (reviewed in ref. *8*). This was confirmed for rats by intraperitoneal application of α-MSH, which suppressed LPS-induced fever via MC receptors in the brain and also inhibited the LPS-induced increase of corticosterone and IL-6 plasma levels via different (peripheral?) action *(139)*. From other experimental models of inflammation, it was concluded that melanocortin peptides may have antiinflammatory activity also via direct interaction with peripheral host cells *(140)*. It has long been claimed that lymphocytes from LPS-sensitive mice can process POMC to ACTH and β-endorphin, whereas LPS-resistant mice cannot process POMC *(141)*. Although macrophages were reported to constitutively produce POMC-derived peptides, ACTH was found in mononuclear leukocytes only after mitogen stimulation *(142)*. ACTH appears to stimulate calcium uptake in rat lymphocytes *(143)*, and α-MSH induces IL-10 production in human monocytes *(144)*. These latter cells express MC1 receptors, which are upregulated after exposure of the cells to endotoxin or mitogen for 3–5 days *(145)*. The modulation of immune responses by α-MSH requires preceding upregulation of MC1-R expression and costimulatory factors. Analysis of mouse pro-B-lymphocytes showed that these cells express MC5-R *(146)*. In a preclinical study with plasma from cardiopulmonary bypass patients, which usually react to surgery with diffuse inflammatory responses and whose plasma hyperstimulates monocytes and granulocytes in culture, pretreatment of the cells with α-MSH significantly diminished the hyperstimulation *(147)*.

3.1.10 Regulation of Exocrine Gland Function

Several exocrine glands are regulated by melanocortin peptides. For example, the activity of sebaceous glands and the size of the preputial gland, a specialized sebaceous gland implicated in pheromone production, is markedly reduced after removal of the neurointermediate lobe of experimental animals (reviewed in ref. *8*). Application of α-MSH restores sebaceous gland function and increases size and sebum content of preputial glands. Behavioral changes induced by melanocortin peptides, including altered sexual attraction of females or aggressive behavior of male animals due to olfactory cues, are linked to changes of preputial gland activity (see ref. *8*). The lacrimal gland where ACTH and α-MSH stimulate protein discharge was shown to express high-affinity melanocortin receptors *(148)*. Similarly, the harderian gland, which secretes lipids and porphyrins, expresses melanocortin receptors *(149)*. Recently, Cone and coworkers *(150)* generated MC5-R-deficient mice and they demonstrated that exocrine gland function was impaired in these animals and hence that melanocortin peptides act through MC5 receptors. Further screening for MC5-R expression in different glands confirmed that this MC receptor subtype is found, in addition to lacrimal, harderian, sebaceous, and preputial glands, also in porstate glands, pancreas, adrenal, esophagus, thymus, and spleen *(151)*.

3.1.11. Cardiovascular System

Melanocortins, predominantly γ-MSH and ACTH fragments, display pressor, cardioaccelerator, and natriuretic activity in rats *(152)*, and γ-MSH also plays a role in the adjustment to high-salt diet *(153)*. Intravenous application of γ_2-MSH to conscious rats causes a dose-dependent increase in blood pressure and heart rate *(154)*. The C-terminus of the γ_2-MSH molecule is crucial for eliciting this effect. Truncation of the first five residues at the N-terminus (i.e., γ_2-MSH[6–12] increased the potency of γ_2-MSH). The shortest fragment with measurable pressure activity is the MSH core peptide His-Phe-Arg-Trp but Asp^9-Arg^{10}-Phe^{11} are important for full intrinsic activity *(154)*. The potency of γ_2-MSH is about 10-fold higher than that of ACTH[4–10] *(155)*; γ_3-MSH, α-MSH and [Nle^4, D-Phe^7]-α-MSH do not induce these responses *(152)*. ACTH[1–24] has a depressor effect, combined with a tachycardiac response. Cerebral hemodynamics in the rat, that is, induction of pressor tachycardic response, was shown to require similar structural elements *(156)*; [Nle^4, D-Phe^7]-α-MSH and ACTH[1–24] were without activity in this test. In humans, plasma γ-MSH levels are elevated in patients with severe congestive heart failure and in primary hyperaldosteronism *(157)*. It appears that γ-MSH-related peptides are involved in sodium homeostasis as well as in certain forms of hypertension and that the effect is mediated via interaction at MC3 receptors, possibly localized in the anteroventral third ventricle region, situated outside the blood–brain barrier *(152)*.

3.1.12. Gonads, Eye, Pituitary Gland

A last area of effects of melanocortin peptides relates to anterior pituitary function and effects in the eye and the gonads. α-MSH was shown to modulate the activity of hypothalamic releasing factors on lactotrophic, gonadotrophic, and somatotrophic cells (reviewed in ref. *8*). In lactotrophs, α-MSH potentiates TRH- or ATP-triggered prolactin release by increasing Ca^{2+}-entry induced by prolacting secretagogues *(158)*.

α-MSH may modulate the neuroactivity of the retina and induce increased permeability of the blood–aqueous barrier of the eye *(8)*. For example, prostaglandin production by bovine retinal pigment epithelium is increased by α-MSH *(159)*. However, chronic stimulation of retinal pigment epithelium in a patient receiving ACTH treatment led to central serous retinopathy, possibly caused by disruption of the outer blood–retinal barrier or leakage from choroidal vessels induced by melanocortins *(160)*.

MSH and ACTH peptides have long been discovered in the testes, ovaries, and placenta of experimental animals (reviewed in ref. *8*) and more recently, MC5 receptor mRNA was also detected in these tissues *(36)*. In the ox, MC5-R is highly expressed in the testes, suggesting an involvement in spermatogenesis in this species *(161)*. In the female, α-MSH stimulates the release of progesterone from prepubertal ovaries *(162)*; the MC receptor subtype involved in ovarial stimulation is not yet known but likely to be MC5-R.

3.2. Assays of Melanocortins

3.2.1. Bioassays

The melanophore and melanoma cell assays used currently for determination of the bioactivity of melanocortin peptides at MC1 receptor subtypes are described in ref. *8* in full detail. The in vivo melanophore assays (e.g., *Xenopus*, *Rana*) are less sensitive and rarely applied but in vitro fish, frog, and lizard skin assays (e.g., *Ctenopharyngodon*, *Synbranchus*, *Oncorhynchus*, *Rana*, *Hyla*, *Xenopus*, *Anolis*, *Bufo*) are still frequently used for testing novel melanocortin analogs. The assay sensitivity is in the picomolar range. It should be noted, however, that there are some marked differences in MC1 receptor–ligand recognition between different species, in particular with respect to agonist/antagonist behavior of melanocortin analogs *(163)*. Therefore, the MC1 receptor assays mainly used at present are based on mammalian cells, in particular mouse and human melanoma cell lines (mouse B16 and Cloudman S91, human D10, HBL, and others) or on HEK-293 or COS cells expressing MC receptor subtypes. Novel melanoma cell lines useful for receptor studies include the MC1-R-deficient mouse B16-G4F and clones derived thereof expressing human MC1-R *(164)* or other types of receptor. Several signals can be assayed with these melanoma cells, the most frequently being receptor binding assays, adenylate cyclase activation, determination of cAMP levels, tyrosinase activity, or melanin formation. For assaying the corticotropic or lipolytic activity of ACTH at MC2 receptors, rat adrenocortical cells or adipocytes are used to quantify receptor binding, adenylate cyclase, cAMP levels or corticosteroid production. Similarly, rabbit adipocytes are applied to study MSH-type ligands interacting with fat cells (all these assays are described with full experimental details in ref. *8*).

A new combinatorial chemistry-based diffusion assay was developed to screen random tripeptides for antagonistic activity and to identify pharmacologic groups responsible for receptor interaction *(165)*. For efficient determination of cAMP, a rapid nonradioactive colorimetric assay was introduced by Chen et al. *(166),* based on β-galactosidase gene fused to five copies of the cAMP response element (CRE). When performed in a 96-well microplate, the activation of CRE-binding protein that results from an increase in intracellular cAMP or Ca^{2+} can be determined directly in a microplate reader.

3.2.2. Immunoassays

There are different commercially available assay kits for the determination of POMC-derived peptides in plasma and other biologic samples. Besides soliδ-phase-based two-step assays for α-MSH *(8)* and ACTH using a single antibody, two-site immunoradiometric assays (IRMAs) were developed for the larger POMC peptides (ACTH, *N*-POMC, β-LPH) with which the intact circulating peptides in man could be determined accurately *(167)*. Basal ACTH ranges from 0.9 to 11.3 pmol/L, whereas α-MSH ranges from <0.5 to 10 pmol/L; levels of circulating POMC precursor or large POMC fragments are higher (5–40 pmol/L) than those of the melanocortin peptides *(167)*. Similar values (up to 50–60 pmol/L) for high molecular weight forms of γ-MSH were determined with a hetero-two-site enzyme immunoassay for γ_2-MSH *(168)*; the small-molecular form of γ-MSH was around 1 pmol/L. Whether circulating POMC represents an additional source for the formation of peripheral melanocortins is not yet clear. Breakdown of ACTH and α-MSH in the circulation is relatively rapid, with a half-life in human plasma of 15–30 minutes at 37°C *(8)*.

3.2.3. Receptor Binding Assays

Receptor binding assays for MC receptor subtypes performed with intact cells *(169)* or cell membranes are usually carried out with either tritiated or radioiodinated MSH or ACTH radioligands (*see* 5.1.). Radioiodinated tracers are generally preferred because of considerably higher specific radioactivity. The most widely used MSH radioligand is [^{125}I]-[Nle^4, D-Phe^7]-α-MSH because, with this peptide oxidation of methionine is eliminated, and the high bioactivity of the D-Phe^7 analog is maintained *(170)*. However, for some human melanoma cell lines, [^{125}I]-α-MSH or [^{125}I]-[D-Phe^7]-α-MSH containing Met^4 are the preferred radioligands as nonspecific binding is lower. Particularly unfavorable for human melanoma cells is [^{125}I]-[Nle^4]-α-MSH because of high nonspecific binding *(8)*; by contrast, the same radioligand has excellent characteristics in mouse melanoma cell assays. ACTH radioligand based on radioiodination of Tyr^2 is difficult to prepare and radioiodination frequently leads to inactive products. As an alternative, [Phe^2, Nle^4, Tyr^{23}]-ACTH[1–38] *(171)* and its shorter, 1–24, form were developed for radioiodination without impairing the bioactivity; this modified ACTH[1–24] radioligand is commercially available. The assay conditions (temperature, time, addition of enzyme inhibitors, and separation of bound from free radioligand) depends on the cell types used; for example, mouse melanoma cells are incubated at 16°C, whereas human melanoma cells require an incubation temperature of 37°C for optimal receptor binding *(172)*. When HEK-293 or COS cells with overexpressed MC receptor subtypes are used, the conditions are less critical than with cells exhibiting only small receptor numbers.

3.3. Signaling of Melanocortins

The main intracellular signal generated by stimulation of melanocortin receptors is cAMP production via MC-R/G_s coupling to adenylate cyclase. Cells (e.g., HEK-293 or COS) transfected with any of the MC receptor subtypes respond to stimulation by MSH agonists usually by a marked increase in cAMP content. Human melanocytes *(173)* or human melanoma cells in culture *(172)* frequently respond with only small increments of cAMP elevation, except for those cell lines that express high numbers of MC1-R (e.g., HBL cells), which produce a high cAMP increase *(172)*. Mouse melanoma cells also respond with marked cAMP production. Thus it appears that MC receptors mainly interact with $G_s\alpha$, possibly with more than one of the four splice variants *(164)*. From a study comparing α-MSH and agouti protein interacting with MC1-R on mouse melanoma cells, Siegrist et al. *(174)* concluded that intracellular signaling of these two ligands is not confined to the adenylate cyclase/cAMP/PKA system, but that additional signals, such as PKC activation *(175,176)*, may be generated. For pigment migration in melanophores, it was recently demonstrated that PKA and PKC activate two different pathways for melanosome dispersion and that protein phosphatase 2A mediates this effect *(177)*. Also, when Cloudman S91 mouse melanoma cells *(178)* or human melanocytes *(179)* are depleted from PKC activity, α-MSH-induced melanogenesis is markedly reduced or abolished. In the adrenal cortex, activation of PKC is an important element in the response of glomerulosa cells to α-MSH *(180)*. These data show, that both PKA and PKC form part of the MC receptor signaling system.

The role of inositol phospholipids and Ca^{2+} in mediating intracellular melanocortin signaling is less well established. For example, Hepa cells transfected with hMC3-R respond to α-MSH or ACTH with a 15-fold elevated cAMP level, but inositol phosphates increase only moderately over basal levels (1.5-fold) and in a biphasic manner *(181)*. Intracellular mobilization of Ca^{2+} after receptor stimulation could not be detected in these cells, which confirms our own findings with various melanoma cell lines. However, inhibition of PKA with its specific inhibitor H-89 produced a $[Ca^{2+}]_i$ and monophasic inositol phosphate signal *(181)*. Human ASP was shown to increase $[Ca^{2+}]_i$ in cells transfected with human MC1 (or MC3) receptor with an EC_{50} of 18 n*M*; Ca^{2+} entry could be blocked by nitrendipine *(182)*. The lack of $[Ca^{2+}]_i$ mobilization potency of melanocortins is contrasted by the fact that Ca^{2+} is required for ligand binding to MC receptors and for the early phase of receptor signaling (receptor coupling) for which evidence was presented many years ago with melanophores *(183)* and melanoma cells *(184)*.

MC3 receptor stimulation on mouse B-lymphocytes leads to activation of the Jak/STAT pathway and cell proliferation as recently demonstrated by Buggy *(146)*. Physiologic concentrations of α-MSH (10 n*M*) induced phosphorylation of Jak2 and STAT1. Whether this type of activation is confined to MC3-R on lymphocytes or is also found in other cell types and for other MC receptor subtypes is not yet known.

4. Synthetic Melanocortin Agonists and Antagonists

4.1. Molecular "Anatomy" of Melanocortins

The molecular "anatomy" of α-MSH, which describes the function of individual residues as inferred from structure–activity studies, is briefly outlined in Table 6; a more detailed description is found in ref. *8*. The minimal structural requirement for eliciting a melanotropic response in most MSH-target cell systems is the tetrapeptide His-Phe-Arg-Trp *(8,185)*; in some assay systems, even shorter peptides of this central region of α-MSH elicit a response. The N-terminal portion with its balanced hydrophilic (Ser-Tyr-Ser)/hydrophobic (Met) nature has the function of a "potentiating" element for the linear central part of the hormone. Increase of hydrophobicity within the 1–3 region or, particularly oxidation of Met^4, considerably lowers the potency of the peptide. The N-terminal acetyl group protects the peptide from aminopeptidase attack and also increases the potency in some of the melanophore and melanoma cell assays. The Glu^5 residue seems to be important for interaction of ACTH with the MC2-R in the adrenocortical cell assay; for the α-MSH-type receptors (i.e., MC1-R, MC4-R, MC5-R), the negative charge at position 5 can be omitted. Within the tetrapeptide core His-Phe-Arg-Trp, change of configuration or modification individual residues may drastically alter the potency of the peptide (agonist/antagonist properties; increase/decrease of affinity). The Gly^{10} has a spacer function for the C-terminal tripeptide but can be replaced by Lys-amide, for example, in α-MSH[4–10]

Table 6

Molecular 'Anatomy' of α-MSH, Stabilized Analogs with Increased Potency, and Radioligands/Affinity Ligands for Receptor Analysis

	Comments[a]
Ac-	protects from degradation; increases affinity/bioactivity
-Ser-Tyr-Ser-	balanced hydropho-/hydrophilicity, potentiates the core
-Met-	oxidation drastically diminishes affinity and activity
-Glu-	important for interaction with MC2 receptor (ACTH)
-His-Phe-Arg-Trp-	crucial for binding and triggering all MC receptors
-Gly-	spacer function, can be replaced by different residues
$-Lys-Pro-Val-NH_2$	increases affinity; elicits responses on its own
$Ac-Ser-Tyr-Ser-Met-Glu-His-Phe-Arg-Trp-Gly-Lys-Pro-Val-NH_2$	
-Nle-	increases affinity and protects peptide from oxidation
-D-Phe-	dramatically increases affinity and bioactivity (MC1-R)
Ac-Nle-Asp- -D-Phe- $-Lys-NH_2$	linear (4-10) sequence is equipotent with α-MSH
-Cys- -Cys-	displays about the same activity as α-MSH at MC1-R
-Cys- -D-Phe- -Cys-	almost as potent with linear [Nle^4, D-Phe^7]-α-MSH
Ac-Nle-Asp- -D-Phe- $-Lys-NH_2$	100-fold more potent than α-MSH; very stable in vivo
Ac-Cys -D-Nal- $-Cys-NH_2$	antagonist activity at MC4-R, agonist activity at MC1-R
Ac-Cys- -D-Nal- $-Nle-Cys-NH_2$	displays MC3-R selectivity, agonist activity at MC1-R

Ac-	Ser-	Tyr-	Ser-	Met-	Glu-	His-	Phe-	Arg-	Trp-	Gly-	Lys-	Pro-	Val-NH_2	Comments[a]
		-Tyr ($^3H_2$)												>50 Ci/mmol; full biological activity; binding studies
													-Nva ($^3H_4$)	>100 Ci/mmol; 100% bioactivity, brain MC-R mapping
				-Nva ($^3H_4$)									-Nva ($^3H_4$)	>200 Ci/mmol; 100% bioactivity, receptor mapping
		-Tyr (^{125}I)												>2000 Ci/mmol; binding studies with human melanoma
		-Tyr (^{125}I)		-Nle-										>2000 Ci/mmol; binding studies with mouse melanoma
		-Tyr (^{125}I)		-Nle-			-D-Phe-							>2000 Ci/mmol; most frequently used radioligand
		-Tyr (^{125}I)					-D-Phe							>2000 Ci/mmol; binding studies with human melanoma
											-Lys (^{125}IB)			dehalogenation-resistant in vivo; high bioactivity in vitro
Ac-D-Ala-		$^3H_2$ Pap-		-Nva-										>30 Ci/mmol; MC1 receptor crosslinking
		-Tyr (^{125}I)		-Nle-					-Trp (Naps) -					>2000 Ci/mmol; MC1 receptor crosslinking
		-Tyr (^{125}I)		-Nle-			-D-Phe-		-Trp (Naps) -					>2000 Ci/mmol; MC1 receptor crosslinking
ApSSpr		-Tyr (^{125}I)		-Nle-			-D-Phe-		-Trp (Naps) -				-Pap-	>2000 Ci/mmol; MC1 receptor crosslinking
DTPA-				-Nle			-D-Phe-							[^{111}In]-DTPA-MSH for in vivo tumor imaging
			DTPA-	Nle-	Asp-		-D-Phe-			-Lys (DHP) -NH_2				[^{111}In]-DTPA-MSH for in vivo tumor imaging
			DOTA-	Nle-	Asp-		-D-Phe-			-Lys (DHP) -NH_2				[^{111}In]-DOTA-MSH for in vivo tumor imaging

ApSSpr, (4-azidophenyl)-1,3'-dithiopropionyl; DOTA, (1,4,7,10-tetraazacyclododecane-1,4,7,10-tetraacetic acid); DTPA, diethylenetriaminopentaacetic acid; IB, iodobenzoate; Naps, 2-nitro,4-azidophenylsulfenyl; Pap, p-azidophenylalanine.

[a]For references *see text*.

fragments without loss of activity. The C-terminal tripeptide again has a potentiating function for the linear central core peptide but may display independent biologic activity in specific assays or tests (e.g., antiinflammatory activity, stimulation of dopaminergic neurons, certain melanophores). The latter has been questioned by Castrucci et al. *(185)*. C-terminal modification of the α-MSH molecule generally affects the potency more than N-terminal modifications *(8)*. Comparative studies between α-MSH and γ-MSH interacting with MC1, MC3, and MC4 receptors *(186,187)* confirmed that Pro^{12} and also Tyr^{2} of α-MSH are important at MC4-R whereas the central portion of α-/γ-MSH is the primary determinant for MC3-R.

The molecular ''anatomy'' of ACTH corresponds with that of α-MSH: the structural elements indispensable for eliciting a steroidogenic response in the adrenal cortex reside within the 4–10 sequence as this fragment is the shortest displaying corticotropic activity *(8)*. The N-terminal tripeptide enhances the potency and oxidation of Met^{4} almost completely abolishes the corticotropic activity. The 11–24 region is a weak antagonist of ACTH without biologic activity. This means that the adrenocortical MC2 receptor recognizes the 11–24 portion of the ACTH molecule. The 25–39 region of ACTH is a 'species label', conveys antigenicity to the hormone and is the site of phosphorylation (Ser^{31}). The importance of free or blocked termini of ACTH was tested with ACTH[4–10] eliciting aldosterone or corticosterone production by zona glomerulosa cells or, respectively, fasciculata/reticularis cells in the rat *(188)*. Whereas free termini of MSH/ACTH fragments were crucial for the latter effect, the former was elicited by fragments with free and blocked peptide ends.

The molecular ''anatomy'' of γ-MSH is less well established. It appears that for interaction with MC3-R the central portion of the molecule is the most relevant, as described for α-MSH. The C-terminal tripeptide adds intrinsic activity to the central core but can be extended by–Gly-OH without loss of activity to form γ_2-MSH. No specific function has been attributed to the N-terminal lysine.

4.2. Melanocortins with Increased Potency

A so-called alanine-scan in the α-MSH molecule, in which each residue individually is replaced by alanine and the peptide tested for binding and tyrosinase activity, confirmed that the sequence 4–9 is crucial for bioactivity as well as Met^{4} whose replacement led to a considerable loss in activity *(189)*. However, Nle^{4} or Nva^{4} do not or only minimally affect the biologic activity of α-MSH and of ACTH *(8)* and are frequently used in peptide analogs for radioiodination. Whereas most other structural modifications or changes of configuration at individual residues along the peptide chain of α-MSH did not yield compounds with markedly increased affinity, the introduction of the D-Phe^7 residue *(190)* led to a class of MSH compounds with very high bioactivity and receptor affinity, combined with increased stability in the circulation. In particular, [Nle^4, D-Phe^7]-α-MSH *(190)* is currently the most universally used agonist of α-MSH with a potency of 5-20-fold higher than that of the native peptide, depending on the assay used *(8)*. [Nle^4, D-Phe^7]-α-MSH exhibits a characteristic "stickiness" to the receptor binding pocket *(8,191)*, resulting in a much slower dissociation than that of α-MSH. Most synthetic MSH peptide analogs developed for receptor pharmacology are based on the D-Phe^7 structure.

With the introduction of cyclic α-MSH analogs either by disulfide bridge *(192)* or lactam ring formation, a new class of MSH agonists and antagonists were introduced that became the basis for the development of highly potent MC receptor subtype-specific ligands. However, the first cyclic MSHs did not show improved characteristics as compared to linear α-MSH, because c[Cys^4, Cys^{10}]-α-MSH displays just about the same potency as α-MSH *(8)*. In certain assays, this analog is a partial agonist and its 4–10 fragment is almost inactive *(193,194)*. Introduction of a D-Phe^7 residue increases both binding and bioactivity. Highly constrained *bicyclic* MSH analogs are all less potent than α-MSH (25 to 400-fold) in the frog and lizard skin assay *(195)* indicating that a certain degree of flexibility is required for the stimulation of MC1 receptors.

From a number of cyclic lactam analogs of α-MSH *(196,197)*, [Nle^4, *c*{Asp^5, D-Phe^7, Lys^{10}}]-α-MSH[4–10] (Melanotan-II, MT-II) was found to be the most potent with a 100-fold higher bioactivity in the lizard skin assay than that of α-MSH. Variation of the ring size of cyclic analogs reduced the potency of the analogs. MT-II displayed high melanogenic activity also in melanoma cells and in human skin in vivo *(198)*. In cells expressing transfected human MC1-R, the dissociation rate of [Nle^4, *c*{Asp^5, D-Phe^7, Lys^{10}}]-α-MSH[4–10] from hMC1-R is much lower than that of α-MSH and even lower than that of [Nle^4, D-Phe^7]-α-MSH [191]. Exchange of the D-Phe^7 residue in [Nle^4, c{Asp^5, D-Phe^7, Lys^{10}}]-α-MSH[4–10] by D-Phe(p*I*) or D-Nal(2) led to compounds with potent antagonist activity at the MC4-R and weak antagonist activity at the MC3-R (*see* 4.4.).

Fatty acid conjugates of α-MSH fragments exhibit prolonged biologic activity *(199)*: When palmitoyl, myristoyl, decanoyl, and hexanoyl chains are coupled to the N-terminus of the cyclic (Asp-Lys) lactam-bridged analog *H*-Asp-His-D-Phe-Arg-Trp-Lys-*NH2*, shorter fatty acid chains do not affect the biologic activity of the analog in the lizard skin assay, the longer fatty acids decrease it, but all peptides displayed markedly prolonged activity. In mouse melanoma cells, the analogs were 10–100 times more potent than α-MSH *(199)*. Whether this increase will lead to a more or less favorable tissue distribution of the compounds in vivo has not yet been examined.

Minimum energy calculations of α-MSH gave some first insights into possible conformations of the linear peptide *(200)*. More recently, the biologically active conformations of the tetrapeptide core His-Phe-Arg-Trp were calculated *(201)*, which demonstrated that the aromatic moieties of the His^6, Phe^7, and Trp^9 side chains form a continuous hydrophobic "surface," presumably interacting with a complementary receptor site. Solution structures of α-MSH determined with two-dimensional NMR spectroscopy and dynamic simulate-annealing calculations show that the α-MSH forms a hairpin loop conformation that includes the His-Phe-Arg-Trp part, whereas [Ahx^4, Asp^5, D-Phe^7, Lys^{11}]-α-MSH prefers a type I β-turn comprising the first four residues *(202)*.

In another approach using the novel high-affinity ligands interacting with mutants of the MC receptors, sites of contact between ligand and receptor could be inferred from the binding data. For example, [Nle^4-*c*{Asp^5, D-Phe^7, Lys^{10}}]-α-MSH[4–10], α-MSH, and γ-MSH were tested with MC1 receptor mutants in which acidic residues in TM 2 and TM3 (e.g., Glu^{94}, Asp^{117}, Asp^{123}) or aromatic residues (e.g., Phe^{175}, Phe^{196}, and Phe^{257} or different tyrosine residues) in TM4, TM5, and TM6 participate in binding of MSH *(203)*. Multiple mutagenesis or introduction of a positive charge instead of a negative charge also changed the ratio of agonist potency vs binding affinity of these peptides. In human MC5, Gln^{235}, and Arg^{272} appear to be responsible for the low affinity of α-MSH to this receptor; if mutated to those residues that are conserved in other MC receptors (i.e., K^{235} and C^{272}), the ligand affinity of α-MSH is increased 10-fold for Q235K or, respectively, 690-fold for R272C *(204)*. In human MC1-R, Asp^{184} is crucial for interaction with [Nle^4, D-Phe^7]-α-MSH because replacement by Ala completely abolishes MSH binding *(205)*; similarly, Ser^6, Glu^{269}, and Thr^{272} are important for MSH binding.

4.3. Small Peptide and Nonpeptide Melanocortin Ligands

Small peptide ligands for frog and lizard melanocortin receptors have been studied extensively *(8)*. More recently, new MC receptor ligands were discovered using a combinatorial chemistry-based diffusion assay for screening of a tripeptide library *(165)*. The tripeptide *H*-D-Trp-Arg-Leu-NH_2 turned out to be a receptor antagonist with a *K* in the micromolar range (Table 7). Systematic modification of the tetrapeptide His-Phe-Arg-Trp produced two analogs with micromolar binding activity at the hMC1-R (Table 7) and two tripeptides, Ac-D-Phe-Arg-Trp-NH_2 and Ac-D-Phe-Arg-D-Trp-NH_2, with similar activity at the MC4-R *(206)*. Nonpeptide molecules were developed by Heizmann et al. *(207)* who identified a number of novel MC1 ligands from a large peptoid library active in the micromolar range (Table 7). Some of these tripeptoids display antagonistic activity in the adenylate cyclase assay and, when integrated into a partial α-MSH sequence, they show mixed agonist/antagonist activity in the nanomolar range.

Table 7
Small Molecular Weight Peptides and Peptoids With hMC1 Receptor Affinity

Peptoid/Peptide	Sequence (R_1-R_2-R_3)	K_D (μmol/l)
Peptoid 1	H-60-21-66-NH_2	1.58 ± 0.11
Peptoid 2	H-60-21-61-NH_2	1.93 ± 0.22
Peptoid 3	H-60-21-60-NH_2	1.99 ± 0.11
Peptoid 4	H-60-20-66-NH_2	2.07 ± 0.55
Peptoid 5	H-66-21-60-NH_2	4.06 ± 0.99
Peptoid 6	H-60-21-35-NH_2	6.48 ± 1.73
Peptoid 7	H-60-21-54-NH_2	7.04 ± 0.90
Peptoid 8	H-60-21-11-NH_2	7.24 ± 1.64
Peptoid 9	H-61-21-66-NH_2	27.2 ± 0.76
Peptide 1	H-D-Trp-Arg-Leu-NH_2	3.69 ± 1.24
Peptide 2	Ac-His-D-Phe-Arg-Trp-NH_2	0.6
Peptide 3	Ac-His-Phe-Arg-D-Trp-NH_2	6.3

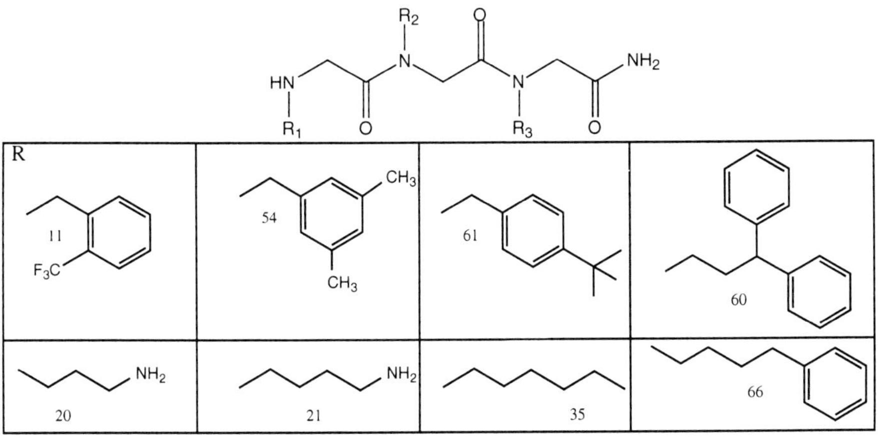

The numbers of the peptoid sequences relate to the structural elements in the box. The general structure of the tripeptoids is given above. All data originate from Heizmann et al. *(206)*, except for peptide 2 and peptide 3 which originate from Haskell-Luevano et al. *(205)*.

4.4. Agonists, Antagonists, Inverse Agonists

Antagonist acitivty of α-MSH analogs has been reported repeatedly in the past *(208)*, but usually such activity was confined to lower vertebrate melanophore MC receptor systems *(8)*. The shortest peptides with demonstrated MSH antagonist activity at amphibian MC receptors are D-Trp-Nle-NH_2 and D-Trp-Arg-NH_2 as well as some structural analogs *(209)*. At the human MC1 receptor, they are devoid of activity. Another potent MSH antagonist at amphibian MC receptors is [Arg^8, D-$Trp^{7,9}$, *N-methyl*-Phe^8]-substance P *(209)*. The cyclic lactam-bridged heptapeptide, [Nle^4-*c*{Asp^5, D-Phe^7, Lys^{10}}]-α-MSH[4–

10], is a potent agonist in both frog and mammalian MC1 receptors (*see above*). Replacement of D-Phe^7 by bulky amino acids such as D-p-iodophenylalanine or D-2'-naphthylalanine leads to potent antagonists in the frog skin assay ($pA_2 \approx 10.3$) *(210)*. Both peptides, [Nle^4-*c*{Asp^5, D-Phe(pI)^7, Lys^{10}}]-α-MSH[4–10] and [Nle^4-*c*{Asp^5, D-Nal(2)^7, Lys^{10}}]-α-MSH[4–10] are potent antagonists at the mammalian MC4 receptor (pA2 ≈ 9.3) and less potent antagonists at the MC3 receptor (pA2 ≈ 8.3) but full agonists at the MC1-R (Table 6). On the other hand, *p*-chloro-and *p*-fluorophenylalanine^7 derivatives are full agonists.

Replacement of the lactam bridge of [Nle^4-c{Asp^5, D-Nal(2)^7, Lys^{10}}]-α-MSH[4–10] by disulfide bridge increased MC4-R selectivity but decreased the overall affinity to MC receptors *(211)*. Enlarging the disulfide ring structure by one residue, that is, bridge formation between position 4 and 11 with incorporation of a D-Nal^7 residue (Table 6) led to a peptide with weak MC1-R agonist but potent MC4-R antagonist activity *(212)*. This MSH antagonist, *c*[Cys^4, D-Nal^7, Cys^{11}]-α-MSH[4–11], increases food intake in free-feeding rats *(213)*. The corresponding compound with a 29-membered ring instead of a 26-membered ring, *c*[Cys^3, D-Nal^7, Nle^{10}, Cys^{11}]-α-MSH[3–11], had highest affinity for the MC3 receptor *(212)*. Further development of this compound produced an analog, *c*[Cys^3, Nle^4, Arg^5, D-Nal^7, Cys^{11}]-α-MSH[3–11], which exerts antagonist activity at all four MC receptor subtypes, MC1-R, MC3-R, MC4-R and MC5-R *(214)*.

From the α-MSH[5–13] sequence, a peptide library consisting of 31,360 structurally different peptides was generated and many of them individually screened for antagonistic activity *(215)*. This led to the identification of a potent antagonist, Met-Pro-D-Phe-Arg-D-Trp-Phe-Lys-Pro-Val-NH_2 with an IC_{50} of 11 ± 7 n*M*. Crucial determinants for antagonistic activity reside in positions 5–6, 7–9, and 10 of the α-MSH molecule, as D-Trp^5 and Phe^6 were shown to be indispensable elements, whereas D-Phe^3 potentiated the effect *(215)*. The tripeptide MSH antagonist, D-Trp-Arg-Leu-NH_2, and the tripeptoid antagonists are briefly described in Section 4.3.

The physiology and mechanism of action of peripherally produced agouti protein (AP) or agouti signaling protein (ASP) as well as its counterpart in the brain, agouti gene-related protein (AGRP) is presented in Chapter 17 of this volume. In the context of MC receptor antagonists, agouti protein does not simply act as MC1-R antagonist of MSH but rather as inverse agonist, as demonstrated by Siegrist et al. *(216,217)*. Whereas agouti suppresses MSH-induced cAMP, tyrosinase, and melanin production and even lowers basal melanogenic activity, it affects cell proliferation and MC1-R downregulation in the same way as α-MSH. Hence agouti does not simply block MC1-R but elicits a dual form of signaling into the cell, through interaction with MC1-R.

4.5. Analogs for Therapeutic Application

Whereas α-MSH has not yet been introduced to the clinic, ACTH is frequently applied in medicine, in particular for functional tests, for example, of the adrenal cortex, or for the treatment of neurologic and rheumatoid disorders, skin diseases, and disorders of the gastrointestinal tract as well as for adjuvant therapy of oncologic patients. Several analogs of ACTH have been introduced, one of the most frequently used being Synacthen, the ACTH[1–24] fragment. The ACTH[1–17] analog, [β-Ala^1, Lys^{17}]-NH-$(CH_2)_4$-NH_2, was reported to be an efficient antiallergic agent.

With regard to MSH peptides, there is a considerable interest to develop specific MC receptor agonists and antagonists for treatment of disorders of weight homeostasis, such as obesity or anorexia, and for the control of the cardiovascular system (see Chapter 15). The agonist peptide Melanotan-II, [Nle^4, *c*{Asp^5, D-Phe^7, Lys^{10}}]-α-MSH[4–10] received attention recently because of its twofold effect in humans: skin tanning and penile erection. Skin tanning is provoked after five low doses of 0.01–0.025 mg/kg given every other day *(218)*. Slightly higher doses (0.03 mg/kg) produce somnolescence and fatigue or yawning and stretching, which correlated with spontaneous penile erections. Indeed, injection of 0.025 mg/kg of Melanotan-II initiated erections in men with psychogenic erectile dysfunction *(219)*; the side effects appear to be manageable.

Future clinical application of α-MSH or ACTH[1–24] agonist peptides may involve conditions of transient cardiac hypoxia and reoxygenation such as occur in coronary artery disease. MSH or ACTH (at 160 μg/kg IV) immediately increased cardiac output, heart rate, mean arterial pressure, and pulse pressure with full recovery of EEC in experimental rats that had been subjected to a 5-min period of ventilation interruption. These rats invariably died from cardiac arrest within 6–9 min of resumption of ventilation *(220)*. Another area of future clinical application of MSH-like peptides is the neurotrophic and neuroprotective potential of melanocortins that may become useful in treating spinal cord injury; in an experimental model the peptide signficantly improved recovery in animals *(221)*.

A new area of potential therapeutic application is the use of MC1 receptor partial sequences for antimelanoma immunization: HLA-A2-restricted CTL epitopes of the human MC1 receptor were selected and some of them were found to induce peptide-specific CTLs from peripheral blood mononuclear cells of healthy HLA-A2$^+$ donors after repeated in vitro stimulation with peptide-pulsed antigen-presenting cells *(222)*.

4.6. Melanocortin-Toxin Conjugates

A recombinant fusion protein containing the peptide sequences of α-MSH, the catalytic (cytotoxic) A-chain and the lipophilic part of the B-chain of diphtheria toxin was shown to exhibit cytotoxic activity on melanoma cells in vitro *(223)*. In vitro binding to mouse and human melanoma tumor sections of the fusion protein was compared with that of [Nle^4, D-Phe^7]-α-MSH and α-MSH: in mouse melanoma sections, the fusion protein was 15-fold/3-fold less potent than the two peptides and in human melanoma specimens, the potency was 6-fold/ 5-fold lower, indicating that the fusion protein retained good MC1-R recognition also in tumors *(224)*. A clinical study with this fusion protein was not yet reported. In another approach, Ghanem et al. *(225)* prepared a covalent conjugate between α-MSH and melphalan which they found to exert receptor-mediated cytotoxicity on human melanoma cells. Similarly, melphalan attached to the 4–10 sequence of α-MSH also appeared to exert its toxic effect via a receptor-mediated mechanism *(226)*. Yet, these MSH-toxin conjugates were not tested in man, most likely because of insufficient in vivo selectivity and/or potency as toxins. A prerequisite for such studies is the availability of MC1-R-selective ligands.

5. Labeled Melanocortins for Receptor Analysis

5.1. Radiolabeled Melanocortins

The first radiolabeled melanocortin peptides were prepared by incorporation of [^{14}C]phenylalanine^7 or [^3H]tyrosine^2 into ACTH or, respectively, α-MSH (reviewed in ref. *8*). Then several preparations of tritiated melanocortin peptides were reported using halogen–tritium exchange at tyrosine or reduction of double bonds of α-MSH- or ACTH-precursor molecules containing either diiodotyrosine or allylglycine residues (*see* ref. *8*). These peptides exhibited specific radioactivities of up to >50 Ci/mmol. Highly tritiated α-MSH molecules with >100 Ci/mmol or even >200 Ci/mmol were obtained by catalytic reduction of acetylenic bonds with tritium gas, using precursor molecules that contained either one propargylglycine (Pra) yielding a radioligand with four ^3H atoms or two Pra residues and producing a radioligand with eight ^3H atoms (Table 6). These compounds were used for receptor mapping in the developing rat brain *(8)*.

Much higher specific radioactivities were obtained with ^{125}I incorporated into Tyr^2 of α-MSH, Tyr^5 of β-MSH or Tyr^{23} of ACTH[1–24], yielding tracer molecules of >2000 Ci/mmol *(8)*. However, the biologic activity of monoiodinated α-MSH (or β-MSH) could only be retained by using a peptide whose Met^4 was either replaced by Nle^4 or which was carefully reduced after partial oxidation during the iodination procedure *(170)*. Radioiodination of Tyr^2 of ACTH[1–24] or 1–39 generally led to inactive radioligands; inactivation could be avoided *(171)* by radioiodination of Tyr^{23} of a precursor whose Tyr^2 was replaced by Phe^2 and its Met^4 by Nle^4 (see above). An important factor in the preparation process of melanocortin radioligands is the careful purification to homogeneity by HPLC. All the different α-MSH radioligands, [^{125}I]-α-MSH, [^{125}I]-[Nle^4]-α-MSH, [^{125}I]-[D-Phe^7]-α-MSH or [^{125}I]-[Nle^4, D-Phe^7]-α-MSH (Table 6), retained the full biologically activity of their parent peptides. [^{125}I]-[Nle^4, D-Phe^7]-α-MSH was the most active and hence most frequently used for binding assays, for the study of MC receptor expression and receptor up- and downregulation (*see* Chapter 17). In mouse melanoma tissue sections, [^{125}I]-[Nle^4]-α-MSH produced excellent autoradiography results and its binding characteristics in the tissue ($K \approx 1$ nmol/L; $B_{max} \approx 15{,}000$) were very similar to those found with isolated cells *(227)*. For human melanoma tissue autoradiography, [^{125}I]-[Nle^4, D-Phe^7]-α-MSH was the preferred radioligand *(228)*, although with isolated human melanoma cells MC1-R studies may lead to different results when either [^{125}I]-α-MSH, [^{125}I]-[D-Phe^7]-α-MSH or [^{125}I]-[Nle^4, D-Phe^7]-α-MSH is used *(170)*.

Iodination of α-MSH or [Nle^4, D-Phe^7]-α-MSH using *N*-succinimidyl-3-[^{125}I]iodobenzoate (SIB), with which the Lys^{11} side-chain of α-MSH is acylated, was reported to lead to tracer molecules which are more resistant to dehalogenation reaction in vivo and which exhibit an up to 10-fold lower dissociation constant in vitro when compared with MSH tracer molecules containing monoiodinated Tyr^2 *(229)*.

5.2. MSH-Radiopharmaceuticals for Potential Clinical Application

Radioactive MSH tracers for in vivo application were developed on the basis of peptide analogs conjugated to chelators for heavy metal radionuclides such as diethylenetriaminopentaactic acid (DTPA) *(230,231)* or 1,4,7,10-tetraazacyclododecane-1,4,7,10-tetraacetic acid (DOTA) *(232)*. These chelators were usually attached to the N-terminus of α-MSH (Table 6), yielding compounds with good bioactivity that represent potential imaging agents. For example, bis-α-MSH-DTPA was found to be equipotent with α-MSH in the in vitro tyrosinase assay using Cloudman S91 cells *(233)*. When tested in vivo with S91-tumor-bearing mice, uptake by the tumor was significantly higher than by other organs *(233)*, and, when injected into 15 patients who were shown to have a total of 46 melanoma lesions, 41 of these lesions (89%) were imaged with the [^{111}In]DTPA-labeled derivative, without any false positives, and in two cases the scan was instrumental for correct diagnosis *(234)*. An improved derivative based on [Nle^4, D-Phe^7]-α-MSH appeared to yield even better results *(235)*. Bagutti et al. *(236)* developed DTPA-α-MSH[4–10] fragment analogs which showed improved tumor : tissue ratios *(236)*, but mono-

MSH-DTPA analogs were superior ligands for in vivo tumor imaging compared to bis-MSH-DTPA.

Labeling of [Nle^4, D-Phe^7]-α-MSH with ^{18}F using *N*-succinimidyl 4-[^{18}F] fluorobenzoate yielded [Nle^4, D-Phe^7, Lys^{11}[^{18}F]PFB]-α-MSH, which retained almost the same potency in the B16-F1 mouse melanoma cell binding assay as the parent [Nle^4, D-Phe^7]-α-MSH (IC_{50} 112 p*M* vs 82 p*M*), suggesting that addition of 4-fluorobenzoate to Lys^{11} did not compromise MSH receptor binding affinity *(237)*. The normal tissue clearance of [Nle^4, D-Phe^7, Lys^{11}[^{18}F]PFB]-α-MSH in mice was quite rapid, with little evidence for defluorination. Other radionuclides such as rhenium or technetium were incorporated into α-MSH fragment analogs either via the peptide chelator *N*-acetyl-Cys-Gly-Cys-Gly attached to the N-terminus of the α-MSH molecule *(238)* or via insertion into the disulfide bridge of cyclic [Cys^4, Cys^{10}]- or [Cys^5, Cys^{10}]-α-MSH derivatives, thus forming a thiolate–metal–thiolate bridge *(239)*. These analogs were chemically stable and biologically active and may, perhaps, become useful for tumor targeting.

5.3. Photoreactive Melanocortins for Receptor Identification

Photocrosslinking of MC1 receptors on frog and lizard melanophores with α-MSH derivatives containing one or two photoreactive groups at positions 1, 7, 9, or 13 of the molecule was shown to induce long-lasting receptor stimulation *(8,240,241)*. It seems therefore that stimulation of MC1-R on frog and lizard melanophores does not lead to receptor downregulation within the time-period of the experiment. In fact, the intracellular signal can be shut down by deprivation of the system from extracellular Ca^{2+} or addition of α_2-adrenergic agonist; addition of normal buffer following such shutdown restores the signal *(8)*. Simultaneous crosslinking of α-MSH to lizard MC1-R via three photoreactive groups can either lead to irreversible receptor activation or receptor inhibition: [ApSSpr-Ser^1, D-Pap^7, Pap^{13}]-α-MSH with the photolabels in position 1, 7, and 13 induced irreversible stimulation, whereas [ApSSpr-Ser^1, Trp(Naps)^9, Pap^{13}]-α-MSH with the photolabels in positions 1, 9, and 13 led to long-lasting inhibition of pigment dispersion *(240)*. If the latter compound after crosslinking to the receptor was treated with β-mercaptoethanol, thereby cleaving the photolabel at position 1, the long-lasting inhibition was transformed to long-lasting stimulation *(241)*. This demonstrates that the 'tight' or altered crosslinking of MC1-R ligands to the receptor may change the characteristics of an MSH agonist into those of an antagonist and vice versa. It should be noted that irreversible MC1-R stimulation could also be obtained by introduction of a phenylalanine mustard into α-MSH fragments *(242)*.

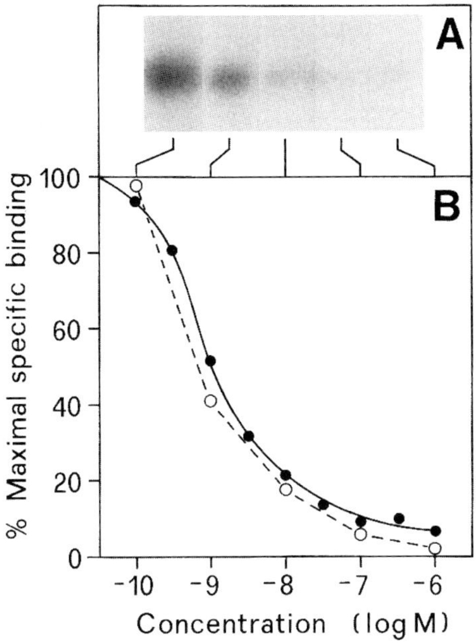

Fig. 2. Quantitative autoradiography of an SDS-polyacrylamide gel analysis of the photolabeled 45 kDa MC1R protein of mouse B16-F1 melanoma cells using [^{125}I]-[Nle^4, D-Phe^7, Trp(Naps)^9]-α-MSH as photoprobe; comparison with the displacement curve in the binding assay. A: Autoradiogram after gradual (prephotolysis) displacement of the label with increasing concentration of α-MSH. B: densitometric quantification of the autoradiograms (———) and displacement curve in the binding assay (----------).

MC1 receptors on mouse and human melanoma cells were biochemically characterized by photocrosslinking *(25–27)*. Using [^{111}I]-[Nle^4, D-Phe^7, Trp(Naps)^9]-α-MSH as photoradioligand, with which the extent of receptor labeling exactly paralleled the liganδ–receptor binding characteristics (Fig. 2), the method proved ideal to compare MC1-R on various mouse and human melanoma cells. The molecular weights of most MC1 receptors determined by sodium dodecyl sulfate (SDS)-polyacrylamide gel electrophoresis were around 45 kDa. The W4 variant of B16 mouse melanoma cells, which have a low sialic acid content, showed a lower molecular weight *(26)* and another variant, the B16-M2R clone, displayed a doublet band of approx 43 and 46 kDa for the MC1-R *(243)*, in contrast to the single bands of B16-F1 and F10 cells *(26)*. On the other hand, MC1-R of Cloudman S91 mouse melanoma or human HBL melanoma cells generally showed higher molecular weights

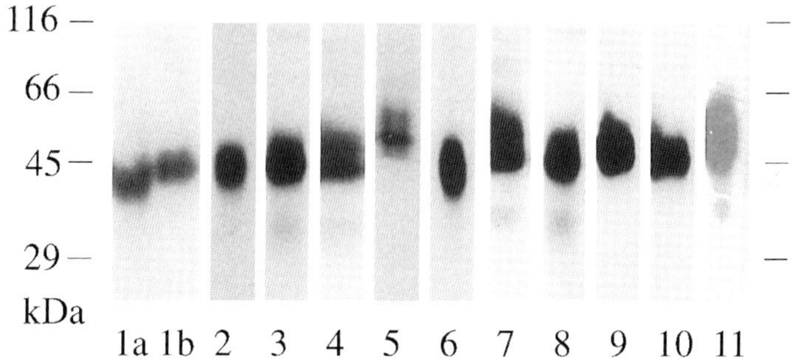

Fig. 3. Comparison of MC1R labeling of different melanoma cells using [^{125}I]-[Nle^4, D-Phe^7, Trp(Naps)^9]-α-MSH as photoprobe. *1a:* B16-W4; *1b:* B16-F1; *2:* B16-F5; *3:* B16-F10; *4:* B16-M2R; *5:* Cloudman S91; *6:* B16-F1 subclone; *7:* human 205; *8:* human D10; *9:* mouse G4F containing human MC1R; *10:* B16-F10 (for comparison); *11:* human HBL.

(Fig. 3). This demonstrates that there is considerable variation of glycosylation of the MC1-R on different melanoma cells and that this has little or no effect on liganδ-receptor binding.

In another approach, radioiodinated α-MSH containing a biotinyl group at the N-terminus and a photoactivatable group on the Lys^{11} side chain, into which a disulfide group had been incorporated, was crosslinked with the MC1-R of B16 mouse melanoma cells. The cells were homogenized and the receptor solubilized in Triton X-100 and bound to magnetic beads containing streptavidin *(244)*. A doublet-band of labeled receptor of approx 43–46 kDa was identified and partially purified. Using Cloudman S91 cells, crosslinking of external and internalized MC1 receptor showed identical molecular weights of 50–53 kDa *(245)*. Intracellular and membrane receptors share common antigenic determinants, which indicates a structural relationship between the two populations of receptor.

5.4. Carrier-bound and Fluorescent Melanocortins

Proteins or other polymers to which bioactive peptides are covalently attached are useful carriers for receptor studies. For example, the possibility of coupling a large number of marker groups to the carrier without harming the attached peptide (and hence the bioactivity of the conjugate) makes such conjugates ideal tools for receptor localization and characterization. The first conjugates between α-MSH and human serum albumin or thyroglobulin were shown to retain biological activity *(8)* and served as model for much larger conjugates. Tobacco mosaic virus to which approx 300 copies of α-MSH had been chemically linked showed dramatically increased potency (>1000-fold) and long-lasting receptor stimulation *(8)*. On the other hand, a 1:1 complex between α-MSH and PEG-5000 retained only 1% of the activity of α-MSH *(8)*. Dimers and trimers of α-MSH displayed similar or slightly higher activity than α-MSH, depending on the type and length of spacer used to link the peptides *(246,247)*. An α-MSH antagonist was changed to an agonist by dimerization *(246)*.

Biotinyl-[β-Ala[1], Lys[17]]-ACTH[1–17]-NH-$(CH_2)_4$-NH_2 *(248)*, which displayed very high potency to MC1 receptors expressed on human melanoma cells (*K* 0.02 ± 0.005 nmol/L for D10 cells and 0.21 ± 0.02 nmol/L for HBL cells) and derivatives of [Nle[4], D-Phe[7]]-α-MSH containing biotinyl and fluorescent groups *(249)* were developed for attachment to carriers suitable for receptor localization and human melanoma targeting. A carrier-α-MSH conjugate was applied to treatment of experimental melanoma *(250)*: a hydroxypropylmethacrylamide copolymer to which doxorubicin and α-MSH were coupled displayed biological activity in vitro, and the conjugate was significantly more effective in vivo when applied to melanoma tumor-bearing animals than polymers containing only doxorubicin or than nonconjugated drug. Yet, undesired deposition of such conjugates may occur and therefore this approach needs further studies.

Multiple copies (10–20) of both [Nle[4], D-Phe[7]]-α-MSH and fluorescein isothiocyanate (FITC) as fluorophore were conjugated to polyvinyl alcohol (PVA) yielding the multivalent macromolecular conjugate (FITC-PVA-MSH) *(251)*. This carrier specifically labeled MC1-R on human epidermal melanocytes and keratinocytes *(252)* and human melanoma cells *(253)* but not cells from other origin. Binding of the conjugate to the cells exhibited a unique cluster pattern (capping) suggesting a receptor internalization related phenomenon. Most importantly, every cell of all melanoma cell lines tested, melanotic or amelanotic, possessed receptors as visualized by fluorescence microscopy. Since the cultivated cells were not synchronized, some binding apparently took place during all phases of the cell cycle. Therefore, receptor expression appears not to be cell cycle-dependent. This shows that fluorescent melanotropin conjugates might prove useful for determining the state of MC1-R expression on human melanoma tumors and hence for melanoma diagnosis.

6. Proopiomelanocortin, the Precursor for the Melanocortins

Proopiomelanocortin was the first polyprotein discovered which serves as precursor for several functionally different peptides. Its existence was demonstrated by a combination of immunoprecipitation and SDS/PAGE-separation experiments of biosynthetic products obtained from AtT-20 mouse pituitary tumor cells *(20)* or by translating mRNA isolated from AtT-20 cells or ectopic ACTH-producing human tumor cells in a reticulocyte cell-free system *(21)* using antibodies against ACTH, β-endorphin and γ-MSH (*see* ref. *8*). POMC was also the first prohormone whose entire mRNA sequence *(23)* and gene structure *(254)* was determined by DNA sequence analysis. More recently, POMC-like proteins were also discovered in nonvertebrate species.

6.1. Gene and Protein Structure of POMC

The structural organization of the human POMC gene has been reviewed by Chang et al. *(255)*. The gene is located on a 7.8 kb segment that consists of three exons separated by a 3.9- and 2.8-kb intron. The three exons consist of 87, 152, and 663 base pairs, which, after transcription and splicing of the pre-mRNA, make up the mature mRNA of 1150–1200 bases from which the corresponding pre-POMC of 267 amino acid residues is translated. Except for the signal sequence and the first 18 amino acid residues of POMC, located on exon 2, the genomic DNA information for POMC is all contained within exon 3.

Table 8 lists the amino acid sequence of 18 different POMC molecules, including seven mammalian, four amphibian and seven fish molecules, as well as proopiocortin (POC) and proopiomelanotropin (POM) from the lamprey. The shortest precursor with 226 residues is POMC-A from the salmon and the longest with 320 residues is POMC from the dogfish. Whereas the mammalian, amphibian, and elasmobranch species express a POMC containing the three sequence types of α-, β-, and γ-MSH, the teleost POMC lacks the γ-MSH region. In several species, two forms of POMC were found in the pituitary gland. Additional forms of POMC were also found in other tissues (*see below*). In Table 8, the sequences of ACTH, MSH, and β-endorphin within the 20 different POMC precursor molecules are specifically visualized.

6.2. POMC Isoforms and Mutants

Human ACTH-secreting pituitary tumors were shown to secrete three forms of POMC-mRNA: the normal size of 1150–1200 bases, a short variant of approx 800 bases, and a long form of approx 1500 bases *(256)*. Longer or shorter forms of POMC have also been reported to occur in nonpituitary ACTH-producing tumors, pancreatic and other peripheral tumors, testis, epididymis, ovaries, placenta, and in the brain (*see* ref. *8*). At least two different nonallelic gene products of POMC were discovered in the mouse and rat pituitary gland *(257)*. Several POMC variants or mutant forms were also found in humans: one of them contains a 9-bp deletion, corresponding to the loss of the Ser-Ser-Gly sequence between residues 67–73 *(258)*. Expression and processing of the variant form in Chinese hamster ovary cells was considerably reduced as compared to normal POMC *(259)*. The existence of another variant of POMC-mRNA was suggested for the human epidermis *(260)*. Severe early-onset obesity, adrenal insufficiency and red hair pigmentation can be caused by POMC mutations which show disturbed processing of the POMC molecule (lack or reduced amount of circulating ACTH) or whose mRNA cannot be translated *(261)*. Finally, a number of different polymorphisms in the POMC gene were discovered which led to insertions or deletions of residues or truncation of POMC *(262)*; however, these mutations could not readily be associated with either obesity of anorexia nervosa.

6.3. Biosynthesis and Processing of POMC

The biosynthesis of POMC includes several posttranslational processing steps (reviewed in ref. *8*), such as

1. *N*-Glycosylation of Asn^{65} in the γ_3-MSH region and Asn^{136} within the CLIP region (the numbering relates to mature human POMC, that is, without signal peptide)
2. Disulfide bond formation between Cys^{2}-Cys^{24} and Cys^{8}-Cys^{20}
3. Phosphorylation of Ser^{142} within the CLIP region
4. *O*-Glycosylation of Thr^{45} in the N-POMC(1–49) region
5. Sulfation of carbohydrate moiety(ies)

POMC is variably glycosylated, phosphorylated or sulfated in different species. These modifications do not affect biosynthesis and processing of the prohormone or release of the mature peptides but lead to heterogeneity of circulating peptides *(8)*.

Cleavage of POMC by paired basic amino acid convertases PC1 and PC2 proceeds in the following way *(263)*: (i) PC1 cleaves POMC into *N*-POMC/ ACTH and β-LPH and *N*-POMC/ACTH into *N*-POMC, JP and ACTH; (ii) PC2 cleaves β-LPH into β-endorphin and γ-LPH as well as ACTH into either α-MSH or desacetyl-α-MSH (Fig. 1). Thus, PC1 catalyzes the early steps of POMC processing, whereas PC2 catalyzes the later steps *(264)*. In general, even though both convertases can cleave POMC in cells devoid of secretory granules, POMC processing is more efficient in cells containing secretory granules *(263)* and the acidic milieu of secretory granules is favorable but not indispensable for processing by PC1 *(265)*. Basic amino acid residues at cleavage sites are then removed by carboxypeptidase H (CPH) and glycine-extended C-terminal ends of γ- and α-MSH are amidated by peptidylglycine α-amidating monooxygenase (PAM) *(266)*. PAM is a bifunctional enzyme containing peptidylglycine α-hydroxylating monooxygenase (PHM) and peptidyl-α-hydroxyglycine α-amidating lyase (PAL). PHM catalyzes the conversion of peptidylglycine termini into α-hydroxyglycine intermediates. This process is

Table 8
Amino Acid Sequence of Proopiomelanocortin From Different Species[a]

Homo sapiens (man)	1	mprsccsrsg	alllalllqa	smevrgwcle	ssqcqdltte	snllecirac	kpdlsaetpm	60
Macaca nemestrina (monkey)	1	mprsccsrsg	alllalllqa	smevrgwcle	ssqcqdltte	snllecirac	kpdlsaetpv	60
Sus scrofa (pig)	1	mprlcgsrsg	allltlllqa	smgvrgwcle	ssqcqdlste	snllacirac	kpdlsaetpv	60
Bos taurus (ox)	1	mprlcssrsg	alllalllqa	smevrgwcle	ssqcqdltte	snllacirac	kpdlsaetpv	60
Rattus norvegicus (rat)	1	mprfcnsrsg	alllalllqt	sidvwswcle	ssqcqdltte	snllacirac	rldlsaetpv	60
Mus musculus (mouse)	1	mprfcysrsg	alllalllqt	sidvwswcle	ssqcqdltte	snllacirac	kldlsletpv	60
Cavia porcellus (guinea pig)	1	mprscysrsg	tlllalllqi	smevrgwcle	ssqcqdltte	rhlleclrac	kpdlsaetpv	60
Rana ridibunda (laughing frog)	1	mlqpvwscil	allgvfifhv	gevrsqcwes	nkctdlssed	gilecikack	mdlsaespvf	60
Rana catesbeiana (bullfrog)	1	mlqpvwhaci	lailgvfifh	vgevrsqcwe	snkctdlsse	dgilecikac	kmdlsaespv	60
Xenopus laevis A (African clawed frog)	1	mfrptggcsl	ailgvfifhi	gevqsqcwes	srcadlssed	gilecikack	mdlsaespvf	60
Xenopus laevis B (African clawed frog)	1	mfrplwgcfl	ailgicifhi	gevqsqcwes	srcadlssed	gvlecikack	tdlsaespvf	60
Oncorhynchus keta A (chum salmon)	1	mvcapwllav	vvvcvcnpgv	ggqcwdsshc	kdlpsedkil	ecthlfrsgl	qdespeprsa	60
Oncorhynchus keta B (chum salmon)	1	gvipfcsvrs	crerqetcgr	tkqesersiw	lnmvcapwll	avvvvcvcnp	gvggqcwdss	60
Oncorhynchus mykiss A (rainbow trout)	1	mlcpawllav	avvgvvrgvk	gqcwenprch	dlssennlle	ciqlcrsdlt	tkspifpvkv	60
Oncorhynchus mykiss B (rainbow trout)	1	mfgtflqnqs	vrlnmvcapw	llavvvvcvc	npgvegqcwd	sshckdlpse	dkilecihlf	60
Acipenser transmontanus (sturgeon)	1	mlhpvwgcvv	avmgvlwfys	sgvqsqcweh	sqcrdlasea	nileciqack	vdlsaesplf	60
Lepisosteus osseus (gar)	1	mlrsvwvysl	glavllqqsg	reqcwehsqc	rdlsseenil	eciqacnsdl	taespifpgn	60
Squalus acanthias (dogfish)	1	mmqqsmwrsv	lvvlcmvwar	ssgqlrecwd	htkcrqltsa	pklmecieac	kvektlespi	60
Petromyzon marinus POC (lamprey)	1	mmgncsrlll	llemlsiisp	sasamcwarl	dqgcftdckk	ycsngtragt	paavlenlla	60
Petromyzon marinus PMC (lamprey)	1	mattssaptr	ssslpctval	llsgliallg	paasrsaplv	clqacescle	paqpeplcwm	60

Homo sapiens (man)	61	fpgngdeqpl	tenprkyvmg	hfrwdrfgrr	nssssgssga	gqkredvsag	edcgplpegg	120
Macaca nemestrina (monkey)	61	fpgngdeqpl	tenprkyvmg	hfrwdrfgrr	nsssgsahqk	redvaagedr	gllpeggpep	120
Sus scrofa (pig)	61	fpgngdaqpl	tenprkyvmg	hfrwdrfgrr	ngsssggggg	gggagqkree	eevaagegpg	120
Bos taurus (ox)	61	fpgngdeqpl	tenprkyvmg	hfrwdrfgrr	ngssssgvgg	aaqkreeeva	vgegpgprgd	120
Rattus norvegicus (rat)	61	fpgngdeqpl	tenprkyvmg	hfrwdrfgpr	nsssaggsaq	rraeeetagg	dgrpepspre	120
Mus musculus (mouse)	61	fpgngdeqpl	tenprkyvmg	hfrwdrfgpr	nsssagsaaq	rraeeeavwg	dgspepspre	120
Cavia porcellus (guinea pig)	61	fpggadeqtp	tesprkyvtg	hfrwgrfgrg	nssgasqkre	eeaaaadpgf	hgdgvepglr	120
Rana ridibunda (laughing frog)	61	pgnghmqpls	enirkyvmsh	frwnkfgrrn	stsndnnngg	ykredianyp	ilnlltgsdn	120
Rana catesbeiana (bullfrog)	61	fpgnghiqpl	senirkyvms	hfrwnkfgrr	nstsndnnnn	nggykredia	nypilnlflg	120
Xenopus laevis A (African clawed frog)	61	pgnghlqpls	esirkyvmth	frwnkfgrrn	ntgndgssgg	ykredisnyp	vlnlfpssdn	120
Xenopus laevis B (African clawed frog)	61	pgnghlqpls	esirkyvmth	frwnkfgrrn	stgndgsntg	ykredissyp	vfslfplsdq	120
Oncorhynchus keta A (chum salmon)	61	aqqsteesls	lgillaalts	geraldadpe	phsdkrhsys	mehfrwgkpi	ghkrrpikvy	120
Oncorhynchus keta B (chum salmon)	61	hckdlpsedk	ilecthlfrs	glqdespepr	saaqqstees	lslgillaal	tsgeraldad	120
Oncorhynchus mykiss A (rainbow trout)	61	hlqppspsds	dspplylpls	llspssplyp	teqqnsvspq	akrsysmehf	rwgkpvgrkr	120
Oncorhynchus mykiss B (rainbow trout)	61	rsglqdespe	prsaaqqste	eslslgilla	altsgerald	adpephsdkr	hsysmehfrw	120
Acipenser transmontanus (sturgeon)	61	pgnghlqpts	ediqnyvmsh	fhwntfgqrm	ngtpggskre	gastalsvll	ealsqprdev	120
Lepisosteus osseus (gar)	61	ghlqppsead	rnyakshfrs	talgrrtngs	vgsskqagen	aalsilfaal	appqaeeeme	120
Squalus acanthias (dogfish)	61	ypgnghtepi	aeslrnyvmg	hfrwnkfgkk	rgnntgfsgn	kredepvraf	lnhlpavvsq	120
Petromyzon marinus POC (lamprey)	61	cvqlkcsddg	ddndddapll	qwiasraesr	sdfdiannkw	wlvrwggqsg	lsgeggesgg	120
Petromyzon marinus PMC (lamprey)	61	qclgecsrla	apsadgseiv	llgggggdea	peggevsadk	rvqesadgyr	mqhfrwgqpl	120

Homo sapiens (man)	121	peprsdgakp	gpregkrsys	mehfrwgkpv	gkkrrpvkvy	pngaedesae	afplefkrel	180
Macaca nemestrina (monkey)	121	rgdgagpgpr	egkrsysmeh	frwgkpvgkk	rrpvkvypng	aedesaeafp	lefkreltgq	180
Sus scrofa (pig)	121	prgdgvapgp	rqdkrsysme	hfrwgkpvgk	krrpvkvypn	gaedelaeaf	plefrrelag	180
Bos taurus (ox)	121	daetgpredk	rsysmehfrw	gkpvgkkrrp	vkvypngaed	esaqafplef	kreltgerle	180
Rattus norvegicus (rat)	121	gkrsysmehf	rwgkpvgkkr	rpvkvypnva	enesaeafpl	efkrelegeq	pdgleqvlep	180
Mus musculus (mouse)	121	gkrsysmehf	rwgkpvgkkr	rpvkvypnva	enesaeafpl	efkreleger	plgleqvles	180
Cavia porcellus (guinea pig)	121	edkrsysmeh	frwgkpvgkk	rrpvkvyang	aeeesaeafp	lefkreltge	rpaaapgpdg	180
Rana ridibunda (laughing frog)	121	qntqqgimed	eavdrqdskr	sysmehfrwg	kpvgkkrrpi	kvfptdaeee	sseifplelr	180
Rana catesbeiana (bullfrog)	121	sdnqntqegi	meddaldrqd	skrsysmehf	rwgkpvgkkr	rpikvfptda	eeessesfpi	180
Xenopus laevis A (African clawed frog)	121	qnaqgdnmee	epmdrqenkr	aysmehfrwg	kpvgrkrrpi	kvypngveee	saesfpmelr	180
Xenopus laevis B (African clawed frog)	121	napgdnmeee	pldrqenkra	ysmehfrwgk	pvgrkrrpik	vypngveees	aesypmelrr	180
Oncorhynchus keta A (chum salmon)	121	assleggdss	egtfplqarr	qlgswedemv	galgnqgaka	qtkvvprtlt	vtglqdkkdg	180
Oncorhynchus keta B (chum salmon)	121	pephsdkrhs	ysmehfrwgk	pighkrrpik	vyassleggd	ssegtfplqa	rrqlgswede	180
Oncorhynchus mykiss A (rainbow trout)	121	rpvkvytngv	eeesseafps	emrrelgtdd	avypsleagt	aeggeaegme	gvfslqekkd	180
Oncorhynchus mykiss B (rainbow trout)	121	gkpighkrrp	ikvyassleg	gdssegtfpl	qarrqlsswe	demvgalgnq	gakaqtkvvp	180
Acipenser transmontanus (sturgeon)	121	ereseeeegl	qqhrrddkrs	ysmehfrwgk	pvgrkrrpvk	vypngveees	aesypaeirr	180
Lepisosteus osseus (gar)	121	esessqqqrr	edkrsysmeh	frwgkpvgrk	rrpvkvypng	veeesaeayp	temrrdlmsd	180
Squalus acanthias (dogfish)	121	tsqmedeeme	tlfprqdgkr	sysmehfrwg	kpmgrkrrpi	kvypnsfede	svenmgpelk	180
Petromyzon marinus POC (lamprey)	121	sprveqvdla	gqvesspass	ssqakrsvss	pkyamghfrw	gspdkatirk	rrpvrpntsd	180
Petromyzon marinus PMC (lamprey)	121	pgkkrqpeqs	qgvplgmgsd	enarvvnggq	awdegwtldd	qanevnarqw	saapskkdst	180

Homo sapiens (man)	181	tgqrlregdg pdgpaddgag aqadlehsll vaaekkdegp yrmehfrwgs ppkdkryggf						240
Macaca nemestrina (monkey)	181	rpragdgpdg paddgagpra dlehsllvaa ekkdegpyrm ehfrwgsppk dkryggfmts						240
Sus scrofa (pig)	181	appepardpe apaegaaara eleyglvaea eaaekkdegp ykmehfrwgs ppkdkryggf						240
Bos taurus (ox)	181	qargpeaqae saaaraeley glvaeaeaea aekkdsgpyk mehfrwgspp kdkryggfmt						240
Rattus norvegicus (rat)	181	dtekadgpyr vehfrwgnpp kdkryggfmt seksqtplvt lfknaiiknv hkkgq						
Mus musculus (mouse)	181	daekddgpyr vehfrwsnpp kdkryggfmt seksqtplvt lfknaiikna hkkgq						
Cavia porcellus (guinea pig)	181	lgfglvaeae aeaaaaekkd aaekkddgsy rmehfrwgtp rkgkryggfm tseksqtplv						240
Rana ridibunda (laughing frog)	181	relslefdyp dtnseedldd gelldgpvkk drkykmhhfr wegppkdkry ggfmtpersq						240
Rana catesbeiana (bullfrog)	181	elrrelslef dypdtnseee ldngellegp vkkgrkykmh hfrwegppkd kryggfmtpe						240
Xenopus laevis A (African clawed frog)	181	relsleldyp eidldedied neveraltkk ngnyrmhhfr wgsppkdkry ggfmtpersq						240
Xenopus laevis B (African clawed frog)	181	elsleldype idldediedn evksaltkkn gnyrmhhfrw gsppkdkryg gfmtpersqt						240
Oncorhynchus keta A (chum salmon)	181	syrmghfrwg sptaikrygg fmkpytkqsh kplitllkhi tlkneq						
Oncorhynchus keta B (chum salmon)	181	mvgalgnqga kaqtkvvprt ltvtglqdkk dgsyrmghfr wgsptaikry ggfmkpytkq						240
Oncorhynchus mykiss A (rainbow trout)	181	gsykmnhfrw sgppaskryg gfmkswders qkplltlfkn viikdgqqkr eqwgreegee						240
Oncorhynchus mykiss B (rainbow trout)	181	rtltvtglqd kkdgsyrmgh frwgsptaik ryggfmkpyt qqshkplitl lkhvtlkneq						240
Acipenser transmontanus (sturgeon)	181	dlslkldypq geeleevfgg endllnlqkk dgsykmnhfr wsgppkdkry ggfmkswder						240
Lepisosteus osseus (gar)	181	ldyplleeve eelggenevl nlqekkdgsy kmhhfrwsrp pkdkryggfm kswdersqkp						240
Squalus acanthias (dogfish)	181	reasvdfdyp vvetseagee emledakkkd gkiykmthfr wgrgpkgsaq swgpdrtqpm						240
Petromyzon marinus POC (lamprey)	181	speipdyafn gvegpaddag dsvfmsrret pdaaghrgvd eaaatgedae vgnkdgvfrv						240
Petromyzon marinus PMC (lamprey)	181	plsmqkenpe lyqmnhfrwg qppthfkqkr yggfmrkspg yahlkplvtf frdvmkndsp						240

Homo sapiens (man)	241	*mtseksqtpl vtlfknaiik naykkge* (267)
Macaca nemestrina (monkey)	241	*eksqtplvtl fknaiiknay kkgq* (264)
Sus scrofa (pig)	241	*mtseksqtpl vtlfknaivk nahkkgq* (266)
Bos taurus (ox)	241	*seksqtplvt lfknaiikna hkkgq* (265)
Rattus norvegicus (rat)	(235)	
Mus musculus (mouse)	(235)	
Cavia porcellus (guinea pig)	241	*tlfknaivkn ahkkgq* (256)
Rana ridibunda (laughing frog)	241	*tplmtlfkna iiknahkkgq* (260)
Rana catesbeiana (bullfrog)	241	*rsqtplmtlf knaiiknahk kgq* (263)
Xenopus laevis A (African clawed frog)	241	*tplmtlfkna iiknthkkgl* (260)
Xenopus laevis B (African clawed frog)	241	*plmtlfknai iknshkkgq* (259)
Oncorhynchus keta A (chum salmon)	(226)	
Oncorhynchus keta B (chum salmon)	241	*shkplitllk hitlkneq* (258)
Oncorhynchus mykiss A (rainbow trout)	241	kralgerkyh fqg (253)
Oncorhynchus mykiss B (rainbow trout)	(240)	
Acipenser transmontanus (sturgeon)	241	*sqkplltlfk nvmikdghek kgq* (263)
Lepisosteus osseus (gar)	241	*lltlfknvii kdghqkkgq* (259)
Squalus acanthias (dogfish)	241	qftnledmlq esmdndlpee evkk***dgddyk fghfrwsvpl kd***kryggfmk swdergqkpl 300
	301	ltlfrnvivk dghekkaqsq (320)
Petromyzon marinus POC (lamprey)	241	pppfkr*yggf mkvmqeidhw plvpvirkvm hkestksl* (278)
Petromyzon marinus PMC (lamprey)	241	*tmlnn* (245)

[a]Complete POMC amino acid sequences, including signal peptides, based on the EMBL protein data bank (Dec 1998). The one-letter symbols are used in the lower case for ease of legibility. Melanocortin and endorphin sequences are represented in italics, MSH peptides in addition in bold. The processing sites at dibasic residue pairs are underlined.

References: *Homo sapiens (99), Macaca nemestrina (89), Sus scrofa (98), Bos taurus (23), Rattus norvegicus (75), Mus musculus (51), Cavia porcellus (52), Rana ridibunda (76), Rana catesbeiana (77), Xenopus laevis* A *(96), Xenopus laevis* B *(96), Oncorhynchus keta* A *(97), Oncorhynchus keta* B *(100), Oncorhynchus mykiss* A *(58), Oncorhynchus mykiss* B *(58), Acipenser transmontanus (59), Lepisosteus osseus (60), Squalus acanthias (24), Petromyzon marinus POC (65), Petromyzon marinus PMC (66).*

dependent on copper, ascorbate, and molecular oxygen *(266)*. PAL catalyzes the conversion of the intermediates into α-amidated products along with the generation of glyoxylate. Like PC1, PC2 and CPH, both enzymes of PAM are dependent on the presence of a divalent metal ion. The last step of the processing of POMC is the N^{α}-acetylation of α-MSH and of β-endorphin in the neurointermediate lobe by N^{α}-acetyltransferase 1 or 2 (NAT-1, NAT-2). As has been pointed out above, there is a difference in this process between mammalian and amphibian intermediate pituitary cells: whereas in the latter cells monoacetylated α-MSH is formed during peptide secretion, in mammalian cells mono-and diacetylated forms of α-MSH are formed in an earlier step and stored until secretion *(267)*. By contrast, in both mammalian and amphibian cells, β-endorphin is stored in acetylated form. Additional information on intracellular targeting and sorting mechanisms can be found in *(268)*. A protein of particular interest for interaction with POMC at acidic pH in maturating secretory granules is carboxypeptidase E, which appears to coaggregate with POMC, thus directing the sorting and retention of secretory granule proteins during granule maturation *(269)*. It is well known that the ionic milieu controls the compartment-specific activation of the POMC processing enzymes *(270)*.

6.4. Occurrence and Regulation of POMC

The occurrence of POMC-producing cells is not confined to the pituitary gland and the brain; it is well documented that POMC-mRNA and POMC-derived peptides are also synthesized in various peripheral tissues such as the skin, testes, ovaries, placenta, adrenal medulla, gastrointestinal tract, cells of the immune system, and in tumor cells. Earlier literature on the occurrence of POMC and its peptides and on the distriubtion of melanotroph and corticotroph cells in the pituitary gland and of POMC neurons in the central nervous system has been summarized in ref. *8*; novel information on corticotroph cells in pituitary adenomas was presented in a recent case report from which it was concluded that a POMC-producing adenoma can originate from a somatotropic adenoma as a result of mutations that occurred during tumor progression *(271)*.

POMC biosynthesis is regulated by a complex network of inhibitory and stimulatory factors which differs for the different POMC-producing cell types. For example, POMC secretion from *Xenopus* intermediate lobe melanotrophs is inhibited by nerves originating from the hypothalamic suprachiasmatic nucleus that make synaptic contact with the intermediate lobe cells and contain different neurotransmitters, such as dopamine, neuropeptide Y, and GABA. Secretion from melanotrophs can be stimulated by sauvagine and TRH which are released in the neural lobe by nerve terminals originating from the magnocellular nucleus *(272)*. CRF primarily stimulates ACTH secretion from corticotrophs and γ_3-MSH secretion *(8)*. Several other

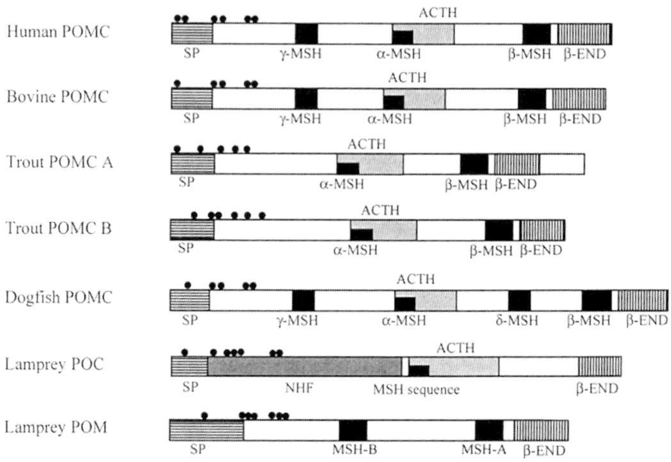

Fig. 4. Schematic diagram and evolution of the POMC family. MSH sequences (black), ACTH sequences (stippled) and β-endorphin sequences (hatched) are shown. SP, signal peptide; NHF, N-terminal glycopeptide of lamprey POMC. Closed circles show Cys residues.

factors exert modulatory effects on corticotrophs and melanotrophs, including oxytocin, vasopressin, acetylcholine, serotonin, steroid hormones, catecholamines, somatostatin, and opiates.

At the level of POMC gene regulation, there is evidence that the mechanism of POMC gene expression in ectopic ACTH-producing tumors differs from that of pituitary cells *(273)*. POMC gene expression is also regulated at the translational level which involves the recognition of the stem-loop by RNA-binding proteins *(274)*. It was also found that POMC stem-loop RNA-binding proteins specifically recognize a predicted stem-loop found in the coding region of CRF, which would suggest a new mechanism of POMC gene regulation.

6.5. Evolution of POMC

A comparison of the structural organization of POMC molecules from different vertebrate species is shown in Fig. 4. Mammalian POMCs all contain the sequence of four melanocortin peptides, that is, α-, β- and γ-MSH and ACTH. Teleost fishes (e.g., trout) lack the γ-MSH region and hence produce only three different melanocortin peptides. By contrast, POMC of the elasmobranch fish *Squalus acanthias* (dogfish) contains four different MSH peptides, that is, an additional δ-MSH besides α-, β- and γ-MSH, and also ACTH *(24)*. This means that dogfish melanotrophs potentially secretes five melanocortin peptides. The lamprey has two distinctly different POMC-like molecules: POM from which MSH-A (corresponding to β-MSH) and MSH-B (corresponding to α-MSH) are derived, and POC from which ACTH is derived. It is likely that ancestral (pre-)vertebrate POMC genes contained only one MSH-like sequence which was duplicated once, twice or three times during evolution of the vertebrates. In the tree of evolution, elasmobranch and teleost fishes had been separated about as early as elasmobranchs from mammals, which explains the differences in POMC structural organization between these classes of animals. It is however surprising that the nonvertebrate POMC from leech also contains α-, β- and γ-MSH-like and ACTH-like peptides (*see below*).

POMC-like peptides were also discovered in several nonvertebrate species, for example, in the mollusc *Planorbarius (275)*, the trematod *Schistosoma (276)*,and in the leech *(277)*. Whereas POMC of the parasite *Schistosoma* may originate from a genetic transfer of vertebrate POMC-DNA to the helminth, the POMC-like protein of the leech is unique in its structure. Leech POMC was isolated from immunocytes and characterized by sequence determination of its cDNA and of its isolated ACTH peptide *(277)*. The linear arrangement of α-, β-, and γ-MSH and of ACTH/CLIP on the POMC molecule does not differ from that of vertebrate POMC, and there is also considerable sequence identity between leech and vertebrate POMC regions. However, in leech POMC the sequence for β-endorphin is replaced by Met-enkephalin sequence. Processed leech α-MSH is biologically as active as mammalian α-MSH *(277)*. Therefore it is likely that MSH peptides exert important regulatory functions also in non-vertebrate species.

Comparison of Tables 1–4 shows that the α-and γ-MSH sequence has been preserved over a very long period of time whereas the β-MSH structure, outside the MSH core sequence has been subjected to various alterations. The same is true for the ACTH sequence in the CLIP region which also underwent a number of changes during evolution. From a comparison of mutations within the α-and the β-MSH region of different species, it was calculated *(8)* that one mutation occurred in the β-MSH structure about every 25 million years, whereas the rate of mutation in the α-MSH structure was only about 0.08 in the same time period. Similarly, ACTH[1–24] was about as stable as α-MSH, whereas CLIP had a similar frequency of mutations as β-MSH. Thus it appears that most likely the MSH core structure was formed at an early phase of evolution and thereafter underwent relatively little change.

Acknowledgments

This work was supported by the Swiss National Science Foundation and the Swiss Cancer League. I thank Dr. A. Miserez for his help with Fig. 3.

References

1. Lerner, A. B. (1993) The discovery of the melanotropins. *Ann. N. Y. Acad. Sci.* **680,** 1–12.
2. Shizume, K., Lerner, A. B., and Fitzpatrick, T. B. (1954) In vitro bioassay for melanocyte stimulating hormone. *Endocrinology* **54,** 553–560.
3. Lerner, A. B. and Lee, T. H. (1955) Isolation of a homogeneous melanocyte-stimulating hormone from hog pituitary gland. *J. Am. Chem. Soc.* **77,** 1066–1067.
4. Lee, T. H. and Lerner A. B. (1956) Isolation of melanocyte-stimulating hormone from hog pituitary gland. *J. Biol. Chem.* **221,** 943–959.
5. Geschwind, I. I., Li, C. H., and Barnafi, L. (1956) Isolation and structure of melanocyte-stimulating hormone from porcine pituitary glands. *J. Am. Chem. Soc.* **78,** 4494–4495.
6. Harris, J. I. and Lerner, A. B. (1957) Amino-acid sequence of the α-melanocyte-stimulating hormone. *Nature* **179,** 1346–1347.
7. Harris, J. I. and Roos, P. (1959) Studies on pituitary polypeptide hormones. I. The structure of α-melanocyte-stimulating hormone from pig pituitary glands. *Biochem. J.* **71,** 434–445.
8. Eberle, A. N. (1988) *The Melanotropins: Chemistry, Physiology and Mechanisms of Action.* Karger, Basel.
9. Bell, P. H. (1954) Purification and structure of β-corticotropin. *J. Am. Chem. Soc.* **76,** 5565–5567.
10. Riniker, B., Sieber, P., Rittel, W., and Zuber, H. (1972) Revised amino acid sequences for porcine and human adenocorticotropic hormone. *Nature New Biol.* **235,** 114–115.
11. Li, C. H., Barnafi, L., Chrétien, M., and Chung, C. (1966) Isolation and amino acid sequence of β-LPH from sheep pituitary glands. *Nature* **208,** 1093–1094.
12. Chrétien, M., Gilardeau, C., and Li, C. H. (1972) Revised structure of sheep β-lipotropic hormone. *Int. J. Pept. Protein Res.* **4,** 263–265.
13. Chrétien, M. and Li, C. H. (1967) Isolation, purification and characterization of γ-lipotropic hormone from sheep pituitary glands. *Can. J. Biochem.* **43,** 1163–1174.
14. Scott, A. P., Ratcliffe, J. G., Rees, L. H., Landon, J., Bennett, H. P. J., Lowry, P. J., and McMartin, C. (1973) Pituitary peptide. *Nature New Biol.* **244,** 65–67.
15. Bradbury, A. F., Smyth, D. G., Snell, C. R., and Birdsall, N. J. (1976) The C-fragment of lipotropin: an endogenous peptide with high affinity for brain opiate receptors. *Nature* **260,** 793–795.
16. Graf, L., Ronai, A. Z., Bajusz, S., Cseh, G., and Szekely, J. I. (1976) Opioid agonist activity of β-lipotropin fragments: a possible biological lipotropin source of morphine-like substances in the pituitary. *FEBS Lett.* **64,** 181–184.
17. Li, C. H. and Chung, D. (1976) Isolation and structure of an untriakontapeptide with opiate activity from camel pituitary glands. *Proc. Natl. Acad. Sci. U.S.A.* **73,** 1145–1148.
18. Yalow, R. S. and Berson, S. A. (1971) Size heterogeneity of immunoreactive human ACTH in plasma and in extracts of pituitary glands and ACTH-producing thymoma. *Biochem. Biophys. Res. Commun.* **44,** 439–445.
19. Yalow, R. S. and Berson, S. A. (1973) Characteristics of 'big ACTH' in human plasma and pituitary extracts. *J. Clin. Endocrinol. Metab.* **36,** 415–423.
20. Mains, R. E., Eipper, B. A., and Ling, N. (1977) Common precursor to corticotropins and endorphins. *Proc. Natl. Acad. Sci. U.S.A.* **74,** 3014–3018.

21. Roberts, J. L. and Herbert, E. (1977) Characterization of a common precursor to corticotropin and β-lipotropin: cell-free synthesis of the precursor and identification of corticotropin peptides in the molecule. *Proc. Natl. Acad. Sci. U.S.A.* **74,** 4826–4830.
22. Bertagna, X., Nicholson, W. E., Sorenson, G. D., Pettengill, O. S., Mount, C. D., and Orth, D. N. (1978) Corticotropin, lipotropin, and β-endorphin production by a human non-pituitary tumor in culture: evidence for a common precursor. *Proc. Natl. Acad. Sci. U.S.A.* **75,** 5160–5164.
23. Nakanishi, S., Inoue, A., Kita, T., Nakamura, M., Chang, A. C. Y., Cohen, S. N., and Numa, S. (1979) Nucleotide sequence of cloned cDNA for bovine corticotropin-β-lipotropin precursor. *Nature* **278,** 423–427.
24. Takahashi, A., Amemiya, Y., Sakai, M., Yasuda, Y., Suzuki, N., Sasayama, Y., and Kawauchi, H. (1998) Occurrence of four MSHs in dogfish POMC and their immunomodulating effects. *Exp. Dermatol.* **7,** 231; Amemiya, Y., Takahashi, A., Suzuki, N., Sasayama, Y., and Kawauchi, H. (1999) A newly characterized melanotropin in proopiomelanocortin in pituitaries of an elasmobranch, *Squalus acanthias*. *Gen. Comp. Endocrinol.*, **114,** 387–395.
25. Scimonelli, T. and Eberle, A. N. (1987) Photoaffinity labelling of melanoma cell MSH receptors. *FEBS Lett.* **226,** 134–138.
26. Solca, F., Siegrist, W., Drozdz, R., Girard, J., and Eberle, A. N. (1989) The receptor for α-melanotropin of mouse and human melanoma cells: application of a potent α-melanotropin photoaffinity label. *J. Biol. Chem.* **264,** 14,277–14,281.
27. Gerst, J. E., Sole, J., Hazum, E., and Salomon, Y. (1988) Identification and characterization of melanotropin binding proteins from M2R melanoma cells by covalent photoaffinity labeling. *Endocrinology* **123,** 1792–1797.
28. Mountjoy, K. G., Robbins, L. S., Mortrud, M. T., and Cone, R. D. (1992) The cloning of a family of genes that encode the melanocortin receptors. *Science* **28,** 1248–1251.
29. Chhajlani, V. and Wikberg, J. E. (1992) Molecular cloning and expression of the human melanocyte-stimulating hormone receptor cDNA. *FEBS Lett.* **309,** 417–420.
30. Roselli-Rehfuss, L., Mountjoy, K. G., Robbins, L. S., Mortrud, M. T., Low, M. J., Tatro, J. B., Entwistle, M. L., Simerly, R. B., and Cone, R. D. (1993) Identification of a receptor for gamma melanotropin and other proopiomelanocortin peptides in the hypothalamus and limbic system. *Proc. Natl. Acad. Sci. U.S.A.* **90,** 8856–8860.
31. Gantz, I., Konda, Y., Tashiro, T., Shimoto, Y., Miwa, H., Munzert, G., Watson, S. J., DelValle, J., and Yamada, T. (1993) Molecular cloning of a novel melanocortin receptor. *J. Biol. Chem.* **268,** 8246–8250.
32. Gantz, I., Miwa, H., Konda, Y., Shimoto, Y., Tashiro, T., Watson, S. J., DelValle, J., and Yamada, T. (1994) Molecular cloning, expression, and gene localization of a fourth melanocortin receptor. *J. Biol. Chem.* **268,** 15,174–15,179.
33. Chhajlani, V., Muceniece, R., and Wikberg, J. E. (1993) Molecular cloning of a novel human melanocortin receptor. *Biochem. Biophys. Res. Commun.* **195,** 866–873.
34. Gantz, I., Shimoto, Y., Konda, Y., Miwa, H., Dickinson, C. J., and Yamada, T. (1994) Molecular cloning, expression, and characterization of a fifth melanocortin receptor. *Biochem. Biophys. Res. Commun.* **200,** 1214–1220.
35. Griffon, N., Mignon, V., Facchinetti, P., Diaz, J., Schwartz, J. C., and Sokoloff, P. (1994) Molecular cloning and characterization of the rat fifth melanocortin receptor. *Biochem. Biophys. Res. Commun.* **200,** 1007–1014.

36. Labbe, O., Desarnaud, F., Eggerickx, D., Vassart, G., and Parmentier, M. (1994) Molecular cloning of a mouse melanocortin-5 receptor gene widely expressed in peripheral tissues. *Biochemistry* **33,** 4543–4549.
37. Fan, W., Boston, B. A., Kesterson, R. A., Hruby, V. J., and Cone, R. D. (1997) Role of melanocortinergic neurons in feeding and the agouti obesity syndrome. *Nature* **385,** 165–168.
38. Huszar, D., Lynch, C. A., Fairchild-Huntress, V., Dunmore, J. H., Fang, Q., Berkemeier, L. R., Gu, W., Kesterson, R. A., Boston, B. A., Cone, R. D., Smith, F. J., Campfield, L. A., and Burn, P. (1997) Targeted disruption of the melanocortin-4 receptor results in obesity in mice. *Cell* **88,** 131–141.
39. Boston, B. A., Blaydon, K. M., Varnerin, J., and Cone, R. D. (1997) Independent and additive effects of central POMC and leptin pathways on murine obesity. *Science* **278,** 1641–1644.
40. Tritos, N. A., Vicent, D., Gillette, J., Ludwig, D. S., Flier, E. S., and Maratos-Flier, E. (1998). Functional interactions between melanin-concentrating hormone, neuropeptide Y, and anorectic neuropeptides in the rat hypothalamus. *Diabetes* **47,** 1687–1692.
41. Bennett, H. P. J. (1986) Use of ion-exchange Sep-Pak cartridges in the batch fractionation of pituitary peptides. *J. Chromatogr.* **359,** 383–390.
42. Bennett, H. P. J. (1986) Biosynthetic fate of the amino-terminal fragment of pro-opiomelanocortin within the intermediate lobe of the mouse pituitary. *Peptides* **7,** 615–622.
43. Van Strien, F. J., Jesperson, S., van der Greef, J., Jenks, B. G., and Roubos, E. W. (1996) Identification of POMC processing products in single melanotrope cells by matrix-assisted laser desorption/ionization mass spectrometry. *FEBS Lett.* **379,** 165–170.
44. Rudman, D., Chawla, R. K., and Hollins, B. M. (1979) *N,O*-Di-acetylserine[1]-α-melanocyte-stimulating hormone, a naturally occurring melanotropic peptide. *J. Biol. Chem.* **254,** 10,102–10,108.
45. Geschwind, I. I. (1959) Species variation in protein and polypeptide hormones, in *Comparative Endocrinology* (Grobman, A., ed.), Wiley, New York pp. 421–443.
46. Dixon, J. S. and Li, C. H. (1960) The isolation and structure of α-melanocyte-stimulating hormone from horse pituitaries. *J. Am. Chem. Soc.* **82,** 4568–4572.
47. Lee, T. H., Lerner, A. B., and Buettner-Janusch, V. (1961) The isolation and structure of α-and β-melanocyte-stimulating hormones from monkey pituitary glands. *J. Biol. Chem.* **236,** 1390–1394.
48. Scott, A. P., Lowry, P. J., Ratcliffe, J. G., Rees, L. H., and Landon, J. (1974) Corticotrophin-like peptides in the rat pituitary. *J. Endocrinol.* **61,** 355–367.
49. Li, C. H., Danho, W. O., Chung, D., and Rao, A. J. (1975) Isolation, characterization, and amino acid sequence of melanotropins from camel pituitary glands. *Biochemistry* **14,** 947–952.
50. Lee, T. H., Lerner, A. B., and Buettner-Janusch, V. (1963) Melanocyte-stimulating hormones from sheep pituitary glands. *Biochim. Biophys. Acta* **71,** 706–709.
51. Notake, M., Tobimatsu, T., Watanabe, Y., Takahashi, H., Mishina, M., and Numa, S. (1983) Isolation and characterization of the mouse corticotropin-β-lipotropin precursor gene and a related pseudogene. *FEBS Lett.* **156,** 67–71.
52. Keightley, M. C., Funder, J. W., and Fuller, P. J. (1991) Molecular cloning and sequencing of a guinea pig pro-opiomelanocortin cDNA. *Mol. Cell. Endocrinol.* **82,** 89–98.

53. Tonon, M. C., Desrues, L., Lazure, C., Jenks, B. G., Chrétien, M., and Vaudry, H. (1989) Melanin concentrating hormone. V. Isolation and characterization of α-melanocyte-stimulating hormone from frog pituitary glands. *Life Sci.* **45,** 1155–1161.
54. Martens, G. J. M., Civelli, O., and Herbert, E. (1985) Nucleotide sequence of cloned cDNA for pro-opiomelanocortin in the amphibian *Xenopus laevis*. *J. Biol. Chem.* **260,** 13,685–13,689.
55. Follenius, E., van Dorsselaer, A., and Meunier, A. (1986) Proportions of mono-and diacetylated forms of α-MSH in individual neurointermediate lobe extracts of *Cyprinus carpio* L. *Comp. Biochem. Phyisol.[B]* **84,** 315–317.
56. Kawauchi, H. and Muramoto, K. (1979) Isolation and primary structure of melanotropins from salmon pituitary glands. *Int. J. Pept. Protein Res.* **14,** 373–374.
57. Kawauchi, H., Adachi, Y., and Tsubokawa, M. (1980) Occurrence of a new melanocyte stimulating hormone in the salmon pituitary gland. *Biochem. Biophys. Res. Commun.* **96,** 1508–1517.
58. Salbert, G., Chauveau, I., Bonnec, G., Valotaire, Y., and Jego, P. (1992) One of the two trout proopiomelanocortin messenger RNAs potentially encodes new peptides. *Mol. Endocrinol.* **6,** 1605–1613.
59. Amemiya, Y., Takahashi, A., Dores, R. M., and Kawauchi, H. (1997) Sturgeon proopiomelanocortin has a remnant of γ-melanotropin. *Biochem. Biophys. Res. Commun.* **230,** 452–456.
60. Smith, T. R., Rubin, D., Dores, R. M., Youson, J., and Marra, L. (1996) GENBANK, accession U59910.
61. Arends, R. J., Vermeer, H., Martens, G. J., Leunissen, J. A., Wnedelaar Bonga, S. E., and Flik, G. (1998) Cloning and expression of two proopiomelanocortin mRNAs in the common carp (*Cyprinus carpo* L.). *Mol. Cell. Endocrinol.* **143,** 23–31.
62. Okuta, A., Ando, H., Ueda, H., and Urano, A. (1996) Two types of cDNAs encoding proopiomelanocortin of sockeye salmon, *Oncorhynchus keta*. *Zoolog. Sci.* **13,** 421–427.
63. Bennett, H. P. J., Lowry, P. J., McMartin, C., and Scott, A. P. (1974) Structural studies of α-melanocyte-stimulating hormone and a novel β-melanocyte-stimulating hormone from the neurointermediate lobe of the pituitary of the dogfish *Squalus acanthias*. *Biochem. J.* **141,** 439–444.
64. Takahashi A., Amemiya, Y., Nozaki, M., Sower, S. A., Joss, J., Gorbman, A., and Kawauchi, H. (1995) Isolation and characterization of melanotropins from lamprey pituitary glands. *Int. J. Pept. Protein Res.* **46,** 197–204.
65. Heinig, J. A., Keeley, F. W., Robson, P., Sower, S. A., and Yourson, J. H. (1995) The appearance of proopiomelanocortin early in vertebrate evolution: cloning and sequencing of POMC from a lamprey pituitary cDNA library. *Gen. Comp. Endocrinol.* **99,** 137–144.
66. Takahashi, A., Amemiya, Y., Sarashi, M., Sower, S. A., and Kawauchi, H. (1995) Melanocortin and corticotropin are encoded on two distinct genes in the lamprey, the earliest evolved extant vertebrate. *Biochem. Biophys. Res. Commun.* **213,** 490–498.
67. Vaudry, H., Jenks, B. G., and Overbeeke, A. P. (1983) The frog pars intermedia contains only the non-acetylated form of α-MSH: acetylation to generate α-MSH occurs during the release process. *Life Sci.* **33,** 97–100.
68. Van Strien, F. J. C., Galas, L., Jenks, B. G., and Roubos, E. W. (1995) Differential acetylation of POMC-derived peptides in the pituitary gland of Xenopus laevis in relation to background adaptation. *J. Endocrinol.* **146,** 159–167.

69. Ellerkmann, E., Porter, T. E., Nagy, G. M., and Frawley, L. S. (1992) *N*-Acetylation is required for the lactotrope recruitment activity of α-melanocyte-stimulating hormone and β-endorphin. *Endocrinology* **131,** 566–570.
70. Bundel, D. T., Conlon, J. M., Chartrel, N., Tonon, M. C., and Vaudry, H. (1992) Isolation and structural characterization of peptides related to α-and γ-melanocyte-stimulating hormone (MSH) from the frog brain. *Brain Res. Mol. Brain Res.* **15,** 1–7.
71. Bertagna, X., Seidah, N., Massias, J. F., Lenne, F., Luton, J. P., Girard, F., and Chrétien, M. (1989) Microsequencing evidence for the maturation of human proopiomelanocortin into an 18 amino acid β-melanocyte stimulating hormone [h β-MSH(5-22)] in nonpituitary tissue. *Peptides* **10,** 83–87.
72. Harris, J. I. and Roos, P. (1956) Amino-acid sequence of a melanophore stimulating peptide. *Nature* **178,** 90.
73. Dixon, J. S. and Li, C. H. (1961) The isolation and structure of β-melanocyte-stimulating hormone from horse pituitary gland. *Gen. Comp. Endocrinol.* **1,** 161–169.
74. Geschwind, I. I., Li, C. H., and Barnafi, L. (1957) The isolation, characterization and amino acid sequence of a melanocyte-stimulating hormone from bovine pituitary glands. *J. Am. Chem. Soc.* **79,** 6394–6401.
75. Drouin, J., Chamberland, M., Charron, J., Jeannotte, L., and Nemer, M. (1985) Structure of the rat pro-opiomelanocortin (POMC) gene. *FEBS Lett.* **193,** 54–58.
76. Hilario, E., Lihrmann, I., and Vaudry, H. (1990) Characterization of the cDNA encoding proopiomelanocortin in the frog Rana ridibunda. *Biochem. Biophys. Res. Commun.* **173,** 653–659.
77. Pan, F. M. and Chang W. C. (1989) Nucleotide sequence of bullfrog proopiomelanocortin cDNA. *Nucleic Acids Res.* **17,** 5843.
78. Kawauchi, H., Adachi, Y., and Ishizuka, B. (1980) Isolation and structure of another β-melanotropin from salmon pituitary glands. *Int. J. Pept. Protein Res.* **16,** 79–82.
79. Love, R. M. and Pickering, B. T. (1974) A β-MSH in the pituitary gland of the spotted dogfish (*Scyliorhinus canicula*): isolation and structure. *Gen. Comp. Endocrinol.* **24,** 398–404.
80. Scott, A. P. and Lowry, P. J. (1974) Adenocorticotrophic and melanocyte-stimulating peptides in the human pituitary. *Biochem. J.* **139,** 593–602.
81. Bertagna, X. (1994) Proopiomelanocortin-derived peptides. *Endocrinol. Metab. Clin. North Am.* **23,** 467–485.
82. Kitahara, N., Nishizawa, T., Iida, K., Okazaki, H., Andoh, T., and Soma, G. I. (1988) Absence of a γ-melanocyte-stimulating hormone sequence in proopiomelanocortin mRNA of chum salmon *Oncorhynchus keta. Comp. Biochem. Physiol. [B]* **91,** 365–370.
83. Böhlen, P., Esch, F., Shibasaki, T., Baird, A., Ling, N., and Guillemin, R. (1981) Isolation and characterization of a γ-melanotropin-like peptide from bovine neurointermediate pituitary. *FEBS Lett.* **128,** 67–70.
84. Rouillé, Y., Michel, G., Chauvet, M. T., Chauvet, J., and Acher, R. (1989) Particular proecessing of pro-opiomelanocortin in *Xenopus laevis* intermediate pituitary: sequencing of α-and β-melanocyte stimulating hormones. *FEBS Lett.* **245,** 215–218.
85. Chauvet, J., Michel, G., Rouillé, Y., Chauvet, M. T., and Acher, R. (1991) Study of frog (*Rana esculenta*) proopiomelanocortin processing in the intermediate pituitary. Identification of α-melanotropin, β-melanotropin, Lys-γ-melanotropin, and corticotropin-like intermediate lobe peptide. *Int. J. Pept. Protein Res.* **37,** 236–240.

86. Van Strien, F. J., Devreese, B., van Beeumen, J., Roubos, E. W., and Jenks, B. G. (1995) Biosynthesis and processing of the N-terminal part of proopiomelanocortin in *Xenopus laevis*: characterization of γ-MSH peptides. *J. Neuroendocrinol.* **7,** 807–815.
87. Salzet, M., Wattez, C., Bulet, P., and Malecha, J. (1994) Isolation and structural characterization of a novel peptide related to γ-melanocyte stimulating hormone from the brain of the leech Theromyzon tessulatum. *FEBS Lett.* **348,** 102–106.
88. Bennett, H. P. J., Seidah, N. G., Benjannet, S., Solomon, S., and Chrétien, M. (1986) Reinvestigation of the disulfide bridge arrangement in human proopiomelanocortin N-terminal segment (hNT 1-76). *Int. J. Pept. Protein Res.* **27,** 306–313.
89. Patel, P. D., Sherman, T. G., and Watson, S. J. (1988) Characterization of proopiomelanocortin cDNA from the Old World monkey, *Macaca nemestrina*. *DNA* **7,** 627–635.
90. Seger, M. A. and Bennett, H. P. J. (1986) Structure and bioactivity of the aminoterminal fragment of proopiomelanocortin. *J. Steroid Biochem.* **25,** 703–710.
91. Gen, K., Hirai, T., Kato, T., and Kato, Y. (1994) Presence of the same transcript of proopiomelanocortin (POMC) genes in the porcine anterior and intermediate pituitary lobes. *Mol. Cell. Endocrinol.* **103,** 101–108.
92. Browne, C. A., Bennett, H. P. J., and Solomon, S. (1981) The isolation and characterization of γ_3-melanotropin from the neurointermediate lobe of the rat pituitary. *Biochem. Biophys. Res. Commun.* **100,** 336–343.
93. Bennett, H. P. J. (1986) Biosynthetic fate of the amino-terminal fragment of proopiomelanocortin within the intermediate lobe of the mouse pituitary. *Peptides* **7,** 615–622.
94. McLean, C. and Lowry, P. J. (1981) Natural occurrence but lack of melanotropic activity of γ-MSH in fish. *Nature* **290,** 341–343.
95. Schwyzer, R. (1977) ACTH: a short introductory review. *Ann. N.Y. Acad. Sci.* **297,** 3–26.
96. Martens, G. J. M. (1986) Expression of two proopiomelanocortin genes in the pituitary gland of *Xenopus laevis*: complete structures of the two preprohormones. *Nucleic Acids Res.* **14,** 3791–3798.
97. Soma, G. I., Kitahara, N., Nishizawa, T., Nanami, H., Kotake, C., Okazaki, H., and Andoh, T. (1984) Nucleotide sequence of a cloned cDNA for proopiomelanocortin precursor of chum salmon, *Onchorynchus keta*. *Nucleic Acids Res.* **12,** 8029–8041.
98. Boileau, G., Barbeau, C., Jeannotte, L., Chrétien, M., and Drouin, J. (1983) Complete structure of the porcine pro-opiomelanocortin mRNA derived from the nucleotide sequence of cloned cDNA. *Nucleic Acids Res.* **11,** 8063–8071.
99. Chang, A. C., Cochet, M., and Cohen, S. N. (1980) Structural organization of human genomic DNA encoding the pro-opiomelanocortin peptide. *Proc. Natl. Acad. Sci. U.S.A.* **77,** 4890–4894.
100. Kitahara, N., Nishizawa, T., Iida, K., Okazaki, H., Andoh, T., and Soma, G. (1988) Absence of γ-melanocyte-stimulating hormone mRNA of chum salmon *Oncorhynchus keta*. *Comp. Biochem. Physiol.* **91,** 365–370.
101. Maier, R., Barthe, P. L., Schenkel-Hulliger, L., and Desaulles, P. A. (1971) The biological activity of (1-D-serine, 17-18-dilysine)-corticotrophin-(1-18)-octadecapeptide-amide. *Acta Endocrinol.* **68,** 458–466.
102. Geiger, R. (1971) Synthese eines Heptadecapeptids mit hoher adrenocorticotroper Wirkung. *Liebigs Ann. Chem.* **750,** 165–171.

103. Roubos, E. W. (1997) Background adaptation by *Xenopus laevis*: a model for studying neuronal information processing in the pituitary pars intermedia. *Comp. Biochem. Physiol. A Physiol.* **118,** 533–550.
104. Desrues, L., Tonon, M. C., Leprince, J., Vaudry, H., and Conlon, J. M. (1998) Isolation, primary structure, and effects on α-melanocyte-stimulating hormone release of frog neurotensin. *Endocrinology* **139,** 4140–4146.
105. Chartrel, N., Conlon, J. M., Danger, J. M., Fournier, A., Tonon, M. C., and Vaudry, H. (1992) Characterization of melanotropin-release-inhibiting factor (melanostatin) from frog brain: homology with human neuropeptide Y. *Proc. Natl. Acad. Sci. U.S.A.* **88,** 3862–3866.
106. Mor, A., Chartrel, N., Vaudry, H., and Nicolas, P. (1994) Skin peptide tyrosine-tyrosine, a member of the pancreatic polypeptide family: isolation, structure, synthesis, and endocrine activity. *Proc. Natl. Acad. Sci. U.S.A.* **91,** 10,295–10,299.
107. Wakamtsu, K., Graham, A., Cook, D., and Thody, A. J. (1997) Characterization of ACTH peptides in human skin and their activation of the melanocortin-1 receptor. *Pigment Cell Res.* **10,** 288–297.
108. Thody, A. J. and Graham, A. (1998) Does α-MSH have a role in regulating skin pigmentation in humans? *Pigment Cell Res.* **11,** 265–274.
109. Furumura, M., Sakai, C., Abdel-Malek, Z., Barsh, G. S., and Hearing, V. J. (1996) The interaction of agouti signal protein and melanocyte stimulating hormone to regulate melanin formation in mammals. *Pigment Cell Res.* **9,** 191–203.
110. Aberdam, E., Bertolotto, C., Sviderskaya, E. V., de Thillot, V., Hemesath, T. J., Fisher, D. E., Bennett, D. C., Ortonne, J. P., and Ballotti, R. (1998) Involvement of microphthalmia in the inhibition of melanocyte lineage differentiation and of melanogenesis by agouti signal protein. *J. Biol. Chem.* **273,** 19,560–19,565.
111. Cone, R. D., Lu, D., Koppula, S., Vage, D. I., Klungland, H., Boston, B., Chen, W., Orth, D. N., Pouton, C., and Kesterson, R. A. (1996) The melanocortin receptors: agonists, antagonists, and the hormonal control of pigmentation. *Recent Progr. Horm. Res.* **51,** 287–317.
112. Smith, R., Healy, E., Siddiqui, S., Flanagan, N., Steijlen, P. M., Rosdahl, I., Jacques, J. P., Rogers, S., Turner, R., Jackson, I. J., Birch-Machin, M. A., and Rees, J. L. (1998) Melanocortin I receptor variants in an Irish population. *J. Invest. Dermatol.* **111,** 119–122.
113. Kijas, J. M., Wales, R., Tornsten, A., Chardon, P., Moller, M., and Andersson, L. (1998) Melanocortin receptor 1 (MC1-R) mutations and coat color in pigs. *Genetics* **150,** 1177–1185.
114. Funasaka, Y., Chakraborty, A. K., Hayashi, Y., Komoto, M. Ohashi, A., Nagahama, M., Inoue, Y., Pawelek, J., and Ichihashi, M. (1998) Modulation of melanocyte-stimulating hormone receptor expression on normal human melanocytes: evidence for a regulatory role of ultraviolet B, interleukin-1α, interleukin-1β, endothelin-1 and tumour necrosis factor-α. *Br. J. Dermatol.* **139,** 216–224.
115. Loir, B., Bouchard, B., Morandini, R., Del Marmol, V., Deraemaecker, R., Garcia-Borron, J. C., and Ghanem, G. (1997) Immunoreactive α-melanotropin as an autocrine effector in human melanoma cells. *Eur. J. Biochem.* **244,** 923–930.
116. Morandini, R., Boeynaems, J. M., Hedley, S. J., MacNeil, S., and Ghanem, G. (1998) Modulation of ICAM-1 expression by α-MSH in human melanoma cells and melanocytes. *J. Cell Physiol.* **175,** 276–282.

117. Scott, G., Cassidy, L., and Abdel-Malek, Z. (1997) α-Melanocyte-stimulating hormone and endothelin-1 have opposite effects on melanocyte adhesion, migration, and pp125FAK phosphorylation. *Exp. Cell Res.* **237,** 19–28.
118. Ghanem, G., Loir, B., Hadley, M., Abdel-Malek, Z., Libert, A., Del Marmol, V., Lejeune, F., Lozano, J., and Garciα-Borron, J. C. (1992) Partial characterization of IR-α-MSH peptides found in melanoma tumors. *Peptides* **13,** 989–994.
119. Schauer, E., Trautinger, F., Kock, A., Schwarz, A., Bhardwaj, R., Simon, M., Ansel, J. C., Schwarz, T., and Luger, T. A. (1994) Proopiomelanocortin-derived peptides are synthesized and released by human keratinocytes. *J. Clin. Invest.* **93,** 2258–2262.
120. Luger, T. A., Scholzen, T., and Grabbe, S. (1997) The role of α-melanocyte-stimulating hormone in cutaneous biology. *J. Invest. Dermatol. Symp. Proc.* **2,** 87–93.
121. Hartmeyer, M., Scholzen, T., Becher, E., Bhardwaj, R. S., Schwarz, T., and Luger, T. A. (1997) Human dermal microvascular endothelial cells express the melanocortin receptor type 1 and produce increased levels of IL-8 upon stimulation with α-melanocyte-stimulating hormone. *J. Immunol.* **159,** 1930–1937.
122. Strand, F. L., Williams, K. A., Alves, S. E., Antonawich, F. J., Lee, T. S., Lee, S. J., Kume, J., and Zuccarelli, LA. (1994) Melanocortins as factors in somatic neuromuscular growth and regrowth. *Pharmacol. Ther.* **62,** 1–27.
123. Zohar, M. and Salomon, Y. (1992) Melanocortins stimulate proliferation and induce morphological changes in cultured rat astrocytes by distinct transducing mechanisms. *Brain Res.* **576,** 49–58.
124. Joosten, E. A., Verhaagh, S., Martin, D., Robe, P., Franzen, R., Hooiveld, M., Doornbos, R., Bar, P. R., and Moonen, G. (1996) α-MSH stimulates neurite outgrowth of neonatal rat corticospinal neurons in vitro. *Brain Res.* **736,** 91–98.
125. Lichtensteiger, W., Hanimann, B., Siegrist, W., and Eberle, A. N. (1996) Region- and stage-specific patterns of melanocortin receptor ontogeny in rat central nervous system, cranial nerve ganglia and sympathetic ganglia. *Brain Res. Dev. Brain Res.* **22,** 93–110.
126. Kistler-Heer, V., Lauber, M. E., and Lichtensteiger, W. (1998) Different developmental patterns of melanocortin MC3 and MC4 receptor mRNA: predominance of MC4 in fetal rat nervous system. *J. Neuroendocrinol.* **10,** 133–146.
127. De Wied, D. and Ree, J. M. (1989) Neuropeptides: animals behaviour and human psychopathology. *Eur. Arch. Psychiatry Neurol. Sci.* **238,** 323–331.
128. Pitsikas, N., Spruijt, B. M., Algeri, S., and Gispen, W. H. (1990) The ACTH/MSH (4-9) analog Org2766 improves retrieval of information after a fimbria fornix transection. *Peptides* **11,** 911–914.
129. Von Frijtag, J. C., Croiset, G., Gispen, W. H., Adan, R. A., and Wiegant, V. M. (1998) The role of central melanocortin receptors in the activation of the hypothalamus-pituitary-adrenal-axis and the induction of excessive grooming. *Br. J. Pharmacol.* **123,** 1503–1508.
130. Steiger, A. and Holsboer, F. (1997) Neuropeptides and human sleep. *Sleep* **20,** 1038–1052.
131. Arborelius, L., Owens, M. J., Plotsky, P. M., and Nemeroff, C. B. (1999) The role of corticotropin-releasing factor in depression and anxiety disorders. *J. Endocrinol.* **160,** 1–12.
132. Lichtensteiger, W. and Monnet, F. (1979) Differential response of dopamine neurons to α-melanotropin and analogues in relation to their endocrine and behavioural potency. *Life Sci.* **25,** 2079–2087.

133. Shimida, M., Tritos, N. A., Lowell, B. B., Flier, J. S., and Maratos-Flier, E. (1998) Mice lacking melanin-concentrating hormone are hypophagic and lean. *Nature* **396,** 670–674.
134. Baker, B. I. (1993) The role of melanin-concentrating hormone in color change. *Ann. N. Y. Acad. Sci.* **680,** 279–289.
135. Krude, H., Biebermann, H., Luck, W., Horn, R., Brabant, G., and Gruters, A. (1998) Severe early-onset obesity, adrenal insufficiency and red hair pigmentation caused by POMC mutations in humans. *Nat. Genet.* **19,** 155–157.
136. Ng, T. B. (1990) Studies on hormonal regulation of lipolysis and lipogenesis in fat cells of various mammalian species. *Comp. Biochem. Physiol. [B]* **97,** 441–446.
137. Liakos, P., Chambaz, E. M., Feige, J. J., and Defaye, G. (1998) Expression of ACTH receptors (MC2-R and MC5-R) in the glomerulosa and the fasciculata-reticularis zones of bovine adrenal cortex. *Endocr. Res.* **24,** 427–432.
138. Lipton, J. M. and Clark, W. G. (1986) Neurotransmitters in temperature control. *Ann. Rev. Physiol.* **48,** 613–623.
139. Huang, Q. H., Hruby, V. J., and Tatro, J. B. (1998) Systemic α-MSH suppresses LPS fever via central melanocortin receptors independently of its suppression of corticosterone and IL-6 release. *Am. J. Physiol.* **275,** R524–R530.
140. Catania, A. and Lipton, J. M. (1998) Peptide modulation of fever and inflammation within the brain. *Ann. N. Y. Acad. Sci.* **856,** 62–68.
141. Harbour, D. V., Galin, F. S., Hughes, T. K., Smith, E. M., and Blalock, J. E. (1991) Role of leukocyte-derived pro-opiomelanocortin peptides in endotoxic shock. *Circ. Shock* **35,** 181–191.
142. Lysons, P. D. and Blalock, J. E. (1997) Pro-opiomelanocortin gene expression and protein processing in rat mononuclear leukocytes. *J. Neuroimmunol.* **78,** 47–56.
143. Clarke, B. L., Moore, D. R., and Blalock, J. E. (1994) Adrenocorticotropic hormone stimulates a transient calcium uptake in rat lymphocytes. *Endocrinology* **135,** 1780–1786.
144. Bhardwaj, R. S., Schwarz, A., Becher, E., Mahnke, K., Aragane, Y., Schwarz, T., and Luger, T. A. (1996) Pro-opiomelanocortin-derived peptides induce IL-10 production in human monocytes. *J. Immunol.* **156,** 2517–2521.
145. Bhardwaj, R. S., Becher, E., Mahnke, K., Hartmeier, M., Schwarz, T., Scholzen, T., and Luger, T. A. (1997) Evidence for the differential expression of functional α-melanocyte-stimulating hormone receptor MC-1 on human monocytes. *J. Immunol.* **158,** 3378–3384.
146. Buggy, J. J. (1998) Binding of α-melanocyte-stimulating hormone to its G-protein-coupled receptor on B-lymphocytes activates the Jak/STAT pathway. *Biochem. J.* **331,** 211–216.
147. Bilfinger, T. V., Hughes, T. K., Rodriguez, M., Glass, R., Casares, F., and Stefano, G. B. (1996) Hyperstimulation of leukocytes by plasma from cardiopulmonary bypass patients is diminished by α-MSH pretreatment. *Int. J. Cardiol.* **53,** Suppl., S47–S53.
148. Leiba, H., Garty, N. B., Schmidt-Sole, J., Piteman, O., Azrad, A., and Salomon, Y. (1990) The melanocortin receptor in the rat lacrimal gland: a model system for the study of MSH (melanocyte-stimulating hormone) as a potential neurotransmitter. *Eur. J. Pharmacol.* **181,** 71–82.
149. Tatro, J. B. and Reichlin, S. (1987) Specific receptors for α-melanocyte-stimulating hormone are widely distributed in tissue of rodents. *Endocrinology* **121,** 1900–1907.

150. Chen, W., Kelly, M. A., Opitz-Araya, X., Thomas, R. E., Low, M. J., and Cone, R. D. (1997) Exocrine gland dysfunction in MC5-R-deficient mice: evidence for coordinated regulation of exocrine gland function by melanocortin peptides. *Cell* **91,** 789–798.
151. Van der Kraan, M., Adan, R. A., Entwistle, M. L., Gispen, W. H., Burbach, J. P., and Tatro, J. B. (1998) Espression of melanocortin-5 receptor in secretory epithelia supports a functional role in exocrine and endocrine glands. *Endocrinology* **139,** 2348–2355.
152. Versteeg, D. H., van Bergen, P., Adan, R. A., and de Wildt, D. J. (1998) Melanocortins and cardiovascular regulation. *Eur. J. Pharmacol.* **360,** 1–14.
153. Mayan, H., Ling, K. T., Lee, E. Y., Wiedemann, E., Kalinyak, J. E., and Humphreys, M. H. (1996) Dietary sodium intake modulates pituitary proopiomelanocortin mRNA abundance. *Hypertension* **28,** 244–249.
154. Van Bergen, P., Janssen, P. M., Hoogerhout, P., De Wildt, D. J., and Versteeg, D. H. (1995) Cardiovascular effects of γ-MSH/ACTH-like peptides: structure-activity relationships. *Eur. J. Pharmacol.* **294,** 795–803.
155. Gruber, K. A. and Callahan, M. F. (1989) ACTH-(4-10) through γ-MSH: evidence for a new class of central autonomic nervous system-regulating peptides. *Am. J. Physiol.* **257,** R681–694.
156. Van Bergen, P., Van der Vaart, J. G., Kasbergen, C. M., Versteeg, D. H., and De Wildt, D. J. (1997) Structure-activity analysis for the effects of γ-MSH/ACTH-like peptides on cerebral hemodynamics in rats. *Eur. J. Pharmacol.* **318,** 357–368.
157. Valentin, J. P., Wiedemann, E., and Humphreys, M. H. (1993) Natriuretic properties of melanocyte-stimulating hormones. *J. Cardiovasc. Pharmacol.* **22,** Suppl. **2,** S114–S118.
158. Nunez, L. and Frawley, L. S. (1998) α-MSH potentiates the responsiveness of mammotropes by increasing Ca^{2+} entry. *Am. J. Physiol.* **274,** E971–E977.
159. Bar-Ilan, A., Savion, N., and Naveh, N. (1992) α-Melanocyte-stimulating hormone (α-MSH) enhances eicosanoid production by bovine retinal pigment epithelium. *Prostaglandins* **43,** 31–44.
160. Zamir, E. (1997) Central serous retinopathy associated with adrenocorticotrophic hormone therapy: a case report and a hypothesis. *Graefes Arch. Clin. Exp. Ophthalmol.* **235,** 339–344.
161. Vanetti, M., Schonrock, C., Meyerhof, W., and Hollt, V. (1994) Molecular cloning of a bovine MSH receptor which is highly expressed in the testis. *FEBS Lett.* **348,** 268–272.
162. Durando, P. E. and Celis, M. E. (1998) In vitro effect of α-MSH administration on steroidogenesis of prepurbertal ovaries. *Peptides* **19,** 667–675.
163. Castrucci, A. M., Almeida, A. L., Al-Obeidi, F. A., Hadley, M. E., Hruby, V. J., Staples, D. J., and Sawyer, T. K. (1997) Comparative biological activities of α-MSH antagonists in vertebrate pigment cells. *Gen. Comp. Endocrinol.* **105,** 410–416.
164. Chluba-de Tapia, J., Bagutti, C., Cotti, R., and Eberle, A. N. (1996) Induction of constitutive melanogenesis in amelanotic mouse melanoma cells by transfection of the human melanocortin-1 receptor gene. *J. Cell Sci.* **109,** 2023–2030.
165. Quillan, J. M., Jayawickreme, C. K., and Lerner, M. R. (1995) Combinatorial diffusion assay used to identify topically active melanocyte-stimulating hormone receptor antagonists. *Proc. Natl. Acad. Sci. U.S.A.* **92,** 2894–2898.

166. Chen, W., Shields, T. S., Stork, P. J., and Cone, R. D. (1995) A colorimetric assay for measuring activation of G_s-and G_q-coupled signaling pathways. *Anal. Biochem.* **226,** 349–354.
167. Gibson, S., Crosby, S. R., Stewart, M. F., Jennings, A. M., McCall, E. , and White, A. (1994) Differential release of proopiomelanocortin-derived peptides from the human pituitary: evidence from a panel of two-site immunoradiometric assays. *J. Clin. Endocrinol. Metab.* **78,** 835–841.
168. Yogi, Y., Hashida, S., Ekman, R., Setoguchi, T., and Ishikawa, E. (1995) Noncompetitive enzyme immunoassay (hetero-two-site enzyme immunoassay) for γ_2-melanocyte-stimulating hormone (γ_2-MSH) and measurement of immunoreactive γ_2-MSH in plasma of healthy subjects. *J. Clin. Lab. Anal.* **9,** 397–406.
169. Siegrist, W., Oestreicher, M., Stutz, S., Girard, J., and Eberle, A. N. (1988). Radioreceptor assay for α-MSH using mouse B16 melanoma cells. *J. Recept. Res.* **8,** 323–343.
170. Eberle, A. N., Jäggin Verin, V., Solca, F., Siegrist, W., Küenlin, C., Bagutti, C., Stutz, S., and Girard, J. (1988). Biologically active monoiodinated α-MSH derivatives for receptor binding studies using human melanoma cells. *J. Recept. Res.* **11,** 311–322.
171. Buckley, D. I., Yamashiro, D., and Ramachandran, J. (1981) Synthesis of a corticotropin analog that retains full biological activity after iodination. *Endocrinology* **109,** 5–9.
172. Siegrist, W., Solca, F., Stutz, S., Giuffrè, L., Carrel, S., Girard, J., and Eberle, A. N. (1989) Characterization of receptors for α-melanocyte-stimulating hormone on human melanoma cells. *Cancer Res.* **49,** 6352–6358.
173. Friedmann, P. S., Wren, F., Buffey, J., and MacNeil, S. (1990) α-MSH causes a small rise in cAMP but has no effect on basal or ultraviolett-stimulated melanogenesis in human melanocytes. *Br. J. Dermatol.* **123,** 145–151.
174. Siegrist, W., Drozdz, R., Cotti, R., Willard, D. H., Wilkison, W. O., and Eberle, A. N. (1997) Interaction of α-melanotropin and agouti on B16 melanoma cells: evidence for inverse agonism of agouti. *J. Recept. Signal Transduct. Res.* **17,** 75–98.
175. Buffey, J., Thody, A. J., Bleehen, S. S., and MacNeil, S. (1992) α-Melanocyte-stimulating hormone stimulates protein kinase C activity in murine B16 melanoma. *J. Endocrinol.* **133,** 333–340.
176. Siegrist, W., Sauter, P., and Eberle, A. N. (1995) A selective protein kinase C inhibitor (CGP 41251) positively and negatively modulates melanoma cell MSH receptors. *J. Recept. Signal Transduct. Res.* **15,** 283–296.
177. Reilein, A. R., Tint, I. S., Peunova, N. I., Enikolopov, G. N., and Gelfand, V. I. (1998) Regulation of organelle movement in melanophores by protein kinase A (PKA), protein kinase C (PKC), and protein phosphatase 2A (PP2A*). J. Cell Biol.* **142,** 803–813.
178. Park, H. Y., Russakovsky, V., Ao, Y., Fernandez, E., and Gilchrest, B. A. (1996) α-Melanocyte stimulating hormone-induced pigmentation is blocked by depletion of protein kinase C. *Exp. Cell Res.* **227,** 70–79.
179. McLeod, S. D., Smith, C., and Mason, R. S. (1995) Stimulation of tyrosinase in human melanocytes by pro-opiomelanocortin-derived peptides. *J. Endocrinol.* **146,** 439–447.
180. Kapas, S., Purbrick, A., and Hinson, J. P. (1995) Role of tyrosine kinase and protein kinase C in the steroidogenic actions of angiotensin II, α-melanocyte-stimulating hormone and corticotropin in the rat adrenal cortex. *Biochem. J.* **305,** 433–438.

181. Konda, Y., Gantz, I., DelValle, J. Shimoto, Y., Miwa, H., and Yamada, T. (1994) Interaction of dual intracellular signaling pathways activated by the melanocortin-3 receptor. *J. Biol. Chem.* **269,** 13,162–13,166.
182. De Graan, P. N. E., Eberle, A. N., and van de Veerdonk, F. C. G. (1982) Calcium sites in MSH stimulation of Xenopus melanophores: studies with photoreactive α-MSH. *Cell. Mol. Endocrinol.* **26,** 327–339.
183. Gerst, J. E., Sole, J., and Salomon, Y. (1987) Dual regulation of β-melanotropin receptor function and adenylate cyclase by calcium and guanosine nucleotides in the M2R melanoma cell line. *Mol. Pharmacol.* **31,** 81–88.
184. Kim, J. H., Kiefer, L. L., Woychik, R. P., Wilkison, W. O., Truesdale, A., Ittoop, O., Willard, D., Nichols, J., and Zemel, M. B. (1997) Agouti regulation of intracellular calcium: role of melanocortin receptors. *Am. J. Physiol.* **272,** E379–E384.
185. Castrucci, A. M., Hadley, M. E., Sawyer, T. K., Wilkes, B. C., al-Obeidi, F., Staples, D. J., de Vaux, A. E., Dym, O., Hintz, M. F., and Riehm, J. P. (1989) α-Melanotropin: the minimal active sequence in the lizard skin bioassay. *Gen. Comp. Endocrinol.* **73,** 157–163.
186. Miwa, H., Gantz, I., Konda, Y., Shimoto, Y., and Yamada, T. (1995) Structural determinants of the melanocortin peptides required for activation of melanocortin-3 and melanocortin-4 receptors. *J. Pharmacol. Exp. Ther.* **273,** 367–372.
187. Peng, P. J., Sahm, U. G., Doherty, R. V., Kinsman, R. G., Moss, S. H., and Pouton, C. W. (1997) Binding and biological activity of C-terminally modified melanocortin peptides: a comparison between their actions at rodent MC1 and MC3 receptors. *Peptides* **18,** 1001–1008.
188. Löw, M., Szalay, K. S., and Kisfaludy, L. (1990) Role of chain termini in selective steroidogenic effect of ACTH/MSH(4-10) on isolated adrenocortical cells. *Peptides* **11,** 29–31.
189. Sahm, U. G., Olivier, G. W., Branch, S. K., Moss, S. H., and Pouton, C. W. (1994) Synthesis and biological evaluation of α-MSH analogues substituted with alanine. *Peptides* **15,** 1297–1302.
190. Sawyer, T. K., Sanfilippo, P. J., Hruby, V. J., Engel, M. H., Heward, C. B., Burnett, J. B., and Hadley, M. E. (1980) 4-Norleucine, 7-D-phenylalanine-α-melanocyte-stimulating hormone: a highly potent α-melanotropin with ultralong biological activity. *Proc. Natl. Acad. Sci. U.S.A.* **77,** 5754–5758.
191. Haskell-Luevano, C., Miwa, H., Dickinson, C., Hadley, M. E., Hruby, V. J., Yamada, T., and Gantz, I. (1996) Characterizations of the unusual dissociation properties of melanotropin peptides from the melanocortin receptor, hMC1-R. *J. Med. Chem.* **39,** 432–435.
192. Sawyer, T. K., Hruby, V. J., Darman, P. S., and Hadley, M. E. (1982) [half-Cys**4,** half-Cys^{10}]-α-melanocyte-stimulating hormone: a cyclic α-melanotropin exhibiting superagonist biological activity. *Proc. Natl. Acad. Sci. U.S.A.* **79,** 1751–1755.
193. Sawyer, T. K., Staples, D. J., de Lauro Castrucci, A. M., and Hadley, M. E. (1989) Discovery and structure-activity relationships of novel α-melanocyte-stimulating hormone inhibitors. *Peptide Res.* **2,** 140–146.
194. Sahm, U. G., Olivier, G. W., Branch, S. K., Moss, S. H. and Pouton, C. W. (1996) Receptor binding affinities and biological activities of linear and cyclic melanocortins in B16 murine melanoma cells expressing the native MC1 receptor. *J. Pharm. Pharmacol.* **48,** 197–200.

195. Haskell-Luevano, C., Shenderovich, M. D., Sharma, S. D., Nikiforovich, G. V., Hadley, M. E., and Hruby, V. J. (1995) Design, synthesis, biology, and conformations of bicyclic α-melanotropin analogues. *J. Med. Chem.* **38,** 1736–1750.
196. Sugg, E. E., Castrucci, A. M., Hadely, M. E., van Binst, G., and Hruby, V. J. (1988) Cyclic lactam analogues of Ac[Nle[4]]α-MSH_{4-11}-NH_2. *Biochemistry* **27,** 8181–8188.
197. Al-Obeidi, F., Castrucci, A. M., Hadley, M. E., and Hruby, V. J. (1989) Potent and prolonged acting cyclic lactam analogues of α-melanotropin: design based on molecular dynamics. *J. Med. Chem.* **32,** 2555–2561.
198. Dorr, R. T., Lines, R., Levine, N., Brooks, C., Xiang, L., Hruby, V. J., and Hadley, M. E. (1996) Evaluation of melanotan-II, a superpotent cyclic melanotropic peptide in a pilot phase-I clinical study. *Life Sci.* **58,** 1777–1784.
199. Al-Obeidi, F., Hruby, V. J., Yaghoubi, N., Marwan, M. M., and Hadley, M. E. (1992) Synthesis and biological activities of fatty acid conjugates of a cyclic lactam α-melanotropin. *J. Med. Chem.* **35,** 118–123.
200. Nikiforovich, G. V., Shenderovich, M. D., and Chipens, G. I. (1981) The space structure of α-melanotropin. *FEBS Lett.* **126,** 180–182.
201. Nikoforovich, G. V., Sharma, S. D., Hadley, M. E., and Hruby, V. J. (1998) Studies of conformational isomerism in α-melanocyte stimulating hormone by design of cyclic analogues. *Biopolymers* **46,** 155–167.
202. Lee, J. H., Lim, S. K., Huh, S. H., and Lee, W. (1998) Solution structures of the melanocyte-stimulating hormones by two-dimensional NMR spectroscopy and dynamical simulated-annealing calculations. *Eur. J. Biochem.* **257,** 31–40.
203. Yang, Y. K., Dickinson, C., Haskell-Luevano, C., and Gantz, I. (1997) Molecular basis for the interaction of [Nle**4**, D-Phe[7]]melanocyte stimulating hormone with the human melanocortin-1 receptor. *J. Biol. Chem.* **272,** 23,000–23,010.
204. Frandberg, P. A., Xu, X., and Chhajlani, V. (1997) Glutamine[235] and arginine[272] in human melanocortin-5 receptor determines its low affinity to MSH. *Biochem. Biophys. Res. Commun.* **236,** 489–492.
205. Chhajlani, V., Xu, X., Blauw, J., and Sudarshi, S. (1996) Identification of ligand binding residues in extracellular loops of the melanocortin 1 receptor. *Biochem. Biophys. Res. Commun.* **219,** 521–525.
206. Haskell-Luevano, C., Hendrata, S., North, C., Sawyer, T. K., Hadley, M. E., Hruby, V. J., Dickinson, C., and Gantz, I. (1997) Discovery of prototype peptidomimetic agonists at the human melanocortin receptors MC1-R and MC4-R. *J. Med. Chem.* **40,** 2133–2139.
207. Heizmann, G., Tanner, H., and Eberle, A. N. (1997) New ligands for the human melanoma MSH receptor identified by a peptoid library (oligo N-substituted glycines), in: *Innovation and Perspectives in Solid-Phase Synthesis and Combinatorial Libraries 1996: Peptides, Proteins and Nucleic Acids,* (Epton, R. ed.) Mayflower Scientific, Birmingham. pp. 391–394.
208. Al-Obeidi, F., Hruby, V. J., Hadley, M. E., Sawyer, T. K., and Castrucci, A. M. (1990) Design, synthesis, and biological activities of a potent and selective α-melanotropin antagonist. *Int. J. Pept Protein Res.* **35,** 228–234.
209. Quillan, J. M. and Sadee, W. (1996) Structure-based search for peptide ligands that cross-react with melanocortin receptors. *Pharm. Res.* **13,** 1624–1630.
210. Hruby, V. J., Lu, D., Sharma, S. D., Castrucci, A. L., Kesterson, R. A., al-Obeidi, F. A., Hadley, M. E., and Cone, R. D. (1995) Cyclic lactam α-melanotropin analogues of Ac-Nle[4]-cyclo[Asp**5,** D-Phe**7,** Lys[10]]α-melanocyte-stimulating hormone-

(4-10)-NH_2 with bulky aromatic amino acids at position 7 show high antagonist potency and selectivity at specific melanocortin receptors. *J. Med. Chem.* **38,** 3454–3461.

211. Schiöth, H. B., Muceniece, R., Mutulis, F., Prussis, P., Lindeberg, G., Sharma, S. D., Hruby, V. J., and Wikberg, J. E. (1997) Selectivity of cyclic [D-Nal^7] and [D-Phe^7] substituted MSH analogues for the melanocortin receptor subtypes. *Peptides* **18,** 1009–1013.
212. Schiöth, H. B., Mutulis, F., Muceniece, R., Prusis, P., and Wikberg, J. E. (1998) Discovery of novel melanocortin-4 receptor selective MSH analogues. *Br. J. Pharmacol.* **124,** 75–82.
213. Kask, A., Rago, L., Mutulis, F., Pahkla, R., Wikberg, J. E., and Schöth, H. B. (1998) Selective antagonist for the melanocortin 4 receptor (HS014) increases food intake in free-feeding rats. *Biochem. Biophys. Res. Commun.* **245,** 90–93.
214. Kask, A., Mutulis, F., Muceniece, R., Pahkla, R., Mutule, I., Wikberg, J. E., Rago, L., and Schiöth, H. B. (1998) Discovery of a novel superpotent and selective melanocortin-4 receptor antagonist (HS024): evaluation in vitro and in vivo. *Endocrinology* **139,** 5006–5014.
215. Jayawickreme, C. K., Quillan, J. M., Graminski, G. F., and Lerner, M. R. (1994) Discovery and structure-function analysis of α-melanocyte-stimulating hormone antagonists. *J. Biol. Chem.* **269,** 29,846–29,854.
216. Siegrist, W., Willard, D. H., Wilkison, W. O., and Eberle, A. N. (1996) Agouti protein inhibits growth of B16 melanoma cells in vitro by acting through melanocortin receptors. *Biochem. Biophys. Res. Commun.* **218,** 171–175.
217. Siegrist, W., Drozdz, R., Cotti, R., Willard, D. H., Wilkison, W. O., and Eberle, A. N. (1997) Interactions of α-melanotropin and agouti on B16 melanoma cells: evidence for inverse agonism of agouti. *J. Recept. Signal Transduct. Res.* **17,** 75–98.
218. Dorr, R. T., Lines, R., Levine, N., Brooks, C., Xiang, L., Hruby, V. J., and Hadley, M. E. (1996) Evaluation of melanotan-II, a superpotent cyclic melanotropic peptide in a pilot phase-I clinical study. *Life Sci.* **58,** 1777–1784.
219. Wessels, H., Fuciarelli, K., Kansen, J., Hadley, M. E., Hruby, V. J., Dorr, R., and Levine, N. (1998) Synthetic melanotropic peptide initiates erections in men with psychogenic erectile dysfunction: double-blind, placebo controlled crossover study. *J. Urol.* **160,** 389–393.
220. Guarini, S., Bazzani, C., and Bertolini, A. (1997) Resuscitating effect of melanocortin peptides after prolonged respiratory arrest. *Br. J. Pharmacol.* **121,** 1454–1460.
221. Van de Meent, H., Hamers, F. P., Lankhorst, A. J., Joosten, E. A., and Gispen, W. H. (1997) Beneficial effects of the melanocortin α-melanocyte-stimulating hormone on clinical and neurophysiological recovery after experimental spinal cord injury. *Neurosurgery* **40,** 122–130.
222. Salazar-Onfray, F., Nakazawa, T., Chhajlani, V., Perssson, M., Karre, K., Masucci, G., Celis, E., Sette, A., Southwood, S., Appella, E., and Kiessling, R. (1997) Synthetic peptides derived from the melanocyte-stimulating hormone receptor MC1-R can stimulate HLA-A2-restricted cytotoxic T lymphocytes that recognize naturally processed peptides on human melanoma cells. *Cancer Res.* **57,** 4348–4355.
223. Murphy, J. R. (1988) Dipththeria-related peptide hormone gene fusions: a molecular genetic approach to chimeric toxin development. *Cancer Treat. Res.* **37,** 123–140.

224. Tatro, J. B., Wen, Z., Entwistle, M. L., Atkins, M. B., Smith, T. J., Reichlin, S., and Murphy, J. R. (1992) Interaction of an α-melanocyte-stimulating hormone-diphtheria toxin fusion protein with melanotropin receptors in human melanoma metastases. *Cancer Res.* **52,** 2545–2548.
225. Ghanem, G. E., Libert, A., Arnould, R., Vercammen, A., and Lejeune, F. (1991) Human melanoma targeting with α-MSH-melphalan conjugate. *Melanoma Res.* **1,** 105–114.
226. Morandini, R., Süli-Vargha, H., Libert, A., Loir, B., Botyanszki, J. Medzihradszky, K., and Ghanem, G. (1994) Receptor-mediated cytotoxicity of α-MSH fragments containing melphalan in a human melanoma cell line. *Int. J. Cancer* **56,** 129–133.
227. Siegrist, W., Girard, J., and Eberle, A. N. (1991). Quantification of MSH receptors on mouse melanoma tissue by receptor autoradiography. *J. Receptor Res.* **11,** 323–331.
228. Tatro, J. B., Atkins, M., Mier, J. W., Hardarson, S., Wolfe, H., Smith, T., Entwistle, M. L., and Reichlin, S. (1990). Melanotropin receptors demonstrated in situ in human melanoma. *J. Clin. Invest.* **85,**1825–1832.
229. Garg, P. K., Alston, K. L., Welsh, P. C., and Zalutsky, M. R. (1996) Enhanced binding and inertness to dehalogenation of α-melanotropic peptides labeled using *N*-succinimidyl 3-iodobenzoate. *Bioconjug. Chem.* **7,** 233–239.
230. Bard, D. R., Wraight, E. P., and Knight, C. G. (1993) Bis-MSH-DTPA. A potential imaging agent for malignant melanoma. *Ann. N.Y. Acad. Sci.* **680,** 451–453.
231. Bagutti, C., Stolz, B., Albert, R., Bruns, C., Pless, J., and Eberle, A. N. (1993) [^{111}In]DTPA-labeled analogues of α-MSH for the detection of MSH receptors in vitro and in vivo. *Ann. N.Y. Acad. Sci.* **680,** 445–447.
232. Heppeler, A., Froidecaux, S., Mäcke, H. R., Jermann, E., Béhé, M., Powell, P., and Hennig, M. (1999) Radiometal labelled macrocyclic chelator derivatised somatostatin analogue with superb tumour targeting properties and potential for receptor mediated internal radiotherapy. *Chem. Eur. J.* **5,** 1974–1981.
233. Bard, D. R., Knight, C. G., and Page-Thomas, D. P. (1990) A chelating derivative of α-melanocyte stimulating hormone as a potential imaging agent for malignant melanoma. *Br. J. Cancer* **62,** 919–922.
234. Wraight, E. P., Bard, D. R., Maughan, T. S., Knight, C. G., and Page-Thomas, D. P. (1992) The use of a chelating derivative of α-melanocyte stimulating hormone for the clinical imaging of malignant melanoma. *Br. J. Radiol.* **65,** 112–118.
235. Bard, D. R. (1995) An improved imaging agent for malignant melanoma, based on [Nle**4,** D-Phe^7]-α-melanocyte stimulating hormone. *Nucl. Med. Commun.* **16,** 860–866.
236. Bagutti, C., Stolz, B., Albert, R., Bruns, C., Pless, J., and Eberle, A. N. (1994) [^{111}In]DTPA-labeled analogues of α-melanocyte-stimulating hormone for melanoma targeting: receptor binding in vitro and in vivo. *Int. J. Cancer* **58,** 749–755.
237. Vaidyanathan, G. and Zalutsky, M. R. (1997) Fluorine-18-labeled [Nle**4,** D-Phe^7]-α-MSH, an α-melanocyte stimulating hormone analogue. *Nucl. Med. Biol.* **24,** 171–17**8,**
238. Giblin, M. F., Jurisson, S. S., and Quinn, T. P. (1997) Synthesis and characterization of rhenium-complexed α-melanotropin analogs. *Bioconjug. Chem.* **8,** 347–353.
239. Giblin, M. F., Wang, N., Hoffman, T. J., Jurisson, S. S., and Quinn, T. P. (1998) Design and characterization of α-melanotropin peptide analogs cyclized through

rhenium and technetium metal coordination. *Proc. Natl. Acad. Sci. U.S.A.* **95,** 12,814–12,818.

240. Eberle, A. N. (1993) Peptides containing multiple photolabels: a new tool for the analysis of ligand-receptor interactions: Reversible long-lasting stimulation and inhibition of MSH receptors by multiple photocrosslinks with α-MSH. *J. Recept. Res.* **13,** 27–37.
241. Eberle, A. N. (1995) Transformation of an irreversible MSH antagonist into an irreversible MSH agonist by differential receptor crosslinking using the photoaffinity technique. *J. Mol. Recognit.* **8,** 47–51.
242. Süli-Vargha, H., Botyanszki, J., Medzihradszky-Schweiger, H., and Medzihradszky, K. (1990) Synthesis of α-MSH fragments containing phenylalanine mustard for receptor studies. *Int. J. Pept. Protein Res.* **36,** 308–315.
243. Solca, F. F., Salomon, Y., and Eberle, A. N. (1991) Heterogeneity of the MSH receptor among B16 murine melanoma subclones. *J. Recept. Res.* **11,** 379–390.
244. Ahmed, A. R., Olivier, G. W., Adams, G., Erskine, M. E., Kinsman, R. G., Branch, S. K., Moss, S. H., Notarianni, L. J., and Pouten, C. W. (1992) Isolation and partial purification of a melanocyte-stimulating hormone receptor from B16 murine melanoma cells: a novel approach using a cleavable biotinylated photoactivated ligand and streptavidin-coated magnetic beads. *Biochem. J.* **286,** 377–382.
245. Chakraborty, A. K., Orlow, S. J., Bolognia, J. L., and Pawelek, J. M. (1991) Structural/functional relationships between internal and external MSH receptors: modulation of expression of Cloudman melanoma cells by UVB radiation. *J. Cell. Physiol.* **147,** 1–6.
246. Carrithers, M. D. and Lerner, M. R. (1996) Synthesis and characterization of bivalent peptide ligands targeted to G-protein-coupled receptors. *Chem. Biol.* **3,** 537–542.
247. Brandenburger, Y., Rose, K., Bagutti, C., and Eberle, A. N. (1999) Synthesis and receptor binding analysis of thirteen oligomeric α-MSH analogs. *J. Recept. Signal Transduct. Res.* **19,** 467–480.
248. Bagutti, C. and Eberle, A. N. (1993) Synthesis and biological properties of a biotinylated derivative of ACTH(1-17) for MSH receptor studies. *J. Recept. Res.* **13,** 229–244.
249. Erskine-Grout, M. E., Olivier, G. W., Lucas, P., Sahm, U. G., Branch, S. K., Moss, S. H., Notarianni, L. J., and Pouton, C. W. (1996) Melanocortin probes for the melanoma MC1 receptor: synthesis, receptor binding and biological activity. *Melanoma Res.* **6,** 89–94.
250. O'Hare, K. B., Duncan, R., Strohalm, J., Ulbrich, K., and Kopeckova, P. (1993) Polymeric drug-carrier containing doxorubicin and melanocyte-stimulating hormone: in vitro and in vivo evaluation against murine melanoma. *J. Drug Target.* **1,** 217–229.
251. Sharma, S. D., Granberry, M. E., Jiang, J., Leong, S. P., Hadley, M. E., and Hurby, V. J. (1994) Multivalent melanotropic peptide and fluorescent macromolecular conjugates: new reagents for characterization of melanotropin receptors. *Bioconjug. Chem.* **5,** 591–601.
252. Jiang, J., Sharma, S. D., Hruby, V. J., Bentley, D. L., Fink, J. L., and Hadley, M. E. (1996) Human epidermal melanocyte and keratinocyte melanocortin receptors: visualization by melanotropic peptide conjugated microspheres (latex beads). *Pigment Cell Res.* **9,** 240–247.

253. Jiang, J., Sharma, S. D., Fink, J. L., Hadley, M. E., and Hruby, V. J. (1996) Melanotropic peptide receptors: membrane markers of human melanoma cells. *Exp. Dermatol.* **5,** 325–333.
254. Chang, A. C. Y., Cochet, M., and Cohen, S. N. (1980) Structural organization of human genomic DNA encoding the pro-opiomelanocortin peptide. *Proc. Natl. Acad. Sci. U.S.A.* **77,** 4890–4894.
255. Chang, A. C. Y., Cochet, M., and Cohen, S. N. (1985) Structural analysis of the gene encoding human proopiomelanocortin: in *Biogenetics of Neurohormonal Peptides.* (Håkanson, R. and Thorell, J. eds.), Academic Press, London, pp. 15–28.
256. Clark, A. J., Lavender, P. M., Besser, G. M., and Rees, L. H. (1989) Pro-opiomelanocortin mRNA size heterogeneity in ACTH-dependent Cushing's syndrome. *J. Mol. Endocrinol.* **2,** 3–9.
257. Crine, P., Lemieux, E., Fortin, S., Seidah, N. G., Lis, M., and Chrétien, M. (1981) Expression of variant forms of proopiomelanocortin, the common precursor to corticotropin and β-lipotropin in the rat pars intermedia. *Biochemistry* **20,** 2475–2481.
258. Morris, J. C., Savva, D., and Lowry, P. J. (1995) Reduced expression of a naturally deleted form of human proopiomelanocortin complementary deoxyribonucleic acid after transfection into Chinese hamster ovary cells. *Endocrinology* **136,** 195–201.
259. Bicknell, A. B., Savva, D., and Lowry, P. J. (1996) Pro-opiomelanocortin and adrenal function. *Endocr. Res.* **22,** 385–393.
260. Can, G., Abdel-Malek, Z., Porter-Gill, P. A., Gill, P., Boyce, S., Grabowski, G. A., Nordlund, J., and Farooqui, J. (1998) Identification and sequencing of a putative variant of proopiomelanocortin in human epidermis and epidermal cells in culture. *J. Invest. Dermatol.* **111,** 485–491.
261. Krude, H., Biebermann, H., Luck, W., Horn, R., Brabant, G., and Grüters, A. (1998) Severe early-onset obesity, adrenal insufficiency and red hair pigmentation caused by POMC mutations in humans. *Nat. Genet.* **19,** 155–157.
262. Hinney, A., Becker, I., Heibult, O., Nottebom, K., Schmidt, A., Ziegler, A., Mayer, H., Siegfried, W., Blum, W. F., Remschmidt, H., and Hebebrand, J. (1998) Systematic mutation screening of the pro-opiomelanocortin gene: identification of several genetic variants including three different insertions, one nonsense and two missense point mutations in probands of different weight extremes. *J. Clin. Endocrinol. Metab.* **83,** 3737–3741.
263. Seidah, N. G., Day, R., Marcinkiewicz, M., and Chrétien, M. (1993) Mammalian paired basic amino acid convertases of prohormones and proproteins. *Ann. N.Y. Acad. Sci.* **680,** 135–160.
264. Marcinkiewicz, M., Day, R., Seidah, N. G., and Chrétien, M. (1993) Ontogeny of the prohormone convertases PC1 and PC2 in the mouse hypophysis and their colocalization with corticotropin and α-melanotropin. *Proc. Natl. Acad. Sci. U.S.A.* **90,** 4922–4926.
265. Tanaka, S., Yora, T., Nakayama, K., Inoue, K., and Kurosumi, K. (1997) Proteolytic processing of pro-opiomelanocortin occurs in acidifying secretory granules of AtT-20 cells. *J. Histochem. Cytochem.* **45,** 425–436.
266. Eipper, B. A., Bloomquist, B. T., Husten, E. J., Milgram, S. L., and Mains, R. E. (1993) Peptidylglycine α-amidating monooxygenase and other processing enzymes in the neurointermediate pituitary. *Ann. N.Y. Acad. Sci.* **680,** 147–160.

267. Dores, R. M., Steveson, T. C., and Price, M. L. (1993) A view of the *N*-acetylation of α-melanocyte-stimulating hormone and β-endorphin from a phylogenetic perspective. *Ann. N.Y. Acad. Sci.* **680,** 161–174.
268. Castro, M. G. and Morrison, E. (1997) Post-translational processing of proopiomelanocortin in the pituitary and in the brain. *Crit. Rev. Neurobiol.* **11,** 35–57.
269. Rindler, M. J. (1998) Carboxypeptidase E, a peripheral membrane protein implicated in the targeting of hormones to secretory granules, co-aggregates with granule content proteins at acidic pH. *J. Biol. Chem.* **273,** 31,180–31,185.
270. Schmidt, W. K. and Moore, H. P. (1995) Ionic milieu controls the compartment-specific activation of pro-opiomelanocortin processing in AtT-20 cells. *Mol. Biol. Cell* **6,** 1271–1285.
271. Kovacs, K., Horvath, E., Stefaneanu, L., Bilbao, J., Singer, W., Muller, P. J., Thapar, K., and Stone, E. (1997) Pituitary adenoma producing growth hormone and adrenocorticotropin: a histological, immunocytochemical, electron microscopic, and in situ hybridization study: case report. *J. Neurosurg.* **88,** 1111–1115.
272. Dotman, C. H., Maia, A., Cruijsen, P. M. J. M., Jenks, B. G., and Roubos, E. W. (1998) Inhibitory and stimulatory control of proopiomelanocortin biosynthesis in the intermediate pituitary of *Xenopus laevis*. *Ann. N.Y. Acad. Sci.* **839,** 472–474.
273. Picon, A., Leblond-Francillard, M., Raffin-Sanson, M. L., Lenne, F., Bertagna, X., and Keyzer, Y. (1995) Functional analysis of the human pro-opiomelanocortin promoter in the small cell lung carcinoma line DMS-79. *J. Mol. Endocrinol.* **15,** 187–194.
274. Spencer, C. M. and Eberwine, J. (1999) Cytoplasmic proteins interact with a translational control element in the protein-coding region of proopiomelanocortin mRNA. *DNA Cell Biol.* **18,** 39–49.
275. Ottaviani, E., Capriglione, T., and Franceschi, C. (1995) Invertebrate and vertebrate immune cells express pro-opiomelanocortin (POMC) mRNA. *Brain Behav. Immunol.* **9,** 1–8.
276. Duvaux-Miret, O., Stefano, G. B., Smith, E. M., Dissous, C., and Capron, A. (1992) Immunosuppression in the definitive and intermediate hosts of the human parasite *Schistosoma mansoni*. *Proc. Natl. Acad. Sci. U.S.A.* **89,** 778–781.
277. Salzet, M., Salzet-Raveillon, B., Cocquerelle, C., Verget-Bocquet, M., Pryor, S. C., Rialas, C. M., Laurent, V., and Stefano, G. B. (1997) Leech immunocytes contain proopiomelanocortin: nitric oxide mediates hemolymph proopiomelanocortin processing. *J. Immunol.* **159,** 5400–5411.

CHAPTER 2

Melanocortins and Pigmentation

Aaron B. Lerner

It all began in 1916, the story not only for the melanocortins and pigmentation but also for the entire field of pituitary endocrinology. Two independent papers appeared in *Science* by two young biologists, Philip E. Smith in California *(1)* and Bennett M. Allen in Kansas *(2)*. They described a way to ablate the pituitary glands of tadpoles without killing the animals, and they observed that the tadpoles so treated were light in color. Soon after these reports it was found that injections of pituitary extracts into tadpoles and frogs would turn them dark *(3)*. Before this apparently simple achievement, it was thought that the pituitary gland was necessary for life. No investigator had been able to destroy or remove that gland from any animal and keep it alive. Ten years later Smith *(4)*, in another major success, reported his procedure for the ablation of the hypophysis in rats. He opened the door for intense research on the role of the pituitary gland in mammalian systems. It should not be a surprise that a change in color of tadpoles marked the beginning of pituitary endocrinology. It was essential that one be able to see and measure a change with visible light. There were no spectrophotometers or other equipment to monitor the metabolic processes that occurred outside the visible range after the destruction or removal of a gland or the injection of extracts from glands into animals.

While impressive advances were being made in the basic biologic and medical sciences, there were numerous questions in clinical medicine regarding disorders of pigmentation. Some people had defects from birth—albinism, piebaldism, large nevi, and so on. Others had conditions that were acquired—local or generalized hyperpigmentation or hypopigmentation, or both. Both acquired conditions occur in adrenal insufficiency or Addison's disease. In this paper I will be concerned mostly with the darkening that occurs in patients with loss of adrenal function from any cause (idiopathic atrophy, tuberculosis, metastatic cancer, removal of the adrenals, etc.). In this disorder there is hyperpigmentation of the exposed areas (face, hands, arms) the body folds and

The Melanocortin Receptors Ed.: R. D. Cone

creases (axillae, groin, palms), pigmented nevi, the oral cavity, and sites of recent scars. People of medium dark color and who tan well can become extremely dark. Replacement therapy with relatively low doses of cortisone, 37.5 mg daily, is usually sufficient to get the patient back to his or her original color.

What caused the darkening? It was generally assumed that the darkening that occurred in tadpoles and frogs minutes after the injection of pituitary extracts was totally unrelated to what happens in human beings. It was assumed—wrongly—that it takes days or weeks for human being to darken following adrenalectomy. It should have been realized that under the proper conditions human beings do have the capacity to darken quickly. For example, some people of medium dark complexion can darken within 24 hours after exposure to strong sunlight. Injection of melanocyte-stimulating hormone (MSH) can darken someone in 2 or 3 days.

In the early 1950s efforts were being made to isolate MSH from the pituitary gland. Smith had also previously identified an adrenocorticotropic principle when he demonstrated that adrenal atrophy following hypophysectomy could be reversed by implantation of pituitaries. At about the same time the Armour Laboratories began to market adrenocorticotropic hormone (ACTH) for clinical use. It was found that Armour ACTH was a potent darkening agent for tadpoles and frogs. In addition, patients receiving ACTH for several weeks were turning dark. Some investigators were beginning to conclude that ACTH was the major darkening peptide in the pituitary. But when α- and β-MSH were isolated, they proved to be more potent than ACTH in darkening frog skin, with no ability to stimulate the adrenal glands. Armour produced their ACTH from whole bovine pituitary glands while we isolated MSH from bovine posterior pituitary glands, which we knew included cells of the intermediate lobe. Armour changed their method and began to separate physically the anterior lobes from the posterointermediate lobes. When their commercial ACTH came only from the anterior lobes the darkening stopped. Something other than ACTH caused the darkening. Their first ACTH product was contaminated with MSH and it was the MSH that was the offending agent. Injections of MSH into human subjects made them dark *(5–9)*.

The five peptides α-, β-, and γ-MSH, ACTH, and β-lipotropin that are part of the precursor molecule proopiomelanocortin (POMC) are referred to as *melanocortins*. They are peptide hormones and neuropeptides that together with their receptors participate in the control of an amazing array of processes, including pigmentation, adrenocortical steroidogenesis, energy homeostasis, inflammation, and others. Most is known about α-MSH and ACTH and their receptors. It appears that γ-MSH has no role in pigmentation. We do not know whether β-MSH can be processed from its parent peptide β-lipotropin or from

POMC, nor do we know much about β-lipotropin and pigmentation. Even though the sequence of the 13 amino acids that make up α-MSH is acetylated via an *N*-acetyltransferase to give α-MSH. The N and C-terminal ends of ACTH are free, but they are blocked in α-MSH—by an acetyl group at the N-terminus and by an amide group at the C-terminus. On the dispersion of melanin granules in frog melanocytes α-MSH is 30 times more potent than ACTH but only about five times more potent than *N*-acetylated ACTH made in the laboratory. We do not know whether or not there is an *N*-acetyltransferase for ACTH. If *N*-acetyl-ACTH were made either under normal or abnormal conditions it would be a potent darkening peptide. We know from limited clinical studies that synthetic α-MSH and an *N*-acetyl 23-amino acid ACTH can produce striking hyperpigmentation *(6,7)*. We also know that there was no darkening of a patient who received 1 mg (~100 units) of ACTH daily for 21 d but marked darkening in another patient who received 24 mg (approx 2400 units) of ACTH daily for 8 d *(8)*. On an equimolar basis α-MSH produced more rapid darkening than ACTH. The darkening of a patient with an unknown disorder *(11)* as well as those with primary biliary cirrhosis *(12)* may have been due to high levels of α-MSH.

Now that POMC has been found to be present in keratinocytes, melanocytes and other cells far from the pituitary gland we need to know whether or not it can be processed in these locations to produce the melanocortins, α-MSH and ACTH. This is particularly important, since hypophysectomy does not produce the depigmentation in mammals that it does in amphibians. What *N*-acetyl and des-*N*-acetyl forms exist? If they are produced, do they affect pigmentation locally? What defects occur in the receptors for α-MSH and ACTH?

Under normal conditions is there a role for MSH-ACTH on pigmentation? Probably yes. The melanocortins may serve to prime pigment cells so that, when needed, the cells can produce pigment. Such is the case of tanning of skin in response to exposure to ultraviolet light to protect against further photo damage.

In the past few years several growth factors have been found to be potent mitogens for melanocytes in culture *(13)*. Most of the pigment cells for these studies came from the foreskins of newborns. The factors include basic fibroblast growth factor (bFGF), endothelin-1, hepatocyte growth factor/scatter factor and stem-cell factor. These growth factors are generally more potent as mitogens for melanocytes in culture than α-MSH and ACTH. We don't know if the MSH-ACTH receptors of neonatal human pigment cells are modified in the culturing process so that they become less active to MSH-ACTH *(14)*. As stated above, we do know that α-MSH and ACTH injected into adults produce marked darkening. It is tempting to give these growth factors to people with vitiligo to see if the loss of melanocytes can be stopped and the proliferation of new cells increased.

Another area of investigation that is opening up concerns melanocytes and nitric oxide (NO). Human melanocytes express the neuronal form of nitric oxide synthase (NOS) and they are sensitive to NO produced via inducible NOS from Langerhans cells (E.A. Lerner, personal communication). We need to understand the interactions of the melanocortins, growth factors, and NO on pigment cells.

Summary

The melanocortins α-,β-,and γ-MSH, ACTH and β-lipotropin are neuropeptides processed from the precursor molecule POMC in different cells of the pituitary gland. POMC is also present in other cells including keratinocytes but the processing in these cells is still unknown. An *N*-acetyltransferase catalyzes the acylation of deacetyl MSH to α-MSH. Both α-MSH and ACTH can bring about the darkening of human beings. If ACTH could be acetylated in vivo it would be much more potent darkening agent than free ACTH but still not as potent as α-MSH.

What started off as a quest to explain the hyperpigmentation seen in patients with adrenal insufficiency led to the isolation of the melanocortins and their receptors. This knowledge together with the advances made on growth factors and nitric oxide will in turn be the basis for explaining the mechanism of many disorders of pigmentation.

References

1. Smith, P. E. (1916) Experimental ablation of the hypophysis in the frog embryo. *Science* **44,** 280.
2. Allen, B. M. (1916) The results of extirpation of the anterior lobe of the hypohysis and of the thyroid of rana pipiens larvae. *Science* **44,** 755.
3. Atwell, W. J. (1919) On the nature of pigmentary changes following hypophysectomy in the frog larvae. *Science* **39,** 48.
4. Smith, P. E. (1926) Ablation and transplantation of the hypophysis in the rat. *Anat. Rec.* **32,** 221.
5. Lerner, A. B. and McQuire, J. S. (1961) Effect of alpha-and beta-melanocyte stimulating hormones on the skin color of man. *Nature* **189,** 176.
6. Lerner, A. B. and Snell, R. S., Chanco-Turner, M. L., and McQuire, J. (1966) Vitiligo and sympathectomy. *Arch. Dermatol.* **94,** 269.
7. McQuire, J. S. and Lerner, A. B. (1963) Effects of tricosapeptide "ACTH" and alpha-melanocyte-stimulating hormone on skin color of man. *Ann. N. Y. Acad. Sci.* **100,** 622–630.
8. Lerner, A. B. and McQuire, J. S. (1964) Melanocyte-stimulating hormone and adrenocorticotropic hormone: Their relation to pigmentation. *N. Engl. J. Med.* **270,** 539–546.
9. Lerner, A. B., Shizume, K., and Bunding, J. (1954) Endocrine control of pigmentation. *J. Clin. Endocrinol. Metab.* **14,** 1463.

10. Dores, R. M., Stevenson, T. C. and Price, M. L. (1993) A view of the *N*-acetylation of α-melanocyte-stimulating hormone and β-endorphin from a phylogenetic perspective. *Ann. N. Y. Acad. Sci.* **680,** 161.
11. Pears, J. S., Jung, R. T., Bartlett, W., Browning, M. C. K., Kenicer, K., and Thody, A. J. (1992) A case of skin hyperpigmentation due to α-MSH hypersecretion. **126,** 286–289.
12. Bergasa, N. V., Vergalla, J., Turner, M. L., Loh, P. Y., and Jones E.A. (1993) α-melanocyte-stimulating hormone in primary biliary cirrhosis. *Ann. N. Y. Acad. Sci.* **680,** 454.
13. Halaban, R., Tyrell, L., Longley, J., Yarden, Y., and Rubn, J. (1993) Pigmentation and proliferation of human melanocytes and the effects of melanocyte-stimulating hormone and ultraviolet B light. *Ann. N. Y. Acad. Sci.* **680,** 290–301.
14. Hunt, G. (1995) Melanocyte-stimulating hormone: a regulator of human melanocyte physiology. *Pathobiology* **63,**12–21.

CHAPTER 3

Melanocortins and Adrenocortical Function

Martine Bégeot and José M. Saez

1. Introduction

Although the existence of a functional relationship between the pituitary gland and the adrenal cortex was revealed by the classic studies of Smith almost seventy years ago *(1)*, the first purified adrenocorticotropin (ACTH) preparation from sheep pituitary was obtained only in 1954 *(2)*, and its structure was determined in the following few years *(3)*. In the 1960s, it was shown that ACTH stimulated cyclic adenosine monophosphate (cAMP) production by bovine adrenocortical slices and that cAMP itself could stimulate steroidogenesis, suggesting the role of cAMP as an obligatory mediator of the effects of ACTH *(4)*. Moreover, several groups presented evidence that the hormones did not have to enter cells to stimulate steroidogenesis, since anti-ACTH antibody added several minutes after ACTH obliterated this effect *(5)* and ACTH[1–24] linked to cellulose was able to stimulate steroidogenesis of Y-1 adrenal tumor cells *(6)*. Finally, the presence of specific binding of ^{125}I-ACTH[1–39] to adrenal cell subcellular fraction, which contained ACTH-sensitive adenylate cyclase activity, was demonstrated in 1971 *(7)*. Taken together, these findings led to the proposition that the initial event in the action of ACTH on adrenal cells was the binding of the hormone with specific receptors on the cell membrane leading to stimulation of adenylate cyclase and an increase in cAMP production, which in turn mediates an increase in steroidogenesis *(8)*. This classical schema of the mechanism of ACTH action has been questioned for several reasons. First, several groups found that the affinity of labeled ACTH[1–39] or ACTH[1–24] for its receptor was a least two orders of magnitude lower that the concentration of the hormone than both in vivo and in vitro stimulated steroidogenesis, an observation that cast doubt on the physiologic significance of these binding studies. Further studies indicated

The Melanocortin Receptors Ed.: R. D. Cone

that the biologic activity of these labeled ACTH preparations was low; however, it resulted from the presence of the large iodine atom on Tyr[2] *(9)* and the oxidation of Met[5] during the iodination reaction *(10)*. Second, the obligatory mediator role of cAMP was also put in question by the fact that no significant changes in cAMP production were observed with low but steroidogenically effective, concentrations of ACTH and some of its analogs (reviewed in ref. *11*).

A successful solution to these apparently contradictory findings has been afforded by the progress in three areas :

1. Utilization of ACTH or ACTH analogs in which only Tyr[23] was monoiodinated and which conserved full biologic activity *(12,13)*.
2. Careful analysis of the activities of cAMP-dependent protein kinase during ACTH-induced steroidogenesis *(14,15)* and of the Y-1 mutants having an alteration of cAMP-dependent protein kinase *(16)*.
3. The cloning of ACTH receptor (ACTH-R) *(17)* and the discovery that some patients with ACTH resistance have mutations of ACTH-R. (*see* Chapter 12).

In this chapter, we present an overview of ACTH-R gene organization, mRNA and protein, regulation of ACTH receptors, structure-function relationships of ACTH and related peptides, effects of ACTH on adrenocortical cells and ACTH receptor mutations in human adrenocortical pathology.

2. Characterization of the ACTH Receptor Gene, mRNA, and Protein

2.1. Characterization of the ACTH Receptors

2.1.1. Binding Studies

Except in the first reports *(18,19)*, the detection of ACTH binding sites with high specificity has been successful by using either the ACTH analog {Phe[2],Nle[4]}ACTH[1–38] or the normal ACTH molecule 1–39 monoiodinated on tyrosine 23, both of which retain the biologic activity of the peptide on adrenocortical cells of several species, as summarized in Table 1. In the different species, high–affinity (K^D varying from $2 \times 10^{-10}M$ to $2 \times 10^{-9}M$) and low capacity (from 900 to 3500 sites per cell) binding sites have been characterized in all studied species. These sites are responsible for the physiologic response of adrenal cells to ACTH. A second binding site of low affinity (K^D varying from $10^{-8}M$ to $10^{-7}M$) and high capacity (>20,000 sites per cell) has been also reported in some species. This may result from binding of contaminating {$_{125}$I-Tyr[2]}ACTH[1–39], which explains why its presence has not been reported by some authors. The physiological significance of these low-affinity binding sites is not well understood. The presence of very high-affinity binding sites has been reported only in the rat *(20)*.

Table 1
ACTH or Analog Binding Characteristics in Different Species

Species	K_D (M)	sites per cell	ACTH or derivative used	Authors
Rat	2.5×10^{-10} 10^{-8}	3,000 30,000	$[^{125}I]ACTH_{1-39}$	McIlhinney and Schulster, 1975
Rat	2.6×10^{-10} 7×10^{-9}	7,500 57,400	$[^{125}I]ACTH_{1-39}$	Yanagibashi et al., 1978
Rat	2.4×10^{-9}	4,000	$[3,5\text{-}^{3}H]Tyr^{2,23}\text{-}ACTH_{1-39}$	Ramachandran et al., 1980
Rat - Fasciculata	1.4×10^{-9}	3,500	$[^{125}I\text{-}Tyr^{23},Phe_2,Nle_4]ACTH_{1-38}$	Buckley and Ramachandran, 1981
Rat - Fasciculata	10^{-11} 3×10^{-9}	7,000 630,000	$[^{125}I\text{-}Tyr^{23},Phe_2,Nle_4]ACTH_{1-38}$	Gallo-Payet and Escher, 1985
Rat - Glomerulosa	7×10^{-11} 1.2×10^{-9}	65,000 10^6	$[^{125}I\text{-}Tyr^{23},Phe_2,Nle_4]ACTH_{1-38}$	Gallo-Payet and Escher, 1985
Domestic fowl	1×10^{-9} 2×10^{-8}	3 fmol/50 µg DNA 7.5 fmol/50 µg DNA	$[^{125}I\text{-}Tyr^{23}]ACTH_{1-39}$	Carsia and Weber, 1988
Bovine - Fasciculata	2.3×10^{-10} 1.6×10^{-7}	2,000 30,000	$[^{125}I\text{-}Tyr^{23}]ACTH_{1-39}$	Penhoat et al., 1989
Ovine - Fasciculata	2.7×10^{-10} 10^{-7}	1,200 >30,000	$[^{125}I\text{-}Tyr^{23}]ACTH_{1-39}$	Rainey et al., 1989
Porcine - Total adrenal cortex	4.2×10^{-10}	maximum 56 fmol /mg protein	$[^{125}I\text{-}Tyr^{23},Phe_2,Nle_4]ACTH_{1-38}$	Klemcke and Pond, 1991
Human - Total	1.6×10^{-9}	3,500	$[^{125}I\text{-}Tyr^{23},Phe_2,Nle_4]ACTH_{1-38}$	Catalano et al., 1986
Human - Fasciculata	5.7×10^{-10} 5×10^{-8}	900 20,000	$[^{125}I\text{-}Tyr^{23}]ACTH_{1-39}$	Lebrethon et al., 1994
Human - Glomerulosa	1×10^{-9}	47 fmol/mg protein	$[^{125}I\text{-}Tyr^{23}]ACTH_{1-39}$	Gallo-Payet et al., 1996

2.1.2. Purification of the ACTH Receptor: ^{125}I-ACTH Crosslinking Studies

Before cloning of the ACTH receptor was achieved, several groups attempted to characterize ACTH receptors by using binding of ACTH or analogs and crosslinking. In the first study, a tritiated photoreactive derivative of ACTH was used: {2-nitro-5-azidophenylsulfenyl Trp^9}ACTH[1–39] on rat adrenal cells, and the presence of a single band of M^r 100 was reported *(21)*. Thereafter, a biotin ACTH analog, ^{125}I{Phe^{2}Nle^4, D TBa t^{25}}ACTH[1–25] amide, was used for binding and crosslinking of the probe with disuccinimidyl suberate (DSS). Purification by succinoylavidin Sepharose followed by SDS-PAGE identified a crosslinked protein of an apparent M^r of 43 in bovine adrenal cell membranes *(22)*. In a similar approach, two groups have been able to characterize ACTH receptors by using the {^{125}I-Tyr^{23}}ACTH[1–39] in covalent crosslinking with DSS. In membranes from bovine adrenal cortex, the radioligand bound specifically a 40-kDa protein as revealed by sodium dodecyl sulfate-polyacrylamide gel electrophoresis (SDS-PAGE) of the solubilized membrane proteins *(23)*. However, in a more recent study, using intact cultured bovine adrenal fasciculata cells, crosslinking resulted in the specific labeling of two specific proteins with apparent M^r of 154 and 43 as measured by SDS-PAGE *(24)*. A third band at 124 kDa could be also observed in some experiments. In the presence of phenylarsine oxide, which prevents internalization of the ACTH receptor complexes, only the 43-kDa protein persisted. Moreover, crosslinking to plasma membrane enriched fractions prepared from human or bovine adrenals resulted in the labeling of the 43-kDa protein only. These data would suggest that the ACTH receptor at the cell surface is a protein of 43-kDa and that during the internalization process, this protein became associated with another macromolecule giving rise to the 154-kDa labeled protein observed in intact cells.

2.2. Molecular Cloning of the ACTH Receptor and mRNA Expression

2.2.1. Cloning of the ACTH Receptor

The first approach to characterize the ACTH receptor mRNA was functional expression from rat adrenal mRNA after microinjection in *Xenopus laevis* oocytes. Size fractionation of rat poly(A$^+$) RNA by sucrose density gradient centrifugation revealed that the mRNA encoding the ACTH receptor was present in the 1.1- to 2-kb fraction, as assessed by the ability of ACTH to induce cAMP production by the injected oocytes *(25)*. In 1992, cloning of a family of genes encoding the human melanocortin receptors using a polymerase chain reaction (PCR) strategy was reported *(17)*. Oligonucleotides used for amplification were based on degenerate sequences as previously described for dopamine receptors. Two different genes were isolated encoding the ACTH and melanocyte stimulating hormone (MSH) receptors, respectively, which permitted the prediction of the amino-acid sequences (297 amino acids for the ACTH receptor) and secondary structure. Alignment with other G protein-coupled receptors revealed that these receptors define a novel subfamily of G protein-coupled receptors; the melanocortin receptor family, with some novel features *(17,26)*. The ACTH receptor has a predicted molecular weight of 33 kDa in its unmodified form with two potential sites for N-linked glycosylation, compatible with the sizes reported previously *(22–24)*. Cloning of the ACTH receptor has been achieved in other species, including mouse and bovine *(27–29)*, and alignment of amino-acid sequences reveal between 81% and 88% identity with the human counterpart. More recently, the baboon ACTH receptor cDNA has been cloned with 97% identity with the human ACTH receptor *(30)*. Further work by several groups has identified three additional members in the melanocortin receptor family two of which are neural receptors. A nomenclature has been suggested with a general term melanocortin receptor (MCR), and the following assignments : MSH-R = MC1-R, ACTH-R = MC2-R, MC3-R, then MC4-R and MC5-R.

All these receptors have been cloned in human *(31–33)*, in mouse *(34,35)* and rat *(36)*. By fluorescence *in situ* hybridization, it has been reported that the ACTH receptor in human maps to 18p11.2 *(37,38)* and the corresponding location of the mouse ACTH receptor is also at a single locus at the distal end of chromosome 18 *(38)*.

2.2.2. Expression of ACTH Receptor mRNA in Adrenals

After cloning the ACTH receptor, the preparation of different cDNA probes has allowed the study of the expression of mRNA encoding the ACTH receptor in different tissues as well as the developmental expression by Northern blot analysis or *in situ* hybridization.

By using a human ACTHR probe, it has been reported that a single transcript at 4 kb was detected in adrenals of rhesus macaque but not in the other tissues tested: pituitary, liver, lung, thyroid, or kidney *(17)*. Localization by *in situ* hybridization was limited to cortex in the zona fasciculata reticularis and in the cortical half of the zona glomerulosa. Further studies have revealed the presence of multiple transcripts encoding the ACTHR in several species. In bovine fasciculata cells, using a bovine probe, a major mRNA transcript of 3.6 kb and three minor ones of 1.3, 1.8, and 4.2 kb were detected *(39)*. The same transcripts were detected in ovine fasciculata cells *(40)*. In human adrenals, by using a human probe, two major RNA transcripts at 1.8 and 3.4 kb were detected as well as three minor ones at 4, 7, and 11 kb *(41,42)*. These observations are illustrated in Fig. 1 for bovine and human cells. Basal expression was lower in cultured human adrenal cells than in cultured bovine cells *(43)*. In human fetal adrenal glands at midgestation (16–24 wk), mRNA transcripts were detected using *in situ* hybridization in higher abundance in definitive cortical zone than in fetal cortical zone, particularly in the more central areas of the fetal zone *(42)*.

The presence of multiple ACTH receptor mRNA transcripts has also been reported in the baboon adrenal glands *(30)*. A major mRNA transcript was observed at 3.4 kb in the fetal adrenal gland. Two lesser, although relatively intense, mRNA transcripts of 4.0 and 1.8 kb and three minor ones at 7, 10, and 11 kb were also observed at midgestation. The size and intensity of these transcripts was very similar to the human adrenal cells *(41)*. There was a biphasic developmental expression of the ACTH receptor mRNA in fetal adrenal gland during pregnancy in the baboon, with a marked increase between early and midgestation and a decline of approximately 70% between mid-and late gestation. However, by using *in situ* hybridization it has been demonstrated that ACTH-R mRNA levels were twofold greater in the fetal zone at mid-than at late gestation and threefold greater in the definitive zone than in the fetal zone in late gestation *(44)*. These results are similar to those previously described in fetal human adrenal glands at midgestation *(42)*. The presence of at least two mRNA transcripts was also reported in human adrenal adenoma tissues at 2.0 and 4.0 kb *(45)*. Similar observations were reported in H-295R cells from human adrenal carcinoma *(46)*. On the contrary, in mouse Y-1 cells, only the mRNA transcript at 2 kb was detected *(46)*, whereas two mRNA transcripts at 2 and 4.5 kb were detected in rat adrenal glands *(45)*.

Analysis of the distribution of melanocortin receptor subtypes in human tissues by reverse transcriptase-polymerase chain reaction (RT-PCR) and hybridization has shown that a single and specific PCR product for MC5-R has been detected in adrenal gland, but not for MC1-R, MC3-R and MC4-R *(47,48)*.

2.2.3. Expression of ACTH Receptor mRNA in Other Tissues

In a recent paper, the first evidence has been provided by using RT-PCR followed by hybridization that mRNA for MC2-R is expressed in normal and pathologic human skin as well as in cultured cells derived from epidermis. There are no data available on the implication of these receptors in the regulation of skin functions *(49)*.

Although several years ago the presence of ACTH receptor in human leukocytes, assessed by binding studies, was reported *(50)*, recent studies using RT-PCR approach indicate that these ACTH binding sites are likely to be MC5-R rather than MC2-R *(47)*.

ACTH receptors have been also characterized in 3T3-L1 cells differentiated into adipocytes by binding of an ^{125}I-ACTH analog (K_D : $4 \times 10^{-9} M$, 3500 sites

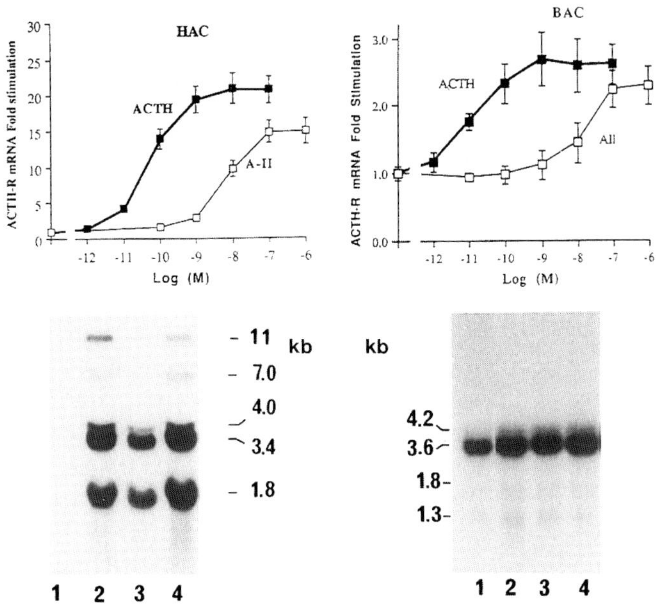

Fig. 1. **Top:** Dose-response of ACTH and AngII on ACTH-R mRNA in human (HAC) and bovine (BAC) adrenal cells. The results, expressed as fold increase over the control, are mean ± SEM. **Bottom:** Effects of ACTH, AngII or both on human and bovine adrenal cells ACTH-R mRNA. 1, Control; 2, ACTH $10_{-9}M$; 3, AngII $10_{-7}M$; 4, ACTH plus AngII.

per cell) *(51)*. There is now evidence that an mRNA transcript of 1.8 kb coding for MC2-R is expressed in mouse adipose tissue even though this expression is limited (only 0.1 of mRNA levels than in adrenals) as well as in 3T3-L1 differentiated in adipocytes *(52,53)*. Expression in adipocytes may be species-specific, however, since RT-PCR studies have shown that human adipocytes did not express MC2-R or MC5-R *(47)*.

2.3. Structure of the ACTH-R Gene and Promoter Characterization

2.3.1. Genomic Organization of the ACTH-R Gene

The entire coding region of the ACTH receptor gene is contained in a single exon *(17)*. For the human gene, by using 5'-rapid amplification of cDNA ends (5'-RACE), it has been demonstrated that a major initiation site of transcription was contained in a 49 bp upstream exon (exon 1) *(54)*. A perfect alignment of the 5'-untranslated region of the ACTH-R mRNA with the previously described genomic sequence *(17)* was obtained until position - 128bp from the ATG codon. The upstream 49 bp of the cDNA end were divergent with a consensus splicing acceptor site found at the point of divergence and corresponded to exon 1. This result was confirmed *(55)* by comparison of the results obtained by primer extension and S1 nuclease protection analysis. The size of the intron separating both exons was determined by long range PCR and is about 18 kb *(56)*. The presence of a major initiation site for transcription could not explain the multiple mRNA transcripts described in human fetal or adult adrenal cells. The presence of multiple function polyadenylation signals has been reported by using 3'-RACE methodology, which explains the size of the major mRNA transcripts at 1.8, 3.4, and 4 kb *(57)*. In fact, two polyadenylation signals at position +1404 and +1578 bp (from the ATG codon) could explain the large band observed at 1.8 kb on Northern blot, assuming a polyA tail, since the initiation site of transcription is located 177 bp upstream of the ATG codon of around 250 nucleotides. The genomic organization of the human ACTH receptor gene is shown in Fig. 2.

The genomic organization of the mouse ACTH receptor gene has been reported by two different laboratories *(58,59)*. The gene comprises at least four exons: exon 1 (between 109 and 113 bp), exon 3 (112 bp), and exon 4 (>1000 bp) containing the whole coding region, 96 bp of the 5'-UTR and 445 bp of the 3'-UTR followed by a single polyadenylation signal at position 1291 from the ATG codon, which explains the unique mRNA transcript of about 2 kb in mouse adrenal cortex. Moreover, a fourth alternative exon (exon 2) of 57 bp between exon 1 and exon 3 has been also described in some clones *(58)*. Exon 3 and exon 4 are separated by an intronic sequence of about 1.6 kb *(58)* and exon 1 is separated from the alternative exon by about 6 kb of intronic sequence (*see* Fig. 2).

2.3.2. Promoter Characterization of the ACTHR Gene

The promoter region of the human ACTH receptor gene has been cloned *(56,60)* (accession number is Y10100 HSACTH PRO). The sequence of this region contains no TATA or CAAT boxes but one sequence resembling an initiator element overlapping the major initiation site of transcription. This region contains a site for the steroidogenic factor 1 (SF1), a specific regulator for the steroidogenic tissues, at position –35 bp and several putative regulatory binding sites like SP1, AP1 sites, and cAMP response element (CRE)-like elements. This promoter is fully functional when ligated to the human GH reporter gene and transfected in Y-1 adrenocortical cells. Moreover, cAMP induced a 2.5-fold increase in stimulation over control which demonstrates the involvement of one or several CRE elements in the cAMP regulation of this

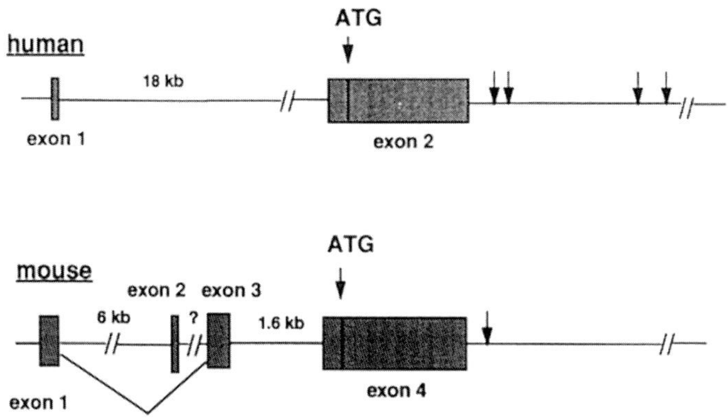

Fig. 2. Structure of the ACTH receptor gene. **Top:** Human **Bottom:** Mouse.

gene. It has been already reported *(41)* that upregulation of the ACTH receptor by ACTH in human adrenal cells is due to an increase in the transcription rate of this gene. The promoter of the mouse ACTH-receptor gene has been cloned (about 1.8 kb) and contains several potential regulatory sites *(59)*. Two SF1 sites at position –25 and –896 bp have been identified. The site located at position –25 binds one or two proteins from adrenal nuclear extracts and is fully functional after transfection in Y-1 cells but not in fibroblasts *(59)*. This promoter region contains also an SP1 site, an AP1 and AP2 site, but no consensus CRE binding sites *(59)*.

3. Regulation of ACTH Receptors

3.1. Regulation by Peptide Hormones: ACTH and Angiotensin II

In contrast to the loss of receptors and desensitization of target cells caused by most polypeptide hormones, ACTH appears to positively regulate its own receptor and the responsiveness of adrenal cells in all species studied. In human, in vivo studies have shown that repeated *(61)* or long-term *(62)* treatment with ACTH enhances the response to further ACTH stimulation. Similarly, in Cushing's disease and in the ectopic ACTH syndrome, glucocorticoid secretion is increased *(63)*. Although initially this enhanced ACTH responsiveness was related to the trophic effects of the hormone on the expression of the genes encoding several enzymes involved in the steroidogenic pathway *(64)*, further studies demonstrated that ACTH also enhances the expression of its own receptor.

In vitro studies using adrenal cells from normal human adult *(41)* and fetal *(42,65,66)*, bovine *(39,67)* and ovine *(66)*, as well as human and mouse adrenocortical tumor cell lines *(46)* have demonstrated that ACTH enhances ACTHR mRNA and/or protein. Concerning the effect of ACTH on ACTHR mRNA levels, they are time- (maximum stimulation between 12 and 24 h) and dose-dependent with an $ED^{50} \approx 10^{-11}M$ (Fig. 1). At maximal concentrations ACTH caused a marked increase in ACTH-R mRNA: 20-fold in human adult *(41)* and fetal *(42,65)* adrenal cells, 3-fold in bovine adrenal cells *(39)*, by 2 to 4-fold in human adrenocortical tumor cell line NCI-H295 and 6-fold in mouse tumor cell line Y-1 *(46)*. In adult human adrenal cells the effects of ACTH on ACTHR mRNA are exerted at both transcriptional and posttranscriptional levels *(41)*, whereas in bovine adrenal cells the effects are mainly posttranscriptional by increasing the mRNA stability *(39)*. Moreover, ACTH treatment also increases in a dose-and time-dependent manner ACTH receptor number *(41,66–68)*. However, whereas in bovine adrenal cells, the stimulatory effect of the hormone on ACTHR mRNA and receptor number was similar *(39,67)*, in both human adult and fetal adrenal cells, the effects on mRNA levels (≈ 20-fold) *(41,42)* were much higher than in receptor number (Fig. 3) *(41,66)*. This discrepancy between mRNA and receptor number has also been observed in transfection studies using human ACTHR cDNA *(69,70)*. These studies have shown that following transient or stable transfection of several cell lines with ACTHR cDNA, all of them expressed receptor mRNA, but the protein was only expressed in cells expressing an endogenous melanocortin receptor. By contrast, following transient *(27)* or stable *(71)* transfection of mouse ACTHR cDNA, a high number of functional receptors is expressed at the cell surface in a cell line that lacks any endogenous receptor.

In addition to ACTH, several other factors have been shown to be able to regulate ACTH-R. In both bovine *(39)* and human (adult and fetal) *(41,65)* adrenal cells, angiotensin-II (AngII) increases in a dose-dependent manner ACTHR mRNA (Fig. 1). The effects of AngII were less than those produced by ACTH, which in turn were less than those produced by the two hormones added together (Fig. 1). In human, but not in bovine adrenal cells, the enhanced ACTH-R mRNA levels were associated with a small increase in receptor number (Fig. 3).

3.2. Regulation by Growth Factors: IGFs and TGFβ

The second factor able to regulate positively ACTH-R expression is insulin-like growth factor I (IGF-I). In bovine adrenal cells IGF-I enhanced ACTH-R number in a dose- and time-dependent manner *(72)*, an effect associated with an increase of ACTH-R mRNA (Fig. 3). Similarly, treatment of human adrenal cells with IGF-I or IGF-II, increases ACTH-R mRNA *(73)* and receptor number (Fig. 4). Interestingly, in bovine adrenal cells the effects of ACTH and IGF-I on receptor number are synergistic *(72)*. Since IGF-I is secreted by bovine adrenal cells and this secretion is increased by ACTH *(74)*, IGF-I may play an autocrine role in ACTH-R expression.

In contrast to IGF-I, transforming growth factor β (TGFβ) has been shown to cause a decrease in ACTH receptor number in both ovine *(68)* and bovine *(75)* adrenal cells, and to block almost completely the stimulatory effect of ACTH on its own receptor. In bovine adrenal cells, the decrease of ACTH-R number was associated with a decrease of ACTH-R mRNA levels (Fig. 4). By contrast, in both adult *(76)* and fetal *(65)* adrenal cells TGFβ had no significant effect on either ACTH-R mRNA or binding sites (Fig. 3), despite the fact that these cells contain both subtypes, I and II, of TGFβ receptors, and that this peptide regulates negatively the expression of P-450c17. Moreover, since both bovine *(77)* and human *(76)* adrenal cells express $TGF\beta_1$, which is negatively regulated by ACTH, it was postulated that TGFβ may play an autocrine role in adrenal cell functions and this was recently confirmed, at least in bovine adrenal cells, by using an antisense approach *(77)*.

4. Structure–Function Relationships of ACTH and Related Peptides

The elucidation of the amino acid sequence of ACTH from different mammals demonstrated that all are composed of 39 amino acids with a serine at the amino-terminal and phenylalanine at the carboxyl terminal. The sequences of the NH2-terminal[1–24] and the COOH-terminal[34–39] are identical, whereas some differences are observed between positions 25 and 34, suggesting that the first 24 amino acids contained the biologically important part of the molecule. This was confirmed by the observation that the synthetic nonadecapeptide corresponding to the first 19 residues of ACTH exhibited all the biologic activities of ACTH *(78)*. Further studies of the structure–activity relationships of ACTH suggest that the C-terminal, in particular the residues 15–18 containing the positive charge Lys-Lys-Arg-Arg could play an important role in the binding of the hormone to its receptor, whereas the NH_2-terminal, in particular residue 4–10 forms the "active core" of the molecule for steroidogenesis (reviewed in refs. *79* and *80*). Thus, the biological activity of ACTH[4–24] is higher than that of ACTH[5–24], which in turn is higher than that of ACTH[6–26], whereas ACTH[7–23] loses all biologic activity, although it still binds to the receptor *(81)*.

Binding studies using fully biologically active labeled ACTH, have allowed a better definition of the structure-binding relationship of ACTH.

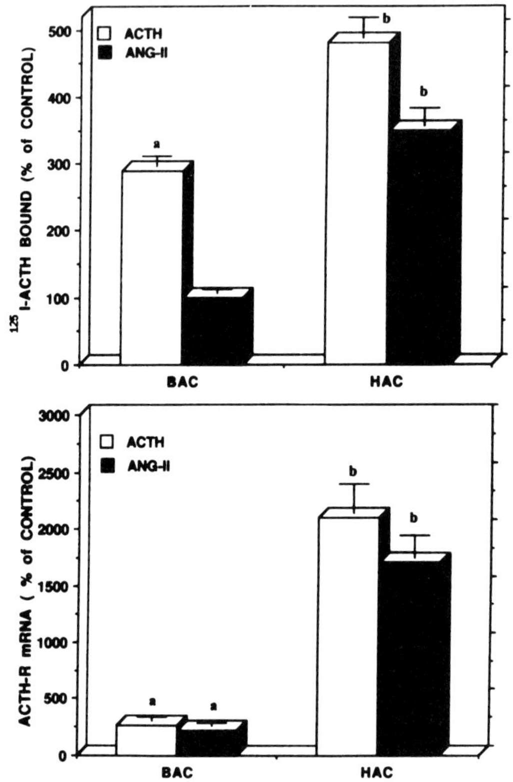

Fig. 3. Effects of ACTH ($10^{-9}M$), AngII ($10^{-7}M$) on ACTHR mRNA and receptor number in human (HAC) and bovine (BAC) adrenal cells. The results are expressed as percent of control are mean ± SEM.

Using the ^{125}I-ACTH analog {Phe2,Nle4}ACTH[1–38], the ability of several analogs, in which the charge profile of residues 15–18 was altered, to compete for the ACTH receptor, demonstrated that the inhibition was proportional to the positive charge in this sequence *(12)*. Thus, the binding affinities were in the following order : ACTH > ACTH[1–19 NH_2] > ACTH[1–17 NH_2] > ACTH[1–19] > ACTH[1–17]. The concentration–response curves of these analogs for cAMP production were superimposable on the binding inhibition curves. In contrast, a maximal steroidogenic response was observed when less than 5% of the binding sites were occupied. Using the same labeled ACTH analog, it has been also shown that the inhibitory potency of ACTH[1–24] and Phe^2,Nle^4 ACTH[1–38] was similar whereas no inhibition was observed with α-MSH at 1µ*M (82)*. More recently, using a stable cell line transfected with

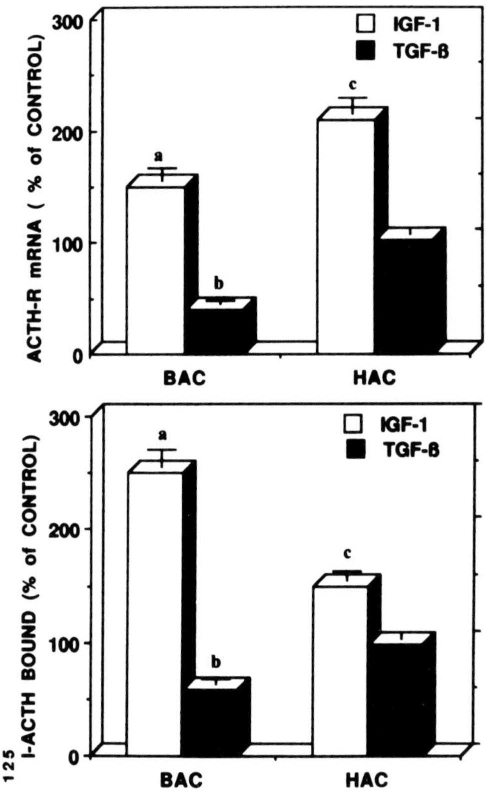

Fig. 4. Effect of IGF-I (50 ng/mL) and TGFβ (2 ng/mL) on ACTHR mRNA and binding sites in human (HAC) and bovine (BAC) adrenal cells. The results expressed as percent of control are mean ± SEM.

mouse ACTH-R cDNA, it has been reported that the binding affinity of ACTH[1–39], ACTH[1–24], ACTH[1–17], ACTH[11–24] and ACTH[7–39] was similar ($K^D \approx 10^{-9}M$), but that the last two analogs were only able to displace 60–70% of the tracer *(71)*. Neither ACTH[18–39] nor α-MSH at $10^{-7}M$ caused a significant displacement of the tracer. In this same study, it was reported that the EC_{50} for cAMP production of ACTH[1–24] ($7.6 \times 10^{-12}M$) was about 10 times lower than those of ACTH[1–39] and ACTH[1–17], whereas ACTH[7–39] and ACTH[11–24] were pure antagonists with an IC_{50} of about $10^{-7}M$. These results are surprising and in discrepancy with the results reported by several groups using adrenal cells. Thus, the EC_{50} for cAMP production of ACTH[1–24] and ACTH[1–17] is about 500 and 10,000 times lower respectively than observed using normal adrenal cells *(12,81,83)*. Whether these marked discrepancies are related to the high number of ACTH binding sites expressed by the transfected cell line compared to normal adrenal cells (26,000 vs 3000 sites per cell) or to another reason is unknown.

Several groups have reported that ACTH[11–24] behaves as a pure antagonist *(71,84–87)*, but two laboratories have reported that at micromolar concentrations, this peptide has a steroidogenic effect on rat adrenal cells *(88,89)*. Whether these effects are mediated by ACTH-R or by another receptor as previously suggested *(87)* remains unknown.

5. Effects of ACTH on Adrenocortical Cells

ACTH has three types of effects on adrenal cells: acute stimulation of the steroidogenesis, stimulation of expression of primary response genes and long-term trophic effects on cell growth and differentiation.

Although at the end of the 1980s, cAMP fulfilled many of the criteria established by Sutherland et al. *(90)* to be considered as the second messenger in the mechanism of action of ACTH, doubts about its role first surfaced at the beginning of the 1970s when it was observed that physiologic concentrations of the hormone could stimulate steroid production in isolated adrenal cells without causing measurable changes in the intracellular concentration of cAMP *(91–93)*. Further studies with analogs of ACTH also emphasized the discrepancy between the peptide concentrations required to stimulate steroid secretion versus those needed to stimulate cAMP formation *(85)*. To explain such a discrepancy, several hypotheses were postulated, either the involvement of other second messengers, cGMP *(94,95)*, Ca^{2+} *(95,96)*, and inositol phosphates *(97)* in the mechanism of action of ACTH or the existence of two populations of ACTH receptors, one which acts through cAMP, the other through a separate unknown mechanism *(87)*.

The availability of labeled ACTHs with full biologic activity has allowed the demonstration that the K_D and the concentrations of ACTH required for half–maximal stimulation of cAMP were an excellent agreement in both rat adrenal cells *(12)* and in bovine adrenal cells (Fig.5) but not in human adrenal cells *(98)*, and that in all cases the ACTH ED_{50} for steroid stimulation was about 40-fold lower. Despite these discrepancies, confirmation of the obligatory role of cAMP on the pleiotropic effects of ACTH on adrenal cells comes from several types of studies. First, ACTH dose–response curves for either protein kinase A (PKA) activation *(14)* or cAMP binding to the regulatory subunit of PKA *(15)* were very close. Second, a family of mutants of the mouse adrenocortical cell line Y-1, which are resistant to the effect of both ACTH and dibutyryl cAMP *(16)* have mutations of the regulatory subunit of PKA *(99)* and the degree of resistance was correlated with the degree of alteration of PKA. Stable transfection of these cells with an expression vector encoding the catalytic subunit of PKA, restored the response to both ACTH and 8-bromo-cAMP *(100)*. Taken together, the above data clearly establish the obligatory mediator role of cAMP on ACTH action.

For many years it has been shown that Ca^{2+} is required for the steroidogenic effect of ACTH *(101)* but a number of controversies still persist as to the exact locus of Ca^{2+} action. In one study, it was reported that Ca^{2+} was obligatory for ACTH binding to its receptor and therefore the lack of steroidogenic effect of ACTH in Ca^{2+}-free medium was due mainly to the absence of binding *(102)*, whereas in two other studies it was reported that extracellular Ca^{2+} did not significantly alter ACTH binding *(103,104)*. Similarly, the role of Ca^{2+} on ACTH-induced cAMP production is also subject to debate *(105,106)*. More recent studies have shown that ACTH increases intracellular Ca^{2+}, through stimulation of L-type Ca^{2+} channels in human adrenal cells *(104)* or T-type in bovine adrenal cells *(107)*. This ACTH-induced intracellular Ca^{2+} increase, which is blocked by PKA inhibitors, is essential for the stimulation of cAMP by ACTH *(104)*. Moreover, calmodulin inhibitors blocked, in a dose-dependent manner, the stimulatory effects of ACTH on both cAMP and steroids in the adrenal tumor cell line Y-1 *(108)*. The above finding leads to the following proposal *(104)* : upon binding to its receptor, ACTH triggers a small increase in intracellular cAMP, activation of PKA and phosphorylation of Ca^{2+} channels, which in turn increase the opening probability and therefore the Ca^{2+} influx, further increasing production of cAMP. However, since extracellular Ca^{2+} also regulates the steroidogenic response to cAMP derivatives, Ca^{2+} must be involved in some step beyond cAMP formation.

5.1. Acute Steroidogenic Response

Both in vivo and in vitro ACTH increases steroid output within the first minutes. This acute stimulation is mainly due to an increased conversion of cholesterol to pregnenolone by the P-450scc which is located within the mitochondrial inner membrane. Many years ago, it was shown that cycloheximide, an inhibitor of protein synthesis, could block ACTH-induced steroid production *(109)*. However, cycloheximide had no effect on either the activity of the P-450scc or the delivery of cholesterol to the outer mitochondrial membrane *(110)*. Therefore, it was concluded that the translocation of cholesterol from the outer to the inner mitochondrial membrane, and therefore the initiation of steroidogenesis, was dependent upon the synthesis of a labile protein. Several candidate proteins have been put forth as the acute regulator

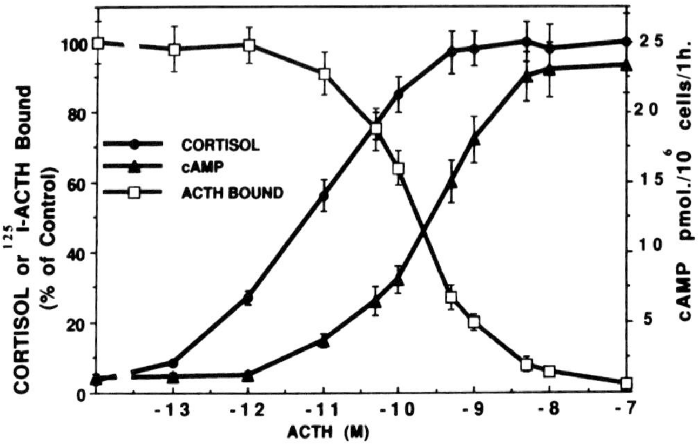

Fig. 5. ACTH dose-response curve for cortisol (●) and cAMP (▲) production and displacement of bound [^{125}I-Tyr^{23}]ACTH[1–39]. The cortisol production and the displacement of bound [^{125}I-Tyr^{23}] are expressed as percent of control, whereas cAMP production is expressed in pmol/10^6 cells/L h.

of steroidogenesis, including steroidogenesis activator protein (SAP), the peripheral benzodiazepine receptor (PBR/DBI) and the steroidogenic acute regulatory (StAR) protein (reviewed in refs. *111* and *112*). From biochemical *(113,114)* and genetic studies *(115)* it appears that StAR is the essential labile protein factor required for acute steroid production in both adrenals and gonads, but not in other steroidogenic tissues as placenta and brain *(114)*. However, the exact mechanism by which ACTH stimulates the transcription and translation of StAR still remains to be elucidated.

Recent studies indicate that gap junction-mediated intercellular communication is one mechanism by which adrenal cells increase their responsiveness to low ACTH concentrations and therefore can explain in part the discrepancy between the EC_{50} for steroid and cAMP production *(83)*. Both in vivo and in vitro, human and bovine adrenal cells express connexin43. Moreover, when cultured at high density, one cell can communicate with as many as 20 other cells, as assessed by the transfer of the fluorescent dye, Lucifer yellow. When the cells were cultured at lower density, the dye transfer was less marked but in both cases the cell-to-cell communication was completely blocked by gap junction inhibitors. Interestingly, although the steroid production per cell at maximal concentrations of ACTH was similar when cells were cultured at high or low densities, the ACTH ED_{50} for cortisol production by cells cultured at high density was significantly lower (6-fold) than that by cells cultured at low density (Fig. 6). However, in the presence of gap junction inhibitors, there was a shift of the ACTH concentration-response curves in the two culture conditions. The ACTH ED_{50} of high and low density culture increased 25- and 5-fold, respectively, and became similar (Fig. 6). Since gap-junction inhibitors did not modify either ACTH binding, ACTH-induced cAMP production, or bromo-cAMP-induced steroid production *(83)*, the most likely explanation for the above findings is that the adrenal cells belong to the threshold response model, in which the threshold concentration of hormone necessary to initiate the response differs among individual cells. Thus, at low concentrations of ACTH, only a few cells respond, but the cAMP response is maximal and can diffuse to neighbouring cells through gap junctions. Considering the volume of adrenal cells, one can calculate that the cAMP produced by a cell can theoretically activate the PKA of 10 to 15 cells.

5.2. *Effects of ACTH on Early Response Genes*

Most nucleated eukaryotic cells constantly respond to a variety of factors that bind to cell surface receptors which induce, in addition to acute effects, long-term transcription-dependent changes in phenotype. One mechanism that has been proposed to link ligand–receptor interactions with long-term effects is the induction of transcriptional regulatory proteins, which are encoded by the "immediate early gene," the induction of which is independent of protein synthesis but requires posttranscriptional modification of preexisting factors *(116)*. Among the very large number of immediate early genes, cellular nuclear protooncogenes, in particular members of the Fos and Jun families, appear to play a crucial role in linking the indirect ligand–receptor interactions to the effects of such ligands on growth, development, and/or differentiation *(116,117)*.

In vivo ACTH has been shown to produce a rapid but transient increase in rat adrenal c-*fos* mRNA levels *(118)* and both acute stress *(119)* and ACTH administration *(120)* enhanced the number of adrenal cells containing c-*fos* immunoreactivity. Furthermore, in vitro ACTH increased c-fos mRNA level in Y-1 adrenocortical cells *(121)* and in cultured adrenal cells of several species *(122–124)*. In both bovine and ovine adrenal fasciculata cells, ACTH stimulated in a dose-dependent manner the mRNA level of both c-*fos* and *Jun*-B, but not of c-*jun* *(124)*. In these cells, as well as in bovine glomerulosa cells *(122)*, angiotensin-II also increased c-*jun* mRNA levels. Although it is clear that ACTH regulates protooncogene expression, and some of the promoters of genes regulated late on by ACTH contain a putative AP-1 like sequence, it is not yet clear whether the protooncogenes are involved in the long-term effects of ACTH on adrenal cells.

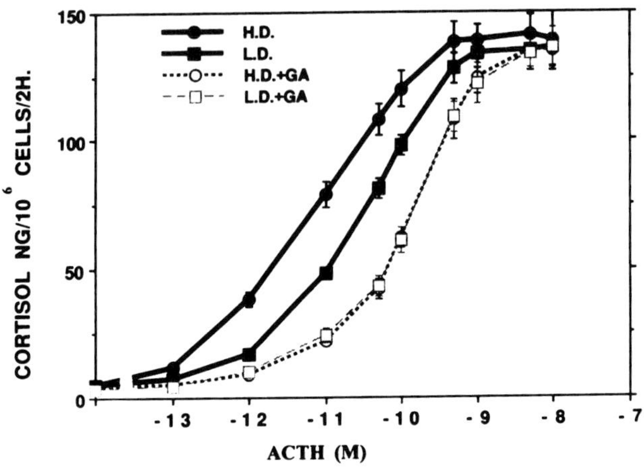

Fig. 6. Dose-dependent effect of ACTH on cortisol production on bovine adrenal culture at high density (HD ●, ○, 2.4×10^5 cells/cm^2) or low density (LD ■, □, 0.25×10^5 cells/cm^2) in the absence (●, ■) or presence (○, □) of 5 µ*M* of 18α-glycyrrhetinic acid) (GA), an inhibitor of gap junctions. When treated by GA, adrenal cell cultures at either high or low density exhibit the same sensitivity to ACTH.

5.3. Effect of ACTH on Adrenal Cell Specific Function

The first indication that ACTH was required for the maintenance of adrenal growth and differentiation was given by the observation that in hypophysectomized rats, both the adrenal weight and the steroid hydroxylase content decreased, and the administration of ACTH to such animals led to recovery of the weight and of the steroidogenic capacity of the gland *(125,126)*.

ACTH is one of the few hormones which, even at high concentrations, do not produce desensitization of its target cells. Thus, both in vivo and in vitro, ACTH enhances the response of adrenal cells to further hormonal stimulation. This increased steroidogenic capacity is due to the positive action of ACTH on the expression of most of the genes encoding proteins involved in adrenal cell function (Table 2). Using bovine adrenal cells, it has been shown that ACTH increased the expression of all steroid hydroxylases (reviewed in refs. *64* and *127*). In addition, as described before, ACTH also regulates the expression of its own receptor and, at least in the mouse, StAR *(128)*. Although cAMP mimics the effects of ACTH on the expression of these genes, only the promoter of P-45011β has a consensus CRE. On the other hand, all promoters of the steroid enzymes *(129)*, human *(56)* and mouse

Table 2

Effects of Peptide Hormones and Growth Factors on the Expression of Specific Genes by Human Adrenal Cells

	ACTH-R	ANGII-R (AT1)	StAR	P-450scc	3ß-HSD	P-450 21	P-450c17	P-450 11ß1	P-450 11ß2
ACTH	↗	→	↗	↗	↗	↗	↗	↗	↗
AngII	↗	↘	ND	↗	↗	ND	↗	↗	↗
IGF-I	↗	↗	↗	↗	↗	ND	↗	ND	ND
TGFß1	→	↗	ND	→	↗	ND	↘	ND	ND

ND : not determined

(59) ACTHR, and mouse StAR *(128)* contain a binding site for the orphan nuclear receptor steroidogenic factor-1 (SF-1). This factor is also required for the development of adrenal and gonads during fetal life *(130,131)*. However, in addition to SF-1, other tissue-and species-specific transcriptional factors or coactivators are required to express the above adrenal specific genes. The presence of these other transcriptional factors helps explain why not all steroidogenic tissues expressing SF-1 have the same type of steroidogenic enzyme expression. Within the same steroidogenic tissue, there are also species-specific differences,for example, in bovine adrenal cells AngII decreases P-450c17 expression *(75)*, whereas in human adrenal cells AngII has opposite effects. These species-specific differences are even more marked when the long-term effects of peptide hormones and growth factors on the steroidogenic response to ACTH and AngII are studied (Fig. 7).

5.4. Effect of ACTH on Adrenal Growth

Although in vivo studies in humans and experimental animals have shown that an excess of endogenous or exogenous ACTH causes adrenal hyperplasia, whereas an atrophy of the gland is observed in the absence of ACTH *(132,133)* or in the syndrome of ACTH resistance (*see below*), most in vitro cell culture studies *(134)* have shown that ACTH is not a direct mitogen for adrenocortical cells. Indeed, in these cell models, ACTH is antimitogenic, rather than mitogenic. This effect is exerted at very low concentrations and appears to be directly correlated with the steroidogenic response, but is not due to an inhibition produced by the accumulation of steroids, and can be mimicked by cAMP derivatives.

Further in vivo studies have shown that the primary response of adrenal to exogenous ACTH is a hypertrophy as revealed by the RNA/DNA ratio and that the proliferative cellular response is secondary to the primary hypertrophic response (reviewed in ref.*135*). These observations lead to the proposal that the mitogenic effect of ACTH is indirect *(134)*. In favor of this hypothesis is the fact that the compensatory adrenal growth observed after unilateral adrenalectomy is inhibited by ACTH *(136)* but not by ACTH antisera that effectively neutralize circulating ACTH *(137,138)*. It has been proposed that the indirect mitogenic effect of ACTH might be mediated by increasing adrenal blood flow *(139)* which would facilitate the access of mitogens and growth stimulation to the adrenal cells *(134)*. It has been reported that N-terminal pro-opiomelanocortin (POMC) peptides produce a mitogenic response in adult, dexamethasone-treated rats *(140)* and in hypophysectomized animals *(141)*. Moreover, N-terminal proopiomelanocortin (POMC) antisera blocked the compensatory adrenal growth following unilateral adrenalectomy *(138)*. Since N-terminal POMC peptides are

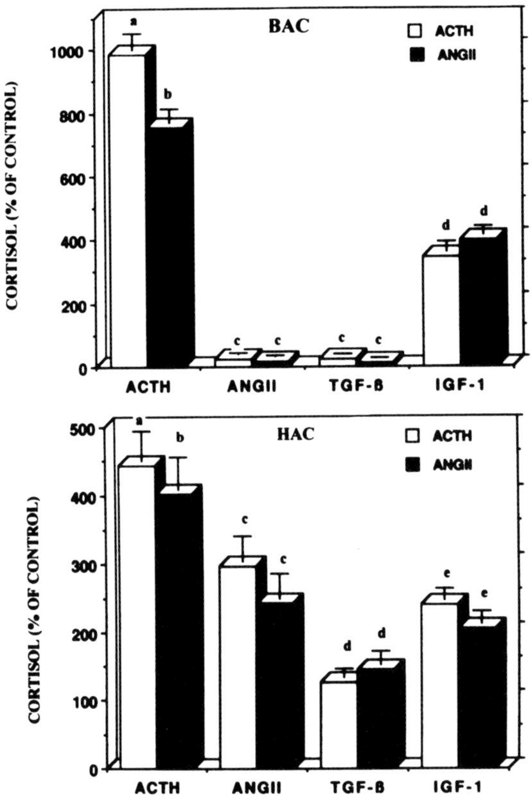

Fig. 7. Effects of a 3-day treatment with ACTH (10^{-9} *M*), AngII (10^{-7} *M*), TGFβ1 (10^{-10} *M*) or IGF-I (7.10^{-9} *M*) on bovine (BAC) and human (HAC) steroidogenic responsiveness. At the end of the experimental period, the media were removed and cells incubated in the presence of either ACTH (10^{-9} *M*) or AngII (10^{-7} *M*) and after 2 h the cortisol in the medium was measured. The results, expressed as percent of untreated cells, are mean ± SEM.

cosecreted with ACTH, the above findings can explain the adrenal cell proliferation observed in all pathologic or experimental situations in which there is an increase of ACTH. However, pituitary-derived N-terminal POMC peptides cannot account for the mitogenic effect observed following administration of synthetic ACTH and the compensatory growth response to unilateral adrenalectomy in hypophysectomized rats.

In addition to the indirect mitogenic action, another mechanism by which ACTH prevents adrenal atrophy might be through inhibiting adrenal cell apoptosis. Following hypophysectomy, apoptotic cells are observed in both zona fasciculata and zona reticularis within the first days and this cell death can be largely prevented by ACTH replacement *(142)*. However, the mechanisms, direct or indirect, by which ACTH prevents apoptosis are unknown.

In summary, while ACTH appears to regulate directly normal cell hypertrophy, its direct effects on adrenal cell proliferation are inhibitory. The mechanisms by which ACTH exerts its indirect mitogenic effects in vivo are still largely unknown. Although several growth factors (IGF-I, IGF-II, fibroblast growth factor [FGF]) can stimulate adrenal cell proliferation in vitro, and are synthesized by adrenal cells, it remains to be demonstrated if these factors acting in an autocrine/paracrine manner are responsible for the indirect mitogenic effects of ACTH.

6. ACTH Receptor in Human Adrenocortical Pathology

6.1. Familial Glucocorticoid Deficiency

Familial glucocorticoid deficiency (FGD) is a rare autosomal recessive disease characterized by a severe glucocorticoid deficiency (less than 10 ng/ml plasma) associated with very high plasma ACTH levels (often more than 1000 pg/ml) but a well-preserved renin–angiotensin–aldosterone axis and requires a treatment with glucocorticoid replacement alone (for a detailed discussion see Chapter 12). This disorder was first described in 1959 *(143)* and about 100 cases have been described in the world. Histologic examination of the adrenal glands from patients dying from this syndrome has revealed a large atrophy of the fasciculata-reticularis zona of the adrenal cortex but a preserved glomerulosa zona *(144)*. These clinical and histological features have led to the conclusion that this syndrome is typically due to a specific ACTH resistance. It was suggested that it may reflect some defects in the ACTH receptor or in the signal transduction system. In 1993, the first mutation in the coding exon of the receptor was described *(145)*. A homozygous point mutation at codon 74 (Ser → Ile) was detected in the second transmembrane domain. Subsequently, other mutations in homozygous or heterozygous state have been described in affected patients *(69,146–148)*. These mutations are scattered throughout the ACTH receptor structure and could affect either binding of the ligand or coupling to the adenylate cyclase.

6.2. Triple A Syndrome

Triple A syndrome was first described in 1978 *(149)* and consists of the triad of achalasia of the cardia, absence of tears and ACTH insensitivity. In many cases, there is a progressive appearance of a polyneuropathy and a deficiency in aldosterone production. It has been also proposed that the ACTH receptor could be partly responsible for this disease. Several laboratories failed to detect any mutation in this receptor in all the studied cases *(69,147,148)*. In a recent paper, it has been demonstrated that Triple A syndrome gene mapped to chromosome 12q13 without any heterogeneity among families, which excludes the ACTH receptor gene in this syndrome *(150)*.

6.3. *Adrenal Tumors*

6.3.1. ACTH Receptor mRNA Expression in Adrenal Tumors

Expression and distribution of ACTH-R mRNA transcripts has been studied by means of Northern blot, RT-PCR or *in situ* hybridization in different benign adrenal tumors or adrenal carcinomas. The highest ACTH-R mRNA levels were found in aldosteronomas and very low levels were reported in carcinomas or non-functioning adenomas. No major differences have been described in adrenal tissue from Cushing's syndrome or adrenal hyperplasia and normal adrenal tissue *(151)*. This result is contradictory with previous data showing in two patients with Cushing's syndrome that the intensity of the mRNA transcripts was much higher in adenoma tissue than in normal tissue *(45)*. It has been also reported that in carcinomas the low levels of expression of the ACTH-R mRNA are associated with a strong expression of P-450 side chain cleavage mRNA. This association could be the reflection of a malignant phenotype as postulated by the authors *(151)*.

6.3.2. ACTH Receptor Mutations and Adrenal Tumors

The presence of activating mutations in G protein-coupled receptors can lead to a gain of function of cells in an agonist independent fashion. This has been described in hyperfunctioning thyroid adenomas, where somatic mutations have been reported in the TSH receptor *(152)*. The mechanism of tumorigenesis in adrenocortical neoplasm remains unknown and ACTH receptor has been proposed as a candidate oncogene. No missense point mutations or silent polymorphisms have been detected within the coding region of the ACTH receptor gene in 16 adrenocortical tumors, including benign adenomas as well as carcinomas *(151,153)*. Similiar observations were obtained in 17 adenomas and 8 carcinomas by another group *(154)*. Apparently, activating mutations are not a common mechanism associated with adrenal neoplasia.

Acknowledgments

This work has been supported by grants from La Fondation de la Recherche Médicale (Paris) and Programme National de la Recherche Clinique (Ministère de la Santé). We thank J Bois and MA Di Carlo for secretarial assistance and John Carew for reviewing the English manuscript.

References

1. Smith, P. E. (1930) Hypophysectomy and a replacement therapy in the rat. *Am. J. Anat.* **45,** 205–273.
2. Li, C. H., Geschwind, I. I., Levy, A. L., Harris, J. I., Dixon, J. S., Pon, N. G., and Porath, J. O. (1954) Isolation and properties of alpha-corticotropin from sheep pituitary glands. *Nature* **173,** 251–254.
3. Li, C. H. and Oelofsen, W. (1967) The chemistry and biology of ACTH and related peptides in *The Adrenal Cortex* Eisenstein, A. B.,ed. Little, Brown, Boston. pp. 185–201.
4. Grahame-Smith, D. G., Butcher, R. W., Ney, R. L., and Sutherland, E. W. (1967) Adenosine 3',5'-monophosphate as the intracellular mediator of the action of adrenocorticotropic hormone on the adrenal cortex. *J. Biol. Chem.* **242,** 5535–5541.
5. Taunton, O. D., Roth, J., and Pastan, I. (1967) The first step in ACTH action, binding to tissue. *J. Clin. Invest.* **46,** 1122–1129.
6. Schimmer, B. P., Ueda, K., and Sato, G. H. (1968) Site of action of adrenocorticotropic hormone. *Biochem. Biophys. Res. Commun.* **32,** 806–810.
7. Lefkowitz, R. J., Roth, J., Pricer, W., and Pastan, I. (1970) ACTH receptors in the adrenal, specific binding of ACTH-125I and its relation to adenyl cyclase. *Proc. Natl. Acad. Sci. U.S.A.* **65,** 745–752.
8. Halkerston, I. D. K. (1975) Cyclic AMP and adrenocortical function. *Adv. Cycl. Nucleot. Res.* **6,** 100–136.
9. Lowry, P. J., McMartin, C., and Peters, J. (1973) Properties of a simplified bioassay for adrneocorticotropic activity using the steroidogenic response of isolated adrenal cells. *J. Endocrinol.* **59,** 43–55.
10. Rae, P. A. and Schimmer, B. P. (1974) Iodinated derivatives of adrenocorticotropic hormone. *J. Biol. Chem.* **249,** 5649–5653.
11. Saez, J. M., Morera, A. M., and Dazord, A. (1981) Mediators of the effects of ACTH on adrenal cells. *Adv. Cycl. Nucleot. Res.* **14,** 563–579.
12. Buckley, D. I. and Ramachandran, J. (1981) Characterization of corticotropin receptors on adrenocortical cells. *Proc. Natl. Acad. Sci. U.S.A.* **78,** 7431–7435.
13. Carsia, R. V. and Weber, H. (1988) Protein malnutrition in the domestic fowl induces alterations in adrenocortical cell adrenocorticotropin receptors. *Endocrinology* **122,** 681–688.
14. Saez, J. M., Evain, D., and Gallet, D. (1978) Role of cAMP and protein kinase on the steroidogenic action of ACTH, prostaglandin E1 and dibutyryl cyclic AMP in normal adrenal cells and adrenal tumor cells from humans. *J. Cycl. Nucleot. Res.* **4,** 311–321.
15. Sala, G. B., Hayashi, K., Catt, K. J., and Dufau, M. L. (1979) Adrenocorticotropin action in isolated adrenal cells. The intermediate role of cyclic AMP in stimulation of corticosterone synthesis. *J. Biol. Chem.* **254,** 3861–3865.
16. Rae, P. A., Guttman, N. S., Tsao, J., and Schimmer, B. P. (1979) Mutations in cyclic AMP-dependent protein kinase and corticotropin-sensitive adenyate cyclase affect adrenal steroidogenesis. *Proc. Natl. Acad. Sci. U.S.A.* **76,** 1896–1900.
17. Mountjoy, K. G., Robbins, L. S., Mortrud, M. T., and Cone, R. D. (1992) The cloning of a family of genes that encode the melanocortin receptors. *Science* **257,** 1248–1251.

18. McIlhinney, R. A. J. and Schulster, D. (1975) Studies on the binding of 125I labelled corticotropin to isolated rat adrenocortical cells. *J. Endocrinol.* **64,** 175–184.
19. Yanagibashi, K., Kamiya, N., Ling, G., and Matsuba, M. (1978) Studies on adrenocorticotropic hormone receptor using isolated rat adrenocortical cells. *Endocrinol. Jpn.* **25,** 545–551.
20. Gallo-Payet, N. and Escher, E. (1985) Adrenocorticotropin receptors in rat adrenal glomerulosa cells. *Endocrinology* **117,** 38–46.
21. Ramachandran, J., Muramoto, K., Kenez–Keri, M., Keri, G., and Buckley, D. I. (1980) Photoaffinity labeling of corticotropin receptors. *Proc. Natl. Acad. Sci. U.S.A.* **77,** 3967–3970.
22. Hofmann, K., Stehle, C. J., and Finn, F. M. (1988) Identification of a protein in adrenal particulates that binds adrenocorticotropin specifically and with high affinity. *Endocrinology* **123,** 1355–1363.
23. Mizuno, T., M, O., Itoh, A., Maruyama, T., Hagiwara, H., and Hirose, S. (1989) Affinity labeling of ACTH receptors in bovine adrenal cortex membranes. *Biochem. Int.* **19,** 695–700.
24. Penhoat, A., Jaillard, C., and Saez, J. M. (1993) Identification and characterization of corticotropin receptors in bovine and human adrenals. *J. Steroid Biochem. Mol. Biol.* **44,** 21–27.
25. Mertz, L. M. and Catt, K. J. (1991) Adrenocorticotropin receptors – Functional expression from rat adrenal messenger RNA in Xenopus-Laevis oocytes. *Proc. Natl. Acad. Sci. U.S.A.* **88,** 8525–8529.
26. Cone, R. D. and Mountjoy, K. G. (1993) Molecular genetics of the ACTH and melanocyte-stimulating hormone receptors. *Trends Endocrinol. Metab.* **4,** 242–247.
27. Cammas, F. M., Kapas, S., Barker, S., and Clark, A. J. L. (1995) Cloning, characterization and expression of a functional mouse ACTH receptor. *Biochem. Biophys. Res. Commun.* **212,** 912–918.
28. Kubo, M., Ishizuka, T., Kijima, H., Kakinuma, M., and Koike, T. (1995) Cloning of a mouse adrenocorticotropin receptor-encoding gene. *Gene* **153,** 279–280.
29. Raikhinstein, M., Zohar, M., and Hanukoglu, I. (1994) cDNA cloning and sequence analysis of the bovine adrenocorticotropic hormone (ACTH) receptor. *Biochim. Biophys. Acta* **1220,** 329–332.
30. Albrecht, E. D., Aberdeen, G. W., Babischkin, J. S., Tilly, J. L., and Pepe, G. J. (1996) Biphasic developmental expression of adrenocorticotropin receptor messenger ribonucleic acid levels in the baboon fetal adrenal gland. Endocrinology **137,** 1292–1298.
31. Chhajlani, V., Muceniece, R., and Wikberg, J. E. S. (1993) Molecular cloning of a novel human melanocortin receptor. *Biochem. Biophys. Res. Commun.* **195,** 866–873.
32. Gantz, I., Konda, Y., Tashiro, T., Shimoto, Y., Miwa, H., Munzert, G., Watson, S. J., Del Valle, J., and Yamada, T. (1993) Molecular cloning of a novel melanocortin receptor. *J. Biol. Chem.* **268,** 8246–8250.
33. Gantz, I., Miwa, H., Konda, Y., Shimoto, Y., Tashiro, T., Watson, S. J., Del Valle, J., and Yamada, T. (1993) Molecular cloning, expression and gene localization of a fourth melanocortin receptor. *J. Biol. Chem.* **268,** 15,174–15,719.
34. Desarnaud, F., Labbe, O., Eggerickx, D., Vassart, G., and Parmentier, M. (1994) Molecular cloning, functional expression and pharmacological characterization of a mouse melanocortin receptor gene. *Biochem. J.* **299,** 367–373.

35. Gantz, I., Shimoto, Y., Konda, Y., Miwa, H., Dickinson, C. J., and Yamada, T. (1994) Molecular cloning, expression, and characterization of a fifth melanocortin receptor. *Biochem. Biophys. Res. Commun.* **200,** 1214–1220.
36. Griffon, N., Mignon, V., Facchinetti, P., Diaz, J., Schwartz, J. C., and Sokoloff, P. (1994) Molecular cloning and characterization of the rat fifth melanocortin receptor. *Biochem. Biophys. Res. Commun.* **200,** 1007–1014.
37. Gantz, I., Tashiro, T., Barcroft, C., Konda, Y., Shimoto, Y., Miwa, H., Glover, T., Munzert, G., and Yamada, T. (1993) Localization of the genes encoding the melanocortin-2 (adrenocorticotropic hormone) and melanocortin-3 receptors to chromosomes 18p11.2 and 20q13.2–q13.3 by fluorescence insitu hybridization: brief report. *Genomics* **18,** 166–167.
38. Magenis, R. E., Smith, L., Nadeau, J. H., Johnson, K. R., Mountjoy, K. G., and Cone, R. D. (1994) Mapping of the ACTH, MSH, and neural (MC3 and MC4) melanocortin receptors in the mouse and human. Mamm. *Genome* **5,** 503–5508.
39. Penhoat, A., Jaillard, C., and Saez, J. M. (1994) Regulation of bovine adrenal cell corticotropin receptor mRNA levels by corticotropin (ACTH) and angiotensin-II (A-II). *Mol. Cell. Endocrinol.* **103,** R7–R10.
40. Picard–Hagen, N., Penhoat, A., Hue, D., Jaillard, C., and Durand, P. (1997) Glucocorticoids enhance corticotropin receptor mRNA levels in ovine adrenocortical cells. *J. Mol. Endocrinol.* **19,** 29–36.
41. Lebrethon, M. C., Naville, D., Begeot, M., and Saez, J. M. (1994) Regulation of corticotropin receptor number and messenger RNA in cultured human adrenocortical cells by corticotropin and angiotensin II. J. Clin. Invest. **93,** 1828–1833.
42. Mesiano, S., Fujimoto, V. Y., Nelson, L. R., Lee, J. Y., Voytek, C. C., and Jaffe, R. B. (1996) Localization and regulation of corticotropin receptor expression in the midgestation human fetal adrenal cortex: implications for in utero homeostasis. *J. Clin. Endocrinol. Metab.* 8**1,** 340–345.
43. Penhoat, A., Lebrethon, M. C., Begeot, M., and Saez, J. M. (1995) Regulation of ACTH receptor mRNA and binding sites by ACTH and angiotensin II in cultured human and bovine adrenal, fasciculata cells. *Endocr. Res.* **21,** 157–168.
44. Aberdeen, G. W., Babischkin, J. S., Davies, W. A., Pepe, G. J., and Albrecht, E. D. (1997) Decline in adrenocorticotropin receptor messenger ribonucleic acid expression in the baboon fetal adrenocortical zone in the second half of pregnancy. *Endocrinology* **138,** 1634–1641.
45. Matsuyama Morita, T., Imai, T., Murata, Y., Kambe, F., Funahashi, H., Takagi, H., and Seo, H. (1995) Adrenocorticotropic hormone (ACTH) increases the expression of its own receptor gene. *Endocr. J.* **42,** 475–480.
46. Mountjoy, K. G., Bird, I. M., Rainey, W. E., and Cone, R. D. (1994) ACTH induces up-regulation of ACTH receptor mRNA in mouse and human adrenocortical cell lines. *Mol. Cell. Endocrinol.* **99,** R17–R20.
47. Cammas, F.M. and Clark, A. J. L. (1995) Tissue distribution of ACTH receptor messenger RNA. *J. Endocrinol.* **147** (Suppl.), P11.
48. Chhajlani, V. (1996) Distribution of cDNA for melanocortin receptor subtypes in human tissues. *Biochem. Mol. Biol. Int.* **38,** 73–80.
49. Slominski, A., Ermak, G., and Mihm, M. (1996) ACTH receptor, CYP11A**1,** CYP17 and CYP21A2 genes are expressed in skin. *J. Clin. Endocrinol. Metab.* **81,** 2746–2749.

50. Smith, E. M., Brosnan, P., Meyer, W. J. I., and Blalock, J. E., (1987) An ACTH receptor on human mononuclear leukocytes: relation to adrenal ACTH receptor activity. *N. Engl. J. Med.* **317,** 1266–1269.
51. Grunfeld, C., Hagman, J., Sabin, E. A., Buckley, D. I., Jones, D. S., and Ramachandran, J. (1985) Characterization of adrenocorticotropin receptors that appear when 3T3–L1 cells differentiate into adipocytes. *Endocrinology* **116,** 113–117.
52. Boston, B. A., and Cone, R. D. (1996) Characterization of melanocortin receptor subtype expression in murine adipose tissues and in the 3T3–L1 cell line. *Endocrinology* **137,** 2043–2050.
53. Kijima, H., Shimizu, C., Koike, T., Kakinuma, M., and Kubo, M. (1997) Different effect of in vivo ACTH administration on ACTH receptor mRNA levels in mouse adrenal and adipose tissue. Preceedings of the 79th Annual Meeting of the Endocrine Society [Abstract]. P3–538.
54. Naville, D., Barjhoux, L., Jaillard, C., Lebrethon, M. C., Saez, J. M., and Begeot, M. (1994) Characterization of the transcription start site of the ACTH receptor gene, presence of an intronic sequence in the 5'-flanking region. *Mol. Cell. Endocrinol.* **106,** 131–135.
55. Clark, A. J. L. and Cammas, F. M. (1996) The ACTH receptor. *Baillières Clin. Endocrinol. Met.* **10,** 29–47.
56. Naville, D., Jaillard, C., Barjhoux, L., Durand, P., and Begeot, M. (1997) Genomic structure and promoter characterization of the human ACTH receptor gene. *Biochem. Biophys. Res. Commun.* **230,** 7–12.
57. Penhoat, A., Naville, D., Jaillard, C., Durand, P., and Bégeot, M. (1997) Presence of multiple functional polyadenylation signals in the 3'-untranslated region of human corticotropin receptor cDNA. *Biochim. Biophys. Acta* **1356,** 249–252.
58. Shimizu, C., Kubo, M., Saeki, T., Matsumura, T., Ishizuka, T., Kijima, H., Kakinuma, M., and Koike, T. (1997) Genomic organization of the mouse adrenocorticotropin receptor. *Gene* **188,** 17–21.
59. Cammas, F. M., Pullinger, G. D., Barker, S., and Clark, A. J. L. (1997) The mouse adrenocorticotropin receptor gene , cloning and characterisation of its promoter and evidence for a role for the orphan nuclear receptor steroidogenic factor 1. *Mol. Endocrinol.* **11,** 867–876.
60. Naville, D., Penhoat, A., Barjhoux, L., Jaillard, C., Fontanay, S., Saez, J., Durand, P., and Begeot, M. (1996) Characterization of the human ACTH receptor gene and in vitro expression. *Endoc. Res.* **22,** 337–348.
61. Oelkers, W. (1985) Prolonged ACTH infusion suppresses aldosterone secretion in spite of high renin activity. *Acta Endocrinol.* **108,** 91–97.
62. Kolanowski, J., Esselinck, W., Nagani, C., and Crabbe, J. (1977) Adrenocortical response upon repeated stimulation with corticotrophin in patients lacking endogenous corticotrophin secretion. *Acta Endocrinol.* **85,** 595–607.
63. Baxter, J. D. and Tyrrel, J. B. (1986) The adrenal cortex *Endocrinology and Metabolism* (Felig, P., Baxter, J. D., Broadus, A. E., and Frohman, L. A., eds.). McGraw–Hill, New York, 511–650.
64. Simpson, E. R. and Waterman, M. R. (1988) Regulation of the synthesis of steroidogenic enzymes in adrenal cortical cells by ACTH. *Annu. Rev. Physiol.* **50,** 427–440.

65. Lebrethon, M. C., Jaillard, C., Naville, D., Begeot, M., and Saez, J. M. (1994) Regulation of corticotropin and steroidogenic enzyme mRNAs in human fetal adrenal cells by corticotropin, angiotensin-II and transforming growth factor β1. *Mol. Cell. Endocrinol.* **106,** 137–143.
66. Rainey, W. E., McAllister, J. M., Byrd, E. W., Mason, J. I., and Carr, B. R. (1991) Regulation of corticotropin responsiveness in human fetal adrenal cells. *Am. J. Obstet. Gynecol.* **165,** 1649–1654.
67. Penhoat, A., Jaillard, C., and Saez, J. M. (1989) Corticotropin positively regulates its own receptors and cAMP response in cultured bovine adrenal cells. *Proc. Natl. Acad. Sci. U.S.A.* **86,** 4978–4981.
68. Rainey, W. E., Viard, I., and Saez, J. M. (1989) Transforming growth factor ß treatment decreases ACTH receptors on ovine adrenocortical cells. *J. Biol. Chem.* **264,** 21,474–21,477.
69. Naville, D., Barjhoux, L., Jaillard, C., Faury, D., Despert, F., Esteva, B., Durand, P., Saez, J. M. and Begeot, M. (1996) Demonstration by transfection studies that mutations in the adrenocorticotropin receptor gene are one cause of the hereditary syndrome of glucocorticoid deficiency. *J. Clin. Endocrinol. Metab.* **81,** 1442–1448.
70. Naville, D., Barjhoux, L., Jaillard, C., Saez, J. M., Durand, P. and Bégeot, M. (1997) Stable expression of normal and mutant human ACTH receptor. Study of ACTH binding and coupling to adenylate cyclase. *Mol. Cell. Endocrinol.* **129,** 83–90.
71. Kapas, S., Hinson, F. M. C. J. P., and Clark, A. J. L. (1996) Agonist and receptor binding properties of adrenocorticotropin peptides using the cloned mouse adrenocorticotropin receptor expressed in a stably transfected HeLa cell line. *Endocrinology* **137,** 3291–3294.
72. Penhoat, A., Jaillard, C., and Saez, J. M. (1989) Synergistic effects of corticotropin and insulin-like growth factor I on corticotropin receptors and corticotropin responsiveness in cultured bovine adrenocortical cells. *Biochem. Biophys. Res. Commun.* **165,** 355–359.
73. L'Allemand, D., Penhoat, A., Lebrethon, M. C., Ardevol, R., Baehr, V., Oelkers, W., and Saez, J. M. (1996) Insulin–like growth factors enhance steroidogenic enzyme and corticotropin receptor messenger ribonucleic acid levels and corticotropin steroidogenic responsiveness in cultured human adrenocortical cells. *J. Clin. Endocrinol. Metab.* **81,** 3892–3897.
74. Penhoat, A., Naville, D., Jaillard, C., Chatelain, P. G., and Saez, J. M. (1989) Hormonal regulation of insulin-like growth factor I secretion by bovine adrenal cells. *J. Biol. Chem.* **264,** 6858–6862.
75. Penhoat, A., Ouali, R., Viard, I., Langlois, D., and Saez, J. M. (1996) Regulation of primary response and specific genes in adrenal cells by peptide hormones and growth factors. *Steroids* **61,** 176–183.
76. Lebrethon, M. C., Jaillard, C., Naville, D., Begeot, M., and Saez, J. M. (1994) Effects of transforming growth factor–β1 on human adrenocortical fasciculata-reticularis cell differentiated functions. *J. Clin. Endocrinol. Metab.* **79,** 1033–1039.
77. Le Roy, C., Leduque, P., Dubois, P. M., Saez, J. M., and Langlois, D. (1996) Repression of transforming growth factor β1 protein by antisense oligonucleotide-induced increase of adrenal cell differentiated functions. *J. Biol. Chem.* **271,** 11,027–11,033.

78. Ramachandran, J. and Li, C. H. (1967) Structure–activity relationships of the adrenocorticotropins and melanotropins , the synthetic approach. *Adv. Enzymol.* **29,** 391–477.
79. Ramachandran, J. (1973) The structure and function of adrenocorticotropin, in *Hormonal Proteins and Peptides* (Li, C. H., ed.) Academic Press, New York. pp. 1–28.
80. Tell, G. P. E., Morera, A. M., and Saez, J. M. (1977) Mechanism of action of adrenocorticotropic hormone in *Congenital Adrenal Hyperplasia* (Lee, P. A., Plotnick, L. P., Kowarski, A. A., and Migeon, C. J., eds.) Baltimore University Park Press pp. 33–76.
81. Sayers, G., Seelig, S., Kumar, S., Karlaganis, G., Schwyzer, R., and Fujimo, M. (1974) Isolated adrenal cortex cells: ACTH4–23 (NH2), ACTH5–2**4,** ACTH6–2**4,** and ACTH7–23 (NH2): cyclic AMP and corticosterone production. *Proc. Soc. Exp. Biol. Med.* **145,** 176–181.
82. Klemcke, H. G. and Pond, W.G. (1991) Porcine adrenal adrenocorticotropic hormone receptors: characterization, changes during neonatal development and response to a stressor. *Endocrinology* **128,** 2476–2488.
83. Munari–Silem, Y., Lebrethon, M. C., Morand, I., Rousset, B. and Saez, J. M. (1995) Gap junction-mediated cell-to-cell communication in bovine and human adrenal cells: a process whereby cells increase their responsiveness to physiological corticotropin concentrations. *J. Clin. Invest.* **95,** 1429–1439.
84. Seelig, S., Sayers, G., Schwyzer, R., and Schiller, P. (1971) Isolated adrenal cells: ACTH11–2**4,** a competitive antagonist of ACTH1–39 and ACTH1–10. *FEBS Lett.* **19,** 232–234.
85. Seelig, S. and Sayers, G. (1973) Isolated adrenal cells: ACTH agonists, partial agonists, antagonists, cyclic AMP and corticosterone production. *Arch. Biochem. Biophys.* **154,** 230–239.
86. Finn, F. M., Johns, P. A., Nishi, N., and Hofmann, K. (1976) Differential response to adrenocorticotropic hormone analogs of bovine adrenal plasma membranes and cells. *J. Biol. Chem.* **251,** 3576–3585.
87. Bristow, A. F., Gleed, C., Fauchere, J. L., Schwyzer, R., and Schulster, D. (1980) Effects of ACTH (corticotropin) analogues on steroidogenesis and cyclic AMP in rat adrenocortical cells: evidence for two different steroidogenically responsive receptors. *Biochem. J.* **186,** 599–603.
88. Goverde, H. J. M. and Smals, A. G. H. (1984) The anomalous effect of some ACTH fragments missing the amino acid sequence 1–10 on the corticosteroidogenesis in purified isolated rat adrenal cells. *FEBS Lett.* **173,** 23–26.
89. Szalay, K. S., De Wied, D., and Stark, E. (1989) Effects of ACTH–(11–24) on the corticosteroid production of isolated adrenocortical cells. *J. Steroid Biochem.* **32,** 259–262.
90. Robison, G. A., Butcher, R. W., and Sutherland, E. W. (1971) Cyclic AMP.
91. Beall, R.J. and Sayers, G. (1972) Isolated adrenal cells: steroidogenesis and cyclic AMP accumulation in response to ACTH. *Arch. Biochem. Biophys.* **148,** 70–76.
92. Mackie, C., Richardson, M. C., and Schulster, D. (1972) Kinetics and dose-response characteristics of adenosine 3',5'-monophosphate production by isolated rat adrenal cells stimulated with adrenocorticotrophic hormone. *FEBS Lett.* **23,** 345–348.
93. Moyle, W. R., Kong, Y. C., and Ramachandran, J. (1973) Steroidogenesis and cyclic adenosine 3',5'-monophosphate accumulation in rat adrenal cells: divergent

effects of adrenocorticotropin and its alpha-nitrophenyl sulfenyl derivative. *J Biol. Chem.* **248,** 2409–2417.
94. Perchellet, J. P., Shanker, G., and Sharma, R. K. (1978) Regulatory role of guanosine 3',5'-monophosphate in adrenocorticotropin hormone-induced steroidogenesis. *Science* **199,** 311–312.
95. Perchellet, J.P., and Sharma, R. K. (1979) Mediatory role of calcium and guanosine 3',5'-monophosphate in adrenocorticotropin-induced steroidogenesis by adrenal cells. *Science* **203,** 1259–1261.
96. Yanagibashi, K. (1979) Calcium ion as "second messenger" in corticoidogenic action of ACTH. *Endocrinol. Jpn.* **26,** 227–232.
97. Farese, R. V., Rosic, N., Babischkin, J., Farese, M. G., Foster, R., and Davis, J. S. (1986) Dual activation of the inositol-triphosphate-calcium and cyclic nucleotide intracellular signaling systems by adrenocorticotropin in rat adrenal cells. *Biochem. Biophys. Res. Commun.* **135,** 742–748.
98. Catalano, R. D., Stuve, L., and Ramachandran, J. (1986) Characterization of corticotropin receptors in human adrenocortical cells. *J. Clin. Endocrinol. Metab.* **62,** 300–304.
99. Williams, S. A. and Schimmer, B. P. (1983) mRNA from mutant Y-1 adrenal cells directs the synthesis of altered regulatory subunits of type 1 cAMP-dependent protein kinase. *J. Biol. Chem.* **258,** 10,215–10,218.
100. Wong, M. and Schimmer, B. P. (1989) Recovery of responsiveness to ACTH and cAMP in a protein kinase-defective adrenal cell mutant following transfection with a protein kinase gene. *Endocr. Res.* **15,** 49–64.
101. Birmingham, M. K., Elliot, F. H., and Valero, P. H. L. (1953) The need for the presence of calcium for the stimulation in vitro of rat adrenal glands by adrenocorticotrophic hormone. *Endocrinology* **53,** 687–689.
102. Cheitlin, R., Buckley, D. I., and Ramachandran, J. (1985) The role of extracellular calcium in corticotropin-stimulated steroidogenesis. *J. Biol. Chem.* **260,** 5323–5327.
103. Lefkowitz, R. J., Roth, J., and Pastan, I. (1970) Effects of calcium on ACTH stimulation of the adrenal , separation of hormone binding from adenyl cyclase activation. *Nature* **228,** 864–866.
104. Gallo-Payet, N., Grazzini, E., Côté, M., Chouinard, L., Chorvatova, A., Bilodeau, L., Payet, M. D., and Guillon, G. (1996) Role of Ca_{2+} in the action of adrenocorticotropin in cultured human adrenal glomerulosa cells. *J. Clin. Invest.* **98,** 460–466.
105. Mahaffee, D. D. and Ontjes, D. A. (1980) The role of calcium in the control of adrenal adenylate cyclase. *J. Biol. Chem.* **255,** 1565–1571.
106. Shima, S., Kawashima, Y., and Hirai, M. (1979) Effects of ACTH and calcium on cyclic AMP production and steroid output by the zona glomerulosa of the adrenal cortex. *Endocrinol. Jpn.* **26,** 219–225.
107. Enyeart, J. J., Mlinar, B., and Enyeart, J. A. (1993) T-type Ca_{2+} channels are required for adrenocorticotropin-stimulated cortisol production by bovine adrenal zona fasciculata cells. *Mol. Endocrinol.* **7,** 1031–1040.
108. Papadopoulos, V., Widmaier, E. P., and Hall, P. F. (1990) The role of calmodulin in the responses to adrenocorticotropin of plasma membranes from adrenal cells. *Endocrinology* **126,** 2465–2474.
109. Garren, L. D., Ney, R. L., and Davis, W. W. (1965) Studies on the role of protein synthesis in the regulation of corticosterone production by adrenocorticotropic hormone in vivo. *Proc. Natl. Acad. Sci. U.S.A.* **53,** 1443–1450.

110. Privalle, C. T., Crivello, J. F., and Jefcoate, C. R. (1983) Regulation of intramitochondrial cholesterol transfer to side-chain cleavage cytochrome P-450 in rat adrenal gland. *Proc. Natl. Acad. Sci. U.S.A.* **80,** 702–706.
111. Papadopoulos, V. (1993) Peripheral-type benzodiazepine/diazepam binding inhibitor receptor: biological role in steroidogenic cell function. *Endocr. Rev.* **14,** 222–240.
112. Stocco, D. M. and Clark, B. J. (1996) Regulation of the acute production of steroids in steroidogenic cells. *Endocrine Rev* **17,** 221–244.
113. Clark, B. J., Wells, J., King, S. R., and Stocco, D. M. (1994) The purification, cloning, and expression of a novel luteinizing hormone-induced mitochondrial protein in MA-10 mouse Leydig tumor cells: characterization of the steroidogenic acute regulatory protein (stAR). *J. Biol. Chem.* **269,** 28,314–28,322.
114. Sugawara, T., Holt, J. A., Driscoll, D., Strauss, J. F., Lin, D., Miller, W. L., Patterson, D., Clancy, K. P., Hart, I. M., Clark, B. J., and Stocco, D. M. (1995) Human steroidogenic acute regulatory protein, functional activity in COS-1 cells, tissue-specific expression, and mapping of the structural gene to 8p11.2 and a pseudogene to chromosome 13. *Proc. Natl. Acad. Sci. U.S.A.* **92,** 4778–4782.
115. Lin, D., Sugawara, T., Strauss, J. F., Clark, B. J., Stocco, D. M., Saenger, P., Rogol, A., and Miller, W. L. (1995) Indispensable role of steroidogenic acute regulatory protein in adrenal and gonadal steroidogenesis. *Science* **267,** 1828–1831.
116. Herschman, H.R. (1992) Primary response genes induced by growth factors and tumor promoters. *Annu. Rev. Biochem.* **60,** 281–319.
117. Angel, P. and Karin, M. (1991) The role of Jun, Fos and the AP-1 complex in cell-proliferation and transformation. *Biochim. Biophys. Acta* **1072,** 129–158.
118. Imai, T., Seo, H., Murata, Y., Ohno, M., Satoh, Y., Funahashi, H., Takagi, H., and Matsui, N. (1990) Adrenocorticotropin increases expression of c-fos and β-actin genes in the rat adrenals. *Endocrinology* **127,** 1742–1747.
119. Yang, G., Koistinaho, J., Zhu, S. H., and Hervonen, A. (1989) Induction of c-fos-like protein in the rat adrenal cortex by acute stress. Immunocytochemical evidence. *Mol. Cell. Endocrinol.* **66,** 163–170.
120. Yang, G., Koistinaho, J., Iadarola, M., Zhu, S. H., and Hervonen, A. (1991) Administration of adrenocorticotropic hormone (ACTH) enhances Fos expression in the rat adrenal cortex. *Regul. Pept.* **30,** 21–31.
121. Kimura, E. and Armelin, H. A. (1990) Phorbol ester mimics ACTH action in corticoadrenal cells stimulating seroidogenesis, blocking cell cycle, changing cell shape, and inducing c-fos proto-oncogene expression. *J. Biol. Chem.* **265,** 3518–3521.
122. Clark, A. J. L., Balla, T., Jones, M. R. and Catt, K. J. (1992) Stimulation of early gene expression by angiotensin II in bovine adrenal glomerulosa cells: role of calcium and protein kinase C. *Mol. Endocrinol.* **6,** 1889–1898.
123. Miyamoto, N., Seo, H., Kanda, K., Hidaka, H., and Matsui, N. (1992) A 3',5'-cyclic adenosine monophosphate-dependent pathway is responsible for a rapid increase in c-fos messenger ribonucleic acid by adrenocorticotropin. *Endocrinology* **130,** 3231–3236.
124. Viard, I., Hall, S. H., Jaillard, C., Berthelon, M. C., and Saez, J. M. (1992) Regulation of c fos, c-jun and Jun-B messenger ribonucleic acids by angiotensin-II and corticotropin in ovine and bovine adrenocortical cells. *Endocrinology* **130,** 1193–1200.
125. Kimura, T. (1969) Effects of hypophysectomy and ACTH administration on the level of adrenal cholesterol side-chain desmolase. *Endocrinology* **85,** 492–499.

126. Purvis, J. L., Canick, J. A., Mason, J. I., Estabrook, R. W., and McCarthy, J. L. (1973) Lifetime of adrenal cytochrome P-450 as influenced by ACTH. *Ann. N.Y. Acad. Sci.* **212,** 319–342.
127. Waterman, M. R. (1994) Biochemical diversity of cAMP-dependent transcription of steroid hydroxylase genes in the adrenal cortex. *J. Biol. Chem.* **269,** 27,783–27,786.
128. Caron, K. M., Ikeda, Y., Soo, S. C., Stocco, D. M., Parker, K. L., and Clark, B. J. (1997) Characterization of the promoter region of the mouse gene encoding the steroidogenic acute regulatory protein. *Mol. Endocrinol.* **11,** 138–147.
129. Parker, K. L. and Schimmer, B. P. (1997) Steroidogenic factor 1 , a key determinant of endocrine development and function. *Endocr. Rev.* **18,** 361–377.
130. Luo, X. R., Ikeda, Y. Y., and Parker, K. L. (1994) A cell–specific nuclear receptor is essential for adrenal and gonadal development and sexual differentiation. *Cell* **77,** 481–490.
131. Sadovsky, Y., Crawford, P. A., Woodson, K. G., Polish, J. A., Clements, M. A., Tourtellotte, L. M., Simburger, K., and Milbrandt, J. (1995) Mice deficient in the orphan receptor steroidogenic factor 1 lack adrenal glands and gonads but express p450 side-chain-cleavage enzyme in the placenta and have normal embryonic serum levels of corticosteroids. *Proc. Natl. Acad. Sci. U.S.A.* **92,** 10,939–10,943.
132. Idelman, S. (1970) Ultrastructure of the mammalian adrenal cortex. *Int. Rev. Cytol.* **27,** 181–281.
133. Nussdorfer, G. G. (1986) Cytophysiology of the adrenal cortex. *Int. Rev. Cytol.* **98,** 1–405.
134. Hornsby, P. I. (1985) The regulation of adrenocortical function by control of growth and structure, in *The Adrenal Cortex* (Anderson, D. C. and Winters, J. S. D., eds.) Butherworths, London pp. 1–31.
135. Dallman, M. F. (1985) Control of adrenocortical growth in vivo. *Endocr. Res.* **10,** 213–242.
136. Dallman, M. F., Engeland, W. C., Holzwarth, M. A., and Scholz, P. M. (1980) Adrenocorticotropin inhibits compensatory adrenal growth after unilateral adrenalectomy. *Endocrinology* **107,** 1397–1404.
137. Rao, A. J., Long, J. A., and Ramachandran, J. (1978) Effects of antiserum to adrenocorticotropin on adrenal growth and proliferation. *Endocrinology* **102,** 371–378.
138. Lowry, P. J., Silas, L., Linton, E. A., McLean, C., and Estivariz, C. (1983) Pro-gamma-melanocyte stimulating hormone cleavage in adrenal gland undergoing compensatory growth. *Nature* **306,** 70–72.
139. Vinson, G.P., Pudney, J. A., and Whitehouse, B. J. (1985) Review, the mammalian adrenal circulation and the relationship between adrenal blood flow and steroidogenesis. *J. Endocrinol.* **105/2,** 285–294.
140. Estivariz, F. E., Iturriza, F., McLean, C., Hope, J., and Lowry, P. J. (1982) Stimulation of adrenal mitogenesis by N-terminal pro-opiocortin peptides. *Nature* **297,** 419–422.
141. Estivariz, F. E., Carino, M., Lowry, P.J., and Jackson, S. (1988) Further evidence that N-terminal pro-opiomelanocortin peptides are involved in adrenal mitogenesis. *J. Endocrinol.* **116,** 201–206.
142. Ceccatelli, S., Diana, A., Villar, M. J., and Nicotera, P. (1995) Adrenocortical apoptosis in hypophysectomized rats is selectively reduced by ACTH. *Neuroreport* **6,** 342–344.

143. Shepard, T. H., Landing, B. H., and Mason, D. G. (1959) Familial Addison's disease. *Am. J. Dis. Child.* **97,** 154–162.
144. Migeon, C. J., Kenny, F. M., Kowarski, A., Snipes, C. A., Spaulding, J. S., Finkelstein, J. W., and Blizzard, R. H. (1968) The syndrome of congenital adrenocortical unresponsiveness to ACTH: report of six cases. *Pediatr. Res.* **2,** 501–513.
145. Clark, A. J. L., McLoughlin, L., and Grossman, A. (1993) Familial glucocorticoid deficiency associated with point mutation in the adrenocorticotropin receptor. *Lancet* **341,** 461–462.
146. Tsigos, C., Arai, K., Hung, W., and Chrousos, G. P. (1993) Hereditary isolated glucocorticoid deficiency is associated with abnormalities of the adrenocorticotropin receptor gene. *J. Clin. Invest.* **92,** 2458–2461.
147. Tsigos, C., Arai, K., Latronico, A. C., Digeorge, A. M., Rapaport, R., and Chrousos, G. P. (1995) A novel mutation of the adrenocorticotropin receptor (ACTH–R) gene in a family with the syndrome of isolated glucocorticoid deficiency, but no ACTH–R abnormalities in two families with the triple a syndrome. *J. Clin. Endocrinol. Metab.* **80,** 2186–2189.
148. Clark, A. J. L. and Weber, A. (1994) Molecular insights into inherited ACTH resistance syndromes. *Trends Endocrinol. Metab.* **5,** 209–214.
149. Allgrove, J., Clayden, G. S., and Grant, D. B. (1978) Familial glucocorticoid deficiency with achalasia of the cardia and deficient tear production. *Lancet* **1,** 1284–1286.
150. Weber, A., Wienker, T. F., Jung, M., Easton, D., Dean, H. J., Heinrichs, C., Reis, A., and Clark, A. J. L. (1996) Linkage of the gene for the triple A syndrome to chromosome 12q13 near the type II keratin gene cluster. *Hum. Mol. Genet.* **5,** 2061–2066.
151. Allolio, B. and Reincke, M. (1997) Adrenocorticotropin receptor and adrenal disorders. *Horm. Res.* **47,** 273–278.
152. Parma, J., Duprez, L., Vansande, J., Cochaux, P., Gervy, C., Mockel, J., Dumont, J., and Vassart, G. (1993) Somatic mutations in the thyrotropin receptor gene cause hyperfunctioning thyroid adenomas. *Nature* **365,** 649–651.
153. Light, K., Jenkins, P. J., Weber, A., Perrett, C., Grossman, A., Pistorello, M., Asa, S. L., Clayton, R. N., and Clark, A. J. L. (1995) Are activating mutations of the adrenocorticotropin receptor involved in adrenal cortical neoplasia? *Life Sci.* **56,** 1523–1527.
154. Latronico, A. C., Reincke, M., Mendonca, B. B., Arai, K., Mora, P., Allolio, B., Wajchenberg, B. L., Chrousos, G. P., and Tsigos, C. (1995) No evidence for oncogenic mutations in the adrenocorticotropin receptor gene in human adrenocortical neoplasms. *J. Clin. Endocrinol. Metab.* **80,** 875–877.

CHAPTER 4

Effects of Melanocortins in the Nervous System

Roger A. H. Adan

Effects of melanocortins on the nervous system have been known since the 1950s. This chapter focuses on the effects of melanocortins that have been described during the last decades and will cover effects on avoidance behavior (which relate to the effects of melanocortins on learning and memory), on grooming behavior, on social and sexual behavior, on inflammation and fever, on neural control of the cardiovascular system, on the interaction with the opioid system, on epilepsy, and on nerve regeneration. Taken together, the present data indicate that the brain melanocortin system has a widespread involvement in neuroendocrine and behavioral responses to the environment.

1. Introduction: Discovery of the Brain Melanocortin System

During the 1950s and 1960s the direct effects of melanocortins on brain and behavior were first described. During the 1970s it became clear that the brain has its own melanocortin system that is anatomically and physiologically distinct from the pituitary gland. Only since the last decade have receptors for melanocortins been identified in brain.

1.1. The First Described Behavioral Effects of Melanocortins

1.1.1. Avoidance Behavior

Mirsky et al. *(1)* were the first to report behavioral effects of adrenocorticotropic hormone (ACTH): peripheral injection of ACTH enhanced the acquisition of avoidance responses in the shuttle box, where a rat is taught to avoid an electric shock by jumping to another compartment. Hypophysec-

The Melanocortin Receptors Ed.: R. D. Cone

tomy retarded acquisition of this avoidance behavior, while administration of ACTH restored this response *(2)*. At that time it was thought that the effects of peripheral administration of ACTH on conditioned avoidance behavior were mediated via the adrenal cortex. However, in adrenalectomized rats, this effect of ACTH was retained *(3)*. Furthermore, De Wied and Miller found that corticosteroids had an effect on avoidance behavior opposite to that of ACTH (reviewed in ref. *4*). Melanocortins devoid of activity on the adrenal gland, such as ACTH[4–10] and ACTH[1–10], were as active as ACTH in delaying extinction of avoidance behavior *(5,6)*. Taken together, the effects of melanocortins on avoidance behavior were demonstrated to be independent from the adrenal gland and suggested a direct effect on the brain.

1.1.2. Stretching, Yawning, and Grooming

In the late 1950s, Ferrari *(7)* demonstrated that central, but not peripheral, application of melanocortins elicited a stretching and yawning response and grooming in various species. This effect was also observed using melanocortins that lack effects on the adrenal cortex, such as α-and β-melanocyte-stimulating hormone (MSH) *(8)*. These data provided a second line of evidence that melanocortins had a direct effect on the brain.

1.2. Early Pharmacologic Evidence for Multiple Brain Melanocortin Receptors

Based on these early findings, it was clear that brain melanocortin (MC) receptors differed from the peripheral MC receptor(s) since there was a difference in melanocortin structure–activity relationships. Pharmacologic studies further indicated the existence of multiple receptor subtypes in the brain *(9,10)*. The structural requirements for ACTH fragments and ACTH analogs to influence distinct behavioral responses were markedly different *(9,10)*. For instance, ACTH[4–10] stimulated the facilitation of retention of active avoidance behavior, whereas it was inactive in inducing excessive grooming behavior *(11)*. Surprisingly, substitution of Phe at position 7 in ACTH[4–10] for its D-enantiomer generated a peptide that could induce excessive grooming behavior, but had an effect opposite to that of ACTH[4–10] on extinction of active avoidance behavior *(12–15)*. Furthermore, effects of ACTH fragments and ACTH peptides on these behavioral assays could be discriminated pharmacologically from effects of α-MSH on nerve regeneration *(16,17)* as well as from stimulatory effects of intracerebroventricularly administered α-MSH on plasma corticosterone levels (11). Thus, there was evidence for the existence of multiple melanocortin receptor subtypes in the nervous system.

1.3. Discovery of the Brain Melanocortin-Producing Neurons

Until the 1970s, it remained unclear whether melanocortins had a physiologic role in the brain and whether melanocortins in the brain originated from the pituitary gland or the brain itself. Melanocortin bioactivity (from brain extracts) and immunoreactivity was ultimately demonstrated in brains from various species (reviewed in ref. *18*). The demonstration of melanocortin peptide immunoreactivity in the limbic system and in the hypothalamus in the late 1970s was the first indication for a melanocortin system in the brain *(19–21)*. Although first it was believed that this immunoreactivity was the result of retrograde transport of melanocortins from the pituitary gland or melanocortins that reached the brain via the bloodstream, the persistence of brain melanocortin-immunoreactivity following removal of the pituitary gland indicated the existence of a melanocortin system in the brain itself *(21–23)*.

The cloning of the proopiomelanocortin (POMC) gene in the late 1970s and the demonstration that it is also expressed in the brain *(24–28)*, further underscored the existence of a brain melanocortin system. The arcuate nucleus of the hypothalamus and the nucleus of the tractus solitarius (NTS) are the main sites of POMC expression in the rodent brain; expression of POMC in the NTS has not yet been demonstrated in primates. The arcuate nucleus has projections to many brain areas which suggests the involvement of this system in diverse brain functions. The nucleus tractus solitarius melanocortin neurons project to catecholaminergic brain stem nuclei and midline structures in the brainstem *(28)*.

The proopiomelanocortin peptide precursor is processed differently in the brain, in which the majority of ACTH is cleaved to form α-melanocyte-stimulating hormone (α-MSH), as compared to the anterior lobe of the pituitary gland, where ACTH is formed *(29)*. This underscored that the brain has its own melanocortin system using α-MSH and γ-MSH as main melanocortin messengers. α-MSH only poorly stimulates corticosteroidogenesis in the adrenal gland. This also indicated that brain MC receptors differed from those in the adrenal gland.

1.4. Discovery of Brain Melanocortin Receptors

The lack of melanocortin antagonists and the absence of binding sites in the brain had raised doubts about the existence of specific brain MC receptors. It took until the late 1980s before binding sites for melanocortins could be demonstrated in brain *(30,31)*.

The cloning of MC receptors and the demonstration that some of the members of the MC receptor family were expressed in the brain were milestones in this field of research (see Chapters *6,13,* and *14*). This has opened new avenues to study behavioral, endocrinologic and neurotrophic effects of melanocortin peptides *(32–39)*.

The effects of melanocortins on the nervous system have been reviewed extensively *(10,18,40–42)*. This chapter describes in vivo effects of melanocortins on the brain concentrating mostly on melanocortin peptides that contain the core sequence essential for activity on MC receptors.

2. Effects of Melanocortins That Contain the Melanocortin Core Sequence

2.1 Stretching and Yawning Syndrome and Grooming

One of the first described behaviors induced by melanocortins is the stretching and yawning response *(7)* and grooming behavior in the rat *(43)*. These effects were collectively called "ACTH induced behavioral syndrome" by Ferrari et al. *(8)* following the observation that dogs, rabbits, cats, and rats displayed frequent yawning, stretching, drowsiness, and grooming behavior following application of ACTH intracisternally. Grooming behavior consists of activities directed to the animal body surface like face washing, body grooming, licking, scratching, and genital grooming. The display of this behavior is often associated with factors other than the condition of the fur and is for instance also displayed following exposure to a novel environment. Although the significance of grooming to animal homeostasis remains to be determined, it was proposed that the display of this behavior may function to regulate body temperature *(44)* and to reduce arousal, elicited by mild stress conditions *(45,46)*. Excessive grooming can be induced by intracerebral injection of different neuropeptides *(46)*. Pharmacologic experiments utilising receptor antagonists suggested that opioidergic, dopaminergic, and serotoninergic brain systems were involved in grooming behavior. However, despite the complexity of neuronal mechanisms underlying this behavior, it seems that the grooming response depends mainly on the melanocortinergic system, since: (i) ACTH and α-MSH are potent activators of excessive grooming when delivered into rat brain ventricles *(43,45)*; (ii) the pattern (the frequency and timing of different grooming operations) of melanocortin (but not other peptide) induced grooming resembled that which appears under physiologic stimulation (i.e., exposure to novelty) *(47)*.

2.2.1. Structure–Activity Relationships of Melanocortin-Induced Grooming Behavior

The structure–activity relationships of melanocortins on grooming fit well with those of peripheral MC receptors *(43,47,48)*. The MC3 and MC4 receptors are the members of the MC receptor family for which mRNA levels could be detected in brain. The reported expression of MC1 and MC5 receptor in the brain remains to be confirmed *(49–51)*. γ-MSH is a full agonist at the MC3 receptor with an activity comparable to α-MSH *(35,37,52)*. This rules out the MC3 receptor as mediator of melanocortin induced excessive grooming behavior, because γ-MSH does not induce excessive grooming behavior *(45)* (Fig. 1). NDP-α-MSH ([Nle^4,D-Phe^7] α-MSH) is a potent agonist for MC receptors, the activity of which had been demonstrated in the frog and lizard skin pigmentation assays *(53)* as well as in the grooming test *(47)*. This analogue is typically 10 to 100 times more potent than α-MSH. Indeed NDP-MSH is the most potent peptide, both on the MC4 receptor in vitro *(38,52)* as well as on the grooming response, followed by α-MSH *(47)*. Removal of the C-terminal three amino acids (as in ACTH[1–10] ; *(47)* reduced excessive grooming behavior as well as MC receptor activity much more dramatically than removal of the three N-terminal amino acids (as in ACTH[4–13] Fig. 1; *(52)*. ACTH[1–10] and ACTH[4–10] did not elicit excessive grooming behavior *(45)* and only poorly activated the MC4 receptor in vitro *(52)*. [D-Phe^7]ACTH[4–10] *(14,45)* and ACTH[4–13] induced excessive grooming behavior, although the response is less than the response to α-MSH. The order of potency of NDP-MSH, α-MSH and ACTH[4–13] on eliciting excessive grooming behavior correlated with that of the activation of the MC4 receptor in vitro *(52)* (Table 1). Both on the grooming response as well as on the MC4 receptor in vitro, [D-Phe^7]ACTH[4–10] was active, whereas ACTH[4–10] had no or little activity respectively. Taken together, based on the efficacy of MC receptor agonists, the MC4 receptor is probably the MC receptor that mediates melanocortin-induced grooming behavior.

Further evidence to suggest the MC4 receptor as mediator of melanocortin induced grooming is that MC4 antagonists block α-MSH-induced excessive grooming behavior *(54)*; (Fig. 2). SHU9119 (Ac-Nle^4-c[Asp^5, D-2-Nal^7, Lys^{10}]-α-MSH[4–10]-NH_2) a potent competitive MC receptor antagonist of human MC3 and MC4 receptors (pA_2 value 8.3 and 9.3, respectively) but not human MC1 and MC5 receptors *(55)* also inhibited melanocortin-induced grooming behavior at a low dose of 150 ng, whereas the other MC4 receptor antagonists, having lower pA_2 values, were effective at a dose of 15 μg (Fig. 2). Furthermore, the MC4 receptor is the only MC receptor for which mRNA is expressed in all areas *(38)* that have been implicated in α-MSH-induced grooming behavior (i.e., periaquaductal gray, substantia nigra, and paraventricular nucleus *(46,56–58)*. Taken together, the MC4 receptor probably mediates the effects of melanocortins on excessive grooming behavior in rats.

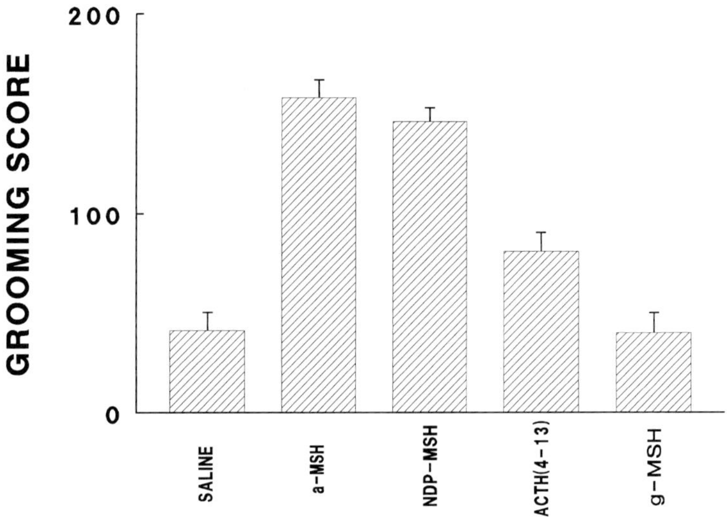

Fig. 1. Melanocortin induced grooming behavior in the rat. Male rats were injected intracerebroventricularly with saline, α-MSH (α-MSH; 1.5 μg), NDP-MSH (15 ng), ACTH[4–13] (3 μg) or γ_2-MSH (g-MSH; 3 μg). Grooming behavior was scored starting 10 min following the injections during 60 min. The data are represented as mean ± standard deviation.

Table 1
Structure–Activity Relationships of Melanocortins in Various Bioassays

	MC-3	MC-4	MC-5	Grooming	Avoidance	Regeneration
α-MSH	+++	+++	+++	++	+	++
γ_{-2}-MSH	+++	+	+	—	—	—
ACTH(4-10)	++	+	+	—	++	+
[D-Phe[7]]-ACTH[4–10]	++	++	++	+	—	—
ORG2766	—	—	—	—	+++	++
NDP-MSH	++++	++++	++++	+++	—	++

The data are representive for activity on cloned MC receptors expressed in cell lines *(35–38,52,166,167)*, grooming behavior *(43,47)*, avoidance behavior *(10,15,41)*,and recovery of sensomotor function in rats following sciatic nerve crushes *(17,155)*.

++++, very potent (on MC receptors EC^{50} values less than 1n*M*), +++, potent (on MC receptors EC^{50} values in nanomolar range), ++, active (on MC receptors EC^{50} values less than 1 μ*M*), +, some activity (on MC receptors EC^{50} values higher than 1μ*M*), -, inactive.

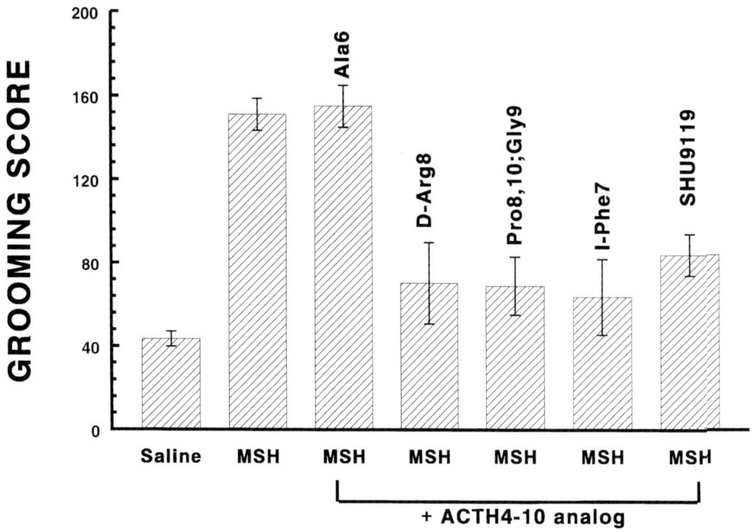

Fig. 2. MC-4 receptor antagonists block α-MSH induced excessive grooming. Male rats were injected intracerebroventricularly with saline, α-MSH (MSH; 1.5 µg) or combinations of α-MSH (1.5 µg) and the MC-4 receptor antagonists [D-Arg^8]ACTH[4–10] (D-Arg$_8$), [paraiodo-Phe^7]ACTH[4–10] (I-Phe$_7$), [Pro$^{8;10}$, Gly^9]ACTH[4–10] (Pro8,10;Gly9) (all three at 15 µg), or SHU9119 (150 ng) or the MC-3 receptor antagonist [Ala^6]ACTH[4–10] (15 µg). Grooming behavior was scored starting 10 min following the injections during 60 min. The data are represented as mean ± standard deviation.

2.1.2. Endogenous Melanocortins Mediate Novelty-Induced Grooming

Grooming behavior is one of the behaviors that animals display spontaneously. It is often observed when a rat is exposed to a stressful situation such as the exposure to a novel environment.

There is good evidence to suggest that α–MSH is the endogenous ligand mediating the grooming behavior as observed following exposure to novelty. First, infusion with ACTH-antiserum into the brain ventricular system blocked novelty induced grooming in the rat *(59)*. Second, downregulation of POMC protein expression in the hypothalamus using antisense oligonucleotides significantly reduced the grooming response to novelty *(60)*. Third, a potent MC receptor antagonist (SHU9119) reduced novelty induced grooming in the rat (Fig. 3). Thus the reaction of a rat to novelty involves activation of the melanocortin system.

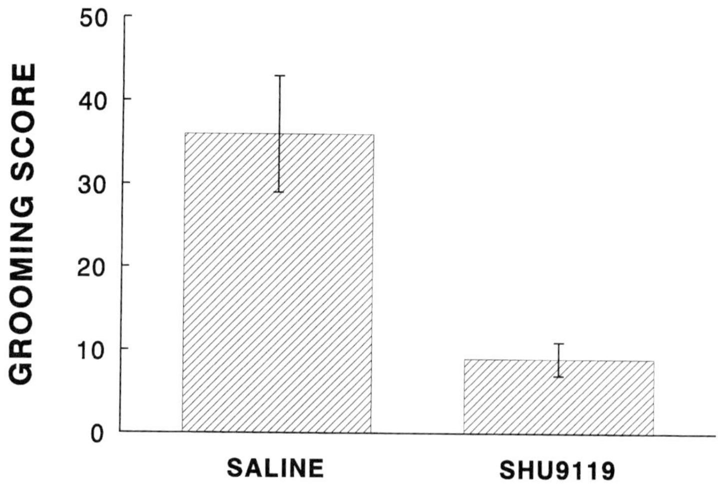

Fig. 3. MC-4 receptor antagonist blocks novelty induced grooming behavior. Male rats were injected intracerebroventricularly with saline or SHU9119 (150 ng) 15 min before exposure to an observation box. Grooming behavior was scored starting 10 min following the injections during 60 min. The data are represented as mean ± standard deviation.

2.1.3. Brain Melanocortin–Induced Activation of the Hypothalamopituitary–adrenal Axis

Exposing a rat to a novel environment induces stress and stimulates the hypothalamopituitary–adrenal (HPA) axis. Central administration of ACTH[1–24] activates the HPA-axis independent from direct effects of ACTH on the adrenal gland *(11)*. Since a novel environment or injection with melanocortins induce grooming behavior, which can be blocked by a MC receptor antagonist, it was tested whether the activation of the HPA axis by central application of melanocortins is also blocked by MC receptor antagonists. Indeed, this effect of ACTH[1–24] is also inhibited by coadministration of the MC4 receptor antagonists [D-Arg^8]ACTH[4–10] and SHU9119 *(60a)* (Fig. 4). It is speculative to suggest that activation of the brain melanocortin system is involved in activation of the HPA axis following mild emotional stress conditions. Therefore, it needs to be determined whether the response of a rat to exposure to stressful situations (activation of the HPA axis) is inhibited by MC receptor antagonists.

In contrast to these findings, a short negative feedback mechanism for melanocortins to inhibit plasma ACTH levels has been proposed *(61)*. However, in this study *(61)* melanocortins were applied chronically, in contrast to the study of Von Frijtag et al. *(60a)* who performed acute administration of melanocortins followed by measurement of plasma ACTH lev-

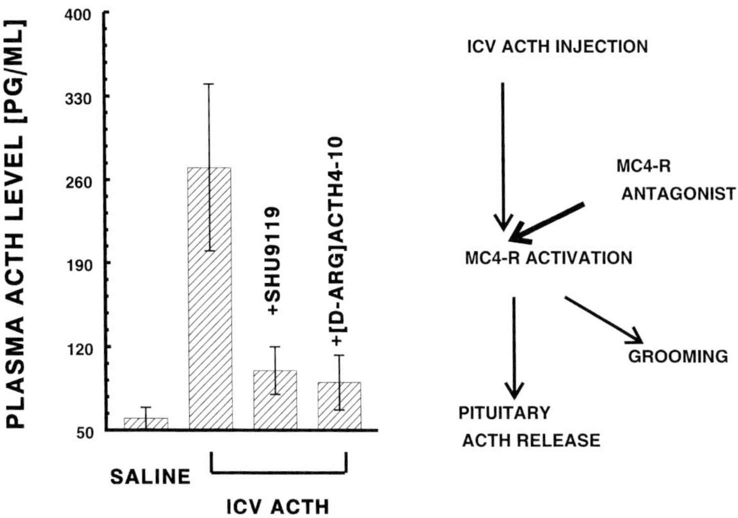

Fig. 4. Brain melanocortin receptor activated stimulation of plasma ACTH levels. Male rats were injected intracerebroventricularly with saline, ACTH[1–24] (1 μg) or a combination of ACTH[1–24] (1 μg) and either SHU9119 or [D-Arg^8]ACTH[4–10] (1 μg). Plasma ACTH levels were determined 30 min after the injection.

els. Taking these data together, chronic high levels of melanocortins in the hypothalamus may inhibit the activity of the HPA axis in a short feedback loop, whereas a single injection with melanocortins applied intracerebroventricularly activates the HPA axis. Corticotrophin-releasing hormone (CRH) release from the hypothalamus in vitro has been demonstrated to be inhibited by melanocortins *(62)*. ACTH may activate the HPA axis in vivo independent from CRH release as described for the hypothalamic culture system, possibly via MC receptors expressed outside the hypothalamus, or via an effect on parvocellular vasopressinergic neurons originating in the paraventricular nucleus.

2.2. Centrally Mediated Melanocortin-Induced Effects on Suppression of Fever and Inflammation

Intracerebroventricular (ICV) administration of α–MSH has a strong antiinflammatory and antipyretic effect in various species *(63–66)*. In fact α–MSH is the most potent antipyretic peptide following exogenous administration *(63)*. Administration of α–MSH as well as NDP-MSH antagonize the effect of cytokines like interleukin-1, interleukin-6, and TNF-α on increasing body temperature and on inducing inflammation *(63,67–70)*. Here these effects of melanocortins and what is known of their mechanism is discussed.

2.2.1. Fever

In rabbits, 100-200 ng α–MSH injected ICV reduced fever induced by ICV injection of either interleukin-1 or TNF *(68,71)*. Structure-activity data on melanocortins able to reduce cytokine-induced fever first suggested that the three C-terminal amino acids (Lys-Pro-Val) were essential *(63)*. However, N-terminal elongation of this tripeptide to α–MSH (9–13) reduced activity whereas further N-terminal elongation, such as α–MSH(8–13) increased potency *(72)*. NDP-MSH is more potent than α–MSH in reducing fever when administered centrally, but it has no, or at best a lower, antipyretic effect than α–MSH when administered peripherally *(73)*. The inactivity of NDP-MSH in the periphery suggests that the peripheral effect of α–MSH and certainly α–MSH(11–13), which lacks the melanocortin core sequence, is not mediated via one of the known MC receptor subtypes. Recently, lipolysaccharide (LPS)-induced fever in rats was inhibited by ICV injection of 300 ng α–MSH, and this effect of α–MSH was antagonized by coinjection of 200 ng SHU9119, suggesting that the antipyretic effect of melanocortins is mediated via MC3 and/or MC4 receptors *(50)*.

In the rabbit, fever has been reported to increase α–MSH levels in the septum *(74–76)*. Antibodies against α–MSH administered intracerebrally aggravated the fever response induced by peripheral injection of interleukin-1 *(77)*. Similarly, ICV injection of 200 ng SHU9119 in LPS-treated rats, aggravates fever *(50)*. Thus the melanocortin system also plays a physiological role in controlling body temperature during fever. MC receptors are indeed expressed in areas involved in regulating body temperature such as the septum, preoptic region, and anterior hypothalamus *(37,38)*.

2.2.2. Inflammation

Intradermal injection of interleukin-1 into the ear of mice elicits an inflammatory response resulting in swelling of the ear which can be measured by determining the thickness of the ear. ICV injection of submicrogram amounts of α–MSH reduced the ear thickness following injection of interleukin-1 into the ear *(64,78)*. When peripheral β2-adrenergic receptors are blocked by propranolol, ICV injection of α–MSH is not effective anymore *(78)*. Furthermore, ICV injection of α–MSH only reduced carrageenan-induced-swelling of the hind paw of mice when the spinal cord is intact *(78)*. These data suggest that activation of the brain melanocortin system by injections of melanocortins regulate the activity of the autonomic system leading to a reduction of the inflammatory response. Indeed MC receptors are expressed in areas that regulate the activity of the autonomic system such as the paraventricular nucleus, the dorsal motor nucleus of the vagus nerve and the nucleus of the tractus solitarius *(38)* as well as in the sympathetic system itself *(79)*. Stimulation of sympathetic postganglionic neurons contributed to increased vascular permeability induced by activation of primary afferents *(80)*. It has been suggested that this melanocortin-induced modulation of inflammation occurs via descending pathways, inhibiting the release of proinflammatory agents such as substance P from sensory neurons *(78)*. Indeed painful cutaneous stimulation is known to induce "neurogenic inflammation" consisting of vasodilatation and hyperalgesia and substance P is a good candidate to mediate both *(81)*. Descending pathways, coming from the periaquaductal gray area and nucleus raphe magnus running in the dorsolateral funiculus, influence pain signaling *(63)*. Both these nuclei in the brain stem express MC receptors *(37,38)*.

The antiinflammatory effects of α-MSH[11–13] are probably not mediated via the same pathway. Much higher doses of the tripeptide α-MSH[11–13] (more than 100-fold on a molar basis as compared to α-MSH) administered either systemically or ICV have a weaker anti- inflammatory effect than α-MSH administered ICV *(65)*. Furthermore, the antiinflammatory effect of α-MSH[11–13] does not depend on an intact spinal cord and sympathetic system *(78)*,suggesting that the C-terminal part of α-MSH has a weak antiinflammatory effect in the periphery, which is independent from the central effect of α-MSH, in line with the lack of the melanocortin core sequence in α-MSH[11–13].

2.3. *Interaction of Melanocortins and Opiates*

Melanocortins injected either ICV or peripherally counteract morphine induced analgesia in mice *(82)*, rabbits *(83)*, and rats *(84)* as assessed using the tail flick assay, the ear withdrawal test and the hot plate assay, respectively. Melanocortins induce hyperalgesia in rats and mice in the tail flick test *(82,85)* at doses that are higher than necessary to counteract opiate-induced analgesia. These effects of melanocortins may also be interpreted by assuming that melanocortins lower the threshold to respond to painful stimuli *(86)*. Although it had been demonstrated that melanocortins displace radiolabeled opioid receptor ligands from opioid receptors *(82,84)* this occurs only at micromolar concentrations, likely to be nonphysiologic. This, together with the fact that α-MSH is only effective when given prior to morphine suggests that melanocortins are not opiate receptor antagonists, but affect some common pathways in the brain in an opposite manner (functional opioid antagonism), consistent with the coproduction a α-MSH, γ-MSH, and β–endorphin in POMC neurons.

Preinjection of morphine a few hours before a second injection of morphine to induce analgesia leads to desensitization, since higher doses of morphine are necessary to obtain analgesia as compared to a single morphine injection, and a lower dose of nalaxon is necessary to block the effect of morphine. This second dose tolerance is counteracted by melanocortins *(82,87)*. Thus melanocortins can also regulate adaptation of the opiate system.

Structure–activity analysis of melanocortin active in antagonizing morphine-induced analgesia are very similar to those active in eliciting excessive grooming behavior *(88)* suggesting that one or more of the cloned MC receptors mediate these effects. For instance [D-Phe7]ACTH[4–10] is active whereas ACTH(4-10) is not. Van Ree et al. *(89)* reported that γ-MSH at a high dose of 50 μg also antagonized β-endorphin-induced analgesia following ICV injections. Thus, based upon these data, the MC3 receptor is a good candidate receptor that may mediate these effects. Another interaction between melanocortins and opioids was described of opioid withdrawal symptoms indicative for opioid dependence *(90)*. Opioid naive rats display opioid withdrawal like symptoms following ICV injection with ACTH(1–24) and γ-MSH *(89,90)*. Also in morphine-dependent rats melanocortins induce opioidlike withdrawal symptoms *(82,86,87)*. Thus, melanocortins counteract morphine-induced analgesia and tolerance, as well as dependence.

An anatomical link between the melanocortin system and the opioid system in the periaquaductal grey area (in this site there is expression of MC receptors as well as POMC projections) may form the substrate for the effects of α-MSH on the opioid system. Recently, it was demonstrated that morphine treatment influenced the expression of the MC4 receptor in periaquaductal gray area and in striatum *(49)*. Thus, long-term adaptive changes following morphine treatment involve alterations in the sensitivity to melanocortins. Furthermore, naloxone inhibited the full expression of excessive grooming behavior observed after intracerebral melanocortin injections *(91,92)*. Thus at the physiologic level, at the biochemical level (since POMC encodes melanocortins and β-endorphin) as well as at the anatomic level, the melanocortin and opioid systems are linked.

3. Effects of Melanocortins Not Mediated via One of the Cloned Brain Melanocortin Receptor Subtypes

3.1. Avoidance Behavior

One of the first reported effects of melanocortins on the brain was the modulation of learning and memory *(1,93)*. The effects of melanocortin fragments on avoidance behavior, attention, and motivation have been reviewed extensively before *(10,94)*. At a low (but not at a high) shock intensity ACTH and α-MSH improved acquisition of shuttle box avoidance behavior *(95)*. The process of extinction of avoidance behavior is more sensitive to the effect of ACTH, especially if ACTH is applied during the extinction period *(10)*. By contrast, ACTH-like peptides facilitate acquisition and retention of passive avoidance behavior. Although in active avoidance behavior [D-Phe[7]]ACTH[4–10] acts opposite to ACTH[4–10], in passive avoidance behavior these peptides both facilitate extinction, the only difference being that the effect of [D-Phe7]ACTH[4–10] was of longer duration *(10)*. Thus, the effect of ACTH-like peptides on avoidance behavior also depends on the experimental device. The bioassays used to demonstrate these effects of melanocortins, involve many skills to be intact such as sensomotor performance, visual signaling, attention, memory storage and retrieval, etc. Thus, melanocortins may have a direct role on memory or an indirect role since melanocortins may stimulate attention *(10,94)*.

The bioassay in which the structure-activity relationship of ACTH peptides had been studied most was the pole jump test *(96)*. This is an active avoidance task in which a rat is taught to avoid an electric footshock by jumping onto a pole. ACTH delays the extinction of this behavior. In this assay, ACTH[4–10] had a potency that equals that of ACTH[1–24] *(9,96)*. In contrast to the induction of excessive grooming behavior in rats, where ICV injection of melanocortins is necessary to evoke the effect, subcutaneous injection of melanocortins are effective in eliciting the avoidance response. Also ACTH[4–9] was active in this assay, which eventually led to the development of ORG2766 (Met(O_2)-Glu-His-Phe-D-Lys-Phe), which is very potent in this test, but was designed to lack activity on steroidogenesis and pigmentation *(96)*. Although even more potent melanocortin analogs were developed using this bioassay, like Met(O_2)-Ala-Ala-Ala-D-Lys- Pro-Val-Gly-Lys-LysNH_2, which is 1000-fold more potent than ORG2766 *(9)*, ORG2766 was studied most extensively since it had been put forward as a drug candidate. Structure–activity relationships of melanocortins active on MC receptors now strongly argue against the involvement of MC receptors in avoidance behavior (Table 1). Typically, [D-Phe[7]]ACTH[4–10], NDP-MSH and γ-MSH had an effect opposite to that of ACTH[4–10] in active avoidance behavior *(10,12,15,97)*. However, NDP-MSH and γ-MSH fully activate the MC3 receptor, whereas [D-Phe[7]]ACTH[4–10] has an activity comparable to ACTH[4–10] on the MC3 receptor *(52)*. The MC4 receptor is activated by NDP-MSH and at higher concentrations, [D-Phe[7]]ACTH[4–10] activates the MC4 receptor, whereas very high doses of ACTH[4–10] are necessary to only partially activate the MC4 receptor *(52)*. Taking into consideration that ACTH[4–9] and ORG2766 were active in delaying the extinction of active avoidance behavior, whereas [D-Phe[7]]ACTH[4–10], NDP-MSH and γ-MSH enhanced the extinction of active avoidance behavior, the involvement of MC3 and MC4 receptors in mediating this behavior should be ruled out. ORG2766 and ACTH[4–9] did not activate, nor bind to these receptors in vitro *(37,38,52)*. Furthermore, the minimal core sequence for melanocortins to delay extinction of active avoidance behavior is ACTH[4–

7] *(98)*, whereas the minimum sequence necessary to stimulate MC receptors is ACTH[6–9] *(99)*. The substitutions in ORG2766 as compared to ACTH[4–9], when introduced separately into ACTH[4–10] resulted in loss of activity on the MC3 or MC4 receptors *(52)*. Therefore, the effects of melanocortins on active avoidance behavior are mediated by another receptor. Since melanocortin antisera injected ICV facilitate extinction of avoidance behavior, melanocortins may be physiologically involved in active avoidance behavior *(10)*. However also antisera against ORG2766 that do not cross-react with ACTH[4–10] are active in facilitating this response *(10)*, suggesting that a structurally related but another signaling molecule than melanocortins is involved physiologically in avoidance behavior. The thalamic parafascicular nucleus has been implicated in mediating the effects of ACTH-like peptides on avoidance behavior *(100)*. Thus, a central target mediates these effects of ACTH-like peptides.

Based upon the potency of ORG2766 in avoidance behavior, ORG2766 had been used widely as a neural specific melanocortin agonist. Since ORG2766 does not act via MC receptors, we now have to consider that ORG2766, although it has structural similarity to melanocortins, is not a melanocortin analog from a functional standpoint of view. Therefore, many effects that reportedly involved the melanocortin system, based on results using ORG2766, should be reconsidered and are not discussed further in this chapter.

3.2. Effects of γ-MSH on Blood Pressure and Cerebral Blood Flow

3.2.1. Pressor Effect

Since the pressor and natriuretic effects of chronic administration of ACTH were not mimicked by gluco- and/or mineralocorticoids, Gruber and Callahan *(101)* started to investigate direct effects of ACTH fragments on blood pressure and heart rate. Peripheral administration of relatively high doses (30–1000 nmol/kg) of ACTH[4–10] and γ_2-MSH increased heart rate and blood pressure, whereas lower doses had a natriuretic effect *(101–105)*. Surprisingly, α-MSH and NDP-MSH do not stimulate heart rate and blood pressure *(105,106)*. The site via which γ_2-MSH elicits this effect is unknown. Injection of ACTH[4–10] or γ_2-MSH ICV only induced a slight pressor effect at very high doses, which suggested that these ligands acted on a structure outside the blood–brain barrier *(107)*. Furthermore, lower doses of γ_2-MSH are needed to evoke the pressor response when γ_2-MSH is injected into the carotic artery instead of intravenously *(108)*. These data suggest that γ_2-MSH acts at a site outside the blood–brain barrier. Indeed, lesions of the AV3V region lead to a right shift of the dose-response curve of γ_2-MSH *(109)*. The pressor effect of γ_2-MSH in rats depends on the arousal potential, since γ_2-MSH has a depressor effect in deeply anesthesized (pentobarbital) rats and a pressor effect in lightly anesthesized (urethane) rats having sufficient sympathetic tone *(105)*. Indeed sympathetic ganglionic blockade eliminated the pressor effect of γ_2-MSH *(110,111)*. Furthermore, application of vasopressin receptor antagonists in the brain reduced the pressor effect of γ_2-MSH, suggesting that circulating (intravenous injection) γ_2-MSH activates the central vasopressinergic system which subsequently activates the sympathetic system *(112)*.

3.2.2. Depressor Effect

Injection of melanocortins into the caudal part of the nucleus of the tractus solitarius leads to bradycardia and a lower blood pressure *(113)*. In this site MC4 receptors are expressed *(38)*. MC3 receptors are not expressed here *(37)* but the high local dose of γ_2-MSH injected in these studies may activate MC4 receptors to an extent high enough to mediate these effects. Depressor effects are also observed when melanocortins (α-MSH) are injected in the dorsal motor nucleus of the vagus nerve *(114)*, where there is abundant expression of the MC4 receptor *(38)*. Injection of a MC4 receptor antagonist (SHU9119) blocks the depressor effect of α-MSH *(115)*. Stimulation of POMC neurons in the arcuate nucleus has a depressor effect which is mediated via the dorsal vagal complex *(116)*, suggesting that the endogenous melanocortin system may play a role in regulating the cardiovascular system.

One may speculate that the MC3 receptor mediates the γ-MSH pressor effect and that the MC4-mediated depressor effect overrules the pressor effect if α-MSH activates both the MC3 and MC4 receptors. However, the pressor effect of γ-MSH is still observed when smaller fragments like γ_2-MSH(6–12) are used, which are devoid of activity on MC receptors *(104)*. Furthermore, an MC3/MC4 receptor antagonist does not antagonize the pressor effect of γ_2-MSH *(115)*. Therefore, the pressor effect of γ-MSH fragments is mediated via another receptor. The effects of γ-MSH on blood pressure may be dependent on a free C-terminal Arg-Phe sequence, which γ_2-MSH has in common with FMRF-amides, that also increase blood pressure in a similar manner *(117)*. γ_3-MSH (the natural γ-MSH peptide in rats and mice) lacks the pressor effect *(102)*, possibly since in γ_3-MSH, C-terminal amino acids may mask the Arg-Phe sequence.

Thus, the depressor effect of melanocortins is probably mediated via MC4 receptors in the nucleus of the tractus solitarius and dorsal motor nucleus of the vagus nerve, whereas the pressor effect of γ_2-MSH is not mediated via one of the identified MC receptors.

3.2.3. *Cerebal Blood Flow*

Intravenous or intracarotic injection with γ_2-MSH also increases intracerebral blood flow *(118,119)*. Therefore, γ_2-MSH or analogs of it, may be used clinically to treat patients suffering from decreased cerebral blood flow, since γ_2-MSH combines an increased blood pressure with an increased cerebral blood flow. Again γ_2-MSH[6–12] is a potent peptide (120), whereas α-MSH and NDP-MSH are inactive, suggesting that this effect is not mediated via one of the cloned MC receptors, but uses the same receptors as the pressor effect of γ_2-MSH.

4. Effects of Melanocortins Partly Mediated via Brain Melanocortin Receptors

4.1. Sexual and Social Behavior

Central effects of melanocortins on sexual and social behavior may be influenced by peripheral effects of melanocortins on for instance the sebaceous and preputial glands *(121–124)*, since α-MSH stimulates pheromone release from these glands, which may in turn influence these behaviors *(125)*. Since most of the these effects of melanocortins were studied following peripheral injections, many of these effects may not be mediated via a direct action of melanocortins on the brain.

4.1.1. *Sexual Behavior*

In male rats, ICV injected α-MSH and ACTH[1–24] stimulates penile erection, ejaculation, sexual posturing, and genital licking *(126,127)*. In female rats α-MSH stimulates lordosis behavior at a low level of receptivity, whereas α-MSH has the opposite effect in receptive females *(128)*. In rabbits, besides stimulating sexual behavior, ICV injection of ACTH[1–24] stimulated plasma LH levels *(129)*. However, in contrast to this ICV injected ACTH[1–24] and ACTH[4–10] delay copulatory behavior in inexperienced rats *(10)* and ACTH[1–24] reduced sexual performance *(130)*. Local injection of α-MSH into the medial preoptic area increased lordosis behavior in the rat *(131)*. In this site MC4 receptors are expressed *(38)*. Depending on whether melanocortins are injected peripherally or centrally and the experimental method used to measure sexual behavior melanocortins appear to be stimulatory or inhibitory (reviewed in refs. *10* and *18*).

4.1.2. *Social Behavior*

Melanocortins generally decrease social interactions *(132–134)*. Many studies reported effects of ORG2766 on social behavior, which are often opposite to the effects of melanocortins such as α-MSH, γ-MSH and

ACTH[4–10] *(133,135,136)*. Again the experimental setup (age of the animals, housing conditions etc) influence the effect that melanocortins have on social behavior. Local injection of ACTH[4–10] into the septum decreased active social interaction and it increased aggressive behavior, while ORG2766 had the opposite effects *(132)*. In the septum there is restricted expression of MC3 and MC4 receptors *(37,38)*. Also injection of α-MSH into the ventromedial nucleus (VMN) of the hypothalamus increased aggressive behavior *(131)*. The VMN expresses MC3 and MC4 receptors *(37,38,137)*.

In these fields of research MC receptor selective ligands have not been used making it difficult to judge whether these effects involve melanocortin receptors.

4.2. Melanocortins and Epilepsy

ACTH is used therapeutically to treat patients suffering from West syndrome, characterized by infantile spasms, hypsarrhythmic electroencephalogram, and mental retardation as well as from Lennox-Gestaut syndrome, characterized by slow spike-wave, atypical absence myoclonus, and frequent ictal falls *(138)*. The mechanism of ACTH being an effective treatment in these disorders is unknown. Corticosteroids are also used therapeutically in these disorders, but there are reports that at least part of the effects of ACTH are independent from its effect on the adrenal gland *(139,140)*. Using magnetic resonance imaging it was demonstrated that ACTH treatment in infants with infantile spasms resulted in a decreased volume of the pons, cerebellar vermis, and corpus callosum *(141)*.

Also in a rat model, ACTH treatment decreased seizure susceptibility in a kindling model *(142)*. Pilocarpine induced epilepsy *(143)*, and convulsive seizures induced by amygdaloid kindling *(144)* in rats is counteracted by subcutanous injections with ACTH and fragments of ACTH. However, the active fragments do not overlap completely with the minimal fragment of ACTH necessary to activate MC receptors. For instance, ORG2766 and ACTH[7–16] as well as ACTH[4–10] and [D-Phe^7]ACTH[4–10] were active, whereas ACTH[1–24] was not. It is unlikely that these effects in rats are mediated via the same neural pathways as the effects of melanocortins applied in for instance West syndrome. The lack of animal models for the various forms of epilepsy and the fact that many different primary defects result in epileptic seizures, make it difficult to determine whether beneficial effects of ACTH in epilepsy are mediated via the adrenal gland or directly on the brain.

4.3. Nerve Regeneration

Perinatal treatment with melanocortins enhanced maturation of the nervous system as indicated by earlier eye opening *(145)* and accelerated maturation of the neuromuscular system (reviewed in ref. *146*). Because of these reasons it was investigated whether melanocortins have trophic effects on damaged nervous tissue. The neurotrophic effects of melanocortins in vitro and in vivo have been reviewed extensively before *(147–150)*.

Strand and Kung *(151)* were the first to describe that ACTH enhanced recovery of neuromuscular function following peripheral nerve crush. The rat sciatic nerve crush model in particular has been used extensively to study neurotrophic effects of melanocortins *(151–153)*. Peripheral injections with microgram quantities of α-MSH accelerated recovery of sensory and motor function and nerve conduction velocity, suggesting that exogenous applied melanocortins stimulate functional recovery from nerve damage. Peptide treatment resulted in more unmyelinated and myelinated newly formed sprouts not affecting the outgrowth rate *(154)*. The rat sciatic nerve crush model has been used for structure-activity studies using melanocortin peptide effects on nerve regeneration (Table 1). These studies demonstrated that the core region of melanocortins necessary for the neurotrophic effect is ACTH[4–9] *(16,155)* suggesting that MC receptors may mediate these effects. However, both α-MSH and ORG2766 are active, suggesting that at least two receptors are involved in mediating effects of melanocortins on stimulating nerve regeneration. α-MSH, NDP-MSH (at lower doses than α-MSH), ACTH[4–10] and ORG2766 are active in recovery of sensibility following sciatic nerve crush, whereas γ-MSH and [D-Phe^7]ACTH[4–10] are not *(17,155)*. Since γ-MSH activated the MC3 receptor in vitro at lower concentrations than ACTH[4–10] *(52)*, the MC3 receptor is probably not involved in mediating neurotrophic effects of melanocortins. Since, in spinal cord MC4 receptor expression was detected *(38)*, the MC4 receptor is a good candidate to mediate these effects. However, since ORG2766 also stimulated functional recovery following sciatic nerve crush, a second receptor outside the melanocortin family may be involved. An ORG2766 receptor has been characterized on rat Schwann cells *(156)*. Binding of ORG2766 to this receptor type is displaced by α-MSH. Activation of the ORG2766 receptor in primary cultures of Schwann cells released a yet unidentified neurotrophic factor that stimulated neurite outgrowth of dorsal root ganglion cells *(156)*.

Melanocortins stimulate neurite outgrowth in primary cultures of dorsal root ganglion and spinal cord *(148)*. NDP-MSH binding has been demonstrated in rat DRG *(157)*. Furthermore, MC4 receptor is expressed in spinal cord *(38)*. Activation of the cAMP signal transduction pathway and transient increases in calcium have been reported following exposure of dorsal root ganglion cultures with melanocortins *(158)*, which is in line with the fact that cloned MC receptors couple to the cAMP signal transduction pathway and in some cases may also activate the IP_3/calcium pathway *(159)*. The endogenously expressed MC4 receptor in Neuro 2A cells stimulates neurite out-

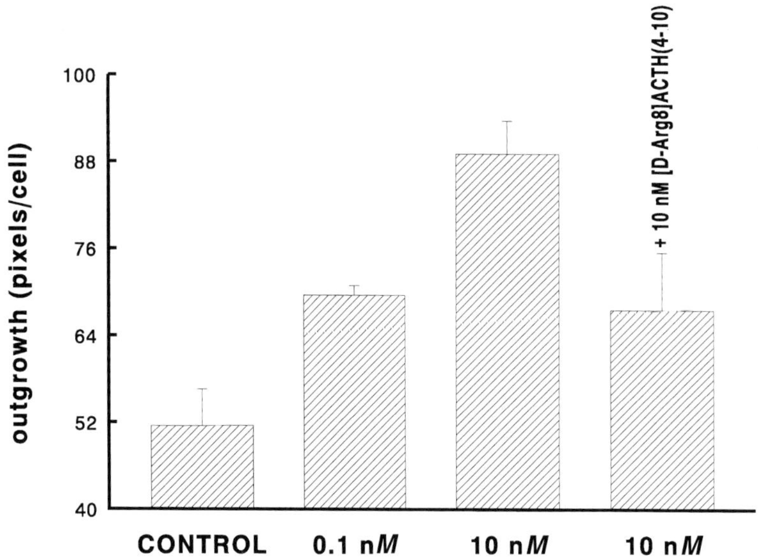

Fig. 5. Melanocortin-stimulated neurite outgrowth in Neuro 2A cells. Neuro 2a cells were incubated with either no peptide added (control), 0.1n*M* α-MSH, 10 n*M* α-MSH or 10 n*M* α-MSH in the presence of 10 n*M* of [D-Arg[8]]ACTH[4–10]. After 24 h, neurite length was measured by a semiautomated procedure. Data are expressed as average with standard error of the mean.

growth *(160)*(Fig. 5). Thus, expression of MC receptors on neurons provides a mechanism for melanocortins to stimulate neurite outgrowth of those neurons.

The effects of ORG2766 and α-MSH on primary cultures of rat spinal cord cells were shown not to overlap completely *(161)*, which suggests that ORG2766 and α-MSH indeed mediate their effect via different receptors.

Following nerve dissection, POMC expression and the immunoreactivity for ACTH/α-MSH has been reported to increase in mouse spinal cord motorneurons *(162)*. In Wobbler mice, which have motorneuron disease and in diabetic neuropathy *(163,164)*, the immunoreactivity for ACTH/α-MSH in nerves is also increased. This is accompanied by an increase in ACTH binding sites in developing, diabetic and dystrophic muscle *(163,165)*, suggesting that both the peptide (melanocortin) as well as probably the MC5 receptor, which is the receptor subtype expressed in muscle *(166,167)* are locally increased. This suggests that melanocortins may act as paracrine factor involved in the innervation of muscle. Following rat sciatic nerve crushes, melanocortin bioactivity (pigment dispersion) in the sciatic nerve is increased *(168,169)*, although POMC mRNA levels are not upregulated *(170)*. Thus, endogenous melanocortin peptides may play a role in recovery from nerve damage. Indeed, a MC receptor antagonist ([D-Trp^{7},Ala^{8},D-Phe^{10}]α-MSH[6–11]amide), which was characterized as a MC receptor antagonist in a lizard skin pigmentation bioassay, decreased the spontaneous recovery after a sciatic nerve crush *(171)*. Since this antagonist also blocked α-MSH-induced excessive grooming *(171)*, which is mediated via MC4 receptors *(54)*, MC4 receptor activation may contribute to the physiologic repair mechanism underlying recovery from nerve damage.

Melanocortins also stimulate recovery following lesions in the central nervous system. These central effects on regeneration have been performed mostly using ORG2766 as melanocortin (reviewed in ref. *172*). Again, there may be an ORG2766 receptor, as well as a MC receptor mediating these effects. For instance, ORG2766 and α-MSH improve functional recovery following 6-hydroxydopamine lesions in the nucleus accumbens *(173)*. α-MSH treatment of rats with spinal cord lesions also enhanced the recovery as judged from motor performance tests and electrophysiologic parameters *(174)*.

5. Conclusions (*see* Fig. 6)

The effects of melanocortins on feeding behavior (*see* Chapter 15), grooming behavior, the depressor effect of melanocortins, the central activation of the HPA axis as well as the antipyretic effect of melanocortins are mediated through activation of brain MC receptors. However, which MC receptor subtype is particularly involved needs further studies, since MC receptor selective ligands have not been used extensively in these studies. For all these effects, the melanocortin core sequence, MSH/ACTH[6–9] (His-Phe-Arg-Trp) is essential. Furthermore, central injections of melanocortins are much more effective as compared to peripheral injections which are usually inactive to elicit these effects. This is in line with the estimate that upon intravenous administration, only 0.01% of α-MSH reaches the brain *(175)*. If one assumes that at least 1 μg of α-MSH applied ICV is necessary to evoke excessive grooming behavior in the rat, this would mean that a dose of 10 mg α-MSH would be required to obtain a similar level of α-MSH in the brain. Indeed, peripheral melanocortin administration of 100 μg per 100 g bodyweight of α-MSH does not stimulate grooming behavior *(43)*.

The effects of melanocortins on grooming behavior and the antipyretic effect may be part of a physiologic repertoire of the melanocortin system in regulation of body temperature as has been suggested before *(44)*. The role of the Harderian gland in producing lipids that are spread over the pelage during grooming in gerbils to prevent heat loss *(176)* and the high density of

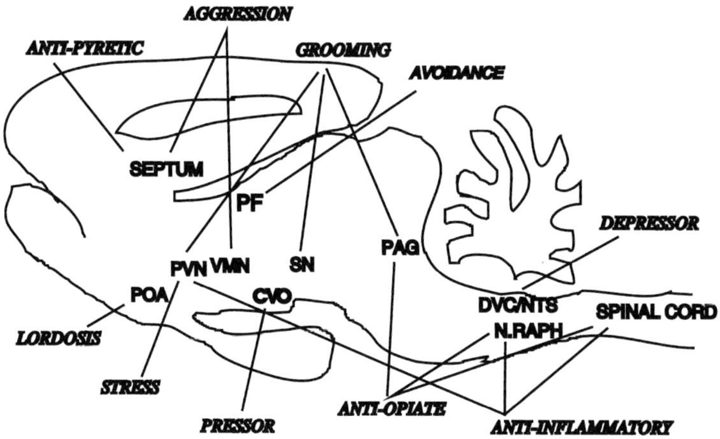

Fig. 6. Neuroanatomy of melanocortin effects. The putative areas in the brain that mediate the effects of melanocortins are depicted. PF, parafascicular nucleus; PVN, paraventricular nucleus; VMN, ventromedial nucleus; POA, medial preoptic area; PAG, paraaquaductal gray; N.Raph, nucleus raphe; NTS, nucleus of the tractus solitarius; DVC, dorsal motor nucleus of the vagus nerve; CVO, circumventricular organs; SN, substantia nigra.

melanocortin binding sites in this gland *(31)*, indicates that also in the periphery melanocortins have this regulatory role, which was confirmed recently in mice in which the MC5 receptor gene was disrupted *(124)*.

Activation of the melanocortin system may primarily stimulate attention (arousal) to peripheral stimuli by modulating the gating of sensory information. Subsequently, this may result in a change in the setpoint for activation of the HPA axis and hyperresponsiveness (as reflected by hyperalgesia) and this may contribute to inhibition of food intake. This involvement of the melanocortin system in gating of sensory information may also be the underlying mechanism for the effects of melanocortins on the suppression of morphine-induced analgesia and the antiinflammatory effect. These effects may involve similar pathways in which activation by melanocortins of descending pathways originating in the periaquaductal grey area and the raphe nucleus, modulate the transmission of afferent nerves (e.g., nociception) at the spinal cord level. The neurotrophic effect of melanocortins may also be explained along this line, since suppression of the inflammatory response by regulating activity of the sympathetic system will be beneficial for the regeneration process. Thus, beneficial effects of melanocortins on stimulation of nerve regeneration may be mediated partially via activation of MC receptors, MC4 and MC5 receptors being the best candidates since these are expressed in the spinal cord and sympathetic system, and in skeletal muscle, respectively. Also the effect of melanocortins counteracting the effects of opioids is a strong candidate effect that may be mediated via MC receptors.

One of the earliest effects of melanocortins (its effects on learning and memory), turned out not to be mediated via one of the identified MC receptor subtypes, since the most potent "melanocortin analogs" on avoidance behavior like ORG2766, were inactive on MC receptors, whereas MC receptor agonists like NDP-MSH and [D-Phe^7]ACTH[4–10] antagonized this response in many experiments measuring avoidance behavior. Subcutaneous injections of melanocortins at microgram quantities are effective in eliciting effects on avoidance behavior. ICV injection of 10 ng ACTH[4–10] delays extinction of active avoidance behavior, which is approximately a thousand-fold lower dose than necessary in the periphery to have the same effect *(177)*. Taking into consideration that 0.01% of a peripherally applied melanocortin passes the blood–brain barrier, the receptor mediating the effects of ACTH-like peptides on avoidance behavior must have a higher affinity for ACTH as compared to the cloned MC receptors. There are many effects of ACTH-like peptides, that do not have the melanocortin core region, like ACTH[4–7] and ACTH[7–16] *(42)*. These effects are mediated via a receptor that probably has little or no sequence homology to the cloned MC receptors.

The pressor effect following intravenous injection with the C-terminal fragments of γ-MSH is not mediated via known MC receptors but may be mediated via FMRF-amide receptors. Similarly the antiinflammatory effect of C-terminal fragments of α-MSH, which lack activity on MC receptors, and which are also effective when given peripherally are not mediated via brain MC receptors. However the antiinflammatory effect of α-MSH, which is most pronounced upon ICV injections is an effect that is probably mediated via MC4-R receptors.

As more selective MC receptor ligands become available, it can be determined which effects of melanocortins are mediated via each of the MC receptor subtypes. These MC receptor selective ligands can be applied locally at sites that express a particular receptor subtype. This will help to clarify the role MC receptors play in behavioral responses and in neuroendocrine control of homeostasis. Furthermore, the relationship to pathology and the role of MC receptors as potential drug targets can now be investigated.

Acknowledgments

I thank D. de Wied, W. H. Gispen and J. P. H. Burbach for introducing me to this field of research, for their continuing support and interest, and for helpful suggestions during the preparation of this chapter.

References

1. Mirsky, R., Miller, R. E., and Stein, M. (1953) Relation of adrenocortical activity and adaptive behavior. *Psychosom. Med.* **15,** 574–588.
2. Applezweig, M. H. and Baudry, F. D. (1955) The pituitary-adrenocortical system in avoidance learning. *Psychol. Rep.* **1,** 417–420.
3. Miller, R. E. and Ogawa, N. (1962) The effect of adrenocorticotropic hormone (ACTH) on avoidance conditioning in the adrenalectomized rat. *J. Comp. Physiol. Psychol.* **55,** 211–213.
4. Klavdieva, M. (1996) The history of neuropeptides III. Front. Neuroendocrinol. **17,** 155–179.
5. De Wied, D. (1966) Inhibitory effect of ACTH and related peptides on extinction of conditioned avoidance behavior in rats. Proc. Soc. Exp. Biol. Med. **122,** 28–32.
6. De Wied, D. (1983) Neuropeptides and behavior. *Psychoneuropharmacology.* **1,** 307–353.
7. Ferrari, W. (1958) Behavioural changes in animals after intracisternal injection with adrenocoticotrophic hormone and melanocyte stimulating hormone. *Nature* 925–926.
8. Ferrari, W., Gessa, G. L., and Vargiu, L. (1963) Behavioral effects inducted by intracisternally injected ACTH and MSH. *Ann. N. Y. Acad. Sci.* **104,** 330–345.
9. Greven, H. M. and De Wied, D. (1977) Influence of peptides structurally related to ACTH and MSH on active avoidance behavior. *Front. Horm. Res.* **4,** 140–152.
10. De Wied, D. and Jolles, J. (1982) Neuropeptides derived form pro-opiomelanocortin, behavioral, physiological, and neurochemical effects. *Physiol. Rev.* **62,** 976–1059.
11. Wiegant, V. M., Jolles, J., Colbern, D. L., Zimmerman, E., and Gispen, W. H. (1979) Intracerebroventricular ACTH activates the pituitary–adrenal system: dissociation from a behavioral response. *Life Sci* **25,** 1791–1796.
12. Fekete, M. and De Wied, D. (1982) Potency and duration of action of the ACTH 4–9 analog (ORG 2766) as compared to ACTH 4–10 and [D-Phe7] ACTH 4–10 on active and passive avoidance behavior of rats. *Pharmacol. Biochem. Behav.* **16,** 387–392.
13. Wiegant, V. M., Colbern, D., van Wimersma Greidanus, T. J., and Gispen, W. H. (1978) Differential behavioral effects of ACTH 4–10 and [D-Phe7] ACTH 4–10. *Brain Res. Bull.* **3,** 167–170.
14. Kobobun, K., O'Donohue, T. L., Handelman, G. E., Sawyer, T. K., Hruby, V. J., and Hadley, M. E. (1983) Behavioral effects of [4-norleucine,7-D-phenylalanine]-α-melanocyte-stimulating hormone. *Peptides* **4,** 721–724.
15. Beckwith, B. E., Tinius, T. P., Hruby, V. J., Al–Obeidi, F., Sawyer, T. K., and Affholter, J. A. (1989) The effects of structure–conformation modifications of melanotropin analogs on learning and memory, D-amino acids substituted linear and cyclic analogs. *Peptides* **10,** 361–368.
16. Bijlsma, W. A., Jennekens, F. G. I., Schotman, P., and Gispen, W. H. (1981) Effects of corticotropin (ACTH) on recovery of sensorimotor function in the rat, structure–activity study. *Eur. J. Pharmacol.* **76,** 73–79.
17. Van Der Zee, C. E. E. M., Brakkee, J. H., and Gispen, W. H. (1991) Putative neurotrophic factors and functional recovery from peripheral nerve damage in the rat. *Br. J. Pharmacol.* **103,** 1041–1046.

18. Eberle, A. N. (1988) *The Melanotropins*. S.Karger, Basel.
19. Watson, S. J., Richard-III, C. W., and Barchas, J. D. (1978) Adrenocorticotropin in rat brain, immunocytochemical localization in cells and axons. *Science* **200,** 1180–1182.
20. Jacobowitz, D. M., and O'Donohue, T. L. (1978) Alfa-melanocyte stimulating hormone, immunohistochemical identification and mapping in neurons of the rat brain. *Proc. Natl. Acad. Sci. U. S. A.* **75,** 6300–6304.
21. Krieger, D. T., Liotta, A., and Brownstein, M. J. (1977) Presence of corticotropin in brain of normal and hypophysectomized rats. *Proc. Natl. Acad. Sci. U. S. A.* **74,** 648–652.
22. Rudman, D., Del Rio, A. E., Chawla, R. K., Houser, D. H., and Sheen, S. (1974) An antimelanotropic protein in human cerebrospinal fluid. *Am. J. Physiol.* **226,** 693–697.
23. O'Donohue, T. L., Holmquist, G. E., and Jacobowitz, D. M. (1979) Effect of hypophysectomy on alpha-melanotropin in discrete regions of the rat brain. *Neurosci. Lett.* **14,** 271–274.
24. Nakanishi, S., Inoue, A., Kita, T., Nakamura, M., Chang, A. C. Y., Cohen, S. N., and Numa, S. (1979) Nucleotide sequence of cloned cDNA for bovine corticotropin-β-lipotropin precursor. *Nature* **278,** 423–427.
25. Drouin, J. and Goodman, H. M. (1980) Most of the coding region of rat ACTH// beta-LPH precursor gene lacks intervening sequences. *Nature* **288,** 610–613.
26. Nakanishi, S., Teranishi, Y., Watanabe, Y., Notake, M., Noda, M., Kakidani, H., jingami, H., and Numa, S. (1981) Isolation and characterization of the bovine corticotropin/beta-lipotropin precursor gene. *Eur. J. Biochem.* **115,** 429–438.
27. Gee, C. E., Chen, C. L. C., Roberts, J. L., Thompson, R., and Watson, S. J. (1983) Identification of proopiomelanocortin neurons in the rat hypothalamus by in situ cDNA-mRNA hybridization. *Nature* **306,** 374–375.
28. Schwartzberg, D. G., and Nakane, P. K. (1983) ACTH-related peptide containing neurons within the medulla oblongata of the rat. *Brain Res.* **276,** 351–356.
29. Gramsch, C., Kleber, G., Hollt, V., Pasi, A., Mehraein, P., and Herz, A. (1980) Proopiocortin fragments in human and rat brain, β-endorphin and α-MSH are the predominant peptides. *Brain Res.* **192,** 109–119.
30. Tatro, J. B. (1990) Melanotropin receptors in the brain are differentially distributed and recognize both corticotropin and α-melnocyte stimulating hormone. *Brain Res.* **536,** 124–132.
31. Tatro, J. B. and Reichlin, S. (1987) Specific receptors for alpha-MSH are widely distributed in tissues of rodents. *Endocrinology* **121,** 1900–1907.
32. Chhajlani, V., Muceniece, R., and Wikberg, J. E. S. (1993) Molecular cloning of a novel human melanocortin receptor. *Biochem. Biophys. Res. Commun.* **195,** 866–873.
33. Low, M. J., Simerley, R. B., and Cone, R. D. (1994) Receptors for the melanocortin peptides in the central nervous system. *Curr. Opin. Endocrinol. Biab.* **1,** 79–88.
34. Chhajlani, V., and Wikberg, J. E. S. (1992) Molecular cloning and expression of the human melanocyte stimulating hormone receptor cDNA. *FEBS Lett.* **309,** 417–420.
35. Gantz, I., Konda, T., Tashiro, T., Shimoto, Y., Miwa, H., Munzert, G., Watson, S. J., DelValle, J., and Yamada, T. (1993) Molecular cloning of a novel melanocortin receptor. *J. Biol. Chem.* **268,** 8246–8250.

36. Gantz, I., Miwa, H., Konda, Y., Shimoto, Y., Tashiro, T., Watson, S. J., DelValle, J., and Yamada, T. (1993) Molecular cloning, expression, and gene localization of a fourth melanocortin receptor. *J. Biol. Chem.* **268,** 15,174–15,179.
37. Rehfuss–Roselli, L., Mountjoy, K. G., Robbins, L. S., Mortrud, M. T., Low, M. J., Tatro, J. B., Entwistle, M. L., Simerly, R. B., and Cone, R. D. (1993) Identification of a receptor for melanotropin and other proopiomelanocortin peptides in the hypothalamus and limbic system. *Proc. Natl. Acad. Sci. U. S. A.* **90,** 8856–8860.
38. Mountjoy, K. G., Mortrud, M. T., Low, M. J., Simerley, R. B., and Cone, R. D. (1994) Cloning and functional characterization of a melanocortin receptor (MC4-R) localized in neuroendocrine and autonomic circuitry in the brain. *Mol. Endocrinol.* **8,** 1298–1308.
39. Barrett, P., MacDonald, A., Helliwell, R., Davidson, G., and Morgan, P. (1994) Cloning and expression of a new member of the melanocyte-stimulating hormone receptor family. *J. Mol. Endocrinol.* **12,** 203–213.
40. Beckwith, B. E., and Sandman, C. A. (1982) Central nervous system and peripheral effects of ACTH, MSH, and related neuropeptides. *Peptides* **3**,411–420.
41. De Wied, D. (1993) Melanotropins as neuropeptides. *Ann. N. Y. Acad. Sci.* **680,** 21–28.
42. De Wied, D., and Wolterink, G. (1988) Structure–activity studies on the neuroactive and neurotropic effects of neuropeptides related to ACTH. *Ann. N. Y. Acad. Sci.* **525,** 130–140.
43. Gispen, W. H., Wiegant, V. M., Greven, H. M., and De Wied, D. (1975) The induction of excessive grooming in the rat by intraventricular application of peptides derived from ACTH, structure–activity studies. *Life Sci.* **17,** 645–652.
44. Colbern, D. L. and Twombly, D. A. (1988) ACTH–induced grooming behaviors and body temperature: temporal effects of neurotensin, naloxone, and haloperidol. *Ann. N. Y. Acad. Sci.* **525,** 180–200.
45. Gispen, W. H. and Isaacson, R. L. (1986) Excessive grooming in response to ACTH. in *Neuropeptides and Behavior, Vol. 1.* (De Wied, D., Gispen, W. H., and Van Wimersma Greidanus, T. B., eds.) Pergamon, Oxford, pp. 273–312.
46. Spruijt, B. M., Van Hooff, J. A. R. A. M., and Gispen, W. H. (1992) Ethology and neurobiology of grooming behavior. *Physiol. Rev.* **72,** 825--852.
47. Spruijt, B. M., De Graan, P. N. E., Eberle, A. N., and Gispen, W. H. (1985) Comparison of structural requirements of αMSH and ACTH for inducing excessive grooming and pigment dispersion. *Peptides* **6,** 1185–1189.
48. Hirsch, M. D., O'Donohue, T. L., Wilson, R., Sawyer, T. K., Hruby, V. J., Hadley, M. E., Cody, W. L., Knittel, J. J., and Crawley, J. N. (1984) Structural and conformational modifications of αMSH/ACTH4–10 provide melanotrpin analogues with high potent behavioral activities. *Peptides* **5,** 1197–1201.
49. Alvaro, J. S., Tatro, J. B., Quillan, J. M., Fogliano, M., Eisenhard, M., Lerner, M. R., Nestler, E. J., and Duman, R. S. (1996) Morphine down–regulates melanocortin-4 receptor expression in brain regions that mediate opiate addiction. *Mol. Pharmacol.* **50,** 583–591.
50. Huang, Q. H., Entwistle, M. L., Alvaro, J. D., Duman, R. S., Hruby, V. J., and Tatro, J. B. (1997) Antipyretic role of endogenous melanocortins mediated by central melanocortin receptors during endotoxin-induced fever. *J. Neurosci.* **17,** 3343–3351.
51. Adan, R. A. and Gispen, W. H. (1997) Brain melanocortin receptors, from cloning to function. *Peptides* **18,** 1279–1287.

52. Adan, R. A. H., Cone, R. D., Burbach, J. P. H., and Gispen, W. H. (1994) Differential effects of melanocortin peptides on neural melanocortin receptors. *Mol. Pharmacol.* **46,** 1182–1190.
53. Sawyer, T. K., Sanfilippo, P. J., Hruby, V. J., Engel, M. H., Heward, C. B., Burnett, J. B., and Hadley, M. E. (1980) 4-Norleucine,7-D-phenylalanine-α-melanocyte-stimulating hormone, a highly potent α–melanotropin with ultralong biological activity. *Proc. Natl. Acad. Sci. U. S. A.* **77,** 5754–5758.
54. Adan, R. A. H., Oosterom, J., Ludvigsdottir, G., Brakkee, J H., Burbach, J. P. H., and Gispen, W. H. (1994) Identification of antagonists for MC**3,** MC4 and MC5 receptors. *Eur. J. Pharmacol.* **269,** 331–337.
55. Hruby, V. J., Lu, D., Sharma, S. D., Castrucci, A. L., Kesterson, R. A., Al-Obeidi, F. A., Hadley, M. E., and Cone, R. D. (19950 Cyclic lactam alfa-melanotropin analogues of Ac-Nle4-cyclo(Asp5,D-Phe7,Lys10)alfa-melanocyte stimulating hormone-(4–10)-NH2 with bulkyaromatic amino acids at position 7 show high potency and selectivity at specific melanocortin receptors. *J. Med. Chem.* **38,** 3454–3460.
56. Bressers, W. M. A., Kruk, M. R., Van Erp, H. H. M., Willekens-Bramer, D. C., Haccou, P., and Meelis, E. (1995) A time-structured analysis of hypothalamically induced increases in self-grooming and activity in the rat. *Behav. Neurosci.* **109,** 1158–1171.
57. Spruijt, B. M. and Gispen, W. H. (1984) The neural substrate involved in ACTH 1–24 excessive grooming. *Neurosci. Lett. Suppl.* **18,** 362.
58. Spruijt, B. M., Cools, A. R., and Gispen, W. H. (1986) The periaquaductal gray, a prerequisite for ACTH-induced excessive grooming. *Behav. Brain. Res.* **20,** 19–25.
59. Dunn, A. J., Green, E. J., and Isaacson, R. L. (1979) Intracerebral adreno-corticotrope hormone mediates novelty-induced grooming in the rat. *Science* **203,** 281–283.
60. Spampinato, S., Canossa, M., Carboni, L., Campana, G., Leanza, G., and Ferri, S. (1994) Inhibition of proopiomelanocortin expression by an oligonucleotide complementary to beta-endorphin mRNA. *Proc. Natl. Acad. Sci. U. S. A.* **91,** 8072–8076.
60a. Von Frijtag, J. C., Croiset G., Gispen, W. H., Adan, R. A., and Wiegant, V. M. (1998) The role of central melanocortin receptors in the activation of the hypothalamus-pituitary-adrenal-axis and the induction of excessive grooming. *Br. J. Pharmacol.* **123 (8)**, 1503–1508.
61. Motta, M., Mangili, G., and Martini, L. (1965) A "short" feedback loop in the control of ACTH secretion. *Endocrinology* **77,** 392–395.
62. Suda, T., Yajima, F., Tomori, N., Sumitomo, T., Nakagami, Y., Ushiyama, T., Demura, H., and Shizume, K. (1986) Inhibitory effect of adrenocorticotropin on corticotropin-releasing factor release from rat hypothalamus in vitro. *Endocrinology* **118,** 459–461.
63. Catania, A. and Lipton, J. M. (1993) α-Melanocyte stimulating hormone in the modulation of host reactions. *Endocr. Rev.* **14,** 564–576.
64. Watanabe, T., Hiltz, M. E., Catania, A., and Lipton, J. M. (1993) Inhibition of Il-1-induced peripheral inflammation by peripheral and cetral administration of analogs of the neuropeptide alpha–MSH. *Brain Res. Bull* **32,** 311–314.
65. Ceriani, G., Macaluso, A., Catania, A., and Lipton, J. M. (1994) Central neurogenic antiinflammatory action of alpha-MSH. *Neuroendocrinology* **59,** 138–143.
66. Lipton, J. M. and Catania, A. (1997) Anti-inflammatory actions of the neuroimmunomodulator α-MSH. *Immunol. Today* **18,** 140–145.

67. Robertson, B., Dostal, K., and Daynes, R. A. (1988) Neuropeptide regulation of inflammatory and immunologic responses. *J. Immunol.* **140,** 4300–4307.
68. Martin, L. W., Catania, A., Hiltz, M. E., and Lipton, J. M. (1991) Neuropeptide alpha-MSH antagonizes IL-6- and TNF-induced fever. *Peptides* **12,** 297–299.
69. Hiltz, M. E., Catania, A., and Lipton, J. M. (1992) Alpha-MSH peptides inhibit acute inflammation induced in mice by rIL-1 beta, rIL-6, rTNF-alpha and endogenous pyrogen but not that caused by LTB4, PAF and rIL-8. *Cytokine* **4,** 320–328.
70. Rajora, N., Boccoli, G., Burns, D., Sharma, S., Catania, A. P., and Lipton, J. M. (1997) alpha-MSH modulates local and circulating tumor necrosis factor-alpha in experimental brain inflammation. *J. Neurosci.* **17,** 2181–2186.
71. Opp, M. R., Obal, F. Jr., and Krueger, J. M. (1988) Effects of alpha-MSH on sleep, behavior, and brain temperature, interactions with IL 1. *Am. J. Physiol.* **255,** R914–922.
72. Deeter, L. B., Martin, L. W., and Lipton, J. M. (1988) Antipyretic properties of centrally administered alpha-MSH fragments in the rabbit. *Peptides* **9,** 1285–1288.
73. Holdeman, M, and Lipton, J. M. (1985) Antipyretic activity of a potent alpha-MSH analog. *Peptides* **6,** 273–275.
74. Samson, W. K., Lipton, J. M., Zimmer, J. A., and Glyn, J. R. (1981) The effect of fever on central alpha-MSH concentrations in the rabbit. *Peptides* **2,** 419–423.
75. Bell, R. C. and Lipton, J. M. (1987) Pulsatile release of antipyretic neuropeptide alpha-MSH from septum of rabbit during fever. *Am. J. Physiol.* 25**2,** R1152–1157.
76. Holdeman, M., Khorram, O., Samson, W. K., Lipton, J. M. (1985) Fever-specific changes in central MSH and CRF concentrations. *Am. J. Physiol.* **248,** R125–129.
77. Shih, S. T., Khorram, O., Lipton, J. M., and McCann, S. M. (1986) Central administration of alfa-MSH antiserum augments fever in the rabbit. *Am. J. Physiol.* **250,** r803–r806.
78. Macaluso, A., McCoy, D., Ceriani, G., Watanabe, T., Biltz, J., Catania, A., and Lipton, J. M. (1994) Antiinflammatory influences of alpha-MSH molecules, central neurogenic and peripheral actions. *J. Neurosci.* **14,** 2377–2382.
79. Mountjoy, K. G., and Wong, J. (1997) Obesity, diabetes and functions for proopiomelanocortin-derived peptides. *Mol. Cell. Endocrinol.* **128,** 171–177.
80. Coderre, T. J., Basbaum, A. I., and Levine, J. D. (1989) Neural control of vascular permeability, interactions between primary afferents, mast cells, and sympathetic efferents. *J. Neurophysiol.* **62,** 48–58.
81. Besson, J. M. and Chaouch, A. (1987) Peripheral and spinal mechanisms of nociception. *Physiol. Rev.* **67,** 167–186.
82. Contreras, P. C. and Takemori, A. E. (1984) Antagonism of morphine-induced analgesia, tolerance and dependence by alpha-melanocyte-stimulating hormone. *J. Pharmacol. Exp. Ther.* **229,** 21–26.
83. Williams, D. W. Jr., Lipton, J. M., and Giesecke, A. H. Jr. (1986) Influence of centrally administered peptides on ear withdrawal from heat in the rabbit. *Peptides* **7,** 1095–1100.
84. Gispen, W. H., Buitelaar, J., Wiegant, V. M., Terenius, L., and De Wied, D. (1976) Interaction between ACTH fragments, brain opiate receptors and morphine–induced analgesia. *Eur. J. Pharmacol.* **39,** 393–397.
85. Sandman, C. A. and Kastin, A. J. (1981) Intraventricular administration of MSH induces hyperalgesia in rats. *Peptides* **2,** 231–233.

86. Bertolini, A., Poggioli, R., and Fratta, W. (1981) Withdrawal symptoms in morphine-dependent rats intracerebroventricularly injected with ACTH1–24 and with beta-MSH. *Life Sci.* **29,** 249–252.
87. Szekely, J. I., Miglecz, E., Dunai–Kovacs, Z., Tarnawa, I., Ronai, A. Z., Graf, L., and Bajusz, S. (1979) Attenuation of morphine tolerance and dependence by alpha-melanocyte stimulating hormone(alpha-MSH). *Life Sci.* **24,** 1931–1938.
88. Wiegant, V. M., Gispen, W. H., Terenius, L., and De Wied, D. (1977) ACTH-like peptides and morphine, interaction at the level of the CNS. *Psychoneuroendocrinology* **2,** 63–70.
89. Van Ree, J. M., Bohus, B., Csontos, K. M., Gispen, W. H., Greven, H. M., Nijkamp, F. P., Opmeer, F. A., de Rotte, G. A., van Wimersma Greidanus T. B., Witter, A., and De Wied, D. (19810 Behavioral profile of gamma-MSH, relationship with ACTH and beta-endorphin action. *Life Sci.* **28,** 2875–2878.
90. Jacquet, Y. F. (1978) Opiate effects after adrenocorticotropin or beta-endorphin injection in the periaqueductal gray matter of rats. *Science* **201,** 1032–1034.
91. Aloyo, V. J., Spruyt, B., Zwiers, H., and Gispen, W. H. (1983) Peptide induced excessive grooming behavior, the role of opiate receptors. *Peptides* **4,** 833–836.
92. Gispen, W. H. and Wiegant, V. M. (1976) Opiate antagonists suppress ACTH1–24 induced excessive grooming in the rat. *Neurosci. Lett.* **2,** 159–164.
93. De Wied, D. (1966) Inhibitory effects of ACTH and related peptides on extinction of conditioned avoidance behavior in rats. *Proc. Soc. Exp. Biol. Med.* **122,** 28–32.
94. Beckwith, B. E. and Sandman, C. A. (1978) Behavioral influences of the neuropeptides ACTH and MSH, A methodological review. *Neurosci. Biobehav. Rev.* **2,** 311–338.
95. Stratton, L. O. and Kastin, A. J. (1974) Avoidance learning at two levels of shock in rats receiving MSH**1,** *Horm. Behav.* **5,** 149–155.
96. Van Nispen, J. W. and Greven, H. M. (1986) Structure–activity relationships of peptides derived from ACTH, β-LPH and MSH with regard to avoidance behavior in rats in *Neuropeptides and Behavior, vol.1.* (De Wied, D., Gispen, W. H., Van Wimersma Greidanus, T. B., eds.) Pergamon, Oxford, pp. 349–384.
97. Flood, J. F., Jarvik, M. E., Bennett, E. L., and Orme, A. E. (1976) Effects of ACTH peptide fragments on memory formation. *Pharmacol. Biochem. Behav.* **5,** 41–51.
98. Fekete. M., Bohus, B., and De Wied, D. (1983) Comparative effects of ACTH-related peptides on acquisition of shuttle– box avoidance behavior of hypophysectomized rats. *Neuroendocrinology* **36,** 112–118.
99. Hruby, V. J., Sharma, S. D., Toth, K., Jaw, J. Y., Al–Obeidi, F., Sawyer, T. K., and Hadley, M. E. (1993) Design, synthesis, and conformation of superpotent and prolonged acting melanotropins. *Ann. N. Y. Acad. Sci.* **680,** 51–63.
100. van Wimersma Greidanus, T. B., Bohus, B., and De Wied, D. (1974) The parafasicular area as the site of action of ACTH analogs on avoidance behavior. *Prog. Brain Res.* **41,** 429–432.
101. Gruber, K. A. and Callahan, M. F. (1989) ACTH (4–10) through gamma–MSH, evidence for a new class of central autonomic nervous system-regulating peptides. *Am. J. Physiol.* 257**,** r681–r694.
102. Klein, M. C., Hutchins, P. M., Lymangrover, J. R., and Gruber, K. A. (1985) Pressor and cardioaccelerator effects of gamma MSH and related peptides. *Life Sci.* **36,** 769–775.

103. Gruber, K. A., Klein, M. C., Hutchins, P. M., Buckalew, V. M. Jr., and Lymangrover, J. R. (1984) Natriuretic and hypertensive activities reside in a fragment of ACTH. *Hypertension* **6,** 468–474.
104. Van Bergen, P., Janssen, P. M., Hoogerhout, P., De Wildt, D. J., and Versteeg, D. H. (1995) Cardiovascular effects of gamma-MSH/ACTH-like peptides, structure–activity relationship. *Eur. J. Pharmacol.* **294,** 795–803.
105. De Wildt, D. J., Krugers, H., Kasbergen, C..M., De Lang, H., and Versteeg, D. H. G. (1993) The hemodynamic effects of gamma2-melanocyte-stimulating hormone and related melanotropins depend on the arousal potential of the rat. *Eur. J. Pharmacol.* **233,** 157–164.
106. Van Bergen, P., Kleijne, J. A., De Wildt, D. J., and Versteeg, D. H. (1997) Different cardiovascular profiles of three melanocortins in conscious rats; evidence for antagonism between gamma 2–MSH and ACTH–(1–24) [In Process Citation]. *Br. J. Pharmacol.* **120,** 1561–1567.
107. Versteeg, D. H., Krugers, H., Meichow, C., De Lang, H., and De Wildt, D. J. (1993) Effect of ACTH-(4–10) and gamma 2-MSH on blood pressure after intracerebroventricular and intracisternal administration. *J. Cardiovasc. Pharmacol.* **21,** 907–911.
108. Kunos, G., Li, S., Varga, K., Archer, P., Kesterson, R. A., Cone, R. D., Hruby, V. J., and Sharma, S. D. (1997) Novel neural pathways of cardiovascular control by alpha- and gamma-MSH. *Fund. Clin. Pharmacol.* **11,** 44s–48s.
109. Callahan, M. F., Cunningham, J. T., Kirby, R. F., Johnson, A. K., and Gruber, K. A. (1988) Role of the anteroventral third ventricle (AV3V) region of the rat brain in the pressor response to gamma 2-melanocyte-stimulating hormone (gamma 2-MSH). *Brain Res.* **444,** 177–180.
110. Callahan, M. F., Kirby, R. F., Wolff, D. W., Strandhoy, J. W., Lymangrover, J. R., Johnson, A. K., and Gruber, K. A. (1985) Sympathetic nervous system mediation of acute cardiovascular actions of gamma 2-melanocyte-stimulating hormone. *Hypertension* **7,** I145–I150.
111. Callahan, M. F., Kirby, R. F., Johnson, A. K., and Gruber, K. A. (1988) Sympathetic terminal mediation of the acute cardiovascular response of gamma 2-MSH. *J. Auton. Nerv. Syst.* **24,** 179–182.
112. Gruber, K. A. and Eskridge, S. L. (1986) Central vasopressin system mediation of acute pressor effect of gamma-MSH. *Am. J. Physiol.* **251,** E134–E137.
113. De Wildt, D. J., Van Der Ven, J. C., Van Bergen, P., De Lang, H., and Versteeg, D. H.G. (1994) A hypotensive and bradycardic action of gamma2-MSH microinjected into the nucleus tractus solitarii of the rat. *Arch. Pharmacol.* **349,** 50–56.
114. Li, S. J., Varga, K., Archer, P., Hruby, V. J., Sharma, S. D., Kesterson, R. A., Cone, R. D., and Kunos, G. (1996) Melanocortin antagonists define two distinct pathways of cardiovascular control by alpha- and gamma-melanocyte-stimulating hormones. *J. Neurosci.* **16,** 5182–5188.
115. Li, S. J., Varga, K., Cone, R. D., Hruby, V. J., and Kunos, G. (1995) Melanocortin-4 receptors (MC4-R) in the dorsal vagal complex (DVC) mediate hypotensive and bradycardic effects of alpha-MSH. *Soc. Neurosci. Abstr.* **21,** 889.
116. Mastrianni, L. A., Palkovits, M., and Kunos, G. (1989) Activation of brainstem endorphinergic neurons causes cardiovascular depression and facilitates baroreflex bradycardia. *Neuroscience* **33,** 559–566.
117. Mues, G., Fuchs, I., Wei, E. T., Weber, E., Evans, C. E., Barchas, J. D., and Chang, J. K. (1982) Blood pressure elevation in rats by peripheral administration of Tyr-

Gly-Gly-Phe-Met-Arg-Phe and the invertebrate neuropeptide, Phe-Met-Arg-Phe-NH2. *Life Sci.* **31,** 2555–2561.

118. De Wildt, D. J., Kasbergen, C. M., and Versteeg, D. H. G. (1995) Effect of gamma-melanocyte stimulating hormone on cerebral blood flow in rats. *J. Cardiovasc. Pharmacol.* **25,** 898–905.
119. Herz, R. C., De Wildt, D. J., and Versteeg, D. H. (1996) The effects of gamma 2-melanocyte-stimulating hormone and nimodipine on cortical blood flow and infarction volume in two rat models of middle cerebral artery occlusion. *Eur. J. Pharmacol.* **306,** 113–121.
120. Van Bergen, P., Van Der Vaart, J. G., Kasbergen, C. M., Versteeg, D. H., and De Wildt, D. J. (1996) Structure–activity analysis for the effects of gamma-MSH/ACTH-like peptides on cerebral hemodynamics in rats. *Eur. J. Pharmacol.* **318,** 357–368.
121. Thody, A. J., and Shuster, S. (1975) Control of sebaceous gland function in the rat by alpha-melanocyte-stimulating hormone. *J. Endocrinol.* **64,** 503–510.
122. Donohoe, S. M., Thody, A. J., and Shuster, S. (1981) Effect of alpha-melanocyte-stimulating hormone and ovarian steroids on preputial gland function in the female rat. *J. Endocrinol.* **90,** 53–58.
123. Krahenbuhl, C., and Desaulles, P. A. (1969) Interaction between alpha-MSH and sex steroids on the preputial glands of female rats. *Experientia* **25,** 1193–1195.
124. Chen, W., Kelly, M. A., Opitz-Araya, X., Thomas, R. E., Low, M. J., and Cone, R. D. (1997) Exocrine gland dysfunction in MC5-R-deficient mice, evidence for coordinated regulation of exocrine gland function by melanocortin peptides. *Cell* **91,** 789–798.
125. Thody, A. J., Donohoe, S. M., and Shuster, S. (1981) alpha-Melanocyte stimulating hormone and the release of sex attractant odors in the female rat. *Peptides* **2,** 125–129.
126. Bertolini, A., Gessa, G. L., Vergoni, W., and Ferrari, W. (1968) Induction of sexual excitement with intraventricular ACTH; permissive role of testosterone in the male rabbit. *Life Sci.* **7,** 1203–1206.
127. Bertolini, A., Vergoni, W., Gessa, G. L. , and Ferrari, W. (1969) Induction of sexual excitement by the action of adrenocorticotrophic hormone in brain. *Nature* **221,** 667–669.
128. Thody, A. J. and Wilson, C. A. (1983) Melanocyte stimulating hormone and the inhibition of sexual behaviour in the female rat. *Physiol. Behav.* **31,** 67–72.
129. Haun, C. K. and Haltmeyer, G. C. (1975) Effects of an intraventricular injection of synthetic ACTH on plasma testosterone, progesterone and LH levels and on sexual behavior in male and female rabbits. *Neuroendocrinology* **19,** 201–213.
130. Spruijt, B. M., Hoglund, U., Gispen, W. H., and Meyerson, B.J. (1985) Effects of ACTH1–24 on male rat behavior in an exploratory, copulatory and socio–sexual approach test. *Psychoneuroendocrinology* **10,** 431–438.
131. Gonzalez, M. I., Vaziri, S., and Wilson, C. A. (1996) Behavioral effects of alpha-MSH and MCH after central administration in the female rat. *Peptides* **17,** 171–177.
132. Clarke, A. and File, S. E. (1983) Social and exploratory behaviour in the rat after septal administration of ORG 2766 and ACTH4–10. *Psychoneuroendocrinology* **8,** 343–350.
133. Niesink, R. J., and Van Ree, J. M. (1984) Neuropeptides and social behavior of rats tested in dyadic encounters. *Neuropeptides* **4,** 483–496.

134. File, S. E. and Clarke, A. (1980) Intraventricular ACTH reduces social interaction in male rats. *Pharmacol. Biochem. Behav.* **12,** 711–715.
135. File, S. E. (1981) Contrasting effects of Org 2766 and alpha-MSH on social and exploratory behavior in the rat. *Peptides* **2,** 255–260.
136. van Rijzingen, I. M., Gispen, W. H., and Spruijt, B. M. (1996) The ACTH(4–9) analog ORG 2766 and recovery after brain damage in animal models—a review. *Behav. Brain. Res.* **74,** 1–15.
137. Xia, Y. and Wikberg, J. E. (1997) Postnatal expression of melanocortin-3 receptor in rat diencephalon and mesencephalon [In Process Citation]. *Neuropharmacology* **36,** 217–224.
138. Snead, O. C. (1995) Other antiepileptic drugs: adrenocorticotrophic hormone (ACTH) in *Antiepileptic Drugs. vol. 4.* (Levy, R. H., Mattson, R. H., and Meldrum, B. S., eds.) Raven Press, New York, pp. 941–947.
139. Crossley, C. J., Richman, R. A., and Thorpy, M. J. (1980) Evidence for cortisol-independent anticonvulsant activity of adrenocorticotropichormone in infantile spasms. *Ann. Neurol.* **8,** 220.
140. Farwell, J., Milstein, J., Opheim, K., Smith, E., and Glass, S. (1984) Adrenocorticotropic hormone controls infantile spasms independently of cortisol stimulation. *Epilepsia* **25,** 605–608.
141. Konishi, Y., Hayakawa, K., Kuriyama, M., Saito, M., Fujii, Y., and Sudo, M. (1995) Effects of ACTH on brain midline structures in infants with infantile spasms. *Pediatr. Neurol.* **13,** 134–136.
142. Holmes, G. L. and Weber, D. A. (1986) Effects of ACTH on seizure susceptibility in the developing brain. *Ann. Neurol.* **20,** 82–88.
143. Croiset, G. and De Wied, D. (1992) ACTH, a structure–activity study on pilocarpine-induced epilepsy. *Eur. J. Pharmacol.* **229,** 211–216.
144. Cottrell, G. A., Nyakas, C., Bohus, B., and De Wied, D. (1983) ACTH and MSH reduce the after–discharge and behavioral depression following kindling. in *Integrative Neurohumoral Mechanisms.* (Endröczi, E., De Wied, D., Angelucci, L., and Scapagnini, U., eds.) Elsevier Biomedical, Amsterdam, pp. 91–97.
145. van der Helm–Hylkema, H. and De Wied, D. (1976) Effect of neonatally injected ACTH and ACTH analogues on eye-opening of the rat. *Life Sci.* **18,** 1099–1104.
146. Strand, F. L., Williams, K. A., Alves, S. E., Antonawich, F. J., Lee, T. S., Lee, S. J., Kume, J., and Zuccarelli, L. A. (1994) Melanocortins as factors in somatic neuromuscular growth and regrowth. *Pharmacol. Ther.* **62,** 1–27.
147. Gispen, W. H., De Koning, P., Kuiters, R. P. F., Van Der Zee, C. E. E. M., and Verhaagen, J. (1987) On the neurotrophic action of melanocortins. *Prog. Brain Res.* **72,** 319–331.
148. Hol, E. M., Gispen, W. H., and Bar, P. R. (1995) ACTH–related peptides, receptors and signal transduction systems involved in their neurotrophic and neuroprotective actions. *Peptides* **16,** 979–993.
149. Strand, F. L., Rose, K. J., Zuccarelli, L. A., Kume, J., Alves, S. E., Antonawich, F. J., and Garrett, L. Y. (1991) Neuropeptide hormones as neurotrophic factors. *Physiol. Rev.* **71,** 1017–1046.
150. Gispen, W. H., Verhaagen, J., and Bar, D. (1994) ACTH/MSH-derived peptides and peripheral nerve plasticity, neuropathies, neuroprotection and repair. *Prog. Brain. Res.* **100,** 223–229.

151. Strand, F. L. and Kung, T. T. (1980) ACTH accelerates recovery of neuromuscular function following crushing of peripheral nerve. *Peptides* **1,** 135–138.
152. Edwards, P. M., Kuiters, R. R. F., Boer, G. J., and Gispen, W. H. (1986) Recovery from peripheral nerve transection is accelerated by local application of αMSH by means of microporous Accurel polypropylene tubes. *J. Neurol. Sci.* **74,** 171–176.
153. Van Der Zee, C. E. E. M., Brakkee, J. H., and Gispen, W. H. (1988) αMSH and Org2766 in peripheral nerve regeneration, different routes of delivery. *Eur. J. Pharmacol.* **147,** 351–357.
154. Verhaagen, J., Edwards, P. M., Jennekens, F. G. I. , and Gispen, W. H. (1987) Pharmacological aspects of the influence of melanocortins on the formation of regenerative peripheral nerve sprouts. *Peptides* **8,** 581–584.
155. Bijlsma, W. A., Schotman, P., Jennekens, F. G. I., Gispen, W. H., and De Wied, D. (1983) The enhanced recovery of sensorimotor fuction in rats is related to the melanotropic moiety of ACTH/MSH neuropeptides. *Eur. J. Pharmacol.* **92,** 231–236.
156. Dyer, J. K., Philipsen, H. L., Tonnaer, J. A., Hermkens, P. H., and Haynes, L. W. (1995) Melanocortin analogue Org2766 binds to rat Schwann cells, upregulates NGF low-affinity receptor p7**5,** and releases neurotrophic activity. *Peptides* **16,** 515–522.
157. Lichtensteiger, W., Hanimann, B., Siegrist, W., and Eberle, A.N. (1996) Region- and stage-specific patterns of melanocortin receptor ontogeny in rat central nervous system, cranial nerve ganglia and sympathetic ganglia. *Brain Res. Dev. Brain Res.* **91,** 93–110.
158. Hol, E. M., Verhage, M., Gispen, W. H., and Bar, P. R. (1994) The role of calcium and cAMP in the mechanism of action of two melanocortins, αMSH and the ACTH 4–9 analogue Org 2766. *Brain Res.* **662,** 109–116.
159. Konda, Y., Gantz, I., DelValle, J., Shimoto, Y., Miwa, H., and Yamada, T. (1994) Interaction of dual intracellular signaling pathways activated by the melanocortin-3 receptor. *J. Biol. Chem.* **269,** 13,162–13,166.
160. Adan, R. A. H. , Kraan van der, M., Doornbos, R. P., Bar, P. R., Burbach, J. P. H., and Gispen, W. H. (1996) Melanocortin receptors mediate α-MSH-induced stimulation of neurite outgrowth in Neuro 2a cells. *Mol. Brain. Res.* **36,** 37–44.
161. Hol, E. M., Hermens, W. T. J. M. C., Verhaagen, J., Gispen, W. H., and Bär, P. R. (1993) αMSH but not ORG 2788 induces expression of c-fos in cultured rat spinal cord cells. *Neuroreport* **4,** 651–654.
162. Hughes, S. and Smith, M. E. (1994) Upregulation of the pro-opiomelanocortin gene in motoneurones after nerve section in mice. *Mol. Brain. Res.* **25,** 41–49.
163. Smith, M. E. and Hughes, S. (1993) Pro-opiomelanocortin neuropeptide receptors on developing and dystrophic muscle fibers. *Mol. Chem. Neuropathol.* **19,** 137–145.
164. Smith, M. E., Hughes, S., Simpson, M. G., and Allen, S. L. (1994) Upregulation of the POMC gene in rats by a neurotoxicant which targets motoneurons. *Neurotoxicology* **15,** 769–772.
165. Hughes, S. and Smith, M. E. (1993) β- Endorphin and ACTH receptors in skeletal muscles in diabetes mellitus. *Ann. N. Y. Acad. Sci.* **680,** 542–544.
166. Labbe, O., Desarnaud, F., Eggerickx, D., Vassart, G., and Parmentier, M. (1994) Molecular cloning of a mouse melanocortin 5 receptor gene widely expressed in peripheral tissues. *Biochemistry* **33,** 4543–4549.

167. Gantz, I., Shimoto, Y., Konda, Y., Miwa, H., Dickinson, C. J., and Yamada, T. (1994) Molecular cloning, expression, and characterization of a fifth melanocortin receptor. *Biochem. Biophys. Res. Commun.* **200,** 1214–1220.
168. Edwards, P. M, Van Der Zee, C. E. E. M., Verhaagen, J., Schotman, P., Jennekens, F. G. I., and Gispen, W. H. (1984) Evidence that the neurotrophic action of αMSH may derive from its ability to mimick the actions of a peptide formed in degenerating nerve stumps. *J. Neurol. Sci.* **64,** 333–340.
169. Verhaagen, J., Edwards, P. M., Schotman, P., Jennekens, F. G. I., and Gispen, W. H. (1986) Characterization of epitopes shared by α-melanocyte-stimulating hormone (αMSH) and the 150-kD neurofilament protein (NF150), relationship to neurotrophic sequences. *J. Neurosci. Res.* **16,** 589–600.
170. Plantinga, L. C., Verhaagen, J., Edwards, P. M., Schrama, L. H., Burbach, J. P.H., and Gispen, W. H. (1992) Expression of the pro-opiomelanocortin gene in dorsal root ganglia, spinal cord and sciatic nerve after sciatic nerve crush in the rat. *Mol. Brain Res.* **16,** 135–142.
171. Plantinga, L. C., Verhaagen, J., Edwards, P. M., Hali, M., Brakkee, J. H., and Gispen, W. H. (1995) Pharmacological evidence for the involvement of endogenous α–MSH-like peptides in peripheral nerve regeneration. *Peptides* **16,** 319–324.
172. Antonawich, F. J., Azmitia, E. C., Kramer, H. K., and Strand, F. L. (1994) Specificity versus redundancy of melanocortins in nerve regeneration. *Ann. N. Y. Acad. Sci.* **739,** 60–73.
173. Wolterink, G., Van Zanten, E., and Van Ree, J. M. (1990) Functional recovery after destruction of dopamine systems in the nucleus accumbens of rats. IV. Delay by intra-accumbal treatment with Org2766- or a α-MSH-antiserum. *Brain Res.* **507,** 115–120.
174. Van de Meent, H., Hamers, F. P., Lankhorst, A. J., Joosten, E. A., and Gispen, W. H. (1997) Beneficial effects of the melanocortin alpha-melanocyte-stimulating hormone on clinical and neurophysiological recovery after experimental spinal cord injury. *Neurosurgery* **40,** 122–30; discussion 130–1.
175. De Rotte, A. A., Bouman, H. J., and van Wimersma Greidanus, T. B. (1980) Relationships between alpha-MSH levels in blood and in cerebrospinal fluid. *Brain Res. Bull.* **5,** 375–381.
176. Thiessen, D. D. (1988) Body temperature and grooming in the mongolian gerbil. *Ann. N. Y. Acad. Sci.* **525,** 27–39.
177. Greven, H. M. and De Wied, D. (1980) Structure and behavioural activity of peptides related to corticotrophin and lipotrophin, in *Hormones and the Brain.* (De Wied, D. and Van Keep, P. A., eds.) MTP Press MA, Lancaster, 115–127.

CHAPTER 5

Peripheral Effects of Melanocortins

Bruce A. Boston

1. Introduction

Peptides derived from the proopiomelanocortin (POMC) prohormore precursor have been implicated in a wide variety of biologic functions since its discovery in 1977 *(1,2)*, and the cloning of the POMC gene in 1979 *(3)*. Some of the peptides derived from POMC are classified as melanocortins because of their ability to stimulate eumelanogenesis in the *melano*cyte or to stimulate steroid production in the adreno*cortic*al cell. Although the two most thoroughly studied of the melanocortin biologic functions are adrenocorticotropic hormone (ACTH) stimulation of adrenal steroidogenesis and melanocyte stimulating hormone (MSH) stimulation of eumelanin production, numerous other effects of these peptides have been reported. Melanocortins have been implicated in the regulation of feeding and grooming behavior, learning and memory, thermogenesis, neural regeneration, metabolism, inflammmation, exocrine gland function, and natriuresis *(54,79,115,127,134)*. Many of these alterations in biologic function are clearly mediated via melanocortin receptors in the central nervous system, while others are mediated at least partially by melanocortin receptors in the periphery. In this chapter, we discuss aspects of melanocortin function other than melanogenesis and steroidogenesis that appear to be at least partially a result of interaction with peripheral melanocortin receptors.

2. Peripheral Melanocortin Receptors and Proopiomelanocortin Binding Sites

Since the cloning of the first melanocortin receptor in 1992 *(4)*, five known melanocortin receptors have been discovered *(4–15)*. All the receptors are small seven transmembrane G protein-coupled receptors that stimu-

The Melanocortin Receptors Ed.: R. D. Cone

late increases in intracellular cAMP by stimulating the activity of adenylyl cyclase. The melanocortin 1 (MC1) receptor is the classic MSH receptor located on melanocytes and is responsible for stimulating the production of brown or black pigment (eumelanin). The melanocortin 2 (MC2) receptor is the classic ACTH receptor located in the adrenal cortex and mediates ACTH stimulation of glucocorticoid production. The MC2 receptor has also been found on the adipocyte *(16)*. Three other melanocortin receptors were recently cloned and have been named the MC3, MC4, and MC5 receptors. The MC3 receptor is primarily located in the central nervous system but has also been detected in the heart and testis *(17)*. Thus far, the data suggest the MC4 receptor is located primarily in the central nervous system. The MC5 receptor is more widely distributed with both peripheral and central expression. Messenger RNA for the MC5 receptor is most abundant in skeletal muscle and brain but detectable levels are also found in lung, testis, spleen, heart, kidney, and liver *(10,13,17,18)*. Very high levels of MCS receptor mRNA have recently been demonstrated in several exocrine glands *(127)*.

In addition to studies detailing the distribution of the known melanocortin receptors, investigators have looked for specific melanocortin binding sites. Tatro and Reichlin *(19)* reported a comprehensive investigation of α-MSH binding sites in both mice and rats using a radioiodinated superpotent α-MSH analog Nle^4, D-Phe^7-α-MSH (NDP-MSH). It should be noted that this technique will reveal potential sites of expression of MC1, MC3, MC4, and MC5 receptors but not MC2, as the MC2 receptor has little affinity for MSH or its analogs. Regardless of the lack of sensitivity for the MC2 receptor, Tatro and Reichlin found significant binding of NDP-MSH in exocrine tissue including the pancreas, lacrimal, Harderian, submandibular, and preputial glands. Significant binding was also found in adrenal glands, white adipose tissue, skin, spleen, bladder, duodenum, and hypothalamus. With the abundance of potential melanocortin receptor sites, it seems quite possible that the melanocortin peptides play a role in peripheral biologic functions other than steroidogenesis and melanogenesis.

3. Distribution of Proopiomelanocortin and Related Peptides

In addition to the location of melanocortin receptors in a multitude of "nonclassical" sites, POMC and POMC-like peptides have been found in many extrapituitary locations. Immunoreactive POMC-like peptides have been detected in the rat in the gastrointestinal and reproductive tracts, heart, liver, kidney, pancreas, and brain *(20–23)*. Furthermore, many human tissues also contain immunoreactive POMC, including the gastrointestinal tract and pancreas, adrenal medulla, brain, and placenta *(24–26)*. More recently, corticotropin was found to be secreted in vitro from rat leukocytes *(27)* and isolated pancreatic islets *(28)*. In addition, POMC mRNA was found to be expressed in keratinocytes *(29)*. Although the expression of melanocortin peptides at these sites is not likely to contribute significantly to serum levels of melanocortins, the wide distribution of expression raises the possibility of paracrine and/or autocrine functions of these peptides. Therefore, not all of the peripheral effects of melanocortin peptides must result from circulating melanocortins in serum. Combined with the wide distribution of melanocortin receptors, the widespread expression of POMC peptides raises multiple possibilities for the role of melanocortin peptides in the regulation of biological systems in vivo.

4. Melanocortins and Metabolism

The ability of melanocortins to directly alter glucose metabolism was first suggested in the early 1930s with the observation that pituitary extracts could cause hypoglycemia *(30)*. Later studies confirmed that the hypoglycemic effect observed with these pituitary extracts could be attributed to pituitary derived ACTH *(31–33)*. It was suspected that the pancreas played a primary role in the ability of corticotropin to induce hypoglycemia. This was confirmed by studies demonstrating that ACTH failed to induce hypoglycemia in mice with alloxan-induced diabetes *(34)*. The first direct evidence that ACTH could induce insulin secretion in vivo was obtained in 1964 by Genuth and Lebovitz *(30)*. Mice administered chromatographically purified corticotropin had a 20-fold rise in plasma insulin, which was accompanied by a significant drop in plasma glucose levels. Furthermore, there was a significant rise in both corticotropin-stimulated and glucose-stimulated insulin secretion in isolated mouse pancreas preparations, confirming that ACTH stimulation of insulin secretion was specific to the pancreas and was not mediated by other humoral or neural factors. The site of ACTH action was further localized to the endocrine portion of the pancreas in later studies using isolated rat islets *(35)*. The ability of ACTH to cause hyperinsulinemia has now been confirmed in vivo in multiple other species including the rat *(36)*, rabbit *(37–39)*, dog *(40)*, and human *(41,42)*.

Although the ability of corticotropin to induce hyperinsulinemia seems to be consistent across species, not all species demonstrate hypoglycemia in response to elevations in serum insulin. Although both the mouse and rat demonstrate hypoglycemia in response to ACTH *(30,36,37,43)*, the rabbit *(37–39)*, dog *(40)* , and human *(41,42)* fail to exhibit hypoglycemia despite an elevation in serum insulin levels. This ability to protect against corticotropin-induced hypoglycemia has been attributed to the adrenal gland as adrenalectomy followed by corticotropin administration in both rabbits *(39)* and humans *(42)* results in profound insulin-induced hypoglycemia. This protective effect is likely mediated through epinephrine released from the adrenal medulla either from the stress of the experimental protocol or as a direct effect of ACTH itself on epinephrine release *(44–46)*. Epinephrine stimulates gluconeogenesis in addition to directly stimulating glucagon secretion, resulting in rapid increases in serum glucose. It is unlikely that the observed hyperglycemia is a direct result of ACTH stimulation of the adrenal cortex as most of the experiments demonstrating hyperglycemia despite hyperinsulinemia were conducted well within the time frame expected for glucocorticoid induction of gluconeogenesis.

Further evidence that the sympathetic nervous system is involved in protecting against corticotropin induced hyperinsulinemia was obtained in experiments using the α-adrenergic blocking agent phentolamine and the β-adrenergic blocker propranolol. In studies using rabbits, phentolamine augmented corticotropin-stimulated insulin secretion, while it blocked the ability of corticotropin to stimulate glucagon secretion (47). Propranolol had no effect on ACTH-induced hyperinsulinemia or hyperglucagonemia. Therefore, the "gluconeogenic" α-adrenergic effects of epinephrine release on glucose metabolism appear to predominate over any β-adrenergic effects. Thus, the presence of intact adrenal glands in these experimental protocols serves to protect the animal from hypoglycemia resulting from corticotropin-induced hyperinsulinemia.

The concentrations of corticotropin needed to induce insulin secretion suggest that this may be a pharmacologic effect rather than a response to circulating basal or stress-induced serum ACTH levels. Although none of the investigators report measurements of circulating ACTH in response to boluses of corticotropin, the serum levels can be estimated. Assuming full absorption of the administered corticotropin, minimum serum concentrations of ACTH needed to produce hyperinsulinemia were approximately 30n*M* in rabbits *(47)*. No information was available for the minimum corticotropin dose needed in rodents. For comparison, circulating concentrations of ACTH in the rat range from 5p*M* in the basal state to approximately 424p*M* during stress, levels far below the minimum concentrations needed to stimulate insulin secretion experimentally *(48)*. With the reports of immunoreactive POMC-like peptides in pancreatic islet cells *(24)*, however, it is possible that ACTH acts as a paracrine factor in the regulation of insulin secretion. This hypothesis has been strengthened by recent experiments by Borelli and colleagues. Using isolated rat islets, these investigators were able to detect corticotropinlike peptide (ACTH-LP) immunoreactive material in the perifusate *(28)*. Furthermore, they were able to demonstrate a significant 17-fold increase in immunoreactive ACTH-LP material in response to an increase in glucose concentrations from 3.3m*M* to 16.7m*M*. As expected, there was also a significant rise in insulin with an increase in glucose in the perifusion media. The increase in insulin in response to glucose was partially inhibited by both corticotropin inhibitory polypeptide (ACTH 7–38) and antiserum directed to the "midportion" of ACTH. Borelli and colleagues went on to demonstrate that the ability of exogenous ACTH to increase insulin secretion in isolated rat islets was itself dependent upon glucose concentration. At 4m*M* glucose, exogenous ACTH[1–39] was unable to stimulate insulin secretion. At concentrations of 8m*M* and 16m*M*, however, ACTH-stimulated insulin secretion 36% and 119% over control, respectively. Taken together, these experiments suggest a paracrine action of islet cell ACTH-LP in the augmentation of glucose stimulated insulin secretion *(28)*.

The physiologic role of ACTH augmentation of insulin secretion is unclear. Previous experiments have demonstrated that expression of POMC mRNA in islet cells appears to be regulated by glucocorticoids. Dexamethasone treatment of rats decreases expression of POMC mRNA in pancreatic islets *(49)*. Other investigators have reported that glucose-stimulated insulin secretion from isolated islets is inhibited by treatment with glucocorticoids *(50,51)*. It is possible that glucocorticoid inhibition of insulin secretion is mediated at least partially through its ability to decrease expression of POMC mRNA and thus decrease the paracrine effect of ACTH on insulin. Both epinephrine, via α-adrenergic receptors, and glucocorticoids inhibit insulin secretion. The combination of α-adrenergic and glucocorticoid inhibition of insulin secretion could function to maintain adequate serum levels of glucose during severe stress. During acute stressful periods, rapid stimulation of gluconeogenesis and inhibition of insulin secretion would occur by activation of the α-adrenergic system, activation of which is partially supported by corticotropin stimulation of epinephrine release from the adrenal medulla *(44–46)*. This would be followed up later by glucocorticoid inhibition of islet cell melanocortin peptide expression which may result in a relative decrease in glucose-stimulated insulin secretion. The shift toward gluconeogenesis initiated by α-adrenergic stimulation would therefore be maintained by glucocorticoids until the stressful period had passed. Once glucocorticoid levels returned to normal, intraislet levels of POMC would increase and the responsiveness of the beta cell to glucose would return to normal.

Several melanocortin peptides other than ACTH[1–39] and ACTH[1–24], the two corticotropin peptides primarily used in the experiments described above, have been reported to alter glucose metabolism. The reported findings, however, are less consistent than those reported with corticotropin. Corticotropinlike intermediate lobe peptide (CLIP or ACTH[18–39] is the peptide remaining after proteolytic cleavage of ACTH[1–39] into α-MSH

(ACTH[1–13]). Early studies demonstrated that CLIP was a potent stimulus of insulin secretion *(52)*. More recent studies, however, have failed to demonstrate any activity of CLIP on insulin, glucagon, or glucose levels *(47, 53)*. Furthermore, CLIP does not contain the core His-Phe-Arg-Trp tetrapeptide sequence, a sequence contained in both MSH and ACTH. This sequence is required for binding and activation of all the known melanocortin receptors. If CLIP does alter glucose metabolism, the mechanism is likely independent of the melanocortin receptors.

Melanocyte-stimulating hormone (MSH) has also been reported to alter glucose metabolism. The evidence that MSH alters glucose metabolism is more limited than that available for corticotropin. In rabbits, α-MSH has been reported to increase glucose, glucagon, and insulin levels in vivo *(47)*. β-MSH was also found to increase glucose, glucagon, and insulin levels while there was no response to γ-MSH. In mice, however, NDP-α-MSH, a potent analog of α-MSH, caused a decrease in insulin levels and an increase in glucose levels *(54 and Boston, unpublished data)*while older studies suggest that α-MSH caused hyperinsulinemia *(37)*. Some of this difference between species might be accounted for by differential activation of the sympathetic nervous system, similar to what is seen with corticotropin. In vitro evidence for MSH-induced insulin-secretion is even more limited. In a study using HIT-T 15 cells, a hamster insulin secreting tumor cell line, α-MSH inhibited insulin secretion *(55)*. It is not clear from this report if the insulin inhibiting action of α-MSH was mediated via a melanocortin receptor or through some other mechanism as the carboxy-terminal tripeptide of α-MSH (α-MSH[11–13]), a peptide lacking the necessary core His-Phe-Arg-Trp sequence, also potently inhibited insulin secretion in this study.

The site and mechanism by which α-MSH alters glucose levels has yet to be proven. Although the in-vitro studies using HIT-T 15 cells suggest a direct inhibitory action on islets, this has yet to be confirmed in isolated primary islet cells. Furthermore, the opposite effects of α-MSH on insulin release in rabbits and mice is confounding. The hyperinsulinemic response in rabbits, however, might be explained by the concurrent rise in serum glucose *(38)* and not a direct result of α-MSH stimulation of the β cell. As is the case with corticotropin, experiments with α-MSH suggest an interaction with the sympathetic nervous system. Pretreatment with phentolamine in rabbits to block α-adrenergic input augments α-MSH stimulated insulin release, while blocking β-adrenergic input with propranolol had no effect on insulin *(38)*. It is likely that the adenergic input to the β cell from α-MSH stimulation is mediated via autonomic neurons as there is no evidence that MSH itself stimulates release of epinephrine from the adrenal medulla. Recent reports support a role of α-MSH in the autonomic regulation of insulin secretion.

Hypothalamic α-MSH containing neurons and the MC4 receptor have been implicated in the central regulation of insulin secretion, however, the data here suggests central α-MSH tonically inhibits insulin release via sympathetic inhibition of islets*(54, 56)*. Paradoxically, the high doses of peripherally administered α-MSH needed to stimulate insulin secretion in vivo in many of these studies also suggests a central mode of action. Since central regulation of insulin secretion has been shown to be regulated by the autonomic nervous system *(57)*, peripherally administered MSH may be acting centrally to alter glucose metabolism. Future investigations should further define the mechanisms and physiologic relevance of MSH modulation of glucose metabolism.

Melanocortin peptides, including both ACTH and α-MSH, also alter lipid metabolism. Melanocortins appear to stimulate lipolysis and increase free fatty acid levels in multiple species including rabbits *(38,47,58,59)* and rats *(60–66)*. Using isolated rabbit adipose cells, Richter and colleagues *(59)* demonstrated a requirement for the core His-Phe-Arg-Trp peptide sequence in the induction of lipolysis in vitro. This is good experimental evidence that the melanocortin receptors are required for this action. It also appears that the lipolytic activity of ACTH and MSH are not mediated via or significantly altered by the autonomic nervous system. Neither phentolamine nor propranolol altered the ability of ACTH and MSH to stimulate release of free fatty acids in rabbits *(38, 47)*. Furthermore, the action of corticotropin in vivo on adipose cells does not appear to be significantly altered by adrenalectomy *(30, 62)* despite early reports that adrenalectomy rendered adipocytes insensitive to corticotropin when tested in vitro *(60, 61)*. Therefore, the accumulated evidence points toward a direct effect of melanocortin peptides on the adipocyte. This hypothesis was further supported with the finding of both the MC2 receptor (ACTH receptor) and the MC5 receptor on adipocytes *(16)*. The ability of melanocortins to stimulate adenylyl cyclase in adipocytes was tested in 3T3-L1 adipocytes, a mouse embryonic fibroblast cell line that can be induced to differentiate into mature adipocytes *(67)*. Incubation with either ACTH or α-MSH resulted in a significant increase in intracellular cAMP levels *(16)*. Since α-MSH does not bind and activate the MC2 receptor, any activation of adenylyl cyclase by α-MSH must be mediated via the MC5 receptor. It is interesting to note that in these experiments, the potent α-MSH analog, NDP-α-MSH, was able to bind to adipocytes but did not activate cyclase. In addition, NDP-α-MSH was able to block the ability of α-MSH to stimulate cyclase and therefore appears to act as an MC5 receptor antagonist in adipoctyes. NDP-α-MSH does not block ACTH-stimulated cyclase activity, however, confirming that MC2 receptors on the adipocyte are also functional *(16)*. Although release of free fatty acids into the medium was not measured in this experiment, previous investigators have shown that lipolysis is mediated by an increase in adenylyl cyclase activity *(64)*. Therefore, it appears that melanocortins exert their effects on lipid metabolism primarily at the adipocyte by increasing adenylyl cyclase activity via stimulation of either the MC2 and/or the MC5 receptor.

Although both the MC2 and MC5 receptors appear to stimulate lipolysis, it is more likely that the MC2 receptor plays the physiologic role in lipid metabolism. For example, only ACTH has significant lipolytic activity in the rat adipocyte with reported EC_{50}'s between 0.15n*M* and 1.34n*M*, whereas α-MSH has relatively little lipolytic activity with an EC_{50} of 1.53μ*M* *(63, 68)*. The concentrations of ACTH that promote lipolysis approximate the levels seen in the rat during stress (up to 0.43n*M*) *(48)*. ACTH stimulation of the MC2 receptor on the adipocyte during stress may serve to augment the release of energy stores into the circulation during times of increased energy demand.

In summary, the melanocortin peptides have multiple effects on metabolism. Corticotropin alters glucose metabolism directly by stimulation of insulin secretion at the pancreatic β cell, and indirectly via the adrenal gland with stimulation of glucocorticoid secretion and possibly the release of epinephrine. Epinephrine then alters glucose levels by stimulating glucagon release and inhibiting insulin secretion. Melanocyte stimulating hormone also alters insulin secretion, although it appears to have different effects in different species. There is some evidence that α-MSH has a direct effect on inhibition of insulin secretion at the islet cell, although it appears most likely that α-MSH acts in the hypothalamus to alter central control of insulin secretion via the autonomic nervous system. Finally, both ACTH and MSH have direct effects on lipolysis in the adipocyte via activation of the MC2 and/or the MC5 receptors.

5. Melanocortins and Inflammation

Regulation of the host immune system in response to infection or injury involves a vast number of inflammatory mediators, or cytokines, and an army of immune cells. There is an accumulating body of evidence that the melanocortin peptides, particularly α-MSH, are involved in the modulation of host immune responses in many animal species. Investigators have discovered that the melanocortins alter the production and/or action of many cytokines and have a profound effect on host responses to infection such as fever and inflammation. One of the earliest observed effects of α-MSH on host respone was the ability to inhibit fever when administered centrally *(69, 70)*. α-MSH potently inhibited fever induced by such agents as endogenous pyrogen, endotoxin, IL-1, IL-6, and tumor necrosis factor *(71–75)*. In addition, intracerebroventricular administration of anti-MSH antibodies increased body temperature in rabbits *(76)*. Intracerebroventricular administration of

SHU-9119, a potent MC3 and MC4 receptor antagonist, also blocks the antipyretic effects of α-MSH *(135)*. Furthermore, α-MSH levels in the hypothalamus are increased during elevations in body temperature caused by administration of endogenous pyrogens *(77)* but not by elevations in body temperature caused by an increase in the ambient temperature of the environment *(78)*. Therefore, α-MSH appears to play a role in the central regulation of fever and serves as an endogenously produced antipyretic agent.

It was the observation that α-MSH altered response to the fever-inducing actions of many of the cytokines that led investigators to study its role in the peripheral actions of the immune system. An elegantly designed experiment led investigators to believe that indeed α-MSH could alter peripheral host responses and had potent antiinflammatory activity. Rabbits were injected with a blue dye and either saline or α-MSH *(79)*. This was followed by intradermal injections of histamine, a potent mediator of inflammation. A blue spot formed at the site of histamine injection in the saline-treated animals indicating leakage of intravascular fluids into the surrounding tissues, a common feature of inflammation. The α-MSH pretreated animals had very little extravasation of the dye into the surrounding subcutaneous tissues, indicating profound acute antiinflammatory activity of the peptide. Later experiments further demonstrated that α-MSH had acute antiinflammatory activity. Intraperitoneal injection of α-MSH prevented mouse paw edema caused by local injection of carrageenan *(80)* and prevented dermal reactions in response to endogenous pyrogens *(79)*. More recently, α-MSH was found to inhibit lipopolysaccharide (LPS) induced hepatitis in mice *(81)*. In addition, α-MSH inhibited systemic models of inflammation including endotoxin-induced adult respiratory distress syndrome and also improved survival in mice with peritonitis and endotoxic shock *(82)*.

α-MSH has also been found to modulate delayed-type hypersensitivity reactions. Delayed-type hypersensitivity reactions are T cell-mediated immune responses that require previous sensitization by an allergen. The reaction to the allergen occurs 24 to 72 h after reexposure to the allergen. Grabbe and collegaues *(83)* demonstrated in a series of experiments that α-MSH can suppress delayed-type hypersensitivity reactions when injected either just prior to initial allergen exposure or just before reexposure. Using trinitrochlorobenzene (TNCB) as a sensitizing agent, mice were pretreated systemically with either α-MSH or saline control injected intravenously before application of TNCB to the abdominal skin. Animals were challenged 7 days later by injection of 10 μL TNCB into the ear. Animals pretreated with α-MSH before sensitization showed a significantly decreased immune response to TNCB with initial and subsequent reexposures, indicating tolerence to this allergen. Furthermore, tolerance was hapten specific as exposure and rechallenge with similar but not identical allergens in TNCB-desensitized animals still produced delayed-type hypersensitivity reactions. Finally, α-MSH was administered just prior to TNCB challenge in animals intially pretreated with saline during sensitization. α-MSH blocked the immune response normally seen with rechallenge to TNCB. Later TNCB challenges without α-MSH pretreatment in these animals, however, resulted in an immune response. Therefore, α-MSH can induce hapten specific tolerance in delayed-type hypersensitivity when administered at the time of initial antigen exposure. α-MSH can also block the immune response to rechallenge with an antigen in previously sensitized animals but cannot induce immune tolerance in animals that have already been exposed to that antigen *(83)*.

Melanocortin peptides also inhibit the immune response in animal models of chronic inflammation. Ceriani and colleagues *(82,84)* used a rat model of inflammatory arthritis to demonstrate the effect of α-MSH on chronic inflammation. Rats develop an inflammatory arthritis when injected with ground and sonicated mycobacterium tuberculosis intradermally. In this experiment, rats were randomly assigned to one of three treatment groups: twice daily intraperitoneal injections of saline, α-MSH, or prednisolone. Animals injected with saline developed arthritis within 11 days and became significantly different from animals treated with α-MSH by day 16. Animals injected with prednisolone were protected from developing arthritis *(84)*. Therefore, α-MSH appeared to inhibit chronic inflammation in this rodent model of arthritis. Although this model is not identical to human forms of autoimmune arthritis such as rheumatoid arthritis, they probably share many of the cytokine inflammatory mediators, some of which may be inhibited by α-MSH. Thus, it remains possible that melanocortins may play a role in the pathogenesis of this clinical disorder. Indeed, α-MSH has been found in the synovial fluid of patients with arthritis in concentrations that seemed to correspond to the degree of inflammation *(85)*.

α-MSH exerts many of its antiinflammatory effects by decreasing the concentrations and/or inhibiting the actions of inflammatory cytokines. α-MSH administration has demonstrated an inhibitory effect on a multitude of cytokines including IL-1, IL-6, IL-8, TNF-α, IFN-γ, and leukotriene B^4 *(82, 86–92)*. Many of these cytokines are important chemotactic agents which promote the migration of immune cells such as neutrophils into local areas of inflammation. These cells produce more cytokines stimulating a cascade reaction that further stimulates the inflammatory process. Intraperitoneal injection of α-MSH, however, was able to block the migration of neutrophils into implanted sponges impregnated with chemotactic agents IL-1 or tumor necrosis factor-alpha (TNF-α) *(93)* , and therefore helped to prevent further activation of the inflammatory cascade. More recently, the effect of α-MSH on the production of nitric oxide, a mediator thought to be common to all forms of inflammation, was investigated. In cultured murine macrophages stimulated with lipopolysaccharide and interferon-gamma (IFN-γ), α-MSH inhibited the production of nitric oxide by inhibiting the transcription of a key enzyme, nitric oxide synthase II *(94)*.

Although in general, α-MSH acts to inhibit cytokine action, this melanocortin peptide actually induced the production of IL-10 in human monocytes. Unlike the cytokines mentioned previously, IL-10 functions as an immunosuppressant and has been shown to down regulate other cytokines such as IL-1β *(95, 96)*. Therefore, α-MSH appears to act by not only inhibiting potent stimulants of inflammation but by increasing the activity of important immunosuppressors. In summary, α-MSH appears to modulate the immune system by counteracting the proinflammatory actions of multiple different cytokines and enhancing the activity of antiinflammatory cytokines.

As is the case with α-MSH and glucose metabolism, it is unclear whether the antiinflammatory actions of this melanocortin peptide are peripheral or mediated via the central nervous system. It is clear that α-MSH exerts its antipyretic effects through central mechanisms. There is ample evidence, however, that many of the antiinflammatory actions of α-MSH are also mediated centrally. Intracerebroventricular administration of α-MSH in mice potently inhibited ear edema induced by either local picryl chloride or IL-1β injections *(97, 98)*. The same dose that was used centrally failed to block IL-1β induced edema when injected peripherally *(98)*. Furthermore, the minimum peripherally administered dose of α-MSH required to inhibit inflammation in this and other mouse models *(81,83,98)* is approximately 1μ*M* in serum, assuming the peptide is fully absorbed after injection and is evenly distributed within the intravascular space. This dose far exceeds the α-MSH concentration observed in vitro to stimulate cAMP production in 293 cells transfected with the MC1 receptor (EC_{50} 2.0n*M*) *(4)*. The concentration of α-MSH required to inhibit TNF-α production from LPS-treated murine brain tissue in vitro (IC_{50} approx 0.1n*M*), however, was more in line with that expected from MSH stimulation of a melanocortin receptor *(99)*. Previous reports have found that systemically administered melanocortins can cross the blood–brain barrier. In experiments using high doses of intravenously injected α-MSH in rats, concentrations of the peptide entering the cerebrospinal fluid are approx 1/1000 the estimated concentrations in the serum *(100)*. Therefore, a melanocortin peptide concentration of at least 0.1 μ*M* in serum would be needed to produce a centrally mediated melanocortin effect on the immune system. This is consistent with the dose of α-MSH required to inhibit inflammation *(81,83,98)*. Although a peripheral site of action of melanocortins can not be completely ruled out, it appears likely that systemically administered peptide is acting through central mechanisms.

The ability of centrally administered melanocortin peptides to alter peripheral inflammation appears to be mediated via the autonomic nervous system. Again, this is remarkably similar to the proposed mechanism of central regulation of glucose metabolism. In a study by Macaluso and colleagues *(101)*, the ability of various sympathetic blocking agents to block α-MSH actions on inflammation in mice were studied. Using intradermal injections of IL-1β to stimulate inflammation, centrally administered α-MSH significantly reduced the IL-1β induced edema. This effect was blocked by systemic administration of propranolol, a nonspecific β-adrenergic blocking agent. The antiinflammatory aspects of central α-MSH, however, were not effected by systemic administration of the anticholinergic atropine or the α-adrenergic blocking agent phentolamine. Furthermore, butoxamine, a selective β_2 antagonist blocked the antiinflammatory effect of α-MSH while there was no effect with administration of atenolol, a selective β_1 antagonist. Therefore, it appears that the central antiinflammatory effect of α-MSH is mediated primarily through activation of the β_{2-}adrenergic receptor. As would be predicted by a central mechanism of melanocortin action on inflammation, transection of the spinal cord and interruption of autonomic neurons to the periphery also blocked the effect of centrally administered α-MSH on peripheral inflammation *(101)*.

If melanocortins do directly effect the immune system by activating peripheral melanocortin receptors, the specific melanocortin receptor that mediates this effect of α-MSH on host immune response is unclear. Numerous reports are now available that have found the MC1 receptor mRNA in cells of the immune system. The MC1 receptor mRNA has been found in monocytes *(102)*, macrophages *(94)*, and neutrophils *(103)* using RT-PCR techniques. Therefore, this receptor has been proposed as the likely peripheral melanocortin receptor that mediates α-MSH antiinflammatory activity. Many of the experiments reporting the antiinflammatory activity of melanocortins, however, provide data that would be inconsistent with the known pharmacology of the MC1 receptor. A major argument against activity at the MC1 receptor is the observation that the terminal tripeptide of α-MSH, α-MSH[11–13], posesses potent peripheral antiinflammatory activity by itself *(79,98,101,102)* and is necessary for the antiinflammatory properties of α-MSH *(104)*. However, melanocortin peptides that stimulate eumelanogenesis and pigment formation, a biologic process known to be mediated via α-MSH stimulation of peripheral MC1 receptors, require the presence of the tetrapeptide His-Phe-Arg-Trp sequence at positions 6 through 9 while the terminal tripeptide antiinflammatory sequence is not required for stimulation of eumelanogenesis *(105)*. Furthermore, β-MSH, a potent stimulator of eumelanogenesis and an agonist of the MC1 receptor, has no apparent antiinflammatory activity *(83)*. β-MSH also lacks the terminal Lys-Pro-Val tripeptide sequence, further suggesting this sequence is neccessary for antiinflammatory activity.

An alternative hypothesis for the peripheral effects of α-MSH and the Lys-Pro-Val tripeptide is suggested in a report by Mugridge and colleagues *(106)*. They found that both α-MSH and the terminal tripeptide α-MSH[11–13] reduced the IL-1β potentiated contractions seen in isolated rat stomach strip preparations. Remarkably, the minimum concentration needed to see a significant inhibition of IL-1β potentiated contractions ($6 \times 10^{-8} M$ for α-MSH; $3 \times 10^{-7} M$ for α-MSH[11–13]), is not far from the minimum estimated in vivo concentrations of peptide neccessary to inhibit inflammation. Furthermore, in experiments using EL4-6.1 T cells, both α-MSH and the tripeptide sequence significantly reduced binding of radiolabeled recombinant IL-1β suggesting direct competition for sites on the IL-1 receptor. This observation could not have been a direct effect of melanocortin receptor mediated down regulation of IL-1 receptor sites as T cells reportedly do not express any of the known melanocortin receptors *(107)*. Therefore, it is likely that many of the observed effects of melanocortin peptides on peripheral inflammation are independent of the known melanocortin receptors or are mediated via melanocortin receptors in the central nervous system.

In summary, melanocortin peptides, predominantly α-MSH, have potent effects on both fever and inflammation. The primary mechanism of action of melanocortin peptides is to inhibit the proinflammatory action of many of the cytokines produced by cells of the immune system. In addition, α-MSH induces the production of cytokines, primarily IL-10, that have an immunosuppressive role. Systemically administered α-MSH is most likely primarily acting centrally via melanocortin receptors in the central nervous system, with the peripheral antiinflammatory signals carried via the autonomic nervous system and stimulation of β_2 adrenergic receptors. Alternatively, α-MSH may act via antagonism of peripheral IL-1 receptors. Although it appears that pharmacologic doses of systemically administered melanocortins are needed to produce antiinflammatory effects, the presence of POMC peptides in both the epidermis and gut suggest a paracrine role for these peptides. The epidermis and the gastrointestinal lining are prime sites for invasion of pathogens and therefore must have potent immune defense mechanisms. Since it would be detrimental to the organism to have these potent immune defense mechanisms proceed unchecked, α-MSH may act locally as a paracrine factor to suppress the inflammatory process. Indeed, both local injury and increases in IL-1 have been shown to increase the transcription of POMC mRNA in human keratinocytes *(108)*. Regardless of whether peripherally injected α-MSH is acting as a pharmacologic agent or reflects a physiologic role of this peptide in regulating inflammation, α-MSH has the ability to profoundly suppress the host immune response. Therefore, α-MSH and other melanocortin analogs may prove useful someday in the treatment of many of the immunologic and infectious diseases that plague mankind.

6. Melanocortins and Natriuresis

One of the more unique actions of peripherally administered melanocortin peptides is the ability to stimulate renal sodium excretion or natriuresis. Early experiments with MSH demonstrated that these melanocortin compounds could stimulate sodium excretion and increase urine volume in rats *(109)*. A wide variety of melanocortins could induce natriuresis including ACTH, α-MSH, β-MSH, and γ-MSH. γ-MSH, a proteolytic cleavage product of the N-terminal fragment of POMC, has a much more limited spectrum of biological activities than does ACTH, α-MSH, or β-MSH, including a complete lack of activity in glucose metabolism and inflammation. Despite this limited spectrum of biological activity, γ-MSH is a potent natriuretic with doses as little as 0.64 pmol in rats resulting in significant increases in renal sodium excretion *(110)*.

A well-studied model of natriuresis is the increase in sodium excretion seen in the remaining kidney after unilateral nephrectomy. It had previously been determined that this reflex natriuresis seen in the contralateral kidney after nephrectomy depended on an intact pituitary gland *(111)* and was mediated through pathways involving the central nervous system *(112,113)*. Since γ-MSH was a potent natriuretic, and measurable levels of γ-MSH had been found in the intermediate lobe of the pituitary and in serum, Lin and colleagues *(114)* hypothesized that γ-MSH was involved in the postnephrectomy reflex natriuresis seen in the contralateral kidney. After unilateral nephrectomy in rats, a significant twofold rise in circulating γ-MSH could be detected. This was accompanied by the expected increase in sodium excretion. If the animals were pretreated with specific γ-MSH antibodies, the reflex natriuresis did not occur. Furthermore, the natriuretic effect of γ-MSH seems to occur at the kidney itself. Infusions of γ-MSH directly into the renal artery produced natriuresis in the ipsilateral kidney while there was no effect at the contralateral kidney *(114)*. Taken together, these observations support a role for γ-MSH in the regulation of natriuresis.

In previous studies, natriuresis was stimulated by γ-MSH infusions directly into the renal artery *(114)*. Since the MC3 receptor is the only known melanocortin receptor with any significant γ-MSH activity, this supports a peripheral γ-MSH receptor. Recent data by Ni and colleagues demonstrates significant inhibition of post nephrectomy reflex natriuresis in the rat by SHU-9119 *(136)*. Since SHU-9119 is a potent MC3 receptor antagonist, this further supports the role of a peripheral MC3 receptor in sodium excretion. Although the MC3 receptor is present in some peripheral sites including both testis and heart *(17)*, it was not detected in the kidney by reverse transcriptase-polymerase chain reaction (RT-PCR) *(R. Kesterson, personal communication)*. The receptor, however, may be located on local renal nerves as deinnervation of the kidney blocks natriuresis resulting from direct infusion of γ-MSH into the renal artery *(115)*.

The physiologic significance of γ-MSH stimulated natriuresis is unclear. A number of early studies had suggested that melanocortins and the pituitary, particularly the pars intermedia in rodents, may be involved with sodium metabolism. Investigators had noticed an increased granularity in cells of the pars intermedia of rats after ingestion of hypertonic saline *(116)* and changes in weight of the pars intermedia after prolonged dehydration *(117)*. Furthermore, studies showed that hypertonic saline solutions decreased pituitary MSH concentrations *(109)*. These studies led to the initial discovery that melanocortins caused natriuresis *(109)*. Recent studies by Mayan and colleagues *(118)* provide further evidence for sodium regulation of intermediate lobe melanocortin production and secretion. After 1 wk of a high sodium diet, there was a twofold increase in plasma γ-MSH levels. There was also an increase in expression of POMC mRNA in the pituitary with most of the change occurring in the pars intermedia. Furthermore, there was a significant increase in the γ-MSH immunoreactivity in the intermediate lobe but not the anterior lobe of the pituitary after the high salt diet *(118)*. Since sodium intake regulates γ-MSH concentrations in plasma at potentially physiologic concentrations, it is possible that release of γ-MSH from the pars intermedia is physiologically important in the regulation of sodium excretion in rodents. Whether this represents a physiologic feedback mechanism that also occurs in humans has yet to demonstrated.

7. Melanocortins and Exocrine Gland Function

The first suggestion that melanocortins could effect exocrine gland function can be traced back to early observations that hypophysectomy could decrease sebum production from sebaceous glands *(119)*. This result was initially attributed to removing gonadotropin stimulation of sex steroid production. Later work confirmed that simply removing the neurointermediate lobe of rats significantly decreased sebum production to levels nearly as low as a total hypophysectomy *(120)*. Therefore, it was implied that pituitary factors other than gonadotropin mediated sex steroid production were necessary for normal sebum production. That α-MSH may be this neurointermediate lobe factor was suggested in further studies which demonstrated that replacement of α-MSH in hypophysectomized rats restored sebum production to near normal levels *(121)*. Simply treating with testosterone alone also restored sebum production to near normal levels. However, treating with both α-MSH and testosterone restored levels to normal *(122,123)*. Furthermore, it appears that α-MSH and testosterone have different mechanisms of action on the sebaceous gland. Testosterone seemed to predominantly effect sebaceous gland growth, while α-MSH stimulated lipogenesis within the gland *(122)*. Therefore, both sex steroids and α-MSH appear to act synergistically in the regulation of sebum production in the rat.

Melanocortins are also potent secretagogues of the lacrimal gland. Both ACTH and α-MSH stimulated protein discharge from rat lacrimal glands in vitro with maximal stimulation occurring at 20n*M* concentrations for both melanocortin peptides *(124,125)*. The lacrimal glands are highly innervated by both sympathetic and parasympathetic neurons but are primarily regulated by parasympathetics, again providing for the possibility of interactions between the melanocortin peptides and the autonomic nervous system. ACTH stimulation of peroxidase secretion was not blocked by phentolamine, atropine, or timolol suggesting that the action of ACTH on the lacrimal gland was independent of activation of parasympathetic or sympathetic neurons *(125)*. Low doses of the cholinergic agonist carbachol, however, potentiated the action of ACTH suggesting an interaction of the parasympathetic nervous system and melanocortins in lacrimal gland function. The physiologic significance of melanocortin stimulation of the lacrimal gland is unknown *(125)*. Although the maximal effect of ACTH and α-MSH was seen with 20n*M* of peptide, effects on lacrimal gland protein secretion were observed with peptide concentrations less than 10n*M*. This minimum concentration is still higher than is seen in circulating physiologic corticotropin levels, even during stress. Recent descriptions of ACTH-like immunoreactivity in the lacrimal gland, however, again raises the possibility of autocrine or paracrine actions of melanocortins within the lacrimal gland itself.

Recent work has helped to define the receptors involved in both lacrimal gland function and sebum production. Since both ACTH and α-MSH stimulate these glands, it is unlikely that the MC2 receptor is exclusively involved as α-MSH has no significant activity at the MC2 receptor. If the MC2 receptor is involved, it would have to be in addition to other melanocortin receptors. Furthermore, there is significant specific binding of radiolabelled NDP-α-MSH in the lacrimal gland which is competed by both ACTH and MSH *(126)*, also good evidence for the presence of melanocortin receptors other than MC2. The best evidence for the melanocortin receptor subtype involved in melanocortin stimulation of lacrimal and sebaceous glands comes with targeted disruption of the MC5 receptor gene in mice by Chen and colleagues *(127)*. The MC5 receptor knockout mouse produces significantly less sebum than wild type mice. This decrease in sebum results in a decreased ability of the mouse to maintain a normal body temperature when wet. In addition, it takes longer for the mouse to dry its fur after submersion indicating a decreased ability of the fur to repel water. Furthermore, Chen and colleagues demonstrated that the MC5 receptor was responsible for the ACTH and α-MSH mediated increase in lacrimal gland protein production. Finally, they also confirmed expression of MC5 receptor mRNA in both cells surrounding the hair follicle and in the lacrimal gland. Expression of MC5 receptor mRNA was obviously absent in the MC5 receptor knockout mice confirming knockout of the MC5 receptor gene *(127)*.

Evidence for another potential melanocortin mediated exocrine gland function in rats is illustrated in experiments that demonstrate production of an "alarm substance" when rats are stressed. When rats are placed in a water tank from which they cannot escape, they initially paddle vigorously but soon stop paddling and simply float. This behavior is termed the immobility response and is probably an adaptation to conserve energy until escape is possible. However, if a rat is placed in soiled water that previously contained a swimming rat, they exhibit almost no immobility response. Previous experiments have established that the substance in the water meets all the characteristics of a pheromone *(128)*. At the present time, the identity of this substance is unknown. Since this substance was released during stress, involvement of the hypothalamic-pituitary-adrenal was suspected. It was initially found that adrenalectomy had no effect on pheromone-induced inhibition of the immobility response *(128)*. Later, however, it was discovered that the pituitary gland mediated the release of this substance as rats swimming in water previously conditioned by hypophysectomized rats continued to exhibit the immobility response *(129)*, a response similar to that expected if the rat was swimming in fresh water. Furthermore, administration of ACTH peripherally to the hypophysectomized rats restored the ability of the rat to secrete this "alarm signal" into the water resulting in no immobility response in rats subsequently exposed to this water. This implies a role of melanocortins in the production of this stress induced alarm signal although it remains unclear if this is mediated via peripheral or central melanocortin receptors.

In summary, melanocortins have been implicated in the regulation of exocrine glands including the sebaceous and lacrimal glands. Additionally, melanocortins also appear to be involved in the release of pheromones that signal stress from one animal to another. The specific mechanism of release of the "stress pheromone" is not known, although it appears that melanocortins regulate other exocrine glands, specifically lacrimal and sebaceous glands, via stimulation of the MC5 receptor. A role for the MC5R in the regulation of human exocrine gland function has not yet been demonstrated.

8. Melanocortins and Testicular Function

Recent studies have started to define a role for melanocortins in the regulation of testicular function. The testis is composed primarily of steroid-producing interstitial cells called Leydig cells and cells lining the seminiferous tubules termed Sertoli cells. Immunoreactive POMC peptides have been found in testicular extracts and appear to be localized to the Leydig cell *(22,130)*. Furthermore, melanocortin peptides have been detected in testicular interstitial fluid in vivo *(131)*. The Leydig and Sertoli cells have separate functions and are controlled by different hormonal feedback systems. There is evidence, however, of considerable crosstalk between these two cells mediated by paracrine factors. With evidence that POMC peptides are produced and secreted from the Leydig cell, the possibility that melanocortin peptides were one of those paracrine factors was raised. In vitro experiments using isolated Sertoli cells demonstrated that both ACTH and MSH caused a dose-dependent increase in cAMP production *(132)*. ACTH, α-MSH, and β-MSH stimulated aromatase activity and inhibited plasminogen activator activity in isolated Sertoli cells *(133)*. These Sertoli cell functions were previously known to be mediated via increases in cAMP.

Multiple melanocortin receptor mRNAs have been detected in the testis including MC1, MC2, MC3, and MC5 *(17)*. Since γ-MSH had little or no effect on aromatase or plasminogen activator activity, it is unlikely that the MC3 receptor is involved in mediating these melanocortin induced functions. In addition, the observation that α-MSH is a potent stimulant of Sertoli cell function rules out an exclusive role of MC2 since α-MSH does not stimulate the MC2 receptor to any significant degree. With detectable levels of both the MC1 and MC5 receptors found in testicular tissue, it is very likely that the paracrine functions of melanocortins released from the Leydig cell interact with one or both melanocortin receptors expressed on Sertoli cells. It remains to be confirmed, however, that the Sertoli cell is indeed the site of melanocortin receptor expression in the testis.

9. Summary

Research into the melanocortin peptides has stretched our knowledge of the biological properties of the peptides far beyond the classic functions of melanogenesis and steroidogenesis. Evidence now implicates their involvement in a vast array of physiologic functions including carbohydrate and lipid metabolism, inflammation and fever, and natriuresis. Furthermore, melanocortins seem to be involved in the regulation of exocrine glands such as the lacrimal and sebaceous glands and endocrine glands such as the testis. Some common themes emerge when these seemingly divergent functions are compared. Many of the effects of peripherally administered melanocortins are mediated via central mechanisms. This is at least partially true in the melanocortin mediated regulation of carbohydrate metabolism and in the regulation of inflammation and fever. Furthermore, the central melanocortin peptide regulation of these peripheral physiologic functions are mediated via the autonomic nervous system. Another common theme is the potential for autocrine and paracrine actions of locally produced melanocortin peptides. The location of melanocortin producing cells near cells that either express melanocortin receptors, or have biological activities known to be altered by melanocortins, supports this hypothesis. Therefore, not all of the demonstrated peripheral activities of melanocortin peptides need to regulated by pituitary derived melanocortins in order to be physiologically important. Finally, not all of the reported peripheral functions of melanocortin peptides may be mediated via the known melanocortin receptors. Reports that demonstrate the lack of requirement for the essential melanocortin peptide core amino acid sequence in some aspects of inflammation and carbohydrate metabolism are prime examples of this principle. As research provides us with better tools for the study of the melanocortin peptides and their receptors, including receptor-specific antagonists and knockout mouse models, investigators will be able to determine the physiologic significance of the peripheral effects of the melanocortin peptides.

References

1. Mains, R. E., Eipper, B. A., and Ling, N. (1977) Common precursor to corticotropins and endorphins. *Proc. Natl. Acad. Sci. U.S.A.* **74,** 3014–3018.
2. Roberts, J. L. and Herbert, E. (1977) Characterization of a common precursor to corticotropin and beta-lipotropin, cell-free synthesis of the precursor and identification of corticotropin peptides in the molecule. *Proc. Natl. Acad. Sci. U.S.A.* **74,** 4826–4830.
3. Nakanishi, S., Inoue, A., Kita, T., Nakamura, M., Chang, A. C., Cohen, S. N., and Numa, S. (1979) Nucleotide sequence of cloned cDNA for bovine corticotropin-beta-lipotropin precursor. *Nature* **278,** 423–427.
4. Mountjoy, K. G., Robbins, L. S., Mortrud, M. T., and Cone, R. C. (1992) The cloning of a family of genes that encode the melanocortin receptors. *Science* **257,** 543–546.
5. Barret, P., MacDonald, A., Helliwell, R., Davidson, G., and Morgan, P. (1994) Cloning and expression of a new member of the melanocyte-stimulating hormone receptor family. *J. Mol. Endocrinol.* **12,** 203–213.
6. Chhajlani, V. and Wikberg, J. E. S. (1992) Molecular cloning and expression of the human melanocyte stimulating hormone receptor cDNA. *FEBS Lett.* **309,** 417–420.
7. Roselli–Rehfuss, L., Mountjoy, K. G., Robbins, L. S., Mortrud, M. T., Low, M. J., and Simerly, R. R. D. C. (1993) Identification of a receptor for γ-MSH and other

proopiomelanocortin peptides in the hypothalamus and limbic system. *Proc. Natl. Acad. Sci. U. S. A.* **90,** 8856–8860.
8. Gantz, I., Konda, Y., Tashiro, T., Shimoto, Y., Miwa, H., Munzert, G., Watson, S. J., DelValle, J., and Yamada, T. (1993) Molecular cloning of a novel melanocortin receptor. *J. Biol. Chem.* **268,** 8246–8250.
9. Gantz, I., Miwa, H., Konda, Y., Shimoto, Y., Tashiro, T., Watson, S. J., DelValle, J., and Yamada, T. (1993) Molecular cloning, expression, and gene localization of a fourth melanocortin receptor. *J. Biol. Chem.* **268,** 15,174–15,179.
10. Gantz, I., Shimoto, Y., Konda, Y., Miwa, H., Dickinson, C. J., and Yamada, T. (1994) Molecular cloning, expression, and characterization of a fifth melanocortin receptor. *Biochem. Biophys. Res. Commun.* **200,** 1214–1220.
11. Chhajlani, V., Muceniece, R., and Wikberg, J. E. S. (1993) Molecular cloning of a novel human melanocortin receptor. *Biochem. Biophys. Res. Commun.* **195,** 866–873.
12. Desarnaud, F., Labbe, O., Eggerickx, D., Vassart, G., and Parmentier, M. (1994) Molecular cloning, functional expression and pharmacological characterization of a mouse melanocortin receptor gene. *Biochem. J.* **299,** 367–373.
13. Labbe, O., Desarnaud, F., Eggerickx, D., Vassart, G., and Parmentier, M. (1994) Molecular cloning of a mouse melanocortin 5 receptor gene widely expressed in peripheral tissues. *Biochemistry.* **33,** 4543–4549.
14. Mountjoy, K. G., Mortrud, M. T., Low, M. J., Simerly, R. B., and Cone, R. D. (1994) Localization of the melanocortin-4 receptor (MC4-R) in neuroendocrine and autonomic control circuits in the brain. *Mol. Endocrinol.* **8,** 1298–1308.
15. Griffon, N., Mignon, V., Facchinetti, P., Diaz, J., Schwartz, J.-C., and Sokoloff, P. (1994) Molecular cloning and characterization of the rat fifth melanocortin receptor. *Biochem. Biophys. Res. Commun.* **200,** 1007–1014.
16. Boston, B. A. and Cone, R. D. (1996) Characterization of melanocortin receptor subtype expression in murine adipose tissues and in the 3T3-L1 cell line. *Enodcrinology* **137,** 2043–2050.
17. Chhajlani, V. (1996) Distribution of cDNA for melanocortin receptor subtypes in human tissues. *Biochem Mol. Biol. Int.* **38,** 73–80.
18. Fathi, Z., Lawerence, G. I., and Parker, E. M. (1995) Cloning, expression, and tissue distribution ofa fifth melanocortin receptor subtype. *Neurochem. Res.* **20,** 107–113.
19. Tatro, J. B. and Reichlin, S. (1987) Specific receptors for α-melanocyte-stimulating hormone are widely distributed in tissues of rodents. *Endocrinology* **121,** 1900–1907.
20. Sanchez-Franco, F., Patel, Y. C., and Reichlin, S. (1981) Immunoreactive adrenocorticotropin in the gastrointestinal tract and pancreatic islets of the rat. *Endocrinology* **108,** 2235–2238.
21. Orwoll, E. S. and Kendall, J. W. (1980) Beta-endorphin and adrenocorticotropin in extrapituitary sites:gastrointestinal tract. *Endocrinology* **107,** 438–442.
22. Tsong, S.-D., Phillips, D. M., Bardin, C. W., Halmi, N., Liotta, A. J., Margioris, A., and Kreiger, D. T. (1982) ACTH and β-endorphin related peptides are present in multiple sites in the reproductive tract of the rat. *Endocrinology* **110,** 2204–2206.
23. Saito, E. Iwasa, S., and Odell, W. D. (1983) Widespread presence of large molecular weight adrenocorticotropin-like substances in normal rat extrapituitary tissues. *Endocrinology* **113,** 1010–1019.

24. Larsson, L. I. (1977) Corticotropin-like peptides in central nerves and in endocrine cells of gut and pancreas. *Lancet* **2,** (8052–8053) 1321–1323.
25. Tanaka, I., Nakai, Y., and Nakao, K., (1982) Presence of immunoreactive γ-melanocyte stimulating hormone, adrenocorticotropin, and β-endorphin in human gastric antral mucosa. *J. Clin. Endocrinol. Metab.* **54,** 392–396.
26. Liotta, A., Osathanondh, R., and Krieger, D. T. (1977) Presence of ACTH in human placenta, Demonstration of in vitro synthesis. *Endocrinology* **101,** 1552–1558.
27. Lyons, P. D. and Blalock, J. E. (1995) The kinetics of ACTH expression in rat leukocyte subpopulations. *J. Neuroimmunol.* **63,** 103–112.
28. Borelli, M. I., Estivariz, F. E., and Gagliardino, J. J. (1996) Evidence for the paracrine action of islet-derived corticotropin-like peptides on the regulation of insulin release. *Metabolism* **45,** 565–570.
29. Wintzen, M., Yaar, M., Burbach, J. P. H., and Gilchrest, B. A. (1996) Proopiomelanocortin gene product regulation in keratinocytes. *J. Invest. Dermatol.* **106,** 673–678.
30. Genuth, S. and Lebovitz, H. E. (1965) Stimulation of insulin release by corticotropin. *Endocrinology* **76,** 1093–1099.
31. Westermeyer, V. W. and Raben, M. S. (1954) Fall in blood sugar from anterior pituitary extract. *Endocrinology* **54,** 173–180.
32. Engel, F. L., Fredericks, J., Lopez, E., and Albertson, T. (1958) Some extra-adrenal actions of corticotropin on carbohydrate metabolism in the rat. *Endocrinology* **63,** 768–777.
33. Engel, F. L. and Engel, M. G. (1945) The influence of corticotropin on ketonemia and glycemia in normal and adrenalectomized rats. *Endocrinology* **55,** 845–856.
34. Miller, W. L. and Krake, J. J. (1963) Effect of corticotropin on exhaled carbon dioxide of mice. *Endocrinology* **72,** 518–522.
35. Schatz, H., Maier, V., Hinz, H., Schleyer, M., Nierle, C., and Pfeiffer, E. F. (1973) Hypophysis and function of pancreatic islets. III. Secretion and biosynthesis of insulin in isolated pancreatic islets of intact and hypophysectomized rats in the presence of growth hormone, corticotrophin and human chorionic somatotrophin in vitro. *Horm. Metab. Res.* **5,** 29–33.
36. Love, T. A., Sussman, K. E., and Timmer, R. F. (1965) The effect of adrenocorticotrophic hormone on plasma insulin and blood glucose in the adrenalectomized rat. *Metabolism* **14,** 632–638.
37. Lebovitz, H. E., Genuth, S., and Pooler, K. (1966) Relationships between the structure and biological activities of corticotropin and related peptides, hyperglycemic action of *N*-acetylated corticotropin and related peptides. *Endocrinology* **79,** 635–642.
38. Knudtzon, J. (1984) Alpha-melanocyte stimulating hormone increases plasma levels of glucagon and insulin in rabbits. *Life Sci.* **34,** 547–554.
39. Lesault, A., Elchinger, B., and Desbals, B. (1991) Circadian variations and extraadrenal effect of ACTH on insulinemia in rabbit. *Horm. Metab. Res.* **23,** 461–464.
40. Ohsawa, N., Kuzuya, T., Tanioka, T., Kanazawa, Y., Ibayashi, H., and Nakao, K. (1967) Effect of administration of ACTH on insulin secretion in dogs. *Endocrinology* **81,** 925–927.
41. Kitabchi, A. E., Jones, G. M., and Duckworth, W. C. (1973) Effect of hydrocortisone and corticotropin on glucose-induced insulin and proinsulin secretion in man. *J. Clin. Endocrinol. Metab.* **37,** 79–84.

42. Kasperlik–Zaluska, A. A. and Krassowski, J. (1980) Synthetic 1–24ACTH-stimulated insulin release in bilaterally adrenalectomized patients. *Horm. Res.* **12,** 10–15.
43. Lundquist, T. and Rerup. C. (1967) Blood glucose level in mice. 3. On the nature of corticotrophin-induced hypoglycemia. *Acta Endocrinol.* **56,** 713–725.
44. Axelrod, J. and Weinshilboum, R. (1972) Catecholamines. *N. Eng. J. Med.* **287,** 237–242.
45. Laborit, H., Baron, C., and Thuret, F. (1976) Action de l'ACTH sur le taux de norepinephrine plasmatique chez le lapin surrenalectomise. *Agressologie* **17,** 27–32.
46. Fenske, M., Fuchs, E., and Probst, R. (1982) Corticosteroid and catecholamine plasma levels in rabbits stressed repeatedly by exposure to a novel environment or by injection of (1–24) ACTH or insulin. *Acta Endocrinol. Suppl.* **246,** 110.
47. Knudtzon, J. (1984) Acute in-vivo effects of adrenocorticotrophin on plasma levels of glucagon, insulin, glucose, and free fatty acids in rabbits, involvement of the alpha-adrenergic nervous system. *J. Endocrinol.* **100,** 345–352.
48. Rees, L. H., Cook, D. M., Kendell, J. W., Allen, C. F., Kramer, R. M., Ratcliffe, J. G., and Knight, R. A. (1971) A radioimmunoassay for rat plasms ACTH. *Endocrinology* **89,** 254–261.
49. Hummel, A., Lendeckel, U., and Hahn, V. (1992) Presence and regulation of a truncated proopiomelanocortin gene transcript in rat pancreatic islets. *Biol. Chem. Hoppe–Seyler* **373,** 1039–1044.
50. Gremlich, S., Roduit, R., and Thorens, B. (1997) Dexamethasone induces post-translational degradation of GLUT2 and inhibition of insulin secretion in isolated pancreatic beta cells. comparison with the effects of fatty acids. *J. Biol. Chem.* **272,** 3216–3222.
51. Lambillotte, C., Gilon, P., and Henquin, J. C. (1997) Direct glucocorticoid inhibition of insulin secretion. An in vitro study of dexamethasone effects in mouse islets. *J. Clin. Invest.* **99,** 414–423.
52. Beloff-Chain, A., Edwardson, J. A., and Hawthorn, J. (1975) Influence of the pituitary gland on insulin secretion in the genetically obese (ob/ob) mouse. *J. Endocrinol.* **65,** 109–116.
53. Bailey, C. J. and Flatt, P. R. (1987) Insulin releasing effects of adrenocorticotropin (ACTH 1–39) and ACTH fragments (1–24 and 18–39) in lean and genetically obese hyperglycaemic (OB/OB) mice. *Int. J. Obesity* **11,** 175–181.
54. Fan, W., Boston, B. A., Kesterson, R. A., Hruby, V. J., and Cone, R. D. (1997) Role of melanocortinergic neurons in feeding and the agouti obesity syndrome. *Nature* **384,** 165–168.
55. Shimizu, H., Tanaka, Y., Sato, N., and Mori, M. (1995) α-melanocyte-stimulating hormone (MSH) inhibits insulin secretion in HIT-T 15 cells. *Peptides* **16,** 605–608.
56. Boston, B. A. Blaydon, K. M. Varnerin, J., and Cone, R. D. (1997) Independent and additive effects of central POMC and leptin pathways on murine obesity. *Science* **278,** 1641–1644.
57. Bloom, S. R., Edwards, A. V., and Hardy, R. N. (1978) The role of the autonomic nervous system in the control of glucagon, insulin and pancreatic polypeptide release from the pancreas. *J. Physiol.* **280,** 9–23.
58. Lafontan, M. and Agid, R. (1979) An extra-adrenal action of adrenocorticotrophin, physiological induction of lipolysis by secretion of adrenocorticotrophin in obese rabbits. *J. Endocrinol.* **81,** 281–290.

59. Richter, W. O. and Schwandt, P. (1983) In vitro lipolysis of proopiocortin peptides. *Life Sci.* **33,** 747–750.
60. Schotz, M. C., Masson, G. M. C., and Page, I. N. (1959) ACTH in vitro on release of nonesterified fatty acids from adipose tissue of adrenalectomized rats. *Proc. Soc. Exp. Biol. Med.* **101,** 159–161.
61. Braun, T. and Hechter, O. (1970) Glucocorticoid regulation of ACTH sensitivity of adenyl cyclase in rat fat cell membranes. *Proc. Natl. Acad. Sci. U. S. A.* **66,** 995–1001.
62. Spirovski, M. Z., Kovacev, V. P., Spasovska, M., and Chernick, S. S. (1975) Effect of ACTH on lipolysis in adipose tissue of normal and adrenalectomized rats in vivo. *Am. J. Physiol.* **228,** 382–385.
63. Oelofsen, W. and Ramachandran, J. (1983) Studies of corticotropin receptors on rat adipocytes. *Arch. Biochem. Biophys.* **225,** 414–421.
64. Ramachandran, J. and Lee, V. (1976) Divergent effects of adrenocorticotropin and melanotropin on isolated rat and rabbit adipocytes. *Biochim. Biophys. Acta* **428,** 339–346.
65. Ramachandran, J., Farmer, S. W., Liles, S., and Li, C. H. (1976) Comparison of the steroidogenic and melanotropic activities of corticotropin, alpha-melanotropin, and analogs with their lipolytic activities in rat and rabbit adipocytes. *Biochim. Biophys. Acta* **428,** 347–354.
66. White, J. E. and Engel, F. L. (1958) Lipolytic action of corticotropin on rat adipose tissue in vitro. *J. Clin. Invest.* **37,** 1556–1563.
67. Cornelius, P., MacDougald, O. A., and Lane, M. D. (1994) Regulation of adipocyte development. *Annu. Rev. Nutr.* **14,** 99–129.
68. Richter, W. O. and Schwandt, P. (1987) Lipolytic potency of proopiomelanocorticotropin peptides in vitro. *Neuropeptides* **9,** 59–74.
69. Glyn, J. R. and Lipton, J. M. (1981) Hypothermic and antipyretic effects of centrally administered ACTH (1–24) and α-melanotropin. *Peptides* **2,** 177–187.
70. Murphy, M. T., Richards, D. B., and Lipton, J. M. (1983) Anitpyretic potency of centrally administered α-melanocyte stimulating hormone. *Science* **221,** 192–193.
71. Martin, L. W. and Lipton, J. M. (1990) Acute phase response to endotoxin, rise in plasma α-MSH and effects of α-MSH injection. *Am. J. Physiol.* **259,** R768–R772.
72. Opp, M. R., Obal, F., Krueger, J. M. (1988) Effects of α-MSH on sleep, behavior, and brain temperature, interactions with IL-1. *Am. J. Physiol.* **255,** R914–R922.
73. Davidson, J., Milton, A. S., and Rotondo, D. (1992) α-Melanocyte stimulating hormone suppresses fever and increases in plasma levels of prostaglandin E2. *J. Physiol.* **451,** 491–502.
74. Robertson, B., Dostal, K., and Daynes, R. (1988) Neuropeptide regulation of inflammatory and immunologic responses. *J. Immunol.* **140,** 4300–4307.
75. Martin, L. W., Cantania, A., Hiltz, M. E., and Lipton, J. M. (1991) Neuropeptide α-MSH antagonizes IL-6 and TNF induced fever. *Peptides* **12,** 297–299.
76. Shih, S. T., Khorram, O., Lipton, J. M., and McCann, S. M. (1986) Central administration of α-MSH antiserum augments fever in the rabbit. *Am. J. Physiol.* **250,** R803–R806.
77. Samson, W. K., Lipton, J. M., and Zimmer, J. A. (1981) The effect of fever on central alpha-MSH concentration in the rabbit. *Peptides* **2,** 419–423.
78. Holdeman, M., Khorram, O., Samson, W. K., and Lipton, J. M. (1985) Fever-specific changes in MSH and CRF concentrations. *Am. J. Physiol.* **248,** R125–R129.

79. Lipton, J. M. (1989) Neuropeptide alpha-melanocyte stimulating hormone in control of fever, the acute phase response and inflammation., in *Neuroimmune Networks, Physiology and Diseases.,* Goetzl, E. and Spector, N. H. (eds.) Liss, New York. pp. 243–250.
80. Hiltz, M. E. and Lipton, J. M. (1990) Alpha-MSH peptides inhibit acute inflammation and contact hypersensitivity. *Peptides* **11,** 972–982.
81. Chiao, H., Foster, S., Thomas, R., Lipton, J., and Star, R. A. (1996) α-Melanocyte-stimulating hormone reduces endotoxin-induced liver inflammation. *J. Clin. Invest.* **97,** 2038–2044.
82. Lipton, J. M., Ceriani, G., Macaluso, A., McCoy, D., Carnes, K., Blitz, J., and Catania, A. (1994) Antiinflammatory effects of the neuropeptide α-MSH in acute, chronic and systemic inflammation. *N. Y. Acad. Sci.* **741,** 137–148.
83. Grabbe, S., Bhardwaj, R. S., Mahnke, K., Simon, M. M., Schwarz, T., and Luger, T. A. (1996) α-Melanocyte-stimulating hormone induces hapten-specific tolerance in mice. *J. Immunology* **156,** 473–478.
84. Ceriani, G., Diaz, J., Murphree, S., Catania, A., and Lipton, J. M. (1994) The neuropeptide α-MSH inhibits experimental arthritis in rats. *Neurooimmunomodulation* **1,** 28–32.
85. Lipton, J. M. and Catania, A. (1997) Anti-inflammatory actions of the neuroimmunomodulator α-MSH. *Immunol. Today* **18,** 140–145.
86. Lipton, J. M. and Catania, A. (1992) α-MSH peptides modulate fever and inflammation., in *Neuro-immunology of Fever* Bartfai, T. and Ottoson, D., (eds.) Pergamon Press, New York. pp. 123–126.
87. Lipton, J. M. (1990) Modulation of host defense by the neuropeptide α-MSH. *Yale J. Biol. Med.* **63,** 173–182.
88. Lipton, J. M. and Catania, A. (1993) Pyrogenic and inflammatory actions of cytokines and their modulation by neuropeptides, Techniques and interpretations., in *Methods in Neuroscience* DeSouza, E. B., (ed.) Academic Press, Orlando. FL, pp. 61–79.
89. Catania, A. and Lipton, J. M. (1993) α-Melanocyte stimulating hormone in the modulation of host reactions. *Endocr. Rev.* **14,** 564–576.
90. Ceriani, G., Macaluso, A., Catania, A., and Lipton, J. M. (1994) Central neurogenic antiinflammatory action of α-MSH, Modulation of peripheral inflammation induced by cytokines and other mediators of inflammation. *Neuroendocrinology* **59,** 138–143.
91. Luger, T. A., Schauer, E., Trautinger, F., Krutmann, J., Ansel, J., Schwarz, A., and Schwarz, T. (1993) Production of immunosuppressing melanotropins by keratinocytes. *Ann. N. Y. Acad. Sci.* **680,** 567–570.
92. Hiltz, M. E., Catania, A., and Lipton, J. M. (1992) α-Melanocyte stimulating hormone peptides inhibit acute inflammation induced in mice by rIL-1β, rIL-**6,** rTNF-α, and endogenous pyrogen but not that caused by LTB**4,** PAF and rIL-8. *Cytokine* **4,** 320–328.
93. Mason, M. J. and Van Epps, D. (1989) Modulation of IL–**1,** tumor necrosis factor, and C5α-mediated murine neutrophil migration by α-melanocyte-stimulating hormone. *J. Immunol.* **142,** 1646–1651.
94. Star, R. A., Rajora, N., Huang, J., Stock, R., Catania, A., and Lipton, J. M. (1995) Evidence of autocrine modulation of macrophage nitric oxide synthase by α-melanocyte-stimulating hormone. *Proc. Natl. Acad. Sci. U. S. A.* **92,** 8016–8020.

95. Fiorentino, D. F., Zlotnik, A., Mosmann, T. R., Howard, M., and O'Garra, A. (1991) IL-10 inhibits cytokine production by activated macrophages. *J. Immunol.* **147,** 3815–3822.
96. Fiorntino, D. F., Zlotnik, A., Vieira, P., Mosmann, T. R., Howard, M., Moore, K. W., and O'Garra, A. (1991) IL-10 acts on the antigen-presenting cell to inhibit cytokine production by Th1 cells. *J. Immunol.* **146,** 3444–3451.
97. Lipton, J. M., Macaluso, A., Hiltz, M. E., and Catania, A. (1992) Central administration of the peptide alpha-MSH inhibits inflammation in the skin. *Peptides* **12,** 795–798.
98. Watanabe, T., Hiltz, M. E., Catania, A., and Lipton, J. M. (1993) Inhibition of IL-1β-induced peripheral inflammation by peripheral and central administration of analogues of the neuropeptide α-MSH. *Brain Res. Bull.* **32,** 311–314.
99. Rajora, N., Boccoli, G., Burns, D., Sharma, S., Catania, A., and Lipton, J. M. (1997) α-MSH modulates local and circulating tumor necrosis factor-α in experimental brain inflammation. *J. Neuroscience* **17,** 2181–2186.
100. De Rotte, A. A., Bouman, H. J., and van Wimersma Greidanus, T. B. (1980) Relationships between α-MSH levels in blood and in cerebrospinal fluid. *Brain Res. Bull.* **5,** 375–381.
101. Macaluso, A., McCoy, D., Ceriani, G., Watanabe, T., Beltz, J., Catania, A., and Lipton, J. M. (1994) Antiinflammatory influences of α-MSH molecules, Central neurogenic and peripheral actions. *J. Neurosci.* **14,** 2377–2382.
102. Bhardwaj, R., Becher, E., Mahnke, K., Hartmeyer, M., Schwarz, T., Scholzen, T., and Luger, T. A. (1997) Evidence for the differential expression of the functional alpha-melanocyte-stimulating hormone receptor MC-1 on human monocytes. *J. Immunol.* **158,** 3378–3384.
103. Catania, A., Rajora, N., Capsoni, F., Minonzio, F., Star, R. A., and Lipton, J. M. (1996) The neuropeptide alpha-MSH has specific receptors on neutrophils and reduces chemotaxis in vitro. *Peptides* **17,** 675–679.
104. Poole, S., Bristow, A. F., Lorenzetti, B. B., Gaines Das, R. E., Smith, T. W., and Ferreira, S. H. (1992) Peripheral analgesic activities of peptides related to α-melanocyte stimulating hormone and interleukin-1β. *Br. J. Pharmacol.* **106,** 489–492.
105. Wilkes, B. C., Sawyer, T. K., Hruby, V. J., and Hadley, M. C. (1983) Differentiation of the structural features of melanotropins important for biological potency and prolonged activity in vitro. Int. J. Peptide. *Protein Res.* **22,** 313–324.
106. Mugridge, K. G., Perretti, M., Ghiara, P., and Parente, L. (1991) α-Melanocyte-stimulating hormone reduces interleukin-1β effects on rat stomach preparations possibly through interference with a type 1 receptor. *Eur. J. Pharmacol.* **197,** 151–155.
107. Bhardwaj, R. S., Schwarz, A., Becher, E., Mahnke, K., Aragane, Y., Schwarz, T., and Luger, T. A. (1996) Pro-opiomelanocortin-derived peptides induce IL-10 production in human monocytes. *J. Immunol.* **156,** 2517–2521.
108. Schauer, E., Trautinger, F., Kock, A., Schwarz, A., Bardwaj, R., Simon, M., Ansel, J. C., Schwarz, T., and Luger, T. A. (1994) Proopiomelanocortin-derived peptides are synthesized and released by human keratinocytes. *J. Clin. Invest.* **93,** 2258–2262.
109. Orias, R. (1970) Natriuretic effect of α-MSH in the water-loaded rat. *Proc. Soc. Exp. Biol.* **133,** 469–474.
110. Lymangrover, J. R., Buckalew, V. M., Harris, J., Klein, M. C., and Gruber, K. A. (1985) Gamma-2 MSH is natriuretc in the rat. *Endocrinol.* **116,** 1227–1229.

111. Lin, S.-Y., Wiedemann, E., and Humphreys, M. H. (1985) Role of the pituitary in reflex natriuresis following acute unilateral nephrectomy. *Am. J. Physiol.* **249,** F282–F290.
112. Lin, S.-Y., Humphreys, M. H. (1985) Centrally administered naloxone blocks reflex natriuresis after acute unilateral nephrectomy. *Am. J. Physiol.* **249,** F390–F395.
113. Ayus, J. C. and Humphreys, M. H. (1982) Hemodynamic and renal functional changes after acute unilateral nephrectomy in the dog, role of carotid sinus baroreceptors. *Am. J. Physiol.* **242,** F181–F189.
114. Lin, S. Y., Chaves, C., Wiedemann, E., and Humphreys, M. H. (1987) A γ-melanocyte stimulating hormone-like peptide causes reflex natriuresis after acute unilateral nephrectomy. *Hypertension* **10,** 619–627.
115. Humphreys, M. H., Wiedemann, E., Valentin, J.-P., Chen, X.-W., and Ying, W.-Z. (1993) Natriuretic actions of g-melanocyte-stimulating hormone. *Ann. N.Y. Acad. Sci.* **680,** 545–548.
116. Howe, A. and Thody, A.J. (1970) The effect of ingestion of hypertonic saline on the melanocyte-stimulating hormone content and histology of the pars intermedia of the rat pituitary gland. *J. Endocrinol* **46,** 201–208.
117. Duchen, L. W. (1968) Changes in the volume of the lobes of the pituitary gland and in the weight and in the weight and water content of organs of rats given hypertonic saline. *Endocrinol.* 4**1,** 593–600.
118. Mayan, H., Ling, K.-T., Lee, E. Y., Wiedemann, E., Kalinyak, J. E., and Humphreys, M. H. (1996) Dietary sodium intake modulates pituitary proopiomelanocortin mRNA abundance. *Hypertension* **28,** 244–249.
119. Ebling, F. J., Ebling, E., and Skinner, J. (1969) The influence of pituitary hormones on the response of the sebaceous glands of the rat to testosterone. *J. Endocrinol.* **45,** 401–406.
120. Thody, A. J. and Shuster, S. (1972) The control of sebum secretion by the posterior pituitary. *Nature* **237,** 346–347.
121. Thody, A. J. and Shuster, S. (1973) A possible role of MSH in the mammal. *Nature* **245,** 207–209.
122. Thody, A. J., Cooper, M. F., Bowden, P. E., Meddis, D., and Shuster, S. (1976) Effect of α-melanocyte-stimulating hormone and testosterone on cutaneous and modified sebaceous glands in the rat. *J. Endocrinol.* **71,** 279–288.
123. Ebling, F. J., Ebling, E., Randall, V., and Skinner, J. (1975) The synergistic action of α-melanocyte stimulating hormone and testosterone on the sebaceous, prostate, preputial, harderian and lachrymal glands, seminal vesicles and brown adipose tissue in the hypophysectomized–castrated rat. *J. Endocrinol.* **66,** 407–412.
124. Jahn, R., Padel, U., Porsch, P.–H., and Soling, H.–D. (1982) Adrenocorticotropic hormone and α-melanocyte-stimulating hormone induce secretion and protein phosphorylation in the rat lacrimal gland by activation of a cAMP-dependent pathway. *Eur. J. Biochem.* **126,** 623–629.
125. Cripps, M. M., Bromberg, B. B., Patchen-Moor, K., and Welch, M. H. (1987) Adrenocorticotropic hormone stimulation of lacrimal peroxidase secretion. *Exp. Eye Res.* **45,** 673–683.
126. Entwistle, M. L., Hann, L. E., Sullivan, D. A., and Tatro, J. B. (1990) Characterization of functional melanotropin receptors in lacrimal glands of the rat. *Peptides* **11,** 477–483.

127. Chen, W., Kelly, M. A., Opitz-Araya, X., and Cone, R. D. (1997) Exocrine gland dysfunction in MC5-R deficient mice: evidence for coordinated regulation of exocrine gland function by melanocortin peptides. *Cell* **91,** 789–798.
128. Abel, E. L. and Bilitzke, P. J. (1992) Adrenal activity does not mediate alarm substance reaction in the forced swim test. *Psychoneuroendocrinology* **17,** 255–259.
129. Abel, E. L. (1994) The pituitary mediates production or release of an alarm chemosignal in rats. *Horm. Behav.* **28,** 139–145.
130. Tsong, S.-D., Phillips, D. M., Halmi, N, Krieger, D. T., and Bardin, C. W. (1982) β-Endorphin is present in the male reproductive tract of five species. *Biol. Reprod.* **27,** 755–764.
131. Valenca, M. M. and Negro-Vilar, A. (1986) Pro-opiomelanocortin-derived peptides in testicular interstitial fluid: characterization and changes in secretion after human chorionic gonadotropin or leuteinizing hormone-releasing hormone analog treatment. *Endocrinology* **118,** 32–37.
132. Boitani, C., Mather, J. P., and Bardin, C. W. (1986) Stimulation of cAMP production in rat Sertoli cells by α-MSH and des-acetyl α-MSH. *Endocrinology* **1986,** 1513–1518.
133. Boitani, C., Farini, D., Canipari, R., and Bardin, C. W. (1988) Estradiol and plasminogen activator secretion by cultured rat Sertoli cells in response to melanocyte-stimulating hormones. *J. Androl.* **10,** 202–209.
134. Eberle, A. N. (1988) The Melanotropins: Chemistry, Physiology and Mechanisms of Action. Karger, Basel, Switzerland. p. 556.
135. Huang, Q. H., Entwistle, M. L., Alvaro, J. D., et al. (1997) Antipyretic role of endogenous melanocortins mediated by central melanocortin receptors during endotoxin-induced fever. *J. Neurosci.* **17,** 3343–3351.
136. Ni, X. P., Kesterson, R. A., Sharma, S. D., et al. (1998) Prevention of reflex natriuresis after acute unilateral nephrectomy by melanocortin receptor antagonists *Am. J. Physiol.* **274,** R931–R938.

Part II

Characterization of the Melanocortin Receptors

CHAPTER 6

Melanocortin Receptor Expression and Function in the Nervous System

Jeffrey B. Tatro

1. Introduction

By the late 1970s, a range of evidence indicated that melanocortins could affect behavioral and visceral functions, neuroendocrine circuits, and the neurochemistry of the brain *(1,2)* in addition to well-characterized roles in pigmentation and adrenocortical steroidogenesis. The discovery of releasable neurosecretory pools of α-MSH in brain tissue *(3)*, and the discovery of an intrinsic POMC (proopiomelanocortin) and melanocortin-containing neuron system in the brain *(4,5)*, began to point to a potential role of endogenous central nervous system (CNS) melanocortins in regulating many of these functions. The facts that similar melanocortinergic systems exist in the brains of lower vertebrate species as primitive as the lungfish *(6)*, and in mammals are predominantly distributed in the phylogenetically ancient visceral neuraxis, suggests that the melanocortin system may subserve highly conserved roles. As discussed below and in Chapters 4 and 13, a fairly extensive literature now supports a fundamental role of melanocortins in diverse CNS functions. Nevertheless, the identification of CNS-associated melanocortin receptors is a fairly recent development. Following the demonstration of MCR in the CNS in 1990 *(7)*, the cloning of a family of MCR-encoding genes (*see* Chapter 7) paved the way for the recent explosive growth in interest in the physiological roles of melanocortins in the nervous system, and the molecular bases of melanocortin actions.

Another major recent development was the remarkable finding that the repertoire of endogenous ligands of MCR present within neurons of the CNS

The Melanocortin Receptors Ed.: R. D. Cone

Table 1
Native Agonist/Antagonist Profiles
of CNS-Associated Melanocortin Receptors

Agonists[a]	
MCR Subtype	*Relative Potency/Binding Affinity*
MC3-R	γ-MSH = ACTH ≥ α-MSH
MC4-R	α-MSH = ACTH >> γ-MSH
MC5-R	α-MSH ≥ ACTH > γ-MSH
Antagonist: AgRP[b]	
	Relative Potency/Binding Affinity
	MC3-R = MC4-R >> MC5-R

Based on assays in MCR-transfected cells. Potency estimated by stimulation of cyclic AMP accumulation (agonists) or inhibition of α-MSH-induced cyclic AMP accumulation (antagonist). Binding affinities estimated by inhibition of ^{125}I-NDP-MSH binding.

[a]ref. *92*

[b]Human MCR; ref. *7c*

includes not only agonists (POMC-derived melanocortins), but also the MCR antagonist, agouti-related protein (AgRP) *(7a,7b)*. Aside from the adrenal ACTH receptor (MC2-R), which is ACTH-selective, all known MCR subtypes recognize multiple forms of native melanocortin agonists (e.g. α-MSH, γ-MSH, ACTH) (summarized in Table 1; *see also* Chapters 1, 7, and 8). AgRP is a selective and essentially equipotent antagonist of the MC3-R and MC4-R (Table 1; *7a,7c*).

The major presumptive source of endogenous MCR agonists in the CNS is the central POMC neuron system. The CNS contains two discrete groups of POMC-synthesizing neurons. The principal POMC neuron group is located in the arcuate nucleus of the medial basal hypothalamus, from which it projects widely to innervate numerous structures in the forebrain, brainstem and spinal cord involved in neuroendocrine and autonomic functions *(4)*. A second, minor POMC neuron group is located in the commissural part of the nucleus of the solitary tract. This group is less well-characterized, but its projections are believed to be distributed mainly in the hindbrain and possibly the spinal cord *(7d,60,61)*. POMC-containing neurons have generally been found to contain multiple forms of melanocortins as well as endorphins *(9,11,12)*. In this review, neurons containing POMC and its derivative peptides are considered to be *melanocortinergic*. In addition to melanocortins of neuronal origin, bloodborne melanocortins appear to be capable of activating MCR within the CNS *(12a)*, as discussed further below. A single AgRP-synthesizing neuron group has been identified in the CNS *(104)*. Like the principal POMC group, it is located in the arcuate nucleus, and lies juxtaposed just medially to the POMC cells *(12b)*. The projections of AgRP-synthesizing neurons are essentially coextensive with those of POMC neurons in the forebrain *(12c,12d)*, but are notably less extensive or absent in the hindbrain and spinal cord *(12d)* (Table 1). Therefore, the working concept of the *melanocortin system* of the CNS is defined here to include the intrinsic central POMC and AgRP neuron systems, and the MCR-bearing target cells of the CNS.

An intriguing property of melanocortins that captured our interest was their pluripotent ability to suppress the actions of proinflammatory cytokines *(13)*. This is particularly true in the CNS, the classical example being the antipyretic action of melanocortins *(14,14a)*. Exogenous melanocortins, including not only α-MSH but also ACTH and several α-MSH analogs, were shown to suppress fevers induced by endogenous leukocytic pyrogens and by cytokine-inducing microbial toxins. They were more potent following intracerebroventricular administration than conventional antipyretics such as aspirin and acetaminophen, were effective at lower doses centrally than peripherally, and did not affect thermoregulation in afebrile animals at low doses. In fact, melanocortins are now known to act as functional antagonists of multiple central actions of proinflammatory cytokines, including interleukin-1 (IL-1), IL-6, and tumor necrosis factor-alpha (TNF-α) *(15)*, suggestive of a conserved, adaptive *cytokine-counterregulatory* role *(14a)*.

The neuroanatomic mapping localization of MCR in brain tissue was first undertaken to identify candidate loci of the cytokine-inhibitory and neuroendocrine functions of α-MSH and ACTH *(7)*. Although our knowledge of the physiologic roles and the molecular pharmacology of central MCR is really still in its infancy, substantial information has been gained concerning the neuroanatomic distribution of MCR and the regional expression of MCR-encoding mRNA subtypes in the CNS. These data provide a rich source of clues to the functional roles of endogenous melanocortins and the neuropharmacologic substrates of exogenous melanocortin actions. Achieving a more detailed understanding of the neuroanatomic distribution of MCR subtypes and the regulation of their expression in the nervous system will be an essential part of future efforts to understand the functional organization and roles of the central melanocortinergic system.

This chapter focuses on MCR expression within the CNS, but the more limited information concerning MCR expression in the peripheral nervous system is also discussed. The objective is to provide an overview of the approaches used and the current state of knowledge concerning the anatomic organization and the regulation of expression of MCR-encoding genes in the nervous system. The impetus for this research is to understand the functional roles and the therapeutic opportunities of the central melanocortinergic system.

In this review, hypotheses are formulated by integrating neuroanatomic, pharmacologic, and genetic lines of evidence.

2. Methodological Considerations

Viewed from a functional anatomic standpoint, achieving a working knowledge of MCR expression is a multilayered problem, entailing the acquisition of several kinds of information at multiple levels of organization. Fundamentally, one wishes to know the cellular sites, and the factors that control the levels, of MCR expression, and the state of activation of functional MCR protein under a given set of conditions. However, because it is not possible to visualize functional MCR directly, one is limited technically to using a number of indirect approaches. Once the localization of the MCR is established, the goal becomes to identify the phenotype and functions of the MCR-expressing cells. This entails knowing the location of the cell body, which may be quite distant if the MCR in question is expressed on a nerve terminal, for example; the neurotransmitters produced by the MCR-bearing cell, its connectivity with other cell types, and its responses to various stimuli. One also needs to know the factors that regulate MCR synthesis and expression in a functional form at the cell surface, which involves understanding the hierarchical control of MCR gene expression and posttranslational processing. There are multiple MCR genes expressed in the nervous system, which further necessitates knowing the functional MCR protein subtype in each case, and its parent gene. Furthermore, because certain evidence suggests the existence of novel MCR subtypes that have yet to be identified, the physiologic or pharmacologic relevance of the particular MCR under study is always in some degree of doubt.

The complexity of this problem obviously demands the use of multiple complementary experimental approaches. Accordingly, several approaches have been used to determine the neuroanatomic localization and molecular identities of MCR expressed in the nervous system. The detection of MCR proteins has been accomplished by ligand-based methods, and MCR subtype gene expression has been assessed by various mRNA hybridization approaches. Ultimately, the use of classical and functional neuroanatomic techniques will be required to determine the cell types, connectivity, and neurochemical coding of cells expressing different MCR subtype mRNA, and the levels and sites of functional expression of MCR subtype proteins under various physiologic conditions.

2.1. MCR Proteins

2.1.1. Neuroanatomic Localization

To visualize the neuroanatomic distribution of putative functional MCR proteins, a classical *in situ* ligand binding and autoradiography approach has been used to localize specific melanocortin binding sites in the rat CNS *(7,16–18)*. Briefly, unfixed CNS tissue is used to prepare slide-mounted, dried cryostat sections, which are then incubated with a radiolabeled MCR ligand. After elution of unbound tracer and rapid drying, the localization of bound tracer is determined autoradiographically, using either direct apposition to X-ray film or by coating of the slides with a liquid photographic emulsion. The specificity of ligand binding is determined by its inhibition in the presence of excess competing unlabeled ligand.

The approach has several advantages. First, it allows the neuroanatomic localization of MCR by a functionally relevant property, with a moderate degree of resolution. That is to say, MCR can be localized to identified nuclei and subnuclei of the CNS based on anatomic features. It can sometimes be determined whether the distribution of binding sites occurs principally as either punctate labeling of individual cell bodies or as labeling of the neuropil surrounding certain cells or cell groups. On the other hand, the technique has several limitations. The anatomic localization of ligand binding is typically not discernable at the cellular level, and it cannot be determined whether the MCR are localized on cellular processes such as axons or terminals, due to diffuse fields of silver grains in photographic emulsions with the use of ^{125}I-labeled ligand. The neurochemical coding of MCR-expressing neuron groups (i.e., the profile of neurotransmitters produced) cannot be identified within the same tissue sections because the tissue fixation required for immunochemical studies is generally not compatible with ligand binding studies.

At the time of writing, the most reliable and widely applied MCR binding assays employ a ^{125}I-labeled derivative of the synthetic agonist, Nle^4, D-Phe^7-α-MSH (NDP-MSH) *(19)*, first developed to determine the tissue distribution of specific MSH binding sites in vivo *(20)* and to characterize MSH receptors in melanoma *(21,22)*. The ability of ^{125}I-NDP-MSH to bind specifically and with high affinity, but nonselectively, to each of the four identified MSH-binding MCR subtypes (MC1-R, MC3-R, MC4-R, MC5-R) was fortuitous in that it allowed the detection of multiple MCR subclasses in vivo and in vitro *(7,18, 20–23)*, and contributed to the rapid characterization of cloned MCR subtypes in 1992–94 *(24–28)*. Theoretically, the use of MCR subclass-selective ligands as probes would permit assessment of the differential neuroanatomic distribution of individual MCR subclasses. However, except for γ-MSH, highly selective ligands are only just beginning to be developed (Chapter 14). Other types of melanocortin probes, including fluoresceinated ligands and other novel melanocortin derivatives, have been developed for receptor studies *(29–32)*, and continue to be the subject of much interest, but these have not yet been widely applied to the study of MCR in the nervous system. Hence the development of ligand-based, subclass-specific MCR probes remains a challenge for future work. It is likely that the development of MCR subtype-directed antibodies for immunocytochemical localization, and studies in mice containing genetic deletions of MCR subtypes, will permit more specific and precise localization of MCR subclasses in the near future.

Another useful application of radiolabeled MCR ligands including NDP-MSH and ACTH[1–24] has been for the localization of brain MCR in vivo *(33,34)*. In this approach, radiolabeled MCR agonists were administered systemically in the presence and absence of excess unlabeled agonists, followed by autoradiographic localization of the radioligand. This allowed the localization of MCR to structures to which systemic melanocortins have direct access, that is, MCR for which the blood–brain barrier does not prevent access from blood *(33,34)*.

2.1.2. Ligand Binding Properties

The in vitro autoradiographic approach has also permitted characterization of the ligand binding properties of *native* MCR populations in brain tissue; a method widely applied to other neurotransmitter receptors (e.g. ref. *35*). Because of the high signal-to-noise ratio (>95% binding specificity in some regions) and sensitivity of this approach, it is possible to characterize the ligand binding properties of MCR populations within individual brain nuclei by combining the method with computerized densitometric image analysis *(7,18,36,37)*. One major applications of this method is to study the regulation of MCR protein expression, by quantifying the effects of experimental treatments on MCR binding levels. Second, the relative binding affinities of localized MCR populations for different ligands can be determined. Of course, the technique has certain limitations. First, because the tissue composition changes within any given series of tissue sections, the number and composition of these MCR populations also vary to some extent, depending partly on the size and complexity of the region sampled. Second, the approach is nonideal for kinetic binding studies because tissue sections are nonuniform with respect to diffusion barriers, tracer access, and proximity of the detection system to the signal. Despite these caveats, this approach is quite robust, and it has provided the only insight to date into the regional ligand-binding properties of native brain MCR populations expressed in vivo (as opposed to those of isolated recombinant MCR expressed in heterologous cells). As more selective MCR ligands become available, this approach will undoubtedly be applied to characterize the regional expression of native MCR in the nervous system, and potentially to identify novel MCR subtypes.

Classical studies of other neurotransmitter receptors have made extensive use of radioligand binding in brain homogenates and membrane preparations containing the receptors of interest. This approach offers the technical advantages of homogeneity of the pooled receptor population under study and amenability to formal kinetic binding analyses. However, it also has the obvious drawback of information loss through pooling of regionally heterogeneous receptor populations, and tends to have a low signal-to-noise ratio. Several groups have characterized specific melanocortin-binding properties of rat brain membrane preparations with mixed success. For years, the very existence of specific MCR was disputed, based in part on the argument that melanocortin analogs could compete for binding to receptors for opioids and other classical neurotransmitters, albeit at very low affinities (K_D in the 10^{-5} to 10^{-4} M range) (reviewed in ref. *38*), whereas specific melanocortin-selective binding in brain was not demonstrated before the 1990s *(7)*. Using brain tissue homogenates, another group described the existence of binding sites having dual selectivity for ACTH and vasoactive intestinal polypeptide (VIP), but the specificity of this binding was low (e.g.,< 40% of ACTH binding was specific), and α-MSH did not compete for binding *(39)*, indicating that this binding was not attributable to currently known MCR subtypes. A major source of potential artifacts in this paradigm is via direct interactions of test substances with G proteins present in the membrane preparations *(40)*. This could be due to alterations in receptor radioligand binding affinity, giving the false appearance of competitive inhibition of radioligand–receptor binding by the test substance. Indeed, such interactions have been demonstrated for a number of cationic amphiphilic peptides, including ACTH *(40)*, and may conceivably occur with VIP as well, potentially accounting for the putative brain ACTH/VIP binding sites *(39)*. Thus, whether the cited findings *(38,39)* have functional relevance awaits further clarification.

Cell membranes prepared from isolated neuronal or glial cell populations should prove useful for biochemical characterization of MCR proteins expressed in the nervous system, but as yet have not been extensively studied. A specific melanocortin binding protein was described in membranes of cultured rat Schwann cells. Crosslinking studies with a photoreactive derivative of NDP-MSH demonstrated a specifically labeled protein doublet band of approximately 42–45 kDa *(42)*, similar to that of a MCR protein detected in mouse melanoma cells by a similar method *(41)*. The Schwann cell MCR protein was not further characterized *(42)*.

2.2. MCR Subtype mRNA

Detection of MCR subtype-encoding mRNA in the CNS has been determined by standard approaches, including Northern blotting, RNAase protection, reverse transcriptase-polymerase chain reaction (RT-PCR), and *in situ* hybridization. These methodologies are well established and need not be reviewed in detail here, but a brief consideration of their salient features will assist the interpretation of existing and future studies of MCR expression in the nervous system.

Northern blotting, RNAase protection, and RT-PCR entail the preparation of RNA extracts from the tissue of interest. Therefore, analysis of regional expression of MCR subtype mRNA in the CNS based on these methods is dependent on dissection of the tissue region of interest. Any further neuroanatomic information is lost upon the disruption and pooling of tissue. In the case of RT-PCR, its extreme sensitivity is advantageous for the detection of rare transcripts, but also presents well-recognized caveats for the interpretation of anatomic studies and studies of the regulation of gene expression. First, the method is susceptible to contamination artifacts from extraneous sources and genomic DNA, necessitating the use of rigorous negative controls. Second, small amounts of mRNA present in non-CNS tissue elements such as vascular cells and passenger leukocytes can give rise to PCR products easily misinterpreted as originating in CNS parenchyma. Therefore, evidence that a particular MCR subtype gene is expressed in the CNS based on qualititative RT-PCR analysis alone, particularly in whole brain preparations, is generally regarded as somewhat preliminary until corroborated by other means. The exponential amplification inherent in the RT-PCR approach, combined with differences in efficiencies of both the RT and PCR reactions also presents some challenges for quantitative studies of mRNA abundance, but modified methods such as competitive RT-PCR *(50)* and other coamplification methods are amenable to quantitative studies.

The use of *in situ* hybridization allows indirect localization of target mRNA transcripts at the cellular level, and is thus an important tool for neuroanatomic studies. The adequacy of its sensitivity depends on a number of factors, most importantly, mRNA abundance, and also technical factors such as tissue preparation, the type and size of nucleic acid probe used, hybridization stringency, and the type of radioisotopic or immunochemical label used *(51)*. The use of *in situ* hybridization of MCR mRNA allows for studies of colocalization with other neuroanatomic markers. The relatively low abundance of MCR mRNA transcripts in CNS tissue, particularly in the case of MC5-R *(36,52)*, renders such colocalization studies technically demanding, but MCR mRNA has been successfully colocalized with other transmitters *(12d)*. The *in situ* hybridization approach is amenable to quantification of even fairly subtle changes in mRNA levels in specific brain regions under various conditions, and has been a major instrument for the study of regulation of other neurotransmitter and receptor genes in the CNS *(53)*.

Regardless of the technical approach used, the detection or quantification of receptor-encoding mRNA alone provides no information on the presence, absence or anatomic localization of functional receptors. Receptor mRNA need not be translated into protein that is processed and expressed on the cell surface at functionally relevant levels. Further, any such translated proteins may be either expressed on the cell body or exported to distant sites on nerve fibers, terminals, or glial processes. Conversely, depending on factors including assay sensitivities, mRNA stability and the kinetics of protein synthesis and turnover, MCR protein may be detectable by ligand-based methods, while the corresponding mRNA is not detectable. Hence MCR-encoding transcripts should not be interpreted as or referred to as *receptors*, as is often practiced in the literature. In fact, the development of methods for the direct detection of MCR-subtype *proteins* presently remains one of the technical challenges facing brain MCR researchers. In this chapter, the term *MCR* is reserved for receptor protein; MCR-encoding mRNA is designated as such.

The localization and quantification of mRNA are essential components of the overall effort to understand the anatomic organization of MCR-expressing cells and the regulation of MCR gene expression. The combined application of approaches for the detection of MCR-encoding transcripts and MCR protein will continue to provide a powerful tool for analysis of MCR expression in the nervous system.

2.3. Functional Assays

Current understanding of MCR function at the cellular level in the nervous system is rather scant. As described in Chapters 7 and 8, most studies of MCR molecular pharmacology at the cellular level have used heterologous expression of plasmids containing MCR subtype-encoding cDNA—in cells that contain appropriate signaling machinery to mount a brisk adenylate cyclase response, but lack endogenous MCR. To understand the nature of MCR cellular actions in the nervous system, it will be essential to develop more specific cellular models, such as isolated neural or glial cell populations that normally express MCR. This would help to ensure that a given MCR protein under study receives appropriate posttranslational processing and has access to the full complement of signaling pathways reflecting its physiologic situation in vivo. The MC1-R-deficient mutant mouse melanoma cell line, B16-G4F *(43)*, has some of these attributes. In many B16 cell sublines not containing this mutation, exogenous melanocortins act via native MC1-R to stimulate adenylate cyclase, and in some cases increased melanogenesis. This long served as a principal line of evidence supporting a melanogenic role of α-MSH in normal mammalian melanocytes *(44)*. Hence the B16-G4F cell subline has cellular machinery presumably reflective of the normal intracellular signaling environment for MCR proteins and is also relevant to the nervous system based on its neural crest origin. This model has been used successfully for the pharmacologic characterization of the neural MC3-R and MC4-R of the rat in heterologous expression *(45)*. Another cellular model reportedly useful for MCR characterization is the Neuro 2A neuroblastoma cell line. These cells respond to melanocortins with neurite outgrowth, potentially mediated by MC4-R since MC4-R transcripts were detected and an MC4-R antagonist inhibited the effect of α-MSH *(46)*. Glial cells, including astrocytes *(47–49)* and Schwann cells *(42)*, may also prove useful because of their favorable growth properties and their putative melanocortin responsiveness.

3. Regional Distribution Of Melanocortin Receptors And Transcripts In The Central Nervous System

It is now recognized that MCR are widely distributed in the CNS. In this section, an overview of current knowledge of the distribution of MCR and the expression of MCR-subtype-encoding genes is presented. Except where otherwise indicated, the available data pertain to rodents. Virtually no data are yet available describing MCR or MCR mRNA distribution in the primate nervous system, but POMC and AgRP neuron systems, apparently organized similarly to those of rodents, exist in humans and other primates *(12c,54–56,56a,56b)*, and human homologs of the primarily CNS-associated rodent MC3-R and MC4-R subtypes have been cloned and characterized *(36,52,57,58)*. The regional distribution of MCR binding sites and MCR subtype mRNA is summarized in Table 2 and Figs. 1 and 2. The functional implications of the structural organization of this system are also discussed.

3.1. MCR Proteins

3.1.1. Neuroanatomic Distribution

Currently, nearly all information concerning the localization of MCR proteins in the nervous system is derived from in vitro ^{125}I-NDP-MSH binding and autoradiography in the rat. Specific melanocortin binding proteins are present at virtually all levels of the neuraxis. They are most densely distributed in the ventral forebrain, and are more restricted, but also prominent, in hindbrain nuclei and in the spinal cord.

By and large, the distribution of central MCR follows fairly closely that of the central POMC neuron system (Figs. 1 and 2; Table 2). MCR are present in the bed nuclei of both of the known POMC neuron groups of the brain, including the arcuate nucleus (Figs. 1D, 2C) *(7)*, seat of the principal POMC neuron group *(59)*; and the commissural part of the nucleus of the solitary tract, located in the caudal medulla (Fig. 1H) *(16)*, seat of the smaller and less extensive POMC group *(60,61)*. In the basal forebrain, MCR are densely distributed throughout most nuclei and at all rostrocaudal levels within the hypothalamus-preoptic region and septal area, which is also the CNS region most densely innervated by POMC neurons (Fig. 2) *(59,62)*. The suprachiasmatic

Table 2
Comparative Distribution of Melanocortinergic Innervation, Agouti-Related Protein Innervation, Specific Melanocortin Binding, and MCR Subtype mRNA in Selected Regions of the Rat CNS

	POMC Innervation	AgRP Innervation	MCR Protein ^{125}I-NDP-MSH Binding	MCR mRNA		
				MC3-R	MC4-R	MC5-R[a]
Forebrain						
Cortical structures						
Neocortex	–		+	–	+	+[a]
Olfactory bulb	–		–	–	+	+[a]
Olfactory tubercle	–	–/+	+	–	+	+[a]
Entorhinal cortex	–		+	–	+	
Hippocampal formation	–		+	+	+	+[a]
Amygdala						
Medial n.	+	+	+	–	+	
Central n.	+	+	+	–	+	
Basolateral n.	+		+	–	+	
Striatum						+[a]
Caudate putamen	–	–	+	–	+	
Globus pallidus	–	–	–	–	–	
Nucleus Accumbens	+	+	+	–	+	–
Epithalamus						
Medial habenular n.	–	–	+	+	–	
Lateral habenular n.	+		+	–	–	
Thalamus						
Midline nuclei	+		+	+	+	
Zona incerta	+	+	+	–	+	
Septal area						
Medial septum	+	–/+	+	–	+	

(continued)

Table 2 *(continued)*

	POMC Innervation	AgRP Innervation	MCR Protein ^{125}I-NDP-MSH Binding	MCR mRNA		
				MC3-R	MC4-R	MC5-R[a]
Lat. septal n., dorsal	+	–/+	+	–	+	
Lat. septal n., intermed.	+		+	+	+	
Lat. septal n., ventral	+	+	+	–	+	
Bed n., stria terminalis	+	+	+	+	+	
Hypothalamus-preoptic region						+[a]
Supraoptic n.	–/+	+	+	–	+	
Suprachiasmatic n.	–/+	+	–	–	–	
Medial preoptic n.	+	+	+	+	+	
Lateral preoptic area	+	+	+	+	+	
Anterior hypoth. n.	+	+	+	+	+	
Periventricular n.	+	+	+	+	+	
paraventricular n.	+	+	–	+	+	
arcuate n.[b]	+	+	+	+	+	
Post. periventricular n.	+	+	+	+	+	
Ventromedial n.	–/+	+	+	+	+	
Dorsomedial n.	+	+	+	–	+	
Post. hypoth. n.	+	+	+	+	+	
Lateral hypoth. area	+	+	+	+	+	
Premammillary nuclei	+		+	+	+	
Circumventricular organs						
OVLT	+	+	+	–	–	
Subfornical organ	–	+	+	–	+	
Median eminence	+	+	+	–	–	
Midbrain				+[a]		
Superior colliculus		–	+	–	+	
Inferior colliculus	+	–	+	+		

Periaqueductal gray	+	+	+	+	+	
ventral tegmental area	+	–/+	+	+	+	
substantia nigra	+	–	+	–	+	+[a]
interpeduncular n.	–	–/+	+	–	–	
interfascicular n.	+	–/+	+	+	–	
Central linear n.	+	+	+	–		
Cerebellum	–	–	–	–	–	+[a]
Hindbrain						+[a]
Raphé nuclei	+	–/+	+	–	+	
Dorsal tegmentum	+	–	+	–	+	
Parabrachial complex	+	+	+	–	+	
Pontine retic. formation	+	–/+	+	–	+	
Medullary retic. formation	+		+	–	+	
Dorsal vagal complex[c]	+	+	+	–	+	
Ventrolateral medulla	+	+	+	–	+	
Spinal Cord						
Substantia gelatinosa	–	–	+	–	+	
Area X	+	–	+	–	–	
Peripheral Nervous System						
Cranial nerve ganglia[d]			+			
Sympathetic ganglia[d]			+		–	–
Spinal ganglia[d]			+		–	–

This list is illustrative, not comprehensive. +, present; –, not detectable; –/+, present at very low levels or literature conflicts; no symbol, not determined

[a]Detected by RT-PCR only.

[b]Bed nucleus of principal POMC neuron group.

[c]Includes minor POMC neuron group.

[d]Prenatal or early postnatal rats.

References: POMC innervation *(9,12d,59,60,62,112)*; AgRP innervation *(12c,12d,56a)*, MSH binding (*7,16,17,37,65*, and *Tatro, unpublished data*); MC3-R mRNA *(12d,26)*; MC4-R mRNA *(67,71)*; MC5-R mRNA (36,52).

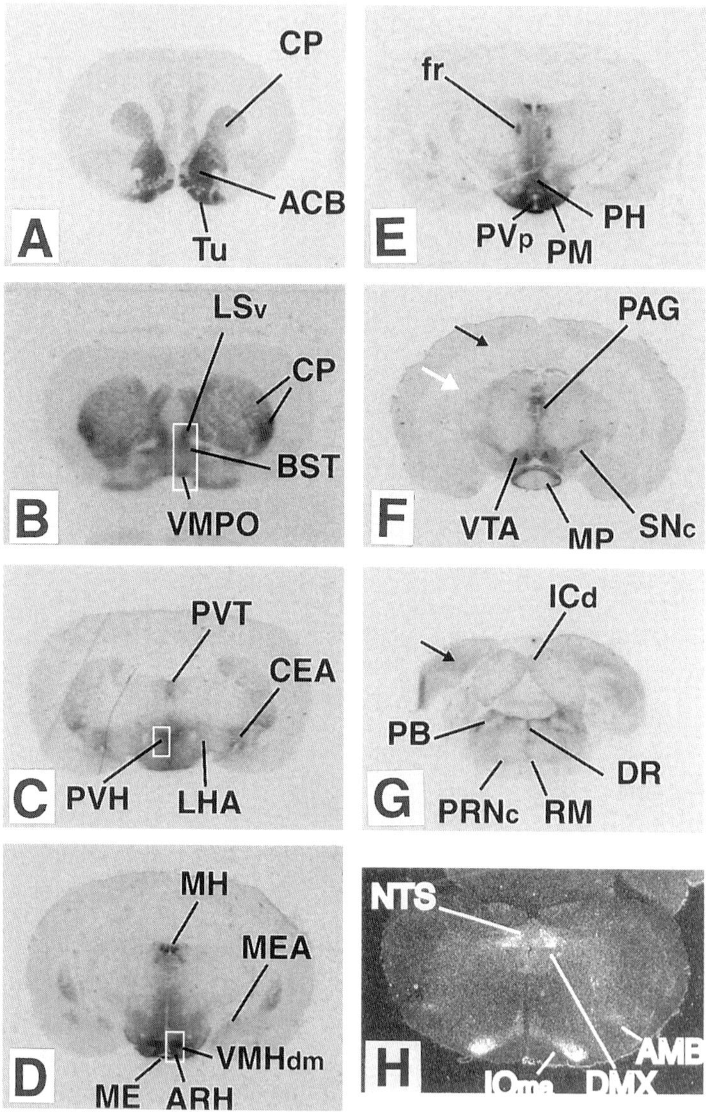

Fig. 1. An overview of the neuroanatomic distribution of melanocortin receptors in the rat brain. Shown are autoradiograms of in vitro ^{125}I-NDP-MSH binding to a rostrocaudal series of coronal tissue sections *(7,16)*. A–G, film autoradiograms of rostral forebrain through rostral pons; H; dark-field photomicrograph of emulsion autoradiogram of caudal medulla. Note the dense MCR distribution in ventral forebrain and hypothalamus (A–E), and the prominent labeling of autonomic centers in the hindbrain (G,H). In F, note faint bands of binding signal (*arrows*) in cortex (*black*) and hippocampal formation (*white*). In G, more intense binding is evident in

nucleus is a notable exception within the hypothalamus, exhibiting little or no MSH binding *(17)* and only sparse POMC innervation *(9,12d,59)*. Some hypothalamic regions in which POMC innervation is very sparse, such as the ventromedial nucleus and the supraoptic nucleus (Fig. 2) *(9,59)*, do contain MCR (Fig. 1, Table 2) *(7,17)*. Among circumventricular organs of the forebrain, MSH binding is present in the vascular organ of the lamina terminalis, subfornical organ *(34)*, and median eminence (Fig. 1D). MCR are also present in the amygdala, and in midline nuclei of the thalamus, including the reuniens, rhomboid, and paraventricular nucleus of the thalamus (Fig. 1C,D), which are also regions receiving POMC innervation *(59,62)*. In the olfactory–striatal complex, intense MSH binding is present throughout the olfactory tubercle and nucleus accumbens, and is distributed more diffusely in the neostriatum, with a prominent density in the ventrolateral neostriatum at the level of the anterior commissure (Fig. 1B). These regions are not innervated by POMC neurons, with the possible exception of part of the accumbens *(59,62)*. In the midbrain, MCR are most prominently distributed throughout the central gray, ventral tegmental area, and the interpeduncular nucleus, and throughout the ventral midline nuclei, but are also present in both superior and inferior colliculi (Fig. 1F, 1G) *(7)*.

In the caudal brainstem, the overall distribution of MCR is much more restricted and is characterized by two general patterns. First, there are focal accumulations of MCR in discrete nuclei, including, among others, the pontine

---

cortical regions including the occipital pole (*arrow*), entorhinal cortex and parasubiculum. To illustrate the extent of MCR codistribution with melanocortinergic innervation, the *boxes* in panels B, C, and D, correspond to the areas shown in Fig. 2, panels A, B, and C, respectively.

ACB, nucleus accumbens; AMB, nucleus ambiguus; ARH, arcuate nucleus of hypothalamus; BST, bed nucleus of stria terminalis; CEA, central nucleus of amygdala; CP, caudate putamen; DMX, dorsal motor nucleus of vagus; DR, dorsal raphe nucleus; fr, fasciculus retroflexus; ICd, inferior colliculus, dorsal nucleus; IOma, inferior olivary complex, medial accessory olive; LHA, lateral hypothalamic area; LSv, lateral septal nucleus, ventral division; ME, median eminence; MEA, medial nucleus of amygdala; MH, medial habenula; MP, mammillary process; PAG, periaqueductal gray matter; PB, parabrachial complex; PH, posterior hypothalamic nucleus; PM, premammillary nuclei; PRNc, pontine reticular nucleus, caudal part; PVH, paraventricular nucleus of hypothalamus; PVp, posterior periventricular nucleus of hypothalamus; PVT, paraventricular nucleus of thalamus; NTS, nucleus of solitary tract; RM, nucleus raphé magnus; SNc, substantia nigra, pars compacta; Tu, olfactory tubercle; VMHdm, ventromedial nucleus of hypothalamus, dorsomedial part; VMPO, ventromedial preoptic nucleus; VTA, ventral tegmental area. Nomenclature and neuroanatomic designations are those of Swanson *(113)*, except for VMPO *(114)*.

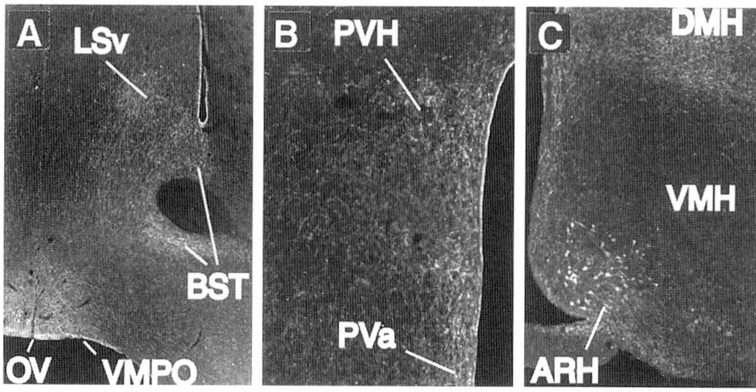

Fig. 2. Distribution of α-MSH-immunoreactive neurons in selected regions of the rat septal region and hypothalamus, for comparison with MCR distribution shown in Fig. 1. Panels A, B, and C correspond to the *boxed* areas shown in autoradiograms of MCR binding in Fig. 1B, C, and D, respectively. Shown are negative images of photomicrographs of tissue sections immunostained with sheep anti-α-MSH antiserum, visualized using a nickel-enhanced avidin–biotin–peroxidase procedure. (A) rostral septal area. Note high density of α-MSH fibers in the BST, LSv, and OVLT/VMPO area, regions of high MCR density. (B) Note dense α-MSH-immunoreactive fiber networks in the parvicellular divisions of PVH and the periventricular nucleus. (C) note α-MSH-containing cell bodies in ARH,the bed nucleus of the major POMC neuron group of the brain, and dense fiber networks in the periventricular region and ventral aspect of the DMH. By contrast, the VMH, including the dorsomedial division which shows intense ^{125}I-NDP-MSH binding (Fig. 1D) is nearly devoid of α-MSH-containing projections. For detailed neuroanatomic distribution of melanocortinergic neuron system, see *(9,59,60,62,112)*. *Abbreviations*: as in Fig. 1, plus DMH, dorsomedial nucleus of hypothalamus; OV, organum vasculosum of lamina terminalis; PVa, periventricular nucleus of hypothalamus, anterior division.

locus ceruleus and dorsal tegmentum, parabrachial complex, various raphé nuclei, and the dorsal vagal complex (Fig. 1G,H) *(16)*. Within the dorsal vagal complex, specific MSH binding is present in parts of the area postrema, in multiple subdivisions of the nucleus of the solitary tract, and is most intense throughout the dorsal motor nucleus of the vagus (DMX); (Fig. 1H) *(16)*, *(J. Tatro, unpublished results)*. Second, MCR are distributed more diffusely and at low to moderate densities in broader regions including the pontine and medullary reticular formation, ventrolateral medulla, dorsal column nuclei, and sensory nucleus of the trigeminal nerve *(16,37)*, and J. Tatro, unpublished results). Excluding the intensely labeled DMX and inferior olivary complex (Fig. 1H) *(16)*, *(J. Tatro, unpublished results)*, which are not known to be innervated by POMC neurons, most or all of these MCR-bearing regions of the midbrain and hindbrain do receive POMC innervation *(59,60,63,64)*. MCR

Table 3
MCR Ligand Affinity Profiles Compared
in Rat Brain Tissue and Recombinant MCR Subtypes

	Affinity Index[a] (K_i or IC_{50}) (n*M*)		Relative Affinity vs α-MSH (ratio)		
	NDP[b]	α-MSH	NDP	γ-MSH[c]	Ref.
Rat brain region					
Bed nucleus, stria terminalis	1.3	56.3	42.7	0.44	*18*
Medial preoptic area	1.7	48.6	28.9	0.35	
Caudate putamen, ventrolat. region	0.76	5.5	7.3	0.033	
Paraventricular n., medial parvo. div.	0.62	4.8	7.7	0.063	
Recombinant MCR subtypes					
MC3R (rat)	10	52	5.2	1.18	*26*
MC4R (human)	2.2	641	291	<0.006	*70*
MC5R (mouse)	1.1	62.5	56.8	0.0492	*28*

[a]Based on potency in inhibiting ^{125}I-NDP-MSH binding.
[b]NDP-MSH.
[c]Lower relative binding affinity of γ-MSH suggests lower proportional content of MC3-R. See also ref. *37*.

are also present in the gray matter of the spinal cord, where they are most densely distributed in the superficial laminae of the dorsal horn (substantia gelatinosa) and the area surrounding the central canal (lamina X), but are also present in other regions of both dorsal and ventral horns *(65)*.

MCR are also present in various cortical regions (Fig. 1, Table 3) *(37)*. Early studies focused on the forebrain showed occasional areas of nonspecific ^{125}I-NDP-MSH binding, but no indication of specific MCR in adult cortex or hippocampus *(7)*. Nevertheless, later studies of more posterior levels clearly indicated the presence of specific NDP-MSH binding in various regions of neocortex, entorhinal cortex, and hippocampal formation (Fig. 1) *(37)*, *(J. Tatro, unpublished data)*. The distribution of MCR in cortical structures has not been determined in detail. In the cerebellum, moderate levels of nonspecific binding have repeatedly been observed in cerebellar white matter *(J. Tatro, unpublished data)*, but specific ^{125}I-NDP-MSH binding has not been detected *(7,17)*.

3.1.2. Ligand Binding Properties

The first demonstration that MSH receptors in the brain had different ligand-binding properties than peripheral (melanoma) MSH receptors was provided by the observation that specific ^{125}I-NDP-MSH binding sites in brain tissue sections showed equivalent relative affinities for α-MSH and ACTH[1–39] *(7)*. This contrasted with mouse melanoma cell MSH receptors, which had substantially higher affinity for α-MSH than for ACTH *(22)*, and adrenal ACTH receptors, which bound ACTH but not α-MSH *(66)*, indicating that the CNS contained a novel MCR population *(7)*. The pharmacology of isolated MCR subtypes is summarized in Table 1 and is covered in detail elsewhere in this volume. Below, the ligand binding profiles of native brain MCR populations *in situ* are briefly compared and contrasted with those of recombinant MCR.

Comparison of the ligand binding profiles of MCR in different brain nuclei provided several insights into the nature and composition of native MCR populations in the brain *(18)*. First, specific NDP-MSH binding in each of 11 different regions of the forebrain was completely blocked by 1 μ*M* α-MSH, desacetyl-α-MSH, β-MSH, or ACTH, indicating that no NDP-MSH-binding MCR subpopulations were present that were capable of discriminating qualitatively between these melanocortin agonists. Second, the MCR receptor populations in different brain regions exhibited different ligand-binding affinity profiles, suggestive of regional heterogeneity of the MCR populations expressed *(18)*. This was consistent with the presence and differential distribution of multiple MCR-encoding genes in the rat CNS *(26,52,67)*. Third, the relative affinity of γ-MSH for binding to MCR in these regions was lower than that of α-MSH *(18,37)*. This indicated that the MC3-R subtype represents only a portion of the total MCR pool present in these rat ventral forebrain regions, because γ-MSH has a binding affinity at least as great as that of α-MSH for the isolated rat MC3-R expressed in heterologous cells (Table 3) *(26)*. The full complement of specific NDP-MSH binding in these regions probably includes MC4-R and possibly MC5-R proteins, which also bind NDP-MSH but have much lower relative affinities for γ-MSH than for α-MSH (Table 3). Based on the available data, the relative affinity for γ-MSH as compared with that of α-MSH in a given brain region may provide a rough indication of the proportion of its total MCR pool represented by MC3-R. Thus, brain regions showing the lowest relative affinities for γ-MSH (e.g., ventrolateral caudate putamen, medial parvocellular region of PVH) presumably contain lower proportions of MC3-R protein than regions such as the BST and MPA, which show about tenfold higher relative affinities for γ-MSH (Table 3). Further, this is consistent with the reported lack of MC3-R mRNA in the CPvl and PVHmp *(26)*. Conversely, those regions showing the highest affinities for γ-MSH likely contain the highest relative proportions of MC3-R as compared with MC4-R and/or MC5-R *(18)*. Finally, none of the native brain MCR studied by *in situ* ligand binding exhibited any ability to bind certain melanocortin analogs having defined melanocortinlike neuropharmacologic activities in vivo, including ACTH[4–9] and its analog, Org2766 *(1)*, and α-MSH[11–13] *(15,68)*. This suggests that if specific receptors exist for these peptides, they either do not bind NDP-MSH, are present at levels insufficient to discriminate in competitive binding studies, or are distributed in brain regions not yet studied *(18)*. Considered together with the fact that that the known MCR subtypes expressed in heterologous cells likewise fail to bind or respond to these peptides *(26,67,69,70)*, this is one line of evidence suggesting that novel receptors for these peptides await discovery.

3.2. Distribution Of MCR Subtype-Encoding mRNA

3.2.1. MC3-R

In the CNS, MC3-R mRNA is distributed mainly in the diencephalon and midbrain, most abundantly in the hypothalamus, but is also present in a few sites in the thalamus and midbrain. MC3-R mRNA was not detected in the hindbrain or spinal cord *(26,58)*. Specific NDP-MSH binding is present (Fig. 1) *(7)* in all or nearly all sites in which MC3-R transcripts have been detected in the rat brain (Table 1), suggesting that the MC3-R may represent some fraction of these sites. MC3-R mRNA is present in a substantial portion of arcuate POMC neurons *(12d)*.

3.2.2. MC4-R

MC4-R is the most widely distributed of the MCR subtype mRNA expressed in the CNS. It was detected at all levels of the neuraxis and in well over 100 specific regions in the rat *(67)*, and is the only MCR subtype gene for which transcripts have been localized in the autonomic centers of the hindbrain, including the parabrachial complex, ventrolateral medulla and the dorsal vagal complex, and in the spinal cord (Table 1) *(67)*. In the spinal cord, MC4-R mRNA appeared to be restricted to the marginal zone (layer I), whereas MCR protein is much more widely distributed *(65)*. The distribution of MC4-R mRNA is more extensive than indicated in Table 1. Like MC3-R mRNA, it is largely codistributed with ^{125}I-NDP-MSH binding sites, suggesting that it may encode many of the MCR proteins present throughout the CNS, at least in the rat *(16,67)*. MC4-R transcripts are also present prenatally in ganglia of the peripheral autonomic nervous system *(71)*, consistent with the presence of NDP-MSH binding in some of these sites *(37)*.

3.2.3. MC5-R

In the nervous system, MC5-R mRNA has been detected only in preparations from whole brain or dissected brain regions. In contrast to its prominent expression in peripheral tissues *(72,73)*, it appears to be very low in abundance in the CNS, and to date has not been localized in neural tissue by in situ hybridization. MC5-R mRNA was detected by RNAase protection assay in cerebral cortex and cerebellum, but not in accumbens/olfactory tubercle, striatum, hypothalamus, or midbrain central gray of the mouse *(28,36,74)*, and was detected by RT-PCR in a dozen regions of the rat brain *(52)* (Table 1). MC5-R mRNA was detected by RT-PCR in an RNA specimen from human brain *(25)*.

3.2.4. MC1-R

Most studies have failed to demonstrate the presence of MC1-R transcripts in the CNS *(36,75)*, but two reports have suggested the presence of MC1-R mRNA in the brain. In one, a few cells in the rat midbrain reportedly hybridized with an oligonucleotide DNA probe directed against the murine MC1-R mRNA *(76)*. The second study reported a MC1-R RT-PCR product generated from whole mouse brain *(77)*. Further studies will help to resolve the significance of these findings.

3.2.5. MC2-R

To date, there are no reports of MC2-R mRNA expression in the nervous system.

3.3. Functional Relationships Of Central MCR To Endogenous Agonists and Antagonists

In order to understand the physiologic or pharmacologic relevance of central MCR, it is critical to assess the relationships of MCR to endogenous and exogenous sources of melanocortins. As discussed above, MCR are prominently and extensively codistributed with the projection fields of melanocortinergic neurons, supporting a functional relationship of postsynaptic MCR with the innervating neurons (Figs. 1 and 2; Table 1). The first direct test of this possibility was reported in a study of the systemic cardiovascular response (hypotension and bradycardia) induced by electrical stimulation of the arcuate nucleus, the bed nucleus of POMC neurons. This response was blocked by microinjection of the MC3-R/MC4-R antagonist, SHU-9119, into the dorsal vagal complex in the rat *(78)*. This finding suggests that the effect of arcuate stimulation was mediated by melanocortins released by POMC neurons innervating the dorsal medullary cardiovascular center *(60)*, which then acted locally on postsynaptic neuronal MCR. Of course, in a number of paradigms, the marked physiologic and metabolic effects of centrally administered MCR antagonists *(45,79)*, targeted genetic MCR ablation *(80)*, and overexpression of endogenous MCR antagonist proteins *(81,82)* support a physiologic role of MCR, but the sites of these interactions have not been determined directly.

Not all central MCR are located in brain regions innervated by melanocortinergic neurons. For example, MCR and/or MCR mRNA are present in cortex, hippocampus, and striatum, regions that do not appear to receive POMC innervation. α-MSH-immunoreactive cortical projections described earlier *(4)* were probably not truly melanocortinergic. The cells of origin were located in the dorsolateral hypothalamus–zona incerta region, a region that contains no POMC-containing cell bodies *(83,84)*, but instead contains a cell group labeled by only certain α-MSH antibodies *(85,86)*. This α-MSH immunoreactivity has been attributed to a cross-reacting peptide contained in the precursor prohormone for melanin-concentrating hormone *(87)*. Other highly specific anti-αMSH antisera do not label these MCH-containing cells *(56a)*. Within cortex and striatum, MCR mRNA could of course be exported to distant nerve terminals. However, unless the MCR detected autoradiographically are themselves intracellular proteins destined for export, or are somehow directly accessible to bloodborne melanocortin hormones, it remains unclear how MCR within these regions may play a functional role in the absence of identified MC innervation. Hence the endogenous ligands for, and the functional significance of, MCR in these areas of anatomic "mismatch" remain to be determined.

It has been recognized for many years that peripherally administered melanocortins can affect CNS functions *(1,2,4)*. However, because the ability of melanocortins to penetrate the blood brain barrier is very limited *(90)*, the mechanisms by which exogenous peripheral melanocortins activate central MCR are unknown. Thus, the MCR present in circumventricular organs, which lack a tight blood–brain barrier *(88)*, comprise one major route by which systemic, pituitary-derived melanocortins may have direct access to central MCR (Fig.1; Table 2). MCR are present in each of four circumventricular organs that function as key autonomic and neuroendocrine regulatory sites: the organum vasculosum of the lamina terminals (OVLT), subfornical organ (SFO), median eminence *(7,17,34)*; and parts of the area postrema *(J. Tatro, unpublished data)*. Indeed, following intravenous injection both ^{125}I-NDP-MSH and ^{125}I-ACTH[1–24] localized specifically in the medial basal hypothalamus, within the median eminence, clearly indicating that bloodborne melanocortins have access to MCR in these sites (20,33,34). Aside from their potential roles in transducing information borne by systemic pituitary-derived melanocortins, MCR expressed in circumventricular organs could play a role in mediating the pharmacologic actions of exogenous melanocortins. For example, we recently showed that the antipyretic effect of *peripherally* administered α-MSH is mediated by MCR located within the CNS, because its effects were blocked by central administration of the MC4/MC3-R antagonist SHU-9119 *(12a)*, at a dose which had no effect on fever when given systemically, and which had no effect in the absence of fever *(45)*. Hence MCR located within circumventricular organs are candidate mediators of these effects. Alternatively, it has been postulated that the ability of MSH peptides to cross the blood–brain barrier, although exceedingly limited *(90)*, may nevertheless suffice to account for the pharmacologic effects of peripherally administered melanocortins on CNS functions *(91)*. These hypotheses, which are independent rather than mutually exclusive alternatives, remain to be tested. Another important, but as yet unresolved, question is whether endogenous circulating melanocortins of pituitary origin are direct physiological activators of MCR located within the CNS.

4. Regulation Of Melanocortin Receptor Expression In The Nervous System

Presently, only limited information is available concerning the regulation of MCR expression in the nervous system. Most of the earlier data concerning MCR regulation derive from studies in melanocytic or adrenocortical MCR-expressing cells (reviewed in ref. *92*), and the regulatory elements responsible for the control of MCR gene expression in vivo are not yet known. Nevertheless, recent studies have indicated that drugs of abuse can alter MCR expression in the CNS. Further, altered MCR expression must be considered as a potential mechanism underlying the development of tolerance to repeated melanocortin administration in certain behavioral paradigms *(93)*, and the physiologic state-dependence of many melanocortin actions in the CNS.

4.1.Ontogeny of MCR Expression

Developmental changes in expression of NDP-MSH binding *(37)* and of MC3-R and MC4-R transcripts *(71,94)* in the rat nervous system have been described. In a comprehensive study, the distribution and intensity of NDP-MSH binding were determined from gestational day 13 (E13) through adulthood in the rat. Prenatal and early postnatal development were characterized by transient, prominent peaks in ^{125}I-NDP-MSH binding density, occurring at different times in different brain regions, suggestive of a role in development of the central and peripheral nervous systems. The patterns of MCR expression *(37)* were related to some extent to the development of central POMC neurons, which appear beginning on E12 *(95)*, and melanocortinergic projections, which appear beginning on E13 *(96)*. The neostriatum was abundantly labeled prenatally, and a pronounced patch-matrix pattern was evident in the early postnatal period, with the more prominent labeling in the patches *(37)*. This pattern did not persist in adults *(37)*, which have a low, diffuse level of NDP-MSH binding throughout most of the neostriatum, with a region of more intense binding in the ventrolateral caudate putamen (Fig. 1) *(7,36)*. A number of regions that do not exhibit NDP-MSH binding in adult rats *(7)*, such as globus pallidus and cerebellar cortex, showed transient peaks of MCR expression that subsided within a few days postnatally *(37)*. Importantly, this study also indicated that MCR are present from a very early stage (E14) in the developing peripheral nervous system, specifically in cranial nerve ganglia, spinal ganglia, and sympathetic ganglia and nerves *(37)*.

The same report included an analysis of ligand affinity profiles designed to assess the relative prevalence of MCR subtypes during ontogeny. Overall, the results for α-MSH and NDP-MSH in developing brain were similar to those reported for adult rats *(18)*, but the relative binding affinity of γ-MSH was generally lower than in adults, suggesting that the relative content of MC3-R in the early postnatal forebrain may be lower than in adults (Table 3) *(37)*. This is consistent with a report that MC3-R mRNA was present at only low levels in rat ventral forebrain up to postnatal day 4 (P4), increasing slightly by P7, but not reaching the substantially higher adult levels of expression until P21 *(94)*. By contrast, MC4-R mRNA was detectable in many regions of the nervous system from E14, the earliest time point studied *(71)*. The distribution of its expression at these developmental stages was generally similar to that described for adults (Table 2) *(67)*, whereas MC5-R mRNA was not detectable by this method. Furthermore, MC4-R mRNA was expressed in sympathetic and vagal trunks and autonomic ganglia *(71)*.

These findings suggest that MC4-R may be the predominant MCR subtype expressed in the developing nervous system. MCR expression seems to be highly regulated during active periods of neurogenesis in the sympathetic nervous system and the thalamus, and a pattern of transient waves of MCR protein and MC4-R mRNA expression occurs during highly active periods of developing neural connectivity. Taken together, these findings suggest a potentially important role of the MC4-R in the development of the CNS and the peripheral autonomic nervous system *(37,71)*.

4.2. Regulation by Addictive Drugs

A number of studies have indicated that melanocortins interact with opiates in the nervous system, generally as functional antagonists of opiate actions in models of analgesia and drug addiction (reviewed in ref. *97*). Therefore, the effects of chronic opiate administration and withdrawal on MCR expression in brain regions believed to be involved in addictive behaviors were determined. As assessed by RNAse protection assay, chronic morphine treatment of rats produced time-dependent decreases in MC4-R transcript levels of 20–40% in striatum, accumbens, and the midbrain periaqueductal gray matter, whereas MC4-R mRNA levels in other brain regions were unaltered. Concomitantly, NDP-MSH binding levels decreased in the ventrolateral striatum, while its apparent binding affinity was unchanged

(36). The decreased striatal binding probably reflected a change in MC4-R protein levels, because the ligand-binding profile of native MCR in the same region was consistent with MC4-R, but not with that of MC3-R (Table 3) *(18,36)*. Moreover, mRNA coding for MC1-R or MC5-R were not detectable in this region, while MC3-R mRNA was barely detectable *(36)*. The morphine-induced decreases in MC4-R transcript levels in neostriatum and periaqueductal gray matter were rapidly reversed by naloxone-precipitated withdrawal *(97)*. By contrast, chronic cocaine administration had an effect opposite to that of morphine, producing a marked increase in striatal MC4-R transcript levels *(98)*. Together, these results suggest that alterations in the central melanocortin system may contribute to the clinical or behavioral manifestations of certain psychoactive drugs.

4.3. Pathophysiologic Regulation

Exogenous melanocortins have neurotrophic actions *(99,100)*, both in an ability to stimulate neurite sprouting and to improve or accelerate functional recovery from CNS lesions. Aside from their obvious relevance to the developmental roles of MCR (above), these have long been a subject of interest with an eye towards novel therapeutic opportunities. To date, only two studies in rats have examined pathophysiologic changes in MCR expression in relevant paradigms. In a model of unilateral hypoxic ischemia, levels of MC4-R transcripts increased selectively in the contralateral striatum *(101)*; whether concomitant changes in MCR protein levels occurred was not determined. In a sciatic nerve crush model, expression of MCR mRNA and MCR protein was unaltered by the nerve crush per se, but subtle changes in MCR protein levels did occur following surgery in both sham-operated and lesioned rats, suggestive of a response of melanocortinergic pathways at the spinal level to surgical trauma or inflammation *(101a)*.

5. Functional Implications

The presence of MCR and MCR-encoding mRNA subtypes in regions of the CNS involved in neuroendocrine and autonomic control, limbic circuitry and sensory processing suggest that endogenous melanocortins may act physiologically at these sites to influence visceral and behavioral homeostatic processes. Within these brain regions, at least two routes of endogenous melanocortin input to central MCR are apparent. First, the generally close association of brain MCR with melanocortinergic and AgRP terminal fields suggests a physiologic relationship of postsynaptic MCR with innervating neurons. Second, in certain circumventricular organs, MCR are positioned to receive hormonal signals from bloodborne pituitary melanocortins, which provides a direct route for functional input from the peripheral (endocrine) melanocortin system into neuroendocrine and autonomic control centers of the CNS. POMC neurons receive melanocortinergic input *(102)*, and a substantial proportion of arcuate POMC cells contain MC3-R mRNA *(12d)*, suggesting a potential role of neuronal melanocortins in the autoregulation of POMC neuron activities.

Two principal factors have previously impeded progress in understanding the physiologic roles of endogenous central melanocortins. These are, first, a lack of knowledge about central MCR, and second, a lack of MCR antagonists. Major progress has been accomplished in identifying, localizing and characterizing pharmacologically the MCR of the nervous system in the past few years. Most recently, the availability of MCR antagonists and genetic models of impaired MCR function have begun to provide significant insights into the roles of endogenous melanocortins and MCR subtypes in the nervous system. For example, a remarkable series of findings followed the discovery that the *agouti* gene product is a MCR antagonist *(103)*. The fact that ubiquitous overexpression of *agouti* produced an obese, hyperphagic, and hypometabolic phenotype in *Ay* mice and related mutants, strongly implicated the MC4-R of the brain as a critical regulator of appetite and energy disposition *(81,103)*. Dramatic support for this hypothesis was provided by the demonstration that genetic ablation of the CNS-associated MC4-R gene in mice produced a phenotype similar to that of *agouti* overexpressors *(80)*. Furthermore, the recent demonstration of an endogenous agoutilike gene product normally expressed in hypothalamus *(104)* raised the possibility that an endogenous MCR antagonist protein may play a physiologic role in the CNS. Indeed, human agouti gene-related protein (AGRP) proved to be a selective antagonist of the human MC3-R and MC4-R in vitro, and its overexpression in vivo in transgenic mice produced obesity *(82)*. Importantly, the obesity-suppressing effects of melanocortins appear not to be simply limited to appetite, but also extend to the central control of metabolism *(104a,104b,104c)*. Furthermore, this is consistent with the presence of MCR protein, MC3-R/MC4-R-encoding transcripts, and extensive POMC and AgRP innvervation in nuclei of the hypothalamus, brainstem and spinal cord which are believed to be involved in regulating these functions (above; and refs. *12c,12d,56a,56b,104d–f*). Together, these findings have stimulated intense interest in the pharmaceutical industry to develop MCR subclass-selective ligands targeted to MCR in the CNS, as novel therapeutic agents for obesity, insulin resistance and diabetes, and at least one central MCR-targeted melanocortin analog (described in refs. *105* and *106*) is presently being tested in human clinical trials in type II diabetes and obesity.

The cytokine-inhibitory actions of melanocortins are of broad interest and potential therapeutic relevance because of the pleiotropic and potent central effects of proinflammatory cytokines, both in coordinating adaptive responses to infectious challenges, and in mediating destructive pathophysiologic processes *(14a,107–109)*. In this context, an antipyretic role of endogenous melanocortins was demonstrated by central MCR blockade using the MC4-R/MC3-R antagonist, SHU-9119 in endotoxin-treated rats *(45)*. The antipyretic effect was mediated by MCR located within the CNS, because the same dose of antagonist that exacerbated fever was without effect when administered intravenously. Further, these findings appeared to rule out a role of MC1-R or MC5-R in mediating the antipyretic effects of α-MSH, at least in the rat. This conclusion is based on the facts that both SHU-9119 and α-MSH are agonists of MC1-R and MC5-R *(110)*, whereas SHU-9119 blocked the antipyretic effect of exogenous α-MSH in endotoxin-treated rats, but had no effect by itself when administered to afebrile rats. Together, these findings indicate that endogenous melanocortins play a *cytokine-counterregulatory* physiologic role that is mediated by central MCR *(45)*. In mice, genetic ablation of the MC4-R dramatically altered both the thermoregulatory response to a low pyrogenic dose of centrally administered interleukin-1, and its modulation by exogenous α-MSH *(110a)*. These findings support a critical role of central MCR, and the MC4-R in particular, in determining the thermoregulatory effects of proinflammatory cytokines. Melanocortins also act centrally to influence behavioral and other aspects of the CNS response to inflammatory stimuli *(14a,15,77,92)*. For example, centrally administered α-MSH potentiated endotoxin-induced anorexia in rats, despite inhibiting fever. Furthermore, central MCR blockade using SHU-9119 reversed endotoxin-induced anorexia, implying that centrally acting endogenous melanocortins contribute to this manifestation of illness behavior *(110b)*. Hence an important goal of future work will be to determine the roles and mechanisms of MCR-mediated inhibition of central cytokine actions.

Within the CNS, the expression of multiple MCR subtypes having differential neuroanatomic distributions and distinct ligand selectivity profiles predicts a complex neuropharmacology of exogenous melanocortins, consistent with long-established observations *(1,78,111)*. Although there are regional variations in the relative proportions of different POMC peptides in the CNS *(8–10)*, there is currently no evidence that POMC neurons can selectively effect the release of certain forms of melanocortins from nerve terminals in response to different types of stimuli. Irrespective of whether differential release of melanocortin peptides occurs, cells may be programmed to respond selectively to certain melanocortins when presented with a complex array of peptides in vivo, through the selective expression of specific MCR subtypes (e.g., MC3-R for responsiveness to γ-MSH) *(26)*. Precise definition of the roles of individual MCR subtypes in the nervous system will continue to rely heavily on both the development of novel MCR-subclass-selective agonists and antagonists, and on targeted genetic ablation studies.

Finally, it is worth emphasizing that several lines of evidence point to the possible existence of as-yet unidentified molecular forms of MCR. For example, several melanocortin analogs having melanocortinlike pharmacologic properties in the nervous system (i.e., actions similar to those of native melanocortins in certain paradigms; e.g., ref. *15,68,115*; *see also* Chapter 4), including the ACTH[4–9] analog Org2766, and α-MSH[11–13], do not appear to bind to native brain MCR, nor do they activate any of the known MCR isoforms *(18,26,67,69,70)*. In addition, the hypertensive and tachycardic effects of intravascular γ-MSH appear to be exerted through a MCR having novel pharmacologic properties, because the effects are not blocked by antagonists of MC3-R, the only MCR subtype for which γ-MSH is a potent agonist, and are not mimicked by α-MSH *(78)*, which is a potent agonist of the other MCR isoforms. It has been postulated that these effects of γ-MSH may be mediated by FMRF amide receptors rather than by MCR *per se* (*see* Chapter 4). Elucidation of potentially novel MCR subtypes responsible for mediating these or other CNS effects of melanocortins will no doubt remain a vigorous area of research.

Acknowledgments

I thank Margaret Entwistle for her expert technical assistance throughout our studies, and Dr. Debbie Beasley for helpful comments on the manuscript. This work was supported by NIH grant no. MH 44694.

References

1. DeWied, D. and Jolles, J. (1982) Neuropeptides derived from pro–opiocortin: behavioral, physiological and neurochemical effects. *Physiol. Rev.* **62,** 976–1059.
2. Dunn, A. J. (1984) Effects of ACTH, β–lipotropin, and related peptides on the central nervous system, in: *Peptides, Hormones, and Behavior* (Nemeroff, C. B. and Dunn, A. J., eds.), Spectrum, New York, pp. 273–347
3. Warberg, J., Oliver, C., Eskay, R. L., Parker, C. R. J., Barnea, A., and Porter, J. C. (1977) Release of α–MSH from a synaptosome–enriched fraction prepaed from rat hypothalamic tissue. *Front. Horm. Res.* **4,** 167–169.
4. O'Donohue, T. L. and Dorsa, D. M. (1982) The opiomelanotropinergic neuronal and endocrine systems. *Peptides* **3,** 353–395.
5. Gee, C. E., Chen, C.–L. C., Roberts, J. L., Thompson, R., and Watson, S. J. (1983) Identification of proopiomelanocortin neurones in rat hypothalamus by *in situ* hybridization. *Nature* **306,** 374–376.
6. Vallarino, M., Tranchand Bunel, D., and Vaudry, H. (1993) Location and identification of α–melanocyte–stimulating hormone in the brain of the lungfish, *Protopterus annectens*. *Ann. N. Y. Acad. Sci.* **680,** 634–637.
7. Tatro, J. B. (1990) Melanotropin receptors in the brain are differentially distributed and recognize both corticotropin and α–melanocyte stimulating hormone. *Brain Res.* **536,** 124–132.

7a. Ollman, M. M., Wilson, B. D., Yang, Y.-K., Kerns, J. A., Chen, Y., Gantz, I., and Barsh, G. S. (1997) Antagonism of central melanocortin receptors in vitro and in vivo by agouti-related protein. *Science* **278,** 135–138.
7b. Broberger, C., Johansen, J., Johansson, C., Schalling, M., and Hökfelt, T. (1998) The neuropeptide Y/agouti related protein (AGRP) brain circuitry in normal, anorectic, and monosodium glutamate-treated mice. *Proc. Natl. Acad. Sci. USA* **95,** 15,043–15,048.
7c. Yang, Y. K., Thompson, D. A., Dickinson, C. J., Wilken, J., Barsh, G. S., Kent, S. B., and Gantz, I. (1999) Characterization of Agouti-related protein binding to melanocortin receptors. *Molecular Endocrinology* **13,** 148–155.
7d. Pilcher, W. H. and Joseph, S. A. (1986) Differential sensitivity of hypothalamic and medullary opiocortin and tyrosine hydroxylase neurons to the neurotoxic effects of monosodium glutamate (MSG). *Peptides* **7,** 783–789.
8. Emeson, R. B. and Eipper, B. A. (1986) Characterization of pro–ACTH/endorphin–derived peptides in rat hypothalamus. *J. Neurosci.* **6,** 837–849.
9. Mezey, E., Kiss, J. Z., Mueller, G. P., Eskay, R., O'Donohue, T. L., and Palkovits, M. (1985) Distribution of the pro–opiomelanocortin derived peptides, adrenocorticotropic hormone, alpha–melanocyte–stimulating hormone and beta–endorphin (ACTH, α–MSH, β–END) in the rat hypothalamus. *Brain Res.* **328,** 341–347.
10. Millington, W. R., Mueller, G. P., and O'Donohue, T. L. (1984) Regional heterogeneity in the ratio of α–MSH:β–endorphin in the rat brain. *Peptides* **5,** 841–843.
11. Nilaver, G., Zimmerman, E. A., Defendini, R., Liotta, A. S., Krieger, D. T., and Brownstein, M. J. (1979) Adrenocorticotropin and β–lipotropin in the hypothalamus. *J. Cell Biol.* **81,** 50–58.
12. Tsou, K., Khachaturian, H., Akil, H., and Watson, S. J. (1986) Immunocytochemical localization of pro–opiomelanocortin–derived peptides in the adult rat spinal cord. *Brain Res.* **378,** 28–35.
12a. Huang, Q.-H., Hruby, V. J., and Tatro, J. B. (1998) Systemic α-MSH suppresses LPS fever via central melanocortin receptors independently of its suppression of corticosterone and IL-6 release. *Am. J. Physiol.* **275,** R524–R530.
12b. Wilson, B. D., Bagnol, D., Kaelin, C. B., Ollmann, M. M., Gantz, I., Watson, S. J., and Barsh, G. S. (1999) Physiological and anatomical circuitry between Agouti-related protein and leptin signaling. *Endocrinology* **140,** 2387–2397.
12c. Haskell-Luevano, C., Chen, P., Li, C., Chang, K., Smith, M. S., Cameron, J. L., and Cone, R. D. (1999) Characterization of the neuroanatomical distribution of agouti-related protein immunoreactivity in the rhesus monkey and the rat. *Endocrinology* **140,** 1408–1415.
12d. Bagnol, D., Lu, X.-Y., Kaelin, C. B., Day, H. W., Ollman, O., Gantz, I., Akil, H., Barsh, G. S., and Watson, S. J. (1999) Anatomy of an endogenous antagonist: Relationship between agouti-related protein and proopiomelanocortin in brain. *J. Neurosci.* **19,** RC26.
13. Cannon, J. G., Tatro, J. B., Reichlin, S. R., and Dinarello, C. A. (1986) α–Melanocyte stimulating hormone inhibits immunostimulatory and inflammatory actions of interleukin–1. *J. Immunol.* **137,** 2232–2236.
14. Lipton, J. M., Glyn, J. R., and Zimmer, J. A. (1981) ACTH and alpha–melanotropin in central temperature control. *Fed. Proc.* **40,** 2760–2764.
14a. Tatro J. B. Endogenous antipyretics. *Clin. Infect. Dis.* (in press).
15. Catania, A. and Lipton, J. M. (1993) α–Melanocyte stimulating hormone in the modulation of host reactions. *Endocr. Rev.* **14,** 564–576.

16. Tatro, J. B. and Entwistle, M. L. (1994) Distribution of melanocortin receptors in the lower brainstem of the rat. *Ann. N.Y. Acad. Sci.* **739,** 311–314.
17. Tatro, J. B. (1993) Melanotropin receptors of the brain. *Methods Neurosci.* **11,** 87–104.
18. Tatro, J. B. and Entwistle, M. L. (1994) Heterogeneity of brain melanocortin receptors suggested by differential ligand binding *in situ. Brain Res.* **635,** 148–158.
19. Sawyer, T. K., Sanfilippo, P. J., Hruby, V. J., Engel, M. H., Heward, C. B., Burnett, J. B., and Hadley, M. E. (1980) [Nle^4,D–Phe^7]α–melanocyte stimulating hormone: A highly potent α–melanotropin with ultralong biological activity. *Proc. Natl. Acad. Sci. U. S. A.* **77,** 5754–5758.
20. Tatro, J. B. and Reichlin, S. (1987) Specific receptors for a–melanocyte–stimulating hormone are widely distributed in tissues of rodents. *Endocrinology***121,** 1900–1907.
21. Tatro, J. B., Atkins, M., Mier, J. W., Hardarson, S., Wolfe, H., Smith, T., Entwistle, M. L., and Reichlin, S. (1990) Melanotropin receptors demonstrated in situ in human melanoma. *J. Clin. Invest.* **85,** 1825–1832.
22. Tatro, J. B., Entwistle, M. L., Lester, B. R., and Reichlin, S. (1990) Melanotropin receptors of murine melanoma characterized in cultured cells and demonstrated in experimental tumors *in situ. Cancer Res.* **50,** 1237–1242.
23. Entwistle, M. L., Hann, L. E., Sullivan, D. A., and Tatro, J. B. (1990) Characterization of functional melanotropin receptors in lacrimal glands of the rat. *Peptides* **11,** 477–483.
24. Chhajlani, V. and Wikberg, J. E. S. (1992) Molecular cloning and expression of the human melanocyte stimulating hormone receptor cDNA. *FEBS Lett.* **309,** 417–420.
25. Chhajlani, V., Muceniece, R., and Wikberg, J. E. S. (1993) Molecular cloning of a novel human melanocortin receptor. *Biochem. Biophys. Res. Commun.* **195,** 866–873.
26. Roselli–Rehfuss, L., Mountjoy, K. G., Robbins, L. S., Mortrud, M. T., Low, M. J., Tatro, J. B., Entwistle, M. L., Simerly, R. B., and Cone, R. D. (1993) Identification of a receptor for POMC peptides in the medial basal hypothalamus and limbic system. *Proc. Natl. Acad. Sci. U. S. A.* **90,** 8856–8860.
27. Desarnaud, F., Labbé, O., Eggerickx, D., Vassar, G., and Parmentier, M. (1994) Molecular cloning, functional expression and pharmacological characterization of a mouse melanocortin receptor gene. *Biochem. J.* **299,** 367–373.
28. Labbé, O., Desarnaud, F., Eggerickx, D., Vassart, G., and Parmentier, M. (1994) Molecular cloning of mouse melanocortin 5 receptor gene widely expressed in peripheral tissues. *Biochemistry* **33,** 4543–4549.
29. Eberle, A. N. (1993) Peptides containing multiple photolabels: a new tool for the analysis of ligand–receptor interactions. *J. Recept. Res.* **13,** 27–37.
30. Bagutti, C. and Eberle, A. N. (1993) Synthesis and biological properties of a biotinylated derivative of ACTH1–17 for MSH receptor studies. *J. Recept. Res.* **13,** 27–37.
31. Hadley, M. E., Hruby, V. J., Jiang, J., Sharma, S. D., Fink, J. L., Haskell–Leuvano, C., Bentley, D. L., Al–Obeidi, F., and Sawyer T. K. (1996) Melanocortin receptors: identification and characterization by melanotropic peptide agonists and antagonists. *Pigment Cell Res.* **9,** 213–234.
32. Erskine–Grout, M.–E., Olivier, G. W. J., Lucas, P., Sahm, U. G., Branch, S. K., Moss, S. H., Notarianni, L. J., and Pouton, C. W. (1996) Melanocortin probes for

the melanoma MC1 receptor: synthesis, receptor binding and biological activity. *Melanoma Res.* **6,** 89–94.
33. Van Houten, M., Khan, M. N., Walsh, R. J., Baquiran, G. B., Renaud, L. P., Bourque, C., Sgro, S., Gauthier, S., Chretien, M., and Posner, B. I. (1985) NH_2–terminal specificity and axonal localization of adrenocorticotropin binding sites in rat median emininc*e*. *Proc. Natl. Acad. Sci. U. S. A.* **82,** 1271–1275.
34. Tatro, J. B. (1993) Brain receptors for central and peripheral melanotropins. *Ann. N. Y. Acad. Sci.* **680,** 621–625.
35. Herkenham, M. and Pert, C. B. (1982) Light microscopic localization of brain opiate receptors: a general autoradiographic method which preserves tissue quality. *J. Neurosci.* **28,** 1129–1149.
36. Alvaro, J. D., Tatro, J. B., Quillan, J. M., Fogliano, M., Eisenhard, M., Lerner, M. R., Nestler, E. J., and Duman, R. S. (1996) Morphine down–regulates melanocortin–4 receptor expression in brain regions that mediate opiate addiction. *Mol. Pharmacol.* **50,** 583–591.
37. Lichtensteiger, W., Hanimann, B., Siegrist, W., and Eberle, A. N. (1996) Region– and stage–specific patterns of melanocortin receptor ontogeny in rat central nervous system, cranial nerve ganglia and sympathetic ganglia. *Dev. Brain Res.* **91,** 93–110.
38. Pranzatelli M. R. (1994) On the molecular mechanism of adrenocorticotrophic hormone in the CNS: neurotransmitters and receptors. *Exper. Neurol.* **125,** 142–161.
39. Li, Z., Queen, G., and LaBella, F. S. (1990) Adrenocorticotropin, vasoactive intestinal polypeptide, growth hormone–releasing factor, and dynorphin compete for common receptors in brain and adrenal. *Endocrinology* **126,** 1327–1333.
40. Mousli, M., Bueb, J.–L., Bronner, C., Rouot, B., and Landry, Y. (1990) G protein activation: a receptor–independent mode of action for cationic amphiphilic neuropeptides and venom peptides. *Trends Pharm. Sci.* **11,** 358–362.
41. Solca, F., Siegrist, W., Drodz, R., Girard, J., and Eberle, A. N. (1989) The receptor for α–melanotropin of mouse and human melanoma cells: Application of a potent photoaffinity label. *J. Biol. Chem.* **264,** 14,277–14,281.
42. Dyer, J. K., Ahmed, A. R. H., Oliver, G. W. J., Poulton, C. W., and Haynes, L. W. (1993) Solubilization partial characterization of the α–MSH receptor on primary rat Schwann cells. *FEBS Lett* **336,** 103–106.
43. Solca, F. F., Chluba–de Tapia, J., Iwata, K., and Eberle, A. N. (1993) B16–G4F mouse melanoma cells: an MSH receptor–deficient cell clone. *FEBS Lett.* **322,** 177–180.
44. Wong, G. and Pawelek, J. (1973) Control of phenotypic expression of cultured melanoma cells by melanocyte stimulating hormones. *Nature New Biol.* **241,** 213–216.
45. Huang, Q.–H., Entwistle, M. L., Alvaro, J. D., Duman, R. S., Hruby, V. J., and Tatro, J. B. (1997) Antipyretic role of endogenous melanocortins mediated by central melanocortin receptors during endotoxin–induced fever. *J. Neurosci.* **17,** 3343–3351.
46. Adan, R. A. H., van der Kraan, M., Doorngos, R. P., Bar, P. R., Burbach, J. P. H., and Gispen, W. H. (1996) Melanocortin receptors mediate α–MSH–induced stimulation of neurite outgrowth in Neuro 2A cells. *Mol. Brain Res.* **36,** 37–44.
47. Cholewinski, A. J. and Wilkin G. P. (1988) Astrocytes from forebrain, cerebellum, and spinal cord differ in their responses to vasoactive intestinal peptide. *J. Neurochem.* **51,** 1626–1633.

48. Van Calker, D., Loffler, F., and Hamprecht, B. (1983) Corticotropin peptides and melanotropins elevate the level of adenosine 3':5'–cyclic monophosphate in cultured murine brain cells. *J Neurochem.* **40,** 418–427.
49. Zohar, M. and Salomon, Y. (1992) Melanocortins stimulate proliferation and induce morphological changes in cultured rat astrocytes by distinct transducing mechanisms. *Brain Res.* **576,** 49–58.
50. Riedy, M. C., Timm, E. A. J., and Stewart, C. C. (1995) Quantitative RT–PCR for measuring gene expression. *BioTechniques* **18,** 70–76.
51. Emson, P. C. (1993) In–situ hybridization as a methodological tool for the neuroscientist. *Trends Neurosci.* **16,** 9–16.
52. Griffon, N., Mignon, V., Facchinetti, P., Diaz, J., Schwartz, J., and Sokoloff, P. (1994) Molecular cloning and characterization of the rat fifth melancortin receptor. *Biochem. Biophys. Res. Commun.* **200,** 1007–1014.
53. Vilaró, M. T., Palacios, J. M., and Mengod, G. (1995) Neurotransmitter receptor histochemistry: the contribution of in situ hybridization. *Life Sci.* **57,** 1141–1154.
54. Désy, L. and Pelletier, G. (1978) Immunohistochemical localization of alpha–melanocyte stimulating hormone (α–MSH) in the human hypothalamus. *Brain Res.* **154,** 377–381.
55. Pilcher, W. H., Joseph, S. A., and McDonald, J. V. (1988) Immunocytochemical localization of pro–opiomelanocortin neurons in human brain areas subserving stimulation analgesia. *J. Neurosurg.* **68,** 621–629.
56. Ibuki, T., Okamura, H., Miyazaki, M., Yanaihara, N., Zimmerman, E. A., and Ibata, Y. (1989) Comparative distribution of three opioid systems in the lower brainstem of the monkey (*Macaca fuscata*). *J. Comp. Neurol.* **279,** 445–456.
56a. Elias, C. F., Saper, C. B., Maratos-Flier, E., Tritos, N. A., Lee, C., Kelly, J., Tatro, J. B., Hoffman, G. E., Ollmann, M. M., Barsh, G. S., Sakurai, T., Yanagisawa, M., and Elmquist, J. K. (1998) Chemically defined projections linking the mediobasal hypothalamus and the lateral hypothalamic area. *J. Comparative Neurol.* **402,** 442–459.
56b. Mihály, E., Fekete, C., Liposits, Z., Stopa, E. G., and Lechan, R. M. (1999) Hypophysiotrophic thyrotropin-releasing hormone-synthesizing neurons of the human hypothalamus are innervated by axons containing neuropeptide Y and agouti-related protein. *Soc. Neurosci. Abstr.* **25,** 1690.
57. Gantz, I., Kondaf, Y., Tashiro, T., Shimoto, Y., Miwa, H., Munzert, G., Watson, S. J., DelValle, J., and Yamada, T. (1993) Molecular cloning of a novel melanocortin receptor. *J. Biol. Chem.* **268,** 8248–8250.
58. Gantz, I., Miwa, H., Konda, Y., Shimoto, Y., Tashiro, T., Watson, S. J., DelValle, J., and Yamada. T. (1993) Molecular cloning, expression, and gene localization of a fourth melanocortin receptor. *J. Biol. Chem.* **268,** 15,174–15,179.
59. O'Donohue, T. L., Miller, R. L., and Jacobowitz, D. M. (1979) Identification, characterization and stereotaxic mapping of intraneuronal α–melanocyte stimulating hormone–like immunoreactive peptides in discrete regions of the rat brain. *Brain Res.* **176,** 101–123.
60. Palkovits, M., Mezey, E., and Eskay, R. L. (1987) Pro–opiomelanocortin–derived peptides (ACTH/β–endorphin/alpha–MSH) in brainstem baroreceptor areas of the rat. *Brain Res.* **436,** 323–328.
61. Bronstein, D. M., Schafer, M. K. H., Watson, S. J., and Akil, H. (1992) Evidence that β–endorphin is synthesized in cells in the nucleus tractus solitarius: detection of POMC mRNA. *Brain Res.* **587,** 269–275.

62. Joseph, S. A. (1980) Immunoreactive adrenocorticotropin in rat brain: a neuroanatomical study using antiserum generated against synthetic $ACTH_{1-39}$. *Am. J. Anat.* **158,** 533–548.
63. Romagnano, M. A. and Joseph, S. A. (1983) Immunocytohemical localization of ACTH 1–39 in the brainstem of the rat. *Brain Res.* **276,** 1–16.
64. Yamazoe, M., Shiosaka, S., Yagura, A., Kawai, Y., Shibasaki, T., Ling, N., and Tohyama, M. (1984) The distribution of α–melanocyte stimulating hormone (α–MSH) in the central nervous system of the rat: An immunohistochemical study. II. Lower brain stem. *Peptides* **5,** 721–727.
65. Tatro, J. B. and Entwistle, M. L. (1997) Distribution of melanocortin receptors in the rat spinal cord. *Soc. Neurosci. Abstr.* **23,** 1492.
66. Klemcke, H. G. and Pond, W. G. (1990) Porcine adrenal adrenocorticotropic hormone receptors: characterization, changes during neonatal development, and response to a stressor. *Endocrinology* **128,** 2476–2488.
67. Mountjoy, K. G., Mortrud, M. T., Low, M. J., Simerly, R. B., and Cone, R. D. (1994) Localization of the melanocortin–4 receptor (MC4-R) in neuroendocrine and autonomic control circuits in the brain. *Mol. Endocrinol.* **8,** 1298–1308.
68. Lichtensteiger, W. and Monnet, F. (1979) Differential response of dopamine neurons to α–melanotropin and analogues in relation to their behavioral potency. *Life Sci.* **25,** 2079–2087.
69. Schiöth, H. B., Muceniece, R., Wikberg, J. E. S., and Chhajlani, V. (1995) Characterisation of melanocortin receptor subtypes by radioligand binding analysis. *Eur. J. Pharmacol.* **288,** 311–317.
70. Schiöth, H. B., Muceniece, R., and Wikberg, J. E. S. (1996) Characterisation of the melanocortin 4 receptor by radioligand binding. *Pharmacol. Toxicol.* **79,** 161–165.
71. Mountjoy, K. G. and Wild, J. M. (1998) Melanocortin-4 receptor mRNA expression in the developing autonomic and central nervous systems. *Brain Res. Dev. Brain Res.* **107,** 309–314.
72. Chen, W., Kelly, M. A., Opitz–Araya, X., Thomas, R. E., Low, M. J., and Cone, R. D. (1997) Exocrine gland dysfunction in MC5–R–deficient mice: evidence for coordinated regulation of exocrine gland function by melanocortin peptides. *Cell* **91,** 789–798.
73. Van der Kraan, M., Adan, R. A. H., Entwistle, M. E., Gispen, W. H., Burbach, J. P. H., and Tatro, J. B. (1998) Expression of melanocortin 5 receptor in secretory epithelia supports a functional role in exocrine and endocrine glands. *Endocrinology* **139,** 2348–2355.
74. Fathi, Z., Iben, L. G., and Parker, E. M. (1995) Cloning, expression, and tissue distribution of a fifth melanocortin receptor subtype. *Neurochem. Res.* **20,** 107–113.
75. Cone, R. D., Mountjoy, K. G., Robbins, L. S., Nadeau, J. H., Johnson, K. R., Roselli–Rehfuss, L., and Mortrud, M. T. (1993) Cloning and functional characterization of a family of receptors for the melanotropic peptides. *Ann. N. Y. Acad. Sci.* **680,** 342–363.
76. Xia, Y., Wikberg, J. E. S., and Chhajlani, V. (1995) Expression of melanocortin 1 receptor in periaqueductal gray matter. *NeuroReport* **6,** 2193–2196.
77. Rajora, N., Boccoli, G., Burns, D., Sharma, S., Catania, A. P., and Lipton, J. M. (1997) α–MSH modulates local and circulating tumor necrosis factor–α in experimental brain inflammation. *J. Neurosci.* **17,** 2181–2186.

78. Li, S.–J., Varga, K., Archer, P., Hruby, V. J., Sharma, S. D., Kesterson, R. A., Cone, R. D., and Kunos, G. (1996) Melanocortin antagonists define two distinct pathways of cardiovascular control by α– and γ–melanocyte–stimulating hormones. *J. Neurosci.* **16,** 5182–5188.
79. Seeley, R. J., Yagaloff, K. A., Fisher, S. L., Burn, P., Thiele, T. E., van Dijk, G., Baskin, D. G., and Schwartz, M. W. (1997) Melanocortin receptors in leptin effects. *Nature* **390,** 349.
80. Huszar, D., Lynch, C. A., Fairchild–Huntress, V., Dunmore, J. H., Fang, Q., Berkemeier, L. R., Gu, W., Kesterson, R. A., Boston, B. A., Cone, R. D., Smith, F. J., Campfield, L. A., Burn, P., and Lee, F. (1997) Targeted disruption of the melanocortin–4 receptor results in obesity in mice. *Cell* **88,** 131–141.
81. Yen, T. T., Gill, A. M., Frigeri, L. G., Barsh, G. S., and Wolff, G. L. (1994) Obesity, diabetes, and neoplasia in yellow $A^{vy/-}$ mice: ectopic expression of the *agouti* gene. *FASEB J.* **8,** 479–488.
82. Ollman, M. M., Wilson, B. D., Yang, Y.–K., Kerns, J. A., Chen, Y., Gantz, I., and Barsh, G. S. (1998) Antagonism of central melanocortin receptors in vitro and in vivo by agouti–related protein. *Science* **278,** 135–138.
83. Shiosaka, S., Shibasaki, T., and Tohyama, M. (1984) Bilateral α–melanocyte stimulating hormonergic fiber system from zona incerta to cerebral cortex: combined retrograde axonal transport and immunohistochemical study. *Brain Res* **309,** 350–353.
84. Kohler, C. and Swanson, L. W. (1984) Acetylcholinesterase–containing cells in the lateral hypothalamic area are immunoreactive for α–melanocyte stimulating hormone and have cortical projections in te rat. *Neurosci. Lett* **49,** 39–43.
85. Watson, S. J. and Akil, H. (1980) α–MSH in rat brain: occurrence within and outside of β–endorphin neurons. *Brain Res.* **182,** 217–223.
86. Khachaturian, H., Akil, H., Brownstein, M. J., Olney, J. W., Voigt, K. H., and Watson, S. J. (1986) Further characterization of the extra–arcuate alpha–melanocyte stimulating hormone–like material in hypothalamus: biochemical and anatomical studies. *Neuropeptides* **7,** 291–313.
87. Nahon, J. L., Presse, F., Bittencourt, J. C., Sawchenko, P. E., and Vale W. (1990) The rat melanin–concentrating hormone messenger ribonucleic acid encodes multiple putative neuropeptides coexpressed in the dorsolateral hypothalamus. *Endocrinology* **125,** 2056–2065.
88. Oldfield, B. J. and McKinley, M. J. (1995) Circumventricular organs. in *The Rat Nervous System*, 2nd ed. (Paxinos, G., ed.), Academic Press, New York, pp. 391–403.
90. Wilson, J. F. (1988) Low permeability of the blood–brain barrier to nanomolar concentrations of immunoreactive alpha–melanotropin. *Psychopharmacology (Berl)* **96,** 262–266.
91. Banks, W. A. and Kastin, A. J. (1995) Permeability of the blood–brain barrier to melanocortins. *Peptides* **16,** 1157–1161.
92. Tatro, J. B. (1996) Receptor biology of the melanocortins, a family of neuroimmunomodulatory peptides. *Neuroimmunomodulation* **3,** 259–284.
93. Jolles, J., Wiegant, V. M., and Gispen, W. H. (1978) Inhibition of behavioral effect of ACTH (1–24) and opioids by repeated administration. *Neurosci. Lett* **9,** 261–266.
94. Xia, Y., and Wikberg, J. E. S. (1997) Postnatal expression of melanocortin–3 receptor in rat diencephalon and mesencephalon. *Neuropharmacology* **36,** 217–224.

95. Khatchaturian, H., Lewis, M. E., Alessi, N. E., and Watson, S. J. (1985) Time of origin of opioid–containing neurons in the rat hypothalamus. *J. Comp. Neurol.* **236,** 538–546.
96. Kawai, Y., Shibasaki, T., Ling, N., and Tohyama, M. (1986) Ontogeny of γ–melanocyte stimulating hormone in the brain and hypohysis of the rat: an immunohistochemical analysis. *Dev. Brain Res.* **28,** 177–193.
97. Alvaro, J. D., Tatro, J. B., and Duman, R. S. (1997) Melanocortins and opiate addiction. *Life Sci.* **61,** 1–9.
98. Alvaro, J. D., Entwistle, M. L., Tatro, J. B., and Duman, R. S. (1996) Chronic cocaine administration increases melanocortin 4–receptor mRNA expression in rat striatum. *Soc. Neurosci. Abstr.* **22,** 80.
99. Strand, F. L., Zuccarelli, L. A., Williams, K. A., Lee, S. J., Lee, T. S., Antonawich, F. J., and Alves, S. E. (1993) Melanotropins as growth factors. *Ann. N. Y. Acad. Sci.* **680,** 29–50.
100. Hol, E. M., Gispen, W. H., and Bär, P. R. (1995) ACTH–related peptides: receptors and signal transduction systems involved in their neurotrophic and neuroprotective actions. *Peptides* **16,** 979–993.
101. Mountjoy, K. G., Guan, J., Elia, C. J., Sirimanne, E. S., and Williams, C. E. (1999) Melanocortin-4 receptor messenger RNA expression is up-regulated in the non-damaged striatum following unilateral hypoxic-ischaemic brain injury. *Neuroscience* **89,** 183–190.
101a. Van der Kraan, M., Tatro, J. B., Entwistle, M. L., Brakkee, J. H., Burbach, J. P. H., Adan, R. A. H., and Gispen, W. H. (1999) Expression of melanocortin 4 receptors and pro-opiomelanocortin in the rat spinal cord in relation to neurotrophic effects of melanocortins. *Mol. Brain Res.* **63,** 276–286.
102. Kiss, J. Z. and Williams, T. H. (1983) ACTH–immunoreactive boutons form synaptic contacts in the hypothalamic arcuate nucleus of rat: evidence for local opiocortin connections. *Brain Res.* **263,** 142–146.
103. Lu, D., Willard, D., Patel, I. R., Kadwell, S., Overton, L., Kost, T., Luther, M., Chen, W., Woychik, R. P., Wilkison, W. O., and Cone, R. D. (1994) Agouti protein is an antagonist of the melanocyte–stimulating–hormone receptor. *Nature* **371,** 799.
104. Shutter, J. R., Graham, M., Kinsey, A. C., Scully, S., Luthy, R., and Stark, K. L. (1997) Hypothalamic expression of ART, a novel gene related to agouti, is up–regulated in obese and diabetic mutant mice. *Genes Dev.* **11,** 593–602.
104a. Cowley, M. A., Pronchuk, N., Fan, W., Dinulescu, D. M., Colmers, W. F., and Cone, R. D. (1999) Integration of NPY, AgRP, and melanocortin signals in the hypothalmic paraventricular nucleus: Evidence of a cellular basis for the adipostat. *Neuron* **24,** 155–163.
104b. Satoh, N., Ogawa, Y., Katsuura, G., Numata, Y., Masuzaki, H., Yoshimasa, Y., and Nakao, K. (1998) Satiety effect and sympathetic activation of leptin are mediated by hypothalamic melanocortin system. *Neurosci. Lett.* **249,** 107–110.
104c. Patterson, T. A., Hedde, J. R., She, L., Newsome, W. P., and Cornelius, P. (1999) Chronic infusion of a melanocortin agonist causes weight loss due to decreased food intake and increased metabolic rate. *Soc. Neurosci. Abstr.* **25,** 618.
104d. Elmquist, J. K., Elias, C. F., and Saper, C. B. (1999) From lesions to leptin: Hypothalamic control of food intake and body weight. *Neuron* **22,** 221–232.

104e. Elias, C. F., Lee, C., Kelly, J., Aschkenasi, C., Ahima, R. S., Couceyro, P. R., Kuhar, M. J., Saper, C. B., and Elmquist, J. K. (1998) Leptin activates hypothalamic CART neurons projecting to the spinal cord. *Neuron* **21,** 1375–1385.

104f. Legradi, G. and Lechan, R. M. (1999) Agouti-related protein containing nerve terminals innervate thyrotropin-releasing hormone neurons in the hypothalamic paraventricular nucleus. *Endocrinol.* **140,** 3643–3652.

105. Abou–Mohamed, G., Papapetropoulos, A., Ulrich, D., Catravas, J. D., Tuttle, R. R., and Caldwell, R. W. (1995) HP–228, a novel synthetic peptide, inhibits the induction of nitric oxide synthase in vivo but not in vitro. *J. Pharmacol. Exp. Ther.* **275,** 584–591.

106. Corcos, I., Thompson, E. E., Omholt, P., Lee, M. D., McPherson, S., McDowell, R., Houghten, R., and Girten, B. (1997) HP–228 is a potent agonist of melanocortin receptor 4, and significantly attenuates obesity and diabetes in Zucker fatty rats. *Soc. Neurosci. Abstr.* **23,** 673.

107. Hopkins, S. J. and Rothwell, N. J. (1995) Cytokines and the nervous system I: expression and recognition. *Trends Neurosci.* **18,** 83–88.

108. Rothwell, N. J. and Hopkins, S. J. (1995) Cytokines and the nervous system II: actions and mechanisms of action. *Trends Neurosci.* **18,** 130–136.

109. Patterson, P. H. (1995) Cytokines in Alzheimer's disease and multiple sclerosis. *Curr. Opin. Neurobiol.* **5,** 642–646.

110. Hruby, V. J., Lu, D., Sharma, S. D., Castrucci, A. L., Kesterson, R. A., Al–Obeidi, F. A., Hadley, M. E., and Cone, R. D. (1995) Cyclic lactam α–melanotropin analogues of Ac–Nle^4–cyclo[Asp^5, D–Phe^7, Lys^{10}]– α–melanocyte–stimulating hormone(4–10)–NH_2 with bulky aromatic amino acids at position 7 show high antagonist potency and selectivity at specific melanocortin receptors. *J. Med. Chem.* **38,** 3454–3461.

110a. Tatro, J. B., Huszar, D., Fairchild-Huntress, V., and Huang, Q.-H. (1999) Role of the melanocortin-4 receptor in thermoregulatory responses to central IL1β. *Soc. Neurosci. Abstr.* **25,** 1558.

110b. Huang, Q. H., Hruby, V. J., and Tatro, J. B. (1999) Role of central melanocortins in endotoxin-induced anorexia. *Am. J. Physiol.* **276,** R864–71.

111. Kobobun, K., O'Donohue, T. L., Handelman, G. E., Sawyer, T. K., Hruby, V. J., and Hadley, M. E. (1983) Behavioral effects of [4–norleucine, 7–D–phenylalanine]–α–melanocyte–stimulating hormone. *Peptides* **4,** 721–724.

112. Kawai, Y., Inagaki, S., Shiosaka, S., Shibasaki, T., Ling, N., Tohyama, M., and Shiotani, Y. (1984) The distribution and projection of γ–melanocyte stimulating hormone in the rat brain: an immunohistochemical analysis. *Brain Res.* **297,** 21–32.

113. Swanson, L. W. (1992) *Brain maps: Structure of the Rat Brain.* Elsevier, New York.

114. Elmquist, J. K., Scammell, T. E., Jacobson, C. D., and Saper, C. B. (1996) Distribution of fos–like immunoreactivity in the rat brain following intravenous lipopolysaccharide administration. *J. Comp. Neurol.* **371,** 85–103.

115. Uehara, Y., Shimizu, H., Sato, N., Tanaka, Y., Shimomura, Y., and Mori, M. (1992) Carboxy-terminal tripeptide of alpha-melanocyte-stimulating hormone antagonizes interleukin-1 induced anorexia. *Eur. J. Pharmacol.* **220,** 119–122.

CHAPTER 7

Cloning of the Melanocortin Receptors

Kathleen G. Mountjoy

1. Introduction

1.1. Biologic Actions of Proopiomelanocortin-Derived Peptides

Melanocortin peptides (adrenocorticotropin [ACTH], α-, β-, and γ-melanocyte stimulating hormone [MSH], and fragments thereof) derived from proopiomelanocortin (POMC) have a diverse array of biologic activities, many of which have yet to be fully elucidated. POMC, produced most abundantly in the pituitary, is also produced in the brain, in the neurons of the arcuate nucleus of the hypothalamus, and the commissural nucleus of the solitary tract of the brainstem; it has also been detected in several peripheral tissues including skin, pancreas, and testis. POMC is differentially processed in the different pituitary lobes, and the processing in the brain differs from that in the pituitary. In the corticotropic cells of the anterior lobe of the pituitary, the major end product is the 39 amino acid, ACTH[1–39]. In the melanotrophs of the intermediate lobe of the pituitary, ACTH[1–39] is the precursor of α-MSH (ACTH[1–13]) and corticotropinlike intermediate lobe peptide (CLIP) (ACTH[18–39]). The major fraction of α-MSH produced by pituitary melanotrophs is acetylated at the amino terminus, while most of brain-derived α-MSH is desacetylated. α, β, γ1, γ2, and γ3-MSH peptides are processed from different regions of the POMC precursor to yield peptides sharing a conserved core of seven amino acid residues. Adult humans lack an intermediate lobe of the pituitary and thus have very little α-MSH in the serum. ACTH[1–39] is the predominant circulating melanocortin peptide in man while α-MSH is the predominant circulating melanocortin in most other species. γ-MSH peptides have been reported to be present in human

The Melanocortin Receptors Ed.: R. D. Cone

skin, are detectable in human adult blood and γ3-MSH is increased in the circulation in patients with cardiac arrest, in sheep blood in response to acute hemorrhagic stress, and is also increased toward the end of gestation. The primary roles of MSH and ACTH are the regulation of pigmentation and adrenal corticosteroid synthesis, respectively. While α-and β-MSH have melanotropic activity, γ-MSH peptides have little, if any, activity when tested in mouse and hamster melanoma cells. ACTH also stimulates proliferation of the adrenal cortex and is crucial for the normal development of this tissue. Numerous other activities for the melanocortin peptides have been demonstrated in the central and peripheral nervous systems, in the immune system, on lipolysis, on pituitary function, parturition, and neuromuscular function. Since the 1950s, a number of biologic responses have been seen on intracerebroventricular introduction of these peptides *(1)*. For example, central administration of melanocortin peptides has been reported to have effects on autonomic controls such as thermoregulation, food intake, cardiovascular function, behavior, and neuroendocrine homeostasis. Retention of learned behaviors, and recovery from nerve damage has also been reported. In addition to their effects on brain, melanocortin peptides exert a neurotrophic action on damaged peripheral nerve tissue *(2)*. ACTH and α-MSH also have antipyretic activity following peripheral or intracerebroventricular administration *(3)*.

1.2. G Protein-Coupled Intracellular Signaling Pathways Associated With Melanocortin Peptides

The receptors for ACTH and α-MSH have long been recognized to be members of the G protein coupled receptor (GPCR) superfamily. ACTH activates adenylyl cyclase and increases cAMP levels in adrenal cells *(4–6)*, while α-MSH activates adenylyl cyclase and increases cAMP levels in melanocytes and melanoma cells *(7)*.

1.3. Melanocortin Peptide Binding to Melanocytes, Adrenal Cells, and Brain Tissue

Melanocortin peptides act by binding to G protein-coupled specific receptors on the cell membrane. Binding sites were initially characterized using radiolabeled α-and β-MSH in cultured mouse and human melanoma and melanocyte cell lines. Scatchard analysis on binding to mouse melanoma cells revealed a single class of binding with a K_d = 1–2 n*M*, while binding of α-MSH to cultured human melanoma cells revealed a higher affinity receptor (K_d = 0.2 n*M*) and lower number of receptors compared with mouse cells. Characterization of MSH binding advanced rapidly following the development of a radiolabeled synthetic superpotent and enzymatically resistant α-MSH analog, Nle^4, D-Phe^7-α-MSH (NDP-α-MSH). Specific receptors were subsequently demonstrated to be present on surgical specimens of human melanoma *(8)* and in lacrimal, Harderian, preputial, submandibular, adrenal glands, pancreas, adipose tissue, bladder, duodenum, skin, spleen, and hypothalamus of mice and rats *(9)*. The relative potency of melanocortin peptides in human melanoma and rodent glandular tissues differed from mouse melanoma, however, and this indicated the existence of multiple melanocortin receptors *(8,9)*. Radioligand binding experiments for ACTH were only made possible following the development of the ACTH derivative, {Phe^2, Nle^4}ACTH[1–38], which retains biologic activity upon iodination of tyrosine at the 23 position. Subsequently it was claimed that the activity of ACTH on the adrenal cortex resulted from the interaction of the hormone with a single type of high-affinity ACTH receptor, present at approximately 3000 copies per cell. It was also claimed by others that there were two receptors for ACTH in the adrenal following the identification of high and low affinity binding sites *(5)*. In the rat brain, radiolabeled ACTH binding was demonstrated in the cortex, hypothalamus, hippocampus, striatum, cerebellum, midbrain, and brainstem *(10)*, and radiolabeled NDP-α-MSH binding was observed in the septum, hypothalamus, thalamus, striatum, and midbrain as well as in the lower brainstem. α-MSH and ACTH were equipotent in competing for [^{125}I]NDP-α-MSH binding in the brain, which suggested that brain melanocortin receptors were distinct from the receptor on mouse melanoma cells *(11)*.

1.4. Search for Melanocortin Receptors

The melanocortin receptors proved to be elusive molecules over many years due to difficulties with receptor solubilization, identifying high-affinity binding sites, and the low abundance of the receptors in different tissues. The melanocortin receptor family were eventually identified in 1992 after applying the molecular biology technique of degenerate oligonucleotide polymerase chain reaction (PCR) to amplify fragments of the melanocortin receptors. This technique utilized the high homology in transmembrane domains three and six of members of the GPCR family cloned before 1990 *(12)*. DNA fragments of the first two members of the melanocortin receptor family (MSH and ACTH receptors) were obtained using PCR and cDNA from a human melanoma tissue shown to have a high number of MSH binding sites *(13)*. The cloning of the MSH and ACTH receptors led to the discovery of three related, but distinct, melanocortin receptors.

1.5. Nomenclature for the Melanocortin Receptors

A family of five melanocortin receptor subtypes has been identified to date and a simple nomenclature devised to distinguish between them *(13,14)*. The MSH receptor on melanocytes was the first to be cloned and is now known as melanocortin 1 receptor, (MC1-R). The second to be identified was the ACTH-R in the adrenal cortex, now known as MC2-R. The other subtypes are similarly identified by the numerical order in which they were discovered; MC3-R, MC4-R, and MC5-R.

2. Cloning of the Melanocyte Stimulating Hormone Receptor

2.1. Degenerate Oligonucleotide PCR on Melanoma Tissue

MSH was known to bind to a receptor that couples to heterotrimeric guanine nucleotide-binding proteins (G proteins) and activates adenylyl cyclase. The receptor was therefore expected to have sequence similarity with other members of the rhodopsinlike GPCR family *(12)*. Degenerate oligonucleotides designed to the highly conserved transmembrane three and six domains of GPCRs were used for the PCR where the template was either cDNA derived from a human melanoma specimen known to express a high number of MSH binding sites *(13)* or human genomic DNA *(15)*. Two PCR subclones resulting from amplification of human melanoma cDNA were determined by DNA sequencing to encode novel GPCRs highly related to one another, and by Northern hybridization analysis to be expressed specifically in melanocytes and adrenal cortex, respectively *(13)*. A PCR subclone resulting from amplification of human genomic DNA was found by Northern hybridization to be specifically expressed in a human melanoma cell line *(15)*.

2.2. Library Screening of Genomic DNA and cDNA

The PCR fragment which was expressed in melanocytes and melanoma was used to screen a human genomic library at high stringency and a mouse cDNA library made from the Cloudman S91 melanoma cell line at moderate stringency *(13)*. Eight independent clones were isolated from the mouse cDNA library and these were determined by restriction mapping and partial sequencing to be derived from the same gene. A single clone of 7.5 kb was isolated from the human genomic library. A human melanoma cDNA library was screened with the PCR fragment obtained from amplification of human genomic DNA *(15)*. One clone was isolated with an insert of 1.8 kb, containing a long open reading frame and the entire coding sequence.

2.3. Peptide Structure and Homology With Other G Protein-Coupled Receptors

The human MC1-R (hMC1-R) amino acid sequence (317 amino acids) (Fig. 1) was 76% identical and colinear with the mouse MC1-R (mMC1-R) cDNA amino acid (315 amino acids) sequence *(13)*. Three hMC1-R clones

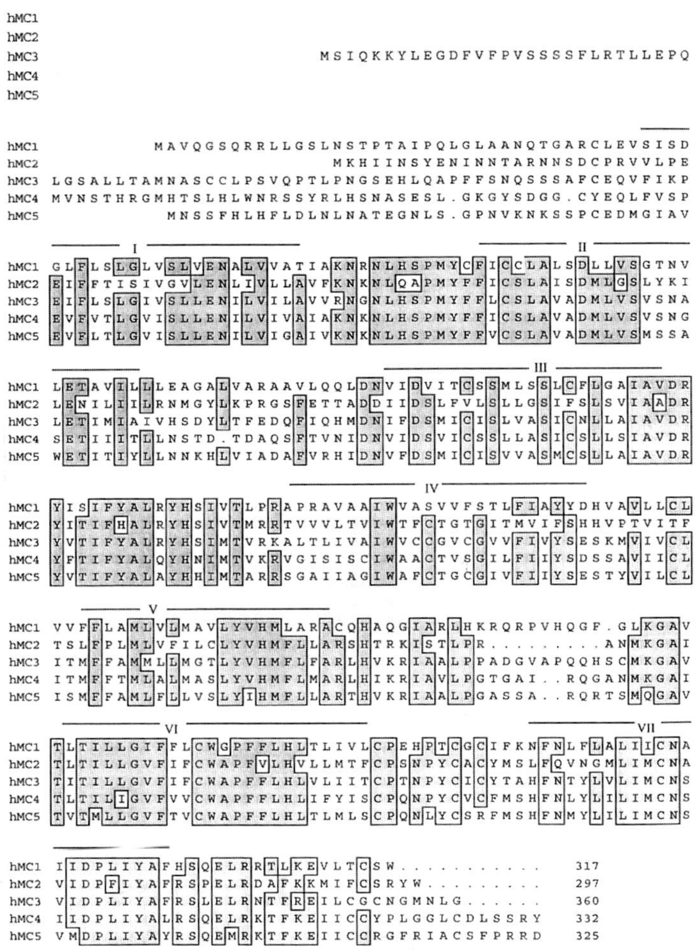

Fig. 1. The amino acid sequence of the human melanocortin receptors is shown using the standard single-letter abbreviations. GenBank accession numbers are hMC1-R (X65634), hMC2-R (X65633), hMC3-R (L06155), hMC4-R (S77415), and hMC5-R (L27080). Residues that are conserved among all the human melanocortin receptors are boxed and shaded.

have been published and these differ from the sequence published by Mountjoy et al. *(13)* at five *(15)* and three *(16)* amino acids. The MC1-R was found not to be highly related to other neuropeptide receptors and has a number of unusual structural features. The MC1-R is among the smallest GPCRs identified and has a short amino terminal extracellular domain, a short carboxy terminal intracellular domain, unusually short fifth transmembrane domain, and a small second extracellular loop *(17)*. The MC1-R lacks several amino acid residues present in most other rhodopsinlike GPCRs *(18)*. These include the proline residues in the fourth and fifth transmembrane domains, which are thought to introduce a bend in the α-helical structure of the transmembrane domains and to participate in the formation of the binding pocket, and one or both of the cysteine residues thought to form a disulfide bond between the first and second extracellular loops. The highly conserved asparagine residue in the seventh transmembrane domain of most rhodopsin like GPCRs is substituted in the MC1-R with an aspartic acid residue. In the N-terminal domain there are potential sites for N-linked glycosylation, while in other regions of the receptor there are potential sites for myristoylation and several sites for phosphorylation by protein kinase C (PKC) and protein kinase A (PKA). The PKC and PKA sites which are located in the third intracellular loop and the C-terminal domain of the receptor may be involved in the regulation of the interaction between the receptors and G proteins. There is one potential PKC site in the third intracellular loop and one in the carboxy terminal domain of the mMC1-R. A potential PKA phosphorylation site is also located in the third intracellular loop of the mMC1-R. The hMC1-R has one potential PKC site in the carboxy terminal domain.

2.4. Gene Structure

The mMC1-R cDNA and the human genomic MC1-R were colinear and therefore there are no introns present in the coding region of this melanocortin receptor. The gene structure of the MC1-R is currently unknown.

2.5. Tissue Expression

Each member of the melanocortin receptor family has a distinct tissue distribution (Table 1). The MC1-R mRNA is expressed in melanocytes and melanoma tissue *(13,15)*. The mMC1-R is encoded predominantly by a single mRNA species of 4-kb, whereas the hMC1-R is encoded predominantly by a 3-kb species in melanoma samples and as both 3-kb and 4-kb mRNA species in melanocytes *(13)*. In addition, the MC1-R mRNA has been found to be expressed in macrophages by RT-PCR *(19)* and in brain periaqueductal gray matter by *in situ* hybridization *(20)*.

2.6. Specificity

The melanocortin receptors have remarkably different pharmacologic properties which were predicted from MSH binding studies (Tables 2 and 3). Not only are there specificity differences between the subtypes but there are also differences between some subtypes in different species. The hMC1-R binds and is potently activated equally by α-MSH and ACTH[1–39] *(15,16,21)* (Fig. 2). α-MSH is as potent as the super potent melanocortin peptide analog,

Table 1
Tissue Expression of Melanocortin Receptor mRNA

Receptor	Tissues	References
MC1-R	Melanocytes, melanoma, macrophage, brain	*13,15,19,20*
MC2-R	Adrenal cortex, adipose tissue	*13,33,34*
MC3-R	Brain, placenta, duodenum, pancreas, stomach	*16,41*
MC4-R	Brain, spinal cord	*45,46,62*
MC5-R	Skin, adrenal cortex, adipose tissue, skeletal muscle	*34,49–53,63*

NDP-α-MSH, at coupling the cloned hMC-1 receptor to the PKA signaling pathway *(21)*. The EC_{50} for α-MSH coupling the hMC1-R is $2 \times 10^{-12}M$. Interestingly, $10^{-10}M$ and lower concentrations of γ2-MSH, γ3-MSH and lys γ3-MSH activate the hMC1-R *(21)*. In contrast to the hMC1-R, the order of potency of melanocortin peptides at the mMC1-R is NDP-α-MSH ($1 \times 10^{-12}M$) > α-MSH ($4 \times 10^{-11}M$) > ACTH ($4 \times 10^{-10}M$) >>γ2-MSH (>$10^{-7}M$) *(21)*. The hMC1-R has therefore evolved to become hyper-sensitive to α-MSH and other melanocortin peptides containing the common MSH core sequence His-Phe-Arg-Trp. This suggests that ACTH may be a potent physiologic melanotropic peptide in man, consistent with the absence of an intermediate lobe of the pituitary in the human. In man, unlike rodents, processing of pituitary or locally derived POMC to yield γ2-or γ3-MSH peptides may have particular relevance to human melanogenesis.

2.7. Chromosomal Mapping

The human chromosomal locations of all five melanocortin receptors have been determined while four of these have been mapped on mouse chromosomes (Table 4). This information has markedly advanced our knowledge about pigmentation. Genetic mapping of the mouse melanocortin receptors was accomplished with an intersubspecific backcross mapping panel. The MC1-R was typed as a *TaqI* restriction fragment length polymorphism (RFLP) and chromosomal mapping placed the mMC1-R near the *extension* locus *(22)*, which had been previously mapped to the distal portion of chromosome 8 in the mouse. Cloning of the MC1-R gene from mice with different *extension* locus alleles demonstrated conclusively that the mouse *extension* locus encodes the MC1-R. The human melanocortin receptors were mapped by fluorescent *in situ* hybridization. In the human, MC1-R maps to 16q24 *(23,24)*. Using a 20-kb genomic fragment of the hMC1-R, three hybridizing sites were identified by *in situ* hybridization to human chromosomes *(24)*. The 16q24 site was the region of

Table 2
Specificity of Recombinant Melanocortin Receptors Coupling to the PKA Signaling Pathway

	Receptor						
Peptide	hMC1-R	mMC1-R	mMC2-R	rMC3-R	hMC4-R	rMC5-R	mMC5-R
NDP-α-MSH	2×10^{-12}	2×10^{-12}	NE	2×10^{-9}	1×10^{-11}	1×10^{-9}	5×10^{-11}
α-MSH	2×10^{-12}	4×10^{-11}	NE	4×10^{-9}	2×10^{-9}	6×10^{-10}	1×10^{-9}
Desacetyl-α-MSH		9×10^{-11}	NE	1×10^{-8}	5×10^{-10}		
$ACTH_{[1-39]}$	8×10^{-12}	4×10^{-10}	6×10^{-11}	4×10^{-9}	7×10^{-10}	6×10^{-9}	6×10^{-9}
$ACTH_{[1-24]}$			8×10^{-12}			5×10^{-10}	
β-MSH	1×10^{-11}	2×10^{-10}	NE			1×10^{-8}	7×10^{-9}
γ1-MSH				4×10^{-9}			9×10^{-9}
γ2-MSH	8×10^{-11}	$>10^{-7}$	NE	4×10^{-9}	$>10^{-7}$	5×10^{-8}	4×10^{-8}
γ3-MSH	$<10^{-10}$	2×10^{-8}		8×10^{-9}	2×10^{-8}		7×10^{-8}
$ACTH_{[4-10]}$	$>10^{-7}$	$>10^{-7}$	NE	$>10^{-7}$	$>10^{-7}$	$>10^{-7}$	NE
CLIP							NE
ORG2766				NE	NE		
β-Lipotropin							NE

EC_{50} *(M)* values obtained from references; hMC1 and mMC1 receptors *(21)*, mMC2-R *(37)*, rMC3-R *(41)*, hMC4-R *(45)*, rMC5-R *(51)*, and mMC5-R *(50)*. NE, no effect.

Table 3
Affinity of Melanocortin Peptides for Recombinant Melanocortin Receptors

	Receptor						
Peptide	hMC1-R	mMC2-R	rMC3-R	mMC3-R	hMC4-R	hMC5-R	mMC5-R
NDP-α-MSH	2×10^{-11}		1×10^{-8}	2×10^{-9}	2×10^{-9}	5×10^{-9}	1×10^{-9}
α-MSH	9×10^{-11}	$>10^{-7}$	5×10^{-8}	3×10^{-8}	6×10^{-7}	9×10^{-7}	6×10^{-8}
desacetyl-α-MSH		$>10^{-7}$	1×10^{-8}		6×10^{-7}	3×10^{-6}	
$ACTH_{[1-39]}$	2×10^{-10}	8×10^{-10}	4×10^{-9}	1×10^{-9}	7×10^{-7}	9×10^{-7}	2×10^{-8}
$ACTH_{[1-24]}$		9×10^{-10}					
β-MSH	5×10^{-10}			2×10^{-8}	4×10^{-7}	1×10^{-6}	2×10^{-8}
γ1-MSH					$>10^{-6}$	4×10^{-5}	7×10^{-7}
γ2-MSH	1×10^{-9}		4×10^{-8}	9×10^{-9}	NE	NE	1×10^{-6}
γ3-MSH			8×10^{-9}		$>10^{-6}$	NE	6×10^{-7}
$ACTH_{[4-10]}$	2×10^{-8}		$>10^{-6}$	2×10^{-6}	$>10^{-6}$	NE	NE
CLIP							NE
ORG2766			NE		NE	NE	

K_i *(M)* values obtained from references; hMC1 *(15)*, mMC2-R *(37)*, rMC3-R *(41)*, mMC3-R *(40)*, hMC4-R *(48)*, hMC5-R *(48)*, and mMC5-R *(50)*. Dispacement curves were performed following binding of [^{125}NDP-α-MSH] for MC1, MC3, MC4, and MC5 receptors and [^{125}I-iodotyrosyl^{23}]ACTH $_{[1-39]}$ for the mMC2-R. NE, no effect.

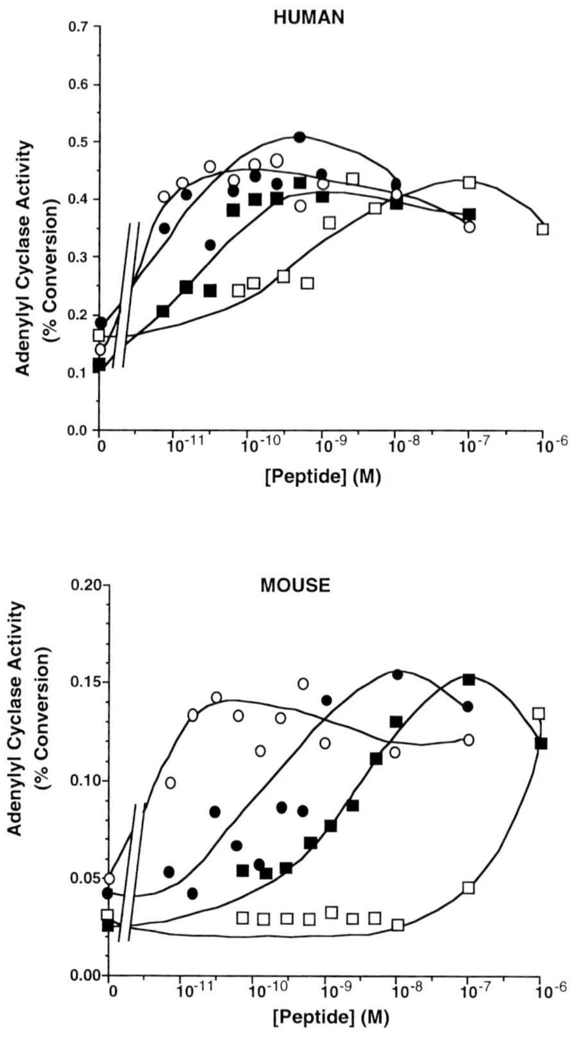

Fig. 2. Species differences in functional coupling of the MC1-R to adenylyl cyclase. The cloned human and mouse MC1 receptors were stably expressed in HEK293 cells. The *x*-axis indicate the concentration of each peptide and the y axis indicate the percent [^{3}H]adenine converted to [^{3}H]cAMP after 1 h incubation at 37°C. Cells were stimulated with NDP-α-MSH *(open circle)*, α-MSH *(closed circle)*, ACTH[1–39] *(closed square)*, γ2-MSH *(open square)*. Curves are representative of two or three separate experiments for each melanocortin peptide and the data are reproduced from ref. *21*.

conserved synteny with the distal tip of chromosome 8 in the mouse *(25,26)*. The two additional sites of hybridization may share homology with the MC1-R coding sequence, or possibly with other gene sequences 5' or 3' of the MC1-R gene.

Table 4
Human and Mouse Chromosomal Map Locations for the Melanocortin Receptors

Receptor	Mouse Chromosome	Human Chromosome
MC1-R	Distal end of 8 identical with the extension locus *(22)*	16q24*(23,24)*
MC2-R	Distal end of 18 near Cdx-1*(24)*	18p11.21*(24,38,39)*
MC3-R	Chr. 2, near E1-2*(24)*	20q13.2*(24,38,43,44)*
MC4-R		18q22*(24,46)*
MC5-R	Distal end of Chr. 18, near D18Mit9 and *sy(64)*	18p11.2*(54)*

References for chromosomal mapping are in parentheses.

3. Cloning of the Adrenocorticotropin Receptor

3.1. Degenerate Oligonucleotide PCR

The hMC2-R was cloned in parallel with the hMC1-R. Degenerate oligonucleotides designed to the highly conserved transmembrane three and six domains of GPCRs were used for the PCR where the template was cDNA derived from a human melanoma specimen known to express a high number of MSH binding sites *(13)*. Two PCR subclones resulting from amplification of human melanoma cDNA were determined by DNA sequencing to encode novel GPCRs highly related to one another, and by Northern hybridization analysis to be expressed specifically in melanocytes and adrenal cortex, respectively *(13)*. The PCR fragment which was specifically expressed in adrenal cortex was the MC2-R and may have been obtained through amplification of contaminating genomic DNA in the melanoma cDNA sample.

3.2. Library Screening of Genomic DNA and cDNA

The PCR fragment which was expressed in adrenal cortex was used to screen a human genomic library at high stringency *(13)* and a bovine adrenal cDNA library at moderate stringency *(17)*. The entire coding sequence of the hMC2-R was found in a 3.9 kb *Hin*dIII fragment and the bovine MC2-R sequence was found in a 1.8-kb cDNA clone. PCR generated mouse and bovine MC2-R probes were generated based on the hMC2-R coding sequence, and used to screen mouse genomic *(27)* and bovine adrenal cortex cDNA *(28)* libraries, respectively.

3.3. Peptide Structure and Homology With the MC1-R

The hMC2-R amino acid sequence (297 amino acids) was 81% identical and colinear with the bovine MC2-R cDNA amino acid (297 amino acids) sequence *(13,17)*. The mMC2-R shares 89% amino acid identity with the hMC2-R, but it lacks the C-terminal tryptophan residue of the hMC2-R thus making it 296 amino acids in total length. The hMC2-R shares 39% amino acid identity with the hMC1-R and is almost colinear (Fig. 1). The MC2-R is the smallest GPCR identified and shares with the MC1-R the unusual structural features compared with other rhodopsinlike GPCRs. The melanocortin receptors are therefore defined as a subfamily of the major G protein-coupled receptor gene family *(13,17)*. The hMC2-R has two potential PKC phosphorylation sites in the third intracellular loop and a third in the second intracellular loop. There is also a potential PKA phosphorylation site in the third intracellular loop of the hMC2-R.

3.4. Gene Structure

The hMC2-R gene consists of two exons; 49 bp of the 5' untranslated region of the hMC2-R is located in exon 1, which is approximately 18 kb upstream of exon 2 *(29)*. Exon one contains one major transcription start site and a minor transcription start site 15 bp downstream of the major start site. Exon 2 contains 128 bp 5' untranslated sequence and the full-length coding sequence. The 5' untranslated region is 177 bp and does not contain CAAT- or TATA-consensus sequences. A consensus Sp1 and near consensus TPA and steroidogenesis factor 1 responsive elements are present. There are seven putative cAMP responsive elements between–107 and–686 *(30)*.

The mMC2-R gene consists of four exons and three of the exons encode the 5' untranslated sequence *(31)*. The second exon (57 bp), which is located approximately 6 kb downstream of exon 1 (109 bp) is only present in some transcripts. Exon 3 (112 bp) is approximately 1.6 kb upstream of exon 4. Exon 4 contains 96 bp of the 5' untranslated region, the full coding sequence, and all of the 3' untranslated region. The 3' untranslated region of the mMC2-R is 445 bp and has a polyadenylation signal AATAAA at 31 bp before the poly A tail. Consensus sites for the steroidogenic cell specific factor, SF1, a glucocorticoid response element, as well as SP1, AP1, and AP2 sites are located within 1.8 kb of the start of transcription *(32)*.

3.5. Tissue Expression

Using *in situ* hybridization, the MC2-R mRNA has been shown to be expressed in all three zones of the adrenal cortex; zona glomerulosa, zona fascicular, and zona reticular *(13)* (Table 1). The hMC2-R mRNA has been shown using RT-PCR to be expressed in skin *(33)* while the mMC2-R has been shown using both RT-PCR and Northern blot hybridization to be abundantly expressed in mouse adipose tissue *(34)*. Northern blot hybridization indicates that the hMC2-R gene is expressed in human and rhesus monkey adrenal as a major mRNA band of 4 kb *(13)*, in human adrenal tissue as two major bands of 1.8 and 3.4 kb and three minor bands at 4, 7, and 11 kb *(29)*. The mMC2-R is expressed in mouse adrenal *(27,35)* and adipose tissue *(34)* as a major band of approx 2 kb and a weaker band of approx 4 kb.

3.6. Specificity

The MC2-R is unique among the melanocortin receptor family in that there have been great difficulties expressing the human and bovine MC2 receptors in heterologous cells. In 1992 the hMC2-R was expressed in the mouse melanoma cell line, M3, and shown to couple to the PKA signaling pathway in response to ACTH *(13)*. This expression system was not suitable however, for specificity studies since the M3 cells express an endogenous MC1-R. The hMC2-R has also been expressed in the mouse OS3 cell line, an ACTH receptor-deficient derivative of the adrenocortical carcinoma cell line Y1, and shown to couple to the PKA signaling pathway in response to ACTH *(36)*. Surprisingly, the mMC2-R expressed in HeLa and HEK293 cells thus allowing the specificity of this receptor to be determined *(27,32,37)* (Tables 2 and 3). The mMC2-R couples to the PKA signaling pathway with EC_{50}s ranging from 7.5 to 57 × $10^{-12}M$ in response to ACTH[1–39], ACTH[1–24] and ACTH[1–17]. α-,β- and γ-MSH peptides do not activate the MC2-R *(37)* (Table 2).

3.7. Chromosomal Mapping

The MC2-R gene maps to a conserved syntenic region on chromosome 18 in the mouse and human (Table 4). The hMC2-R maps to the short arm of human chromosome 18 (18p11.2) *(24,38,39)*. Familial ACTH resistance, a rare endocrine disorder, maps to this locus in approximately half of the individuals with the disease and these people have mutations in the coding region of the hMC2-R *(61)*.The mMC2-R was typed as a *MspI* RFLP and mapped to the distal end of mouse chromosome 18 near *Pdea (24)*.

4. Cloning of the Melanocortin-3 Receptor

4.1. Degenerate Oligonucleotide PCR

A third melanocortin receptor, hMC3-R, was identified using degenerate oligonucleotides based on conserved amino acid sequences of the MC1 and MC2 receptors and cDNA made from human RNA was used as template *(16)*. An mMC3-R gene was also cloned using degenerate oligonucleotide PCR and human genomic DNA as template where the primers were designed to amplify members of the superfamily of rhodopsin like GPCRs *(40)*.

4.2. Library Screening of Genomic and cDNA

The hMC3-R PCR fragment was used as a probe to screen a human genomic library *(16)* to obtain a full-length hMC3-R gene. A rat MC3-R (rMC3-R) gene was identified when the 400-bp hMC1-R PCR fragment originally obtained from human melanoma mRNA was used to screen a rat hypothalamic cDNA library under low stringency *(41)*. A 2-kb clone was identified which encoded the full length coding rMC3-R. The mMC3-R was cloned when the hMC3-R PCR fragment was used as a probe to screen a mouse genomic library to obtain the full-length mMC3-R gene *(40)*.

4.3. Peptide Structure and Homology With MC1 and MC2 Receptors

The hMC3-R gene encodes a 360 amino acid protein which shares 46% amino acid identity with the MC1 and MC2 receptors (Fig. 1). The rMC3 and mMC3 receptors (323 amino acids each) share 90% amino acid identity with the hMC3-R, respectively. The MC3-R also shares the unusual structural features of the MC1 and MC2 receptors, including the absence of a conserved cysteine residue in the first extracellular loop, thought to form a disulfide bond with the second extracellular loop in most GPCRs. The rMC3-R cDNA amino acid sequence is almost colinear with the hMC3-R but the N-terminus of the hMC3-R has an additional 37 amino acids compared with the rat and mouse MC3 receptors. The hMC3-R has two translation initiation codons; Met^{38} aligns with Met^{1} of the rat, and mouse MC3 receptors. Three potential sites for N-linked glycosylation are found in the N-terminal extracellular domain of the human, rat and mouse MC3 receptors. Two potential sites for phosphorylation by PKC are present in the hMC3-R; one in the second intracellular loop and the other in the carboxy terminal domain of the receptor. The second intracellular loop phosphorylation site is shared with the MC2-R, while the carboxy terminal phosphorylation site is shared with the MC1-R. The MC3-R does not contain sites that could be phosphorylated by PKA, in contrast with the MC1 and MC2 receptors.

4.4. Gene Structure

The rat MC3-R cDNA and the human genomic MC3-R were colinear and therefore there are no introns present in the coding region of this melanocortin receptor. The gene structure of the MC3-R is currently unknown. However, the hMC3-R has a longer (37 amino acids) N-terminal extracellular domain than the rat and mouse MC3 receptors. It is possible that there is an intron in the 5' untranslated region. The human and mouse MC3-R genomic sequences and the rMC3-R cDNA sequence are almost identical up to 50 bp 5' of the mouse and rat translational start codon *(40)*. The mouse genomic sequence does not contain the in frame ATG codon corresponding to the first methionine codon of the human sequence. In both the human and mouse genomic sequences there are segments compatible with intron/exon boundaries 50 bp 5' of the mouse translational start codon. The hMC3-R coding region may therefore start at the second methionine of the sequence reported by Gantz et al. *(16)* (Fig. 1).

4.5. Tissue Expression

The MC3-R is predominantly expressed in brain. Using *in situ* hybridization, neuroanatomical mapping in adult rat brain showed that the MC3-R mRNA containing neurons had a rather restricted distribution with the greatest density of labeled neurons present in the hypothalamic cell groups, such as the arcuate nucleus, and in regions such as the anteroventral periventricular nucleus and posterior hypothalamic area *(16,41)*. Northern blot hybridization indicates that the MC3-R gene is expressed in rat hypothalamus as two mRNA bands of 2 and 2.5 kb *(41)*. In addition to its expression in brain, the MC3-R was also found to be expressed in placenta by Northern analysis and in stomach, duodenum and pancreas by RT-PCR *(16)* (Table 1).

4.6. Specificity

The MC3-R is unique in that it does not appear to distinguish between melanocortin peptides, NDP-α-MSH, ACTH[1–39], α-MSH, desacetyl-α-MSH, β-MSH, γ1-MSH, γ2-MSH and γ3-MSH *(41)* (Tables 2 and 3). The EC_{50}s for these peptides coupling the rMC3-R to the PKA signaling pathway range from 1 to $14 \times 10^{-9} M$. ORG2766, an ACTH[4–9] analog (methionylsulfone-Glu-His-Phe-D-Lys-Phe) and ACTH[4–10], synthetic melanocortin peptides with little adrenocorticotropic activity but potent activity in various assays involving the central and peripheral nervous systems, either did not stimulate the MC3-R (ORG2766) or had very reduced ability to activate adenylyl cyclase (ACTH[4–10], $EC_{50} \sim 10^{-7} M$). ORG2766 had no activity as an antagonist either *(41)*. The human, rat and mouse MC3 receptors have similar pharmacologic properties. Melanocortin binding was no different between the full length hMC3-R and a mutated hMC3-R in which the N-terminus is the same length as the mouse and rat MC3 receptors, indicating that both translation initiation sites in the hMC3-R produce receptors with similar binding affinities *(42)*.

4.7. Chromosomal Mapping

The MC3-R maps to a region of conserved synteny between mouse chromosome 2 and human chromosome 20 (Table 4). The hMC3-R maps to the long arm of human chromosome 20 (20q13.2) *(24,38)*. Before the identification of MC3-R as a melanocortin receptor, this gene sequence was reported as an orphan receptor and was shown to be genetically linked to non-insulin-dependent diabetes *(43)*. Furthermore, a dinucleotide repeat polymorphism was used to map the gene, with a somatic cell panel to human chromosome 20q

(44). The mMC3-R was typed as a *PstI* RFLP and mapped to the distal half of mouse chromosome 2 near *El-2 (24)*.

5. Cloning of the Melanocortin-4 Receptor

5.1. Degenerate Oligonucleotide PCR

Degenerate oligonucleotides to the highly conserved transmembrane domains two and seven of the known melanocortin receptors were used to amplify a fragment of the MC4-R by PCR using as template either rat brain cDNA *(45)* or mouse genomic DNA *(46)*.

5.2. Library Screening of Genomic and cDNA

The rat and mouse PCR fragments were used as probes to screen human genomic libraries *(45,46)* to obtain full length hMC4-R genes. A full length rat MC4-R (rMC4-R) gene was identified when a PCR fragment amplified from bovine locus ceruleus cDNA using degenerate oligonucleotides designed for the third and sixth transmembrane domains of cloned G protein coupled receptors, was used as a probe to screen a rat brainstem/spinal cord cDNA library *(47)*. A 3-kb clone containing an open reading frame and a stop codon but lacking several hundred base pairs of 5' sequence was isolated. 5' RACE was conducted to generate the 5' sequence of the rMC4-R cDNA.

5.3. Peptide Structure and Homology With MC1, MC2, and MC3 Receptors

The hMC4-R gene encodes a 332 amino acid protein (Fig. 3) that shares 47% and 46% amino acid identity with the MC1 and MC2 receptors, respectively, and 55% amino acid identity with the MC3-R (Fig. 1). The rMC4-R shares 93% amino acid identity with the hMC4-R. The MC4-R also shares the unusual structural features of the MC1, MC2, and MC3 receptors, including the absence of a conserved cysteine residue in the first extracellular loop, thought to form a disulfide bond with the second extracellular loop in most GPCRs and an aspartic acid in transmembrane seven in the place of an asparagine residue which is highly conserved in most GPCRs. There are two potential PKC phosphorylation sites in the hMC4-R; one in the second intracellular loop and another in the carboxy terminal domain. A potential PKA phosphorylation site is also present in the second intracellular loop of the hMC4-R.

5.4. Gene Structure

The rat MC4-R cDNA and the human genomic MC4-R were colinear up to 27 bp upstream of the translation start codon and therefore there are no introns present in the coding region of this melanocortin receptor. Although the human genomic hMC4-R gene and the rat cDNA MC4-R sequences

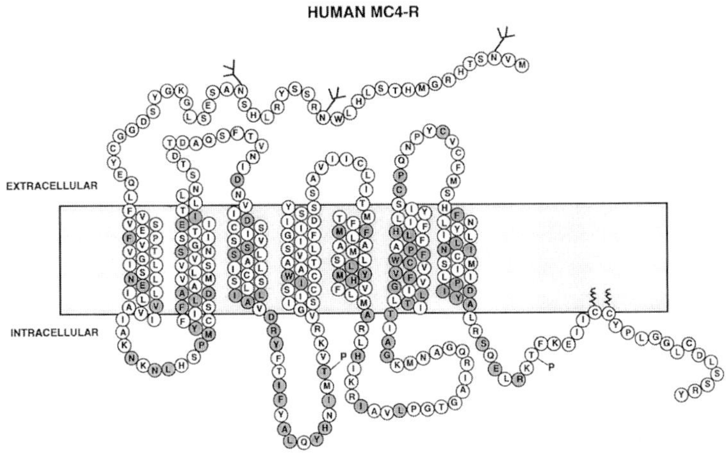

Fig. 3. A pseudostructural plot of the human MC4 receptor is shown with the hydrophobic domains and cellular orientation predicted by hydropathy analysis and by comparison with other known G protein-coupled receptors. Amino acid residues that are identical among the human melanocortin receptors are shaded. Potential sites for N-linked glycosylation in the amino terminal domain are identified. Potential sites for phosphorylation by protein kinase A (Thr162) and protein kinase C (Thr162, Thr314) are shown as is palmitoylation involving Cys320 and Cys321.

diverge 5' of the 27 bp upstream from the translation start codon, there are no consensus exon-intron boundary sequences present in this region. The gene structure of the MC4-R is currently unknown.

5.5. Tissue Expression

The MC4-R has only been shown to be expressed in brain, autonomic nervous system, and spinal cord. (Table 1). Neuroanatomic mapping of the MC4-R in adult rat brain showed that the MC4-R mRNA is more widely expressed than MC3-R mRNA and is found in multiple sites in virtually every brain region, including the cortex, thalamus, hypothalamus, hippocampus, brainstem, and spinal cord *(45,46)*. Unlike the MC3-R, the MC4-R is found in both parvicellular and magnocellular neurons of the paraventricular nucleus of the hypothalamus, suggesting a role in the central control of pituitary function. The MC4-R mRNA is also unique in its expression in numerous cortical and brainstem nuclei, spinal cord, and the developing autonomic nervous sytem *(62)*.

5.6. Specificity

The hMC4-R has a different pharmacological profile from mMC1, MC2 and MC3 receptors but has a similar profile to that of the hMC1-R. The order of potency of melanocortin peptides on this receptor is NDP-α-MSH > desacetyl α-MSH > α-MSH = ACTH[1–39] = β-MSH = lys γ3-MSH > γ2-MSH *(21,45,46)* (Tables 2 and 3). The EC_{50}s for these peptides coupling the hMC4-R to the PKA signaling pathway are: NDP-α-MSH, $1 \times 10^{-11} M$; desacetyl-α-MSH, $5 \times 10^{-10} M$; ACTH[1–39], 7×10^{-10}M; α-MSH, $2 \times 10^{-9} M$; lys γ3-MSH, $1 \times 10^{-9} M$; γ2-MSH, > $10^{-7} M$; ACTH[4–10], > $10^{-7} M$. The order of potency is similar to that of the hMC1-R but these EC_{50} values are all greater than those for the hMC1-R. ORG2766 does not couple the hMC4-R to the PKA signaling pathway. The order of potency for competing for [^{125}I]NDP-α-MSH binding at the hMC4-R is: NDP-α-MSH > β-MSH > desacetyl-α-MSH > α-MSH > ACTH[1–39] > ACTH[4–10] > γ1-MSH > γ2-MSH *(48)*. The K_i values for α-MSH, desacetyl-α-MSH and ACTH[1–39] displacing ^{125}I-NDP-α-MSH binding are ~100-fold greater than the EC_{50} values for these peptides coupling the hMC4-R to the PKA signaling pathway. The rMC4-R has a similar pharmacologic profile to the hMC4-R, but the EC_{50}s for both α-MSH and ACTH[1–39] are approximately $2 \times 10^{-8} M$, an order of magnitude greater than for the hMC4-R *(47)*.

5.7. Chromosomal Mapping

The MC4-R maps to the long arm of human chromosome 18; one report maps it to 18q22 *(24)* while another maps the MC4-R to 18q21.3 *(46)* (Table 4). The mouse MC4-R has not yet been mapped because no RFLP has been identified *(24)*.

6. Cloning of the Melanocortin-5 Receptor

6.1. Degenerate Oligonucleotide PCR

The human and mouse MC5 receptors were identified using genomic DNA as template and PCR with degenerate oligonucleotides designed against conserved sequences in the MSH and ACTH receptors *(46,49)*. The ovine MC5-R (oMC5-R) was identified using PCR with cDNA derived from ovine pars tuberalis as template and oligonucleotides designed to conserved sequences in the GPCR superfamily *(50)*.

6.2. Library Screening of Genomic and cDNA

The products from the human and mouse PCR were used to screen human and mouse genomic libraries respectively to identify full length coding MC5 receptors *(46,49)*. The human clone was originally designated MC2-R by Chhajlani et al. *(49)*. The PCR fragment derived from ovine pars tuberalis was used to screen an ovine genomic library to isolate a full length oMC5-R *(50)*. A full-length mMC5-R was also obtained after homology screening a mouse genomic library with a hMC3-R probe *(51)*. The rMC5-R was cloned when a dopamine D3 receptor probe was used to screen a rat genomic library *(52)*.

6.3. Peptide Structure and Homology With MC1, MC2, MC3, and MC4 Receptors

The hMC5-R gene encodes a 325 amino acid protein that shares 44% and 43% amino acid identity with the MC1 and MC2 receptors, respectively, and 59% and 61% amino acid identity with the MC3 and MC4 receptors, respectively (Fig.1). The MC3, MC4, and MC5 receptors are therefore more closely related to one another than they are to the MC1 and MC2 receptors. The rat and mouse MC5 receptors share 81% amino acid identity with the hMC5-R. The rat and mouse MC5 receptors are more closely related to one another sharing 95% amino acid identity. The oMC5-R shares 83% amino acid identity with the hMC5-R, 78% amino acid identity with the rMC5-R and 77% amino acid identity with the mMC5-R. The MC5-R also shares the unusual structural features of the MC1, MC2, MC3, and MC4 receptors, including the absence of a conserved cysteine residue in the first extracellular loop, thought to form a disulfide bond with the second extracellular loop in most GPCRs. The hMC5-R has four potential PKC phosphorylation sites; one in transmembrane 2, two in transmembrane three and one in the carboxy-terminal domain. These four potential PKC phosphorylation sites are also present in the rMC5-R along with one additional site in transmembrane three and another in the carboxy terminal domain. The mMC5-R has five potential PKC phosphorylation sites with only one of these in the carboxy terminal domain. In contrast to the human and rodent MC5 receptors, the oMC5-R has three potential PKC phosphorylation sites; one in transmembrane two, one in transmembrane three, and another in the carboxy terminal domain.

6.4. Gene Structure

The gene structure of the MC5-R is currently unknown but as with the other four members of the melanocortin receptor family, there are no introns in the coding region. To date no MC5-R cDNA clones have been isolated.

6.5. Tissue Expression

The MC5-R mRNA has been reported to be expressed by PCR and RNase protection assay, in a broad spectrum of tissues including skin, brain (cortex and cerebellum), skeletal muscle, lung, spleen, thymus, bone marrow, testis, ovary, uterus, adrenal gland, pituitary, and thyroid gland *(49–52)* (Table 1). Northern blot analysis indicated the presence of a single 4-kb mRNA species which was expressed in rat adrenal, stomach, lung, and spleen but with much higher levels seen in adrenal and stomach than elsewhere *(52)*. Northern analysis was also used by Gantz et al. *(53)* to demonstrate the presence of an approximately 4-kb mRNA species in mouse skeletal muscle, lung, spleen and brain with higher levels expressed in skeletal muscle than elsewhere.

Using *in situ* hybridization, mMC5-R mRNA has been localized to the submaxillary gland (salivary gland) and all three layers of the adrenal cortex *(52)*. MC5-R mRNA was detected using RT-PCR in mouse adipose tissue and skeletal muscle. By Northern blot analysis the MC5-R hybridized to a single 4.3-kb RNA band in axillary adipose tissue, skeletal muscle and in differentiated 3T3L1 cells, but not to subcutaneous or dorsal adipose tissue *(34)*. MC5-R mRNA was also found by Northern blot analysis to be extremely abundant in harderian, lacrimal, and preputial glands *(63)*. In summary, it appears that the major sites of MC5-R mRNA expression are multiple exocrine tissues, skin, skeletal muscle, adipose tissue, submaxillary gland, adrenal gland, and stomach.

6.6. Specificity

MC5-R from four different species have been expressed heterologously in mammalian cells and they each have different specificities (Tables 2 and 3). The hMC5-R binds NDP-α-MSH with a K_i = 5 n*M*. However, α-MSH, ACTH[1–39] and γ-MSH displaced [^{125}I]NDP-α-MSH binding very poorly (K_is > 300n*M*), and this suggests that the natural ligand for hMC5-R has not yet been identified *(49)*. The relatively high concentrations of α-MSH required to produce a half-maximal response for the hMC5-R coupling to the PKA pathway *(36)* support the suggestion that the natural ligand for the hMC5-R is presently unknown or that the human receptor couples to a novel signaling pathway. The mouse and rat receptors distinguish between α-MSH and ACTH with α-MSH being more potent. The EC_{50}s for melanocortin peptides coupling the mMC5-R to the PKA signaling pathway are: NDP-α-MSH, $5 \times 10^{-11}M$; α-MSH, $1 \times 10^{-9}M$; ACTH[1–39], $6 \times 10^{-9}M$; β-MSH, $6 \times 10^{-9}M$; γ1-MSH, $9 \times 10^{-9}M$; γ2-MSH, $4 \times 10^{-8}M$; γ3-MSH, $7 \times 10^{-8}M$ *(51)* (Table 2). ACTH[4–10] and β-lipotropin did not couple the mMC5-R to increases in cAMP. The response of the mMC5-R coupling to the PKA pathway was similar for both the nonacetylated form of α-MSH (ACTH[1–13]) and α-MSH *(53)*. The rMC5-R differs from the mMC5-R in that α-MSH is twice as potent as NDP-α-MSH on the rat receptor *(52)*. In contrast to the rat and mouse MC5 receptors, the oMC5-R does not distinguish between α-MSH and ACTH. The oMC5-R is, however, similar to the hMC5-R, since the oMC5-R binds NDP-α-MSH with an IC_{50} = 2n*M*, but α-MSH and ACTH[1–39] displaced [^{125}I] NDP-α-MSH binding very poorly (IC_{50}s ~$10^{-6}M$) *(50)*. Furthermore, relatively high concentrations ($10^{-8}M$) of NDP-α-MSH, α-MSH, β-MSH, and ACTH were required to produce a half-maximal response for the oMC5-R coupling to the PKA signaling pathway and γ-MSH had no effect *(50)*. β-Endorphin had no effect on the functional coupling of any species of MC5-R to the PKA signaling pathway.

6.7. *Chromosomal Mapping*

The hMC5-R has been localized to the short arm of human chromosome 18 (18p11.2) by fluorescence *in situ* hybridization *(54)* (Table 4). This is the same locus that the hMC2-R maps to. The mMC5-R was mapped using an intersubspecific mapping panel and mapped to the distal end of chromosome 18 near D18Mit9 and *sy (64).*

7. Are There Further Melanocortin Receptors to be Identified?

Following the identification of the MC1 and MC2 receptors, Southern blots were performed and probed with PCR fragments of the MC1 and MC2 receptors using low-stringency hybridization. These Southern blots indicated the existence of as many as five or six members of the melanocortin receptor family *(13)*. It is currently unclear as to whether there are any further members of the melanocortin receptor family not yet identified; this is primarily because there are many physiologic responses to melanocortin peptides that cannot be explained by the specificity of the five known melanocortin receptors. In addition, not all of the melanocortin peptides that have biologic activity in vivo , such as ORG2766 and β-lipotropin, have been shown to couple to any of the five known melanocortin receptors.

The most potent melanocortin peptides in various assays involving the central and peripheral nervous systems are ACTH[4–10] and ORG2766. Relatively high concentrations of ACTH[4–10], compared with other melanocortin peptides, are required to couple MC3 and MC4 receptors to the PKA signalling pathway, and ORG2766 fails to activate either MC3 or MC4 receptors. Although it is possible that ORG2766 produces its effects in vivo through a melanocortin receptor not yet identified, recent behavioral studies have demonstrated that ORG2766 responses may be mediated via the NMDA receptor *(55,56)*.

Recently, the administration of MC3 and MC4 receptor antagonists either into regions of rat brain involved in cardiovascular regulation or via intracarotid injection, failed to inhibit the pressor and tachycardiac effects of γ-MSH and ACTH[4–10] but did inhibit the α-MSH induced hypotension and bradycardia *(57)*. Similarly, Van Bergen et al. *(58)* showed that whereas the administration of γ2-MSH to conscious rats had a pressor response, ACTH[1–24] decreased blood pressure and the superpotent α-MSH analog, NDP-α-MSH (potent agonist on MC3 and MC4 receptors), was without effect on blood pressure. Furthermore, Huang et al. *(59)* have shown that intracerebroventricular injection of the MC4/MC3 receptor antagonist, SHU9119, alone significantly increased endotoxin-induced fever in rats, whereas α-MSH alone significantly decreased fever.

The results of these physiologic studies cannot be explained by the known pharmacology of the central MC3 and MC4 receptors and therefore the existence of further, yet unknown, melanocortin receptors in the central nervous system has been suggested *(57)*.

Melanocortin peptide activity is complex at many levels. First, endogenous melanocortin peptides can activate more than one melanocortin receptor with almost equal potency. Second, many of the physiologic responses have been studied using superpotent synthetic agonists and antagonists which may not have the same effects in vivo as they do in vitro. Third, expression of MC3 and MC4 receptors overlaps in some regions of the brain and indeed, some neurons may express both receptors. If this is the case, there may be cross talk and even antagonism between these receptors. Finally, the recent identification of an agouti-related transcript (ART) encoding the agouti-related protein (AGRP) *(60)* indicates that there are endogenous antagonists for the melanocortin receptors in the central nervous system. Regardless of whether there are more than five melanocortin receptors, it is clear that melanocortin peptides and melanocortin receptors present a very sophisticated neuropharmacology that may take many years to unravel.

8. Summary and Conclusions

The melanocortin receptors are members of the large superfamily of rhodopsinlike GPCRs. The five identified melanocortin receptors form a unique subfamily of the GPCRs because they are not closely related to any other members of the superfamily and they have a number of unique structural characteristics. They are among, and include, the shortest GPCR due to having short amino-and carboxy-terminal ends, as well as short second extracellular and third intracellular loops. They lack several highly conserved amino acid residues in most rhodopsinlike GPCRs, which include the cysteine in the first extracellular loop, prolines in transmembranes 4 and 5, and the asparagine in transmembrane seven is substituted by an aspartic acid. The five melanocortin receptors exhibit distinct tissue specific expression and specificities. All five receptors couple to the PKA signaling pathway and four respond to multiple different melanocortin peptides including α-MSH. These four all respond to α-MSH and desacetyl-α-MSH with EC_{50}s ≤ 1 n*M*. The MC2-R is unique in that it responds to ACTH only. There are species differences in specificity for the MC1 and MC5 receptors. Three of the melanocortin receptors, MC3, MC4, and MC5, are more closely related to one another than to the MC1 and MC2 receptors. The most highly conserved amino acid sequence across species is the MC4-R (93%) and the least highly conserved across species is the MC1-R (76%). Chromosomal mapping has been completed on all five melanocortin receptors in the human and on four (MC1, MC2, MC3, and MC5) of the mouse.

Three melanocortin receptors (MC2, MC4, and MC5) map to human chromosome 18 and two of these (MC2 and MC5) map to exactly the same locus.

The melanocortin peptide and receptor system is complex due to various ligands having similar effects on different receptors, expression of more than one melanocortin receptor in some tissues, and the presence of endogenous antagonists for at least some of the melanocortin receptors. Some of the physiologic responses to melanocortin peptides are difficult to explain using the known specificity of the melanocortin receptors in vitro and therefore it is possible that there are more than five melanocortin receptors. The complexity of the melanocortin neuropharmacology in vivo, however, will no doubt continue to make the matching of melanocortin peptide physiologic effects to specific melanocortin receptors a challenge over many years.

Acknowledgments

This work is supported by a Wellcome Trust Senior Research Fellowship of NZ and The Health Research Council of NZ. I thank Dr. Laurence Dumont for her generous assistance with the preparation of the figures.

References

1. DeWied, D. and Jolles, J. (1982) Neuropeptides derived from pro-opiocortin: behavioral, physiological, and neurochemical effects. *Physiol. Rev.* **62,** 977–1059.
2. Strand, F. L., Rose, K. J., Zuccarelli, L. A., Kume, J., Alves, S. E., Antonawich, F. J. and Garrett, L. Y. (1991) Neuropeptide hormones as neurotrophic factors. *Physiol. Rev.* **71,** 1017–1046.
3. Catania, A. and Lipton, J. M. (1993) α-Melanocyte stimulating hormone in the modulation of host reactions. *Endocri. Rev.* **14,** 564–576.
4. Buckley, D. I. and Ramachandran, J. (1981) Characterization of corticotropin receptors on adrenocortical cells. *Proc. Natl. Acad. Sci. U. S. A.* **78,** 7431–7435.
5. Gallo–Payet, N. and Escher, E. (1985) Adrenocorticotropin receptors in rat adrenal glomerulosa cell. *Endocrinology.* **117,** 38–46.
6. Gallo–Payet, N. and Payet, M. D. (1989) Excitation–secretion coupling: involvement of potassium channels in ACTH-stimulated rat adrenocortical cells. *J. Endocrinol.* **120,** 409–421.
7. Pawelek, J. (1976) Factors regulating growth and pigmentation of melanoma cells. *J. Invest. Dermatol.* **66,** 201–209.
8. Tatro, J. B., Atkins, M., Mier, J. W., Hardarson, S., Wolfe, H., Smith, T., Entwistle, M. C. and Reichlin, S. (1990) Melanotropin receptors demonstrated *in situ* in human melanoma. *J. Clin. Invest.* **85,** 1825–1832.
9. Tatro, J. B. and Reichlin, S. (1987) Specific receptors for α-melanocyte-stimulating hormone are widely distributed in tissues of rodents. *Endocrinology.* **121,** 1900–1907.
10. Hnatowich, M. R., Queen, G., Stein, D., and LaBella, F. S. (1989) ACTH receptors in nervous tissue. High affinity binding-sequestration of $[^{125}I][Phe^2,Nle^4]$ ACTH 1–24 in homogenates and slices from rat brain. *Can. J. Physiol. Pharmacol.* **67,** 568–576.

11. Tatro, J. B. (1990) Melanotropin receptors in the brain are differentially distributed and recognize both corticotropin and α-melanocyte stimulating hormone. *Brain Res.* **536,** 124–132.
12. Libert, F., Parmentier, M., Lefort, A., Dinsart, C., Van Sande, J., Maenhaut, C., Simons, M. J., Dumont, J. E., and Vassart, G. (1989) Selective amplification and cloning of four new members of the G protein-coupled receptor family. *Science* **244,** 569–572.
13. Mountjoy, K. G., Robbins, L. S., Mortrud, M. T., and Cone, R. D. (1992) The cloning of a family of genes that encode the melanocortin receptors. *Science* **257,** 543–546.
14. Cone, R. D., Mountjoy, K. G., Robbins, L. S., Nadeau, J. H., Johnson, K. R., Roselli-Rehfuss, L., Mortrud, M. T., and Robinson, K. R. (1993) Cloning and functional characterization of a family of receptors for the melanotropic peptides. *Ann. N. Y. Acad. Sci.* **680,** 342–363.
15. Chhajlani, V. and Wikberg, E. S. (1992) Molecular cloning and expression of the human melanocyte stimulating hormone receptor cDNA. *FEBS Lett.* **309,** 417–420.
16. Gantz, I., Konda, K., Tashiro, T., Shimoto, Y., Miwa, H., Munzert, G., Watson, S. J., DelValle, J., and Yamada, T. (1993) Molecular cloning of a novel melanocortin receptor. *J. Biol. Chem.* **265,** 8246–8250.
17. Cone, R. D. and K. G. Mountjoy. (1993) Molecular genetics of the ACTH and melanocyte-stimulating hormone receptors. *Trends Endocrinol. Metab.* **4,** 242–247.
18. Probst, W. C., Snyder, L. A., Schuster, D. I., Brosius, J., and Sealfon, S. C. (1992) Sequence alignment of the G-protein coupled receptor superfamily. *DNA Cell Biol.* **11,** 1–20.
19. Star, R. A., Rajora, N., Huang, J., Stock, R. C., Catania, A., and Lipton, J. M. (1995) Evidence of autocrine modulation of macrophage nitric oxide synthase by α-melanocyte-stimulating hormone. *Proc. Natl. Acad. Sci U. S. A.* **92,** 8016–8020.
20. Xia, Y., Wikberg, J. E. S., and Chhajlani, V. (1995) Expression of melanocortin 1 receptor in periaqueductal gray matter. *Neuroreport* **6,** 2193–2196.
21. Mountjoy, K. G. (1994) The human melanocyte stimulating hormone receptor has evolved to become "super-sensitive" to melanocortin peptides. *Mol. Cell Endocrinol.* **102,** R7–R11.
22. Robbins, L. S., Nadeau, J. H., Johnson, K. R., Kelly, M. A., Roselli–Rehfuss, L., Bacck, E., Mountjoy, K. G., and Cone, R. D. (1993) Pigmentation phenotypes of variant extension locus alleles result from point mutations that alter MSH receptor function. *Cell* **72,** 827–834.
23. Gantz, I., Tashiro, T., Barcroft, C., Konda, Y., Shimoto, Y., Miwa, H., and J. M. Trent, H. (1994) Mapping of the gene encoding the melanocortin-1 (α-melanocyte stimulating hormone) receptor (MC1-R) to human chromosome 16q24.3 by fluorescence *in situ* hybridization. *Genomics* **19,** 394–395.
24. Magenis, R. E., Smith, L., Nadeau, J. H., Johnson, K. R., Mountjoy, K. G., and Cone, R. D. (1994) Mapping of the ACTH, MSH, and neural (MC3 and MC4) melanocortin receptors in the mouse and man. *Mamm. Genome* **5,** 503 –508.
25. Copeland, N. G., Jenkins, N. A., Gilbert, D. J., Eppig, J. T., Maltais, L. C., Miller, J. C., Dietrich, W. F., Weaver, A., Lincoln, S. E., Steen, R. G., Stein, L. D., Nadeau, J. H., and Lander, E. S. (1993) A genetic map of the mouse: current applications and future prospects. *Science* **262,** 57–66.

26. Nadeau, J. H., Davisson, M. T., Doolitlle, D. P., Grant, P., Hillyard, A. P., Kosowsky, M. R., and Roderick, T. H. (1992) Comparative map for mice and humans. *Mamm. Genome* **3,** 480–536.
27. Cammas, F. M., Kapas, S., Barker, S., and Clark, A. J. L. (1995) Cloning, characterization and expression of a functional mouse ACTH receptor. *Biochem. Biophys. Res. Commun.* **211,** 912–918.
28. Raikhinstein, M., Zohar, M. and Hanukoglu, I. (1994) cDNA cloning and sequence analysis of the bovine adrenocorticotropic hormone (ACTH) receptor. *Biochim. Biophys. Acta.* **1220,** 329–332.
29. Naville, D., Barjhous, L., Jaillard, C., Lebrethon, M. C., Saez, J. M., and Begeot M. (1994) Characterization of the transcription start site of the ACTH receptor gene: presence of an intronic sequence in the 5'-flanking region. *Mol. Cell Endocrinol.* **106,** 131–135.
30. Naville, D., Jaillard, C., Barjhoux, L., Durand, P., and Begeot, M. (1997) Genomic structure and promoter characterization of the human ACTH receptor gene. *Biochemi. Biophys. Res. Communi.* **230,** 7–12.
31. Shimizu, C., Kubo, M., Saeki, T., Matsumura, T., Ishizuka, T., Kijima, H., Kakinuma, M., and Koike, T. (1997) Genomic organization of the mouse adrenocorticotropin receptor. *Gene* **188,** 17–21.
32. Clark, A. J. L., Cammas, F. M., and Kapas, S. (1996) Expression of the mouse ACTH receptor gene and characterisation of its promoter. *Endocr. Res.* **22,** 333–335.
33. Slominski, A. J., Ermak, G., and Mihm, M. (1996) ACTH receptor, CYP11A1, CYP17 and CYP21A2 genes are expressed in skin. *J. Clin. Endocrinol. and Metab.* **81,** 2746–2749.
34. Boston, B. A. and Cone, R. D. (1996) Characterization of melanocortin receptor subtype expression in murine adipose tissues and in the 3T3-L1 cell line. *Endocrinology* **137,** 2043–2050.
35. Mountjoy, K. G., Bird, I. M., Rainey, W. E., and Cone, R. D. (1994) ACTH induces up-regulation of ACTH receptor mRNA in mouse and human adrenocortical cell lines. *Mol. Cell Endocrinol.* **99,** R17–R20.
36. Yang, Y.-K., Ollmann, M. M., Wilson, B. D., Dickinson, C., Yamada, T., Barsh, G. S., and Gantz, I. (1997) Effects of recombinant agouti-signalling protein on melanocortin action. *Mol. Endocrinol.* **11,** 274–280.
37. Kapas, S., Cammas, F.M., Hinson, J. P., and Clark, A. J. L. (1996) Agonist and receptor binding properties of adrenocorticotropin peptides using the cloned mouse adrenocorticotropin receptor expressed in a stably transfected HeLa cell line. *Endocrinology* **137,** 3291–3294.
38. Gantz, I., Tashiro, T., Barcroft, C., Konda, Y., Shimoto, Y., Miwa, H., Glover, T., Munzert, G., and Yamada, T. (1993) Localization of the genes encoding the melanocortin-2 (adrenocorticotropic hormone) and melanocortin-3 receptors to chromosome 18p11.2 and 20q13.2–q13.3 by fluorescence *in situ* hubridization. *Genomics* **18,** 166–167.
39. Vamvakppoulos, N. C., Rojas, K., Overhauser, J., Durkin, A. S., Nierman, W. C., and Chrousos, G. P. (1993) Mapping the human melanocortin 2 receptor (adrenocorticotrophic hormone receptor; ACTHR) gene (MC2-R) to the small arm of chromosome 18 (18p11.21–pter). *Genomics* **18,** 454–455.
40. Desarnaud, F., Labbe, O., Eggerickx, D., Vassart, G., and Parmentier, M. (1994) Molecular cloning, functional expression and pharmacological characterization of a mouse melanocortin receptor gene. *Biochem. J.* **299,** 367–373.

41. Roselli–Rehfuss, L., Mountjoy, K. G., Robbins, L. S., Mortrud, M. T., Low, M., Tatro, J. B., Entwistle, M. L., Simerly, R., and Cone, R. D. (1993) Identification of a receptor for γ melanotropin and other proopiomelanocortin peptides in the hypothalamus and limbic system. *Proc. Natl. Acad. Sci. U. S. A.* **90,** 8856–8860.
42. Schioth, H. B., Muceniece, R., Wikberg, J. E. S., and Szardenings, M. (1996) Alternative translation initiation codon for the human melanocortin MC3 receptor does not affect the ligand binding. *Eur. J. Pharmacol.* **314,** 381–384.
43. Bell, G. I., Xiang, K., Newman, M. V., Wu, S., Wright, L. G., Fajans, S. S., Spielman, R. S., and Cox, N. J. (1991) Gene for non–insulin–dependent diabetes mellitus (maturity–onset diabetes of the young subtype) is linked to DNA polymorphism on human chromosome 20q. *Proc. Natl. Acad. Sci. U. S. A.* **88,** 1484–1488.
44. Yamada, Y., Xiang, K., Bell, G. I., Seino, S., and Nishi, M. (1992) Dinucleotide repeat polymorphism in a gene on chromosome 20 encoding a G-protein coupled receptor (D20S32e). *Nucleic Acids Res.* **19,** 2519.
45. Mountjoy, K. G., Mortrud, M. T., Low, M. J., Simerly, R. B., and Cone, R. D. (1994) Localization of the melanocortin-4 receptor (MC4-R) in neuroendocrine and autonomic control circuits in the brain. *Mol. Endocrinol.* **8,** 1298–1308.
46. Gantz, I., Miwa, H., Konda, Y., Shimoto, Y., Tashiro, T., Watson, S. J., DelValle, J., and Yamada, T. (1993) Molecular cloning, expression, and gene localization of a fourth melanocortin receptor. *J. Biol. Chem.* **268,** 15,174–15,179.
47. Alvaro, J. D., Tatro, J. B., Quillan, J. M., Fogliano, M., Eisenhard, M., Lerner, M. R., Nestler, E. J., and Duman, R. S. (1996) Morphine down–regulates melanocortin-4 receptor expression in brain regions that mediate opiate addiction. *Mol. Pharmacol.* **50,** 583–591.
48. Schioth, H. B., Muceniece, R., and Wikberg, J. E. S. (1996) Characterisation of the melanocortin 4 receptor by radioligand binding. *Pharmacol. Toxicol.* **79,** 161–165.
49. Chhajlani, V., Muceniece, R., and Wikberg, J. E. S. (1993) Molecular cloning of a novel human melanocortin receptor. *Biochem. Biophys. Res. Commun.* **195,** 866–873.
50. Barrett, P., MacDonald, A., Helliwell, R., Davidson, G., and Morgan, P. (1994) Cloning and expression of a new member of the melanocyte-stimulating hormone receptor family. *J. Mol. Endocrinol.* **12,** 203–213.
51. Labbe, O., Desarnaud, F., Eggerickx, D., Vassart, G., and Parmentier, M. (1994) Molecular cloning of a mouse melanocortin 5 receptor gene widely expressed in peripheral tissues. *Biochemistry* **33,** 4543–4549.
52. Griffon, N., Mignon, V., Facchinetti, P., Diaz, J., Schwartz, J.-C., and Sokoloff, P. (1994) Molecular cloning and characterisation of the rat fifth melanocortin receptor. *Biochem. Biophys. Res. Commun.* **200,** 1007–1014.
53. Gantz, I., Shimoto, Y., Konda, Y., Miwa, H., Dickson, C. J., and Yamada, T. (1994) Molecular cloning, expression, and characterization of a fifth melanocortin receptor. *Biochem. Biophys. Res. Commun.* **200,** 1214–1220.
54. Chowdhary, B. P., Gustavsson, I., Wikberg, J. E. S., and Chhajlani, V. (1995) Localization of the human melanocortin-5 receptor gene (MC5-R) to chromosome band 18p11.2 by fluorescence in situ hybridization. *Cytogenet. Cell Genet.* **68,** 79–81.
55. Gilchrist, D. P. D., Darlington, C. L., and Smith, P. F. (1994) Org 2766 treatment prevents disruption of vestibular compensation by an NMDA receptor antagonist. *Eur. J. Pharmacol.* **252,** R1–R2.

56. Spruijt, B., Losephy, M., Van Rijzingen, I., and Maaswinkel, H. (1994) The ACTH(4–9) analog Org 2766 modulates the behavioural changes induced by NMDA and NMDA receptor antagonist AP5. *J. Neurosci.* **14,** 3225–3230.
57. Li, S.–J., Varga, K., Archer, P., Hruby, V. J., Sharma, S. D., Kesterson, R. A., Cone, R. D., and Kunos, G. (1996) Melanocortin antagonists define two distinct pathways of cardiovascular control by α- and γ-melanocyte-stimulating hormones. *J. Neurosci.* **16,** 5182–5188.
58. Van Bergen, P., Kleijne, J. A., De Wildt, D. J., and Versteeg, D. H. G. (1997) Different cardiovascular profiles of three melanocortins in conscious rats; evidence for antagonism between γ2-MSH and ACTH-(1–24). *Br. J. Pharmacol.* **120,** 1561–1567.
59. Huang, Q.-H., Entwistle, M. L., Alvaro, J. D., Duman, R. S., Hruby, V. J., and Tatro, J. B. (1997) Antipyretic role of endogenous melanocortins mediated by central melanocortin receptors during endotoxin-induced fever. *J. Neurosci.* **17,** 3343–3351.
60. Shutter, J. R., Graham, M., Kinsey, A. C., Scully, S., Luthy, R., and Stark, K. L. (1997) Hypothalamic expression of ART, a novel gene related to agouti, is up-regulated in obese and diabetic mutant mice. *Genes Dev.* **11,** 593–602.
61. Clark, A. J. L. and Weber, A. (1994) Molecular insights into inherited ACTH resistance syndromes. *Trends Endocrinol Metab.* **5,** 209–214.
62. Mountjoy, K. G. and Wild, J. M. (1998) Melanocortin-4 receptor mRNA expression in the developing autonomic and central nervous systems. *Developmental Brain Research* **107,** 309–314.
63. Chen, W., Kelly, M. A., Opitz-Araya, X., Thomas, R. E., Low, M. J., and Cone, R. D. (1997) Exocrine gland dysfunction in the MC5-R-deficient mice: Evidence for coordinated regulation of exocrine gland function by melanocortin peptides. *Cell* **91,** 789–798.
64. Cone, R. D., Lu, D., Koppula, S., Inge Vage, D., Klungland, H., Boston, B., Chen, W., Orth, D., Pouton, C., and Kesterson, R. A. (1996) The melanocortin receptors: Agonists, antagonists, and the hormonal control of pigmentation. *Recent Progress Hormone Research* **51,** 287–318.

Part III

Biochemical Mechanism of Receptor Action

CHAPTER 8

The Molecular Pharmacology of Alpha-Melanocyte Stimulating Hormone

Structure–Activity Relationships for Melanotropins at Melanocortin Receptors

Victor J. Hruby and Guoxia Han

1. Introduction

α-Melanotropin (α-melanocyte stimulating hormone, α-MSH), Ac-Ser-Tyr-Ser-Met-Glu-His-Phe-Arg-Trp-Gly-Lys-Pro-Val-NH_2, is one of the first peptide hormones isolated from the pituitary gland. The biologic functions of α-MSH and structure–activity relationships of α-MSH analogues before 1993 have been extensively researched and reviewed *(1–12)*. This hormone plays a well-known and principal role in pigmentation. Other effects ascribed to this hormone include:

1. Regulation of the release of pituitary and peripheral hormones, such as somatostatin and corticotropin
2. Sebotrophic effects, adrenal steroidogenesis, immune response, and cardiovascular and metabolic effects
3. Important roles in the nervous system, particularly in the central nervous system (CNS)

Examples of the latter proposed functions include:

A. Effects on attention, learning, memory, motivation, locomotor, sleep, and numerous behavioral changes in grooming and stretching–yawning syndrome (SYS), sexuality, food intake (associated with obesity)
B. Effects on neurotransmission such as cholinergic and dopaminergic systems
C. Neurochemical effects

The Melanocortin Receptors Ed.: R. D. Cone

D. Hypothermic and antipyretic effects
E. Effects on nerve regeneration
F. Interactions with other neuroactive compounds, such as opiates

All of these effects presumably are associated with melanocortin receptors (MCRs). To date, five such receptors *(13–19)*, the MC1-R (pigmentation receptor), MC2-R* (ACTH receptor; recent research has revealed that the MC2-R binds ACTH with high affinity, but that MC2-R does not bind MSH peptides *[20]*), MC3-R, MC4-R, and MC5-R, have been identified. It has been well established that the effects of α-MSH on pigmentation are specifically mediated by the MC1-R. However, given the complexity of expression of the MC3-R, MC4-R, and MC5-R, a simple correlation between these receptors and the biologic effects of their ligands mentioned above have not been established yet. Hence research has been focused on developing potent ligands with specific receptor binding. Many exciting results, particularly in the discovery of potent antagonists, have been reported since our last review in 1993 *(10)*. The progress since then is our focus in this paper.

2. Agonists

2.1. MT-I

As already mentioned, α-MSH has been the subject of extensive structure–activity studies. We provided an overview in 1993 *(10)*, and it is not our intention to discuss again the results presented therein. However, a brief summary is worthwhile as a starting point for a discussion of more recent results. Some of the key general early findings are listed in Table 1 *(10)*. These findings led to the discovery of a series of conformationally constrained and potent α-MSH agonist analogues (Table 2) for the MC1-R.

The first major discovery of enhanced potency for D-Phe^7 analogues suggested the importance of a reverse β-turn *(21)*, which is formed in the active core sequence (His-Phe-Arg-Trp) of α-MSH. By replacing methionine with norleucine (Nle, which has a side chain group pseudoisosteric to Met), the potency of the resulting melanotropin also was increased significantly as a result of the chemically inert and hydrophobic side chain of Nle. The side chain of methionine is rather easily oxidized to its sulfone form, which increased the hydrophilicity dramatically and decreased the bioactivity of the corresponding derivative in the amphibian skin pigmentation assays. These findings led to the discovery of the first-generation superpotent α-MSH agonist, [Nle^4, D-Phe^7]-α-MSH (NDP-MSH, MT-I, Ac-Ser-Tyr-Ser-Nle^4-Glu-His-D-Phe^7-Arg-Trp-Gly-Lys-Pro-Val-NH_2) *(21)*.

*The MC2-R (ACTH receptor) is beyond the scope of this review.

Table 1
Summary of Classical Structure–Activity Relationship for α-MSH

α-MSH: Ac-Ser^1-Tyr^2-Ser^3-Met^4-Glu^5-*His^6-Phe^7-Arg^8-Trp^9*-Gly^{10}-Lys^{11}-Pro^{12}-Val^{13}-NH_2
1. *N*-Acetylation and C-terminal amidation important for potency
2. Core sequence (minimum active sequence): His-Phe-Arg-Trp for frog skin bioassay
3. D-Configuration in position 7 enhances β-turn, resistant to proteolytic hydrolysis
4. Residues in positions 1, 2, 3, 12, and 13 are relatively less important
5. Hydrophobic and unnatural Nle replaces proteolytic Met in position 4 reduced problematic oxidation of Met
6. Truncation of Gly in position 10 increases the potency for frog skin bioassay
7. Bridges (S-S) between positions 4 and 10 increased constraints and potency
8. Lactam bridges between positions 5 and 10 increased constraints and potencies dramatically

Data taken in part from ref. *10*.

Table 2
Relative Biologic Potencies of α-MSH[a] and its Analogues

α-MSH Analogues	Frog Skin	Lizard Skin	Melanoma Tyrosinase
[Nle^4, *D*-Phe^7]-α-MSH, NDP-MSH (MT-I)	22.0	2.0	100
c[Cys^4, Cys^{10}]-α-MSH	20.0	4.0	1.0
Ac-[Nle^4, Asp^5, *D*-Phe^7, Lys^{10}]-α-MSH[4–10]-NH_2	1.0	5.0	—
Ac-Nle^4, *c*[Asp^5, *D*-Phe^7, Lys^{10}]-α-MSH[4–10]-NH_2(MT-II)	0.5	90.0	100

Adapted from ref. *10*.
[a]All potencies relative to α-MSH = 1.0.

2.2. Development of MT-II

In order to enhance the reverse turn effects in the design of α-MSH analogues, we investigated an approach using a pseudoisosteric Cys^4, Cys^{10} cyclic bridge (Table 2) *(22,23)* in α-MSH which made the resulting analogue more constrained than the linear α-MSH analogues.

Later, with the aid of computer modeling and nuclear magnetic resonance (NMR) studies, various analogues were analyzed to determine the relationship between their conformation and potency. This led to the discovery of a new generation of potent α-MSH agonists (for the MC1-R) in which Gly^{10} was replaced by a Lys residue and Glu^5 was reduced one carbon atom in its side chain to Asp^5. Further removal of the relatively unimportant residues in positions 1, 2, 3, 11, 12, and 13, provided analogues such as Ac-[Nle^4, Asp^5, D-Phe^7, Lys^{10}]α-MSH[4–10]-NH_2, which were very potent. By cyclizing *via*

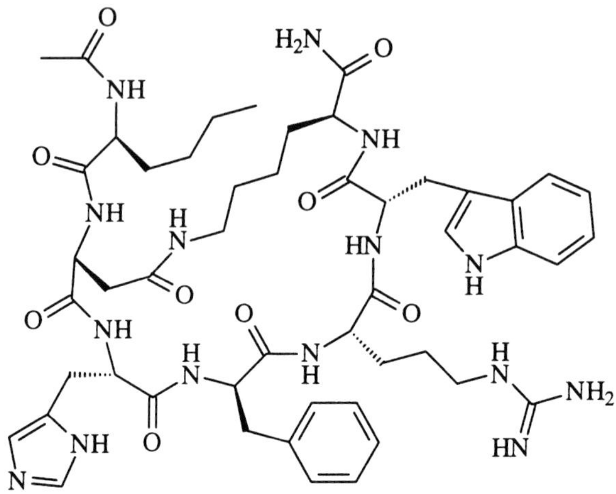

Ac-Nle-Asp-His-*D*Phe-Arg-Trp-Lys-NH_2

Fig. 1. The structure of MT-II.

the side chain group in Asp^{5} and Lys^{10} to form a lactam bridge, a superpotent (with prolonged activity) α-MSH agonist MT-II (Ac-Nle,^{4} *c*[Asp^{5}, D-Phe^{7}, Lys^{10}]α-MSH[4–10]-NH_2, Fig. 1) was obtained *(24,25)*.

The recent discovery of MT-II erectogenic activity in humans *(M. E. Hadley and V. J. Hruby, unpublished result)* and other chemical studies demonstrated that this α-MSH analogue not only is very resistant to proteolysis, but also may pass the blood–brain barrier (BBB). This demonstrates also that this compound may work effectively at more than one melanocortin receptor. MT-II, together with previously reported fatty acid conjugates of α-MSH analogues *(26,27)* and biotin-labeled NDP-MSH *(28)*, are being examined for passage through the BBB.

2.3. MT-II Derivatives

Previous structure–activity and conformation–activity studies have led our group and others to propose bioactive conformations for α-MSH at the classical MC1-R *(23–25,29,30)*. These studies indicate that the side-chain residues in the β-turn region (His^{6}, Phe^{7}, Arg^{8}, Trp^{9}) are critical for agonist activity. In the past, we have found that position 7 is very critical for the potencies of all the α-MSH analogues studied. As discussed above, MT-II is one of the smallest and most constrained melanotropins with potent and prolonged activity in the frog (*Rana pipiens*) skin bioassay (FSB). Given the significance of these side chain groups, it is reasonable to expect that substitution of the phenyl group of D-Phe^{7} in MT-II will affect the bioactivity. Three

Table 3
Activities of MT-II Derivatives at Various MC1-Rs

	EC_{50} values (n*M*)		
Compound	Frog Skin	mMC1-R	hMC1-R
α-MSH	0.10	1.3	0.091
[D-Phe(pF)^7]-MT-II	0.10	0.026	0.016
[D-Phe(pCl)^7]-MT-II	2.0	0.0095	0.0050
[D-Phe(pI)^7]-MT-II	Antagonist pA_2 = 10.3	0.19	0.055
[D-Nal(2')^7]-MT-II	Antagonist pA_2 = 10.5	0.039	0.036

Data adapted from ref. *31*.
MT-II: Ac-Nle-*c*[Asp-His-D-Phe-Arg-Trp-Lys]-NH_2.

different halogen substituents (Table 3) in the *para* position have been examined *(31)*. For all the receptors (frog skin, mMC1-R, hMC1-R, hMC3-R, hMC4-R and mMC5-R) assayed, the fluoro and chloro MT-II analogues showed potent agonist activities. In the FSB, [D-Phe(pF)^7]-MT-II (EC_{50} = 0.10 ± 0.035n*M*) is 20 times more potent than [D-Phe(pI)^7]-MT-II (EC_{50} = 2.0 ± 0.80n*M*). However, for the mMC1-R and hMC1-R, the results are quite different. [D-Phe(pF)^7]-MT-II (EC_{50} = 0.026 ± 0.010n*M*) is one third as potent as [D-Phe(pCl)^7]-MT-II (EC_{50} = 0.0095 ± 0.0053n*M*) for the mMC1-R. [D-Phe(pF)^7]-MT-II (EC_{50} = 0.016 ± 0.003n*M*) is also one third as potent as [D-Phe(pCl)^7]-MT-II (EC_{50} = 0.005 ± 0.004n*M*) for the mMC1-R. There is no good correlation among these two analogues for the various MCRs. If the substituent is iodine, the resulting MT-II derivative, [D-Phe(pI)^7]α-MSH[4–10]-MT-II, displayed no agonist activity at all in the FSB but is a potent antagonist. However, this compound exhibited potent agonist activity in the mMC1-R and hMC1-R bioassays. It is hard to draw any clear structure–activity relationships, just because the [D-Phe(pCl)^7]-MT-II is more potent than [D-Phe(pF)^7]-MT-II and [D-Phe(pI)^7]-MT-II for the hMCR-1 and mMCR-1. Neither steric nor electronic effects can explain these activities adequately. Currently we are investigating the [D-Phe(pBr)^7]-MT-II analogue and the use of other novel aromatic amino acids in this position.

Other ways to incorporate substituents into the phenyl ring of D-Phe^7 include fusing a phenyl ring at the *ortho* and *meta* positions of the phenyl group to another aromatic ring (Nal(1'), Fig. 2). The analogue ([D-Nal(1')^7]-MT-II or SL1-12) showed full agonist activity in all the receptors tested *(S. Lim and V. J. Hruby, unpublished results)*. However, by fusing this phenyl ring at the *meta* and *para* positions (Nal(2'), Fig. 2) of D-Phe^7 in MT-II, the resulting analogue, [D-Nal(2')^7]-MT-II (SHU9119), showed a unique bioactivity profile as an agonist with EC_{50} values ranging from subnanomolar to picomolar

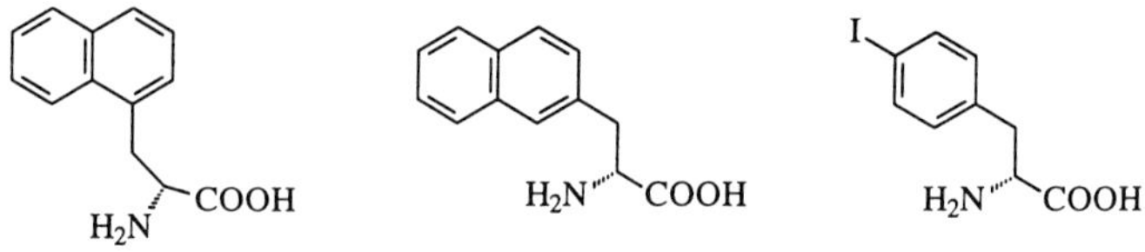

DNal(1')　　DNal(2')　　DPhe(pI)

Fig. 2. Special D-phenylalanine-related analogues with substituted phenyl rings.

Table 4
Bioactivities for D-Phe[7] Substituted MT-II Analogues at the Different MCRs

	EC_{50} values (p*M*)			
Compound	hMC1-R	hMC3-R	hMC4-R	mMC5-R
α-MSH	91	669	210	807
NDP-MSH	23	132	17	ND[a]
[D-Phe(pF)[7]]-MT-II	16	191	19	1360
[D-Phe(pCl)[7]]-MT-II	5.0	63	18	117
[D-Phe(pI)[7]]-MT-II	55	1130 pA_2 = 8.3[b]	573 pA_2 = 9.7[b]	684 Partial agonist
[D-Nal(2')[7]]-MT-II	36	2810 pA_2 = 8.3[b]	No activity pA_2 = 9.3	434 Full agonist

Data from ref. *31*.
MT-II: Ac-Nle-*c*[Asp-His-*D*-Phe-Arg-Trp-Lys]-NH_2.
[a]Not determined.
[b]Partial agonist.

(Table 3) for the mMC1-R and hMC1-R, but as an antagonist in the FSB. As to the agonist activities for the mMC1-R and the hMC1-R, SHU9119 is slightly better than [D-Phe(pI)[7]]-MT-II. No agonistic activity for SHU9119 was found in the FSB.

We are currently investigating the cause of these distinct differences in the bioactivities for [D-Nal(1')[7]]-MT-II (SL-1-12) and [D-Nal(2')[7]]-MT-II (SHU9119). Though the spatial differences between D-Nal(1') and D-Nal(2') are small, they interact in uniquely different ways with the MCRs.

The substituted D-Phe[7] MT-IIs have been further tested using other cloned human MCRs, including the most recently discovered mMC5-R *(19–31)* (Table 4). There is virtually no difference in bioactivities between NDP-MSH and [D-Phe(pF)[7]]-MT-II in the receptors tested. On the other hand, the *p*-chlorophenylalanine[7]-substituted MT-II derivative displayed potent agonist activities for all the receptors, while the bulky *p*-iodo-substituted MT-II derivative was a partial agonist for the hMC3-R, hMC4-R and mMC5-R. The potencies of D-Phe(pI)[7] at these receptors are much lower than that of the *p*-fluoro-substituted MT-II.

Again, SHU9119 displayed almost the same pattern in activities as [D-Phe(pI)[7]]-MT-II at the hMC1-R and the hMC3-R. Though SHU9119 is twice as potent as [D-Phe(pI)[7]]-MT-II at the hMC1-R, the result is just the opposite for the hMC3-R. For the hMC4-R, SHU9119 has no agonist activity, while [D-Phe(pI)[7]]-MT-II is a weak partial agonist with subnanomolar antagonist activity. However, for the mMC5-R, SHU9119 is a full agonist, while [D-Phe(pI)[7]]-MT-II maintains its partial agonist activity with an agonist EC_{50} value of 684 ± 227p*M*. All these results demonstrate that the D-*p*-iodophenylalanine and 2'-naphthylalanine residues may share common spatial features for binding to MCRs. On the other hand, Ac-Nle-*c*[Asp-His-D-Nal(1')-Arg-Trp-Lys]-NH_2 (SL-1-12) is an agonist for all these receptors.

Based on these observations, we believe that slight changes in the side chain conformations of the aromatic residues in position 7 of MT-II can have significant effects on the bioactivities of MT-II derivatives for the MCRs.

Other MT-II derivatives have been reported recently. However, the cyclized analogues were composed with larger rings, which potentially made the analogues more flexible than MT-II. Neither the potencies nor the selectivities (at hMCRs) of these analogues were improved compared with the results from MT-II *(32)*.

2.4. Incorporated of β-Branched Amino Acids into MT-II

Another approach to study structure–activity relationships is by incorporation of novel constrained amino acids into α-MSH analogues. Recently β-methyl–substituted tryptophans (Fig. 3) have been incorporated into MT-II (Table 5) *(33)*. By methyl substitution of the β-carbon in tryptophan, χ space will be constrained. Hence the side chain (indole) cannot rotate as freely as in Trp. As a result, the peptide will have certain restricted conformations that in turn can increase receptor binding preferences.

[(2*S*, 3*R*)-β-Me-Trp[9]]-MT-II showed much higher potency than the other three stereoisomers. Presumably this derivative possesses a better conformation for binding at the MC1-Rs. Comparison of the potency of [(2*S*, 3*S*)-β-Me-Trp[9]]-MT-II with [(2*S*, 3*R*)-β-Me-Trp[9]]-MT-II shows that the former analogue is 60-fold more potent in the FSB. These compounds showed differential prolonged biologic activity at different MCRs. Examination of the results in Table 5 suggest that incorporation of constrained amino acids into MT-II or other melanotropins can further improve bioactivities, provided that these new analogues possess the right conformation for binding to their target receptors.

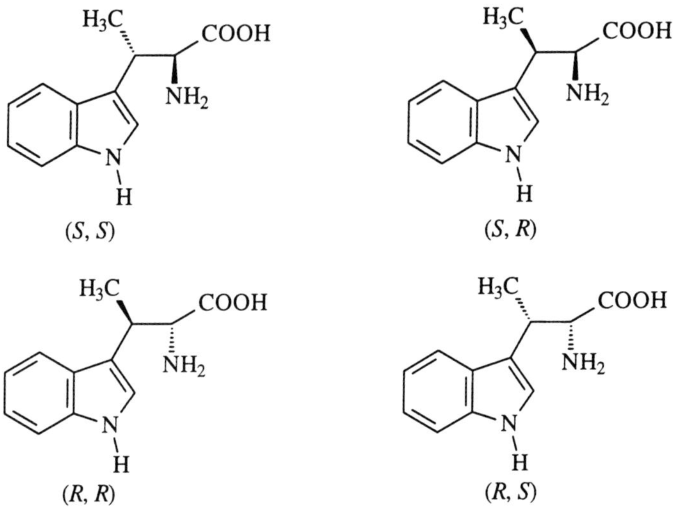

Fig. 3. Structures of β-methyltryptophans.

Table 5
Comparative Biologic Activities of [β-Me-Trp[9]]-MT-II

	EC^{50} (n*M*)		
Compound	Frog Skin	hMC1-R Binding	hMC1-R cAMP
MT-II	0.10	0.50	0.15
[(2*S*, 3*S*)-β-Me-Trp[9]]-MT-II	0.44	0.50	0.30
[(2*S*, 3*R*)-β-Me-Trp[9]]-MT-II	29	15	3.0
[(2*R*, 3*R*)-β-Me-Trp[9]]-MT-II	0.30	2.0	0.40
[(2*R*, 3*S*)-β-Me-Trp[9]]-MT-II	0.060	3.0	1.0

Data adapted from ref. *33*.
MT-II: Ac-Nle-*c*[Asp-His-D-Phe-Arg-Trp-Lys]-NH_2.

3. Antagonists

Despite the great achievements in structure–activity of α-MSH agonist analogues in the past decades, the search for modestly potent α-MSH specific competitive antagonists was not successful until 1989 *(9,34–36)*. Though a few early studies suggested weak or partial inhibition of α-MSH by α-MSH analogues or α-MSH-related fragment derivatives *(8,37–40)*, these were not confirmed in later research. Since then, the design and synthesis of α-MSH antagonists has been slow due to a lack of understanding of antagonist structure–activity relationships.

MSH(6-11): --**His**—**Phe**—**Arg**—Trp—Gly—**Lys**—Pro--

[His[1], Lys[6]]-GHRP: --**His**—DTrp—Ala—**Trp**—**DPhe**—**Lys**—$\mathbf{NH_2}$

⇓

MSH-[His[1], Lys[6]]-GHRP: --**His**—**Phe**—**Arg**—**Trp**—**DPhe**—**Lys**—$\mathbf{NH_2}$

Fig. 4. Hybridized Analogues from MSH[6–11] and [His[1], Lys[6]]GHRP.

Table 6
Comparative Melanotropic Activities of MSH-GHRP Hybrid Analogues

Compound	EC_{50} (μM)[a] or pA_2[b]	
H-His-XXX[7]-YYY[8]-Trp-D-Phe-Lys-NH_2	*R. pipiens*	*A. carolinensis*
D-Trp-Ala	4.7[b]	Inactive[c]
D-Phe-Ala	1.0[a]	1.0[a]
D-Ala-Ala	Inactive[c]	Inactive[c]
D-Arg-Ala	5.0[b]	6.0[b]
D-Trp-Arg	5.5[b]	5.6[b]
Phe-Arg	5.8[b]	Inactive[c]

See ref. *34*.
[a]Agonist.
[b]Antagonist.
[c]No activity was observed in a compound tested at a concentration ≥ $10^{-5}M$ *(34)*.

3.1. α-MSH and GHRP Hybrid Analogues

The first generation α-MSH antagonist (for MC1-R) was designed based upon the structural similarities between MSH[6–11] and [His[1], Lys[6]]GHRP (somatotropin growth hormone releasing peptide) (Fig. 4), the latter of which was discovered to be an antagonist (Table 6) in the frog (*R. pipiens*) melanocyte assay. By hybridizing residues in positions 2 and 3 of [His[1], Lys[6]]GHRP with residues in positions 7 and 8 of α-MSH, the resulting α-MSH-GHRP hybrid analogues (Table 6) were achieved.

A few of the designed hybrids were antagonists in both the frog (*R. pipiens*) and lizard (*Anolis. carolinesis*) bioassay, though all the pA_2 values were around 5~6 *(34)*. One of the most intriguing results is that the analogue *H*-His-Phe-Arg-Trp-D-Phe-Lys-NH_2, which has the core active sequence (His-Phe-Arg-Trp) for agonist activity, is an antagonist in the FSB, but is inactive in the lizard assay. Another interesting point is that only two antagonists, *H*-His-D-Arg-Ala-Trp-D-Phe-Lys-NH_2 and *H*-His-D-Trp-Arg-Trp-D-Phe-Lys-

Table 7
Relative in vitro Potencies of α-MSH Analogues
in Lizard *(A. carolinesis)* Skin Bioassay

α-MSH Fragment	Potency Relative to α-MSH
α-MSH 1.0	
Ac-α-MSH[4–12]-NH_2	5.0
Ac-α-MSH[4–10]-NH_2	0.05
Ac-α-MSH[5–10]-NH_2	0.0007
Ac-α-MSH[6–10]-NH_2	0.0007
Ac-α-MSH[6–9]-NH_2	0.000014
Ac-α-MSH[6–8]-NH_2	0.0
Ac-α-MSH[6–7]-NH_2	0.0
Ac-α-MSH[7–10]-NH_2	0.0
Ac-α-MSH[7–9]-NH_2	0.0
Ac-α-MSH[7–8]-NH_2	0.0
Ac-α-MSH[11–13]-NH_2	0.0

Data adapted from ref. *42*.

NH_2 were found for the lizard, and both of them have either a D-Arg or L-Arg in the 7 or 8 position. This means that positive charges in the side chain may be a requirement for antagonist activity in the lizard, although this requirement is generally not necessary for the frog assay.

One of these analogues (H-His-D-Arg-Ala-Trp-D-Phe-Lys-NH_2), was further tested in two lizards, *Sceloporus jarrovii* and *Urosaurus ornatus (41)*. In *U. ornatus* melanocytes, this analogue displayed potent antagonist activity with a pA_2 value of 7.0. However, in the *S. jarrovii* melanocytes, maximal responses to α-MSH were not achieved in the presence of the antagonist at concentrations higher than $10^{-6}M$. Therefore the pA_2 value could not be determined. This analogue was also tested in vivo assays (at both types of lizards); again the antagonist activity was higher (more than tenfold) in *U. ornatus* than in *S. jarrovii*.

These in vitro and in vivo results demonstrated that despite interspecies variation in potency, this α-MSH antagonist exhibited competitive inhibition of α-MSH on a variety of melanocyte receptors such as the MC1-Rs in *A. carolinesis, S. jarrovii* and *U. ornatus*.

3.2. Truncated α-MSH Analogues

By systematically truncating α-MSH from both the N-terminus and the C-terminus, it was previously found that certain fragments of α-MSH, such as Ac-α-MSH[7–10]-NH_2, expressed no activity in the lizard skin bioassay (*A. carolinesis*, Table 7) *(42)*. Later it was found that this sequence—Ac-α-MSH

Table 8
Melanotropic Activities of Ac-[XXX^7, YYY^{10}]-α-MSH[7–10]-NH_2

Ac-XXX^7-Arg^8-Trp^9-YYY^{10}-NH_2		*R. pipiens*		*A. carolinesis*	
		EC_{50} *(M)*	pA_2	EC_{50} *(M)*	pA_2
XXX	YYY	Agonist	Antagonist	Agonist	Antagonist
Phe	Gly	Inactive[a]	NI[b]	Inactive[a]	4.3
Phe	D-Phe	1.3×10^{-6}	NA	Inactive[a]	5.0
D-Trp	D-Phe	Inactive[a]	4.8	Inactive[a]	5.7
D-Phe	D-Phe	1.5×10^{-4}	NA	Inactive[a]	4.8
D-Phe	Gly	1.4×10^{-5}	NA	1.5×10^{-6}	NA

NA, not applicable.
Data adapted from ref. *43*.
[a]Concentration tested is up to 10^{-5} to 10^{-4}*M*.
[b]No inhibition at concentration of 10^{-5} to 10^{-4}*M*.

[7–10]-NH_2 [Ac-Phe^7-Arg^8-Trp^9-Gly^{10}-NH_2]—is actually a weak antagonist (pA_2 = 4.3) in the same lizard bioassay *(43)*. Other inactive analogues truncated from α-MSH did not possess any inhibitory activities even at very high concentrations (10^{-5} to 10^{-4}*M*). With the earlier discovery of antagonist activities of analogues from MSH-GHRP hybrids, such as H-His-D-Trp-Arg-Trp-D-*Phe*-Lys-NH_2, it was proposed that the Arg^8-Trp^9 sequence could be the part of active core sequence for antagonist activity. Consequently, a few analogues with modified sequences at position either 7 or 10 of Ac-Phe-Arg-Trp-Gly-NH_2 were designed (Table 8). Among all the analogues, only one, Ac-D-Trp^7-Arg-Trp-D-Phe^{10}, is an antagonist with a pA_2 value of 4.8 in FSB (*R. pipiens*). The rest of the analogues (Table 8) displayed either no activity at all or weak agonist activity in this assay. However, in the lizard (*A. carolinesis*) skin bioassay, most of them are antagonists except one, Ac-D-Phe-Arg-Trp-Gly-NH_2, which is an agonist (EC_{50} = 1.5 μ*M*).

These results further confirm our hypothesis that Arg^8-Trp^9 is necessary for antagonist activity. Two analogues, Ac-Phe-Arg-Trp-D-Phe-NH_2 and Ac-D-Phe-Arg-D-Phe-NH_2, are agonists in the FSB (*R. pipiens*). However, both compounds were virtually full antagonists with moderate pA_2 values in the lizard skin bioassay (*A. carolinesis*). This means that lack of antagonist activities in frog skin bioassays are not a predictor for those compounds that will display antagonist activities in lizard skin bioassays (*A. carolinesis*).

3.3. MT-II Analogues

Efforts to design more potent antagonists have been examined by further modifications of potent agonists. Linear MT-II (Ac-Nle^4-Asp^5-His^6-D-Phe^7-

Table 9
Structure–Activity Relationships of α-MSH Analogues

Structure	Potency relative to α-MSH: Frog Skin (*R. pipiens*)	Lizard Skin (*A. carolinesis*)
α-MSH	1.0	1.0
Ac-Nle-Asp-His-D-Phe-Arg-Trp-Lys-NH_2	0.7	9.0
Ac-Nle-Asp-Trp-D-Phe-Arg-Trp-Lys-NH_2	0.1	0.09
Ac-Nle-Asp-His-D-Phe-Nle-Trp-Lys-NH_2	0.0005	0.002
Ac-Nle-Asp-Trp-D-Phe-Nle-Trp-Lys-NH_2	pA_2 = 8.4	Inactive[a]
Ac-Nle-*c*(Asp-Trp-D-Phe-Nle-Trp-Lys)-NH_2	0.001	0.3
Ac-Nle-*c*(Asp-His-D-Phe-Nle-Trp-Lys)-NH_2	0.06	3.0
Ac-Nle-Asp-Trp-D-Phe-Ala-Trp-Lys-NH_2	0.001	0.0009
Ac-Nle-Asp-Trp-D-Phe-Pro-Trp-Lys-NH_2	0.001	0.0008
Ac-Nle-Asp-Trp-Phe-Nle-Trp-Lys-NH_2	Inactive[a]	Inactive[a]
Ac-Nle-Asp-D-Trp-D-Phe-Nle-Trp-Lys-NH_2	0.001	0.001
Ac-Nle-Asp-Trp-D-Phe-Nle-D-Trp-Lys-NH_2	0.001	0.001
Ac-Nle-Asp-D-Trp-D-Phe-Nle-D-Trp-Lys-NH_2	0.0001	0.0009

Data adapted from ref. *35*.
[a]Concentration tested is up to 10^{-4} to $10^{-5}M$.

Arg[8]-Trp[9]-Lys[10]-NH_2) was selected as one of the prototypes (Table 9). Residues in positions of 6, 8, and 9 were modified while residues in positions 4, 5, 7, and 10 were retained.

Among all of these designed analogues *(35)*, only one of them, Ac-Nle-Asp-Trp-D-Phe-Nle-Trp-Lys-NH_2, displayed antagonist activity (pA_2 = 8.4) in the FSB (*R. pipiens*). However, this analogue did not possess any bioactivity in the lizard skin bioassay (*A. carolinesis*). This is the first highly selective and potent antagonist for the FSB. By cyclizing the side chains between Asp and Lys, the resulting cyclic analogue showed no antagonist activity, but is a relatively potent agonist in both the FSB and lizard skin bioassays. The rest of the modifications of either linear or cyclic MT-II analogues, displayed relatively weak activities or no activity.

3.4. ACTH[2][4–10] Analogues

A series of ACTH[4–10]* derivatives (Table 10) have been identified as antagonists in the mMC3-R, hMC4-R and hMC5-R by screening derivatives of

*(α-MSH and ACTH[1–13] have the exact same sequence. However, the N-terminus of α-MSH is acetylated and C-terminus is amidated, while ACTH[1–13] has free amino and carboxyl terminal groups.

Table 10
pA_2 Values of ACTH[4–10] Analogues in the MC3-R, MC4-R and MC5-R[a]

ACTH[4–10] Analogues	mMC3-R	hMC4-R	hMC5-R
[Phe(pI)[7]]- ACTH[4–10]	7.4	8.4	7.9
[Ala[6]]- ACTH[4–10]	6.5	< 6.0	8.7
[Pro[8], Gly[9], Pro[10]]- ACTH[4–10]	—	8.6	6.5
[D-Arg[8]]- ACTH[4–10]	—	8.2	8.1

[a]Data adapted from ref. *44*.
ACTH[4–10]: Met-Glu-His-Phe-Arg-Trp-Gly.

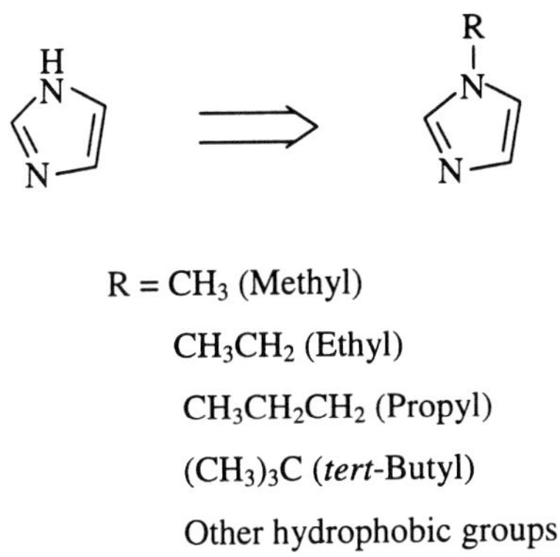

Fig. 5. Alkylated imidazole ring in histidine.

ACTH[4–9] and ACTH[4–10] *(44)*. No antagonists were found for the ACTH[4–9] derivatives, presumably because position 10 is critical for antagonistic activity. This is consistent with the previous finding *(43)*, though only frog MC1-R and lizard MC1-R assays were used at that time. Hence there may be common features for antagonists derived from ACTH[4–10] and ACTH[6–10].

Substitution in the *para* position of the side chain of Phe[7] with a bulky substituent—I (iodine)—converted ACTH[4–10] into an antagonist in all receptors tested. This is consistent with the finding that [D-Phe(pI)[7]]-MT-II is a potent antagonist in mMC1-R *(31)*.

By replacing His[6] with Ala[6], the analogue [Ala[6]]-ACTH[4–10] was obtained which is a weak antagonist at the hMC3-R ($pA_2 = 6.5$) and hMC4-R ($pA_2 < 6$), but a reasonably potent antagonist ($pA_2 = 8.7$) at the hMC5-R (Table 10). This result is also consistent with the recent discovery that elimination of the hydrogen bonding in His[6] by methylation of imidazole ring (Fig. 5) converts MT-II analogues into antagonists *(W. Yuan and V. J. Hruby, unpublished results)*. Comparison of both results suggests that potent antagonists could be

Table 11
IC_{50} Values of the Designed Analogues in Frog *(Xenopus laevis)*

Sequence		IC_{50} (μ*M*)
	Leu	0.62
	Nle	0.93
D-Trp-Arg-XXX-NH_2	Nva3.3	
	Met	5.6
	D-Nle	9.9
D-Trp-YYY-Nle-NH_2	Not available	Not available
ZZZ-Arg-Nle-NH_2	D-Phe	4.4

Data adapted from ref. *45*.

achieved by substituting the imidazole ring (in His[6]) with other hydrophobic groups (Fig. 5). This hypothesis is currently under investigation.

Interestingly, the analogue [Pro[8], Gly[9], Pro[10]]-ACTH[4–10] is a potent antagonist for the hMC4-R but a weaker antagonist for the hMC5-R and apparently inactive at the mMC3-R. Modification of this lead compound could lead to highly selective and potent antagonists for the hMC4-R.

By simply changing the configuration at position 8 from L- to D-(L-Arg to D-Arg) in ACTH[4–10], the resulting analogue—[D-Arg[8]]-ACTH[4–10]—is a reasonably potent antagonist at both hMC4-R and hMC5-R, but apparently this compound is not active at the mMC3-R. These studies provided important insights into melanotropin antagonist activity.

3.5. Combinatorial Screening

3.5.1. D-Trp-Arg-Nle-NH_2 Derivatives

By random screening of a combinatorial library, a series of antagonists have been reported for the in vivo frog (*Xenopus laevis*) skin bioassay. These analogues are designed based upon D-Trp-Arg-Nle-NH_2 *(45)*. By varying all three positions systematically (Table 11), only antagonists with micromolar or weaker values of IC_{50} were obtained. No studies were reported at other melanocortin receptors. However, the most active antagonist in this series (D-Trp-Arg-Nle-NH_2) did not possess any activity in frog (*R. pipiens*) skin bioassay at concentrations up to 10^{-5}*M (M. E. Hadley, S. Hendrata, and V. J. Hruby, unpublished result).*

3.5.2. α-MSH(5–13) Analogues

Another combinatorial library was designed from α-MSH(5–13) (Glu[5]-His[6]-Phe[7]-Arg[8]-Trp[9]-Gly[10]-Lys[11]-Pro[12]-Val[13]-NH_2) which varied sequences between positions 5 and 10 while residues between 11 and 13 remained the same. One lead compound (*Met*[5]-*Pro*[6]-D-*Phe*[7]-Arg[8]-D-*Trp*[9]-*Phe*[10]-Lys[11]-Pro[12]-

Table 12
Structure–Activity Relationship Studies for Positions 5-6, 7-9, and 10 in α-MSH[5–13]

Analogue	IC_{50} (n*M*)
Met[5]-Pro[6]-D-Phe-Arg-D-Trp-Phe-Lys-Pro-Val-NH_2	11
Phe[5]	11
Glu[5]	22
Arg[5]	28
Phe[6]	140
Gly[6]	180
Lys[6]	220
Glu[6]	440
Met-Pro-D-Phe[7]-Arg[8]-*D*Trp[9]-Phe-Lys-Pro-Val-NH_2	11
D-Phe[7]-D-Arg[8]-D-Trp[9]	55
Phe[7]-Arg[8]-D-Trp[9]	60
Phe[7]-D-Arg[8]-D-Trp[9]	470
D-Phe[7]-Arg[8]-Trp[9]	1280
D-Phe[7]-D-Arg[8]-Trp[9]	2800
Phe[7]-Arg[8]-Trp[9]	5100
Phe[7]-D-Arg[8]-Trp[9]	5500
Met-Pro-D-Phe-Arg-D-Trp-Phe[10]-Lys-Pro-Val-NH_2	11
Pro[10]	0.21
Trp[10]	0.26
Ile[10]	0.66
Ala[10]	0.90
Met[10]	1.2
Lys[10]	1.3
His[10]	1.6
Ser[10]	1.7
D-Phe[10]	2.6
Gln[10]	3.1
Glu[10]	4.4

Data adapted from ref. *46*.
α-MSH[5–13]: Glu[5]-His[6]-Phe[7]-Arg[8]-Trp[9]-Gly[10]-Lys-Pro-Val-NH_2 (N-terminus is not acetylated in the reported analogues).

Val[13]-NH_2) *(46)* was identified as the most potent antagonist (IC_{50} = 11 (7n*M*) in the frog (*Xenopus laevis*) skin bioassay used in these studies. These compounds were not examined at other melanocortin receptors.

This lead analogue with Pro[6] is very similar to the Pro[6]-ACTH[4–10] analogue *(44)* discussed above. Structure–activity relationship studies (Table 12) revealed that Met[5] is not critical for antagonist activities. However, replacement of Pro[6] caused more than a 10-fold decrease in antagonist poten-

Table 13
Antagonistic Activities of MT-II Derivatives at the Different MCRs

MT-II Derivative	pA_2 fMC1-R[a]	EC_{50} (p*M*) hMC1-R	pA_2 hMC3-R	pA_2 hMC4-R	EC_{50} (p*M*) mMC5-R
D-Phe(pI)[7]	10.3	55	8.3[b]	9.7[b]	684
D-Nal(2')[7]	≥10.5	36	8.3[b]	9.3[b]	437

Data adapted from ref. *31*.
MT-II: Ac-Nle-*c*[Asp-His-D-Phe-Arg-Trp-Lys]-NH_2.
[a]Frog MC1-R *(R. pipiens)*.
[b]Partial agonist.

cies. Modification of the configuration of residues of positions 7 through 9 revealed that the D-configurations for phenylalanine[7] and tryptophan[9] were very important for potency in the frog (*Xenopus laevis*) skin bioassay (Table 12).

Further structure–activity relationship studies demonstrated that Phe[10] is critical for potent antagonist activity (Table 12). The replacement of Phe[10] with various residues caused a dramatic decrease in potency by orders of magnitude. Though one very potent analogue (IC_{50} = 0.44, Phe-His-D-Phe-Arg-Trp-Gln-Lys-Pro-Val-NH_2 was reported, it is hard to associate the structure–activity relationships due to the lack of data provided *(46)*.

3.6. MT-II Analogues with Bulky Substituents

Though several α-MSH analogues with potent α-MSH antagonist activities were designed in our laboratory several years ago as discussed above, only recently have we obtained highly potent derivatives. An important lead came from examining the radioactive iodolabeled MT-II for pharmacologic investigations, and it was found that the [D-Phe(pI)[7]]-analogue had no agonist activity. This led us to substitute the *para* position of D-Phe[7] with *iodine* in MT-II (Table 14). The resulting analogue, [D-Phe(pI)[7]]-MT-II was a very potent antagonist in the FSB (*R. pipiens*) *(31)*, and this suggested that the bulky iodine in the *para* position of phenyl group in D-Phe[7] might be the cause of antagonist activity. Consequently, we replaced the phenyl group in D-Phe[7] by more bulky groups, for example, D-Nal(2'), to give a more potent antagonist, [D-Nal(2')[7]]-MT-II (Table 13). Further bioassays were conducted for other MCRs. Both of these analogues were potent antagonists for the hMC3-R and hMC4-R *(31)*. For the mMC5-R, [D-Phe(pI)[7]]-MT-II was a weak partial agonist. However, for the hMC1-R, both compounds were very potent agonists. It is interesting to point out that [D-Nal(1')[7]]-MT-II is a full agonist at all of these receptors as discussed earlier *(S. Lim, and V. J. Hruby, unpublished results)*.

Table 14
IC_{50} Values (n*M*) for MT-II Derivatives

MT-II Derivatives	hMC3-R	hMC4-R
NDP-MSH	3.8	3.6
[D-Phe(pI)^7]-MT-II	1.8	2.5
[D-Nal(2')^7]-MT-II	3.3	1.8

Data from ref. *31*.
MT-II: Ac-Nle-*c*[Asp-His-*D*-Phe-Arg-Trp-Lys]-NH_2.

Table 15
Observed Dissociation Characteristics of α-MSH and Its Analogues on hMC1-R Over a Time Period of 6 Hours

Peptide	$T_{1/2}$	k_{-1} (h^{-1})	Relative dissociation rate
α-MSH	4.00	0.17	1.0
MT-I	8.50	0.080	0.50
MT-II	19.5	0.040	0.24

Data from ref. *49*.
MT-I: Ac-Ser-Tyr-Ser-Nle-Glu-His-D-Phe-Arg-Trp-Gly-Lys-Pro-Val-NH_2.
MT-II: Ac-Nle-*c*[Asp-His-D-Phe-Arg-Trp-Lys]-NH_2.

4. Ligand Binding Affinities

To date, various peptides including α-MSH, α-MSH analogues, radiolabeled [^{125}I]NDP-MSH, and [^{125}I]Tyr^{23}ACTH have been used to study binding affinities at all known MCRs (including MC2-R) *(20,31,47–51)*.

By the displacement of [^{125}I]Tyr^{23}-ACTH, it was found that α-MSH and NDP-MSH acted on the MC3-R and MC4-R equally well *(51)*. It was also discovered that sequence 11–13 of α-MSH was very important for binding to the MC3-R and MC4-R. In additional experiments, it was determined that *H*-Met^4-Glu^5-His^6-Phe^7-Arg^8-Trp^9-NH_2 (ACTH[4–9]-NH_2 or *H*-α-MSH[4–9]) was the core sequence for activation of MC3-R and MC4-R where C-terminal amidation was essential for the activity.

For hMC3-R and hMC4-R, the binding affinities of NDP-α-MSH, [D-Phe(pI)^7]-MT-II and [D-Nal(2')^7]-MT-II (SHU-9119) are very similar to each other. This is consistent with their antagonist activity profiles at hMC3-R and hMC4-R (Table 14) *(31)*.

Recent studies in the prolonged (residual) biologic activities of the superpotent α-MSH analogues, MT-I and MT-II, have revealed that the differences reside in their relative dissociation rates from the hMC1-R (Table 15). The

Table 16
K_i and K_d Values (means ± SEM) Obtained From Competition and Saturation Curves, Respectively, for Melanocortin Peptides on Melanocortin (MC1-R, MC3-R, MC4-R, and MC5-R) DNA Transfected COS Cells

Ligand	MC1-R K_i (n*M*)	MC3-R K_i (n*M*)	MC4-R K_i (n*M*)	MC5-R K_i (n*M*)
[^{125}I]NDP-MSH[a]	0.0851	0.396	3.84	5.05
NDP-MSH	0.0231	0.224	2.16	2.39
[Nle[4]]-MSH	0.102	9.85	122	4610
α-MSH	0.0334	20.7	641	8240
Desacetyl-α-MSH	0.0432	3.68	569	3260

Data adapted from ref. *50*.
[a]K_d values (n*M*).

studies have found that α-MSH remained 25% bound, MT-I 65% bound and MT-II 86% bound 6 h after these ligands had been washed away from the assay medium *(49)*. The relative dissociation rate of MT-II was 4 times slower than that for α-MSH and twice slower than that for MT-I, which also was twice slower than that for α-MSH. These data suggested that slow dissociation kinetics may contribute to the prolonged biologic activities observed for both MT-I and MT-II in vitro and in vivo *(24,25,52)*.

It was discovered later that the binding affinities between [Nle[4]]α-MSH and NDP-MSH ([Nle[4], D-Phe[7]]α-MSH) at each of MCRs are quite different. Affinities of α-MSH, NDP-MSH and MT-II for the MC1-R are very similar *(49)*. In addition, the binding for NDP-MSH is only slightly (~five-fold) stronger than that of [Nle[4]]α-MSH *(48)*. However, at the MC3-R and MC4-R, the difference is ~50-fold, while at the MC5-R the difference is more than 200-fold (Table 16) *(50)*. These results confirm our early suggestions that a D configuration in position 7 is critical. They also suggest that a D configuration in position 7 may enhance binding dramatically at the MC5-R. This suggests that variations in position 7 could lead to selective ligands for the MC5-R. There is virtually no difference in binding affinities between α-MSH and [Nle[4]]α-MSH. However, for [Nle[4]]α-MSH, the binding affinity at MC1-R is 3 orders of magnitude greater than at the MC4-R. For α-MSH, the binding affinity at the MC1-R is 4 orders of magnitude greater than at the MC4-R, and the binding affinity at the MC1-R is 620-fold greater than at the MC3-R. Surprisingly, for α-MSH, the binding affinity at the MC1-R is over 5 orders of magnitude greater than at the MC4-R, while there is a difference of 4 orders of magnitude for [Nle[4]]α-MSH for these two receptors. From all these results, it is reasonable to predict that finding highly potent and selective ligands for the MC1-R will be possible. Further studies of ligand binding at all the other MCRs with MSH core peptides will provide the information needed with regard to binding specificity at these receptors.

Clearly, there still is much to learn in order to find ligands that specifically bind at the MC3-R, MC4-R or MC5-R.

5. Perspectives

Considerable progress in the design of selective melanocortin receptor agonists and antagonists, has been achieved in the past few years. The structure–activity relationships have been slowly unveiled by efforts throughout the world. The data reviewed strongly suggest that various useful technologies, such as macromolecular conjugates *(53,54)* which can be used to identify cellular and receptor types, and that fluorescent ligands and combinatorial library approaches *(44,45)*, and other techniques, can be used to accelerate the pace of α-MSH research.

Clearly, many mysteries concerning melanocortin action remain to be solved. For example, ligands for all MCRs (MC1-R, MC3-R, MC4-R, and MC5-R) which are selective and specific; the essential active sequences for each of the MCRs; structural differences for agonists versus antagonists; possible yet unknown new types of MCRs *(51,55)*; and the functions of each of the MCRs. There is much to pursue in this fascinating field.

Finally we would like to cite two distinguished remarks about MSH research as our epilogue for this review. In 1977, it was concluded that "We have about reached the limits of insight that can reasonably be provided by structure–activity studies" *(56)*. However, 20 years later, another outstanding scientist and his colleagues stated that "This field (MSH) is still in its infancy, particularly in consideration of its vast potential." *(57)*.

Acknowledgments

This work is supported by grants from USPHS and NIDA. We thank our many colleagues who have collaborated with us during the past three decades. This review is dedicated to them.

References

1. Tilders, F. J. H., Swaab, D. F., and van Wimersma Greidanus, T. B. (1977) Melanocyte stimulating hormone: control, chemistry and effects, in *Frontiers Hormone Research*, Volume 4 (T. B. van Wimersma Greidanus, ed.) S. Karger AG, Basel, New York.
2. Thody, A. J. (1980) *The MSH Peptides*. Academic Press, New York.
3. Hadley, M. E., Heward, C. B., Hruby, V. J., Sawyer, T. K., and Yang, Y. C. (1981) Biologic actions of melanocyte–stimulating hormone. *Ciba Found. Symp.* **81,** 244–262.

4. Medzihradszky, K. (1982) The bio–organic chemistry of alpha–melanotropin. *Med. Res. Rev.* **2,** 247–270.
5. Hruby, V. J., Mosberg, H. I., Sawyer, T. K., Knittel, J. J., Rockway, T. W., Ormberg, J., Darman, P., Chan, W. Y., and Hadley, M. E. (1983) Conformational and dynamic considerations in the design of peptide hormone analogs. *Biopolymers* **22,** 517–530.
6. Hruby, V. J., Wilkes, B. C., Cody, W. L., Sawyer, T. K., and Hadley, M. E. (1984) Melanotropins: structural, conformational and biologic considerations in the development of superpotent and superprolonged analogs. *Pept. Protein Rev.* **3,** 1–64.
7. Hadley, M. E. (1988) *The Melanotropic Peptides.* CRC Press, Boca Raton, FL.
8. Eberle, A. N. (1988) *The Melanotropins. Chemistry, Physiology and Mechanisms of Action.* S. Karger, Basel, Switzerland.
9. Castrucci, A. M. L., Sawyer, T. K., Al–Obeidi, F., Hruby, V. J., and Hadley, M. E. (1990) Melanotropic peptide antagonists: recent discoveries and biomedical implication. *Drugs Future* **15,** 41–55.
10. Hruby, V. J., Sharma, S. D., Toth, K., Jaw, J. Y., Al–Obeidi, F., Sawyer, T. K., and Hadley, M. E. (1993) Design, synthesis, and conformation of superpotent and prolonged acting melanotropins. *Ann. N. Y. Acad. Sci.* **680,** 51–63.
11. Scharrer, B., Smith, E. M., and Stefano, G. B. (1994) *Neuropeptides and Immunoregulation.* Springer–Verlag, New York.
12. Hadley, M. E. (1996) The melanotropic hormones, in *Endocrinology*, Prentice–Hall, Upper Saddle River, NJ, pp. 153–176.
13. Mountjoy, K. G., Robbins, L. S., Mortrud, M. T., and Cone, R. D. (1992) The cloning of a family of genes that encode the melanocortin receptors. *Science* **257,** 1248–1251.
14. Chhajlani, V. and Wikberg, J. E. (1992) Molecular cloning and expression of the human melanocyte stimulating hormone receptor cDNA. *FEBS Lett.* **309,** 417–420.
15. Chhajlani, V., Muceniece, R., and Wikberg, J. E. (1993) Molecular cloning of a novel human melanocortin receptor. *Biochem. Biophys. Res. Commun.* **195,** 866–873.[published erratum appears in *Biochem. Biophys. Res. Commun.* 1996 Jan 17; **218:** 638].
16. Cone, R. D., Mountjoy, K. G., Robbins, L. S., Nadeau, J. H., Johnson, K. R., Roselli–Rehfuss, L., and Mortrud, M. T. (1993) Cloning and functional characterization of a family of receptors for the melanotropic peptides. *Ann. N. Y. Acad. Sci.* **680,** 342–363.
17. Gantz, I., Konda, Y., Tashiro, T., Shimoto, Y., Miwa, H., Munzert, G., Watson, S. J., DelValle, J., and Yamada, T. (1993) Molecular cloning of a novel melanocortin receptor. *J. Biol. Chem.* **268,** 8246–8250.
18. Gantz, I., Miwa, H., Konda, Y., Shimoto, Y., Tashiro, T., Watson, S. J., Delvalle, J., and Yamada, T. (1993) Molecular cloning, expression, and gene localization of a fourth melanocortin receptor. *J. Biol. Chem.* **268,** 15,174–15,179.
19. Gantz, I., Shimoto, Y., Konda, Y., Miwa, H., Dickinson, C. J., and Yamada, T. (1994) Molecular cloning, expression, and characterization of a fifth melanocortin receptor. *Biochem. Biophys. Res. Commun.* **200,** 1214–1220.
20. Schioth, H. B., Chhajlani, V., Muceniece, R., Klusa, V., and Wikberg, J. E. (1996) Major pharmacological distinction of the ACTH receptor from other melanocortin receptors. *Life Sci.* **59,** 797–801.
21. Sawyer, T. K., Sanfilippo, P. J., Hruby, V. J., Engel, M. H., Heward, C. B., Burnett, J. B., and Hadley, M. E. (1980) 4–Norleucine, 7–D–phenylalanine–alpha–melanocyte–stimulating hormone: a highly potent alpha–melanotropin with ultralong biologic activity. *Proc. Natl. Acad. Sci. U. S. A.* **77,** 5754–5758.

22. Sawyer, T. K., Hruby, V. J., Darman, P. S., and Hadley, M. E. (1982) [half–Cys4,half–Cys10]–alpha–Melanocyte–stimulating hormone: a cyclic alpha–melanotropin exhibiting superagonist biologic activity. *Proc. Natl. Acad. Sci. U. S. A.* **79,** 1751–1755.
23. Nikiforovich, G. V., Rozenblit, S. A., Shenderovich, M. D., and Chipens, G. I. (1984) Possible bioactive conformations of alpha–melanotropin. *FEBS Lett.* **170,** 315–320.
24. Al–Obeidi, F., Hadley, M. E., Pettitt, B. M., and Hruby, V. J. (1989) Design of a new class of superpotent α–melanotropins based on quenched dynamic simulations. *J. Am. Chem. Soc.* **111,** 3413–3416.
25. Al–Obeidi, F., Castrucci, A. M. L., Hadley, M. E., and Hruby, V. J. (1989) Potent and prolonged acting cyclic lactam analogues of alpha–melanotropin: design based on molecular dynamics. *J. Med. Chem.* **32,** 2555–2561.
26. Al–Obeidi, F., Hruby, V. J., Yaghoubi, N., Marwan, M. M., and Hadley, M. E. (1992) Synthesis and biologic activities of fatty acid conjugates of a cyclic lactam alpha–melanotropin. *J. Med. Chem.* **35,** 118–123.
27. Hadley, M. E., Al–Obeidi, F., Hruby, V. J., Weinrach, J. C., Freedberg, D., Jiang, J. W., and Stover, R. S. (1991) Biologic activities of melanotropic peptide fatty acid conjugates. *Pigment Cell Res.* **4,** 180–185.
28. Chaturvedi, D.N., Knittel, J. J., Hruby, V. J., Castrucci, A. M. L., and Hadley, M. E. (1984) Synthesis and biologic actions of highly potent and prolonged acting biotin–labeled melanotropins. *J. Med. Chem.* **27,** 1406–1410.
29. Sugg, E. E., Cody, W. L., Abdel–Malek, Z., Hadley, M. E., and Hruby, V. J. (1986) D–Isomeric replacements within the 6–9 core sequence of Ac–[Nle^4]–alpha–MSH4–11–NH_2: a topological model for the solution conformation of alpha–melanotropin. *Biopolymers* **25,** 2029–2042.
30. Hruby, V. J., Kazmierski, W., Pettitt, B. M., and Al–Obeidi, F. (1988) Conformational constraints in the design of receptor selective peptides: conformational analysis and molecular dynamics, in *Molecular Biology of Brain and Endocrine Peptidergic Systems* (M. Chretien, M. and McKerns, K. W. eds.), Plenum, pp. 13–27.
31. Hruby, V. J., Lu, D., Sharma, S. D., Castrucci, A. M. L., Kesterson, R. A., Al–Obeidi, F. A., Hadley, M. E., and Cone, R. D. (1995) Cyclic lactam alpha–melanotropin analogues of Ac–Nle^4–cyclo[Asp^5, D–Phe^7,Lys^{10}] alpha–melanocyte–stimulating hormone–(4–10)–NH_2 with bulky aromatic amino acids at position 7 show high antagonist potency and selectivity at specific melanocortin receptors. *J. Med. Chem.* **38,** 3454–3461.
32. Haskell–Luevano, C., Nikiforovich, G., Sharma, S. D., Yang, Y.–Y., Dickinson, C., Hruby, V. J., and Gantz, I. (1997) Biologic and conformational examination of stereochemical modifications using the template melanotropin peptide, Ac–Nle–c[Asp–His–Arg–Trp–Ala–Lys]–NH_2, on human melanocortin receptors. *J. Med. Chem.* **40,** 1738–1748.
33. Haskell–Luevano, C., Boteju, L. W., Miwa, H., Dickinson, C., Gantz, I., Yamada, T., Hadley, M. E., and Hruby, V. J. (1995) Topographical modification of melanotropin peptide analogues with beta–methyltryptophan isomers at position 9 leads to differential potencies and prolonged biologic activities. *J. Med. Chem.* **38,** 4720–4729.
34. Sawyer, T. K., Staples, D. J., Castrucci, A. M. L., and Hadley, M. E. (1989) Discovery and structure–activity relationships of novel alpha–melanocyte–stimulating hormone inhibitors. *Pept. Res.* **2,** 140–146.

35. Al–Obeidi, F., Hruby, V. J., Hadley, M. E., Sawyer, T. K., and Castrucci, A. M. L. (1990) Design, synthesis, and biologic activities of a potent and selective alpha–melanotropin antagonist. *Int. J. Pept. Protein Res.* **35,** 228–234.
36. Sawyer, T. K., Staples, D. J., Castrucci, A. M. L., Hadley, M. E., Al–Obeidi, F. A., Cody, W. L., and Hruby, V. J. (1990) Alpha–melanocyte stimulating hormone message and inhibitory sequences: comparative structure–activity studies on melanocytes. *Peptides* **11,** 351–357.
37. McCormack, A. M., Carter, R. J., Thody, A. J., and Shuster, S. (1982) Des–acetyl MSH and gamma–MSH act as partial agonists to alpha–MSH on the Anolis melanophore. *Peptides* **3,** 13–16.
38. Yajima, H. and Kubo, K. (1965) Studies on peptides. II. Synthesis and physiological properties of D–histidyl–D–phenylalanyl–D–arginyl–D–tryptophanylglycine, an optical antipode of an active fragment of α–melanocyte–stimulating hormone. *J. Am. Chem. Soc.* **87,** 2039–2044.
39. Yajima, H., Kawasaki, K., Okada, Y., and Lande, S. (1965) Studies on peptides. V. Color–lightening action of histidylphenylalanylarginyl–5–methoxytryptamine on frog skin in vitro. *Biochim. Biophys. Acta* **107,** 141–143.
40. Yajima, H. and Kawasaki, K. (1968) Studies on peptides. XVII. Synthesis of *N*–alpha–acetyl–seryl–tyrosyl–seryl–methionyl–glutamyl–histidyl–phenylalanyl–arginyl)–5–methoxy–tryptamine and its physiological properties on the frog melanocyte in vitro. *Chem. Pharm. Bull. (Tokyo)* **16,** 1379–1382.
41. Castrucci, A. M. L., Sherbrooke, W. C., Sawyer, T. K., Staples, D. J., Tuma, M. C. B., and Hadley, M. E. (1994) Discovery of an alpha–melanotropin antagonist effective in vitro. *Peptides* **15,** 627–632.
42. Castrucci, A. M. L., Hadley, M. E., Sawyer, T. K., Wilkes, B. C., Al–Obeidi, F., Staples, D. J., de Vaux, A. E., Dym, O., Hintz, M. F., Riehm, J. P., Rao, K. R., and Hruby, V. J. (1989) alpha–Melanotropin: the minimal active sequence in the lizard skin bioassay. *Gen. Comp. Endocrinol.* **73,** 157–163.
43. Sawyer, T. K., Staples, D. J., Castrucci, A. M. L., Hadley, M. E., Al–Obeidi, F. A., Cody, W. L., and Hruby, V. J. (1990) alpha–Melanocyte stimulating hormone message and inhibitory sequences: comparative structure–activity studies on melanocytes. *Peptides* **11,** 351–357.
44. Adan, R. A., Oosterom, J., Ludvigsdottir, G., Brakkee, J. H., Burbach, J. P., and Gispen, W.H. (1994) Identification of antagonists for melanocortin MC3, MC4 and MC5 receptors. *Eur. J. Pharmacol.* **269,** 331–337.
45. Quillan, J. M., Jayawickreme, C. K., and Lerner, M. R. (1995) Combinatorial diffusion assay used to identify topically active melanocyte–stimulating hormone receptor antagonists. *Proc. Natl. Acad. Sci. U. S. A.* **92,** 2894–2898.
46. Jayawickreme, C. K., Quillan, J. M., Graminski, G. F., and Lerner, M. R. (1994) Discovery and structure–function analysis of alpha–melanocyte–stimulating hormone antagonists. *J. Biol. Chem.* **269,** 29,846–29,854.
47. Haskell–Luevano, C., Miwa, H., Dickinson, C., Hruby, V. J., Yamada, T., and Gantz, I. (1994) Binding and cAMP studies of melanotropin peptides with the cloned human peripheral melanocortin receptor, hMC1-R. *Biochem. Biophys. Res. Commun.* **204,** 1137–1142.
48. Schioth, H. B., Muceniece, R., Wikberg, J. E., and Chhajlani, V. (1995) Characterisation of melanocortin receptor subtypes by radioligand binding analysis. *Eur. J. Pharmacol.* **288,** 311–317.

49. Haskell–Luevano, C., Miwa, H., Dickinson, C., Hadley, M. E., Hruby, V. J., Yamada, T., and Gantz, I. (1996) Characterizations of the unusual dissociation properties of melanotropin peptides from the melanocortin receptor, hMC1-R. *J. Med. Chem.* **39,** 432–435.
50. Schioth, H. B., Muceniece, R., and Wikberg, J. E. (1996) Characterisation of the melanocortin 4 receptor by radioligand binding. *Pharmacol. Toxicol.* **79,** 161–165.
51. Adan, R. A., Cone, R. D., Burbach, J. P., and Gispen, W. H. (1994) Differential effects of melanocortin peptides on neural melanocortin receptors. *Mol. Pharmacol.* **46,** 1182–1190.
52. Hadley, M. E., Anderson, B., Heward, C. B., Sawyer, T. K., and Hruby, V. J. (1981) Calcium–dependent prolonged effects on melanophores of [4–norleucine, 7–D–phenylalanine]–alpha–melanotropin. *Science* **213,** 1025–1027.
53. Hadley, M. E., Hruby, V. J., Jiang, J., Sharma, S. D., Fink, J. L., Haskell–Luevano, C., Bentley, D. L., Al–Obeidi, F., and Sawyer, T. K. (1996) Melanocortin receptors: identification and characterization by melanotropic peptide agonists and antagonists. *Pigment Cell Res.* **9,** 213–234.
54. Sharma, S. D., Jiang, J., Hadley, M. E., Bentley, D. L., and Hruby, V. J. (1996) Melanotropic peptide–conjugated beads for microscopic visualization and characterization of melanoma melanotropin receptors. *Proc. Natl. Acad. Sci. U. S. A.* **93,** 13,715–13,720.
55. Li, S. J., Varga, K., Archer, P., Hruby, V. J., Sharma, S. D., Kesterson, R. A., Cone, R. D., and Kunos, G. (1996) Melanocortin antagonists define two distinct pathways of cardiovascular control by alpha–and gamma–melanocyte–stimulating hormones. *J. Neurosci.* **16,** 5182–5188.
56. Schwyzer, R. (1977) ACTH: a short introductory review. *Ann. N. Y. Acad. Sci.* **297,** 3–26.
57. Kastin, A. J., Zadina, J. E., Olson, R. D., and Banks, W. A. (1996) The history of neuropeptide research: version 5.a. *Ann. N. Y. Acad. Sci.* **780,** 1–18.
58. Wessells, H., Fuciarelli, K., Hansen, J., Hadley, M. E., Hruby, V. J., Dorr, R., and Levine, N. (1998) Synthetic melanotropic peptide initiates erections in men with psycogenic erectile dysfunction: double-blind, placebo controlled crossover study. *J. Urology* **160,** 389–393.

CHAPTER 9

In Vitro Mutagenesis Studies of Melanocortin Receptor Coupling and Ligand Binding

Carrie Haskell-Luevano

1. Introduction

The melanocyte stimulating hormone receptor (MSH-R); melanocortin 1 receptor (MC1-R) and the adrenocorticotropin hormone (ACTH) receptor (MC2-R) were the first melanocortin receptors cloned and characterized *(1,2)*. Subsequently, three other melanocortin receptor subtypes have been cloned and designated the MC3-R, MC4-R, and MC5-R. The MC1-R has been clearly demonstrated to be involved in pigmentation and animal coat coloration *(3,4)*. The efficacy of melanocortin peptides at the MC1-R can be summarized as 4-norleucine, 7-D-phenylalanine (NDP-MSH) > α-MSH > ACTH>γ-MSH. With the availability of the cloned melanocortin receptors, several questions can now be studied. In vitro investigations using these cloned receptors may include identifying critical ligand features resulting in receptor selectivity, ligand residues responsible for differing efficacies, and how these ligand residues interact with the receptor for recognition and activation. In lieu of X-ray crystal structures, three-dimensional (3D) homology receptor modeling has become a tool to attempt to identify noteworthy ligand and receptor features. Furthermore, knowledge of the molecular mechanism responsible for the initial intracellular signal transduction cascade would be potentially important for the design of antagonists. This chapter is designed to attempt to address these issues from the available literature.

As discussed in detail in other chapters of this book, the melanocortin peptides are derived by posttranslational processing of the proopiomelanocortin (POMC) preprohormone. The melanocortin peptides (Table 1) contain a common tetrapeptide sequence (His-Phe-Arg-Trp) which, up until 1997, had

The Melanocortin Receptors Ed.: R. D. Cone

Table 1
Primary Sequence of the Melanocortin Peptides With the Core "His-Phe-Arg-Trp"*

ACTH[1–39]	NH_2-SYSME ***HFRW*** GKPVGKKRRPVKVYPNGAEDESAEAFPLEF-OH
α-MSH	Ac-SYSME ***HFRW*** GKPV-NH_2
β-MSH	NH_2-AEKKDEGPYRME ***HFRW*** GSPPKD-OH
γ-MSH	NH_2-YVMG ***HFRW*** DRF-OH

*Emphasized in Bold Italics.

been proposed to be the putative "message" sequence responsible for melanocortin receptor selectivity and activation *(3,5–12)*,with the exception of the MC2-R, which only responds to ACTH *(2)*. This premise, however, has been experimentally demonstrated by two independent laboratories *(13,14)* to be incorrect with regard to the cloned human receptors. In one study, the tetrapeptide His-Phe-Arg-Trp was examined on the human melanocortin receptors (hMC1-R, hMC3-R, hMC4-R, and hMC5-R), and was only able to competitively displace [^{125}I]NDP-MSH at the hMC1-R, albeit possessing 814,336-fold less affinity than the radiolabel *(14)*. In a second study *(13)*, stereochemical modifications of the Ac-His-D-Phe-Arg-Trp-NH_2 tetrapeptide and the Ac-D-Phe-Arg-Trp-NH_2 tripeptide, which were identified previously as the minimal sequence of NDP-MSH to stimulate activity on the frog skin *(15)*, were examined on the hMC1-R, hMC3-R, hMC4-R, and hMC5-R. These studies revealed that only the hMC1-R (0.6 µ*M*) and hMC4-R (1.1 µ*M*) were able to recognize this Ac-His-D-Phe-Arg-Trp-NH_2 peptide. The tripeptides Ac-D-Phe-Arg-Trp-NH_2 and Ac-D-Phe-Arg-D-Trp-NH_2 possessed selective hMC4-R binding, albeit at micromolar concentrations. Additionally, the tetrapeptide Ac-His-Phe-Arg-D-Trp-NH_2 (6.3 µ*M*) was selective for the hMC1-R. These studies demonstrate that the classical "message" sequence of the melanocortin ligands is not the only ligand component necessary for ligand binding and receptor activation at the melanocortin receptors, as previously predicted.

The diagram in Fig. 1 summarizes signal transduction pathways which have been reported to be activated by the melanocortin receptors. All of the melanocortin receptors identified to date activate the cyclic adenosine monophosphate (cAMP) signal transduction pathway. α-MSH has been reported to stimulate protein kinase C (PKC) activity in murine B16 melanoma cells *(16)*, and the human MC3-R has been reported to activate inositol phospholipid/Ca^{2+} -mediated signaling pathways *(17)*. Several signal transduction intermediates are available, such as adenylate cyclase (AC) activity, intracellular cAMP accumulation, inositol 1,4,5, - triphosphate (IP_3), tyrosinase *(18)*, and cAMP response element binding protein (CREB) mediated transcription [detected by a β-galactosidase bioassay *(19)*] to monitor

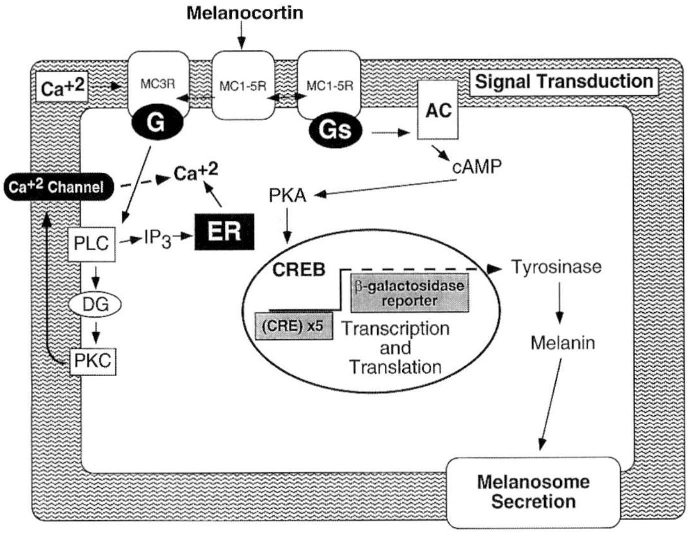

Fig. 1. Diagram of the signal transduction pathways identified for the melanocortin receptor system. Multiple second messenger intermediates such as adenylate cyclase (AC), intracellular cAMP, and CREB-induced nuclear transcription (monitored by the β-galactosidase reporter) are substrates that have been used to monitor changes of ligand efficacy at the MC1-R mutations discussed herein.

melanocortin receptor stimulation and examine ligand–receptor structure–activity relationships (SAR). The role of calcium and its importance for melanocortin biologic activity has been examined and summarized in *(7,8,10,12)* with the importance of this cation in signal transduction demonstrated in references *(20,21)*.

2. Homology Molecular Modeling of the MC1-R

The overall goal of three-dimensional homology molecular modeling is to add a pseudostructure based design strategy and provide further insight for rationale drug design. Fig. 2 illustrates the concept of "pseudostructure-based drug design." The "classical" design approach focuses on ligand (peptide in the case of the melanocortin system) SAR information. Modifications of the peptide ligand are performed to identify the following:

1. Functionally important amino acid residues
2. Residue positions that favor aromatic, constrained, hydrophobic, hydrophilic, or functional moieties
3. Residues that enhance potency or receptor subtype selectivity
4. Minimal peptide length

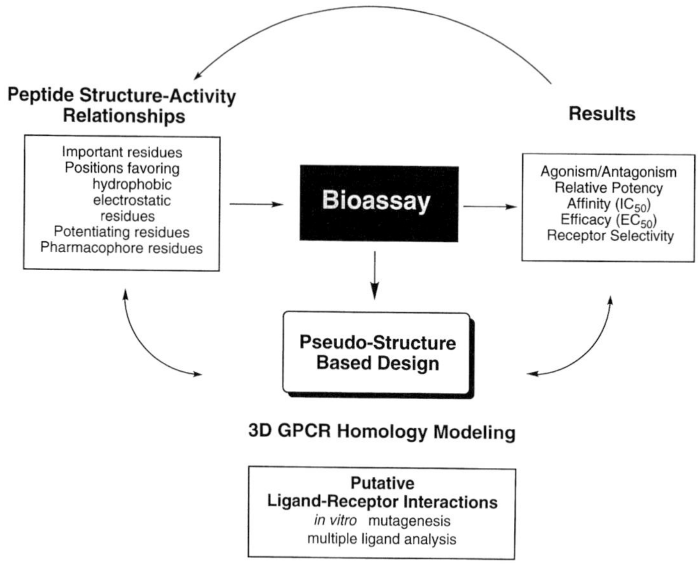

Fig. 2. A summary of the "classical" peptide structure–activity relationship (SAR) studies and how the addition of 3D homology molecular modeling adds a new dimension to the "rational" design approach. In this schematic, the peptide ligand is modified in a variety of ways and tested at a G protein-coupled receptor (GPCR) for its characteristics. With the addition of "pseudo-" structure-based design, the proposed structure of the receptor and its putative binding pocket become a tool to further design and test hypotheses of ligand–receptor interactions.

5. Potential amino acid functional moieties that may participate in the peptide pharmacophore (the three-dimensional orientation of key functional moieties which are the necessary requirements for molecular recognition and/or receptor activation)

Once these modifications are performed, these analogs are then evaluated on the G-protein-coupled receptor (GPCR) of choice, with bioassay information being generated. This approach is similar to a "black box," in that the ligand is being "mapped" experimentally versus "guided" by some sort of putative ligand–receptor structural information. The benefit that 3D molecular homology modeling may offer is to generate working and testable hypotheses by visualizing putative ligand–receptor interactions. Due to the lack of GPCR structural crystallographic information, other than the low-resolution structures of bacteriorhodopsin and rhodopsin, homology modeling has proven to be a valuable tool. This homology modeling technique has been utilized for many proteins (e.g., protease families) and has been routinely used successfully by X-ray crystallographers for several decades.

Several groups have reported *de novo* 3D homology molecular models of the melanocortin receptors *(3,22,23)*. All these models were based to some extent on the non-G protein-coupled seven transmembrane (TM) spanning bacteriorhodopsin protein electron cryomicroscopic structure *(24)* available in the protein data bank (1BRD), and the lower resolution "footprint" of the G protein-coupled rhodopsin receptor *(25)*. However, several other putative seven TM spanning protein rhodopsin and bacteriorhodopsin structures have been subsequently reported *(26–35)* and possess modified transmembrane densities. Concerns and discussions regarding the validity of using the original electron structures for 3D homology modeling have been reviewed *(36–39)*.

2.1. Primary Sequences of the MC1-R in Different Species

Although the receptor residues that are highly conserved in the majority of GPCRs are thought to be important structurally or functionally (discussed below) residues that are conserved in a particular receptor family are proposed to be important for ligand specificity. Fig. 3 summarizes the primary amino acid sequence of the MC1 receptors cloned and include the human *(1,2)*, horse *(40)*, mouse *(2)*, cow *(41,42)*, panther *(43)*, fox *(44)*, and chicken *(45)*. Several highly conserved residues identified by the "Baldwin" alignment *(46)* in the superfamily of GPCRs are indicated in Fig. 4. The residues that are identical in a receptor subtype or family may be indicative of receptor amino acids that may participate in complementary ligand–receptor interactions. However, notable pharmacologic differences were observed between the human and mouse MC1-R *(47)*, thus this may, or may not, be the case for the MC1-Rs. The primary MC1-R sequence of different species diverges significantly in the N-terminal region. Several notable modifications of important residue functional moieties are also observed throughout the various species, particularly in the chicken receptor.

A genetic approach using coat color observations led investigators to clone the mMC1-R from the homozygous *recessive yellow e/e* mouse which identified a framshift mutation at position 183 *(4)*. Cloning of the MC1-R from the *Eso-3J sombre* mouse identified an E92K mutation which resulted in a constitutively active receptor *(4)*. Cloning of the MC1-R from the *Etob tobacco* mouse identified a S69L mutation in the first intracellular loop and possessed increased basal cAMP activity *(4)*. Further examination of other animal species such as cows *(41)*, humans *(48,49)*, guinea pigs *(3)*, foxes *(44)*, and sheep *(50,51)* identified residues important for putative ligand-receptor interactions and MC1-R activation (Fig. 4). Additionally, these mutations localized TMs 2 and 3 as "hot" areas important for putative ligand-induced "active" receptor conformation.

```
                                                                  {           TM1
Human    1    MAVQGSQRRLLGSLNSTPTAIPQLGLAANQTGARCLEV SISDGLFLSLGLVSLVENALVVA
Horse                        TLP*T**L**TT***EPP**** **P*******************T
Mouse    1    *ST*EP*KS*V*****--N*TSH****T**SEPW**Y* **P****************V***I
Cow      1    *PAL**********C**P*TLPFT**P*R**PQ**** *SL***************V****
Panther  1    *SV**P-*X****X***SP*A*R*********P***** *VP**********V***V****
Fox      1    *SG**P*******P*A*SPTT*HFK******P***** **P**********V***V****
Chicken  1    *SMLAPL*-*VREPWNASEGNQSNAT*GAGGAWCQGLD -*PNE***T*********L****

                }       {           TM2           }                       {
Human    62   TIA KNRNLH SPMYCFICCLALSDLLVSGTNVLETA VILLLEAGALVARAAVLQQLD NVIDV
Horse         A** ****** ****Y******V******MS****M* IL******V*ATQ*S****** *I***
Mouse    60   A*T ****** ****Y**********M**VSI****T I*****V*I****V*L***** *L***
Cow      62   A** ****** ****Y******V******VS****** *M******V*ATQ******** *****
Panther  61   A** ****** ****Y******V******VSS***** *M******T*AG********* D****
Fox      62   A** ****** ****Y**G***V******VT****** *M**V*****A*Q******** DI***
Chicken  60   A*L ****** **T*Y******V**N***VSNLAK*L FM**M*H*V**I**SI*RHM* ****M

                    TM3           }                   {           TM4
Human    123  ITCSSMLSSLCFLGAIAVDRY ISIFYALRYHSIVTL PRAPRAVAAIWVASVVFSTLFIAYY
Horse         LI*G**V*******S****** ************MM* **VW**IV****V**LS********
Mouse    121  LI*G**V*******I**I*** *************** ***R***VG**MV*I*S*****T**
Cow      123  LI*G**V************** ***********V*** ***W*II*******ILT*L***T**
Panther  122  LV*GA*V************** *************** ***W**IS*******LS********
Fox      123  LI*G**V************** *************** ***W**IS*******LS********
Chicken  121  LI***VV***S**GV****** *T**********M** Q**VVTM*SV*L**T*S**VL*T**

              }         {            TM5          }                          {
Human    184  D HVAVLL CLVVFFLAMLVLMAVLYVHMLARACQ HAQGIARLHKRQRPVHQGFGLK GAVTLT
Horse         N *T**** ***T**V********************* **R*********H*I******* **A***
Mouse    182  K *T**** ***T******A***I**A**FT**** *V****Q****R*SIR***S** **A***
Cow      184  N *KVI** ***GL*I***A**************** **R*****Q*****I******* **A***
Panther  183  * *T**** ***T**V**************CST** **X*T*************L*** **A***
Fox      184  N *T**** ***S**V*****************R* **R*****R***HS******** **A***
Chicken  182  R NN*I** **IG***F*****L***I**F*L**H *VRS*SSQQ*-*PTIYRTSS** ******

                    TM6           }                    {       TM7               }
Human    245  ILLGIFFLCWGPFFLHLTLI VLCPEHPTCGCIFK NFNLFLALIICNAIIDPLIYAFHSQ EL
Horse         ****V*************S*L I***Q******V** **K***T**L*S**V*******R** **
Mouse    243  ********************* ****Q****S**** ******L**VLSSTV*******R** **
Cow      245  ****V*************S** ****Q********* **************V*******R** **
Panther  244  M*****************S*M ****R**I***V** ******T*****S*V*******R** **
Fox      245  ******************S*M ****Q**I***V*Q ******T*****S****F****R** **
Chicken  243  ****V**I******F**IL* *T**TN*F*T*F*S Y*****I*****SVV*******R** **

Human    306  RRTLKEVLTCSW
Horse         *K*
Mouse    304  *M******L***
Cow      306  *K**Q***Q***
Fox      306  *K**Q***L***
Chicken  304  ****R**VL***
```

Fig. 3. MC1-R primary sequence comparison in different species. The transmembrane regions indicated are predicted based on the "Baldwin" alignment *(46)*. An asterisk (*) signifies identical amino acid conservation as compared with the hMC1-R. A dash (–) signifies a residue deletion at this position.

2.2. Highly Conserved GPCR Residues

The superfamily of GPCRs possesses a few highly conserved amino acid residues which are predominately located in the TM regions *(46,52)*. Due to the functional and positional conservation throughout these receptors, it is thought that they play a role structurally or functionally *(37,53,54)*. Several of these residues, which are conserved throughout the majority of GPCRs, have

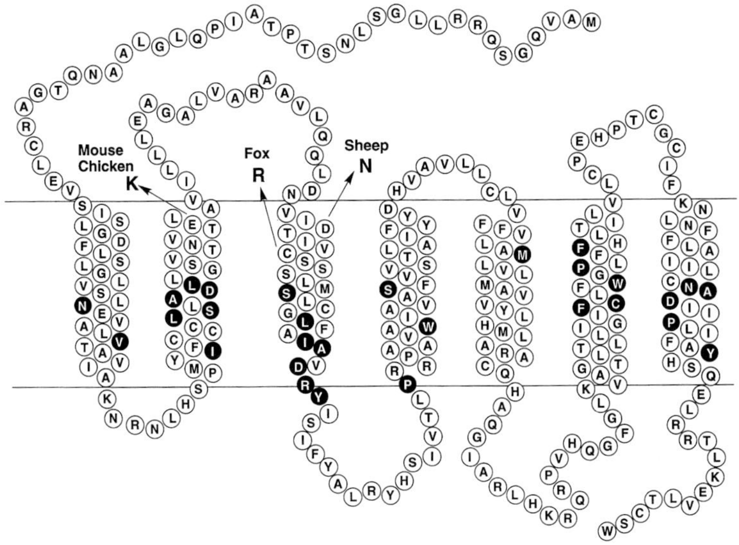

Fig. 4. A schematic illustration of the hMC1-R putative transmembrane regions. Amino acids shown as white text in black circles represent residues found in nearly all GPCR's identified to date. These conserved residues are predicted to be important for GPCR structure and/or functional activation, and are all located within the putative transmembrane domains. The arrows indicate naturally occurring mutations of the MC1-R identified in different species which are located in TMs 2 and 3 and potentially important for ligand–receptor interactions.

been mutated and examined. The highly conserved Pro residues homologous to the hMC1-R P159 (TM4), P256 (TM6), and P295 (TM7) residues (Fig. 4) were mutated to Ala *(55,56)*. The mutated residues homologous to P159 (TM4) and P256 (TM6) were reported not to significantly modify hormone binding, second messenger generation or surface expression *(55,56)*. P295 in TM7, however, was reported to retain high ligand binding affinity but dramatically disrupt functional activity *(55)*, and, further, it is implicated as critical for receptor surface expression *(56)*. Highly conserved Trp residues in the GPCR superfamily homologous to the hMC1-R W169 (TM4) and W254 (TM6) residues were mutated to Phe's and displayed reduced hormone binding affinities but retained functional activities *(55)*. The conserved GPCR amino acid sequence $NPX_{2\text{-}3}Y$ sequence in TM7 was individually mutated to Ala's. The Asn to Ala mutation demonstrated a significant impairment of functional activity *(57,58)*. Additionally, the Pro and Tyr mutations also possessed reduced functional activities *(58)*. These data confirm the importance of the Asn, Pro, and Tyr residues for hormone-induced receptor signaling. A change

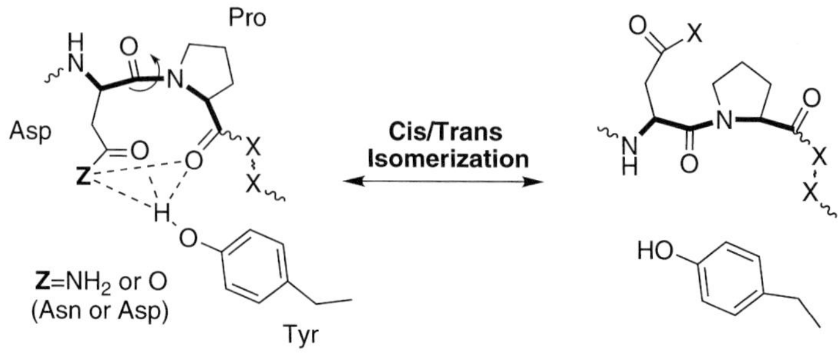

Fig. 5. Illustration of a possible *cis/trans* isomerization mechanism involving the $DPX_{2-4}Y$ conserved GPCR "motif" in TM7. This mechanism may potentially change the electrostatic nature in this receptor region and result in ligand-induced receptor activation. The Z represents either the NH_2 or O functional moiety of Asn or Asp, respectively, depending on which residue is found at this position. The Xs illustrate the residues located between the Pro and Tyr residues in the helices. The dashed lines represent hydrogen bonding interactions.

in the electrostatic characteristics of this region *(59)* could be easily modified by the *cis/trans* isomerization of the Pro residue (Fig. 5) *(60)* and may result in ligand induced signal transduction.

At the time of the original receptor homology modeling studies of the MC1-R *(3,22,23)*, very little information was available about peptide SAR, other than the naturally occurring and commercially available ligands. Receptor mutagenesis information (naturally occurring or induced) was also not available, with the exception of two reports *(4,61)*, making MC1 receptor modeling particularly challenging. This is evident in the different melanocortin receptor models that have been reported. These models diverge significantly in most cases, but were similar in others, that is, some putative ligand Arg^8–receptor interactions. However, in each case, further modifications of these models, based on the emerging receptor mutagenesis experiments and ligand SAR, will help refine these models to make them better potential predictive tools for future drug design.

2.3. GPCR Model Development

As discussed previously, the importance of identifying putative receptor residues important for ligand binding is a key component to GPCR homology modeling goals. A couple of different approaches are available to use for the alignment and orientation of conserved receptor residues. The first and most common approach is to place the conserved receptor residues in the same 3D spatial location for all the GPCRs, with the rationale that these residues are important for all GPCRs. This 3D spatial alignment may be obtained by using the "Baldwin" alignment *(46)* or other alignments *(37,62,63)* that define both the putative TM regions and 3D positioning of these residues within the TM helix bundle. A second method is to position the residues that are conserved for a particular GPCR family into the putative binding pocket (whether this is suggested to be in the TM helix bundle or the extracellular loops appears to be "family"-dependent). If the modeler is fortunate, both approaches will converge and result in the same alignment and spatial placement of predicted key receptor residues.

The conformation and tertiary structure of the ligand—and even which ligand(s)—to dock into the proposed binding pocket is an enigma in itself. The aid of structural information derived from ligand molecular modeling and NMR studies can be beneficial, but one important caveat is that these structures are derived in solution, in the case of nuclear magnetic resonance (NMR) studies, and can produce different conformational families depending on the solvent (NMR) or dielectric constant (modeling). In any event, it essentially comes down to the intuition of the individual(s) developing the ligand–receptor molecular model complex.

In the reported MC1-R models, each author's approach has been outlined *(3,22,23)* in the appropriate reference. In lieu of X-ray structural information, one approach *(23)* proposed the ligand–receptor interactions based upon SAR results of agonist ligands on human and lizard melanocortin receptor assays, as well as a plethora of previous studies performed on MC1 receptors of non-human melanocytes *(6,10,15,64–71)*.

2.4. NDP-MSH Ligand Docking

NDP-MSH is a tridecapeptide (*see* Fig. 7a) that possesses increased efficacy and prolonged biologic activity at the melanocortin receptors *(20)*. This peptide has also been demonstrated to be stable to enzymatic degradation *(72)* and is the ligand of choice for radiolabeling and competitive binding experiments, due to its chemical stability and long duration of action. Previous SAR and computational studies of melanocortin peptides suggested that a β-turn existed in the His-D-Phe-Arg-Trp region *(10,73)*. A proposed bioactive conformational model, consisting of the ligand amino acids, His, D-Phe, and Trp being oriented on one face and the Arg side chain extending on the opposite face, have been reported *(68,69)*. This initial peptide conformation was further manipulated to produce a conformation with a type II' reverse-turn about the His-D-Phe-Arg-Trp residues with the N-and C-terminal portions of the peptide folding back toward the extracellular region of the receptor (Fig. 6). Specific receptor residues that were able to interact with both the hydrophobic portions of the reverse turn and with the Arg^8 side chain were identified in a putative binding pocket between 8 Å and 15 Å from the extracellular portion

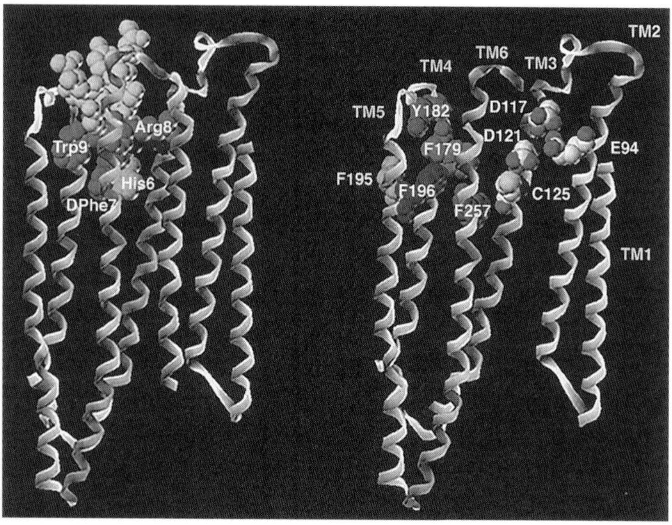

Fig. 6. Proposed interactions of the NDP-MSH melanocortin ligand docked into a model of the hMC1-R. The figure on the left illustrates NDP-MSH docked into the proposed TM domain with the core His[6], D-Phe[7], Arg[8], and Trp[9] melanocortin ligand residues positioned in a type II' β-turn. The remaining NDP-MSH residues are shown. The figure on the right shows the putative hMC1-R resides proposed to interact with the core His-D-Phe-Arg-Trp melanocortin ligand amino acids. The TM domains are numbered as illustrated and TM7 has been removed for clarity of the hydrophilic receptor residues in TMs 2 and 3 which are proposed to interact with the ligand Arg[8] residue. An extensive hydrophobic–aromatic network is proposed to consist of the ligand D-Phe[7] and Trp[9] with the receptor residues illustrated in TMs 4, 5, and 6 (F175 has been omitted for clarity). Both the N-and C-terminal sequences are absent from the model.

of the receptor. NDP-MSH was docked manually into the binding pocket, with the key ligand residues D-Phe[7] and Trp[9] projected toward the hydrophobic aromatic binding region, and the Arg[8] side chain projected toward a hydrophilic binding site consisting of negatively charged residues on TM2, TM3, and TM7 (Fig. 6). Fig. 8 shows a schematic representation of the proposed ligand–receptor interactions of the NDP-MSH peptide residues D-Phe[7]-Arg[8]-Trp[9] docked into the hMC1-R. A network of hydrophobic and aromatic interactions were predicted to be formed between D-Phe[7] and Trp[9] of the ligand with several Phe and Tyr residues of the hMC1 receptor. Additionally, an extensive network of electrostatic interactions (ionic, hydrogen bonding, and Van der Waals) involving the ligand Arg[8] side chain and putative receptor residues in TMs 2 and 3 (the inclusion of water molecules may also be relevant *[32]*) was also proposed. Specific receptor side chains that may interact with the ligand in this region include E94 TM2, D117 TM3, D121 TM3, F175 TM4, F179 TM4, Y182 TM4, Y183 TM4, F196 TM5, F257 TM6, F280 TM7, and N281 TM7. Furthermore, since the orientation of the TM α-helices is speculative, rotation and translation of individual helices allow for additional potential receptor–ligand interactions as previously reported *(23)*.

2.5. MTII Docking

The cyclic MTII analog Ac-Nle^4-*c*[Asp^5-His^6-D-Phe^7-Arg^8-Trp^9-Lys^{10}]-NH_2 *(74,75)* (Fig. 7b), has been docked into the receptor with the lactam bridge oriented toward the extracellular region of the receptor. This alignment was proposed based on ancillary site modifications (regions of the peptide that can be modified by the addition of a large bulky group(s) where these groups do not change the SAR of the original template peptide) of MTII. These ancillary sites have been identified by fatty acid conjugates connected to the N-terminus of MTII *(66)* and incorporation of one or two amino acids into cyclic lactam bridged α-melanocortins *(68,69)*. The previously reported bioactive conformation model of MTII utilized for docking had the His and D-Phe residues positioned on one face, while the Arg and Trp residues are positioned on an opposite face *(74)*. When docked in the hMC1-R model, the receptor–peptide interactions of D-Phe^7 and Arg^8 residues were similar to those found with NDP-MSH, but the Trp^9-receptor interactions differed (*see* Fig. 9). The new proposed location of the Trp^9 residue of MTII with hMC1-R suggests intermolecular interactions between Trp^9 and N281, as well as amino-aromatic interactions *(76–80)* between Arg^8 and F45.

2.6. α-MSH Docking

To account for the differences observed in binding affinity and physiologic response between NDP-MSH and α-MSH, the docked NDP-MSH ligand was extracted from the model, and inversion of the chiral alpha carbon of D-Phe^7 to L-Phe^7 was performed. The Nle^4 of NDP-MSH was converted to Met by replacement of the methylene (CH_2) with sulfur. This ligand was then positioned back into the modeled receptor maintaining the interactions of all the ligand residues except for the Phe^7. This inversion positioned the phenyl ring of L-Phe^7 to interact with the H260 receptor residue, in agreement with a hMC1-R mutagenesis report that was published during the course of those studies *(61)*.

The hMC1-R 3D receptor homology model *(23)* predicted the putative ligand core residues (D-Phe-Arg-Trp) of NDP-MSH and MTII to interact somewhat differently with hMC1-R residues, as summarized in Figs. 8 and 9. α-MSH was only proposed to differ from NDP-MSH by having L-Phe^7 (α-MSH)

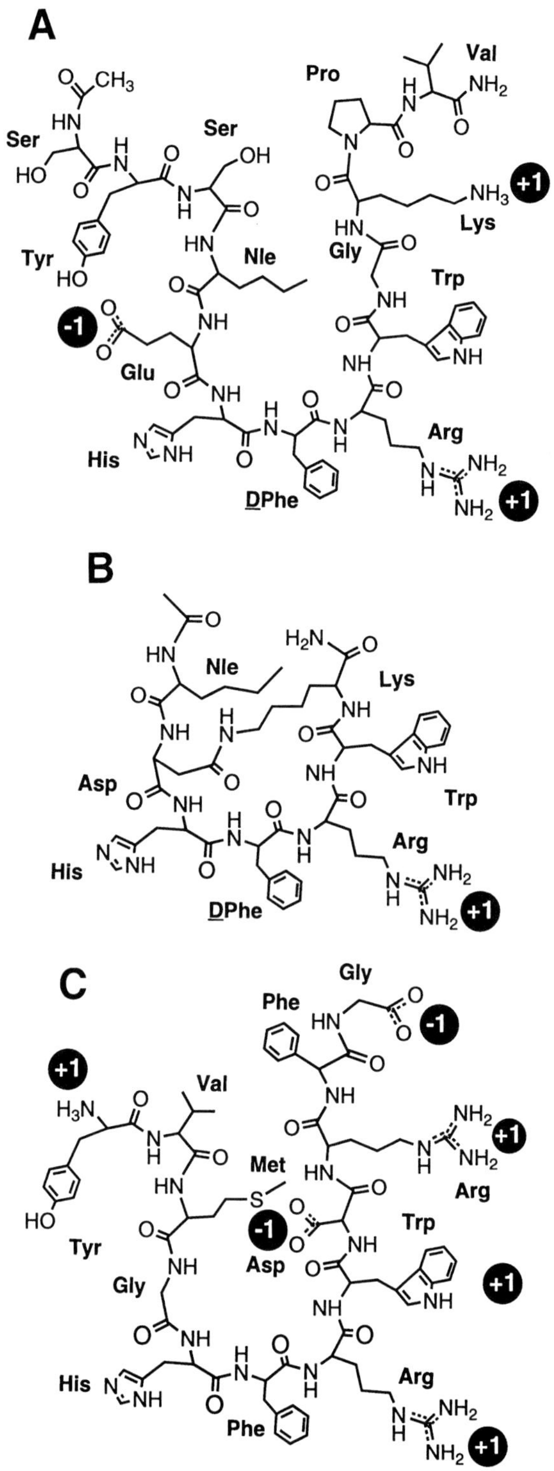
A
Ser
Tyr
Ser
Nle
Glu
His
DPhe
Arg
Trp
Gly
Lys
Pro
Val
-1
+1
+1
B
Nle
Asp
His
DPhe
Arg
Trp
Lys
+1
C
Tyr
Val
Met
Gly
His
Phe
Arg
Trp
Asp
Arg
Phe
Gly
+1
-1
+1
-1
+1
+1

interacting primarily with H260 (TM6) of the hMC1-R as suggested by one of the first hMC1-R mutagenesis report *(61)*, although these premises no longer appear to be correct. γ-MSH (Fig. 7c) was not docked into the hMC1-R model at that time due to a lack of fragment SAR studies to provide necessary functional and structural information.

3. In Vitro Mutagenesis of the hMC1-R and mMC1-R

Ligand–receptor interactions depend on the ability of the hormone to bind and discriminate one receptor from another. The formation of noncovalent complexes between these molecules is ubiquitous and essential for biologic functions. The availability of structural analysis (X-ray crystallography and NMR) of biologic molecules has helped to outline interacting complexes at the molecular level and provide details of protein–protein interactions. Binding and the stimulation of the receptor resulting in signal transduction by ligands can be extremely sensitive to subtle differences in structure as is the case for the melanocortin system. Therefore, a quantitative knowledge of the molecular recognition and binding events are essential. This information includes detailed understanding of the physical forces involved in this process and the extent to which these forces participate in the overall reaction complex (*[81]*, and references therein). Thermodynamics is the overall driving force of the majority of systems in the universe, and is equally important for ligand–receptor systems. The components in a ligand-receptor complex may include solvent molecules, ionic strength, pH, and concentrations of both the ligand and receptor.

Classical ligand–receptor interactions *(81–85)* can be summarized as

$$R + L \underset{k_{-1}}{\overset{k_{+1}}{\leftrightarrow}} RL \tag{1}$$

Fig. 7. *(opposite page)* Structure of the melanocortin ligands used to test in vitro MC1-R mutants for changes in affinity and efficacy. α-MSH (Ac-Ser-Tyr-Ser-Met^4-Glu-His-Phe^7-Arg-Trp-Gly-Lys-Pro-Val-NH_2) differs from (A) NDP-MSH (Ac-Ser-Tyr-Ser-Nle^4-Glu-His-D-Phe^7-Arg-Trp-Gly-Lys-Pro-Val-NH_2) by the isosteric replacement of the S in Met to CH_2 in Nle, and inversion of chirality of L-Phe^7 to D-Phe^7, respectively. Both α-MSH and NDP-MSH contain the same charged residues [Glu^4 (–), Arg^8 (+), Lys^{11} (+)], whereas MTII (B) only possesses the Arg^8 (+) residue, and γ-MSH (C) possesses a free N-terminal (+), C-terminal (–), Arg^8 (+), Asp^{10} (–), and Arg^{11} (+). All these ligands are linear (possessing more conformational flexibility and rotational freedom) except for MTII which possesses a 23-membered ring cyclized by an amide bond between the Asp^5 and Lys^{11} side chains.

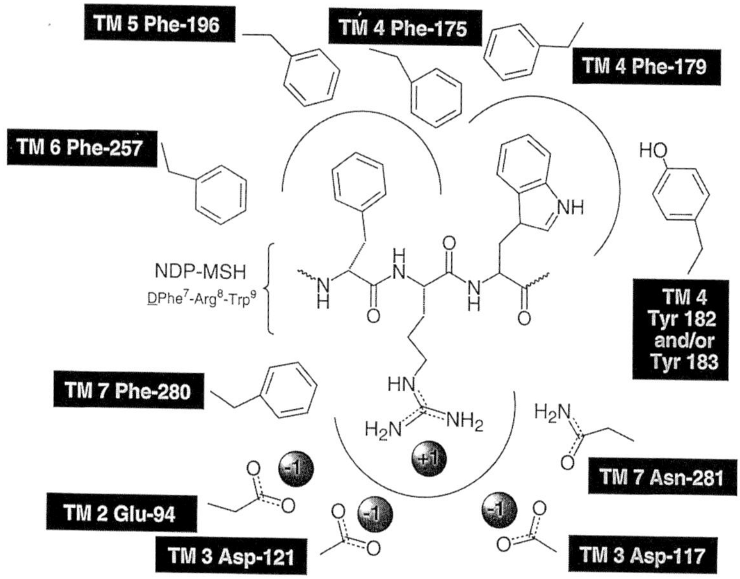

Fig. 8. Summary of the proposed NDP-MSH ligand D-Phe[7]-Arg[8]-Trp[9] residues interacting with the specific hMC1-R residues indicated.

where R is the receptor, L is the ligand, RL is the ligand–receptor complex, k_{+1} is the association rate constant, and k_{-1} is the dissociation rate constant. At equilibrium, when steady–state kinetics are reached (i.e., ligand–receptor association rates equal the dissociation rates), the equilibrium constant (dissociation equilibrium constant) K_d, can be defined as

$$K_d = \frac{[R][L]}{[RL]} \tag{2}$$

where K_d is in units of moles per liter.

The fundamental thermodynamic equation relating the free energy change of a system to changes in enthalpy (energy) ΔH and entropy (disorder) ΔS is

$$\Delta G = \Delta H - T\,\Delta S \tag{3}$$

where ΔG is the change in free energy and T is the absolute temperature (K) of the system. The ligand–receptor equilibrium constant can be related to the free-energy change of the dissociation of the RL complex as

$$\Delta G = \Delta G^\circ - RT \ln K_d \tag{4}$$

where G is the free energy, ΔG is the change in free energy of the ligand–receptor interaction, R is the gas constant, and T is the absolute temperature.

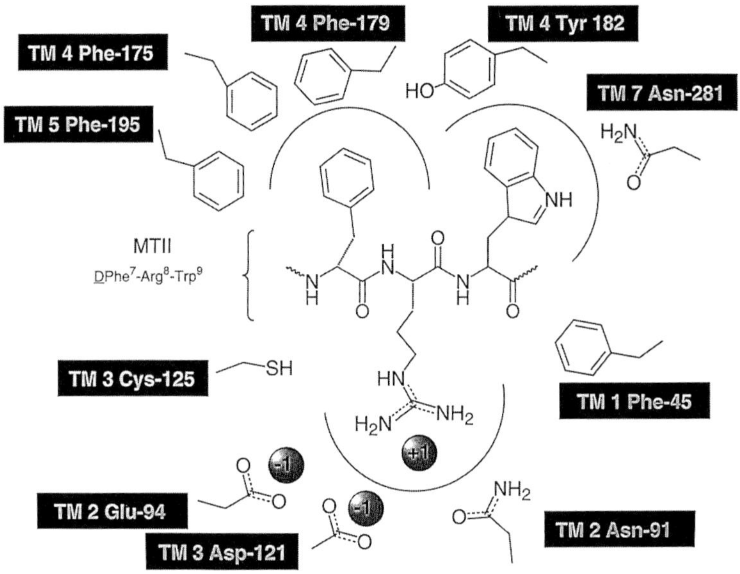

Fig. 9. Summary of the proposed MTII ligand D-Phe[7]-Arg[8]-Trp[9] residues interacting with the specific hMC1-R residues indicated. These interactions differ from those of NDP-MSH (Fig. 8) by the presence of N281 (TM7) and the absence of F257 (TM6) interacting with the ligand Trp[9] and D-Phe[7], respectively. The ligand Trp[9] residue was proposed to interact with F45 (TM1) and the ligand Arg[8] residue via amino-aromatic interactions. Additionally, the ligand Arg[8] putative receptor interactions has been modified to include C125 (TM3), N91 (TM2), F45 (TM1), and exclude D117 (TM3).

At equilibrium and standard conditions (all reactants and products are present at 1M concentration, T = 298 K, and the pressure is 1 atm), $\Delta G = 0$ and

$$\Delta G^\circ = RT \ln K_d \tag{5}$$

This equation can be further extrapolated to relate IC_{50} and Ki values by the equation:

$$Ki = \frac{[IC_{50}]}{1 + \frac{[L]}{K_d}} \tag{6}$$

Either the IC_{50} or Ki values are reported for biological results affiliated with competitive displacement binding experiments and ligand affinity. Although multiple premises are built into this analysis, nevertheless, it is now possible to pseudoquantitate the energy change associated with ligand bind-

Table 2
Theoretical Effect of Changes in Binding Energy (kcal/mol) on Binding Constant (K_d) Values at Room Temperature

Change in Binding Energy	Change in Binding Constant
0.5	2×
1.0	5×
1.5	13×
2.0	29×
2.5	68×
3.0	158×

As summarized by Ajay and Murcko *(81)*.

ing to the receptor with a theoretical binding constant (K_d, which can be defined as the ligand concentration at which 50% of the receptor sites are occupied in a 1:1 complex, at equilibrium *(81)*, or further indirectly using the experimental IC_{50} value) associated with the ligand–receptor intermolecular processes. Table 2 summarizes the previously reported theoretical changes in binding energy which predict the corresponding changes in the binding constant K_d *(81)*. Factors that contribute significantly to the change in free energy *(G)* associated with ligand binding include the following

1. Hydrophobic energy (the entropy gain of water due to ligand binding)
2. Interaction energy between the ligand and receptor
3. Changes in steric interaction on binding (Van der Waals)
4. Changes in conformational energy of the ligand and receptor upon binding

All these parameters are modified when point mutations are introduced into the receptor protein. Changes in these parameters may be observed by differences in ligand binding affinity or efficacy, however, the exact modified characteristic can only be approximated, depending on the amino acid substitution and other modifications introduced.

Theoretically, a linear peptide ligand can possess a large number of different conformations (three-dimensional structures) in the extracellular milieu. However, upon binding to the receptor, a subset of ligand conformations are thought to exist for the necessary ligand–receptor complementarity to be achieved *(10)*. Thus the rationale in the development of the cyclic compounds such as MTII (Fig. 7b) was to limit the conformational flexibility of the ligand to the proposed "bioactive" conformation and thus, ultimately decreasing the overall system energy *(74,75)*. Table 3 is a compilation of multiple studies summarizing interatomic distances between different types of noncovalent interactions that may exist and be important for peptide–receptor interactions *(79,86–88)*. The change in binding values, or IC_{50}'s, associ-

Table 3
Summary of Noncovalent Interactions of Peptide–Protein Interactions

Type of Contact		Nonbonded Contact Distance Å	Binding Energy (kcal/mol)
Salt bridge	-COO$^-$......H_3N^+-	2.4	–5.0
Hydrogen bond	-NH........O= (Amide-carbonyl)	2.9	–6.0
	-OH.....OH- (Hydroxyl-hydroxyl)	2.8	
	-OH.....O= (Hydroxyl-carbonyl)	2.8	
	-NH.....OH- (Amide-hydroxyl)	2.9	
	-NH...N= (Amide-imidazole)	3.1	
	-NH.....S- (Amide-sulfer)	3.7	
Aromatic	π-π stacking	4.5 to 7.5	–2.5 to –5.0
	π-NH (Hydrogen bond)	3.0 to 6.0	–3.0
	π-O (Aromatic-oxygen)	5.1	–1.0
	π-S (Aromatic-sulfer)	5.6	<–1.0
Hydrophobic	Entropically driven	—	–1.0 to –5.0

Data from refs. *79, 86-88.*

ated with these interactions are difficult to predict due to the overall contribution of a multitude of such interactions to the overall system, including the presence or absence of water molecules. Additionally, as eloquently discussed by Schwartz et al. *(89)*, interactions that contribute to the overall binding, either directly or indirectly, are difficult to differentiate experimentally and may be discerned fully only upon crystallization of the ligand–receptor complex.

3.1. hMC1-R Mutagenesis

Figure 10 summarizes the point mutations of the MC1-R discussed in this chapter. Mutations of the hMC1-R are illustrated in black circles with white text, mutations of the mMC1-R are shown in black squares with white text, and mutations performed on both the human and mouse MC1-Rs are shown in shadowed squares with black text. The difference in the numbering nomenclature between the human and mouse MC1-R results from a deletion

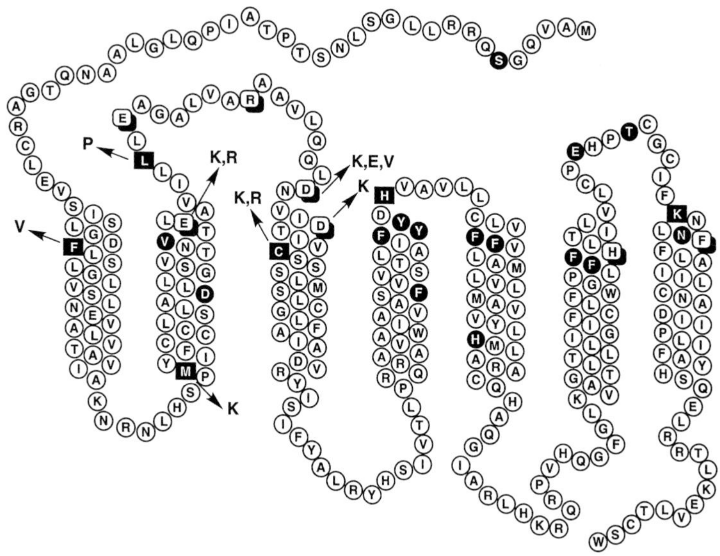

Fig. 10. A summary of point mutations which have been reported in the MC1-Rs. The schematic receptor illustrated is the hMC1-R. Mutations performed solely on the hMC1-R are illustrated in black circles with white text, mutations performed solely on the mMC1-R are shown in black squares with white text, and mutations performed on both the human and mouse MC1-Rs are shown in shadowed squares with black text. The letters indicated by the arrows represent changes in the mMC1-R, which resulted in constitutively active receptors. A difference in residue numbering between the human and mouse MC1-R is due to a deletion of 2 amino acid residues in the N-terminal of the mMC1-R.

of two amino acid residues in the N-terminal of the mMC1-R. The first mutational analyses of the hMC1-R are reported in Tables 4 and 5 *(61,90)*. These data provided some insight into residues potentially involved in ligand binding, such as D117 and H260. Further analysis of the D117 and H260 hMC1-R mutant receptors using multiple ligands has more recently been reported (Table 6) *(91)*.

To test the putative ligand–receptor interactions identified by a hMC1-R 3D model *(23)*, several single, double, triple point mutations, and a quadruple point mutant receptor were made *(92)*. Each of the mutant receptors were examined for both changes in ligand affinity (competitive displacement binding) and efficacy (intracellular cAMP accumulation), using the endogenous ligands α-MSH and γ-MSH, and two synthetic peptides with enhanced potencies at the hMC1-R *(93)*, NDP-MSH *(73)* and MTII *(74,75)* (Fig. 7). α-MSH (Ac-Ser-Tyr-Ser-Met^4-Glu-His-Phe^7-Arg-Trp-Gly-Lys-Pro-Val-NH_2) differs

Table 4
Binding Results of Point Mutations in the hMC1-R

		Binding *K*i (n*M*)									
Mutation	TM	α-MSH	Fold Difference	NDP-MSH	Fold Difference	Nle⁴-α-MSH	Fold Difference	β-MSH	Fold Difference	γ-MSH	Fold Difference
hMC1-R		0.130±0.025	1.0	0.021±0.006	1.0	0.047±0.011	1.0	0.88±0.14	1.0	1.20±0.21	1.0
D117A	3	34.8±8.8	267	0.056±0.005	2.7	8.06±0.99	171	357±32	406	>30000	>25000
F179A	4	0.100±0.014	–1.3	0.058±0.010	2.8	ND	—	ND	—	ND	—
H209A	5	0.040±0.090	–3.3	0.023±0.005	1.1	ND	—	ND	—	ND	—
H260A	6	17.3±2.7	133	0.027±0.010	1.3	0.80±0.09	17	45.8±4.8	52	106±10	88

Reported by Frändberg et al. *(61)*.

Table 5
Summary of the hMC1-R point mutations in the Extracellular Loops

	Binding *K*i (n*M*)			
Mutation	α-MSH	Fold Difference	NDP-MSH	Fold Difference
hMC1-R	0.102±0.019	1.0	0.027±0.004	1.0
S6A	3.78±0.76	37	0.695±0.071	26
E102A	0.033±0.005	–3.1	0.009±0.001	–3.0
R109A	0.119±0.024	1.2	0.012±0.002	–2.3
E269A	0.986±0.181	6.8	0.221±0.036	8.2
T272A	0.749±0.099	7.3	0.102±0.017	3.8

Reported by Chhajlani et al. *(90)*.

from NDP-MSH (Ac-Ser-Tyr-Ser-Nle^4-Glu-His-D-Phe^7-Arg-Trp-Gly-Lys-Pro-Val-NH_2) by the isosteric replacement of the S in Met to CH_2 in Nle, and inversion of chirality of L-Phe^7 to D-Phe^7, respectively. These modifications resulted only in a 4- to 11-fold difference in binding at the hMC1-R *(92,93)*, apparently depending on the cell line in which the hMC1-R is expressed. Both α-MSH and NDP-MSH contain the same charged residues [Glu^4 (–), Arg^8 (+), Lys^{11} (+)], whereas MTII only possesses the Arg^8 (+) residue, and γ-MSH possesses a free N-terminal (+) and C-terminal (–), Arg^8 (+), Asp^{10} (–), and Arg^{11} (+). All these ligands are linear (possessing more conformational flexibility and rotational freedom) except for MTII, which possesses a 23-membered ring cyclized by an amide bond between the Asp^5 and Lys^{11} side chains. This synthetic ligand was identified as possessing increased potencies only at the lizard *(74,75)*, hMC1-R *(64,93)*, and hMC4-R *(93)*, as compared to NDP-MSH.

To examine the putative ligand interactions involving the polar and aromatic receptor amino acids, the selected receptor residues were mutated to Ala. The bioassay results of these mutations are summarized in Tables 7 and 8. Figure 11 summarizes the ligand affinity results at the hMC1-R hydrophilic mutant receptors and illustrates two interesting trends. First, at the wild-type hMC1-R, MTII has a slightly increased affinity compared to NDP-MSH, however, at all the mutations shown, NDP-MSH possessed an increased affinity as compared with MTII, albeit to varying degrees. This supports the hypothesis that the Arg^8 residue of the melanocortin ligand appears to be important for ligand affinity. Second, γ-MSH lost all ability to competitively displace the radiolabel ([^{125}I]NDP-MSH) at these mutations. These data clearly demonstrate that the difference in ligand-charged residues may be an important determinant for different ligand–receptor interactions of these peptides at

Table 6
Summary of Multiple Ligand Analysis on the hMC1-R Mutant Receptors D117A and H260A

Ligand	Sequence	hMC1-R *K*i (n*M*)	D117A *K*i (n*M*)	Fold Difference	H260A *K*i (n*M*)	Fold Difference
[125I]-NDP	Ac-S(I125)YSMEHFRWGKPV-NH_2	0.183	0.478	2.6	0.401	2.2
β-MSHp	Ac-DEGPYRMEHFRWGSPPKD-NH_2	2.26	458	202	51.7	23
Y6-β-MSHp	Ac-DEGPYRMEYFRWGSPPKD-NH_2	634	>300,000	>473	>300,000	>473
	NH-D-Phe-RWG-NH_2	3.57	2,490	697	62.2	17
	MNH-D-Phe-RWG-NH_2	633	31,200	49	2,000	3.2
(1-13)D	Ac-SYS-*c*[CEH-D-Phe-RWC]KPV-NH_2	0.037	0.500	14	0.480	13
(1-13)L	Ac-SYS-*c*[CEHFRWC]KPV-NH_2	0.570	7.80	14	9.55	17
(4-13)D	Ac-*c*[CEH-D-Phe-RWC]KPV-NH_2	0.033	5.10	155	4.94	150
(4-10)D	Ac-*c*[CEH-D-Phe-RWC]-NH_2	197	410	2.1	217	1.1
HS9510	Ac-*c*[CEH-D-Nal(2')-RWC]-NH_2	76.2	>100,000	>1312	600	7.9
MTII	Ac-Nle-*c*[DH-D-Phe-RWK]-NH_2		0.741	251	339	2.37
3.2						
SHU9119	Ac-Nle-*c*[DH-D-Nal(2')-RWK]-NH_2	0.666	444	667	0.225	–3.0

The amino acids listed in the peptide sequences consist of the one-letter abbreviations except where noted by the presence of unnatural residues. The prefix *c* indicates a side chain cyclization between the amino acids designated at the start and end of the indicated brackets. The fold difference is calculated by dividing the *K*i value of the mutant receptor by the corresponding peptide value at the hMC1-R.

From ref. *91*.

Table 7
Summary of the Binding Data of Different Melanocortin Ligands on Point Mutations of the hMC1-R

		Binding IC_{50} (n*M*)							
Mutation	TM	α-MSH	Fold Difference	NDP-MSH	Fold Difference	MTII	Fold Difference	γ-MSH	Fold Difference
hMC1-R		2.58±0.33	1.0	0.67±0.09	1.0	0.24±0.02	1.0	11.5±0.76	1.0
E94A	2	268±13	104	2.15±0.46	3.2	15.5±3.3	65	>1000	>87
D117A	3	125±6	48	5.20±0.35	7.8	42±3	176	>1000	>87
D121A	3	235±9	91	7.10±0.80	11	86±15	360	>1000	>87
D121K	3	>1000	>387	31.2±1.8	47	>1000	>4167	>1000	>87
D121N	3	>1000	>387	27.5±3.8	41	>1000	>4167	>1000	>87
D117A/D121A	3	176±11	68	9.2±2.5	14	97±32	404	>1000	>87
E94A/D117A/D121A	2/3	293±15	113	10.6±1.2	16	123±10	512	>1000	>87
F175A	4	4.45±0.42	1.7	1.75±0.47	2.6	1.10±0.44	4.6	16.4±1.3	1.4
F179A	4	2.78±0.21	1.1	0.79±0.06	1.2	0.34±0.08	1.3	10.2±0.1	0.9
Y182A	4	2.40±0.50	0.9	0.78±0.12	1.2	0.32±0.09	1.3	13.6±0.4	1.2
Y183A	4	2.65±0.21	1.0	0.54±0.08	0.8	0.19±0.01	0.8	12.7±0.3	1.1
F195A	5	3.20±0.32	1.2	0.87±0.11	1.3	0.43±0.05	1.8	19.0±1.2	1.7
F196A	5	2.87±0.43	1.1	0.67±0.13	1.0	0.97±0.10	4.0	32.8±0.8	2.9
F175A/F196A	4/5	11.5±2.9	4.5	1.23±0.10	1.8	1.39±0.36	5.8	57.0±8.5	4.9
F179A/F196A	4/5	3.19±0.68	1.2	1.03±0.20	1.5	0.54±0.15	2.3	375.8±25.9	32.7
Y182A/F196A	4/5	13.7±0.6	5.3	0.99±0.08	1.5	1.67±1.00	6.9	46.0±5.4	4.0
F175A/Y182A/F196A	4/5	12.3±1.8	4.8	1.37±0.08	2.0	1.89±0.90	7.9	>1000	>87
F175A/F179A/ Y182A/ F196A	4/5	22.8±1.3	8.8	1.74±0.26	2.6	2.60±0.20	10.8	>1000	>87
F257A	6	8.10±0.23	3.1	1.32±0.11	2.0	1.78±0.08	7.4	35.0±6.4	3.0
F257A/F258A	6	10.8±0.9	4.2	1.98±0.21	2.9	6.27±0.43	26.1	41.0±7.4	3.6
H260A	6	14.8±2.6	5.7	0.79±0.11	1.2	0.63±0.05	2.6	81.0±3.3	7.0
F280A	7	3.20±0.50	1.2	0.74±0.10	1.1	0.29±0.05	1.2	14.2±0.67	1.2
N281A	7	13.3±1.9	5.1	2.90±0.43	4.3	4.4±0.8	18	>1000	>87

From ref. *92*.

Table 8
Summary of the Intracellular Accumulation cAMP Data of the Melanocortin Ligands on Point Mutations of the hMC1-R

		Intracellular cAMP Accumulation EC_{50} (n*M*)							
Mutation	TM	α-MSH	Fold Difference	NDP-MSH	Fold Difference	MTII	Fold Difference	γ-MSH	Fold Difference
hMC1-R		1.34±0.11	1.0	0.24±0.06	1.0	0.13±0.02	1.0	8.1±0.4	1.0
E94A	2	537±79	400	0.45±0.16	1.8	6.1±0.6	47	-	-
N281A	7	5.43±0.40	4.1	0.55±0.07	2.3	0.38±0.03	2.9	-	-
D117A	3	391±63	291	3.1±0.3	13	98±9	753	>1000	>123
D121A	3	508±34	379	1.2±0.2	5.0	72±9	554	>1000	>123
D121K	3	-	-	>1000	>4167	-	-	-	-
D121N	3	>1000	>746	123±5.8	513	>1000	>7692	-	-
D117A/D121A	3	833±109	621	198±66	825	938±135	7215	>1000	>123
E94A/D117A/D121A	2/3	-	-	782±35	3258	>1000	>7692	-	-
F175A	4	1.25±0.17	-1.0	0.73±0.10	3.0	0.27±0.10	2.0	7.6±1.2	-1.1
F179A	4	0.90±0.10	-1.5	0.65±0.10	2.7	0.29±0.10	2.2	7.8±0.8	-1.0
Y182A	4	1.38±0.09	1.0	0.45±0.07	1.9	0.35±0.06	2.7	9.5±0.4	1.2
Y183A	4	1.49±0.07	1.1	0.38±0.05	1.6	0.46±0.07	3.5	8.6±0.4	1.1
F195A	5	2.38±0.67	1.8	0.63±0.08	2.6	0.32±0.07	2.5	21±3	2.6
F196A	5	2.1±0.59	1.6	0.71±0.10	2.9	0.29±0.05	2.2	42±4	5.2
F175A/F196A	4/5	1.20±0.14	-1.1	0.55±0.12	2.3	0.45±0.07	3.5	>1000	>123
F179A/F196A	4/5	1.46±0.34	1.1	0.53±0.09	2.2	0.38±0.09	2.9	213±12	26
Y182A/F196A	4/5	2.06±0.12	1.5	0.77±0.10	3.2	1.0±0.1	7.7	12±1	1.5
F175A/Y182A/F196A	4/5	1.77±0.09	1.3	0.74±0.04	3.0	0.78±0.07	6.0	>1000	>123
F175A/F179A/Y182A/ F196A	4/5	7.90±0.81	5.9	0.69±0.07	2.9	1.0±0.1	7.7	-	-
F257A	6	5.90±0.64	4.4	0.82±0.13	3.4	2.26±0.37	17	281±52	35
F257A/F258A	6	7.1±0.8	5.3	1.1±0.1	4.6	3.2±0.5	25	519±34	64
H260A	6	10.4±1.2	7.8	0.24±0.10	1.0	0.19±0.04	1.5	169±23	21
F280A	7	1.92±0.10	1.4	0.34±0.08	1.4	0.32±0.03	2.5	13±1	1.6

A dash (-) signifies that no stimulation was detected up to 1μ*M* concentrations of ligand. From ref. *92*.

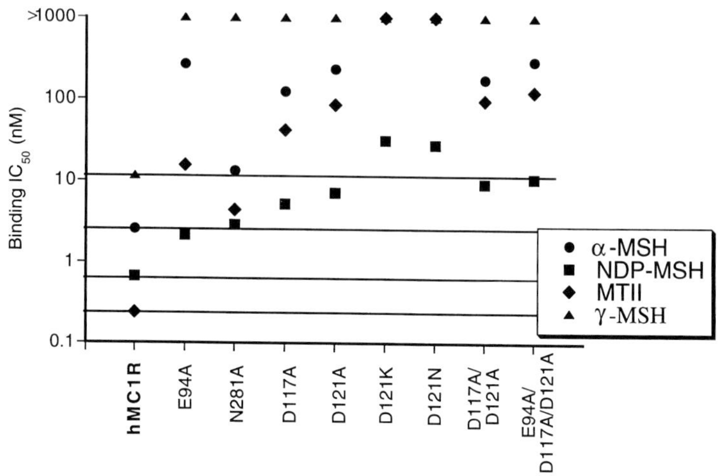

Fig. 11. Competitive displacement binding studies of hMC1-R potentially involved in ligand–receptor electrostatic interactions. The mutant receptor is plotted versus the corresponding ligand binding IC_{50} (n*M*) value, and compared to the wild type hMC1-R.

the hMC1-R and help to account for the differences in efficacy of these ligands at the melanocortin receptors.

Due to the apparent importance of the Phe^7 and Trp^9 residues of the melanocortin ligand *(94,95)* at the MC1-R, an extensive hydrophobic network of receptor residues (F175, F179, Y182, Y183, F195, F196, F257, and F280) were identified as potentially providing complementary aromatic (π–π) interactions with the aforementioned ligand residues (Figs. 8 and 9) *(23)*. These residues were mutated to Ala and the competitive binding and intracellular cAMP accumulation results summarized in Tables 7 and 8. Unexpectedly, these single-point mutations resulted in a maximal affinity difference of 5-fold, but for the majority of ligand–receptor mutant combinations, no significant differences in binding affinity were observed (Fig. 12). It was then rationalized that, since potentially up to 7 aromatic residues may be involved in the aromatic network (including ligand and receptor), the modification of one receptor aromatic residue may be compensated for by the others. Precedent had been found in the neurokinin receptor where only a double aromatic mutation identified significant differences in ligand affinity *(96)*. Additionally, aromatic mutations have been observed to result in small differences in ligand affinity or efficacy as compared with electrostatic residues, which potentially result in larger observed differences in binding energy (Table 3) *(79,86–88)*. This led to the examination of double and triple aromatic residue

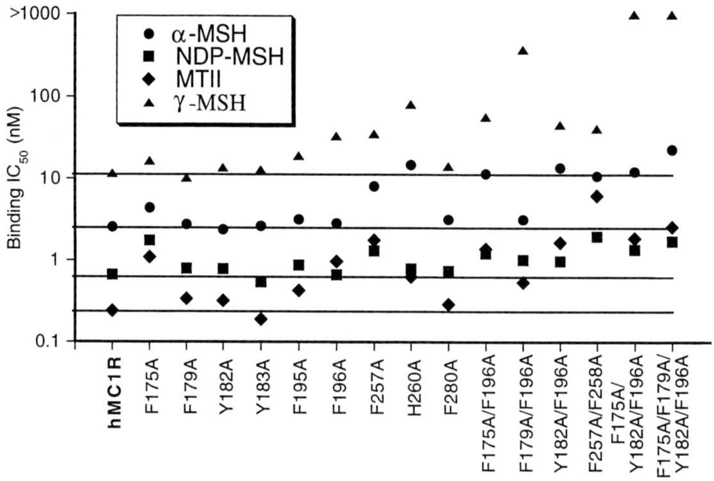

Fig. 12. Competitive displacement binding studies of hMC1-R residues potentially involved in ligand–receptor hydrophobic–aromatic interactions. The mutant receptor is plotted versus the corresponding ligand binding IC_{50} (n*M*) value, and compared to the wild-type hMC1-R.

mutations (Tables 7 and 8 and Fig. 12). The most dramatic observations were that γ-MSH lost the ability to competitively displace the radiolabled NDP-MSH at the mutant receptors F175A/Y182A/F196A and F175A/F179A/Y182A/F196A. α-MSH ligand affinity was most affected by the single F257A and H260A (Tables 4 and 7) mutant receptors, 3-and 5-fold, respectively (within experimental error), and up to 9-fold by the multiple mutant containing receptors. NDP-MSH was apparently not significantly affected by any of these mutant receptors as indicated by up to a 3-fold difference in binding affinity and up to a 4-fold difference in ligand efficacy. The single F257A mutant receptor resulted in a 7-fold difference in binding affinity of MTII. The F175A and F196A mutations also resulted in 4-fold difference in MTII affinity. Overall, these aromatic hMC1-R mutations provided surprisingly indirect results in regards to changes in melanocortin ligand affinity, with multiple simultaneous mutations providing some information about potentially different ligand–receptor interactions of α-MSH, NDP-MSH, MTII and γ-MSH with the hMC1-R.

Ligand efficacy was also examined on these mutations to study the effect of ligand stimulation on the mutant receptors and possibly identify receptor residues that are important for signal transduction and not ligand binding. Theoretically, if a 10-fold decrease in ligand binding affinity was observed, the intracellular cAMP should also demonstrate a 10-fold decrease and cor-

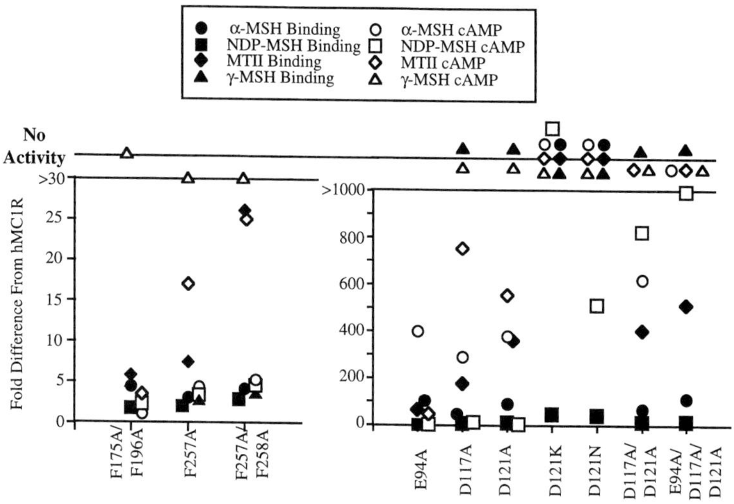

Fig. 13. hMC1-R mutant receptors where notable differences between ligand affinity and efficacy were observed. The mutant receptors are plotted against the fold-difference observed (Tables 7 and 8) and compared to the wild type hMC1-R. Both the fold difference from the wild type receptor of ligand binding affinity and efficacy are included for comparison. It is predicted that for a change in ligand binding affinity, i.e., 10-fold, that a corresponding change in ligand efficacy (10-fold), within experimental error, should also be observed. For the mutant receptors summarized in this figure, this is not the case for one or more of the ligands examined.

relate nicely with the affinity. This was the case for the majority of mutations of the hMC1-R, with the exception of a few notable mutations summarized in Fig. 13. The fold differences for these mutations are summarized in Tables 7 and 8, with the corresponding ligand value (IC_{50} or EC_{50}) defined as 1 on the wild-type hMC1-R. The mutant receptor containing F175A/F196A modifications possessed only a 5-fold decrease in γ-MSH binding affinity while this ligand was unable to stimulate any intracellular cAMP accumulation (Fig. 11). Separately, the F175A and F196A mutant receptors possessed approximately equal γ-MSH affinity and efficacy as compared with the wild-type receptor, albeit a 5-fold decrease in efficacy of the F196A mutant receptor was observed (Table 8). These two receptor residues were proposed to be spatially located between the Phe[7] and Trp[9] ligand residues of NDP-MSH *(23)*, and participate in an aromatic-hydrophobic network that would be continuous with the presence of these ligand residues. The aromatic mutation F257A resulted in approx 12-fold difference between γ-MSH affinity and efficacy. The double mutation F257A/F258A (TM6) also possessed nearly an 18-fold difference in affinity and efficacy. These data suggest that F257 (TM6) appears to be important for receptor activation stimulated by γ-MSH, but not necessarily for the other ligands tested. γ-MSH possesses an aromatic residue (Phe), which is a Pro residue in the corresponding position of α-MSH and NDP-MSH and absent in MTII (Fig. 7). Thus, it can be proposed that this Phe residue of γ-MSH may interact (directly or indirectly) with the hMC1-R residues F175, F196, and F257 as part of the receptor activation process involving this ligand.

More dramatic differences between ligand binding affinity and efficacy were observed for some hydrophilic mutations summarized in Fig. 13. The D117A mutant receptor possessed activities which correlated for NDP-MSH, but α-MSH possessed a 6-fold difference between affinity and efficacy, MTII possessed a 4-fold difference between affinity and efficacy, while γ-MSH could neither bind or stimulate this mutant receptor. The D117A/D121A double mutant receptor resulted in a 9-fold difference in α-MSH affinity and efficacy, 59-fold difference in NDP-MSH affinity and efficacy, MTII was able to bind this mutant receptor with 404-fold decreased affinity, but was unable to generate any intracellular cAMP accumulation, and γ-MSH was unable to either bind or stimulate the receptor. The triple mutant receptor E94A/D117A/D121A was able to bind α-MSH, NDP-MSH, and MTII, albeit with 113-,16-, and 512-fold decrease in binding for these ligands, respectively, but no intracellular cAMP accumulation was detected for α-MSH and MTII. NDP-MSH effected a weak functional response on the triple mutant receptor, which was 3258-fold less efficacious than on the wild-type receptor. Again, γ-MSH was unable to bind or stimulate any activity at this E94A/D117A/D121A mutant receptor. The D121A mutation resulted in 379-, 5-, and 554-fold decreased efficacies of α-MSH, NDP-MSH, and MTII respectively, compared with the wild-type receptor. These aforementioned decreased efficacies correlated with the decreased affinities observed for the corresponding ligands. However, when D121 was substituted with a Lys or Asn residue, α-MSH, MTII, and γ-MSH lost all ability to competitively displace the [^{125}I]NDP-MSH radiolabel. NDP-MSH ligand affinity was only 40-fold less potent on the D121K and D121N mutant receptors. No ligand-induced cAMP accumulation was observed for the D121K mutant and 12-fold difference of NDP-MSH between affinity and efficacy resulted. Surprisingly, NDP-MSH was able to stimulate the triple mutant (E94A/D117A/D121A), albeit with a 3258-fold difference compared to the wild-type receptor. This latter discrepancy may be attributed to a difference in cell surface receptor expression. Affinity constants determined from radiolabeled competitive displacement binding studies are not effected by receptor number, whereas functional activity, such as adenylate cyclase, is affected by receptor number. Receptor number can be quantitated by a variety of techniques including the use of specific antibodies, and the recent development of antibodies against the receptors should help in this regard *(90)*.

In an attempt to discriminate the melanocortin ligand Trp^9 side chain-hMC1-R receptor interactions, cyclic melanocortin ligands containing the L-and D-Trp^9 stereoisomers were examined *(93)*. The hMC1-R 3D model implicated F175 to be located in the putative binding pocket, and specifically proposed to interact with the Trp^9 residue of the MTII ligand (Fig. 9) *(23)*. The absence of this residue was proposed to disrupt the continuity of the previously discussed hydrophobic-aromatic network and result in decreased ligand binding affinity of cyclic peptides containing D-Trp^9. Additionally, this Phe residue is only present in the human MC1-R and is replaced with a Ser or Thr residue in the horse *(40)* mouse *(2)*, cow *(42)*, panther *(43)*, fox *(44)*, and chicken *(45)* (Fig. 3). This is particularly important because the mouse and human MC1-Rs possess different pharmacologic profiles in response to the melanocortin peptides *(47)*. The F175 residue was a likely candidate for the differences observed between these two receptors. Additionally, this aromatic residue has been substituted by other functional moieties in the hMC3-R, hMC4-R, and hMC5-R subtypes. In the analog Ac-Nle-*c*[Asp-His-D-Phe-Arg-D-Trp-Ala-Lys]-NH_2, the Trp residue had been inverted to the D-configuration and resulted in a 78-fold selectivity for the hMC1-R over hMC4-R, whereas the L-Trp containing peptide Ac-Nle-*c*[Asp-His-D-Phe-Arg-L-Trp-Ala-Lys]-NH_2, only possessed 3-fold selectivity (within experimental error). Precedent for stereochemical specificity of the melanocortin ligand (L-Phe^7 in α-MSH versus D-Phe^7 in NDP-MSH) at the hMC1-R has been demonstrated by modifications of residues D117 (TM3) and H260 (TM 6) to Ala's, Tables 4, 6, 7, and 10 (mMC1-R) *(50,61,91,92)*. These data demonstrated that α-MSH binding affinity was significantly affected by these two mutations (up to 267-fold), as compared with the wild-type receptor, whereas NDP-MSH did not possess a difference in binding affinities at these mutated receptors. These observations and the 3D hMC1-R modeling initiated the hypothesis that the receptor F175 of hMC1-R may be specifically interacting with the ligand D-Trp^9 residue. To test this hypothesis, the ligand was predicted to possess differential binding affinities between wild-type hMC1-R and the hMC1-R F175A mutant receptor, with the latter modification resulting in a loss of affinity mutation (approx 70-fold). Table 9 summarizes the ligand binding data of cyclic melanocortin ligands containing L-Trp and D-Trp on the wild-type and hMC1-R F175A mutant receptors. Although the proposed hypothesis of ligand (D-Trp)-receptor (F175) interaction appears to be incorrect based on the fact that no significant differences in binding were observed. The process of studying ligand–receptor complementary interactions has been performed successfully in other GPCR systems *(97–99)*.

Table 9
Binding Affinities of Two Sets of Cyclic Melanotropin Peptides Containing L- and D-Trp[9] Stereoisomer Modifications on the hMC1-R and F175A Mutant Receptor

	Binding IC_{50} (n*M*)		
Peptide Structure	hMC1-R	F175A	Fold Difference[a]
Ac-Nle-*c*[Asp-His-D-Phe-Arg-Trp-Lys]-NH_2	0.25 ± 0.03	0.12 ± 0.01	2.1
Ac-Nle-*c*[Asp-His-D-Phe-Arg-D-Trp-Lys]-NH_2	0.40 ± 0.17	0.69 ± 0.04	0.5
Ac-Nle-*c*[Asp-His-D-Phe-Arg-Trp-Ala-Lys]-NH_2	0.35 ± 0.05	0.34 ± 0.01	1.0
Ac-Nle-*c*[Asp-His-D-Phe-Arg-D-Trp-Ala-Lys]-NH_2	0.91 ± 0.01	0.59 ± 0.05	1.5

[a]The fold difference is calculated by the IC_{50} value of the wild type receptor divided by the IC_{50} value of the F175A mutant receptor.
Fom ref. *93*.

3.2. mMC1-R Mutagenesis

Simultaneously to the mutational analysis studies of the hMC1-R described above, mutational analyses of the mMC1-R were undertaken *(50,100)*. These studies were initiated by the finding of naturally occurring mutations of the mMC1-R, which resulted in constitutively active receptors producing dark coat coloration *(4)*. Further genetic analyses of several different species identified receptor mutations which result in constitutively active MC1-Rs (Fig. 4). These mutations were induced in the mMC1-R and analyzed for the ability to competitively displace [^{125}I]NDP-MSH, functional efficacy, and constitutive activity (β-galactosidase activity). The β-galactosidase assay consists of a colorimetric endpoint measurement based on a β-galactosidase *(lacZ)* gene fused to five copies of the cAMP response element (CRE) that detects the activation of CRE-binding protein (CREB) resulting from an increase in intracellular cAMP or Ca^{+2} (Fig. 1) *(19)*. Tables 10 and 11 summarize these mMC1-R mutational bioassay results. For the majority of experiments NDP-MSH was the ligand used to analyze the effect on ligand affinity and efficacy of these mutations. Several mutations were similar to those mutated in the hMC1-R. These point mutations include E92 (TM2), E100 (EL1), R107 (EL1), D115 (TM3), D119 (TM3), H258 (TM6), F278 (TM7) (mouse MC1-R numbering). The E92A, H258X, (X = A, E, I, W) F278A mutant receptors maintained similar changes between the mouse and human receptors. Mutations in the mouse which resulted in constitutive activation (increased basal levels of β-galactosidase activity above the wild type receptor) include F43V, M71K, E92K/,R, L98P, D115E/K/V, D119K, and C123R/K (Fig. 10). Figure 14 illustrates the constitutive activity of the M71, E92, D115, D119, C123, E92K/D115K, D115K/D119K, and E92K/D115K/D119K

Table 10
Summary of Melanocortin Ligand Binding on mMC1-R Point Mutations

Mutation	TM	Binding IC_{50} (n*M*) α-MSH	NDP-MSH	Fold Difference
mMC1-R		3.68±1.69	0.79±0.48	1.0
F43A	1		11.3±3.3	14.3
F43V	1		17.0±9.4	21.5
M71K	2	1.01±0.05		–3.6
M71K/D119N	2,3		—	
E92A	2		1.07±0.53	1.3
E92D	2		8.05±4.34	10
E92K	2		423±226	535
L98P	EL1	301±55		82
E100P	EL1		6.38±0.48	8.1
R107L	EL1		1.01±0.36	1.3
R107D	EL1		3.08±1.79	3.9
D115E	3		3.52±3.15	4.5
D115K	3	187±15		51
D115V	3		9.17±0.45	12
D119K	3	211±170		57
D119N	3		16.1±7.3	20
D119V	3		179±121	227
C123A	3		3.38±2.07	4.3
C123E	3		1.50±0.42	1.9
C123K	3		7.94±3.27	10
C123R	3		1.14±0.24	1.4
H183E	4	3.55±2.01		–1.0
H258E	6	351±94		95
H258I	6	656±185		178
H258W	6	1360±570		370
K276A	7		2.83±1.24	3.6
K276E	7		3.20±0.85	4.0
K276L	7		2.80±2.40	3.5
F278A	7		3.85±0.07	4.9
F278V	7		3.95±1.48	5
F278Y	7		3.95±2.05	5

Data from refs. *50* and *100*.

mutant receptors as compared to the wild-type mMC1-R. The M71K, E92K, D115E, D115K, D115V, D119K, C123K, and C123R constitutive active receptors can be further stimulated in the presence of NDP-MSH, while the E92R constitutively active receptor cannot.

Table 11
Summary of the β-Galactosidase Activity of mMC1-R Point Mutations

		β-Galactosidase Activity EC_{50} (n*M*)			
Mutation	TM	α-MSH	Fold Difference	NDP-MSH	Fold Difference
mMC1-R		0.20±0.11	1.0	0.02±0.005	1.0
F43A	1	20±15	100	0.05±0.03	2.5
F43V	1	4.45±0.17	22	0.03±0.005	1.5
M71K	2	1.41±0.96	7	5.26±0.40	263
M71K/D119N	2,3	—		1.30±0.52	65
E92A	2	28200±32500	141000	0.55±0.28	27
E92D	2	14.5±11.4	72	0.17±0.23	8.5
E92K	2	—		0.71±0.18	36
L98P	EL1	—		3.27±0.19	163
E100P	EL1	0.91±3.81	4.6	0.02±0.007	1.0
R107L	EL1	1.09±0.61	5.4	0.21±0.08	10
R107D	EL1	0.12±0.01	–1.7	0.01±0.005	–2.0
D115E	3	5.74±2.14	29	0.009±0.008	–2.2
D115K	3	23±7	115	0.02±0.01	1.0
D115V	3	—		4.20±1.60	210
D119K	3	—		1.77±1.02	89
D119N	3	—		1.70±0.86	85
D119V	3	—		152±130	7600
C123A	3	0.23±0.15	1.2	0.01±0.008	–2.0
C123E	3	39±26	195	0.04±0.02	2.0
C123K	3	—		0.19±0.11	9.5
C123R	3	—		—	
H183E	4	0.09±0.01	–2.2	0.02±0.01	1.0
H258E	6	5.44±1.68	27	0.02±0.005	1.0
H258I	6	—		0.09±0.03	4.5
H258W	6	—		0.27±0.16	14
K276A	7	3.3±2.5	17	0.06±0.05	3.0
K276E	7	0.32±0.14	1.6	0.06±0.03	3.0
K276L	7	1.3±0.07	6.5	0.06±0.006	3.0
F278A	7	4.09±1.97	20	0.02±0.007	1.0
F278V	7	4.52±1.53	23	0.02±0.01	1.0
F278Y	7	0.39±0.33	2.0	0.02±0.02	1.0

EL1 is an abreviation for the first extracellular loop. A dash (—) signifies that the value was not determined.

Data from refs. *50* and *100*.

Originally, a mechanism of mMC1 receptor constitutive activation was suggested to mimic the activation of rhodopsin *(101,102)*. In rhodopsin, a salt bridge between K296 and E133 was identified as constraining the receptor in

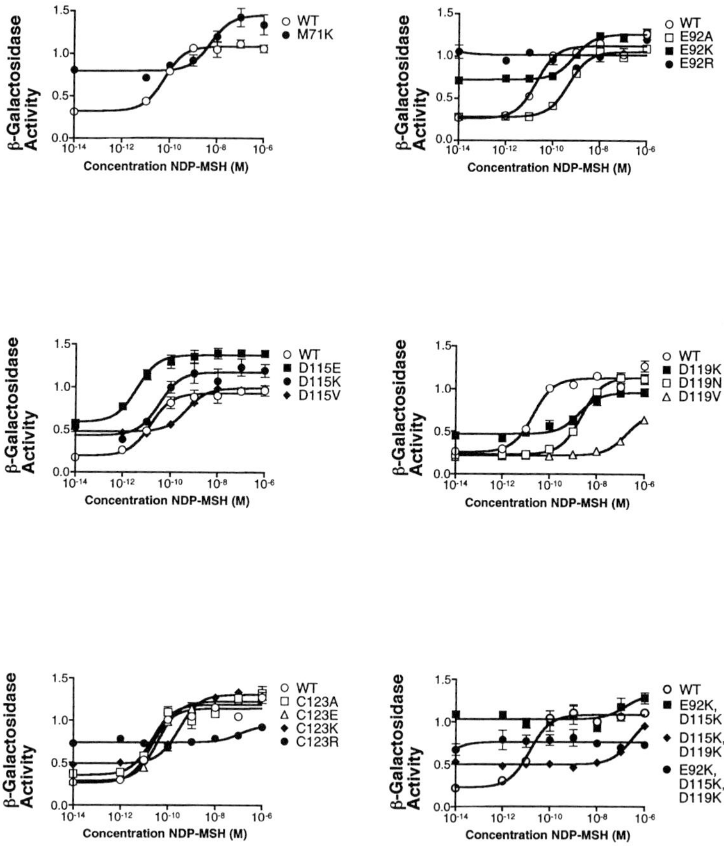

Fig. 14. mMC1-R mutant receptors which resulted in constitutive activity (shown in black symbols), as determined by the β-galactosidase bioassay and normalized for both protein and transfection efficiency.

an "inactive" conformation. When this Lys-Glu interaction was disrupted by retinal, an active (R*) receptor complex resulted. Counter ions for the Glu 92 mMC1-R residue (TM2) were potentially identified and mutated *(3)*. These include H183 (extracellular loop 2), H258 (TM6), and K276 (either extracellular loop 3 or TM7) *(50)*. These mutant receptors, and receptors containing mutations of E92A/D/Q (TM2), did not result in constitutive activity, therefore this hypothesis is not supported by the experimental evidence.

3.3. Implications for General Activation of MCI-R

Two general theories which attempt to explain the comprehensive mechanisms of signal transduction include conformational induction "which involves a receptor conformation never found in the absence of agonist," and

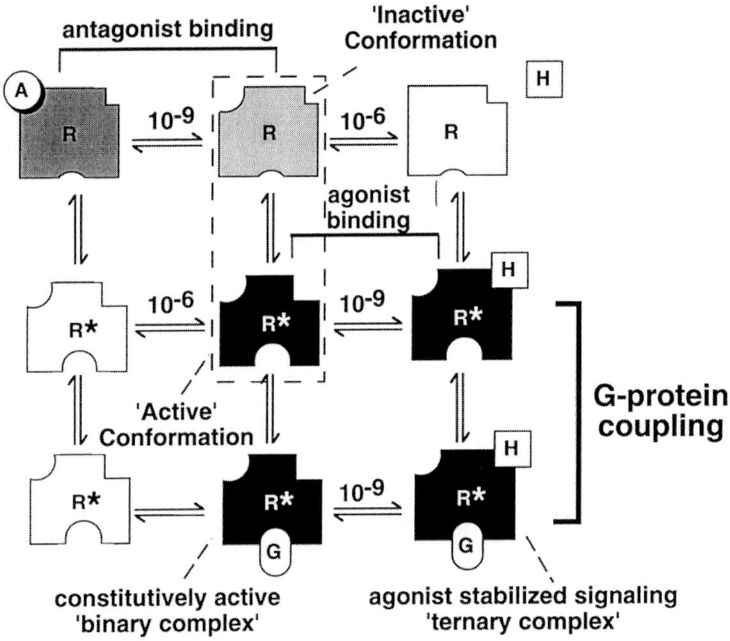

Fig. 15. Ternary complex 2D model for G protein-coupled receptor activation modified from references *(89,103–108)*. The receptor is proposed to exist in a multitude of populations with the R population predominating in the absence of ligand. In the presence of ligand (agonist), the "inactive (R)" receptor conformation if proposed to shift the equilibrium to the "active (R*)" state which possess a higher agonist ligand affinity. Multiple "active (R*)" receptor populations are proposed to exist both in the presence and absence of ligand. Upon agonist ligand stimulation, the predominant receptor population is proposed to be the "agonist stabilized signaling ternary complex."

conformational selection which "involves a choice from a library of conformations" *(103–105)*. The latter theory is also referred to as the "ternary complex, or two-state" model (see *(89,106)* and references therein). In this model the receptor exists in two major populations. One receptor population is considered an "inactive (R)" conformation and the second an "active (R*)" conformation, with the latter state coupled to a G protein in the absence of ligand. Figure 15 summarizes the multiple receptor populations in the ternary complex model as modified from references *(89,103–108)*. This model accounts for a "high-affinity" antagonist ligand binding to the "inactive (R)" receptor population, whereas a "low-affinity" binding state of an agonist results for this receptor population. However, when the receptor is in the "active (R*)" population, the agonist possess a higher affinity for the receptor, and the antagonist possesses a lower affinity for this receptor population. Based on this model, a constitutively active receptor consists of the "active" receptor precoupled to the G protein in the absence of ligand. This receptor population has been proposed to possess the receptor conformation which can be stabilized by binding of the agonist ligand *(107,108)*.

In the case of the mouse MC1-R, the residues E92, D115, and D119 (homologous to the human E94, D117, and D121 residues), have been proposed to interact directly/or indirectly with the Arg^8 residue of the ligand (Figs. 8 and 9). In the case of the E92K mutant receptor, constitutive activation was observed (Fig. 14), however, when E92 was mutated to Arg, nearly maximal basal activity was observed for this mutant receptor. Additionally and importantly, the ability of the ligands to further stimulate these mutant receptors were substantially decreased. This is in contrast to previous data, which demonstrated the constitutively active adrenergic receptors possessed enhanced ligand affinity and efficacy *(109–111)*. Based on 3D homology modeling *(23)*, it is therefore possible to hypothesize that in the case of E92K, electrostatic interactions of the Lys side chain in TM2 may interact with either the D119 or D115 side chains in TM3, but not both (Fig. 16). However, in the case of E92R, it is possible for the Arg side chain to interact with both Asp 115 and 119 in TM3, thus obtaining maximal basal stimulation in the absence of ligand. These interpretations further suggest that if this is the case, then it is possible that the constitutively active receptors resulting from these particular mutations may be mimicking the "agonist-stabilized signaling ternary complex" by obtaining the critical receptor perturbations the ligand (possibly the Arg^8 residue) induces in the receptor.

Double mutant mMC1-R receptors consisting of E92K(TM2)/D115K(TM3), D115K(TM3)/D119K(TM3), and the triple mutant E92K(TM2)/D115K(TM3)/D119K(TM3) all resulted in enhanced basal activities (Fig. 14). Apparent maximal stimulation in the absence of ligand resulted in the mutant receptor E92K/D115K. The triple mutant receptor (E92K/D115K/D119K) possessed decreased basal activity compared to the aforementioned double mutations, with the exception of the D115K/D119K double mutant receptor which possessed the lowest basal activity of these multiple mutant receptors. The ligand NDP-MSH was able to increase β-galactosidase activity on the E92K/D115K and D115K/D119K double mutant receptors, albeit at $10^{-7} M$ concentrations. This would suggest that some receptor component important for maximal stimulation was still present in these double mutant receptors and absent in the E92K/D115K/D119K mutant receptor. A model for receptor activation has been proposed based on these data *(50,100)*. This model is reported as the insertion of one or more basic amino acids in TMs 2 and 3, being responsible for a "vertical" movement of these TM domains and results in the activation of the receptor

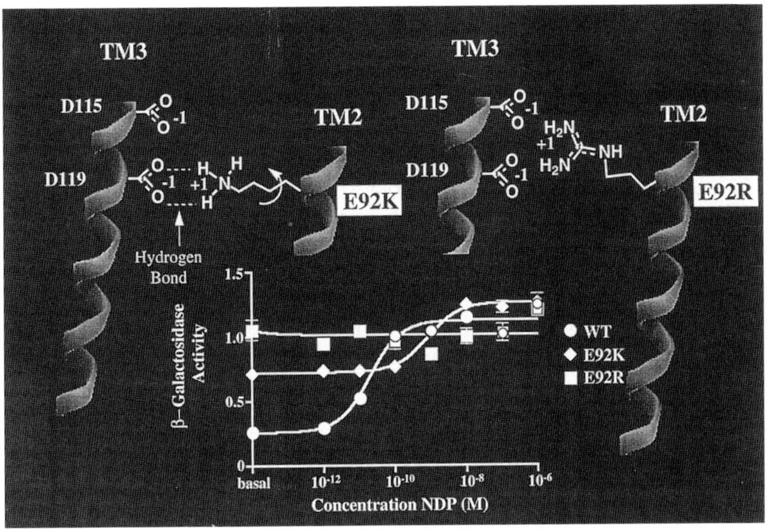

Fig. 16. Proposed molecular interactions of the E92K and E92R constitutively active mMC1-R receptors, based on 3D homology modeling. The figure on the left illustrates how the E92K side chain can interact with either the D119 or D115 (upon rotation of the torsion angle illustrated), but not both. This E92K receptor is constitutively active, but not maximally and can be further stimulated by NDP as shown in the insert. The E92R mutant receptor, however, obtains nearly maximal stimulation in absence of ligand, as compared to the wild-type mMC1-R, and is not further stimulated by ligand. The figure on the right illustrates how the E92R side chain can interact with both the D115 and D119 residues simultaneously, and potentially mimic the ligand Arg^8 side chain interacting with this triad (E92, D115, D119) of electrostatic receptor residues.

in the absence of ligand. Furthermore, perhaps these modifications mimic the ligand Arg^8 residue-induced receptor changes resulting in constitutive activation.

In the mMC1-R model *(43)*, the C123 receptor residue is predicted to be located one helical turn below D119, which is located one helical turn below D115 in TM3. Due to its location in the TM region, this residue may be interacting with the His^6 ligand side chain residue, although the C123 mutations to Ala and Glu did not result in increased basal activity and possessed 4- to 2-fold differences in NDP-MSH binding affinity, respectively, which does not support this hypothesis. However, the C123K constitutively active receptor did possess a decrease (10-fold) in NDP-MSH affinity and efficacy, whereas the C123R mutant receptor was unable to be further stimulated above its inherent basal activity by NDP-MSH. It is possible to predict, due to its location, that the C123K and C123R side chains may be participating in another aspect of "receptor activation, " besides those involving the ligand. Proposals for particular electrostatic interactions between GPCR conserved polar residues have been hypothesized to differentiate the inactive (R) state versus the active (R*) state *(57,112,113)*. Although many different combinations of interactions and hydrogen bonding patterns are possible for these conserved residues which consists of Asn (TM1), Asp (TM2), Asp-Arg-Tyr (DRY, TM3), Asn (TM7), and Tyr (TM7), the fact remains that these residues are highly conserved throughout the entire superfamily of GPCRs and appear to be important, as identified by mutagenesis studies, which lends credibility to this general hypothesis. Unfortunately, with the lack of any X-ray structures, the exact combinations and changes that occur between different receptor populations can only be speculated upon at this time.

4. Use of In Vitro Receptor Mutagenesis Studies for Iterative 3D Model Refinement and Future Directions

The hMC1-R model(s) discussed herein, have been shown by the mutagenesis data to potentially contain correct ligand–receptor interactions, but it is also incorrect in several aspects. The loss of function mutations (hMC1-R D121K, D117A/D121A, and E94A/D117A/D121A mutant receptors) (Fig. 13), and gain of function mutations (mMC1-R E92K/R, D115E/K/V, D119K, E92K/D115K, D115K/D119K, and E92K/D115K/D119K mutant receptors) (Fig. 14) involving these potential MC1-R residues and the ligand Arg^8 amino acid appears to be in agreement with the 3D receptor model. However, the ligand Phe^7 and Trp^9 interactions with the predicted hMC1-R aromatic residues appears to be inconclusive as the changes observed for ligand affinity and efficacy at these mutant receptors appear to be within experimental error (Fig. 12). A couple of notable exceptions to the aforementioned statement are summarized in Fig. 13, where differences between ligand affinity and efficacy implicate the potential role of F257 and F257/F258 in γ-MSH ligand-induced receptor activation. The specific MTII ligand-hMC1-R interactions proposed in Fig. 9 needs to be refined to include putative interactions with D117, F257, and F258. It does appear that the differential interaction of N281 with MTII (18-fold) and not NDP-MSH (4-fold) or α-MSH (5-fold) (Table 7) also needs to be incorporated into the model. Additionally, enough mutational information is now available to develop a specific γ-MSH-hMC1-R 3D molecular model which can be further tested by ligand modifications and further receptor mutational analysis.

The emerging melanocortin ligand SAR and receptor mutations resulting in both loss of function and gain of function provide data to refine the 3D homology models developed in the absence of this information. The results now available regarding the previously predicted "His-Phe-Arg-Trp" message sequence is critical for identifying "other" ligand residues which consist of the ligand pharmacophore for each melanocortin receptor. Additionally, it appears from the emerging ligand SAR, that a different pharmacophore of the melanocortin ligands exists for each of the five melanocortin receptor subtypes. For example, the His[6] ligand residue of the tetrapeptide Ac-His-D-Phe-Arg-Trp-NH_2 appears to be very important at the hMC1-R, as the tripeptide Ac-D-Phe-Arg-Trp-NH_2 was unable to bind or transduce a signal at the hMC1-R. Additionally, the observation that NDP-MSH possesses greater ligand affinity than MTII at the hMC3-R and hMC5-R is further experimental evidence supporting the hypothesis of different pharmacophore models for each MCR subtype. The aromatic mutations of the hMC1-R are also an enigma to be sorted out in regards to hydrophobic-aromatic receptor interactions with the ligand, with a further challenge being to specifically identify which receptor residue(s) the ligand Phe[7] and Trp[9] amino acids are interacting with. Towards this end, refined 3D homology melanocortin receptor modeling may aid in the design of future experiments to address these questions.

5. Summary and Conclusions

The MC1-R 3D modeling *(3,22,23)* and mutagenesis studies *(50,51,61, 90,92,100)* discussed herein, provide new insights into aspects of melanocortin ligand molecular recognition, receptor residues important for ligand affinity and efficacy, and receptor residues important for signal transduction. Mutagenesis of the hMC1-R has identified receptor residues which are important for differentiating melanocortin ligand (α-MSH, γ-MSH, NDP-MSH, and MTII)-receptor interactions, ligand binding affinities, and ligand efficacy (loss of function mutations). Mutagenesis studies of the mMC1-R have verified several of the above observations, and additionally identified receptor mutations which result in constitutively active receptors (gain of function mutations). The role that 3D GPCR homology modeling played was twofold. First, in the case of some hMC1-R mutagenesis, modeling predicted which receptor residues participate in ligand–receptor interactions. Second, modeling helped to generate working hypothesis as to possible (and experimentally testable) mechanism(s) behind some of the mMC1-R mutations which resulted in constitutively active receptors. Together these theoretical and experimental techniques complement each other favorably to propose, explain, and design further experimental studies.

Acknowledgments

This monograph was supported in part by the U.S. Public Health Service grant DK09231 (CHL). Carrie Haskell-Lueravo is a recipient of a Burroughs Wellcome Fund Career Award in the Biomedical Sciences.

References

1. Chhajlani, V. and Wikberg, J. E. S. (1992) Molecular cloning and expression of the human melanocyte stimulating hormone receptor cDNA. *FEBS Lett* **309,** 417–420.
2. Mountjoy, K. G., Robbins, L. S., Mortrud, M. T., and Cone, R. D. (1992) The cloning of a family of genes that encode the melanocortin receptors. *Science* **257,** 1248–1251.
3. Cone, R. D., Lu, D., Kopula, S., Vage, D. I., Klungland, H., Boston, B., Chen, W., Orth, D. N., Pouton, C., and Kesterson, R. A. (1996) The melanocortin receptors: agonists, antagonists, and the hormonal control of pigmentation. *Recent Progr. Horm. Res.* **51,** 287–318.
4. Robbins, L. S., Nadeau, J. H., Johnson, K. R., Kelly, M. A., Roselli–Rehfuss, L., Baack, E., Mountjoy, K. G., and Cone, R. D. (1993) Pigmentation phenotypes of variant extension locus alleles result from point mutations that alter MSH receptor function. *Cell* **72,** 827–834.
5. Castrucci, A. M. L., Hadley, M. E., Sawyer, T. K., Wilkes, B. C., Al–Obeidi, F., Staples, D. J., DeVaux, A. E., Dym, O., Hintz, M. F., Riehm, J., Rao, K. R., and Hruby, V. J. (1989) α–Melanotropin: the minimal active sequence in the lizard skin bioassay. *Gen. Comp. Endocrinol.* **73,** 157–163.
6. Haskell–Luevano, C., Shenderovich, M. D., Sharma, S. D., Nikiforovich, G. V., Hadley, M. E., and Hruby, V. J. (1995) Design, synthesis, biology and conformations of bicyclic α–melanotropin peptide analogues. *J. Med. Chem.* **38,** 1736–1750.
7. Eberle, A. N. (1988)*The Melanotropins: Chemistry, Physiology and Mechanisms of Action*, Karger, Basel.
8. Hadley, M. E. (1989)*The Melanotropic Peptides: Source, Synthesis, Chemistry, Secretion and Metabolism*; Vols. I–III. CRC Press, Boca Raton, FL.
9. Hruby, V. J., Wilkes, B. C., Hadley, M. E., Al–Obeidi, F., Sawyer, T. K., Staples, D. J., DeVaux, A., Dym, O., Castrucci, A. M., Hintz, M. F., Riehm, J. P., and Rao, K. R. (1987) α–Melanotropin: the minimal active sequence in the frog skin bioassay. *J. Med. Chem.* **30,** 2126–2130.
10. Hruby, V. J., Wilkes, B. C., Cody, W. L., Sawyer, T. K., and Hadley, M. E. (1984) Melanotropins: structural, conformational and biological considerations in the development of superpotent and superprolonged analogs. *Pept. Protein Rev.* **3,** 1–64.
11. Medzihradszky, K. (1982) The bio–organic chemistry of α–melanotropin. *Medicinal Res. Rev.* **2,** 247–270.
12. Vaudry, H. and Eberle, A. N. (1993)The melanotropic peptides. *Ann. N. Y. Acad. Sci.* **680**.
13. Haskell–Luevano, C., Hendrata, S., North, C., Sawyer, T. K., Hadley, M. E., Hruby, V. J., Dickinson, C., and Gantz, I. (1997) Discovery of prototype peptidomimetic agonists at the human Melanocortin receptors MC1-R and MC4-R. *J. Med. Chem.* **40,** 2133–2139.
14. Schiöth, H. B., Muceniece, R., Larsson, M., Mutulis, F., Szardenings, M., Prusis, P., Lindeberg, G., and Wikberg, J. E. S. (1997) Binding of cyclic and linear MSH core peptides to the melanocortin receptor subtypes. *Eur. J. Pharm.* **319,** 369–373.
15. Haskell–Luevano, C., Sawyer, T. K., Hendrata, S., North, C., Panahinia, L., Stum, M., Staples, D. J., Castrucci, A. M., Hadley, M. E., and Hruby, V. J. (1996) Truncation studies of α–melanotropin peptides identifies tripeptide analogues exhibiting prolonged agonist bioactivity. *Peptides* **17,** 995–1002.

16. Buffy, J., Thody, A. J., Bleehen, S. S., and Mac Neil, S. (1992) α–MSH stimulates protein kinase c activity in murine b16 melanoma. *J. Endocrinol* **133,** 333–340.
17. Konda, Y., Gantz, I., DelValle, J., Shimoto, Y., Miwa, H., and Yamada, T. (1994) Interaction of dual signal trandsuction pathways activated by the melanocortin–3 receptor. *J. Biol. Chem.* **269,** 13,162–13,166.
18. Abdel–Malek, Z. A., Kreutzfeld, K. L., Marwan, M. M., Hadley, M. E., Hruby, V. J., and Wilkes, B. C. (1985) Prolonged stimulation of S91 melanoma tyrosinase by [Nle^4, D–Phe^7]–substituted α–Melanotropins. *Cancer Res.* **45,** 4735–4740.
19. Chen, W., Shields, T. S., Stork, P. J. S., and Cone, R. D. (1995) A colorimetric assay for measuring activation of Gs–and Gq–coupled signaling pathways. *Anal. Biochem.* **226,** 349–354.
20. Hadley, M. E., Anderson, B., Heward, C. B., Sawyer, T. K., and Hruby, V. J. (1981) Calcium–dependent prolonged effects on melanophores of [4–norleucine, 7–D–Phenylalanine]–α–melanotropin. *Science* **213,** 1025–1027.
21. Haskell–Luevano, C., Miwa, H., Dickinson, C., Hadley, M. E., Hruby, V. J., Yamada, T., and Gantz, I. (1996) Characterizations of the unusual dissociation properties of melanotropin peptides from the melanocortin receptor, hMC1-R. *J. Med. Chem.* **39,** 432–435.
22. Prusis, P., Frändberg, P.–A., Muceniece, R., Kalvinsh, I., and Wikberg, J. E. S. (1995) A three dimensional model for the interaction of MSH with the melanocortin–1 receptor. *Biochem. Biophys. Res. Commun.* **210,** 205–210.
23. Haskell–Luevano, C., Sawyer, T. K., Trumpp–Kallmeyer, S., Bikker, J., Humblet, C., Gantz, I., and Hruby, V. J. (1996) Three–dimensional molecular models of the hMC1-R melanocortin receptor: complexes with melanotropin peptide agonists. *Drug Des. Discov.* **14,** 197–211.
24. Henderson, R., Baldwin, J. M., Ceska, T. A., Zemlin, F., Beckmann, E., and Downing, K. H. (1990) Model for the strucuture of bacteriorhodopsin based on high–resolution electron cryo–microscopy. *J. Mol. Biol.* **213,** 899–929.
25. Schertler, G. F. X., Villa, C., and Henderson, R. (1993) Projection structure of rhodopsin. *Nature* **362,** 770–772.
26. Schertler, G. F. X. and Hargrave, P. A. (1995) Projection structure of frog rhodopsin in two crystal forms. *Proc. Natl. Acad. Sci. U. S. A.* **92,** 11,578–11,582.
27. Schertler, G. F. X., Hargrave, P. A., and Unger, V. M. (1996) Three dimentional structure of rhodopsin obtained by electron cryomicroscopy. *Invest. Ophthalmol. Visual Sci.* **37,** S805.
28. Schertler, G. F. X., Unger, V. M.., and Hargrave, P. A. (1995) The Structure of rhodopsin obtained by cryo-electron microscopy to 7 Å resolution. *Biophys. J.* **68**, A.21
29. Unger, V. M. and Schertler, G. F. X. (1995) Low resolution structure of bovine rhodopsin determined by electron cryo–microscopy. *Biophys. J.* **68,** 1776–1786.
30. Davies, A., Schertler, G. F. X., Gowen, B. E., and Saibil, H. R. (1996) Projection structure of an inverterate rhodopsin. *J. Struct. Biol.* **117,** 36–44.
31. Unger, V. M., Hargrave, P. A., and Schertler, G. F. X. (1995) Localization of the transmembrane helices in the three–dimentsonal structure of frog rhodopsin. *Biophys. J.* **68,** A330.
32. Grigorieff, N., Ceska, T. A., Downing, K. H., Baldwin, J. M., and Henderson, R. (1996) Electron–crystallographic refinement of the structure of bacteriorhodopsin. *J. Mol. Biol.* **259,** 393–421.

33. Pebay–Peyroula, E., Rummel, G., Rosenbusch, J. P., and Landau, E. M. (1997) X–ray structure of bacteriorhodopsin at 2.5 angstroms from microcrystals grown in lipidic cubic phases. *Science* **277,** 1676–1681.
34. Unger, V. M., Hargrave, P. A., Baldwin, J. M., and Schertler, G. F. X. (1997) Arrangement of rhodopsin transmembrane α–helices. *Nature* **389,** 203–206.
35. Kimura, Y., Vassylyev, D. G., Miyazawa, A., Kidera, A., Matusuhima, M., Mitsuoka, K., Murata, K., Hirai, T., and Fujiyoshi, Y. (1997) Surface of bacteriorhodopsin revealed by high–resoluation electron crystallography. *Nature* **389,** 206–211.
36. Hutchins, C. (1994) Three–dimensional models of the D1 and D2 dopamine receptors. *Endoc. J.* **2,** 7–23.
37. Hibert, M. F., Trumpp–Kallmeyer, S., Bruinvels, A., and Hoflack, J. (1991) Three–dimensional models of neurotransmitter G–binding protein–coupled receptors. *Mol. Pharmacol.* **40,** 8–15.
38. Hoflack, J., Trumpp–Kallmeyer, S., and Hibert, M. (1994) Re–evaluation of bacteriorhodopsin as a model for G protein–coupled receptors. *Trend Pharmacol. Sci.* **15,** 7–9.
39. Trumpp–Kallmeyer, S., Hoflack, J., Bruinvels, A., and Hibert, M. (1992) Modeling of G–protein–coupled receptors: application to dopamine, adrenaline, serotonin, acetylcholine, and mammalian opsin receptors. *J. Med. Chem.* **35,** 3448–3462.
40. Marklund, L., Johansson Moller, M., Sandberg, K., and Anderson, L. (1996) A missense mutation in the gene for melanocyte–stimulating hormone receptor (MC1-R) is associated with the chestnut coat color in horses. *Mamm. Genome* **7,** 895–899.
41. Kungland, H., Väge, K. I., Gomez–Raya, L., Adelsteinsson, S., and Lien, S. (1995) The role of melanocyte–stimulating hormone (MSH) receptor in bovine coat color determination. *Mamm. Genome* **6,** 636–639.
42. Vanetti, M., Schonrock, C., Meyerhof, W., and Hollt, V. (1994) Molecular cloning of a bovine MSH receptor which is highly expressed in the testis. *FEBS Lett* **348,** 268–272.
43. Lu, D., Haskell–Luevano, C., Väge, D. I., and Cone, R. D. (1999) Functional variants of the MSH receptor (MC1–R), agouti, and their effects on mammalian pigmentation, in Humana Press, Totowa, N. J. pp. 231–259.
44. Vage, D. I., Lu, D., Klungland, H., Lien, S., Adalsteinsson, S., and Cone, R. D. (1997) A non–epistatic Interaction of *agouti* and *extension* in the fox, *Vulpes Vulpes. Nat. Genet.* **15,** 311–315.
45. Takeuchi, S., Suzuki, S., Hirose, S., Yabuuchi, M., Sato, C., Yamamoto, H., and Takahashi, S. (1996) Molecular cloning and sequence analysis of the chick melanocortin 1–receptor gene. *Biochem. Biophys. Acta* **1306,** 122–126.
46. Baldwin, J. M. (1993) The probable arrangement of the helices in G protein–coupled receptors. *EMBO J.* **12,** 1693–1703.
47. Mountjoy, K. G. (1994) The human melanocyte stimulating hormone receptor has evolved to become "super–sensitive" to melanocortin peptides. *Mol. Cell. Endocrinol.* **102,** R7–R11.
48. Valverde, P., Healy, E., Jackson, I., Rees, J. L., and Thody, A. J. (1995) Variants of the melanocyte–stimulating hormaone receptor gene are associated with red hair and fair skin in humans. *Nat. Genet.* **11,** 328–330.
49. Koppula, S. V., Robbins, L. S., Lu, D., Baack, E., White, C. R., Swanson, N. A., and Cone, R. D. (1997) Identification of common polymorphisms in the coding

sequence of the human MSH receptor (MC1-R) with possible biological effects. *Hum. Mut.* **9,** 30–36.
50. Lu, D., Väge, D. I., and Cone, R. D. (1999) A ligand–mimetic model for constitutive activation of the melanocortin–1 receptor. *Mol. Endocrinol.* **12,** 592–604.
51. Vage, D. I., Klungland, H., Lu, D., and Cone, R. D. (1999) Molecular and pharmacological characterization of dominant black coat color in sheep. *Mamm. Genome* **10,** 39–43.
52. Probst, W. C., Snyder, L. A., Schuster, D. I., Brosius, J., and Sealfon, S. C. (1992) Sequence alignment of the G–protein coupled receptor superfamily. *DNA Cell Biol.* **11,** 1–20.
53. Findlay, J. and Eliopoulos, E. (1990) Three–dimensional modelling of G protein–linked receptors. *TiPS* **11,** 492–499.
54. Savarese, T. M. and Fraser, C. M. (1992) In vitro mutagenesis and the search for structure–function relationships among G protein–coupled receptors. *Biochem. J.* **283,** 1–19.
55. Wess, J., Nanavati, S., Vogel, Z., and Maggio, R. (1993) Functional role of proline and tryptophan residues highly conserved among G–protein–coupled receptors studied by mutational analysis of the m3 muscarinic receptor. *EMBO J.* **12,** 331–338.
56. Hong, S., Ryu, K.–S., Oh, M.–S., Ji, I., and Ji, T. H. (1997) Roles of transmembrane prolines and proline–induced kinks of the lutropin/choriogonadotropin receptor. *J. Biol. Chem.* **272,** 4166–4171.
57. Perlman, J. H., Colson, A.–O., Wang, W., Bence, K., Osman, R., and Greshengorn, M. C. (1997) Interactions between conserved residues in transmembrane helices 1, 2, and 7 of the thyrotropin–releasing hormone receptor. *J. Biol. Chem.* **272,** 11,937–11,942.
58. Hunyady, L., Bor, M., Baukal, A. J., Balla, T., and Catt, K. J. (1996) A conserved nplfy sequence contributes to agonist binding and signal trandsuction but is not an internalization signal for the type 1 angiotensin ii receptor. *J. Biol. Chem.* **270,** 16,602–16,609.
59. Strosberg, A. D., Camoin, L., Blin, N., and Maigret, B. (1993) In receptors coupled to GTP–binding proteins, ligand binding and G–protein activation is a multistep dynamic process. *Drug Des. Discov.* **9,** 199–211.
60. Berlose, J.–P., Convert, O., Brunissen, A., Chassaing, G., and La Vielle, S. (1994) Three–dimensional structure of the highly conserved seventh transmembrane domain of G–protein–coupled receptors. *Eur. J. Biochem.* **225,** 827–843.
61. Frändberg, P.–A., Muceniece, R., Prusis, P., Wikberg, J., and Chhajlani, V. (1994) Evidence for alternate points of attachment for α–MSH and its stereoisomer [Nle^4, D–Phe^7]–α–MSH at the melanocortin–1 receptor. *Biochem. Biophys. Res. Commun.* **202,** 1266–1271.
62. Ballesteros, J. A. and Weinstein, H. (1995) Integrated methods for the construction of three dimensional models and computational probing of structure–function relations in G–protein coupled receptors. *Methods Neurosci.* **25,** 366–428.
63. van Rhee, A. M. and Jacobson, K. A. (1996) Molecular architecture of G protein–coupled receptors. *Drug Devel. Res.* **37,** 1–38.
64. Haskell–Luevano, C., Miwa, H., Dickinson, C., Hruby, V. J., Yamada, T., and Gantz, I. (1994) Binding and cAMP studies of melanotropin peptides with the cloned human peripheral melanocortin receptor, hMC1-R. *Biochem. Biophys. Res. Commun.* **204,** 1137–1142.

65. Sawyer, T. K., Castrucci, A. M., Staples, D. J., Affholter, J. A., DeVaux, A. E., Hruby, V. J., and Hadley, M. E. (1993) Structure–activity relationships of [Nle[4], D–Phe[7]] α–MSH: discovery of a tripeptidyl agonist exhibiting sustained bioactivity *Ann. N. Y. Acad. Sci.* **680**, 597–599.
66. Al–Obeidi, F., Hruby, V. J., Yaghoubi, N., Marwan, M. M., and Hadley, M. E. (1992) Synthesis and Biological activities of fatty acid conjugates of a cyclic lactam α–melanotropin. *J. Med. Chem.* **35,** 118–123.
67. Sahm, U. G., Olivier, G. W. J., Branch, S. K., Moss, S. H., and Pouton, C. W. (1994) Influence of α–MSH terminal amino acids on binding affinity and biological activity in melanoma cells. *Peptides* **15,** 441–446.
68. Sharma, S. D., Nikiforovich, G. V., Jiang, J., Castrucci, A. M., Hadley, M. E., and Hruby, V. J. (1992) A new class of positively charged melanotropin analogs: a new concept in peptide design. In *Peptides*, (Schneider, C. H. and Eberle, A. N., eds.) Escom, Leiden, pp. 95,96.
69. Sharma, S. D., Nikiforovich, G. V., Jiang, J., Castrucci, A. M., Hadley, M. E., and Hruby, V. J. (1994) Cationized melanotropin analogues: structure–function relationships, in *Peptides: Chemistry and Biology*, (Hodges, R. A. and Smith, J. A. eds.), ESCOM, Leiden, pp. 398–399.
70. Chaturvedi, D. N., Hruby, V. J., Castrucci, A. M., Kreutzfeld, K. L., and Hadley, M. E. (1985) Synthesis and biological evaluation of the superagonist [Nα–chlorotriazinylaminofluorescein–Ser[1], Nle[4], D–Phe[7]]–α–MSH. *J. Pharm. Sci.* **74,** 237–240.
71. Chaturvedi, D. N., Knittel, J. J., Hruby, V. J., de L. Castrucci, A. M., and Hadley, M. E. (1984) Synthesis and biological actions of highly potent and prolonged acting biotin–labeled melanotropins. *J. Med. Chem.* **27,** 1406–1410.
72. Castrucci, A. M. L., Hadley, M. E., Sawyer, T. K., and Hruby, V. J. (1984) Enzymological studies of melanotropins. *Comp. Biochem. Physiol.* **78B,** 519–524.
73. Sawyer, T. K., Sanfillippo, P. J., Hruby, V. J., Engel, M. H., Heward, C. B., Burnett, J. B., and Hadley, M. E. (1980) 4–Norleucine, 7–D–phenylalanine–α–melanocyte–stimulating hormone: a highly potent α–melanotropin with ultra long biological activity. *Proc. Natl. Acad. Sci. U. S. A.* **77,** 5754–5758.
74. Al–Obeidi, F., Hadley, M. E., Pettitt, B. M., and Hruby, V. J. (1989) Design of a new Class of superpotent cyclic α–melanotropins based on quenched dynamic stimulations. *J. Am. Chem. Soc.* **111,** 3413–3416.
75. Al–Obeidi, F., Castrucci, A. M., Hadley, M. E., and Hruby, V. J. (1989) Potent and prolonged acting cyclic lactam analogues of α–melanotropin: design based on molecular dynamics. *J. Med. Chem.* **32,** 2555–2561.
76. Mitchell, J. B. O., Nandi, C. L., McDonald, I. K., Thornton, J. M., and Price, S. L. (1994) Amino/aromatic interactions in proteins: is the evidence stacked against hydrogen bonding? *J. Mol. Biol.* **239,** 315–331.
77. Levitt, M. and Perutz, M. (1988) Aromatic rings act as hydrogen bond acceptors. *J. Mol. Biol.* **201,** 751–754.
78. Burley, S. K. and Petsko, G. A. (1986) Amino–aromatic interactions in proteins. *FEBS Lett.* **203,** 139–143.
79. Burley, S. and Petsko, G. (1988) Weakly polar interactions in proteins. *Adv. Protein Chem* **39,** 125–189.
80. Flocco, M. and Mowbray, S. (1994) Planar stacting interactions of arginine and aromatic side–charins in proteins. *J. Mol. Biol.* **235,** 709–717.

81. Ajay and Murcko, M. A. (1995) Computational methods to predict binding free energy in ligand–receptor complexes. *J. Med. Chem.* **38,** 4953–4967.
82. Cuatrecasas, P. and Hollenberg, M. D. (1976) Membrane receptors and hormone action. In *Advances in Protein Chemistry* (Anfinsei, C. B., Edsall, J. T., and Richards, F. M., eds.) Academic Press, New York, pp. 251–451.
83. Yamamura, H. I., Enna, S. J., and Kuhar, M. J., *Methods in Neurotransmitter Receptor Analysis* , Raven Press: New York, (1990).
84. Williams, M., Glennon, R. A., and Timmermans, P. B. M. W. M. *Receptor Pharmacology and Function*; Marcel Dekker, New York, (1989).
85. Hulme, E. C. *Receptor–Ligand Interactions: A Practical Approach*; IRL Press: New York, (1992).
86. Ramachandran, G. N. and Sasisekharan, V. (1968) Conformation of polypeptides and proteins. *Adv. Protein. Chem.* **23,** 283–437.
87. Pimentel, G. C. and McClellan (1960)*The Hydrogen Bond*. Freeman, London. pp. 282–288.
88. Schulz, G. E. and Schirmer, R. H. (1979)*Principles of Protein Structure*. Springer–Verlag, New York, pp. 20–28.
89. Schwartz, T. W., Gether, U., Schambye, H. T., and Hjorth, S. A. (1995) Molecular mechanism of action of non–peptide ligands for peptide receptors. *Curr. Pharm. Des.* **1,** 325–342.
90. Chhajlani, V., Xu, X. L., Blauw, J., and Sudarshi, S. (1996) Identification of ligand binding residues in extracellular loops of the melanocortin 1 receptor. *Biochem. Biophys. Res. Commun.* **219,** 521–525.
91. Schiöth, H. B., Muceniece, R., Szardenings, M., Prusis, P., Lindeberg, G., Sharma, S. D., Hruby, V. J., and Wikberg, J. E. (1997) Characterisation of D117A and H260A mutations in the melanocortin 1 receptor. *Mol. Cell. Endocrinol.* **126,** 213–219.
92. Yang, Y.–K., Dickinson, C., Haskell–Luevano, C., and Gantz, I. (1997) Molecular basis for the interaction of [Nle^4, D-Phe^7] melanocyte stimulating hormone with the human moleanocortin–1 receptor (melanocyte α–MSH receptor). *J. Biol. Chem.* **272,** 23000–23010.
93. Haskell–Luevano, C., Nikiforovich, G. V., Sharma, S. D., Yang, Y.–K., Dickinson, C., Hruby, V. J., and Gantz, I. (1997) Biological and conformational evaluation of stereochemical modifications using the template melanotropin peptide, Ac–Nle–c[Asp–His–Phe–Arg–Trp–Ala–Lys]–NH_2, on human melanocortin receptors. *J. Med. Chem.* **40,** 1738–1748.
94. Haskell–Luevano, C., Boteju, L. W., Miwa, H., Dickinson, C., Gantz, I., Yamada, T., Hadley, M. E., and Hruby, V. J. (1995) Topographical modifications of melanotropin peptide analogues with β–methyltryptophan isomers at position 9 leads to differential potencies and prolonged biological activities. *J. Med. Chem.* **38,** 4720–4729.
95. Hruby, V. J., Lu, D., Sharma, S. D., Castrucci, A. M. L., Kesterson, R. A., Al–Obeidi, F. A., Hadley, M. E., and Cone, R. D. (1995) Cyclic lactam α–melanotropin analogues of Ac–Nle^4–c[Asp^5, D-Phe^7, Lys^{10}]–α–MSH(4–10)–NH_2 with bulky aromatic amino acids at position 7 show high antagonist potency and selectivity at specific melanocortin receptors. *J. Med. Chem.* **38,** 3454–3461.
96. Huang, R.–R. C., Vicario, P. P., Strader, C. D., and Fong, T. M. (1995) Identification of residues involved in ligand binding to the neurokinin–2 receptor. *Biochemistry* **34,** 10,048–10,055.

97. Chini, B., Mouillac, B., Ala, Y., Balestre, M.–N., Trumpp–Kallmeyer, S., Hoflack, J., Elands, J., Hibert, M., Manning, M., Jard, S., and Barberis, C. (1995) Tyr115 is the key residue for determining agonist selectivity in the v1α vasopressin receptor. *EMBO J.* **14,** 2176–2182.
98. Kaupmann, K., Bruns, C., Raulf, F., Weber, H. P., Mattes, H., and Lübbert, H. (1995) two amino acids, located in transmembrane domains vi and vii, determine the selectivity of the peptide agonist sms 201–995 for the sstr2 somatostatin receptor. *EMBO J.* **14,** 727–735.
99. Mouillac, B., Chini, B., Balestre, M.–N., Elands, J., Trumpp–Kallmeyer, S., Hoflack, J., Hibert, M., Jard, S., and Barberis, C. (1995) The binding site of neuropeptide vasopressin VIa receptor: evidence for a major localization within transmembrane regions. *J. Biol. Chem.* **270,** 25,771–25,777.
100. Lu, D. Doctoral Thesis, Oregon Health Science University, 1997.
101. Robinson, P. R., Cohen, G. B., Zhukovsky, E. A., and Oprian, D. D. (1992) Constitutively active mutants of rhodopsin. *Neuron* **9,** 719–725.
102. Cohen, G. B., Oprian, D. D., and Robinsin, P. R. (1992) Mechanism of activation and inactivation of opsin: role of Glu113 and Lys296. *Biochemistry* **31,** 12,592–12,601.
103. Kenakin, T. (1996) Receptor conformation induction versus selection: all part of the same energy landscape. *Trends Pharmacol. Sci.* **17,** 190–191.
104. Kenakin, T. (1995) Agonist–receptor efficacy i: mechanisms of efficacy and receptor promiscuity. *Trends Pharmacol. Sci.* **16,** 188–192.
105. Kenakin, T. (1995) Agonist–receptor efficacy ii: atonist trafficking of receptor signals. *Trends Pharmacol. Sci.* **16,** 232–238.
106. Bond, R. A., Leff, P., Johnson, T. D., Milano, C. A., Rockman, H. A., McMinn, T. R., Apparsundaram, S., Hyek, M. F., Kenakin, T. P., Allen, L. F., and Lefkowitz, R. J. (1995) Physiological effects of inverse agonists in transgenic mice with myocardial overexpression of the b2–adrenoceptor. *Nature* **374,** 272–276.
107. Elling, C. E., Nielsen, S. M., and Schwartz, T. W. (1995) conversion of antagonist–binding site to metal–ion site in the tachykinin NK–1 receptor. *Nature* **374,** 74–77.
108. Thristrup, K., Elling, C. E., Hjorth, S. A., and Schwartz, T. W. (1996) Construction of a high affinity zinc switch in the κ–opioid receptor. *J. Biol. Chem.* **271,** 7875–7878.
109. Cotecchia, S., Exum, S., Caron, M. G., and Lefkowitz, R. J. (1990) Regions of the α1–adrenergic receptor involved in coupling to phosphatidylinositol hydrolysis and enhanced sensitivity of biological function. *Proc. Natl. Acad. Sci. U. S. A.* **87,** 2896–2900.
110. Ren, Q., Kurose, H., Lefkowitz, R. J., and Cotecchia, S. (1993) Constitutively active mutants of the α2–adrenergic receptor. *J. Biol. Chem.* **268,** 16,483–16,487.
111. Samama, P., Cotecchai, S., Costa, T., and Lefkowitz, R. J. (1993) A mutation–induced activated state of the β2–adrenergic receptor. extending the ternary complex model. *J. Biol. Chem.* **268,** 4625–4636.
112. Scheer, A., Fanelli, F., Costa, T., De Benedetti, P. G., and Cotecchia, S. (1996) Constitutively active mutants of the alpha 1β–adrenergic receptor: role of highly conserved polar amino acids in receptor activation. *EMBO J.* **15,** 3566–3578.
113. Oliveira, L., Paiva, A. C. M., Sander, C., and Vriend, G. (1994) A common step for signal transduction in G protein–coupled receptors. *TiPS* **15,** 170–172.

Part IV

Receptor Function

CHAPTER 10

The Melanocortin-1 Receptor

Dongsi Lu, Carrie Haskell-Luevano, Dag Inge Vage, and Roger D. Cone

1. Role of the MC1-R in Mammalian Pigmentation

The melanocyte-stimulating hormone (MSH) receptor, recently renamed the melanocortin-1 receptor (MC1-R), is a 7 transmembrane domain receptor in the rhodopsin superfamily that plays an important role in the regulation of mammalian pigmentation. The study of the MC1-R has introduced at least two novel paradigms to the G protein signaling field: constitutively active receptors *(1)* and endogenous receptor antagonists *(2)*. To elaborate on these and other findings, it is first necessary to briefly review mammalian pigmentation and the role of the MC1-R in its regulation. Findings specific to the human MC1-R and its role in human pigmentation are discussed in Chapter 11.

1.1. The Melanocyte

The complex biopolymer known as melanin is the key determinant of mammalian pigmentation. Melanin in skin and hair is produced by neural crest-derived melanocytes that migrate from the neural crest to populate the epidermis and hair follicles early during gestation. The melanocytes act as unicellular exocrine glands, since melanin is secreted via specialized endoplasmic reticulum (ER)-derived vesicles known as melanosomes. The absorption of melanin by surrounding keratinocytes or by the growing hair shaft is what causes the pigmentation of hair and skin.

Genetics has long played an important role in the study of pigmentation and melanocyte function. Since the beginning of animal husbandary, man has bred animals for the retention of identifiable traits, and pigmentation has, naturally, been one of the most common traits analyzed. In the mouse, an animal long bred both by hobbyists and scientists alike, there are now more

The Melanocortin Receptors Ed.: R. D. Cone

than 60 genes identified that affect pigmentation (reviewed in refs. *3–5*) Classically, these have been divided into 3 *(6)*, and more recently 6 *(4)* categories of genes, affecting

1. Melanocyte development and migration (*steel, piebald*)
2. Melanocyte gene expression (*microphthalmia*)
3. Melanocyte morphology (*dilute, leaden*)
4. Melanosome structure and function (*silver, pink-eyed dilution*)
5. Melanogenic enymes (*albino, brown, slaty*)
6. Regulators of melanogenesis (*extension, agouti, mahogany, mahoganoid, umbrous*)

The MC1-R, encoded by the *extension* locus *(1)*; falls into this last category.

1.2. Biochemistry of Melanin Synthesis

The melanin polymers synthesized by the melanocyte can be divided into two major categories: the sulfer-containing yellow-red pheomelanins, and the brown-black eumelanins (Fig. 1). The synthesis of both classes are completely dependent on the rate-limiting enzyme, tyrosinase, which catalyzes two steps in the conversion of tyrosine to the common precursor dopaquinone. Albinism, or the absence of any melanin pigment, results when tyrosinase activity is lacking. Dopaquinone can spontaneously form high molecular weight melanins, although many enzymatic activities are also known to catalyze reactions downstream from the formation of dopaquinone. For example, tyrosinase also has dihydroxyindole (DHI) oxidase activity, specifically required for the synthesis of black eumelanins.

Less is known about the synthesis of pheomelanins, and no enzymes specific to this pathway have yet been identified. The only requirements for pheomelanin synthesis known to date are tyrosinase and a thiol donor for the conversion of dopaquinone to cysteinyldopa. It is likely that there are multiple enzymes operating along this branch of the melanin synthetic pathway given the diversity of pigment seen in animals lacking eumelanin—from the red coat of the Irish setter to the cream or bright yellow colors of the Labrador retriever, to the orange of the calico cat.

In addition to tyrosinase, three other melanogenic enzymes are known, DHICA oxidase (tyrosinase-related protein [TRP1]), Dopachrome tautomerase (TRP2), and DHICA polymerase (Pmel17). The TRP1 and TRP2 proteins are highly related to tyrosinase, and are encoded by the pigmentation loci *brown (7)* and *slaty (8)*. Pmel 17 has some limited homology to tyrosinase, and maps to a pigmentation locus known as *silver (9)*. Less is known regarding the enzymatic activities of this protein. All three enzymes appear to be primarily involved in eumelanogenesis, and as their associated genetic names imply, these enzymes are modulatory of eumelanin synthesis.

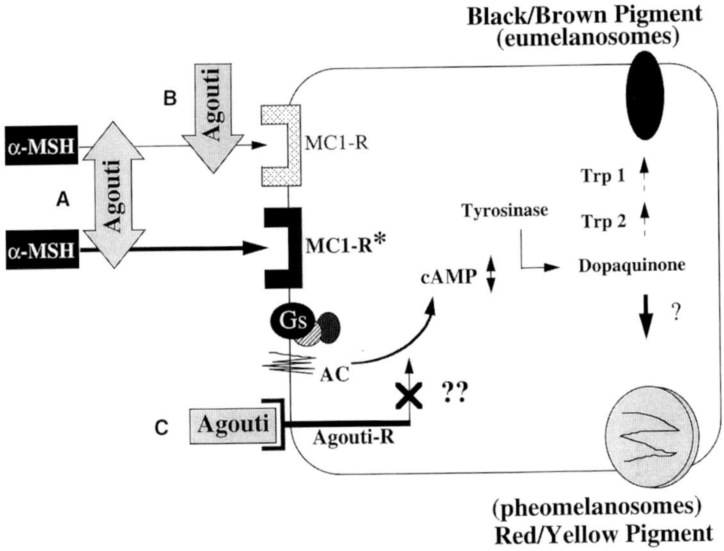

Fig. 1. The eumelanin/pheomelanin switch. α-MSH and agouti stimulate or block MC1-R activation, respectively to control tyrosinase, the rate limiting enzyme in melanogenesis. Basal tyrosinase activity leads to pheomelanin synthesis while α-MSH stimulated levels lead to eumelanin synthesis. From *(9a)* with permission.

1.3. The Eumelanin/Pheomelanin Switch

The switch regulating the mode of melanin synthesis seems to be linked to the rate-limiting enzyme tyrosinase. The level of tyrosinase expression is significantly lower during pheomelanogenesis versus eumelanogenesis *(10,11)*, and stimulation of tyrosinase with a variety of treatments leads to eumelanogenesis *(12,13)*. Thus, low basal levels of tyrosinase lead to default synthesis of pheomelanin, while higher levels lead to eumelanin production; the mechanism by which substrate is routed along one pathway or another on the basis of the levels of expression of the common rate-limiting enzyme is not understood. Other enzymes involved specifically in eumelanogenesis, TRP1, TRP2, and Pmel 17, are undetectable in pheomelanic hair bulbs *(14)*. Tyrosinase, in turn, is regulated both transcriptionally *(15,16)*, and posttranslationally *(17,18)* by cyclic adenosine monophosphate (cAMP). The primary hormonal stimulator of tyrosinase is α-melanocyte-stimulating hormone (α-MSH), which potently elevates intracellular cAMP in the melanocyte via its Gsα-coupled receptor, the MC1-R *(19)*.

Genetic investigations of pigmentation in the mouse (reviewed in refs. *4* and *5*), and a large number of other mammalian species (reviewed in ref. *6*), has led to the identification, primarily, of two loci specifically involved in regulation of the eumelanin/pheomelanin switch, *agouti* and *extension* (Fig. 2). These loci have diametrically opposed actions. Recessive *extension* alleles result in pheomelanization, or yellow-red coat colors, and dominant alleles result in the "extension" of dark black across the coat of the animal; dominant *agouti* alleles cause yellow-red coats while homozygosity for null alleles causes dark black coat colors. As mentioned above, the *extension* locus encodes the MC1-R, while cloning of *agouti* demonstrated the locus to encode a 108 amino acid secreted peptide *(20,21)*, subsequently demonstrated to be a high-affinity antagonist of the MC1-R *(2)*. *Extension* alleles act within the hair follicle melanocyte to regulate the eumelanin/pheomelanin switch *(22–24)*, whereas the *agouti* gene product is made by the surrounding hair follicle cells to regulate the switch both temporally and spatially *(25,26)*. The wild-type allele of *agouti* induces a temporary suppression of eumelanin synthesis during hair growth to produce the subterminal pheomelanin band resulting in the "*agouti*" pigmentation pattern seen in most mammalian coats.

1.4. Structure and Function of the MC1-R

α-MSH and other proopiomelanocortin (POMC)-derived melanotropic peptides (Fig. 3) stimulate eumelanogenesis by binding to a single class of membrane receptor, of approximately 45 kDa, found specifically on the surface of the melanocyte *(27,28)*. Cloning of the murine and human MC1-Rs demonstrated that this receptor is a member of the large superfamily of seven membrane spanning receptors *(27–30)*. MC1-R sequences are now known from the fox *(31)*, cow *(32,33)*, chicken *(34)*, sheep *(35)*, and panther *(R. D. Cone., unpublished observations)* as well (Fig. 4). The MC1 receptor also shares 39-61% amino acid identity with a family of G protein-coupled receptors that all bind melanocortin peptides. This family includes the MC2-R (adrenal ACTHR) *(30)*, MC3-R *(36,37)*, MC4-R *(38,39)*, and MC5-R *(40–44)*. The MC1-R, and related melanocortin receptors, do not appear to be closely related to any other particular G protein coupled receptors, although an initial alignment study suggested some distant relationship with the cannabinoid receptors *(30)*. Recently, a cDNA with sequence properties of a hybrid cannabinoid/melanocortin receptor has been reported from the leech central nervous system (CNS), providing some support for a potential evolutionary relationship between the two receptor families *(45)*.

The MC1-R is somewhat unusual, from a structural point of view, in that hydrophobicity analysis suggests the absence of the second extracellular loop. Furthermore, the disulfide bond present in many GPCRs between the first and second extracellular loops *(46,47)* is absent, due to the loss of the relevant cysteines residues. A structural model of the MC1-R, based on (i) primary sequence, (ii) naturally occuring functional variants, (iii) studies of the

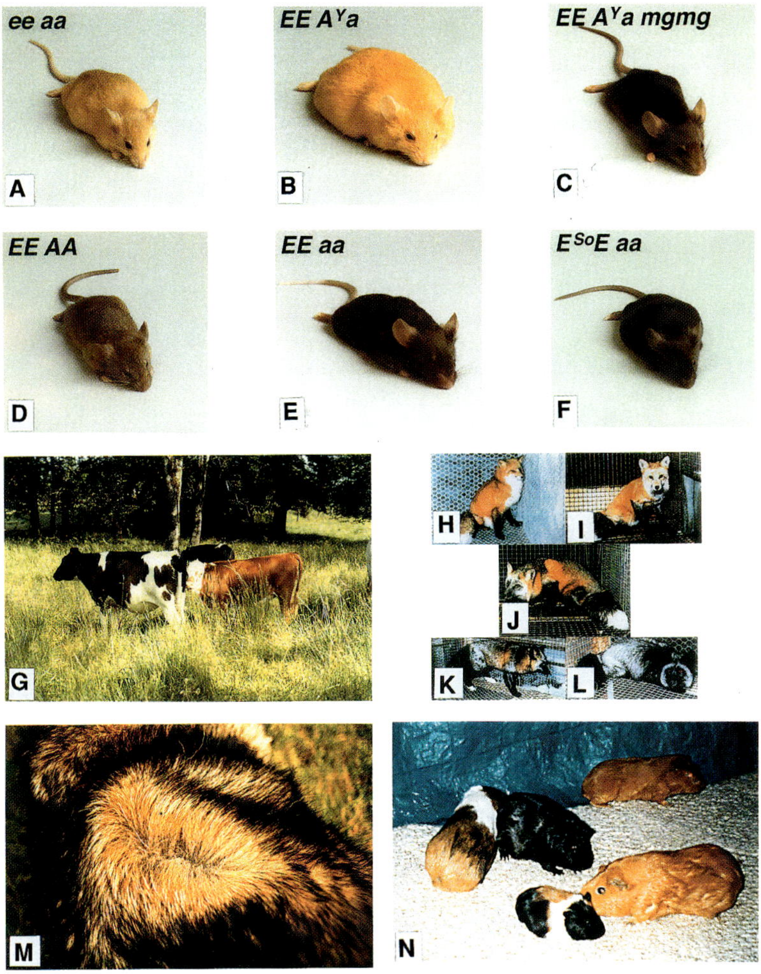

Fig. 2. Mammalian *extension* and *agouti* phenotypes. (a–f) Phenotypic effects of *agouti*, *extension*, and *mahogany* genes in the C57Bl/6J mouse. When homozygous, *mahogany* suppresses both the coat color and obesity phenotypes of the dominant A^y allele of *agouti (113)*. (g) Dominant black (E^D) and recessive red (*ee*) coat colors seen in Holstein and Hereford cattle. (h–l) Coat colors resulting from the non-epistatic interaction of *extension* and *agouti* in the fox, *Vulpes vulpes*. In order, animals are the Red (*EEAA*), Smoky Red (*EEAa*), Gold Cross (EE^AAA), Silver Cross (EE^AAa), and Silver fox (*EEaa*, E^AEaa, E^AE^Aaa, E^AE^AAa, *or* E^AE^AAA). (m) My dog, Coda. She is not pure-bred, but has a marked agouti banding pattern indicative of the wild-type A allele. (n) Black (*E*), red (*ee*), and tricolor (e^p) coat patterns in the guinea pig, *Cavia porcellus*. Portions (h–l) of this figure are reprinted with permission from *Nature Genetics*.

A

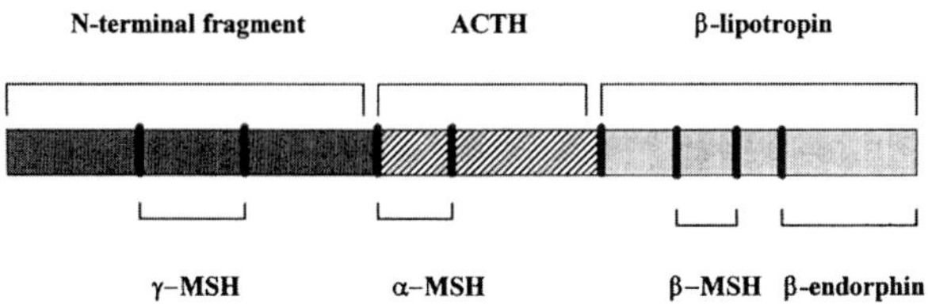

B

ACTH	H2N-SYSMEHFRWGKPVGKKRRPVKVYPNGAEDESAEAFPLEF-OH
α-MSH	Ac-SYSMEHFRWGKPV-NH2
β-MSH	H2N-AEKKDEGPYRMEHFRWGSPPKD-OH
γ-MSH	H2N-YVMGHFRWDRFG-OH
NDP-α-MSH	Ac-SYSXEHZRWGKPV-NH2

Fig. 3. The melanotropic peptides. Peptides with melanotropic activity are cleaved from three regions of the proopiomelanocortin prohormone precursor (A). Retention of melanotropic activity correlates with the presence of the H-F-R-W pharmacophore sequence. The synthetic melanotropic peptide NDP-α-MSH is shown for comparison (X, norleucine, Z, D-phenylalanine).

α-MSH pharmacophore, and (iv) in vitro mutagenesis studies is described in Subheading 3 below.

An extensive body of work exists describing the pharmacologic properties of the MC1-R, and has been reviewed elsewhere *(48,49)*. A key component for recognition of the MC1-R by a peptide ligand is the core pharmacophore His-Phe-Arg-Trp. The MC1-Rs bind most melanocortin peptides containing this pharmacophore, but generally do not recognize γ-MSH-derived peptides cleaved from the amino terminal portion of the POMC precursor. There can be significant pharmacologic variation in the MC1-R from species to species. For example, the relative preference for α-MSH over adrenocorticotropin hormone (ACTH), a 39 amino acid peptide containing α-MSH[1–13] at its amino terminus, varies widely. ACTH and α-MSH are equipotent at the human MC1-R *(50–52)*, while α-MSH is fivefold more potent in activation of the mouse MC1-R and 1000-fold more potent in activation of the MC1-R from *Rana pipiens* and *Anolis carolinensis (48,53)*.

It is interesting to speculate that the increase in ACTH sensitivity of the human MC1-R may be due to the altered biology of the processing of ACTH to α-MSH in man. In most mammals, ACTH is known to be processed to α-MSH in the intermediate lobe of the pituitary, from where it is then secreted. Humans lack this division of the pituitary, and hence have undectable levels of circu-

```
                                                              {--------TM-1------}
MMC1R    1 MSTQEPQKSLVGSLNS--NATSHLGLATNQSEPWCLYVSIPDGLFLSLGLVSLVENVLVVI
HMC1R    1 MAVQGSQRRLLGSLNSTPTAIPQLGLAANQTGARCLEVSISDGLFLSLGLVSLVENALVVA
BMC1R    1 MPALGSQRRLLGSLNCTPPATLPFTLAPNRTGPQCLEVSSLDGLFLSLGLVSLVENVLVVA
PMC1R    1 MSVQGP-RXLLGSXNSTSPAAPRLGLAANQTGPRCLEVSVPDGLFLSLGLVSVVENVLVVA
FMC1R    1 MSGQGPQRRLLGSPNATSPTTPHFKLAANQTGPRCLEVSIPDGLFLSLGLVSVVENVLVVA
CMC1R    1 MSMLAPLR-LVREPWNASEGNQSNATAGAGGAWCQGLD-IPNELFLTLGLVSLVENLLVVA

                  {-------TM-2-------}                                  {
MMC1R   60 AITKNRNLHSPMYYFICCLALSDLMVSVSIVLETTIILLLEVGILVARVALVQQLDNLIDV
HMC1R   62 TIAKNRNLHSPMYCFICCLALSDLLVSGTNVLETAVILLLEAGALVARAAVLQQLDNVIDV
BMC1R   62 AIAKNRNLHSPMYYFICCLAVSDLLVSVSNVLETAVMLLLEAGVLATQAAVVQQLDNVIDV
PMC1R   61 AIAKNRNLHSPMYYFICCLAVSDLLVSVSSVLETAVMLLLEAGTLAGRAAVVQQLDDVIDV
FMC1R   62 AIAKNRNLHSPMYYFIGCLAVSDLLVSVTNVLETAVMLLVEAGALAAQAAVVQQLDDIIDV
CMC1R   60 AILKNRNLHSPTYYFICCLAVSDNLVSVSNLAKTLFMLLMEHGVLVIRASIVRHMDNVIDM


           ---TM-3---------}                      {--------TM-4---------}
MMC1R  121 LICGSMVSSLCFLGIIAIDRYISIFYALRYHSIVTLPRARRAVVGIWMVSIVSSTLFITYY
HMC1R  123 ITCSSMLSSLCFLGAIAVDRYISIFYALRYHSIVTLPRAPRAVAAIWVASVVFSTLFIAYY
BMC1R  123 LICGSMVSSLCFLGAIAVDRYISIFYALRYHSVVTLPRAWRIIAAIWVASILTSLLFITYY
PMC1R  122 LVCGAMVSSLCFLGAIAVDRYISIFYALRYHSIVTLPRAWRAISAIWVASVLSSTLFIAYY
FMC1F. 123 LICGSMVSSLCFLGAIAVDRYLSIFYALRYHSIVTLPRAWRAISAIWVASVLSSTLFIAYY
CMC1R  121 LICSSVVSSLSFLGVIAVDRYITIFYALRYHSIMTLQRAVVTMASVWLASTVSSTVLITYY
XMC1R                FKLLCVIAVDSYITIFYALRYHSIMTLRRVVIAIAVIWSVSLVCAPIFIVYH

                {-------TM-5-------}                                          {--
MMC1R  182 KHTAVLLCLVTFFLAMLALMAILYAHMFTRACQHVQGIAQLHKRRRSIRQGFSL----KGAATLT
HMC1R  184 DHVAVLLCLVVFFLAMLVLMAVLYVHMLARACQHAQGIARLHKRQRPVHQGFGL----KGAVTLT
BMC1R  184 NHKVILLCLVGLFIAMLALMAVLYVHMLARACQHARGIARLQKRQRPIHQGFGL----KGAATLT
PMC1R  183 DHTAVLLCLVTFFVAMLVLMAVLYVHMLCSTCQHAXGTARLHKRQRPVHQGLGL----KGAATLT
FMC1R  184 NHTAVLLCLVSFFVAMLVLMAVLYVHMLARARQHARGIARLRKRQHSVHQGFGL----KGAATLT
CMC1R  182 RNNAILLCLIGFFLFMLVLMLVLYIHMFALACHHVRSISSQQK-QPTIYRTSSL----KGAVTLT
XMC1R      ESGAVILCLIVFFLFMSALMVALYIĦMFALARQHARSISALQKGKRRRITPHQARANMKGAITLT

           ----TM-6--------}                        {-------TM-7-------}
MMC1R  243 ILLGIFFLCWGPFFLHLLLIVLCPQHPTCSCIFKNFNLFLLLIVLSSTVDPLIYAFRSQEL
HMC1R  245 ILLGIFFLCWGPFFLHLTLIVLCPEHPTCGCIFKNFNLFLALIICNAIIDPLIYAFHSQEL
BMC1R  245 ILLGVFFLCWGPFFLHLSLIVLCPQHPTCGCIFKNFNLFLALIICNAIVDPLIYAFRSQEL
PMC1R  244 MLLGIFFLCWGPFFLHLSLMVLCPRHPICGCVFKNFNLFLTLIICNSIVDPLIYAFRSQEL
FMC1R  245 ILLGIFFLCWGPFFLHLSLMVLCPQHPICGCVFQNFNLFLTLIICNSIIDPFIYAFRSQEL
CMC1R  243 ILLGVFFICWGPFFFHLILIVTCPTNPFCTCFFSYFNLFLILIICNSVVDPLIYAFRSQEL
XMC1R  102 LLLGIFMLCRLPLLNS


MMC1R  304 RMTLKEVLLCSW
HMC1R  306 RRTLKEVLTCSW
BMC1R  306 RKTLQEVLQCSW
FMC1R  306 RKTLQEVLLCSW
CMC1R  304 RRTLREVVLCSW
```

Fig. 4. Alignment of known MC1-R sequences. Amino acid sequences are from the mouse, human, bovine, fox, and chicken receptors (references indicated in the text), or from *Xenopus laevis* and *Panthera pardus* (R. D. Cone., unpublished observations).

lating α-MSH in the serum. This highlights the troublesome issue of the source of melanotropic peptide involved in the regulation of melanogenesis in dermal and follicular melanocytes.

High circulating melanotropins clearly induce eumelanogenesis. Injection of α-MSH into mice induces the synthesis of dark black hair *(22,23)*, while injection of α-MSH in man results in eumelanization, or tanning, of the skin *(54,55)*. Furthermore, elevation of endogenous circulating ACTH in endocrine disorders such as Cushing's or Addison's disease can often result in hyperpigmentation *(56)*. Despite the fact that high circulating melanotropins induce eumelanogenesis in skin and hair, it has been clearly demonstrated that the pituitary is unneccesary for maintenance of eumelanogenesis. For example, hypophysectomy in the mouse does not affect resynthesis of the dark black coat in the C57Bl/6J mouse *(23)*. POMC expression has now been demonstrated in a number of sites such as keratinocytes *(57)* and it is possible that this cell type and perhaps the hair follicle cell is the primary site of melanotropin synthesis for the regulation of dermal and hair pigmentation, respectively.

1.5. Structure and Function of *agouti*

Classic genetic studies have demonstrated in a number of species that *agouti* and *extension* alleles interact to produce the final distribution of eumelanin and phaeomelanin pigments both spatially, across the coat of the animal, as well as temporally across the length of each individual hair shaft (Fig. 2m). For example, the *agouti* banding pattern results from temporary inhibition of the wild type allele of *extension*, but the action of *agouti* can be overridden, in most species, by the presence of dominant *extension* alleles. In most species, *extension* is epistatic to *agouti* (e.g., mouse see *[58]*), meaning that when an animal contains a dominant *agouti* and a dominant *extension* allele, the *extension* phenotype prevails, implying that *extension* acts downstream of *agouti*. Another example familiar to most is the coat color variation seen in the German Shepherd dog, where the variable distribution of tan and black results from the interaction of at least two *extension* alleles (*E*, *e*), and three *agouti* alleles (a^y, a^w, and a^t) *(59)*.

Agouti has long been studied in the mouse, where approximately 20 alleles have been identified, beginning with *non-agouti* (*a*) and dominant yellow mutations (A^y) first identified by mouse fanciers (reviewed in ref. *4*). Genetic evidence led to the hypothesis that *agouti* was an antagonist of α-MSH or α-MSH signaling; identification of the MC1-R as *extension* and the cloning of the *agouti* gene supported and allowed a direct test of the hypothesis. Cloning of *agouti* demonstrated the gene to encode a 131 amino acid peptide with a putative 22 amino acid signal peptide *(20,21)* (Fig. 5). The peptide contains a basic amino acid-rich domain followed by a unique cysteine repeat motif that has homology to the cysteine repeats observed in the conotoxins and agatoxins. Agouti was the first example of a mammalian protein to contain this motif. The agouti gene encoding the wild-type allele was shown to be expressed in a developmentally regulated fashion peaking at postnatal day 3, corresponding well to the time period during which the pheomelanin band begins to be deposited in the developing hair shaft in the mouse. Subsequent analysis of additional alleles has demonstrated that the variable distribution of pheomelanin due to those alleles results from various promoter mutations that restrict *agouti* gene expression to the pheomelanized regions (reviewed in ref. *4*).

A

```
                 Agouti Alignment

                 {-Signal Peptide-}     {-----Basic
M-agouti   1     MDVTRLLLATLVSFLCFFTVHSHLALEEKLGDDRSLRSNSSMNS
H-agouti  1     MDVTRLLLATLLVFLCFFTANSHLPPEEKLRDDRSLRSNSSVNL
F-agouti   1     MNIFRLLLATLLVSLCFLTAYSHLA-EEKPKDDRSLRSNSSVNL
D-agouti   1     MNIFRLLLATLLVSLCFLTAYSHLA-EEKPKDDRSLRSNSSVNL

                 Domain-------------------}
M-agouti   45    LDFSSVSIVALNKKSKKISRKEAEKRKRSSKKKASMKKVA--RP
H-agouti   45    LDVPSVSIVALNKKSKQIGRKAAEK-KRSSKKEASMKKVV--RP
F-agouti   45    LDFPSVSIVALNKKSKKISRKEAEK-KRSSKKKASMKNKARPRP
D-agouti   45    LDFPSVSIVALNKKSKKISRKEAEK-KRSSKKKASMKNVARPRP

                      {----Cysteine Rich Domain-----
M-agouti   87    PPPS--PCVATRDSCKPPAPACCDPCASCQCRFFGSACTCRVLN
H-agouti   87    RTPLSAPCVATRNSCKPPAPACCDPCASCQCRFFRSACSCRVLS
F-agouti   87    PPPN--PCVATRNSCKSPAPACCDPCASCQCRFFRSACTCR...
D-agouti   87    PPPT--PCVATRNSCKSPAPACCDPCASCQCRFFRSACTCRVLS

                 --}
M-agouti   129   PNC
H-agouti   129   LNC
F-agouti   129   ...
D-agouti   129   PRC
```

B

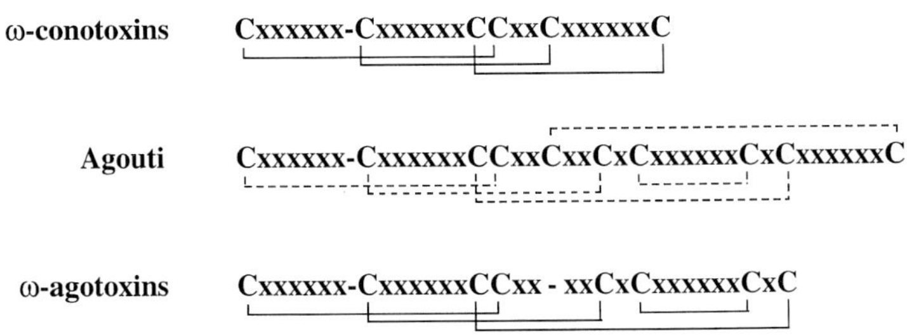

Fig. 5. (A) Amino acid sequence of the *agouti* signaling protein from mouse *(21,114)*, man *(115,116)*, fox, *Vulpes vulpes (31)*, and dog (Daniela Dinulescu and R. D. Cone., unpublished data). (B) Alignment of the conserved cysteine motifs in agouti, conotoxin, and agatoxin.

As mentioned above, the cloning of *agouti* and the MC1-R allowed a direct test of the hypothesis that agouti is an antagonist of α-MSH. Indeed, a 108 amino acid recombinant agouti protein, produced in insect cells, was demonstrated to be a high-affinity competitive antagonist ($K^i = 6.6 \times 10^{-10}$) of the MC1-R *(2)* (Fig. 6). Parenthetically, the mechanism of agouti action held interest for those outside the pigmentation field, because ectopic expression

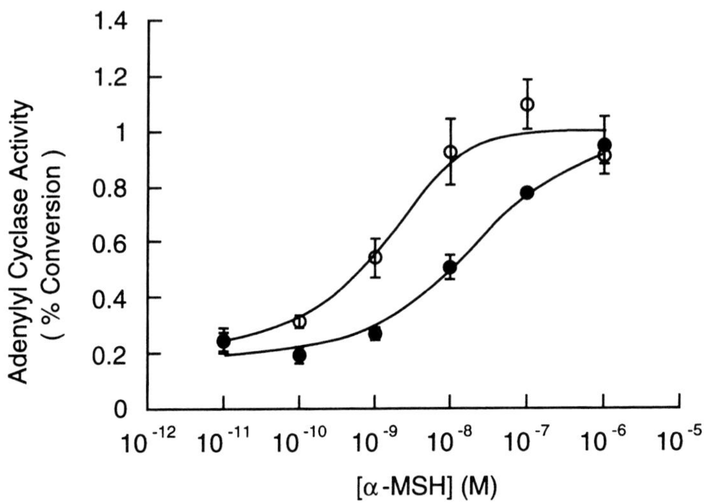

Fig. 6. Functional antagonism of the mMC1-R by agouti. Agouti inhibits activation of the mMC1-R by α-MSH in stably transfected 293 cells as monitored by stimulation of adenylyl cyclase activity. Measurement of adenylyl cyclase activity was performed as described *(117)*. Data represent means and standard deviations from triplicate data points. (Reprinted with permission from *Nature,* ref. *2* [1994] Macmillan Magazines Ltd.)

of agouti resulting from some dominant alleles (A^y, A^{vy}, reviewed in refs. *60* and *61*) produces one of the five monogenic obesity syndromes known in the mouse (compare the mouse in Fig. 2b with that in Fig. 2e).

Initial characterization of baculovirus-produced agouti demonstrated that the peptide was also a high-affinity antagonist of a related melanocortin receptor in the hypothalamus, called the MC4-R *(2)* (*see* Chapter 14). This receptor had been demonstrated to be present in brain regions known to be involved in the regulation of feeding and metabolism *(39)*. However, a number of groups hypothesized that a unique agouti receptor must exist, and could potentially contribute to the action of agouti both in obesity and pigmentation, arguing that endogenous peptide antagonists of the G protein-coupled receptors were not known to exist *(62)*, and that agouti also has effects on intracellular Ca^{2+} that are not likely to be mediated via melanocortin receptors; *(63,64)*. At least in the case of *agouti*-induced obesity this issue seems to have been resolved. A small peptide antagonist of the MC4-R, Ac-Nle^4-c[Arg^5, D-Nal(2')^7, lys^{10}]α-MSH[4–10]-NH_2 *(65)*, that mimics agouti pharmacologically at the MC4-R has been demonstrated to stimulate feeding upon intracerebroventricular administration *(66)*. This finding demonstrates that melanocortinergic neurons exert an inhibitory tone on feeding behavior. Furthermore, ablation of the MC4-R by gene knockout produces an animal that virtually duplicates the obesity phenotype seen in the A^y mouse *(67)*. Together, these studies strongly argue that inhibition of MC4-R signaling is the only alteration required for the *agouti* obesity syndrome.

While it is now generally agreed upon that agouti blocks MSH binding to the MC1-R *(68–70)*, there is still some debate concerning a target for agouti action on the melanocyte in addition to the MC1-R. This derives first from a simple observation: the quality of the pheomelanic pigment in the A^y and *ee* animals, though both have a disruption of MC1-R signaling, is not the same. In the same C57Bl/6J background the A^y animal has a bright yellow coat while the *ee* animal has a more dusty yellow coat (compare the mouse in Fig. 2a with that shown in Fig. 2b). Secondly, several more recent studies have argued that agouti has various actions on melanocytes in the absence of α-MSH. For example, recombinant agouti has been demonstrated not only to block α-MSH-stimulated melanogenesis, but to further reduce basal melanogenesis in B16 F1 murine melanoma cells in the absence of exogenous α-MSH *(69)*. Agouti has also been demonstrated to inhibit forskolin and dibutyryl cAMP (dbcAMP) stimulated proliferation and tyrosinase activity in primary human melanocytes *(71)*. Finally, as mentioned previously, long term exposure to agouti has been demonstrated to produce a rise in intracellular Ca^{2+} in a skeletal muscle cell line *(64)*, and the homology of agouti to the agatoxin/conotoxin family of proteins has been used to argue that agouti must interact with a Ca^{2+} channel. Of course, this family of proteins *(63)* is known to bind to many different proteins other than Ca^{2+} channels *(72)*, and in any event, when they do interact with Ca^{2+} channels they generally act as channel blockers, inhibiting Ca^{2+} entry.

An equally likely hypothesis to explain the action of agouti on basal melanogenesis is that agouti is an inverse agonist of the MC1-R, binding to the receptor in the absence of ligand and downregulating its basal signaling activity *(73,74)*. Support for this hypothesis comes from a recent study with the B16-F1 melanoma cell line, and a subclone, G4F, lacking MC1-R expression *(75)*. The inhibition of cell growth induced by recombinant agouti in the absence of α-MSH was shown to occur in the B16 line but not the MC1-R minus subclone *(70)*.

2. Allelic Variants of the MC1-R

The possibility that the *extension* locus might encode an α-MC1-R was posited in 1984, when Tamate and Takeuchi *(13)* showed that "…the *e* locus controls a mechanism that determines the function of an α-MSHR." This elegant study demonstrated that dbcAMP could induce eumelanin synthesis in hairbulbs from both A^y and *ee* mice, while α-MSH could only due so in A^y mice, demonstrating a primary defect in the α-MSH response mapping to the *e* locus.

The cloned MC1-Rs were used to map the chromosomal location of the receptors in man and mouse to determine if the receptors mapped to a known pigmentation locus such as *extension*, or possibly a melanoma susceptibility locus. Fluorescent *in situ* hybridization to metaphase chromosomes demonstrated that the human receptor maps to 16q2 *(76,77)*, a region not linked to melanoma susceptibility. The murine *extension* locus was previously mapped near the distal end of chromosome 8 in the mouse *(78–80)*, and an intersubspecific mapping panel was used to place the MC1-R near the *Es-11* locus, in this same region *(81)*.

Definitive evidence that *extension* encoded the MC1-R came from a study by Robbins in which the MC1-R was cloned from mice containing four different *extension* locus alleles, *e*, E^+, E^{so}, $E^{so\text{-}3J}$, and E^{tob} *(1)*. At the time of this finding, no functional variants of the G protein coupled receptors had yet been reported. The possible existence of literally hundreds of functionally variant *extension* locus alleles, identified in most domesticated mammals during the past century by classical breeding, raised some exciting research possibilities. First, the data suggested the possibility of finding receptors that had been somehow constitutively activated by naturally occurring mutations in among the dominant *extension* alleles, and second, suggested that perhaps the best in vitro mutagenesis studies of MC1-R structure and function had already been laboriously performed by random mutagenesis, followed by generations of careful trait selection by animal breeders. The functional variants that are known as of this writing can be seen in Fig. 7.

2.1. Mouse

Four *extension* phenotypes are found in the mouse. Wild-type (E^+), *sombre* (encoded by two independently occurring alleles, E^{so} and $E^{so\text{-}3J}$, *tobacco* (E^{tob}), and *recessive yellow* (*e*). *Recessive yellow* (*e*) arose spontaneously in the C57BL inbred strain and is almost entirely yellow due to an absence of eumelanin synthesis in the hair follicles *(82)*. A small number of dark hairs can be found dorsally in *e*/*e* animals, and the animals have black eyes as well. E^{tob} is a naturally occurring *extension* allele present in the tobacco mouse, *Mus poschiavinus (83)*. This wild mouse is confined primarily to the Val Poschiavo region of southeastern Switzerland. The E^{tob} allele in the *Mus poschiavinus* background produces a darkening of the back, which is only visible after the 8th week when the flanks become *agouti*. The E^{tob} allele is epistatic to *agouti*, producing a darkened back when crossed to yellow (A^y) or black (*aa*) mice.

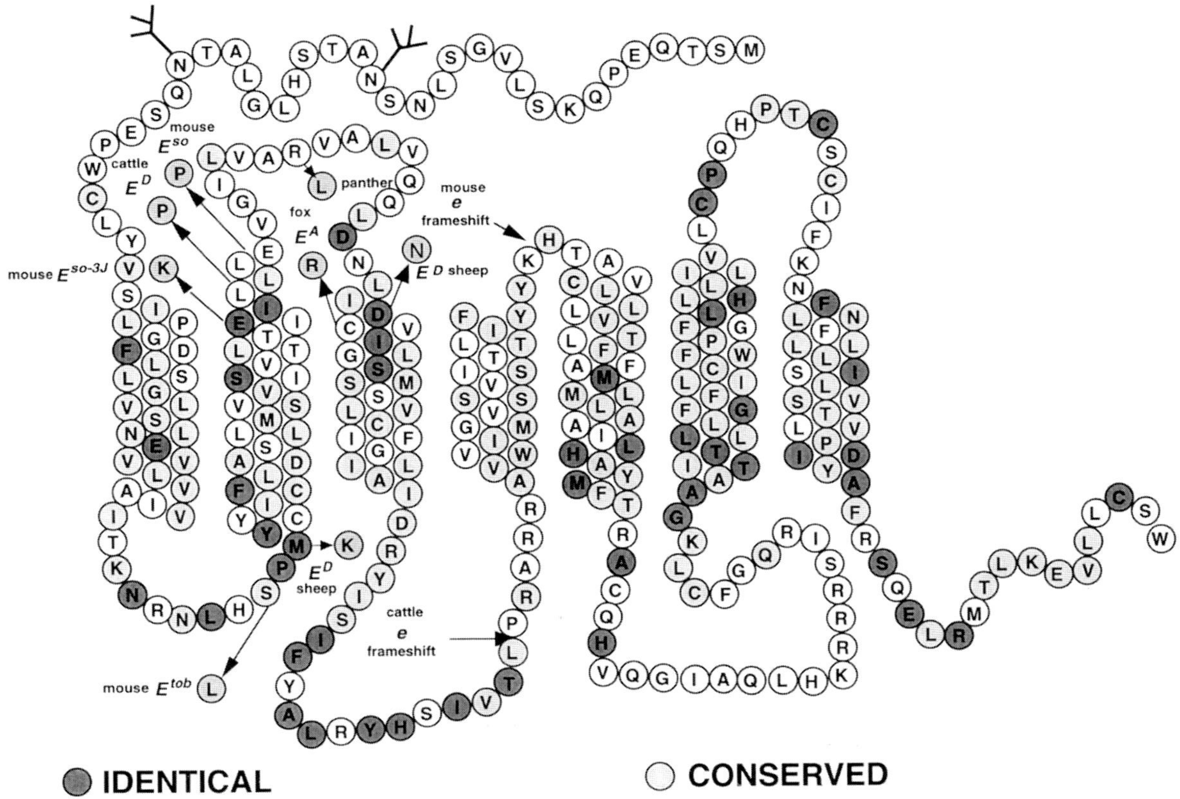

Fig. 7. Naturally occurring functional variants of the MC1-R. Functional mutations are illustrated using the sequence of the mouse MC1-R for reference. Shading indicates residues identical or conserved among all melanocortin receptor sequences. References provided in the text, except for changes seen in the panther (R. D. Cone., unpublished observations).

Consequently, unlike the *sombre* phenotype described below, the dominant melanizing effect of E^{tob} is incompletely expressed. The E^{so} allele arose spontaneously in 1961 in the C3H strain *(58)*. E^{so} homozygotes are, with the exception of a few yellow hairs entirely black, and have darkened skin as well, resembling extreme non-*agouti* mice (a^e/a^e). As heterozygous $E^{so/+}$ animals mature, yellow hairs appear on the flanks and the bellies become grey, clearly distinguishing them from homozygotes, and resembling the non-*agouti* mouse (*a*/*a*). Like E^{tob}, the E^{so} allele is also epistatic to *agouti*. $E^{so\text{-}3J}$ arose spontaneously in 1985 at the Jackson Laboratory in the CBA/J strain and is phenotypically similar to the original E^{so} allele. No evidence has been presented for phenotypic effects of variant *extension* alleles, outside of their effects on pigmentation.

Robbins et. al. *(1)* demonstrated that the murine *extension* locus encodes the murine MC1-R, and the different pigmentation phenotypes of these alleles result from point mutations in the receptor that altered its functional properties *(1)*. In the *recessive yellow* mouse, a frameshift mutation at position 183 between the fourth and the fifth transmembrane domains results in a prematurely terminated nonfunctional MC1-R. In the *sombre* mice, there is a glu-to-lys change at position 92 in the $E^{so\text{-}3J}$ allele, a leu-to-pro change at position 98 in the E^{so} allele, with both of these mutations located in the putative exterior portion of the second transmembrane domain of the receptor. When expressed in the heterologous 293 cell line, both *sombre-3J* and *sombre* receptors are constitutively activated up to 30% to 50% of the maximal stimulation levels of the wild-type receptor, even in the absence of the α-MSH (Fig. 8). Though the phenotype of the *tobacco* and the *sombre* mice are similar, the receptor of the *tobacco* allele, which has a ser-to-leu change at position 69 of the first intracellular loop of the receptor, has different pharmacologic features from the *sombre* receptor. The *tobacco* receptor only has a slightly elevated basal activity but can be further stimulated by α-MSH and has a much higher maximal adenylyl cyclase level than the wild-type receptor *(1)*.

2.2. Cattle

Three *extension* alleles were postulated in the cattle based on the genetic studies, E^D for dominant black, *e* for recessive red, and E^+, the only allele in cattle and mice that allows phenotypic expression of *agouti (84)*. A leu-to-pro change at position 99, homologous to position 97 of the mouse, has been found in the E^D allele, and a frameshift mutation resulting from a single-base deletion at position 104 has been found in the *e* allele of the cattle *(32)*. The red and black pigments that result are represented by the colors seen, for example, in the Hereford and Holstein breeds (Fig. 2g). The E^D allele has not yet been pharmacologically characterized, but is likely to have functionally similar

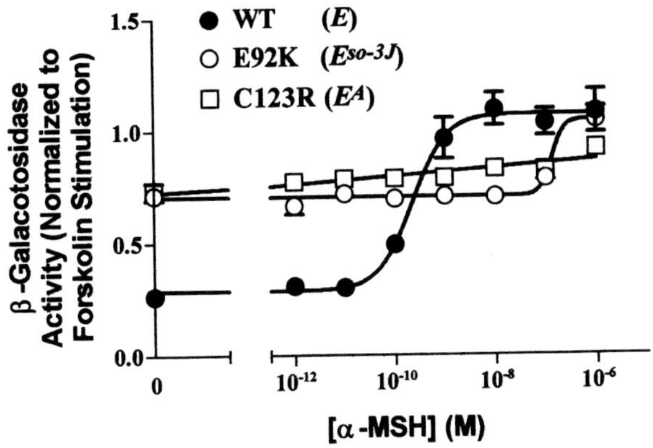

Fig. 8. Pharmacology of the mouse *sombre-3J* and mouse C123R receptors. The wild-type MC1-R, $E^{so\text{-}3J}$ allele, andin vitro -generated C123R mutation were cloned into the pcDNA Neo expression vector (Invitrogen), and transfected stably into the HEK 293 cell line. G418^r cell populations were selected and assayed for intracellular cAMP levels following hormone stimulation using a cAMP-dependent β-galactosidase reporter construct as described previously *(118)*. Data points are the average of triplicate determination with error bars indicating the standard deviation. Data is normalized to cell number and 10μ*M* forskolin-stimulated activity level for each individual cell population. The forskolin-stimulated activities did not vary significantly among cell populations. Reprinted from *(9a)* with permission.

consequences to the leu98pro change that occurs just two amino acids away in the mouse. This change constitutively activates the MC1-R similarly to the glu92lys change.

2.3. *Fox*

As mentioned above, in many species, including the mouse, dominant alleles at *extension* are epistatic to *agouti*. On the molecular level, this translates to the observation that once receptors have been made constitutively active by mutation, they can no longer be inhibited by agouti. However, in the fox, *Vulpes vulpes*, the proposed *extension* locus is not epistatic to the *agouti* locus *(85,86)*. Both the MC1-R and *agouti* genes were recently cloned from this species to attempt to understand this novel relationship between the receptor and its antagonist *(31)*.

A constitutively activating cys125arg mutation in the MC1-R was found specifically in darkly pigmented animals carrying the Alaska Silver allele (E^A). This mutation was introduced by in vitro mutagenesis into the same position (aa123) of the highly conserved mouse MC1 receptor (85% amino acid identity) for pharmacologic analysis. MC1-R (cys123arg), when expressed in the 293 cell system, was found to activate adenylyl cyclase to levels from 25% to 90% maximal levels, in the absence of any hormone stimulation (Fig. 8).

The full-length wild-type fox MC1-R was transiently expressed in Cos-1 cells, and appeared to couple normally to adenylyl cyclase, as measured by analysis of intracellular cAMP concentrations, with an EC_{50} of $1.6 \times 10^{-9} M$ (not shown), comparable to the value reported for the mouse MC1-R *(30)*, ($2.0 \times 10^{-9} M$).

A deletion in the first coding exon of the *agouti* gene was found associated with the proposed recessive allele of *agouti* in the darkly pigmented Standard Silver fox (*aa*). This deletion removes the start codon and the signal sequence, and thus is likely to ablate the production of functional agouti. Thus, as in the mouse, dark pigmentation can be caused by a constitutively active MC1-R, or homozygous recessive status at the *agouti* locus.

These findings allow a detailed interpretation of fox coat color phenotypes resulting from *extension* and *agouti*. Red coat color in cattle and the red guinea pig (*see* 2.4. below) result from homozygosity of defective alleles of the MC1-R. In contrast, no deletions or deleterious mutations in the MC1-R were observed in DNA from the Red fox (*EEAA*). This allele of the receptor appeared normal in functional expression assays in tissue culture (not shown) demonstrating that, in this species, red coat color results from inhibition of the MC1-R by the product of the *A* allele of *agouti*.

When two constitutively active MC1-Rs are found, such as in the Alaskan Silver fox (E^AE^AAA), primarily eumelanin is found. In striking contrast to the mouse, however, heterozygosity of the dominant *extension* allele E^A is not sufficient to override inhibition of eumelanin production by *agouti*. One wild-type *agouti* allele produces significant red pigment around the flanks, midsection, and neck in the Blended Cross fox (E^AEAa) Fig. 2k.). This strongly suggests an interaction between *extension* and *agouti* distinct from the epistasis seen in the mouse. One possible model to explain this interaction is that in the fox, the agouti is an inverse agonist of the MC1-R.

In the recently proposed allosteric ternary complex model *(73)*, G protein-coupled receptors are in equilibrium between the inactive (R) and active (R*) states, even in the absence of ligand. In contrast to the classical competitive antagonist which binds equally well to R and R* and acts by blocking ligand binding, inverse agonists, recently verified experimentally *(74)*, bind preferentially to R and thus shift the receptor equilibrium in the direction of the inactive state. While the mouse agouti behaves like a classical competitive antagonist, it is possible that the fox protein is a inverse agonist and can inhibit the constitutively active E^A allele of the MC1-R.

2.4. Guinea Pig

Variegated pigment patterns, that is coats containing an irregular patchwork of two or more colors, have often been associated with heterozygosity of X-linked pigment genes in the female animal. A classic example is the orange locus (*O*) in the cat, resulting from X chromosome inactivation in the female as proposed by Lyon *(87)*. Males and homozygous females containing this allele are yellow-orange while heterozygous females (+/*O*) have the tortoiseshell or calico coat consisting of irregularly distributed patches of yellow and brown pigment. Yet variegated brindle and tortoiseshell coat color patterns map to the autosomal *extension* locus in a variety of mammals, including the rabbit, dog, cattle, pig, and guinea pig *(6)*.

Preliminary results are available from a study of the *extension* locus in the guinea pig, in which an allele, e^p, produces the tortoiseshell coat pattern in homozygous male or female animals. Our initial hypothesis was that such a phenotype might result from variable MC1-R gene expression that could be easily detectable as a gene rearrangement. Analysis of the MC1-R gene locus by Southern hybridization has not confirmed this, and additional work needs to be done *(88)*. However, after probing DNA from the black, tortoiseshell, and red guinea pig with a small coding sequence fragment of the mouse MC1-R a large deletion in this gene was observed the red guinea pig. This confirms the observations in the mouse and in cattle that absence of functional MC1 receptor does not affect melanocyte development or migration into the skin and hair follicle, but simply ablates expression of eumelanin in the coat of the animal. Further work will be required to understand the mechanism of variegated function of the MC1-R in tortoiseshell and brindle animals.

2.5. Panther

A coat color phenotype that has always fascinated viewers is the melanized coat seen in a number of the large felines. In the leopard, *Panthera pardus*, the classic spotting seen in the wild-type tan and brown animal can actually still be seen beneath the sleek black coat of the eumelanic variant. The absence of a defined *extension* locus in domestic felines further compounds the problem of analyzing the role of the MC1-R in feline pigmentation. Nonetheless, the gene that produces the dark black coat in several of the large cats is reported to be dominant acting, and this laboratory was fortunate to obtain blood samples from Chewy and Boltar, tan and black *Panthera pardus*, respectively, residing at the Octagon Wildlife Sanctuary in Florida. These animals have been bred twice, throwing both black and tan offspring. Cloning and sequence analysis of the MC1-R from both animals demonstrated Boltar to be heterozygous for an arg106leu change, while Chewy was arg106 at both alleles. Given the proximity of this mutation to the constitutively activating mutations in the mouse, cow, and fox, it is tempting to speculate that this change represents a dominant allele of the MC1-R in *Panthera pardus*. The allele has not yet been characterized pharmacologically.

3. Structure of the MC1-R

In addition to studies of naturally occuring variants of the MC1-R, other approaches have been taken to better understand the structure and function of this receptor. Two discussed here include in vitro mutagenesis studies and computer modeling of the receptor.

3.1. In Vitro Mutagenesis Studies

To further understand the structure and function of the MC1-R, in vitro mutagenesis studies have been performed on both murine and human MC1-R. Based on residues conserved across the entire melanocortin receptor family, several residues, including asp117, phe179, his209, and his260, were mutated to alanine in the human MC1-R *(89)*. These mutants were examined for binding of both α-MSH and NDP-α-MSH. For NDP-α-MSH, binding affinities were all similar to the wild type, but for α-MSH, binding affinities were significantly altered in some cases. Affinities were reduced about 267-fold for asp117ala, about 132-fold for his260ala, and were similar to the wild type for phe179ala and his209ala. Although the data clearly show a different interaction of NDP-α-MSH with the receptor compared to the native ligand, it is likely to result from variations on binding to the same binding pocket, with the mutations either directly or indirectly affecting the specific NDP-α-MSH contacts only.

The charged residues in the extracellular loop of the human MC1-R, including ser6, glu102, arg109, arg184, glu269, and thr272, were also mutated to alanine to investigate whether these residues are involved in ligand binding to the receptor *(90)*. The binding affinity to either α-MSH or NDP-α-MSH were reduced for ser6, arg184, glu269 and thr 272, but similar to the wild type for glu102 or arg109. These results suggest that certain extracellular residues are important in the ligand-receptor interaction, although the data do not prove a direct interaction between these residues and the ligand. Nevertheless, it has been known for some time that the residues of α-MSH flanking the core H-F-R-W pharmacophore contribute importantly to the affinity of the interaction between ligand and receptor. These flanking sequences are not neccesary for full agonist activity, and are perhaps the residues interacting with extracellular residues to enhance ligand affinity (reviewed in refs. *48* and *49*).

Studies from our laboratory have focused on mutated residues found to be responsible for constitutive activation of the MC1-R in naturally occurring variants of the MC1-R in mice *(1)*, foxes *(31)*, cattle *(32)*, and sheep *(35)*. Inititially we proposed that, as in the case of the rhodopsin receptor, constitutively activating mutations were likely to be acting by disrupting internal molecular constraints that acted to favor the inactive, R, conformation of the receptor *(1)*. Thus, we fully expected that the E92K change in the mouse sombre receptor constitutively activated the MC1-R by removing acidic residues essential for an electrostatic bridge with an as yet unidentified basic receptor residue. In vitro mutagenesis studies of this residue, as well as others, demonstrated, however, that only insertion of basic residues at the E92, D119, and C125 positions caused constitutive activation of the receptor *(91)*. This led us to propose that many of the activating mutations of the MC1-R are acting by mimicking insertion of the arginine residue of the ligand ion precisely in the pocket where this residue normally inserts to stabilize the active, R*, receptor conformation.

3.2. Computer Modeling of the Receptor

Identification of chemical and structural ligand interactions with receptor proteins may provide insights to designing receptor subtype selective agonists and antagonists and understanding naturally occurring mutations. Determination of true three-dimensional structure at high resolution requires X-ray diffraction techniques. Unfortunately, the members of the G protein-coupled receptor (GPCR) superfamily are resistant thus far to crystallization techniques. Lacking this direct structural information, computer assisted molecular modeling of these receptors has become a common approach to try to predict receptor structure and probable ligand–receptor interactions. This approach is based upon the low resolution electron-microscopy structure of the non-G protein-coupled seven transmembrane spanning protein, bacteriorhodopsin *(92,93)*, with further refinements that include the footprint of the mammalian G protein-coupled rhodopsin receptor *(94)*. Transmembrane region alignment of the sequences that constitute the α-helical regions may be determined by hydrophobicity plots, such as Kyte-Doolittle analysis *(95)*, or more consistently using the "Baldwin" alignment *(96)*, which accommodates similar positioning of the GPCR superfamily conserved amino acid residues.

Several melanocortin receptor models have been developed by different groups *(97–100)* to propose receptor residues that may be interacting with regions of the melanotropin ligands. Figure 9 illustrates the mMC1-R interacting with the NDP-α-MSH peptide. Figure 9A shows a side view of the ligand-docked receptor with flanking residues of the ligand proposed to increase affinity via interaction with extracellular loops shown in yellow. Residues of the pharmacophore are labeled, with charged residues shown in red and blue, and hydrophobic residues in other colors. Examination of the receptor (Fig. 9B) shows two domains that are proposed to interact with the charged and hydrophobic domians of the ligand. A highly charged domain, shown in red and blue, is made of the residues in TMII and TMIII, glu92, arg115, and arg119, while a domain containing multiple phenylalanines and

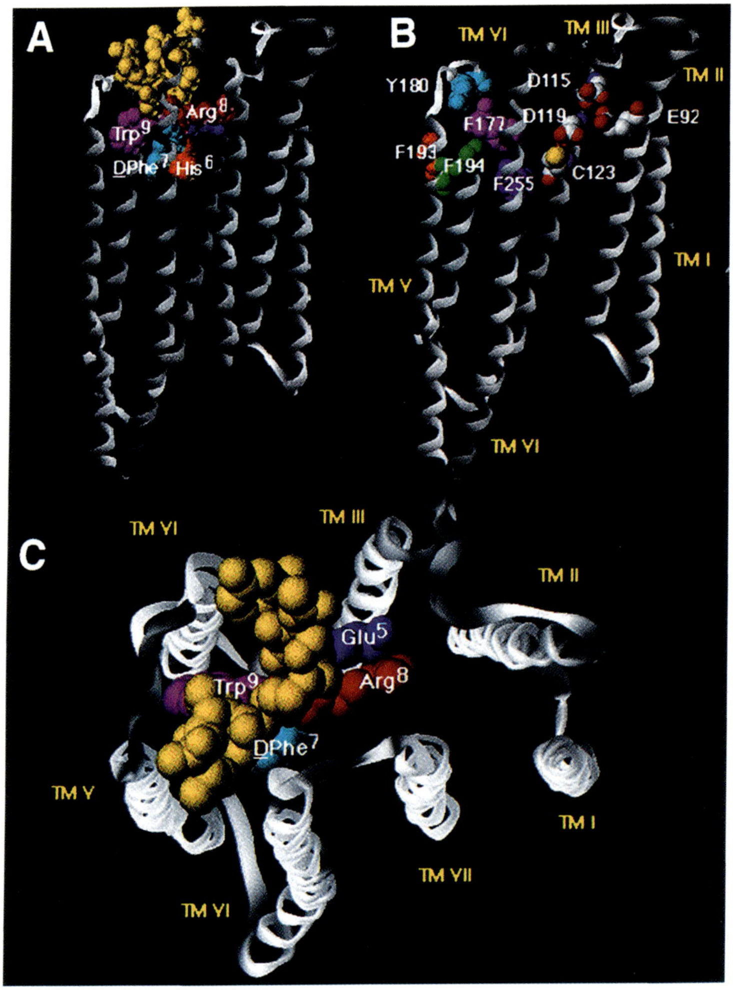

Fig. 9. Molecular model of NDP-α-MSH docked into the mMC1-R. The α-helical backbone of the receptor is denoted in various shades of gray. The ligand residues highlighted in yellow consist of the regions of NDP-MSH which flank the "message" residues. The "message" residues His^6 (orange), D-Phe^7 (aqua), Arg^8 (red), and Trp^9 (magenta) are docked into the putative binding pocket of the receptor, and labeled in Panel A. (A and B) Side-on views of the ligand–receptor complex, with TM I located on the far right, and TM V located on the far left. mMC1-R receptor residues which are proposed to interact with the ligand "message" residues, are labeled in panel B. C Space-filled model of NDP-α-MSH docked into mMC1-R looking down toward the intracellular portion of the ligand–receptor complex. This model was generated based originally upon the bacteriorhodopsin structure (BR1) obtained from the Protein Data Bank *(93)*, modified manually to fit the helical packing arrangement of rhodopsin *(94)*, and based on homology with the hMC1-R *(97)*.

a tyrosine residue is proposed to interact via the aromatic phenylalanine and histidine residues of the pharmacophore. Figure 9C illustrates a view of NDP-α-MSH docked with the mMC1-R looking down on the surface of the membrane. These interactions for the mMC1-R are predicted based on structural and functional homology, albeit with minor differences *(52)*, with the hMC1-R, which has been extensively studied *(97–100)*.

Additionally, based on this type of receptor molecular modeling, we can propose specific residue interactions which may explain the basis for the constitutive activation observed for naturally occurring mutations and can be extrapolated to a general mechanism for melanocortin receptor activation. For the glu92lys mutation, an acidic negatively charged residue in TM II is replaced by a longer basic positively charged residue. The Arg residue of the pharmacophore is proposed to interact with complementary negatively charged receptor residues *(88,97)*. With this in mind, it can be postulated that the lys92 residue can possess similar functional properties as the ligand Arg residue, and therefore, mimic the ligand-induced receptor conformation in the absence of ligand. A report of point mutations of the hMC1-R identified arg117ala (Arg115 in the mouse MC1-R), as significantly (267-fold) decreasing α-MSH binding affinity *(89)*. Thus with this supporting information, the lys92 residue (TM II) can be proposed to interact with the conserved asp residue(s) *(115,119)* seen in Fig. 9. Specifically, these interactions may include complementary electrostatic (charge-charge) interactions and up to two hydrogen bonds. In a hydrophobic environment, a salt bridge such as this may generate up to 10 kcal/mol stabilization energy *(101)*. Furthermore, asp115, located one helical turn above Asp119, is also conserved within the melanocortin receptor family. Rotation around the lys side-chain torsion angles would allow for nearly identical interactions with asp115 as proposed for asp119. This is important as some ambiguity is present as to which particular asp residue, or combination of, may be an acceptable complementary acidic residue. Once these interactions have formed, a receptor conformation may be formed in which the highly conserved DRY sequence in TM III, proposed to be important for signal transduction *(102,103)*, can obtain the necessary conformational and spatial orientation important for signal transduction. The exact mechanism may involve a change in TM spanning α-helical packing of TMs II and III, thus modifying the packing orientation of the entire receptor.

Two additional mutations of the MC1-R, leu98pro (E^{so}) in the mouse *(1)* and leu99pro in bovine *(32)*, resulted in constitutive activation and black coat color. Interestingly, both mutations are at the borderline of the transmembrane spanning helical interface of TM II on the extracellular surface. As alluded to previously, proline residues possess a variety of structural implications in transmembrane helices. This particular amino acid can modify α-helices 20°–30° from helices lacking the proline residue at the interface between the extracellular surface and TM regions, and can affect the packing of entire transmembrane helical spanning regions *(104)*. The hydrophobicity of the wild-type region (mouse, Ile-Ile-Leu-Leu-Leu) is modified dramatically by the leu to pro mutation. Leucine possesses a value of 3.8 on the hydropathy index, whereas proline possesses a value of –1.6 *(95)*. Therefore this particular modification at this location in the receptor may not only modify helical packing of TM II, but also modify the position of the helical secondary structure and orientation in the membrane, thus likely disrupting the normal interaction(s) of glu92.

A mutation found in the fox TM III, cys125arg, has also been demonstrated to result in a constitutively active melanocortin receptor *(31)*. This arg residue is likely to interact electrostatically with the arg residues one and two turns above it on the helix, and an interaction with glu92 of TM II is also probable. In the latter interaction, a similar mechanism described for the glu92lys mutation may be applicable in that these ionic interactions modify the helical packing arrangement in TMs II and III and therefore, generate a receptor population that can couple to the G protein in the absence of ligand.

All the constitutively activating mutations identified to date for the melanocortin receptors are located in the TM II and TM III region. This concentration of mutations allows us to propose a general mechanism for this biological phenomenon for the melanocortin receptors. The direct structural changes, in the case of ser69leu, leu98pro, and leu99pro, or indirect changes in the case of glu92lys and *c*ys125arg result in modifying the overall helical packing of the receptor by possibly modifying important interactions between TM II, TM III, and TM VII (discussed above), leading to a shift in receptor population that is able to couple to the G protein in the absence of ligand, resulting in a dark coat. Although these speculations remain to be experimentally confirmed, molecular modeling has provided new hypotheses that may account for the constitutive activities which result from naturally occurring melanocortin mutations, and can be tested experimentally.

4. Roles for the MC1-R Outside the Regulation of Pigmentation

The only phenotype reported for the MC1-R-null *recessive yellow* (*e*/*e*) mouse is the absence of eumelanin in the coat. Nevertheless, this does not preclude physiologic role(s) for the receptor outside of regulation of the eumelanin-pheomelanin switch. MC1-R mRNA has been reported in the periaquaductal gray region of the brainstem by *in situ* hybridization *(105)*, and in a variety of peripheral sites by polymerase chain reaction, including pituitary, testis, ovary and placenta *(106)*. Expression in human placenta, ovary, and testis has also been reported using a monoclonal antibody to the MC1-R *(107)*. Several reports suggest an antiinflammatory role for α-MSH mediated via MC1-R expression in macrophages *(108,109)*, neutrophils *(110)* and microvascular endothelial cells *(111)*. In these systems, it has been suggested that the antiinflammatory activity of α-MSH appears to derive from its ability to regulate the production of various cytokines. For example, the inhibitory cytokine IL-10 is upregulated in monocytes by α-MSH treatment *(112)*. Additional work will be necessary to definitively demonstrate that the MC1-R is necessary for mediating the peripheral antiinflammatory effects of α-MSH, and to demonstrate a physiologic role for the MC1-R in regulating immune function and inflammation.

5. Conclusions and Future Prospects

A great deal of new information has been learned over the past few years regarding the MC1-R and the regulation of mammalian pigmentation. Thus far, it appears that normal variations in the eumelanin/pheomelanin switch directly involving the MC1-R result primarily from genetic variation in the coding sequence of this receptor. In contrast to many of the constitutively activating mutations in other G protein-coupled receptors, activating mutations of the MC1-R are localized to TMII and TMIII, and appear in some way to mimic ligand binding. Alterations in the expression levels of the receptor or its ligand as a mechanism for genetic diversity remain to be demonstrated. As an alternative mechanism, the eumelanin/pheomelanin switch may also be regulated by the novel G protein-coupled receptor antagonist, agouti. In this case, nearly all variation characterized thus far in the mouse, fox, and cow results from alterations in temporal, spatial, or quantitative aspects of expression of the *agouti* gene. Naturally occurring pharmacologic variants of agouti do not appear to be common.

The high frequency of hMC1-R variants, their regional localization in the human receptor coding sequence, and their association with red hair and fair skin in humans remains a mystery (*see* Chapter 11). Additional work will be required to determine the value of these polymorphisms in relation to melanoma and other disorders of pigment cells or the pigmentation process. Likewise, the role of the conserved human *agouti* in pigmentation, or other physiologic processes, remains to be determined.

Finally, the recent cloning of the mahogany gene *(119,120)* a suppressor of agouti, may ultimately lead to a deeper understanding of agouti action (*see* Chapter 14, Section 5.4.).

Acknowledgments

This chapter is revised and reprinted from *G Proteins and Disease*, with permission from Humana Press.

References

1. Robbins, L. S., Nadeau, J. H., Johnson, K. R., Kelly, M. A., Roselli–Rehfuss, L., Baack, E., Mountjoy, K. G., and Cone, R. D. (1993) Pigmentation phenotypes of variant extension locus alleles result from point mutations that alter MSH receptor function. *Cell* **72,** 827–834.
2. Lu, D., Willard, D., Patel, I. R., Kadwell, S., Overton, L., Kost, T., Luther, M., Chen, W., Woychik, R. P., Wilkison, W. O., and Cone, R. D. (1994) Agouti protein is an antagonist of the melanocyte–stimulating hormone receptor. *Nature* **371,** 799–802.
3. Barsh, G. S. (1996) The genetics of pigmentation: from fancy genes to complex traits. *TIG* **12,** 299–305.
4. Jackson, I. J. (1994) Molecular and developmental genetics of mouse coat color. *Annu. Rev. Genet.* **28,** 189–217.
5. Silvers, W. K. (1979). *The Coat Colors of Mice: A Model for Mammalian Gene Action and Interaction.* Springer–Verlag, New York.
6. Searle, A. G. (1968). *Comparative Genetics of Coat Colors in Mammals.* Logos Press London.
7. Jackson, I. J. (1988) A cDNA encoding tyrosinase–related protein maps to the mouse brown locus. *Proc. Natl. Acad. Sci. U. S. A.* **85,** 4392–4396.
8. Jackson, I. J., Chambers, D. M., Tsukamoto, K., Copeland, N. G., Jenkins, N. A., and Hearing, V. (1992) A second tyrosinase–related protein, TRP–2, maps to and is mutated at the mouse slaty locus. *EMBO J.* **11,** 527–535.
9. Kwon, B. S., Chintamaneni, C.D., Kozak, C.A., Copeland, N.G., Gilbert, D.J., Jenkins, N. A., Barton, D.E., Francke, U., Kobayashi, Y., and Kim, K.K. (1991) A melanocyte–specific gene, Pmel 17, maps near the silver coat color locus on mouse chromosome 10, and is a syntenic region on human chromosome 12. *Proc. Natl. Acad. Sci. U. S. A.* **88,** 9228–9232.

9a. Lu, D., Chen, W., and Cone, R. D. (1988) Regulation of melanogenesis by the MSH receptor. In: *The Pigmentary System and its Disorders.* Norlund, J.J. (ed.) Oxford University Press, New York.

10. Burchill, S. A., Thody, A. J., and Ito, S. (1986) Melanocyte–stimulating hormone, tyrosinase activity and the regulation of eumelanogenesis and phaeomelanogenesis in the hair follicular melanocytes of the mouse. *J. Endocrinol.* **109,** 15–21.
11. Burchill, S. A., Virden, R., and Thody, A. J. (1989) regulation of tyrosinase synthesis and its processing in the hair follicular melanocytes of the mouse during eumelanogenesis and phaeomelanogenesis. *J. Invest. Dermatol.* **93,** 236–240
12. Takeuchi, T., Kobunai, T., and Yamamoto, H. (1989) Genetic control of signal transduction in mouse melanocytes. *Soc. Invest. Dermatol.* **92,** 239S–242S.
13. Tamate, H. B. and Takeuchi, T. (1984) Action of the e locus of mice in the response of phaeomelanic hair follicles to alpha–melanocyte stimulating hormone in vitro. *Science* **224,** 1241–1242..

14. Kobayashi, T., Vieira, W. D., Potterf, B., Saka, C., Imokawa, G., and Hearing, V. (1995) Modulation of melanogenic protein expression during the switch from eutopheomelanogenesis. *J. Cell Sci.* **108,** 2301–2309.
15. Hoganson, G. E., Ledwitz–Rigby, F., Davidson, R. L., and Fuller, B. B. (1989) Regulation of tyrosinase mRNA levels in mouse melanoma cell clones by melanocyte–stimulating hormone and cyclic AMP. *Som. Cell. Mol. Gen.* **15,** 255–263.
16. Kwon, B. S., Wakulchik, M., Haq, A.Q., Halaban, R., Kestler, D. (1988) Sequence analysis of mouse tyrosinase cDNA and the effect of melanotropin on its gene expression. *Biochem. Biophys. Res. Commun.* **153,** 1301–1309.
17. Halaban, R., Pomerantz, S. H., Marshall, S., and Lerner, A. B. (1984) Tyrosinase activity and abundance in cloudman melanoma cells. *Arch. Biochem. Biophys.* **230,** 383–387.
18. Wong, G., and Pawelek, J. (1975) Melanocyte stimulating hormone promotes activation of preexisting tyrosinase molecules in Cloudman S91 melanoma cells. *Nature* **255,** 644–646.
19. Pawelek, J. (1976) Factors regulating growth and pigmentation of melanoma cells. *J. Invest. Dermatol.* **66,** 201–209.
20. Bultman, S. J., Michaud, E. J., and Woychik, R. P. (1992) Molecular characterization of the mouse agouti locus. *Cell* **71,** 1195–1204.
21. Miller, M. W., Duhl, D. M. J., Vrieling, H., Cordes, S. P., Ollmann, M. M., Winkes, B. M., and Barsh, G. S. (1993) Cloning of the mouse *agouti* gene predicts a novel secreted protein ubiquitously expressed in mice carrying the *lethal yellow* (A^y) mutation. *Genes Dev.* **7,** 454–467.
22. Geschwind, I. I. (1966) Change in hair color in mice induced by injection of α–MSH. **79,** 1165–1167.
23. Geschwind, I. I., Huseby, R. A., and Nishioka, R. (1972) The effect of melanocyte–stimulating hormone on coat color in the mouse. *Recent Prog. Horm. Res.* **28,** 91–130.
24. Lamoreux, M. L. and Mayer, T. C. (1975) Site of gene action in the development of hair pigment in recessive yellow (e/e) mice. *Dev. Biol.* **46,** 160–166
25. Silvers, W. K. (1958) An experimental approach to action of genes at the agouti locus in the mouse. III. Transplants of newborn A^w–, A–, and a^t– skin to A^y–, A^w–, A–, and *aa* hosts. *J. Exp. Zool.* **137,** 189–196.
26. Silvers, W. K. and Russel, E. S. (1955) An experimental approach to action of genes at the agouti locus in the mouse. *J. Exp. Zool.* **130,** 199–220.
27. Gerst, J. E., Sole, J., Hazum, E., and Salomon, Y. (1988) Identification and characterization of melanotropin binding proteins from M2R melanoma cells by covaler photoaffinity labeling. *Endocrinology* **123,** 1792–1797.
28. Solca, F., Siegrist, W., Drozdz, R., Girard, J., and Eberle, A. N. (1989) The receptor for α–melanotropin of mouse and human melanoma cells. *J. Biol. Chem.* **264,** 14,277–14,280.
29. Chhajlani, V. and Wikberg, J. E. S. (1992) Molecular cloning and expression of the human melanocyte stimulating hormone receptor cDNA. *FEBS Lett.* **309,** 417–420.
30. Mountjoy, K. G., Robbins, L. S., Mortrud, M. T., and Cone, R. D. (1992) The cloning of a family of genes that encode the melanocortin receptors. *Science* **257,** 543–546.
31. Vage, D. I., Lu, D., Klungland, H., Lien, S., Adalsteinsson, S., and Cone, R.D. (1997) A non–epistatic interaction of *agouti* and *extension* in the fox, *Vulpes vulpes*. *Nat. Genet.* **15,** 311–315.

32. Klungland, H., Vage, D. I., Gomez–Raya, L., Adelsteinsson, S., and Lien, S. (1995) The role of melanocyte–stimulating hormone (MSH) receptor in bovine coat color determination. *Mamm. Genome* **6,** 636–639.
33. Vanetti, M., Schonrock, C., Meyerhof, W., Hollt, V. (1994) Molecular cloning of a bovine MSH receptor which is highly expressed in the testis. *FEBS Lett.* **348,** 268–272
34. Takeuchi, S., Suzuki, S., Hirose, S., Yabuuchi, M., Sato, C., Yamamoto, H., and Takahashi, S. (1996) Molecular cloning and sequence analysis of the chick melanocortin 1–receptor gene. *Biochem. Biophys. Acta* **1306,** 122–126.
35. Vage, D., Klungland, H., Lu, D., and Cone, R. (1999) Molecular and pharmacological characterization of dominant black coat color in sheep. *Mamm. Genome* **10,** 39–43.
36. Gantz, I., Konda, Y., Tashiro, T., Shimoto, Y., Miwa, H., Munzert, G., Watson, S. J., DelValle, J., and Yamada, T. (1993) Molecular cloning of a novel melanocortin receptor. *J. Biol. Chem.* **268,** 8246–8250.
37. Roselli–Rehfuss, L., Mountjoy, K. G., Robbins, L. S., Mortrud, M. T., Low, M. J., Tatro, J. B., Entwistle, M. L., Simerly, R., and Cone, R. D. (1993) Identification of a receptor for γ–MSH and other proopiomelanocortin peptides in the hypothalamus and limbic system. *Proc. Natl. Acad. Sci. U. S. A.* **90,** 8856–8860.
38. Gantz, I., Miwa, H., Konda, Y., Shimoto, Y., Tashiro, T., Watson, S. J., DelValle, J., and Yamada, T. (1993) Molecular cloning, expression, and gene localization of a fourth melanocortin receptor. *J. Biol. Chem.* **268,** 15,174–15,179
39. Mountjoy, K. G., Mortrud, M. T., Low, M. J., Simerly, R. B., and Cone, R. D. (1994) Localization of the melanocortin–4 receptor (MC4–R) in neuroendocrine and autonomic control cirsuits in the brain. *Mol. Endocrinol.* **8,** 1298–1308.
40. Barret, P., MacDonald, A., Helliwell, R., Davidson, G., and Morgan, P. (1994) Cloning and expression of a new member of the melanocyte–stimulating hormone receptor family. *J. Mol. Endocrinol.* **12,** 203–213.
41. Chhajlani, V., Muceniece, R., and Wikberg, J. E. S. (1993) Molecular cloning of a novel human melanocortin receptor. *Biochem. Biophys. Res. Commun.* **195,** 866–873.
42. Gantz, I., Shimoto, Y., Konda, Y., Miwa, H., Dickinson, C. J., and Yamada, T. (1994) Molecular cloning, expression, and characterization of a fifth melanocortin receptor. *Biochem. Biophys. Res. Commun.* **200,** 1214–1220.
43. Griffon, N., Mignon, V., Facchinetti, P., Diaz, J., Schwartz, J.–C., and Sokoloff, P. (1994) Molecular cloning and characterization of the rat fifth melanocortin receptor. *Biochem. Biophys. Res. Commun.* **200,** 1007–1014.
44. Labbe, O., Desarnaud, F., Eggerickx, D., Vassart, G., and Parmentier, M. (1994) Molecular cloning of a mouse melanocortin 5 receptor gene widely expressed in peripheral tissues. *Biochemistry.* **33,** 4543–4549.
45. Elphick, M. R. (1998) An invertebrate G–protein coupled receptor is a chimeric cannabinoid/melanocortin receptor. *Brain Res.* **780,** 170–173.
46. Dixon, R. A. F., Sigal, I. S., Candelore, M. R., Register, R. B., Scatergood, A., Rands, E., and Strader, C. D. (1987) Structural features required for ligand binding to the β–adrenergic receptor. *EMBO J.* **6,** 3269–3275.
47. Karnik, S. S., Sakmann, J. P., Chen, H. B., and Khorana, H. G. (1988) Cysteine residues 110 and 187 are essential for the formation of correct structure in bovine rhodopsin. *Proc. Natl. Acad. Sci., U. S. A.* **85,** 8459–8463.
48. Eberle, A. N. (1988). *The Melanotropins. Chemistry, Physiology and Mechanisms of Action.* (S. Karger, Basel).

49. Hadley, M. E., Hruby, V. J., Jiang, J., Sharma, S. D., Fink, J. L., Haskell–Leuvano, C., Bentley, D. L., Al–Obeidi, F., and Sawyer, T. K. (1996) Melanocortin receptors: identification and characterization by melanotropic peptide agonists and antagonists. *Pigment Cell Res.* **9,** 213–234.
50. Abdel–Malek, Z., Swope, V. B., Suzuki, I., Akcali, C., Harriger, M. D., Boyce, S. T., Urabe, K., and Hearing, V. J. (1995) Mitogenic and melanogenic stimulation of normal human melanocytes by melanotropic peptides. *Proc. Natl. Acad. Sci.U. S. A.* **92,** 1789–1793.
51. Hunt, G., Donatien, P. D., Lunec, J., Todd, C., Kyne, S., and Thody, A. J. (1994) Cultured human melanocytes respond to MSH peptides and ACTH. *Pigment Cell Res.* **7,** 217–21.
52. Mountjoy, K. G. (1994) The human melanocyte stimulating hormone receptor has evolved to become "super–sensitive" to melanocortin peptides. *Mol. Cell. Endocrinol.* **102,** R7–R11.
53. Eberle, A. N., de Graan, P. N. E., Baumann, J. B., Girard, J., van Hees, G., and van de Veerdonk, F. C. G. (1984) Structural requirements of α–MSH for the stimulation of MSH receptors on different pigment cells. *Yale J. Biol. Med.* **57,** 353–354.
54. Lerner, A. B., and McGuire, J. S. (1961) Effect of alpha–and beta–melanocyte stimulating hormones on the skin color of man. *Nature* **189,** 176–177.
55. Levine, N., Sheftel, S. N., Eytan, T., Dorr, R. T., Hadley, M. E., Weinrach, J. C., Ertl, G. A., Toth, K., McGee, D. L., and Hruby, V. J. (1991) Induction of skin tanning by subcataneous administration of a potent synthetic melanotropin. *JAMA* **266,** 2730–2736.
56. Orth, D. N., Kovacs, W. J., and DeBold, C. R. (1992). The adrenal cortex, in *Williams Textbook of Endocrinology*, (Wilson, J. D. and Foster, D. W. eds.) Saunders, Philadelphia. pp. 523.
57. Schauer, E., Trautinger, F., Kock, A., Schwarz, A., Bhardwaj, R., Simon, M., Ansel, J. C., Schwarz, T., and Luger, T. A. (1994) Proopiomelanocortin–derived peptides are synthesized and released by human keratinocytes. *J. Clin. Invest.* **93,** 2258–2262.
58. Bateman, N. (1961) Sombre, a viable dominant mutant in the house mouse. *J. Hered.* **52,** 186–189.
59. Carver, E. A. (1984) Coat color genetics of the German shepherd dog. *J. Hered.* **75,** 247–252.
60. Siracusa, L. D. (1994) The *agouti* gene: turned on to yellow. *TIG* **10,** 423–428.
61. Yen, T. T., Gill, A. M., Frigeri, L. G., Barsh, G. S., and Wolff, G. L. (1994) Obesity, diabetes, and neoplasia in yellow A(vy)/– mice: ectopic expression of the agouti gene. *FASEB J.* **8,,** 479–88.
62. Conklin, B. R., and Bourne, H.R. (1993) Mouse coat color reconsidered. *Nature* **364,** 110.
63. Manne, J., Argeson, A. c., and Siracusa, L. D. (1995) Mechanisms for the pleiotropic effects of the agouti gene. *Proc. Natl. Acad. Sci. U. S. A.* **92,** 4721–4724.
64. Zemel, M. B., Kim, J. H., Woychik, R. P., Michaud, E. J., Kadwell, S. H., Patel, I. R., and Wilkison, W. O. (1995) Agouti regulation of intracellular calcium: Role in the insulin resistance of viable yellow mice. *Proc. Natl. Acad. Sci. U. S. A.* **92,** 4733–4737.
65. Hruby, V. J., Lu, D., Sharma, S. D., Castrucci, A. L., Kesterson, R. A., Al–Obeidi, F. A., Hadley, M. E., and Cone, R. D. (1995) Cyclic lactam α–melanotropin analogues of Ac–Nle^4–c[Asp^4,D–Phe^7,Lys^{10}]α–MSH(4–10)–NH_2 with bulky aromatic

amino acids at position 7 show high antagonist potency and selectivity at specific melanocortin receptors. *J. Med. Chem.* **38,** 3454–3461.

66. Fan, W., Boston, B. A., Kesterson, R. A., Hruby, V. J., and Cone, R. D. (1997) Role of melanocortinergic neurons in feeding and the agouti obesity syndrome. *Nature* **385,** 165–168.
67. Huszar, D., Lynch, C. A., Fairchild–Huntress, V., Dunmore, J. H., Fang, Q., Berkemeier, L. R., Gu, W., Kesterson, R. A., Boston, B. A., Cone, R. D., Smith, F. J., Campfield, L. A., Burn, P., and Lee, F. (1997) Targeted disruption of the melanocortin–4 receptor results in obesity in mice. *Cell* **88,** 131–141.
68. Blanchard, S. G., Harris, C. O., Ittoop, O. R. R., Nichols, J. S., Parks, D. J., Truesdale, A. T., and Wilkison, W. O. (1995) Agouti antagonism of melanocortin binding and action in the B16F10 murine melanoma cell line. *Biochemistry.* **34,** 10,406–10,411.
69. Hunt, G., and Thody, A. J. (1995) Agouti protein can act independently of melanocyte–stimulating hormone to inhibit melanogenesis. *J. Endocrinol.* **147,** R1–R4.
70. Siegrist, W., Willard, D. H., Wilkison, W. O., and Eberle, A. N. (1996) Agouti protein inhibits growth of B16 melanoma cells in vitro by acting through melanocortin receptors. *Biochem. Biophys. Res. Commun.* **218,** 171–175.
71. Suzuki, I., Ollmann, M., Barsh, G. S., Im, S., Lamoreux, M. L., Hearing, V. J., Nordlund, J. J., and Abdel–Malek, Z. (1997) Agouti signalling protein inhibits melanogenesis and the response of human melanocytes to a–melanotropin. *J. Invest. Dermatol.* **108,** 838–842.
72. Olivera, B. M., Miljanich, G. P., Ramachandran, J., and Adams, M. E. (1994) Calcium channel diversity and neurotransmitter release: The w–Conotoxins and w–Agatoxins. *Annu. Rev. Biochem.* **63,** 823–867.
73. Lefkowitz, R. J., Cotecchia, S., Samama, P., and Costa, T. (1993) Constitutive activity of receptors coupled to guanine nucleotide regulatory proteins. *TiPS* **14,** 303–308.
74. Samama, P., Pei, G., Costa, T., Cotecchia, S., and Lefkowitz, R. J. (1994) Negative antagonists promote an inactive conformation of the β_2–adrenergic receptor. *Mol. Pharm.* **45,** 390–394.
75. Solca, F. F., Chluba–de Tapia, J., Iwata, K., and Eberle, A. N. (1993) B16–G4F mouse melanoma cells: an MSH receptor–deficient clone. *FEBS Lett.* **322,** 177–180.
76. Gantz, I., Yamada, T., Tashiro, T., Konda, Y., Shimoto, Y., Miwa, H., and Trent, J. M. (1994) Mapping of the gene encoding the melanocortin–1 (alpha–melanocyte stimulating hormone receptor (MC1–R) to human chromosome 16q24.3 by fluoresence in situ hybridization. *Genomics* **19,** 394–395.
77. Magenis, R. E., Smith, L., Nadeau, J. H., Johnson, K. R., Mountjoy, K. G., and Cone, R. D. (1994) Mapping of the ACTH, MSH, and neural (MC3 and MC4) melanocortin receptors in the mouse and human. *Mamm. Genome* **5,** 503–508.
78. Falconer, D. S. (1962) Sombre (So) on LG XVIII. *Mouse News Lett.* **27,** 30.
79. Meredith, R. (1971) Linkage of am and e. *Mouse News Lett.* **45,** 31.
80. Searle, A. G. and Beechey, C. V. (1970) Linkage of Os and Eso. *Mouse News Lett.* **42,** 27.
81. Cone, R. D., Mountjoy, K. G., Robbins, L. S., Nadeau, J. H., Johnson, K. R., Roselli–Rehfuss, L., and Mortrud, M. T. (1993) Cloning and functional characterization of a family of receptors for the melanotropic peptides. *Ann. N. Y. Acad. Sci.* **680,** 342–363.

82. Hauschka, T. S., Jacobs, B. B., and Holdridge, B. A. (1968) Recessive yellow and its interaction with belted in the mouse. *J. Hered.* **59,** 339–341.
83. von Lehmann, E. (1973) Coat color genetics of the tobacco–mouse (*Mus poschiavinus* Fatio). *Mouse News Lett.* **48,** 23.
84. Adalsteinsson, S., Bjarnadottir, S., Vage, D. I., and Jonmundsson, J. V. (1995) Brown coat color in Icelandic cattle produced by the loci agouti and extension. *J. Hered.* **86,** 395–398.
85. Adalsteinsson, S., Hersteinsson, P., and Gunnarsson, E. (1987) Fox colors in relation to colors in mice and sheep. *J. Hered.* **78,** 235–237.
86. Ashbrook, F. G. (1937) The breeding of fur animals. *Year. Agric.* 1379–1395.
87. Lyon, M. F. (1961) Gene action in the X–chromosome of the mouse (*Mus musculus* L.). *Nature* **190,** 372–373.
88. Cone, R. D., Lu, D., Chen, W., Koppula, S., Vage, D. I., Klungland, H., Boston, B., Orth, D. N., Pouton, C., and Kesterson, R. A. (1996) The melanocortin receptors: agonists, antagonists, and the hormonal control of pigmentation. *Recent Prog. Horm. Res.* **51,** 287–318.
89. Frandberg, P.–A., Muceniece, R., Prusis, P., Wikberg, J., and Chhajlani, V. (1994) Evidence for Alternate Points of Attachment for α–MSH and its Stereoisomer [Nle^4,D–Phe^7]–α–MSH at the Melanocortin–1 Receptor. *Biochem. Biophys. Res. Commun.* **202,** 1266–1271.
90. Chhajlani, V., Xu, X., Blauw, J., and Sudarshi, S. (1996) Identification of ligand binding residues in extracellular loops of the melanocortin 1 receptor. *Biochem. Biophys. Res. Commun.* **219,** 521–525.
91. Lu, D., Vage, D. I., and Cone, R. D. (1998) A ligand–mimetic model for the constitutive activation of the melanocortin–1 receptor. *Mol. Endocrinol.* **12,** 592–604.
92. Grigorieff, N., Ceska, T. A., Downing, K. H., Baldwin, J. M., and Henderson, R. (1996) Electron–crystallographic refinement of the structure of bacteriorhodopsin. *J. Mol. Biol.* **259,** 393–421.
93. Henderson, R., Baldwin, J. M., Ceska, T. A., Zemlin, F., Beckmann, E., and Downing, K. H. (1990) Model for the strucuture of bacteriorhodopsin based on high–resolution electron cryo–microscopy. *J. Mol. Biol.* **213,** 899–929.
94. Schertler, G. F. X., C., V. and R., H. (1993) Projection structure of rhodopsin. *Nature* **362,** 770–772.
95. Kyte, J. and Doolittle, R. F. (1982) A simple method for displaying the hydrophobic character of a protein. *J. Mol. Biol.* **157,** 105–132.
96. Baldwin, J. (1993) The probable arrangement of helices in the G protein–coupled receptors. *EMBO J.* **12,** 1693–1703.
97. Haskell–Luevano, C., Sawyer, T. K., Trumpp–Kallmeyer, S., Bikker, J. A., Humblet, C., Gantz, I., and Hruby, V. J. (1996) Three–dimensional molecular models of the hMC1R melanocortin receptor: complexes with melanotropin peptide agonists. *Drug Des. Discov.* **14,** 197–211.
98. Lu, D., Haskell–Luevano, C., Vage, D. I., and Cone, R. D. (1998). Functional variants of the MSH receptor (MC1–R), agouti, and their effects on mammalian pigmentation, in *G Proteins, Receptors, and Disease*, (Spiegel, A. M. ed.) Humana Press, Totowa pp. 231–260.
99. Prusis, P., Frändberg, P.–A., Muceniece, R., Kalvinsh, I., and Wikberg, J. E. S. (1995) A three dimensional model for the interaction of MSH with the melanocortin–1 receptor. *Biochem. Biophys. Res. Commun.* **210,** 205–210.

100. Prusis, P., Schioth, H. B., Muceniece, R., Herzyk, P., Afshar, M., Hubbard, R. E., and Wikberg, J. E. (1997) Modeling of the three–dimensional structure of the human melanocortin 1 receptor, using an automated method and docking of a rigid cyclic melanocyte–stimulating hormone core peptide. *J. Mol. Graph. Model.* **15,** 307–317.
101. Strader, C. D., Fong, T. M., Tota, M. R., Underwood, D., and Dixon, R. A. F. (1994) Structure and function of G protein–coupled receptors. *Annu. Rev. Biochem.* **63,** 101–132.
102. Savarese, T. M. and Fraser, C. M. (1992) In Vitro Mutagenesis and the Search for Structure–function relationships among G protein–coupled receptors. *Biochem. J.* **283,** 1–19.
103. Zhu, S. Z., Wang, S. Z., Hu, J., and El–Fakahany, E. E. (1994) An arginine residue conserved in most g protein–coupled receptors is essential for the function of the m1 muscarinic receptor. *Mol. Pharm.* **45,** 517–523.
104. Williams, K. A. and Deber, C. M. (1991) Proline residues in transmembrane helicies: structural or dynamic role? *Biochemistry* **30,** 8919–8923.
105. Xia, Y., Wikberg, J. E. S., and Chhajlani, V. (1995) Expression of melanocortin 1 receptor in periaqueductal gray matter. *NeuroReport* **6,** 2193–2196.
106. Schioth, H. B., Muceniece, R., Wikberg, J. E. S., and Chhajlani, V. (1995) Characterisation of melanocortin receptor subtypes by radioligand binding analysis. *Eur. J. Pharmacol. (Mol. Pharmacol. Sec.)* **288,** 311–317.
107. Thornwall, M., Dimitriou, A., Xu, X., Larsson, E., and Chhajlani, V. (1997) Immunohistochemical detection of the melanocortin 1 receptor in human testis, ovary and placenta using specific monoclonal antibody. *Horm. Res.* **48,** 215–218.
108. Rajora, N., Ceriani, G., Catania, A., Star, R. A., Murphy, M. T., and Lipton, J. M. (1996) a–MSH production, receptors, and influence on neopterin in a human monocyte/macrophage cell line. *J. Leukoc. Biol.* **59,** 248–253.
109. Star, R. A., Rajora, N., Huang, J., Stock, R. C., Catania, A., and Lipton, J. M. (1995) Evidence of autocrine modulation of macrophage nitric oxide synthase by α–melanoctye–stimulating hormone. *Proc. Natl. Acad. Sci. U. S. A.* **92,** 8016–8020.
110. Catania, A., Rajora, N., Capsoni, F., Minonzio, F., Star, R. A., and Lipton, J. M. (1996) The neuropeptide α–MSH has specific receptors on neutrophils and reduces chemotaxis in vitro. *Peptides* **17,** 675–679.
111. Hartmeyer, M., Scholzen, T., Becher, E., Bhardwaj, R. S., Schwarz, T., and Luger, T. A. (1997) Human dermal microvascular endothelial cells express the melanocortin receptor type 1 and produce increased levels of IL–8 upon stimulation with a–melanocyte–stimulating hormone. *J. Immunol.* **159,** 1930–1937.
112. Bhardwaj, R. S., Schwarz, A., Becher, E., Mahnke, K., Aragane, Y., Schwarz, T., and Luger, T. A. (1996) Pro–opiomelanocortin–derived peptides induce IL–10 production in human monocytes. *J. Immunol.* **156,** 2517–2521.
113. Lane, P. W. and Green, M. C. (1960) Mahogany, a recessive color mutation in linkage group V of the mouse. *J. Hered.* **51,** 228–230.
114. Bultman, S. J., Klebig, M. L., Michaud, E. J., Sweet, H. O., Davisson, M. T., and Woychik, R. P. (1994) Molecular analysis of reverse mutations from nonagouti (*a*) to black–and–tan (a^t) and white–bellied agouti (A^w) reveals alternative forms of agouti transcripts. *Genes Dev.* **8,** 481–490.
115. Kwon, H. Y., Bultman, S. J., Loffler, C., Chen, W.–J., Furdon, P. J., Powell, J. G., Usala, A.–L., Wilkison, W., Hansmann, I., and Woychik, R. P. (1994) Molecular

structure and chromosomal mapping of the human homolog of the agouti gene. *Proc. Natl. Acad. Sci., U. S. A.* **91,** 9760–9764.

116. Wilson, B. D., Ollmann, M. M., Kang, L., Stoffel, M., Bell, G. I., and Barsh, G. S. (1995) Structure and function of ASP, the human human homnologue of the mouse agouti gene. *Hum. Mol. Genet.* **4,** 223–230.
117. Johnson, R. A. and Salomon, Y. (1991) Assay of adenylyl cyclase catalytic activity. *Methods Enzymol.* **195,** 3–21.
118. Chen, W., Shields, T. S., Stork, P. J. S., and Cone, R. D. (1995) A colorimetric assay for measuring activation of Gs and Gq coupled signaling pathways. *Anal. Biochem.* **226,** 349–354.
119. Gunn, T. M., Miller, K. A., He, L., Hyman, R. W., Davis, R. W., Azarani, A., Schlossman, S. F., Duke-Cohan, J. S. and Barsh, G. S. (1999) The mouse mahogany locus encodes a transmembrane form of human attractin. *Nature* **398,** 1521–1526.
120. Nagle, D. L., McGrail, S. H., Vitale, J., Woolf, E. Z., Dussault, B. J., Jr., DiRocco, L., Holmgren, L., Montagno, J., Bork, P., Huszar, D., Fairchild-Huntress, V., Ge, P., Keilty, J., Ebeling, C., Baldini, L., Gilchrist, J., Burn, P., Carlson, G. A., and Moore, K. J. (1999) The mahogany protein is a receptor involved in suppression of obesity. *Nature* **398,** 148–152.

CHAPTER 11

The Human Melanocortin-1 Receptor

Eugene Healy, Mark Birch-Machin, and Jonathan L. Rees

1. Introduction

One of the more obvious features that distinguishes one human from another are the pigmentatory characteristics (including skin type, hair and eye color) of the individual. Although it had been suspected (as a result of investigations into murine coat color) that several genes were likely to be involved in human pigmentation, and, although it had been known for some time that proopiomelanocortin (POMC) peptides such as alpha-melanocyte stimulating hormone (α-MSH) and adrenocorticotropic hormone (ACTH) can alter cutaneous pigmentation, it has only been during the past 10 years that molecular biologic/genetic approaches have offered some insight into the complexities of human pigmentation *(1–3)*. The detection of mutations within the genes responsible for type I and type II oculocutaneous albinism and piebaldism provided evidence for genotypic/phenotypic relationships in a subset of individuals with pigmentatory disorders, but did little to explain the wide variability in the pigmentatory characteristics of the vast majority of individuals *(4–6)*. However, a basis for understanding ''normal'' human pigmentation became possible with the initial cloning of the human melanocortin 1 receptor (*MC1R*) gene by three separate groups who isolated this gene on the basis of its similarity to other G protein-coupled receptors *(7,8)*, and the subsequent identification of variant alleles within the murine homolog of this gene (*mc1r*) which could differentially activate adenylyl cyclase and which were associated with various coat colors in mice *(9)*.

2. Expression of *MC1R*

The human *MC1R* gene is an intronless gene which has been mapped to chromosome 16q24.3, and which encodes for a seven pass transmembrane

The Melanocortin Receptors Ed.: R. D. Cone

G protein-coupled receptor of 317 amino acids *(7,8,10)*. The human receptor is closely homologous at the amino acid sequence level to that in other mammals and in the chicken, consistent with conservation of function between species *(8,11–15)*. Expression studies originally indicated that transcription of *MC1R* is predominantly confined to certain cell types within the skin and the central nervous system, but recent investigations have suggested that the gene may be transcribed in a wider variety of tissues *(36)*. Previous work had initially shown that α-MSH had effects on mouse melanoma cells and subsequently on human melanoma cells in vitro, suggesting that these cells contained a receptor for this hormone *(16,17)*. Tatro et al. *(18)* later demonstrated that binding sites for [^{125}I]-Nle^{4}-D-Phe^{7}-α-MSH existed on melanoma cells in vivo, and following the cloning of the gene, *MC1R* transcripts were identified in human melanoma cell lines by Northern blot analysis *(7,8)*. However, the number of *MC1R* transcripts and of cell membrane *MC1R* receptors varies greatly between different melanoma cell lines *(19–21)*. Low levels of *MC1R* mRNA have also been detected in cultured human melanocytes by Northern hybridisation and reverse transcriptase polymerase chain reaction (RT-PCR) *(19,22)*; *(Healy, Birch-Machin, and Rees, unpublished observations)*; although Cone et al. *(19)* detected two *MC1R* mRNA species of approximately 3 kb and 4 kb in primary human melanocyte cultures, Suzuki et al. *(22)* could only detect a single 3-kb mRNA species in several human melanocyte strains.

In addition to α-MSH binding to melanocytes in vivo, there is evidence that the hormone also binds to immortalized and normal keratinocytes *(23,24)*, and low levels of *MC1R* mRNA have been detected in human keratinocyte cell lines and normal cultured human keratinocytes by RT-PCR *(25)* *(Healy, Birch-Machin, and Rees, unpublished observations)*. The receptor protein has also been identified in melanoma by immunohistochemistry employing a polyclonal antibody, but this antibody did not detect the receptor on normal keratinocytes and melanocytes *(26)*. Preliminary results from our laboratory using *in situ* hybridization on human skin suggests that the expression of *MC1R* is greater in follicular melanocytes than in interfollicular melanocytes. Expression of the murine *mc1r* on Cloudman S91 melanoma cells is upregulated by ultraviolet (UV) radiation, and UV radiation similarly increases the number of *MC1R* receptors in immortalised human epidermal keratinocytes; despite the upregulation of *MC1R* following exposure to UV radiation, the relevance of this pathway to UV-induced pigmentation in human skin is unknown, with some evidence that other pathways may be more important *(23,27–31)*. *MC1R* mRNA has also been found in a human monocyte/macrophage cell line, in human microvascular endothelial cells *(32–34)*, and low levels of transcripts and receptor in the periaquaductal gray matter in the brain by *in situ* hybridization and immunohistochemistry. Although the receptor is also highly expressed in the bovine testis, *MC1R* transcripts have not been detected in human testis by Northern hybridization *(78,35)*, but Chhajlani *(36)* using RT-PCR followed by detection with a radioactive probe has suggested that the mRNA is present in testis as well as in several other tissues including pituitary, adrenal gland, uterus, ovary, placenta, spleen, lymph node, leukocytes, and lung.

3. Ligand Binding to *MC1R*

Transfection of the human *MC1R* into cultured cells has facilitated investigations on ligand binding to the receptor. Whereas the activation of adenylyl cyclase and levels of cAMP have been used as endpoints for investigations on the murine *mc1r*, most studies to date on the human *MC1R* have relied on ligand binding/displacement of a radiolabeled MSH analog alone *(7–9,22,37)*. Several ligands (both natural and synthetic) are capable of binding to the receptor, with Nle^4-D-Phe^7-α-MSH > α-MSH > β-MSH > γ-MSH in order of potency; ACTH has been reported to bind with a similar potency to α-MSH, but Schioth et al. *(37)* point out that this may be due to degradation of ACTH to α-MSH in the binding assay, because the presence of phosphoramidon in the assay reduces the affinity of ACTH for the receptor *(7,22,37)*. Although the transfection experiments provide an opportunity to investigate several aspects of ligand receptor interactions, the conclusions that can be drawn from transfection studies, especially with regards to cutaneous pigmentation, are limited for the following reasons. First, POMC and its breakdown products are produced in human skin, but there is evidence that the relative amounts of the various POMC breakdown products may differ between melanocytes and keratinocytes in vivo, and the relative amounts of ligand available for binding and activation of the receptor in vivo may differ to the concentrations used in the in vitro studies *(38–42)*. Second, based on adenylyl cyclase activation and bioavailability of ACTH[1–17], it is possible that this may be the more relevant ligand in vivo, at least for *MC1R* on melanocytes *(42)*. Third, α-MSH may stimulate human pigmentation via other pathways, for example via protein kinase C, in addition to the adenylyl cyclase pathway *(43–45)*. Fourth, there are potential protein kinase A and protein kinase C phosphorylation sites on the *MC1R*, but the effect of phosphorylation on the subsequent signal transduction, and whether this alters ligand binding is not known *(21)*.

4. Effects of α-MSH on Human Pigmentation

There are two types of pigment in human (and mammalian) hair and skin that account for their visible coloring, eumelanin (black/brown) and phaeomelanin (red/yellow) *(46,47)*. Although it had been recognized early on (from the numerous pigmented strains of mice gathered together by mouse fanciers) that many genes were involved in cutaneous pigmentation, the relative insolubility and difficulties with purification of the natural melanins prohibited their complete characterization and the elucidation of the enzymatic pathways involved in their synthesis for a long time *(3,46)*. Following the discovery in 1961 by Lerner and McGuire that purified α-MSH could alter cutaneous pigmentation in humans, numerous studies have been carried out to investigate the effects of this hormone on the behavior of melanoma cells and melanocytes with regards to dendricity, growth, attachment to extracellular matrix, and the production of pigment. However, whereas α-MSH has effects under certain conditions in vitro on the proliferation and attachment of pigment cells, the relevance of α-MSH in the control of growth and attachment of normal melanocytes in human skin in vivo is not known *(20,48,49)*. By contrast, α-MSH has effects on human pigmentation in vivo, but the serum levels of the hormone do not vary greatly between individuals of different skin types, suggesting that the hormone itself may not be a physiologic determinant of human pigmentation *(50)*. Despite this, the identification of several of the key enzymes (and their corresponding genes) which are involved in melanogenesis, including the genes for tyrosinase, tyrosinase-related protein 1/5,6-dihydroxyindole-2-carboxylic acid (TRP-1/DHICA) oxidase, and TRP-2/dopachrome tautomerase, has allowed for a better (if still incomplete) understanding of the pathway(s) through which α-MSH has its effects *(51–53)*.

4.1. *Signaling Pathway of* MC1R

The binding of α-MSH to *MC1R* initially causes activation of the relevant heterotrimeric $G\alpha_s$-protein, which in turn activates adenylyl cyclase resulting in an increase in intracellular cAMP *(7,8)*. Although at present the mechanism through which the increased intracellular cAMP causes upregulation of tyrosinase is not entirely known, there is evidence from murine B16 melanoma cells that the higher concentrations of cAMP permit binding of the microphthalmia protein to the promotor region of the tyrosinase gene resulting in transcription of tyrosinase *(54)*; *MITF*, the human homolog of microphthalmia, has also been shown to be capable of binding to and upregulating the tyrosinase promoter *(55)*. Whether α-MSH also causes an increase in the levels of TRP-1 and/or TRP-2 mRNA in murine and human pigment cells is not entirely clear, but there is evidence that α-MSH alters posttranscriptional events which result in increased expression of tyrosinase, TRP-1 and TRP-2 proteins, with the upregulation of these three proteins by α-MSH responsible for the preferential production of eumelanin over pheomelanin *(44,56,57)*. However, this is unlikely to be the complete story, because several groups have reported on the stimulation of protein kinase C activity by α-MSH, and work by Park and colleagues has shown that the abrogation of protein kinase C upregulation by α-MSH in the S91 Cloudman mouse melanoma model does not affect the activity of adenylyl cyclase yet pigmentation is abolished *(43–45)*. There is evidence that another human melanocortin receptor acts via more than one intracellular signaling pathway, in that the human MC3R is coupled to both cAMP and inositol phospholipid / Ca^{2+}-mediated postreceptor signaling systems *(58)*.

5. Evolutionary and Physiologic Aspects of Pigmentation in Humans

There has been some debate on the evolutionary and physiologic relevance of human interfollicular pigmentation, with support for a protective role against the damaging effects of ultraviolet radiation coming from studies on sunburn and skin cancer. At first glance, the higher incidence of melanoma and nonmelanoma skin cancer in white Caucasians in comparison with more pigmented races seems a convincing argument *(59–61)*, and the increased development of squamous cell carcinomas and precursor lesions in albinos in Africa during childhood and early adulthood supports the argument for the sun-protective aspect of melanin against nonmelanoma skin cancer, but the relatively infrequent development of melanoma in this same population raises questions about the protection by pigment against cutaneous melanoma *(62–64)*. The higher doses of ultraviolet B required to produce sunburn in more pigmented individuals also lends support for protection by melanin, and this may be a more physiologically relevant endpoint because intense sunburn would have a more acute effect on viability as a result of fluid and electrolyte imbalance and secondary infection *(65)*. The distribution of more fair-skinned individuals in Northern Europe is thought to have resulted from the dependence of cutaneous vitamin D metabolism on ambient ultraviolet radiation; in areas of low sunshine the development of rickets (with consequences such as deformation of the pelvic bones, and problems during childbirth (with effects on population survival)) is more likely in individuals whose diet is poor in vitamin D, such as the cereal-based diet of the ancestors of Northern European populations *(66,67)*. By contrast, this problem seems to have been avoided by the Eskimos, whose skin is more pigmented, because of their dependence on a fish-based diet, which includes a greater provision of fat-soluble vitamins (including vitamin D). On the other hand, what is the biologic relevance of hair and eye color, and why does secondary sexual hair often differ in color from scalp hair? Is attraction between the sexes (and the resulting reproductive advantage) a sufficient reason, or is it the fact that alterations in hair color do not have more far-reaching biologic consequences permissive enough to allow propagation of this trait?

6. *MC1R* Variants and Human Pigmentation

Although human pigmentation is a complex issue, the cloning of the *MC1R* gene has provided a genetic handle by which various aspects of human pigmentation can be more readily investigated *(7,8,10)*. The initial identification by Robbins et al. *(9)* that *recessive yellow*, *tobacco,* and *sombre* mice contained sequence alterations in the murine *mc1r* gene, and that these alterations in the *mc1r* were likely to be responsible for the different coat colors of these mice because of their reduced or increased ability to activate adenylyl cyclase, provided evidence for the importance of this gene in regulating mammalian pigmentation. In order to investigate whether the human *MC1R* gene had a similar responsibility in the physiologic control of human pigmentation, we initially investigated for alterations of this gene in 30 unrelated Caucasian individuals with red hair (who also tanned poorly) and for comparison in 30 unrelated Caucasian individuals with dark hair who tanned well on exposure to ultraviolet radiation *(68)*. This initial comparison of these more extreme Caucasian phenotypes allowed the identification of a high frequency (70%) of *MC1R* variants in the red-headed subjects, whereas none of the dark-headed group contained a variant. Of the 30 individuals with red hair, 8 contained more than one variant, and cloning of the PCR products followed by sequencing of the clones showed that seven of these were compound heterozygotes. Extension of the study to include individuals with an intermediate phenotype demonstrated that variants were not restricted to subjects with red hair, with germline *MC1R* variants present in 33% of fair/blonde, 10% of brown, 11.5% of black, 22% of auburn and 82% of light-red/deep-red haired individuals, however, everyone with variants on both *MC1R* alleles (i.e., homozygotes or compound heterozygotes) had red hair. In addition, *MC1R* variants were almost always confined to people with fair skin type (59.7% of skin types I and II) who either did not tan or tanned poorly following exposure to ultraviolet radiation, whereas only two individuals with darker skin type (3.4% of skin type III and IV) contained variants. All of the variants that had been identified in the 21 individuals of the original 30 subjects with red hair clustered in and around the second transmembrane and in the seventh transmembrane domains; this prompted us to concentrate on these two areas of the gene when investigating the subjects with intermediate pigmentation, however, subsequent work by our group and others has indicated that the area adjacent to and within the second intracellular loop is also a variant hot spot *(69,94)*. *See* Figs. 1 and 2.

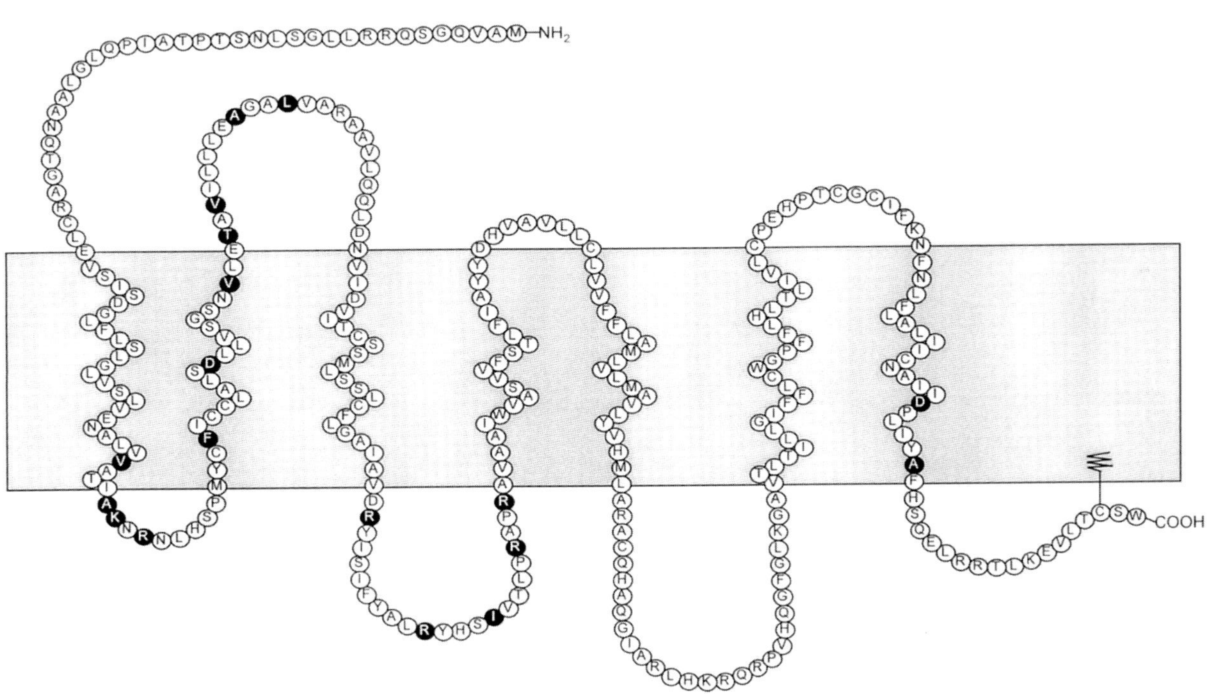

Fig. 1. *MC1R* variants (represented by black circles) identified to date are taken from refs. *68,69,79*, and *Smith, unpublished observations*. The cell membrane is depicted by the shaded area.

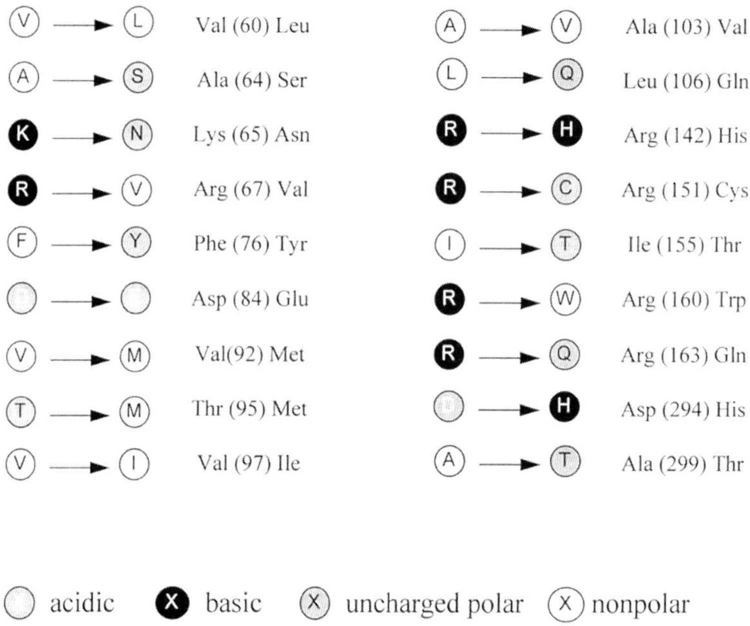

Fig. 2. *MC1R* variants identified to date with an emphasis on the biochemical nature of the amino acid changes. *MC1R* variants taken from refs. *68,69,79*, and *Smith et al., manuscript in preparation.*

The results of this study suggest that the issue of human pigmentation is perhaps more complex than the situation in several animals *(11–15)*. The fact that all subjects with two *MC1R* variant alleles were red-headed would be consistent with inactivation of both copies of this gene being aetiologically associated with red hair in humans, which would be similar to the situation in the *recessive yellow* mouse where the *mc1r* is homozygous for a frameshift mutation that produces a prematurely terminated nonfunctioning receptor *(9)*; yellow mice might seem to be more akin to blonde-haired humans, but analysis of the melanin content has shown that the hair of yellow mice is similar to human red hair, with both containing predominantly pheomelanin pigment *(70)*. On the other hand, no variants were detected in a number of subjects with red hair, and indeed the majority of red-headed subjects contained only one variant allele. It is possible that these individuals contained alterations outside the coding region, perhaps affecting gene expression, but at present this remains speculative. It is also possible that not all cases of red hair are due to alterations in the *MC1R*, and that other genes involved in this pathway are responsible in certain cases. Agouti (an antagonist of *MC1R*) is one such candidate, however, Barsh *(71)* has argued that mutations in agouti are unlikely to account for red hair because mutations are unusual in genes encoding for a ligand. Equally important is the situation regarding the association of *MC1R* variant alleles with other hair colours. Of course, all *MC1R* variants may not be equal in their effects on pigmentation, and some variants may be neutral polymorphisms, such that the receptor still functions adequately and maintains the drive toward eumelanin synthesis. Conversely, as in the case of the sombre and tobacco mice, some variants may constitutively activate the receptor *(9)*. Yet, the relation between the *MC1R* gene and hair and skin pigmentation is not straightforward, as can be evidenced from the case of black-haired fair-skinned individuals who are commonplace in certain populations such as in Ireland *(72)*.

6.1. Assessment of the Pigmentation Phenotype

Difficulties in explaining the exact relationship between pigmentation and *MC1R* variants may also arise from problems inherent with the clinical determination/validity of skin type. The skin typing system as put forward by Fitzpatrick was initially devised to predict the likelihood of burning from psoralen plus ultraviolet A (PUVA) therapy for psoriasis *(73)*. Although an improvement on previous predictors for burning following the administration of PUVA, the Fitzpatrick classification (which relies on a subjective patient history) is a crude measure of two separate responses to a single semistandardized dose of UVR. Both the erythemal and pigmentation responses are combined into single skin type categories, and not all individuals will fall neatly into the proposed groups *(74)*. For the purpose of assessing the role of *MC1R* in skin pigmentation, it might be preferable to employ the ultraviolet-induced tanning response alone following chronic exposure to sunlight (according to the subject's history) or following quantitated chronic exposure to artificial ultraviolet radiation sources in the investigator's institution; temporal and financial constraints with the use of artificial ultraviolet radiation sources would prohibit investigations on large groups of individuals. Neither is the assessment of hair color without its difficulties. Hair color changes throughout life, and secondary sexual hair is often different in color to that on the scalp. In addition, hair color is in reality part of a continuum, and although investigators are likely to agree on the extremes of hair color (e.g., red and black), where does "fair" end and "brown" begin, and should "strawberry blonde" be included under red or blonde? On a molecular basis, problems also exist because hair colors do not simply contain eumelanin or pheomelanin, but are generally a mixture of both pigments *(70,75)*.

6.2. Transfection Experiments in the Analysis of the Function of MC1R Variant Alleles

Transfection of the human wild-type and variant *MC1R* alleles into COS and HEK-293 cells is likely to aid our understanding of the cellular effects of variant alleles, but as mentioned above may be limited by both the ligand used and the assumption that activation of adenylyl cylcase is a suitable assay for predicting the likely effects on eumelanin/pheomelanin synthesis. Investigators studying the function of the *MC1R* receptor have constructed mutant receptors by site-directed mutagenesis, and have identified certain amino acids, including D117A and H260A, which are important in ligand binding *(76)*, but little work has been carried out to date on the variants which are present in vivo in humans. However, Xu et al. *(77)* who detected the Val92Met variant in 7 of 11 individuals with skin type 1, have reported that α-MSH has approximately five times lower potency in displacing a radiolabeled analogue of α-MSH from the Val92Met variant receptor transfected into COS-1 cells as compared to wild-type receptor. By contrast, Koppula et al. *(78)* found no pharmacologic consequences of this polymorphism, and further investigations on the presence of variants in different populations and in individuals with different skin type suggests that the Val92Met variant is likely to be a neutral polymorphism *(69,79) (Healy, Birch-Machin, and Rees, unpublished observations)*. Evidence for the fact that the wild-type human *MC1R* gene is important in the control of pigmentation has been provided by Chluba-de Tapia et al. *(80)* who transfected the gene into amelanotic mouse melanoma cells that did not express the murine *mc1r*. In this system melanogenesis occurred without the addition of exogenous α-MSH, suggesting that the *MC1R* receptor was constitutionally active, although Loir et al. *(81)* have more recently reported on a role for this receptor in the release of α-MSH from melanoma cells, making it possible that constitutional pigmentation in the transfected mouse melanoma cells was due to an autocrine effect. Despite this and the fact that amelanotic mouse melanoma cells are obviously atypical, this might be a preferable system in which to investigate the functional activity of human *MC1R* variants. Hunt et al. *(82)* have also reported on the unresponsiveness of cultured human epidermal melanocytes from individuals with red hair to MSH, suggesting that the *MC1R* receptor in redheads is functionally compromised, but it is not known whether these melanocytes were from individuals with variant *MC1R* alleles.

7. *MC1R* Variants in Celtic Individuals

The association of *MC1R* variants with red hair and fair skin was also examined in a randomly selected group of individuals from a Celtic population, in which these phenotypic characteristics are more frequently observed *(72)*. In addition, several individuals in this group had the classical Celtic phenotype [according to Beirn et al. *(72)*] with dark brown/black hair and fair skin type. In the overall population, 75% of people had a variant *MC1R* allele, with 38% of these (28% of total) subjects containing two or more variants (including some individuals homozygous for *MC1R* variants) *(Smith, unpublished observations)*. Three variants in particular (Arg151Cys, Arg160Trp, and Asp294His) showed an association with red hair, whereas the associations with fair skin type were strongest for Arg151Cys and Arg160Trp. The same three variants are associated with red hair in an Australian twin pair study, and Box et al. *(69)* have commented on an association between Val60Leu and fair/blonde/light brown hair in the same group of Australians, and in the Irish population, a similar association was also observed with the Val60Leu variant *(Smith, unpublished observations)*. Interestingly, a greater number of Irish individuals with darker skin type contained a variant than was detected in our original study *(68)*.

8. Inheritence of Red Hair

The aetiological association of certain *MC1R* variants (especially the Arg151Cys, Arg160Trp and Asp294His variants) with red hair has implications for the study of the inheritance of this trait, and indeed family studies may provide further information on the likely function of these variants. Previous studies on the inheritance of red hair have shown that, although it can segregate as a simple mendelian recessive trait with variable expression, this does not seem to be the case for many families *(83,84)*. Linkage with the MNS locus on chromosome 4 has also been documented in certain families, in accordance with other genes being important in this phenotype *(84)*. Box et al. *(69)* in looking at the *MC1R* gene in Australian red-headed twins have reported that variants in this gene are necessary but not always sufficient for the production of red hair, presumably because of the existence of additional modifier genes. We have gone on to look at the association of red hair with variants in kindreds with a predominance of red hair, and have identified several families where the vast majority of red-headed individuals have inherited two variant alleles, whereas almost all subjects with other hair colours have not (our unpublished results), suggesting that in these families red hair is inherited as a recessive trait.

9. *MC1R* Variants and Skin Cancer

Despite the limitations in our understanding of the association of *MC1R* variants with different pigmentary characteristics, investigations on this gene may provide information on which individuals in the general population are prone to the development of cutaneous neoplasia. The reported incidence of skin cancer is increasing worldwide, with substantial evidence that ultraviolet radiation exposure is a major etiologic factor *(85)*. The pigmentation phenotype of the individual has been identified as an important risk factor for the development of both melanoma and nonmelanoma skin cancer, although the exact mechanism of the greater susceptibility to these tumors in Caucasians as compared with individuals of asian or african descent is not entirely clear *(59–61)*. The more widely accepted viewpoint is that (eu)melanin is photoprotective, and by absorbing incident photons prevents damage to DNA and subsequent mutation of relevant oncogenes and tumor suppressor genes in melanocytes and keratinocytes. This is likely to be the case for squamous cell carcinoma of the skin because African albinos (with little or no pigment) develop this tumour frequently and at an early age, but the fact that this group of people seldom develop melanoma suggests that protection by (eu)melanin is not the complete story *(62,64)*. Conversely, some investigators have documented the increased ability of pheomelanin to generate free radicals in vitro following exposure to ultraviolet radiation, and have proposed that this phenomenon may occur in vivo and might account for the fact that Caucasians (whose skin generally contains a higher phaeomelanin/eumelanin ratio) are more susceptible to skin cancer *(86–88)*. It could be argued that the more costly and time-consuming investigation for *MC1R* variants in order to determine which individuals are susceptible to skin tumours may not offer information over that which can be readily obtained clinically from the assessment of pigmentation, including skin type, eye and hair color, and the presence of freckles. However, not all fair-skinned subjects develop skin cancer, and cutaneous neoplasms also arise in a significant number of more pigmented individuals (i.e., skin types III and IV). Furthermore, in the case of melanoma which can produce α-MSH, the addition of this hormone to cultured melanoma cells and melanocytes in vitro can stimulate proliferation and can promote the ability of melanoma cells to metastasise following transfer of the cells into mice *(20,48,89,90)*. For these reasons we investigated for *MC1R* variants in a group of individuals with sporadic melanoma and in a group of subjects with psoriasis for comparison; the control group of psoriatics was chosen because there is no association between psoriasis and skin type, and no increased risk of melanoma in patients with psoriasis (apart from the possible risk attributable to long-term treatment with PUVA) *(79,91,92)*.

In this study, because of the previous observed clustering of *MC1R* variants around the second and seventh transmembrane domains, we concentrated on these two areas of the gene, and detected a greater number of variants in the subjects with melanoma than in the control group. Although, as expected, there was a mild preponderence of individuals with fair skin type in the melanoma group, the association of variants with melanoma (relative risk 3.91, 95% confidence intervals 1.48 – 10.35) was largely independent of skin type, suggesting that investigation for *MC1R* variants might offer an advantage over conventional examination of pigment phenotype for the identification of individuals susceptible to sporadic melanoma. The population risk attributable to carriers was 34.6%, consistent with this gene being associated with approximately one third of all melanomas. The Asp84Glu variant was detected in over 20% of melanoma cases, with two individuals homozygous for this alteration, whereas only two of 135 subjects (including 77 with fair skin type) in our previous study contained this variant *(68)*. Although the Asp84Glu variant does not seem to be more frequent in all populations with melanoma *(Healy, Birch-Machin, and Rees, unpublished results)*, at present it still seems likely that the presence of this variant does convey a increased risk for the development of melanoma. Future work may identify whether other variants similarly are a risk factor for melanoma, and what is the mechanism of the association between *MC1R* variants and melanoma, that is whether the protective effects of eumelanin, the potentially damaging effects of phaeomelanin, or the mitogenic/metastatic effects of α-MSH are responsible. In addition, attempts at the development of chemotherapeutic agents for melanoma which are bound to *MC1R* receptor ligands (in order to make the drug target the melanoma cells specifically) will require investigations on their ability to bind not only to the wild-type receptor but also to the variant receptors that are present in individuals with melanoma *(24)*.

10. Function of *MC1R* in Other Cell Types

MC1R is also expressed in other human cell types as well as in pigment cells. The function of this receptor in keratinocytes remains unknown, although it is possible that binding by ligand to the *MC1R* receptor may be part of an autocrine loop which further stimulates the production of proopiomelanocortin similar to that observed in pigmented cells *(81)*. Although speculative, the receptor in keratinocytes, as well as in monocytes/macrophages and microvascular endothelial cells, may function as part of the cutaneous immune system, or in the case of monocytes and endothelial cells in inflammation in general. α-MSH has been shown to modulate contact hypersensitivity responsiveness in mice, and stimulation of the *MC1R* receptor on human microvascular endothelial cells in vitro results in the release of interleukin-8, whereas α-MSH inhibits the production of neopterin by a human monocyte/macrophage cell line in vitro following stimulation with interferon-γ and tumor necrosis factor-α *(32,33,93)*. The function of the restricted *MC1R* expression in the periaquaductal gray has not yet been investigated, and the relevance of its expression in other tissues including testis, ovary, adrenal gland, is unknown *(34)*.

11. Conclusion

Similar to the case in other animals and birds, the *MC1R* receptor is an important determinant of human pigmentation. Future work will help establish which *MC1R* variants are functionally relevant, and the mechanism by which variants alter hair color and / or skin type (i.e., through altered ligand binding or altered activation of the intracellular signaling pathway). Variants also seem to convey a risk for the development of cutaneous melanoma, but whether this is via their effects on pigmentation or through effects of the MSH signaling pathway on proliferation of melanocytes and melanoma cells requires further investigation.

References

1. Nelson, D. H., Meakin, J. W., and Thorn, G. W. (1960) ACTH–producing pituitary tumors following adrenalectomy for Cushing's syndrome. *Ann. Intern. Med.* **52,** 561–569.
2. Lerner, A. B. and McGuire, J. S. (1961) Effect of alpha– and beta–melanocyte stimulating hormones on the skin colour of man. *Nature* **189,** 176–179.
3. Jackson, I. J. (1991) Mouse coat colour mutations: a molecular genetic resource which spans the centuries. *BioEssays* **13,** 439–446.
4. Tomita, Y., Takeda, A., Okinaga, S., Tagami, H., and Shibahara, S. (1989) Human oculocutaneous albinism caused by single base insertion in the tyrosinase gene. *Biochem. Biophys. Res. Commun.* **164,** 990–996.
5. Giebel, L. B. and Spritz, R. A. (1991) Mutation of the KIT (mast / stem cell growth factor receptor) protooncogene in human piebaldism. *Proc. Natl. Acad. Sci. U. S. A.* **88,** 8696–8699.
6. Rinchik, E. M., Bultman, S. J., Horsthemke, B., Lee, S. –T., Strunk, K. M., Spritz, R. A., Avidano, K. M., Jong, M. T. C., and Nicholls, R. D. (1993) A gene for the mouse pink–eyed dilution locus and for human type II oculocutaneous albinism. *Nature* **361,** 72–76.
7. Chhajlani, V. and Wikberg, J. E. S. (1992) Molecular cloning and expression of the human melanocyte stimulating hormone receptor cDNA. *FEBS Lett.* **309,** 417–420.
8. Mountjoy, K. G., Robbins, L. S., Mortrud, M. T., and Cone, R.D. (1992) The cloning of a family of genes that encode the melanocortin receptors. *Science* **257,** 1248–1251.
9. Robbins, L. S., Nadeau, J. H., Johnson, K. R., Kelly, M. A., Roselli–Rehfuss, L., Baack, E., Mountjoy, K. G., and Cone, R. D. (1993) Pigmentation phenotypes of variant extension locus alleles result from point mutations that alter MSH receptor function. *Cell* **72,** 827–834.
10. Gantz, I., Yamada, T., Tashiro, T., Konda, Y., Shimoto, Y., Miwa, H., and Trent, J. M. (1994) Mapping of the gene encoding the melanocortin–1 (α–melanocyte stimulating hormone) receptor (MC1R) to human chromosome 16q24.3 by fluorescence *in situ* hybridization. *Genomics* **19,** 394–395.
11. Marklund, L., Johansson, M., Sandberg, K., and Andersson, L. (1996) A missense mutation in the gene for melanocyte–stimulating hormone receptor (*MC1R*) is associated with the chestnut coat color in horses. *Mamm. Genome* **7,** 895–899.
12. Klungland, H., Vage, D. I., Gomezraya, L., Adalsteinsson, S., and Lien, S. (1995) The role of melanocyte–stimulating hormone (MSH) receptor in bovine coat colour determination. *Mamm. Genome* **6,** 636–639.
13. Joerg, H., Fries, H. R., Meijerink, E., and Stranzinger, G.F. (1996) Red coat color in holstein cattle is associated with a deletion in the MSHR gene. *Mamm. Genome* **7,** 317–319.

14. Takeuchi, S., Suzuki, H., Yabuuchi, M., and Takahashi, S. (1996) A possible involvement of melanocortin 1–receptor in regulating feather color pigmentation in the chicken. *Biochim. Biophys. Acta* **1308,** 164–168.
15. Vage, D. I., Lu, D., Klungland, H., Lien, S., Adalsteinsson, S., and Cone, R. D. (1997) A non–epistatic interaction of agouti and extension in the fox, *Vulpes vulpes. Nat. Genet.* **15,** 311–315.
16. Wong, G., Pawelek, J., Sansone, M., and Morowitz, J. (1974) Response of mouse melanoma cells to melanocyte stimulating hormone. *Nature* **248,** 351–354.
17. Fuller, B. B. and Meyskens, F. L. Jr. (1981) Endocrine responsiveness in human melanocytes and melanoma cells in culture. *J. Natl. Cancer Inst.* **66,** 799–802.
18. Tatro, J. B., Atkins, M., Mier, J. W., Hardarson, S., Wolfe, H., Smith, T., Entwistle, M. L., and Reichlin, S. (1990) Melanotropin receptors demonstrated in situ in human melanoma. *J. Clin. Invest.* **85,** 1825–1832.
19. Cone, R. D., Mountjoy, K. M., Robbins, L. S., Nadeau, J. H., Johnson, K. R., Roselli–Rehfuss, L., and Mortrud, M. T. (1993) Cloning and functional characterization of a family of receptors for the melanotropic peptides. *Ann. N.Y. Acad. Sci.* **680,** 342–363.
20. Eberle, A. N., Siegrist, W., Bagutti, C., Chluba–de Tapia, J., Solca, F., Wikberg, J. E. S., and Chhajlani, V. (1993) Receptors for melanocyte–stimulating hormone on melanoma cells. *Ann. N.Y. Acad. Sci.* **680,** 320–341.
21. Siegrist, W., Stutz, S., and Eberle, A. N. (1994) Homologous and heterologous regulation of α–melanocyte stimulating hormone receptors in human and mouse melanoma cell lines. *Cancer Res.* **54,** 2604–2610.
22. Suzuki, I., Cone, R. C., Im, S., Nordlund, J., and Abdel–Malek, Z. A. (1996) Binding of melanotropic hormones to the melanocortin receptor MC1R on human melanocytes stimulates proliferation and melanogenesis. *Endocrinology* **137,** 1627–1633.
23. Chakraborty, A. and Pawelek, J. (1993) MSH receptors in immortalized human epidermal keratinocytes: a potential mechanism for coordinate regulation of the epidermal–melanin unit. *J. Cell. Physiol.* **157,** 344–350.
24. Sharma, S. D., Jiang, J., Hadley, M. E., Bentley, D. L., and Hruby, V. J. (1996) Melanotropic peptide–conjugated beads for microscopic visualization and characterization of melanoma melanotropin receptors. *Proc. Natl. Acad. Sci. U. S. A.* **93,** 13715–13720.
25. Bhardwaj, R.S., Becher, E., Mahnke, K., Hartmeyer, M., Scholzen, T., and Luger, T.A. (1996) Evidence of the expression of a functional melanocortin receptor 1 by human keratinocytes. *J. Invest. Dermatol.* **106,** 817.
26. Xia, Y., Skoog, V., Muceniece, R., Chhajlani, V., and Wikberg, J. E. S. (1995a) Polyclonal antibodies against human melanocortin MC_1 receptor: preliminary immunohistochemical localisation of melanocortin MC_1 receptor to malignant melanoma cells. *Eur. J. Pharmacol.* **288,** 277–283.
27. Friedmann, P. S., and Gilchrest, B. A. (1987) Ultraviolet radiation directly induces pigment production by cultured human melanocytes. *J. Cell. Physiol.* **133,** 88–94.
28. Friedmann, P. S., Wren, F., Buffey, J., and MacNeil, S. (1990a) α–MSH causes a small rise in cAMP but has no effect on basal or ultraviolet–stimulated melanogenesis in human melanocytes. *Br. J. Dermatol.* **123,** 145–151.
29. Friedmann, P.S., Wren, F.E., Matthews, J.N. (1990b) Ultraviolet stimulated melanogenesis by human melanocytes is augmented by di–acyl glycerol but not TPA. *J. Cell. Physiol.* **142,** 334–341.

30. Eller, M. S., Yaar, M., Gilchrest, B. A. (1994) DNA damage and melanogenesis. *Science* **372,** 413–414.
31. Chakraborty, A., Slominski, A., Ermak, G., Hwang, J., and Pawelek, J. (1995) Ultraviolet B and melanocyte–stimulating hormone (MSH) stimulate mRNA production for αMSH receptors and proopiomelanocortin–derived peptides in mouse melanoma cells and transformed keratinocytes: *J. Invest. Dermatol.* **105,** 655–659.
32. Rajora, N., Ceriani, G., Catania, A., Star, R. A., Murphy, M. T., and Lipton, J. M. Alpha–MSH production, receptors, and influence on neopterin in a human monocyte/macrophage cell line. *J. Leukocyte Biol.* **59,** 248–253.
33. Hartmeyer, M., Scholzen, T., Becher, E., Bhardwaj, R. S., Fastrich, M., Schwarz, T., and Luger, T. A. Human microvascular endothelial cells (HMEC–1) express the melanocortin receptor type 1 and produce increased levels of IL–8 upon stimulation with α–MSH. *J. Invest.Dermatol.* **106,** 809.
34. Xia, Y., Wikberg, J. E. S., and Chhajlani, V. (1995b) Expression of the melanocortin 1 receptor in periaqueductal grey matter. *Neuroreport* **6,** 2193–2196.
35. Vanetti, M., Schonrock, C., Meyerhof, W., and Hollt, V. (1994) Molecular cloning of a bovine MSH receptor which is highly expressed in the testis. *FEBS Lett.* **348,** 268–272.
36. Chhajlani, V. (1996) Distribution of cDNA for melanocortin receptor subtypes in human tissues. *Biochem. Mol. Biol. Int.* **38,** 73–80.
37. Schioth, H. B., Muceniece, R., Wikberg, J.E.S., and Chhajlani, V. (1995) Characterisation of melanocortin receptor subtypes by radioligand binding analysis. *Eur. J. Pharmacol.* **288,** 311–317.
38. Thody, A. J., Ridley, K., Penny, R. J., Chalmers, R., Fisher, C., and Shuster, S. (1983) MSH peptides are present in mammalian skin. *Peptides* **4,** 813–816.
39. Farooqui, J. Z., Medrano, E. E., Abdel–Malek, Z., and Nordlund, J. (1993) The expression of proopiomelanocortin and various POMC–derived peptides in mouse and human skin. *Ann. N. Y. Acad. Sci.* **680,** 508–510.
40. Slominski, A., Ermak, G., Hwang, J., Chakraborty, A., Mazurkiewicz, J. E., and Mihm, M. (1995) Proopiomelanocortin, corticotropin releasing hormone and corticotropin releasing hormone receptor genes are expressed in human skin. *FEBS Lett.* **374,** 113–116.
41. Wintzen, M., Yaar, M., Burbach, J. P. H., and Gilchrest, B. A. (1996) Proopiomelanocortin gene product regulation in keratinocytes. *J. Invest. Dermatol.* **106,** 673–678.
42. Kazumasa, W., Graham, A., Cook, D., and Thody, A. J. (1997) Characterisation of ACTH peptides in human skin and their activation of the melanocortin–1 receptor. *Pigment Cell Res.* **10,** 288–297.
43. Buffey, J., Thody, A. J., Bleehen, S. S., and MacNeil, S. (1992) α–Melanocyte–stimulating hormone stimulates protein kinase C activity in murine B16 melanoma. *J. Endocrinol.* **133,** 333–340.
44. Kuzumaki, T., Matsuda, A., Wakamatsu, K., Ito, S., and Ishikawa, K. (1993) Eumelanin biosynthesis is regulated by coordinate expression of tyrosinase and tyrosinase–related protein–1 genes. *Exp. Cell Res.* **207,** 33–40.
45. Park, H. –Y., Russakovsky, V., Ao, Y., Fernandez, E., and Gilchrrest, B. A. (1996) α–Melanocyte stimulating hormone–induced pigmentation is blocked by depletion of protein kinase C. *Exp. Cell Res.* **227,** 70–79.
46. Prota, G. and Thomson, R. H. (1976) Melanin pigmentation in mammals. *Endeavour* **35,** 32–38.

48. Halaban, R., Tyrrell, L., Longley, J., Yarden, Y., and Rubin, J. (1993) Pigmentation and proliferation of human melanocytes and the effects of melanocyte–stimulating hormone and ultraviolet B light. *Ann. N. Y. Acad. Sci.* **680,** 290–301.
47. Thody, A. J., Higgins, E. M., Wakamatsu, K., Ito, S., Burchill, S. A., and Marks, J. M. (1991) Pheomelanin as well as eumelanin is present in human epidermis. *J. Invest. Dermatol.* **97,** 340–344.
49. Hunt, G., Donatien, P. D., Cresswell, J. E., and Thody, A. J. (1993) The effect of alpha–MSH on the attachment of human melanocytes to laminin and fibronectin. *Ann. N. Y. Acad. Sci.* **680,** 549–551.
50. Spiro, J., Parker, S., Oliver, I., Fraser, C., Marks, J. M., and Thody, A. J. (1987) Effect of PUVA on plasma and skin immunoreactive alpha–melanocyte stimulating hormone concentrations. *Br. J. Dermatol.* **117,** 703–707.
51. Kwon, B. S., Haq, A. K., Pomerantz, S. H., and Halaban, R. (1987) Isolation and sequence of a cDNA clone for human tyrosinase that maps at the mouse c–albino locus. *Proc. Natl. Acad. Sci. U. S. A.* **84,** 7473–7477.
52. Tsukamoto, K., Jackson, I. J., Urabe, K., Montague, P. M., and Hearing, V. J. (1992) A second tyrosinase–related protein, TRP–2, is a melanogenic enzyme termed DOPAchrome tautomerase. *EMBO J.* **11,** 519–526.
53. Jimenez–Cervantes, C., Solano, F., Kobayashi, T., Urabe, K., Hearing, V. J., Lozano, J. A., and Garcia–Borron, J. C. (1994) A new enzymatic function in the melanogenic pathway. The 5,6–dihydroxyindole–2–carboxylic acid oxidase activity of tyrosinase–related protein–1 (TRP–1). *J. Biol. Chem.* **269,** 17,993–18,000.
54. Bertolotto, C., Bille, K., Ortonne, J.–P., and Ballotti, R. (1996) Regulation of tyrosinase gene expression by cAMP in B16 melanoma cells involves two CATGTG motifs surrounding the TATA box: implication of the microphthalmia gene product. *J. Cell Biol.* **134,** 747–755.
55. Yasumoto, K., Mahalingam, H., Suzuki, H., Yoshizawa, M., Yokoyama, K. (1995) Transcriptional activation of the melanocyte-specific genes by the human homolog of the mouse Microphthalmia protein. *J. Biochem.* **118,** 874–881.
56. Aroca, P., Urabe, K., Kobayashi, T., Tsukamoto, K., and Hearing, V.J. (1993) Melanin biosynthesis patterns following hormonal stimulation. *J. Biol. Chem.* **268,** 25,650–25,655.
57. Hunt, G., Donatien, P. D., Lunec, J., Todd, C., Kyne, S., and Thody, A. J. (1994) Cultured human melanocytes respond to MSH peptides and ACTH. *Pigment Cell Res.* **7,** 217–221.
58. Konda, Y., Gantz, I., DelValle, J., Shimoto, Y., Miwa, H., and Yamada, T. (1994) Interaction of dual intracellular signaling pathways activated by the melanocortin–3 receptor. *J. Biol. Chem.* **269,** 13,162–13,166.
59. Chuang, T. Y., Reizner, G. T., Elpern, D. J., Stone, J. L., and Farmer, E. R. (1995) Nonmelanoma skin cancer in Japanese ethnic Hawaiians in Kauai, Hawaii: an incidence report. *J. Am. Acad. Dermatol.* **33,** 422–426.
60. Elder, D.E. (1995) Skin cancer: melanoma and other specific nonmelanoma skin cancers. *Cancer* **75(Suppl. 1),** 245–256.
61. Halder, R. M. and Bridgeman–Shah, S. (1995) Skin cancer in African Americans. *Cancer* **75(Suppl. 2),** 667–673.
62. Luande, J., Henschke, C. I., and Mohammed, N. (1985) The Tanzanian human albino skin. *Cancer* **55,** 1823–1828.
63. Diffey, B.L., Healy, E., Thody, A.J., and Rees, J.L. (1995) Melanin, melanocytes, and melanoma. *Lancet* **346,** 1713.

64. Lookingbill, D. P., Lookingbill, G. L., and Leppard, B. Actinic damage and skin cnacer in albinos in northern Tanzania: findings in 164 patients enrolled in an outreach skin care program. *J. Am. Acad. Dermatol.* **32,** 653–658.
65. Kaidbey, K. H., Agin, P. P., Sayre, R. M., and Kligman, A. M. (1979) Photoprotection by melanin: a comparison of black and Caucasin skin. *J. Am. Acad. Dermatol.* **1,** 249–260.
66. Bodmer, W. F. and Cavalli–Sforza, L. L. (1976) Racial differentiation, in *Genetics, Evolution and Man.* (Bodmer, W. F. and Cavalli–Sforza, L. L., eds.) Freeman, New York, pp. 559–604.
67. Kingdon, J. (1993) *Self–Made Man and His Undoing.* Simon & Schuster, London.
68. Valverde, P., Healy, E., Jackson, I., Rees, J. L., and Thody, A. J. (1995) Variants of the melanocyte–stimulating hormone receptor gene are associated with red hair and fair skin in humans: *Nat. Genet.* **11,** 328–330.
69. Box, N. F., Wyeth, J. R., O'Gorman, L. E., Martin, N. G., and Sturm, R. A. (1997) Characterisation of melanocyte–stimulating hormone receptor variant alleles in twins with red hair. *Hum. Mol. Genet.* **11,** 1891–1897.
70. Prota, G., Lamoreux, M. L., Muller, J., Kobayashi, T., Napolitano, A., Vincensi, M. R., Sakai, C., and Hearing, V. J. (1995) Comparative analysis of melanins and melanosomes produced by various coat color mutants. *Pigment Cell Res.* **8,** 153–163.
71. Barsh, G.S. (1996) The genetics of pigmentation: from fancy genes to complex traits. *Trends Genet.* **12,** 299–305.
72. Beirn, S. F., Judge, P., Urbach, F., MacCon, C. F., and Martin, F. (1970) Skin cancer in County Galway, Ireland. *Proc. Natl. Cancer Conf.* **6,** 489–500.
73. Fitzpatrick, T.B. (1988) The validity and practicality of sun–reactive skin types I through VI. *Arch. Dermatol.* **124,** 869–871.
74. Rampen, F. H., Fleuren, B. A., de Boo, T. M., and Lemmens, W. A. (1988) Unreliability of self–reported burning tendency and tanning ability. *Arch. Dermatol.* **124,** 885–888.
75. Jimbow, K., Ishida, O., Ito, S., Hori, Y., Witkop, C.J., and King, R.A. (1983) Combined chemical and electron microscopic studies of pheomelanosomes in human red hair. *J. Invest. Dermatol.* **81,** 506–511.
76. Schioth, H. B., Muceniece, R., Szardenings, M., Prusis, P., Lindeberg, G., Sharma, S. D., Hruby, V. J., and Wikberg, J. E. S. (1997) Characterisation of D117A and H260A mutations in the melanocortin 1 receptor. *Mol. Cell. Endocrinol.* **126,** 213–219.
77. Xu, X., Thornwall, M., Luhdin, L. G., Chhajlani, V. (1996) Val92Met variant of the melanocyte stimulating hormone receptor gene. *Nat. Genet.* **14,** 384.
78. Koppula, S. V., Robbins, L. S., Lu, D., Baack, E., White, C. R.Jr., Swanson, N. A., and Cone, R. D. (1997) Identification of common polymorhpisms in the coding sequence of the human MSH receptor (MC1R) with possible biological effects. *Hum. Mutat.* **9,** 30–36.
79. Valverde, P., Healy, E., Sikkink, S., Haldane, F., Thody, A. J., Carothers, A., Jackson, I. J., and Rees, J. L. (1996) The Asp84Glu variant of the melanocortin 1 receptor (*MC1R*) is associated with melanoma. *Hum. Mol. Genet.* **5,** 1663–1666.
80. Chluba–de Tapia, J., Bagutti, C., Cotti, R., and Eberle, A. N. (1996) Induction of constitutive melanogenesis in amelanotic mouse melanoma cells by transfection of the human melanocortin–1 receptor gene. *J. Cell. Sci.* **109,** 2023–2030.

81. Loir, B., Bouchard, B., Morandini, R., Del Marmol, V., Deraemaecker, R., Garcia–Borron, J. C., and Ghanem G. (1997) Immunoreactive α–melanotropin as an autocrine effector in human melanoma cells. *Eur. J. Biochem.* **244,** 923–930.
82. Hunt, G., Todd, C., and Thody, A. J. (1996) Unresponsiveness of human epidermal melanocytes to melanocyte–stimulating hormone and its association with red hair. *Mol. Cell. Endocrinol.* **116,** 131–136.
83. Singleton, W. R. and Ellis, B. (1964) Inheritance of red hair for six generations. *J. Hered.* **55,** 261 + 266.
84. Eiberg, H. and Mohr, J. (1987) Major locus for red hair color linked to MNS blood groups on chromosome 4. *Clin. Genet.* **32,** 125–128.
85. Armstrong, B. K. and Kricker, A. (1996) Epidemiology of sun exposure and skin cancer, in *Cancer Surveys* Skin Cancer, Vol. 26 (Leigh, I. M., Newton Bishop, J. A., and Kripke, M. L., eds.). Cold Spring Harbor Laboratory Press, New York, pp. 133–153.
86. Menon, I. A., Persad, S., Haberman, H. F., and Kurian, C. J. (1983) A comparative study of the physical and chemical properties of melanins isolated from human black and red hair. *J. Invest. Dermatol.* **80,** 202–206.
87. Persad, S., Menon, I. A., and Haberman, H. F. (1983) Comparison of the effects of UV–visible irradiation of melanins and melanin–hematoporphyrin complexes from human black and red hair. *Photochem. Photobiol.* **37,** 63–68.
88. Hunt, G., Kyne, S., Ito, S., Wakamatsu, K., Todd, C., and Thody, A. J. (1995) Eumelanin and phaeomelanin contents of human epidermis and cultured melanocytes. *Pigment Cell Res.* **8,** 202–208.
89. Lunec, J., Pieron, C., Sherbet, G.V., and Thody, A.J. (1990) Alpha–melanocyte–stimulating hormone immunoreactivity in melanoma cells. *Pathobiology* **58,** 193–197.
90. Lunec, J., Pieron, C., and Thody, A.J. (1992) MSH receptor expression and the relationship to melanogenesis and metastatic activity in B16 melanoma. *Melanoma Res.* **2,** 5–12.
91. Bhate, S. M., Sharpe, G. R., Marks, J. M., Shuster, S., Ross, W.M. (1993) Prevalence of skin and other cancers in patients with psoriasis. *Clin. Exp. Dermatol.* **18,** 401–404.
92. Stern, R. S., Nichols, K. T., and Vakeva, L. H. (1997) Malignant melanoma in patients treated for psoriasis with methoxsalen (psoralen) and ultraviolet A radiation (PUVA). *N. Engl. J. Med.* **336,** 1041–1045.
93. Saunder, D. N., and Nordlund, J. J. (1989) Alpha–melanocyte stimulating hormone modulates contact hypersensitivity responsiveness in C57/Bl6 mice. *J. Invest. Dermatol.* **93,** 511–517.
94. Smith, R., Healy, E., Siddiqui, S., Flanagan, N., Steijlen, P., Rosdahl, I., Rogers, S., et al. (1998) Melanocortin 1 receptor variants in an Irish population. *J. Invest. Dermatol.* **111,** 101–104.

CHAPTER 12

The Melanocortin-2 Receptor in Normal Adrenocortical Function and Familial Adrenocorticotropic Hormone Resistance

Adrian J. L. Clark

The physiologic role of adrenocorticotropic hormone (ACTH) and its part in the pituitary–adrenal axis is one of the most intensively studied systems in endocrinology. ACTH was one of the first hormones that was found to stimulate cAMP production by the adrenal gland *(1)*, and the notion that this effect was mediated via a specific cell surface receptor was confirmed by the elegant studies of Lefkowitz and colleagues *(2)* in work that set a standard for receptor characterization. Nevertheless, progress on the understanding of the ACTH receptor has been relatively slow. It is now clear that the MC2-R is synonymous with the ACTH receptor, and both terms are used in this chapter. In general, the term ACTH receptor is used to describe the functional entity for example, ligand binding to adrenal cells, whereas the term MC2-R is used to describe aspects that can clearly be related to this gene.

1. Physiologic Role of ACTH

The role of ACTH on the adrenal cortex is described in detail in Chapter 3; These actions are reviewed here.

1.1. Steroidogenesis

The adrenal cortex is divided into three histologically distinct zones: an outer zona glomerulosa that synthesizes the mineralocorticoid aldosterone, an internal zona fasciculata that synthesizes glucocorticoids (corticosterone in rodents, cortisol in higher mammals), and the innermost zona reticularis that synthesizes glu-

The Melanocortin Receptors Ed.: R. D. Cone

cocorticoids and the adrenal "androgens," including dihydroxyepiandrosterone and androstenedione. Production of aldosterone by the zona glomerulosa is principally under the control of angiotensin II, which acts through a well characterized G protein-coupled angiotensin receptor that signals mainly through phospholipase C *(3)*. ACTH also has a discernible action in stimulating aldosterone production by these cells, but this action is of less significance than stimulation by angiotensin II.

Production of glucocorticoids by the fasciculata cells is under the control of ACTH. Although several other mediators may have modulatory actions on glucocorticoid production, no other factor is as potent as ACTH. Hypophysectomy to remove the source of natural ACTH results in life threatening glucocorticoid deficiency, which can be restored by replacement of ACTH. Pituitary ACTH secretion normally fluctuates in a circadian rhythm, and consequently circulating glucocorticoid also fluctuates in the same rhythm, lagging behind ACTH by 1 – 2 h, and therefore exhibiting peak plasma cortisol between 6 and 9 A.M. and a nadir of undetectable cortisol at midnight in the human. Synthesis of steroids by the zona reticularis cells is also partly controlled by ACTH. ACTH deficiency or resistance is associated with absent secretion of adrenal androgens as normally occurs in human adrenarche *(4)*.

1.2. ACTH in Adrenal Growth and Development

The role of ACTH in the growth of the adrenal cortex is still debated. While hypophysectomy leads to adrenal atrophy (e.g., *5*), replacement with ACTH does not fully restore adrenal cortex size *(6)*. In primary cultures of adrenocortical cells, ACTH has often been observed to have an antimitogenic action *(7–9)*, although recent data suggest that ACTH has a delayed mitogenic effect after an initial antimitogenic phase *(10)*. There is evidence that the N-terminal peptide from proopiomelanocortin (POMC), *N*-POMC[1–28], has a major role in stimulation of adrenal growth in this situation *(11,12)*. Nevertheless, ACTH resistance resulting from a defect in the MC2-R is associated with marked adrenal atrophy. A possible explanation for these apparently conflicting findings is that ACTH stimulates fasciculata cell differentiation initially *(10,13)*, and after a differentiation phase, is able to exert a more typical mitogenic stimulus. This area remains one of significant uncertainty, however.

1.3. Extraadrenal Actions of ACTH

There is evidence for the existence of a short feedback loop by means of which ACTH secreted by the pituitary acts on the hypothalamus to impair further corticotropin release *(14,15)*. This phenomenon has not been extensively studied, and its functional significance is still uncertain. Furthermore, it is not clear whether this action is mediated through a MC2-R, or some other receptor.

Effects of ACTH in vitro on adipocyte lipolysis have long been recognized, and appear to be mediated through an MC2-R *(16,17)*. The physiologic importance of this action is not entirely clear, and merits further study. In terms of human pathology, one of the well-recognized actions of ACTH excess is in stimulating skin pigmentation. This is almost certainly mediated through the MC1-R, although there has been some recent evidence for the expression of the MC2-R in skin *(18)*. A possible role of ACTH on the immune system is arguable, and there are claims for the identification of receptors for ACTH on peripheral blood mononuclear cells *(19)*. Such actions of ACTH are in all probability mediated through a melanocortin receptor other than the MC2-R.

2. Characterization of the ACTH Receptor In Vivo and In Cells

2.1. Ligand Binding

The ACTH receptor was one of the first to be characterized both in terms of its signal transduction characteristics and its ligand binding properties. However, ligand binding studies with ACTH have been difficult, and consequently there has been some variation in the available published data.

In their original paper on this subject, Lefkowitz and colleagues *(2)* prepared iodinated ACTH[1–39]using the chloramine T method, and they purified a fraction that they believed was monoiodinated on tyrosine 2. Using this material they demonstrated specific binding sites for ACTH on adrenal cell membranes that could be competed off by ACTH but not by a variety of other peptide hormones. They demonstrated its distribution in a cell membrane fraction that exhibited ACTH-dependent adenylate cyclase activity, and its absence in fractions lacking this activity.

Subsequent workers had difficulty in obtaining such effective tracer for similar studies. This is partly due to the tendency for ACTH to bind nonspecifically to cellular and other particulate matter, and partly due to the effect of the large iodine atom on tyrosine 2 *(20)*. Furthermore it was shown that methionine 4 was liable to become oxidized during the iodination process *(21)*. These problems led to the use of the "Ramachandran analogue" — ACTH in which Tyr 2 was substituted by a phenylalanine, and methionoine 4 was substituted by a norleucine. This peptide could therefore only be monoiodinated on Tyr 23, and resulted in successful studies by this and other research groups *(22)*. More recent methodology has allowed successful preparation of normal sequence ACTH[1–39] that is monoiodinated on Tyr 23 and not significantly oxidized at Met 4 *(86)*.

Using these tracers and relatively standard binding assay protocols, several groups have reported ACTH binding data with cells or membrane preparations from the adrenal cortex of various species. These results are summarized in Table 1. As can be seen, most of the more recent studies indicate the presence of adrenal receptors with subnanomolar affinities in a wide variety of species. The density of binding sites is variable, but most workers report several thousand high affinity sites per cell. Of particular interest is the vast excess of sites in rat adrenal glomerulosa cells when compared with fasciculata cells, and the evidence that numbers of sites can be increased by ACTH itself, angiotensin II and dexamethasone *(23–25)*. Also notable is the mouse Y1 corticoadrenal cell line originally described by Yasumura et al. *(26)*, which expresses ACTH receptors, the binding characteristics of which have been characterized recently *(27)*. The use of this cell line and its derivatives will be referred to later.

2.2. Signal Transduction

ACTH was originally shown to stimulate cyclic adenosine monophosphate (cAMP) production in adrenal cells by Haynes *(1)*, and this observation has been widely accepted. Indeed, it may be that stimulation of adenylate cyclase is sufficient for generation of the entire ACTH signal in the adrenal, and the qualitative effects of ACTH can be mimicked by addition of dibutyryl cAMP or forskolin. A number of mouse Y1 cell lines possessing mutations of adenylate cyclase are unable to synthesize steroids in response to ACTH despite possessing the ability to make steroids in response to exogenous cAMP *(28,29)*.

The targets of ACTH-stimulated cAMP generation and protein kinase A activation are numerous, and many remain to be identified. These targets, however, include induction of transcription of several of the key genes whose products are enzymes involved in steroidogenesis, as well as the STAR protein involved in mitochondrial cholesterol import, and apparently the MC2-R gene itself. Some of these established targets are listed in Table 2. The mechanisms by which protein kinase A stimulates expression of these genes is not entirely clear in many cases, and it seems that activation of the cAMP response element binding protein (CREB) by phosphorylation is not used, and alternative cAMP signal transduction pathways are active (reviewed in ref. *30*).

One of the persisting idiosyncrasies of ACTH signaling, however, is the discrepancy between the sensitivity of steroidogenesis to ACTH which is usually found to be in the tens of picomolar range, and the sensitivity of cAMP generation. It is often argued that very small and transient amounts of cAMP are sufficient to activate the steroidogenic process, while much more extensive stimulation is necessary to obtain measurable elevations of cAMP. This may indeed be the case, and it may be that a more complex adenylate cyclase assay is needed to relate ACTH dose responses to cAMP signal transduction processes. Alternatively, it may be that cAMP leaking through gap junctions

Table 1
Summary of Published Ligand Binding Studies Using Iodinated ACTH

Cell Type/Species	Tracer	K^d/IC^{50} (M)	BMAX	Author	Comments
Mouse adrenal particles	^{125}I-ACTH[1–39]	$\sim 10^{-6}$	ND	Lefkowitz et al. *(2)*	
Rat, human, sheep adrenal membranes	^{125}I-ACTH[1–24]	$\sim 5 \times 10^{-7}$	18 – 41 pmol/mg	Saez et al. *(82)*	
Sheep adrenocortica cells	1	5.9×10^{-10}	1038 sites/cell	Darbeida & Durand *(24)*	Sites increased by dexamethasone
Human adrenocortical cells	1	5.7×10^{-10}	850 sites/cell	Lebrethon et al. *(25)*	Sites increased by ACTH and AII
Bovine adrenal fasciculata cells	1	2.3×10^{-10}	1910 sites/cell	Penhoat et al. *(23)*	Sites increased by ACTH
Human adrenocortical cells	2	1.6×10^{-9}	3560 sites/cell	Catalano et al. *(83)*	Calcium essential for binding
Rat adrenocortical cells	2	1.4×10^{-9}	3840 sites/cell	Buckley & Ramachandran *(84)*	
Rat fasciculata cells	2	1.1×10^{-11}	7200 sites/cell	Gallo-Payet & Escher *(85)*	
Rat glomerulosa cells	2	7.6×10^{-11}	65,000 sites/cell	Gallo-Payet & Escher *(85)*	
Mouse 3T3-L1 adipocytes	2	4.3×10^{-9}	21 fmol/50 µg DNA	Grunfeld et al. *(16)*	No binding when undifferentiated
Chicken adrenocortical cells	1	1.1×10^{-9}	3.2 fmol/50 µg DNA	Carsia & Weber *(86)*	Affinity reduced by protein malnutrition
HeLa cells expressing MC2-R	1	0.8×10^{-9}	26,400 sites/cell	Kapas et al. *(58)*	Stably transfected cell line

Tracer: 1 = ^{125}I-Tyr23-ACTH[1–39], 2 = ^{125}I-Tyr^{23},Phe^{2},Nle^{4}] ACTH[1–38].
Data summarized for high affinity ACTH binding sites only.

Table 2
Genes Known to be Regulated by ACTH Stimulation of cAMP in the Adrenal

Gene Name	Alternative or Common Name
CYP11A	P450 side chain cleavage enzyme
CYP17	17α-Hydroxylase
CYP21	21 Hydroxylase
CYP11B1	11β-Hydroxylase
CYP11B2	Aldosterone synthase
MC2-R	ACTH receptor
STAR	
c-FOS	
c-JUN	
JUN-B	
c-MYC	

from a small number of especially sensitive cells activates surrounding cells, and this has been proposed as an explanation for this discrepancy *(31)*. An alternative view was, however, originally proposed by Kojima et al. *(32)* who demonstrated in adrenal glomerulosa cells that the ACTH effect on aldosterone production was enhanced by an additional action in opening membrane calcium channels. This action was absent when forskolin was used alone, but forskolin plus an ionophore mimicked the ACTH dose response curve for aldosterone. More recently, Enyeart et al. *(33)* have shown in bovine adrenal fasciculata cells that ACTH activated T-type calcium channels, and that these channels and ACTH-stimulated cortisol production could be blocked by specific T-channel blockers. However, these findings could not be reproduced in Y1 cells *(34)*.

2.3. Ligand Preference

Most binding studies have reported experiments in which ligands unrelated to ACTH have been shown to be ineffective in displacing bound ACTH. Of particular interest in understanding the determinants of receptor recognition of its natural ligand are those experiments in which ACTH analogues and related peptides have been used to displace bound ACTH. These are summarized in Table 3. It is apparent that truncation of ACTH[1–39] progressively from the C-terminus results in only a small reduction in affinity for the receptor until peptides shorter than ACTH[1–17] are used. Free acid forms are significantly less active than C-terminally amidated peptides. α-MSH and β-MSH are almost without ligand binding activity on these cell preparations.

Understanding these findings is enhanced by the study of ACTH analogs on cAMP production by adrenal cells. As before, C-terminal truncation results

Table 3
Ligand Preference by the ACTH Receptor

Author	Cell type	ACTH Analog 1–39	ACTH Analog 1–19	ACTH Analog 1–17	α-MSH	β-MSH
Buckley & Ramachandran *(22)*	Rat adrenocortical cells	1.6×10^{-9}	4.7×10^{-9}	15.4×10^{-9}	ND	ND
Gallo-Peyet & Escher *(85)*	Rat fasciculata cells	43×10^{-9}	ND	ND	$>>10^{-6}$	$>>10^{-6}$
Grunfeld et al. *(16)*	3T3-L1 cells	4.3×10^{-9}	5.8×10^{-9}	115×10^{-9}	1380×10^{-9}	ND
Kapas et al. *(58)*	MC2-R/HeLa cells	0.8×10^{-9}	ND	1.2×10^{-9}	$>>10^{-6}$	ND

ND = not tested.
Results are the IC_{50} for displacement of ACTH tracer binding by various ACTH anaolgs and related peptides.

in dramatic loss of agonist activity with peptides shorter than ACTH[1–18] (e.g., *35* and *36*). It seems likely that the main function of amino acids C-terminal to this have a role in protecting the shorter peptide from degradation. However, when used in very high concentrations, shorter peptides such as ACTH[1–10] and the N-terminally truncated ACTH[4–10] also had steroidogenic activity with isolated rat adrenal cells *(37,38)*. This has led to the suggestion that a second ACTH receptor exists on adrenal and perhaps other cells *(39)*. Further truncation from the N-terminal end as in ACTH[11–24] leads to peptides that lack cAMP generating or steroidogenic activity, but which have been shown to act as competitive antagonists for ACTH[1–24] *(38,40,41)*.

2.4. Receptor Purification

Several groups have attempted to purify the receptor protein using conventional biochemical methods. Ramachandran et al. *(8)* used an ACTH analog [(2-nitro-5-azidophenylsulfenyl)-Trp^9]ACTH to crosslink to the receptor. On sodium dodecyl sulfate-polyacrylamide gel electrophoresis (SDS PAGE) material of 100 kDa was detected. Hoffman et al. *(42)* used a generally similar strategy, but with the advantage that the crosslinked ligand was biotinylated, and therefore purifiable using an avidin column. They were able to demonstrate the presence of a 43-kDa protein. Penhoat et al. *(43)* also used a crosslinking strategy and identified two complexes of 43 and 154 kDa. These findings are summarized in Table 4.

There have also been attempts to use the concept of receptor–ligand coding complementarity to design an epitope based on the opposite coding strand to that for ACTH *(44)*. This led to the identification of a putative ACTH receptor on immunoblotting, although these substances were never conclusively shown to be ACTH receptors.

Mertz & Catt *(45)* reported the use of a *Xenopus* oocyte expression system for identifying the size of the mRNA encoding the rat ACTH receptor. Maximum ACTH stimulated cAMP generation lay within the RNA fraction of 1.1 – 2 kb in size, but this approach was not successful in cloning the cDNA encoding this receptor.

3. The MC2 Receptor Gene

3.1. Gene Cloning

3.1.1. Human MC2-R Gene

The original cloning of the MC2-R gene is described in detail elsewhere and will not be reviewed here. In brief, Mountjoy et al. *(46)* identified a λ phage clone from a human genomic library that encoded a 297 residue polypeptide that was 39% identical to the human MC1-R coding sequence.

Table 4
Purification of the ACTH Binding Complex

Authors	Strategy	Reagent/ligand	Size of complex (kDa)
Ramachandran et al. *(8)*	Crosslinking	[(2-nitro-5-azidophenylsulphenyl)-Trp^{9}]ACTH	100
Hoffman et al. *(42)*	Crosslinking	[Phe^{2},Nle^{4},$DTBct^{25}$]ACTH-[1–25]amide	43
Penhoat et al. *(43)*	Crosslinking	ACTH[1–39]	43 & 154

This gene was shown using *in situ* hybridization and northern blot analysis to be expressed almost exclusively in the adrenal cortex of the rhesus macaque monkey *(46)*. Subsequently, Lebrethon et al. *(25)* demonstrated its expression in human adrenal cells as major mRNA species of 1.8 and 3.4 kb, and lesser species of 4, 7, and 11 kb. Mountjoy et al. *(46)* also presented limited expression studies after transfection into Cloudman S91 cells — cells that have the disadvantage of expressing the MC1-R.

3.1.2. Bovine MC2-R cDNA

Following the cloning of the human MC2-R gene, Raikhinstein et al. *(47)* were able to use this sequence to clone the bovine MC2-R cDNA from a bovine adrenal cDNA library. This identified several 3 kbp cDNAs encoding a 297- residue polypeptide having 81% identity to the human receptor. Expression studies of this cDNA have not been published.

3.1.3. Mouse MC2-R Gene

We used the human MC2-R sequence to design primers for the polymerase chain reaction with which we amplified a 661-bp fragment from murine genomic DNA, which was then used to screen a mouse genomic library. The λ phage clones so identified encoded a receptor that had 84% amino acid identity to the human MC2-R, and 81% identity to the bovine receptor *(48)*. The mouse receptor is a single amino acid shorter at the C-terminus than the others, but contained the same two N-linked glycosylation sites and two extracellular cysteine residues believed to be involved in disulfide bridging. This gene is expressed in mouse adrenal and in Y1 cells as a major transcript of 1.8 kb and a minor transcript of 4.5 kb. The Y1 cell signal is markedly weaker than that in mouse adrenals. *In situ* hybridization studies have confirmed that this expression in the mouse is limited mainly to the zona glomerulosa and fasciculata cells, with a few scattered MC2-R positive cells in the zona reticularis and adrenal medulla *(49)*.

As with the human gene, the entire coding region of the mouse receptor was contained in a single exon. The 5'-rapid amplification of cDNA ends (5' RACE) technique was used to try to identify the extent and nature of the 5' untranslated region of this receptor, and revealed a fragment that extended 241 bp upstream of the initiator methionine. This sequence diverged from the genomic sequence, implying the existence of one or more exons upstream of the coding exon. Screening of a mouse genomic library revealed the presence of two further constant exons of 113- and 112-bp length *(50)*. There is also evidence for a further alternatively spliced exon of 57 bp lying between exons 1 and 2 in about 2 – 5 % of mouse MC2-R transcripts *(51)*.

Unusually, this gene structure is not entirely maintained in the human. Using a similar strategy for isolating the 5' end of the human cDNA, Naville

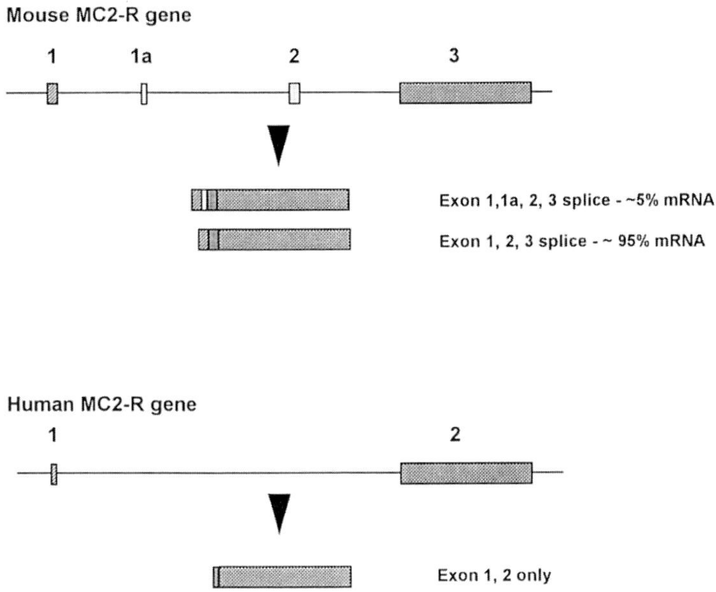

Fig. 1. Diagrammatic representation of the mouse and human MC2-R gene structures and their mature mRNA transcripts. See text for details.

et al. *(52)* found evidence of only a single exon of 49 bp upstream of the coding exon, and no evidence of alternative splicing. However, it is notable that there is some sequence similarity between the human exon 1 and that of the mouse (*see* Fig. 1).

3.2. MC2-R Promoter

The identification of the 5' ends of the cDNAs has allowed identification of the nature of the mouse and human MC2-R promoters *(50–52)*. Both genes are atypical promoters lacking conventional features such as TATA boxes, CAAT boxes, and GC-rich sequences, yet both drive the expression of luciferase reporter genes in mouse Y1 cells. Both genes contain consensus sites for the orphan nuclear receptor, steroidogenic factor 1 (SF1) close to the transcription initiation site. This has been shown in the case of the mouse promoter to be important, but not essential for MC2-R gene expression *(50)*. Other consensus elements in the promoter include several putative cAMP response elements in the human gene which are notably missing in the mouse gene.

3.3. Regulation of the MC2-R

The availability of probes for the MC2-R enabled the study of the regulation of this gene in response to various stimuli in appropriate cells. Thus

Mountjoy et al. *(53)* demonstrated that ACTH or forskolin stimulated the expression of the MC2-R gene 6-fold after 24 h in mouse Y1 cells and 2-to 4-fold in human NCI-H295 cells. They also demonstrated a marked increase of MC2-R mRNA after exposure to angiotensin II for 24 h.

Similar findings were reported by Lebrethon et al. *(25)* with human primary adrenal cultures. In these cells a 21-fold increase in mRNA expression and 4-fold increase in receptor number was found after exposure to ACTH alone. Angiotensin II had effects of a similar magnitude, and the combination of ACTH and angiotensin II were additive. By contrast, the same group reported that transforming growth factor beta-1 (TGFβ-1) was without effect *(54)*, although cycloheximide alone stimulated gene expression *(55)*.

3.4. Expression of the Cloned MC2-R

Expression of the cloned MC2-R after transfection into heterologous cells has provided major experimental difficulties. Using the human MC2-R gene cloned into a variety of well characterized eukaryotic expression vectors with varying amounts of 5' and 3' untranslated region there has been widespread failure to obtain functional expression in a range of well characterized cell lines that are readily used for this purpose with other receptors. Using a highly optimized transfection protocol, we were able to detect weak evidence of receptor expression in Cos 7 cells, but these results were confounded by the presence of an endogenous melanocortin receptor that generated cAMP in response to ACTH *(56)*. Others have found that the Cloudman M3 cell line which expresses an MC1-R is capable of expressing the human MC2-R, but again, the presence of the endogenous receptor confounds the reliable characterization of the MC2-R *(57)*.

Using the mouse MC2-R either in transient expression in human HeLa cells or in a stable HeLa cell line, we have been able to obtain background-free expression of this receptor. These studies revealed a receptor that was highly sensitive to ACTH as indicated by cAMP generation, having an EC_{50} for ACTH[1–24] and ACTH[1–39] of 7.5 and 57×10^{-12} *M*, respectively. The receptor had no significant response to peptides shorter than ACTH[1–17], and ACTH[11–24] and ACTH[7–39] behaved as antagonists. The receptor bound ^{125}I-ACTH[1–39] and had dissociation constants of 0.84 and 0.94×10^{-9} *M* for the 24 and 39 residue biologically active peptides respectively *(58)*. These findings lend strong credibility to the proposal that the MC2-R is the ACTH receptor. However, the explanation as to why the mouse receptor could be expressed so readily in these cells is not clear.

The observations that the human MC2-R could be expressed in cells with endogenous melanocortin receptors led to the speculation that a cofactor for expression is needed. Hypothetically, such a cofactor would be present in cells that expressed endogenous melanocortin receptors, although the relative ease with which other melanocortin receptors can be expressed in various cell types implies that this cofactor is not required for other melanocortin receptors. This reasoning led to studies in which the MC2-R was expressed in mutant Y1 cell lines known as Y6 and OS3 which fail to express the endogenous mouse MC2-R for reasons that are not clear *(59)*. These cells have no response to ACTH before transfection, but appear to express the human MC2-R with some success. Because these cells are relatively difficult to transfect transiently, it is necessary to make stable cell lines for these studies. This system now provides a means of characterizing the normal and mutated forms of the human MC2-R. Recent work has reported the use of this system for characterizing the antagonism of the agouti protein for the MC2-R *(60)*, and the characterization of a naturally occurring mutation of the human MC2-R *(61)*.

4. ACTH Insensitivity

Support for the view that the MC2-R was indeed the receptor for ACTH came from the finding that mutations in this gene were found in patients with a rare autosomal recessive insensitivity to ACTH known as familial glucocorticooid deficiency (FGD), isolated glucocorticoid deficiency, or hereditary unresponsiveness to ACTH *(62,63)*. This syndrome is distinct from a related disorder known as the triple A syndrome or Allgroves syndrome *(64)*; which has recently been shown to be linked to an unidentified gene on human chromosome 12q13 *(65)*.

4.1. Clinical Presentation

The typical presentation of FGD is the result of glucocorticoid deficiency. Thus, in the neonatal period, most patients will exhibit hypoglycemia. This may not be profound and often responds to more frequent feeding. Less commonly in the neonatal period a picture of hepatitis with mild jaundice can be found which reverses after glucocorticoid replacement. Excessive pigmentation of the skin resulting from elevated ACTH levels takes longer to be manifest and is usually first noted after 4 or 5 months of life. Children in whom the diagnosis is not made by this time tend to be especially prone to infection and take longer to recover from relatively minor infective episodes. In some cases this may result in profound and sometimes fatal septic events at any time in the childhood years.

4.2. Diagnosis and Differential Diagnosis

The salient feature of FGD is the finding of subnormal or undetectable plasma cortisol in combination with an elevated plasma ACTH. Frequently, the cortisol values at 9 A.M. are between 100 and 300 nmol/L (normal

≥ 300 nmol/L which may not in itself be very remarkable, but these values respond poorly or not at all to injection of 250 μg of synthetic ACTH[1–24]. In other cases 9 a.m. cortisol values are clearly undetectable and fail to respond to stimulation. Plasma ACTH is usually markedly elevated with values usually over 1000 pg/mL (normal ≤ 80 pg/mL). These features could be found in other adrenal disorders such as autoimmune Addison's disease, but can clearly be distinguished by demonstrating normal renin and aldosterone concentrations and normal electrolytes. Other inherited adrenal disorders that should be excluded include the triple A syndrome, adrenoleucodystrophy, congenital adrenal hyperplasia and congenital adrenal hypoplasia. The first of these is usually associated with deficient tear production from early life, and achalasia of the esophagus which may be detected only on barium swallow. Adrenoleucodystrophy can usually be excluded by demonstrating normal long-chain fatty acids, and congenital adrenal hyperplasia is characterized by elevated 17α-hydroxyprogesterone. Congenital adrenal hypoplasia is associated with failure of both adrenal and gonadal development.

4.3. Pathogenesis

A number of hypotheses had been put forward over the years to explain the origin of FGD. These included proposals of a defect in the receptor for ACTH (e.g., *63*), although a defect in the ACTH signal transduction system, or a defect in adrenal gland development had also been postulated. Evidence favoring the first of these came from Smith et al. *(66)* who demonstrated defective ACTH binding to peripheral blood mononuclear cells in a patient with FGD, in contrast to normal binding characteristics in cells from a control subject. However, Yamaoka et al. *(67)* demonstrated a failure of cAMP generation by ACTH in mononuclear cells and concluded that the disease resulted from a postreceptor defect.

Following the cloning of the human MC2-R *(46)*, we were able to demonstrate a homozygous missense point mutation in two affected siblings *(68)*. This mutation converted Ser74 which lies in the second transmembrane domain to Ile (S74I), and segregated with the disease in the family. Subsequently we and others have reported a number of different missense and nonsense mutations in this gene which occur in homozygous or compound heterozygous form in patients with the disorder *(57,69–71)*. (*See* Fig. 2.) In all cases these mutations co-segregate with the disease in the family. The current status of these published mutations is summarized in Table 5.

Confirmation that these mutations cause the disease depends on expression studies in which the mutant receptor gene is introduced into cells that lack endogenous MC2-R. As already discussed, this has been exceptionally difficult to do, and conventional methods of expression have had limited useful-

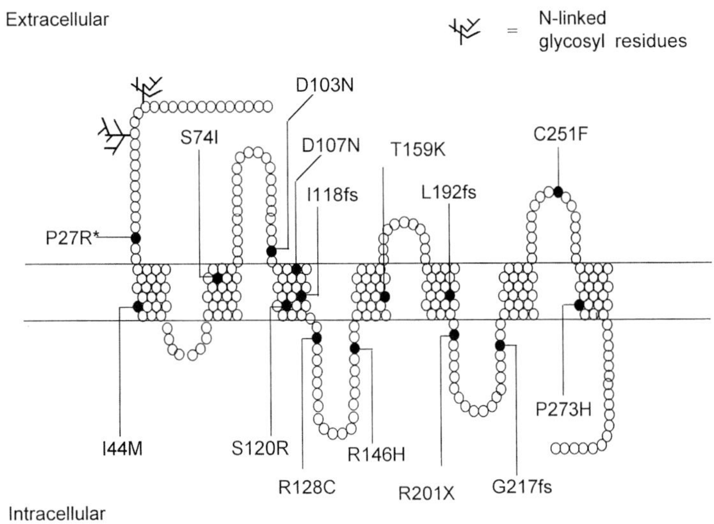

Fig. 2. Two-dimensional model of the MC2-R to demonstrate the location of the polymorphisms and mutations found in patients with familial glucocorticoid deficiency. Further details are in Table 5.

ness *(56)*. Naville et al. *(57)* used the M3 melanoma cell line to express mutant MC2-R, but this data is also confounded by the endogenous MC1 receptors. In this work these authors showed that the D107N, C251 F, and G217frameshift mutations lacked all cAMP generating function in this system in contrast to the normal sequence receptor.

Using the mouse Y6 cell line *(59)*, we find it possible to express transfected human MC2-R in the absence of any background signal. In this system the S74I mutation appears to markedly reduce the ability of the receptor to generate cAMP at doses of ACTH[1–24] or ACTH[1–39] up to $10^{-6}M$. The receptor can still bind ligand with a reduced affinity. This implies that this mutation results mainly in a loss of the ability to transduce the signal *(61)*. Studies of other naturally occurring human MC2-R mutations are in progress.

As these expression studies suggest, different mutations are likely to disable the receptor to different degrees making phenotype–genotype comparisons difficult. However the S74I mutation that we originally described has proved to be the most prevalent of these mutations and we have now identified it in 10 individuals from 6 families in homozygous form and as a compound heterozygote with a more severe mutation in two cases. Many of these cases have a Scottish family background, and it seems highly likely that the founder mutation occurred in this region probably in the past two centuries.

Table 5
ACTH-R Mutations That Have Been Reported in Patients With Familial Glucocorticoid Deficiency

Mutation	Authors
S74I	Clark et al. *(68)*
S120R	Tsigos et al. *(69)*
P27R (polymorphism)	Weber and Clark *(76)*
R128C	Weber et al. *(70)*
I44M	Weber et al. *(70)*
R146H	Weber et al. *(70)*
L192frameshift	Weber et al. *(70)*
R201X	Tsigos et al. *(71)*
T159K	Elias et al. *(61)*
D107N	Naville et al. *(57)*
C251F	Naville et al. *(57)*
G217frameshift	Naville et al. *(57)*
I118frameshift	Elias et al. *(61)*
P273H	Stratakis et al. *(87)*
D103N	Elias *(89)*

MC2-R mutations allow for some interesting physiologic inferences. Since a point mutation in both alleles of this gene can result in complete failure to secrete cortisol, there can be little doubt that the MC2-R gene is the ACTH receptor gene, and that it is the only ACTH receptor gene. There is evidence that adrenocortical cells also express the MC5 receptor *(72)* which has been shown to respond in transfected cells to ACTH [1–39],(K^i= 929 n*M* *[88]*) at high concentrations. However, it appears that the high concentrations of ACTH found in untreated FGD are probably not sufficiently high to recruit this receptor for stimulation of cortisol production.

A second interesting finding is that patients with FGD fail to develop an adrenarche — the prepubertal secretion of adrenal "androgens" from the zona reticularis cells. This implies either that these cells that express relatively few MC2-R *(49)* require ACTH at some stage in their development, or that ACTH acting on the small number of receptors is critically important for the initiation of adrenarche *(4)*. Perhaps the most unexpected finding in many of these patients with FGD is that many are unusually tall despite having a bone age that is appropriate for their height *(73)*. An adequate explanation for this finding is not apparent at present.

4.4. Normal Receptor FGD

Not all cases of FGD are associated with mutations within the coding region of the MC2-R. Of 37 families that we have studied to date, in only 14 families are affected cases associated with homozygous or compound heterozygous mutations. One possibility is that there are mutations in regions of the gene apart from the coding region such as the promoter region, although it is generally true that mutations in promoters are not a common cause of genetic disease

As a result of the human genome mapping project it is now relatively straightforward to identify highly polymorphic microsatellite repeat sequences at given locations in the genome. The MC2-R was mapped to the short arm of chromosome 18 (18p11.2) *(74,75)* and so we investigated the proximity of a number of repeats in this region by performing linkage analysis in the families with MC2-R mutations. This approach revealed that the markers *D18S40* and *D18S44* were positioned on either side of the MC2-R gene at distances of 3 and 4 cM, respectively *(76)*. Such a distance, although large in physical terms, is satisfactory for the segregation studies proposed.

The results of this analysis indicate that in the case of several of the families without MC2-R mutations the segregation analysis was not compatible with an etiologic role for this gene *(76)*. This result is important since it makes it clear that the clinical phenotype of FGD can be caused by a second genetic defect. For ease of reference we have adopted the term FGD type 2 for this syndrome that is not linked to the MC2-R locus. It is hoped that ultimately the identity of the causative gene for this syndrome will be identified, which may allow a more descriptive distinction between the etiologies for this disease.

4.5. Constitutively Activating Mutations

There has been considerable interest in the last few years in the identification of mutations causing constitutive activation of G protein-coupled receptors. Notably, familial pseudoprecocious puberty and toxic nodules of the thyroid gland have been shown to be associated with, and presumably result from, constitutively activated forms of the luteinizing hormone (LH) and thyrotropin stimulating hormone (TSH) receptor, respectively *(77,78)*. Moreover, as discussed in Chapter 10, a variety of coat color mutants have been shown to result from constitutively activating MC1-R mutations *(79)*. It has therefore been reasonable to consider what phenotype might be associated with constitutive activation of the MC2-R. It seems likely that a sporadic mutation of this type could result in focal adrenal hyperplasia - that is, an adrenal adenoma, and that a germline mutation might result in bilateral adrenal hyperplasia. Two groups have sought such mutations in a combined total of 41 cases of varied adrenal pathology and failed to discover any mutation *(80,81)*. Thus if this does occur, it is not a common cause of adrenal disease.

5. Summary

The ACTH receptor is a receptor that has been the subject of extensive study over many years. Investigation has undoubtedly been hindered by technical difficulties in performing ligand binding studies, and the limited sites of expression of this receptor. Despite these problems, it has become apparent that this receptor is unique among the melanocortin receptor family in that it shows no significant response to any of the melanocortin stimulating hormone (MSH) peptides. It seems that the basic residues lying between positions 15 and 18 in ACTH have an important and essential role in permitting interaction of ACTH with its receptor.

Mountjoy et al. *(46)* were the first to succeed in cloning the MC2-R gene, which clearly encodes the ACTH receptor. The evidence for this identity consists of (i) tissue distribution studies, (ii) receptor expression studies, and (iii) evidence for defects in the MC2-R in patients with ACTH resistance or FGD. The second of these pieces of evidence has been especially hard to establish, and the reasons for this are not yet clear. Future research efforts on this interesting gene and its translation product are likely to focus on the determinants of ligand receptor interaction and signal transduction, and on the tissue specific restrictions on expression of the receptor.

References

1. Haynes, R. C. (1958) The activation of adrenal phosphorylase by the adrenocorticotropic hormone. *J. Biol. Chem.* **233,** 1220–1222.
2. Lefkowitz, R. J., Roth, J., Pricer, W., and Pastan, I. (1970) ACTH receptors in the adrenal, specific binding of ACTH–^{125}I and its relation to adenyl cyclase. *Proc. Natl. Acad. Sci. U. S. A.* **65,** 745–752.
3. Bernstein, K. E. and Alexander R. W. (1992) Counterpoint, molecular analysis of the angiotensin II receptor. *Endocr. Rev.* **13,** 381–386.
4. Weber, A. Clark, A. J. L., Perry, L. A., Honour. J. W., and Savage, M. O. (1997) Diminished adrenal androgen secretion in familial glucocorticoid deficiency implicates a significant role for ACTH in the induction of adrenarche. *Clin. Endocrinol.* **46,** 431–437.
5. Robinson, P. M., Comline, R. S., Fowden, A. L., and Silver, M. (1983) Adrenal cortex of fetal lamb, changes after hypophysectomy and effects of synacthen on cytoarchitecture and secretory activity. *Q. J. Exp. Physiol.* **68,** 15–27.
6. Payet, N. and Lehoux, J. G. (1980) A comparative study of the role of vasopressin and ACTH in the regulation of growth and function of rat adrenal glands. *J. Steroid Biochem.* **12,** 461–467.
7. Masui, H. and Garren, L. D. (1970) On the mechanism of action of adrenocorticotropic hormone: stimulation of deoxyribonucleic acid polymerase and thymidine kinase activities in adrenal glands. *J. Biol. Chem.* **245,** 2627–2632.
8. Ramachandran, J., Muramoto, K., Kenez–Keri, M., Keri, G., and Buckley, D. I. (1980) Photoaffinity labelling of corticotropin receptors. *Proc. Natl. Acad. Sci. U. S. A.* **77,** 3967–3970.
9. Hornsby, P. J. and Gill, G. N. (1977) Hormonal control of adrenocortical cell proliferation. *J. Clin. Invest.* **60,** 342–352.
10. Arola, J., Heikkila, P., and Kahri, A. I. (1993) Biphasic effect of ACTH on growth or rat adrenocortical cells in primary culture. *Cell Tissue Res.* **271,** 169–176.

11. Estivariz, F. E., Iturriza, F., McClean, C., Hope, J., and Lowry, P. J. (1982) Stimulation of adrenal mitogenesis by N–terminal proopiocortin peptides. *Nature* **297,** 419–422.
12. Estivariz, F. E., Morano, M. I., Carino, M., Jackson, S., and Lowry, P. J. (1988) Adrenal regeneration in the rat is mediated by mitogenic N–terminal pro–opiomelanocortin peptides generated by changes in precursor processing in the anterior pituitary. *J. Endocrinol.* **116,** 207–216.
13. Heikkila, P., Arola, J., Salmi, A., and Kahri, A. I. (1995) ACTH–induced c–myc proto–oncogene expression precedes antimitogenic effect during differentiation of fetal rat adrenocortical cells. *J. Endocrinol.* **145,** 379–385.
14. Boscaro, M., Sonino, M., Paoletta, A., Rampazzo, A., and Mantero, F. (1988) Evidence for ultra–short loop autoregulation of adrenocorticotropin secretion in man. *J. Clin. Endocrinol. Metab.* **66,** 255–257.
15. Calogero, A. E., Gallucci, W. T., Gold, P. W., and Chrousos, G. P. (1988) Multiple feedback regulatory loops upon rat hypothalamic corticotropin–releasing hormone secretion: potential clinical implications. *J. Clin. Invest.* **82,** 767–774.
16. Grunfeld, C., Hagman, J., Sabin, E. A., Buckley, D. I., Jones, D. S., and Ramachandran, J. (1985) Characterization of adrenocorticotropin that appears when 3T3–L1 cells differentiateinto adipocytes. *Endocrinology* **116,** 113–117.
17. Izawa, T., Mochizuki, T., Komabayashi, T., Suda, K., and Tsuboi, M. (1994) Increase in cytosolic free Ca^{2+} in corticotropin–stimulated white adipocytes. *Am. Physiol.* **266,** E418–E426.
18. Slominski, A., Ermak, G., and Mihm, M. (1996) ACTH receptor, CYP11A1, CYP17 and CYP21A2 genes are expressed in skin. *J. Clin. Endocrinol. Metab.* **81,** 2746–2749.
19. Clarke, B. L. and Bost, K. L. (1989) Differential expression of functional adrenocorticotropic hormone by subpopulations of lymphocytes. *J. Immunol.* **143,** 464–469.
20. Lowry, P. J., McMartin, C., and Peters, J. (1973) Properties of a simplified bioassay for adrenocorticotropic activity using the steroidogenic response of isolated adrenal cells. *J. Endocrinol.* **59,** 43–55.
21. Rae, P. A. and Schimmer, B. P. (1974) Iodinated derivatives of adrenocorticotropic hormone. *J. Biol. Chem.* **249,** 5649–5653.
22. Buckley, D. I., Yamashiro, D., and Ramachandran, J. (1981) Synthesis of a corticotropin analogue that retains full biological activity after iodination. *Endocrinology* **109,** 5–9.
23. Penhoat, A., Jaillard, C., and Saez, J. M. (1989) Corticotropin positively regulates its own receptors and cAMP response in cultured bovine adrenal cells. *Proc. Natl. Acad. Sci. U. S. A.* **86,** 4978–4981.
24. Darbeida, H. and Durand, P. (1990) Mechanism of glucocorticoid enhancement of the responsiveness of ovine adrenocortical cells to adrenocorticotropin. *Biochem. Biophys. Res. Commun.* **166,** 1183–1191.
25. Lebrethon, M. C., Naville, D., Begeot, M., and Saez, J. M. (1994a) Regulation of corticotropin receptor number and messenger RNA in cultured human adrenocortical cells by corticotropin and angiotensin II. J. *Clin. Invest* **93,** 1828–1833.
26. Yasumura, Y., Bunonassissi, V., and Sato, G. (1966) Clonal analysis of differentiated function in animal cell cultures. I. Possible correlated maintenance of differentiated function and the diploid karyotype. *Cancer Res.* **26,** 529–535.
27. Schioth, H. B., Chhajlani, V., Muceniece, R., Klusa, V., and Wikberg, J. E. S. (1996) Major pharmacological distinction of the ACTH receptor from other melanocortin receptors. *Life Sci.* **59,** 797–801.

28. Rae, P. A., Gutmann, N. S., Tsao, J., and Schimmer, B. P. (1979) Mutations in cyclic AMP–dependent protein kinase and corticotropin (ACTH)–sensitive adenylate cyclase affect adrenal steroidogenesis. *Proc. Natl. Acad. Sci. U. S. A.* **76,** 1896–1900.
29. Wong, M., Krolczyk, A. J., and Schimmer, B. P. (1992) The causal relationship between mutations in cAMP–dependent protein kinase and the loss of adrenocorticotropin–regulated adrenocortical functions. *Mol. Endocrinol.* **6,** 1614–1624.
30. Waterman, M. R. and Bischof, L. J. (1996) Mechanisms of ACTH (cAMP)–dependent transcription of adrenal steroid hydroxylases. *Endocr. Res.* **22,** 615–620.
31. Munari–Silem, Y., Lebrethon, M. C., Morand, I., Rousset, B., and Saez, J. M. (1995) Gap junction–mediated cell–to–cell communication in bovine and human adrenal cells: a process whereby cells increase their responsiveness to physiological corticotropin concentrations. *J. Clin. Invest.* **95,** 1429–1439.
32. Kojima, I., Kojima, K., and Rasmussen, H. (1985) Role of calcium and cAMP in the action of adrenocorticotropin on aldosterone secretion. *J. Biol. Chem.* **260,** 4248–4256.
33. Enyeart, J. J., Mlinar, B., and Enyeart, J. A. (1993) T–type Ca^{2+} channels are required for adrenocorticotropin–stimulated cortisol production by bovine adrenal zona fasciculata cells. *Mol. Endocrinol.* **7,** 1031–1040.
34. Coyne, M. D., Wang, G., and Lemos, J. R. (1996) Calcium channels do not play a role in the steroid response to ACTH in Y1 adrenocortical cells. *Endocr. Res.* **22,** 551–556.
35. Seelig, S. and Sayers, G. (1973) Isolated adrenal cortex cells, ACTH agonists, partial agonists, antagonists; cyclic AMP and corticosterone production. *Arch. Biochem. Biophys.* **154,** 230–239.
36. Goverde, H. J. M. and Smals, A. G. H. (1984) The anomalous effect of some ACTH–fragments missing the amino acid sequence 1–10 on the corticosteroidogenesis in purified isolated rat adrenals. *FEBS Lett.* **173,** 23–26.
37. Schyzer, R., Schiller, P., Seelig, S., and Sayers, G. (1971) Isolated adrenal cells: log dose response curves for steroidogenesis induced by $ACTH_{1-24}$, $ACTH_{1-10}$, $ACTH_{4-10}$ and $ACTH_{5-10}$. *FEBS Lett.* **19,** 229–231.
38. Seelig, S., Sayers, G., Schyzer, R., and Schiller, P. (1971) Isolated adrenal cells, $ACTH_{11-24}$, a competitive antagonist of $ACTH_{1-39}$ and $ACTH_{1-10}$. *FEBS Lett.* **19,** 232–234.
39. Bristow, A. F., Gleed, C., Fauchere, J.–L., Schwyzer, R., and Schulster, D. (1980) Effects of ACTH (corticotropin) analogues on steroidogenesis and cyclic AMP in rat adrenocortical cells. *Biochem. J.* **186,** 599–603.
40. Finn, F. M., Johns, P. A., Nishi, N., and Hoffman, K. (1976) Differential response to adrenocorticotropic hormone analogs of bovine adrenal plasma membranes and cells. *J. Biol. Chem.* **251,** 3576–3585.
41. Szalay, K. S., De Wied, D., and Stary, E. (1989) Effects of ACTH–(11–24) on the corticosteroid production of isolated adrenocortical cells. *J. Steroid Biochem.* **32,** 259–262.
42. Hoffman, K., Stehle, C. J., and Finn, F. M. (1988) identification of a protein in adrenal particulates that binds adrenocorticotropin specifically and with high affinity. *Endocrinology* **123,** 1355–1363.
43. Penhoat, A., Jaillard, C., and Saez, J. M. (1993) Identification and characterization of corticotropin receptors in bovine and human adrenals. *J. Steroid Biochem. Mol. Biol.* **44,** 21–27.

44. Bost, K. L., Smith, E. M., and Blalock, J. E. (1985) Similarity between the corticotropin (ACTH) receptor and a peptide encoded by an RNA that is complementary to ACTH mRNA. **82,** 1372–1375.
45. Mertz, L. M. and Catt, K. J. (1991) Adrenocorticotropin receptors: functional expression from rat adrenal mRNA in *Xenopus laevis* oocytes. *Proc. Natl. Acad. Sci. U. S. A.* **88,** 8525–8529.
46. Mountjoy, K. G., Robbins, L. S., Mortrud, M. T., and Cone, R. D. (1992) The cloning of a family of genes that encode melanocortin receptors. *Science* **257,** 1248–1251.
47. Raikhinstein, M., Zohar, M., and Hanukoglu, I. (1994) cDNA cloning and sequence analysis of the bovine adrenocorticotropic hormone (ACTH) receptor. *Biochem. Biophys. Acta* **1220,** 329–332.
48. Cammas, F. M., Kapas, S., Barker, S., and Clark, A. J. L. (1995) Cloning, characterisation and expression of a functional mouse ACTH receptor. *Biochem. Biophys. Res. Commun.* **212,** 912–918.
49. Xia, Y. and Wikberg, J. E. S. (1996) Localization of ACTH receptor messenger RNA by in situ hybridization in mouse adrenal gland. *Cell Tissue Res.* **286,** 63–68.
50. Cammas, F. M., Pullinger, G. D., Barker, S., and Clark, A. J. L. (1997) The mouse adrenocorticotropin receptor gene, characterization of its promoter and evidence for a role for the orphan nuclear receptor steroidogenic factor 1. *Mol. Endocrinol.* **11,** 867–876.
51. Shimizu, C., Kubo, M., Saeki, T., Matsumura, T., Ishizuka, T., Kijima, H., Kakinuma, M., and Koike, T. (1997) Genomic organization of the mouse adrenocorticotropin receptor. *Gene* **188,** 17–21.
52. Naville, D., Jaillard, C., Barjhoux, L., Durand, P., and Begeot, M. (1997) Genomic structure and promoter characterization of the human ACTH receptor gene. *Biochem. Biophys. Res. Commun.* **230,** 7–12.
53. Mountjoy, K. G., Bird, I. M., Rainey, W. E., and Cone, R. D. (1994) ACTH induces upregulation of ACTH receptor mRNA in mouse and human adrenocortical cell lines. *Mol. Cell. Endocrinol.* **99,** R17–R20.
54. Lebrethon, M. C., Jaillard, C., Naville, D., Begeot, M., and Saez, J. M. (1994) Effects of transforming growth factor–beta 1 on human adrenocortical fasciculata–reticularis cell differentiated functions. *J. Clin. Endocrinol. Metab.* **79,** 1033–1039.
55. Penhoat, A., Jaillard, C., Begeot, M., Durand, P., and Saez, J. M. (1996) Cycloheximide enhances ACTH–receptor messenger RNA through transcriptional and post-transcriptional mechanisms in bovine adrenocortical cells. *Mol. Cell. Endocrinol.* **121,** 57–63.
56. Weber, A., Kapas, S., Hinson, J., Grant, D. B., Grossman, A., and Clark, A. J. L. (1993) Functional characterization of the cloned human ACTH receptor, impaired responsiveness of a mutant receptor in familial glucocorticoid deficiency. *Biochem. Biophys. Res. Commun.* **197,** 172–178.
57. Naville, D., Barjhoux, L., Jaillard, C., Faury, D., Despert, F., Esteva, B., Durand, P., Saez, J. M., and Begeot, M. (1996) Demonstration by transfection studies that mutations in the adrenocorticotropin receptor gene are one cause of the hereditary syndrome of glucocorticoid deficiency. *J. Clin. Endocrinol. Metab.* **81,** 1442–1448.
58. Kapas, S., Cammas, F. M., Hinson, J. P., and Clark, A. J. L. (1996) Agonist and receptor binding properties of adrenocorticotropin peptides using the cloned mouse adrenocorticotropin receptor expressed in a stably transfected HeLa cell line. *Endocrinology* **137,** 3291–3294.

59. Schimmer, B. P., Kwan, W. K., Tsao, J., and Qiu, R. (1995) Adrenocorticotropin–resistant mutants of the Y1 adrenal cell line fail to express the adrenocorticotropin receptor. *J. Cell. Physiol.* **163,** 164–171.
60. Yang, Y.–K,, Ollmann, M. M., Wilson, B. D., Dickinson, C., Yamada, T., Barsh, G. S., and Gantz, I. (1997) Effect of recombinant agouti–signalling peptide on melanocortin action. *Mol. Endocrinol.* **11,** 274–280.
61. Elias, L. L. K., Huebner, A., Metherell, L. A., Canas, A., Warne, G. L., Bitti, M. L. M., Cianfirani, S., Clayton, P. E., Savage, M. O., and Clark, A. J. L. Tall stature in familial glucocorticoid deficiency (submitted).
62. Shepherd, T. H., Landing, B. H., and Mason, D. G. (1959) Familial Addison's disease. *Am. J. Dis. Child.* **97,** 154–162.
63. Migeon, C. J., Kenny, F. M., Kowarski, A., Snipes, C. A., Spaulding, J. S., Finkelstein, J. W., and Blizzard, R. M. (1968) The syndrome of congenital adrenocortical unresponsiveness to ACTH. report of six cases. *Pediatr. Res.* **2,** 501–513.
64. Allgrove, J., Clayden, G. S., Grant, D. B., and Macaulay, J. C. (1978) Familial glucocorticoid deficiency with achalasia of the cardia and deficient tear production. *Lancet* **1,** 1284–1286.
65. Weber, A., Wienker, T. F., Jung, M., Easton, D., Dean, H. J., Heinrichs, C., Reis, A., and Clark, A. J. L. (1996) Linkage of the gene for the triple A syndrome to chromosome 12q13 near the type II keratin gene cluster. *Hum. Mol. Genet.* **5,** 2061–2066.
66. Smith, E. M., Brosnan, P., Meyer, W. J., and Blalock, J. E. (1987) An ACTH receptor on human mononuclear leukocytes, relation to adrenal ACTH–receptor activity. *N. Engl. J. Med.* **317,** 1266–1269.
67. Yamaoka, T., Kudo, T., Takuwa, Y., Kawakami, Y., Itakura, M., and Yamashita, K. (1992) Hereditary adrenocortical unresponsiveness to adrenocorticotropin with a postreceptor defect. *J. Clin Endocrinol. Metab* **75,** 270–274.
68. Clark, A. J. L., McLoughlin, L., and Grossman, A. (1993) Familial glucocorticoid deficiency caused by a point mutation in the ACTH receptor. *Lancet* **341,** 461–462.
69. Tsigos, C., Arai, K., Hung, W., and Chrousos, G. P. (1993) Hereditary isolated glucocorticoid deficiency is associated with abnormalities of the adrenocorticotropin receptor gene. *J. Clin. Invest.* **92,** 2458–2461.
70. Weber, A., Toppari, J., Harvey, R. D., Klann, R. C., Shaw, N. J., Ricker, A. T., Nanto–Salonen, Bevan, J. S., and Clark, A. J. L. (1995) Adrenocorticotropin receptor gene mutations in familial glucocorticoid deficiency, relationships with clinical features in four families. *J. Clin. Endocrinol. Metab.* **80,** 65–71.
71. Tsigos, C., Arai, K., Latronico, A. C., DiGeorge, A. M., Rapaport, R., and Chrousos, G. P. (1995) A novel mutation of the adrenocorticotropin receptor (ACTH–R) gene in a family with the syndrome of isolated glucocorticoid deficiency, but no ACTH–R abnormalities in two families with the triple A syndrome. *J. Clin. Endocrinol. Metab.* **80,** 2186–2189.
72. Labbe, O., Desarnaud, F., Eggerickx, D., Vassart, G., and Parmentier, M. (1994) Molecular cloning of a mouse melanocortin 5 receptor gene widely expressed in peripheral tissues. *Biochemistry* **33,** 4543–4549.
73. Clark, A. J. L., Cammas, F. M., Watt, A., Kapas, S., and Weber, A. (1997) Familial glucocorticoid deficiency, one syndrome, but more than one gene. *J. Mol. Med.* **75,** 394–399.
74. Gantz, I., Tashiro, T., Barcroft, C., Konda, Y., Shimoto, Y., Miwa, H., Glover, T., Munzert, G., and Yamada, T. (1993) Localization of the genes encoding the

melanocortin–2 (adrenocorticotropic hormone) and melanocortin–3 receptors to chromosomes 18p11.2 and 20q13.2 – q13.3 by fluorescent in situ hybridization. *Genomics* **18,** 166–167.
75. Magenis, R. E., Smith, L., Nadeau, J. H., Johnson, K. R., Mountjoy, K. G., and Cone, R. D. (1994) Mapping of the ACTH, MSH, and neural (MC3 & MC4) melanocortin receptors in the mouse and human. *Mamm. Genome* **5,** 503–508.
76. Weber, A. and Clark, A. J. L. (1994) Mutations of the ACTH receptor gene are only one cause of familial glucocorticoid deficiency. *Hum. Mol. Genet.* **3,** 585–588.
77. Shenker, A., Laue, L., Kosugi, S., Merendino, J. J., Minegishi, T., and Cutler, G. B. (1993) A constitutively activating mutation of the luteinizing hormone receptor in familial male precocious puberty. *Nature* **365,** 652–654.
78. Parma, J., Duprez, L., Van Sande, J., Cochaux, P., Gervy, C., Mockel, J., Dumont, J., and Vassart, G. (1993) Somatic mutations in the thyrotropin receptor gene causing hyperfunctioning thyroid adenomas. *Nature* **365,** 649–651.
79. Robbins, L. S., Nadeau, J. H., Johnson, K. R., Kelly, M. A., Roselli–Rehfuss, L., Baack, E., Mountjoy, K. G., and Cone, R. D. (1993) Pigmentation phenotypes ofvariant extension locus alleles result from point mutations that alter MSH receptor function. *Cell* **72,** 827–834.
80. Latronico, A. C., Reincke, M., Mendonca, B. B., Arai, K., Mora, P., Allolio, B., Wajchenberg, B. L., Chrousos, G. P., and Tsigos, C. (1995) No evidence for oncogenic mutations in the adrenocorticotropin receptor gene in human adrenal neoplasms. *J. Clin. Endocrinol. Metab.* **80,** 875–877.
81. Light, K., Jenkins, P. J., Weber, A., Perrett, C., Grossman, A., Pistorello, M., Asa, S. L., Clayton, R. N., and Clark, A. J. L. (1995) Are activating mutations of the adrenocorticotropin receptor involved in adrenal cortical neoplasia? *Life Sci.* **56,** 1523–1527.
82. Saez, J. M., Dazord, A., Morera, A. M., and Bataille, P. (1975) Interactions of adrenocorticotropic hormone with its adrenal receptors. *J. Biol. Chem.* **250,** 1683–1689.
83. Catalano, R. D., Stuve, L., and Ramachandran, J. (1986) Characterization of corticotropin receptors in human adrenocortical cells. *J. Clin. Endocrinol. Metab.* **62,** 300 – 304.
84. Buckley, D. I. and Ramachandran, J. (1981) Characterization of corticotropin receptors on adrenocortical cells. *Proc. Natl. Acad. Sci. U. S. A.* **78,** 7431–7435.
85. Gallo–Payet, and Escher, E. (1985) Adrenocorticotropin receptors in rat adrenal glomerulosa cells. *Endocrinology* **117,** 38–46.
86. Carsia, R. V. and Weber, H. (1988) Protein malnutrition in the domestic fowl induces alterations in adrenocortical cell adrenocorticotropin receptors. *Endocrinology* **122,** 681–688.
87. Wu, S.-M., Stratakis, C. A., Chan, C. H. Y., Hallermeier, K. M. Bourdony, C. J. Rennert, O. M., and Chan, W. Y. (1998) Genetic heterogeneity of adrenocorticotropin (ACTH) resistance syndromes: Identification of a novel mutation of the ACTH receptor gene in hereditary glucocorticoid deficiency. *Mol Genet. Metab.* **64,** 256.
88. Chhajlani, V., Muceniece, R., and Wikberg, J. E. S. (1993) Molecular cloning of a novel human melanocortin receptor. *Biochem. Biophys. Res. Commun.* **195,** 866–873.
89. Elias, L. L. K., Weber, A., Pullinger, G. D., Mirtella, A., and Clark, A. J. L. (1999) Functional characterization of naturally occuring mutations of the human adrenocorticotropin receptor: poor correlation of phenotype and genotype *J. Clin. Endocrinol. Metabol.* **84,** 2766–2770.

CHAPTER 13

The Melanocortin-3 Receptor

Robert A. Kesterson

1. Introduction

This chapter deals with perhaps the least understood receptor for the melanocortin peptides, that is the melanocortin-3 receptor (MC3-R). Although naturally occurring and genetically engineered mutations have provided us with insight into the function of the other known melanocortin receptors, little is known about the physiologic role of the MC3-R. Therefore, in order to further our understanding and potentially ascribe a function to the MC3-R, I will review the literature which describes the cloning and tissue-specific expression of the MC3-R gene. Particular attention will be paid to the neural expression of the MC3-R, as well as the pharmacological characterization of this receptor in vitro as a "γ-MSH" melanocortin receptor. Additionally, I will review the recent data, which describes the pharmacologic interaction of agouti and agouti-related peptide with the MC3-R. Finally, I will describe in vivo data which convincingly demonstrates one physiologic role of γ-MSH in mediating the response of reflex natriuresis. Since introduction of antagonists of the MC3-R potently block the natriuretic response induced by γ-MSH, one likely physiologic function for the MC3 receptor has thereby been identified.

2. Cloning and Genomic Localization of the MC3-R Gene

After the successful cloning of MC1-R (MSHR) and MC2-R (ACTH-R) cDNAs using degenerate oligonucleotide primers designed to recognize known G protein-coupled receptors *(1,2)*, the MC3-R was the first new member of the melanocortin receptor gene family isolated using polymerase chain reaction (PCR) and low-stringency hybridization techniques based upon MC1-R and MC2-R sequences. In 1993, two groups reported the cloning and

The Melanocortin Receptors Ed.: R. D. Cone

characterization of both the rat *(3)* and human *(4)* MC3-R genes, while in 1994, a third group independently isolated the mouse MC3-R gene while screening a genomic library with a G protein-coupled receptor PCR fragment *(5)*. Previously, the sequence of an orphan G protein-coupled receptor genetically linked to non-insulin-dependent diabetes mellitus on human chromosome 20q was identified, which we now know is the MC3-R gene *(6,7)*. The genbank accession numbers for the cloned MC3-R genes are X70667, X74983, and L06155 for the rat, mouse, and human genes, respectively.

The genomic localization of the human MC3-R gene was mapped to position 20q13.2 by fluorescent *in situ* hybridization (FISH), while chromosomal mapping with intersubspecific panels localized the mouse MC3-R gene to a syntenic region on the distal half of chromosome 2 *(8)*. Although initial mapping suggested that the MC3-R gene was linked to the generalized epilepsy disorder known as benign familial neonatal convulsions (BFNC) *(9)*, cloning of the MC3-R gene from a family with BFNC failed to demonstrate mutations within the coding region of the human MC3-R gene *(Kesterson and Cone, unpublished observations)*. Furthermore, probands from several families with BFNC have recently been identified as having mutations in a novel potassium channel gene (KCNQ2) *(10,11)*. Therefore, the human MC3-R gene has yet to be associated with a known disease state.

3. Structure of MC3-R

The predicted primary structures and sequence similarities of the human, mouse, and rat MC3-R gene products are depicted in Fig. 1. Inspection of the MC3-R sequences reveals seven highly conserved putative transmembrane domains, potential protein kinase C (PKC) phosphorylation sites in the second intracellular loop and in the carboxy-terminus, along with a conserved cysteine residue found in the carboxy-terminal tail which may function as a membrane anchoring site if palmitoylated *(12)*. Additionally, there are three potential N-linked glycosylation sites located in the amino-terminal extracellular domain. Biochemical data supporting the latter comes from photoaffinity labeling studies of the rat MC3-R using an analog of α-MSH in which sodium dodecyl sulfate-polyacrylamide gel electrophoresis (SDS-PAGE) followed by autoradiography showed a single band at 53–56 kDa for the native receptor, but a 35-kDa band after deglycosylation with peptide N glycosidase F *(13)*.

The sequence of the rat MC3-R cDNA predicts an open reading frame of 323 amino acids encoding a 35,800-Da protein *(3)*, as does the mouse MC3-R genomic sequence *(5)*. By contrast, the human MC3-R gene apparently encodes for a protein of 360 amino acids in length *(4)* due to an extended N-terminal extracellular domain. The additional 37 residues do not appear to

```
                 MsiqkkylegdfvfpvssssflrtllepqlgsalltaMNaSCCLpSVqPTLPNgSEHlqAPffSNqSsSaFCEQVFIKPE   80 hMC3-R
                                                      MNSSCCLSSVsPmLPNLSEHPAAPpASNRSGSGFCEQVFIKPE   43 mMC3-R
                                                      MNSSCCpSSsyPTLPNLSqHPAAPsASNRSGSGFCEQVFIKPE   43 rMC3-R

 ------ I ------     ------ II ------                            ------ III ------
iFLsLGIVSLlENILVILAVVRNGNLHSPMYFFLCSLAvADMLVSvSNaLETIMIAivhSDdyTfEDQFIQHMDNIFDSMICISLVASICNLLAIAV  177 hMC3-R
VFLALGIVSLMENILVILAVVRNGNLHSPMYFFLCSLAaADMLVSLSNSLETIMIAVINSDSLTLEDQFIQHMDNIFDSMICISLVASICNLLAIAi  140 mMC3-R
VFLALGIVSLMENILVILAVVRNGNLHSPMYFFLlSLlqADMLVSLSNSLETIMIvVINSDSLTLEDQFIQHMDNIFDSMICISLVASICNLLAIAV  140 rMC3-R

              ------ IV ------      ------ V ------
DRYVTIFYALRYHSIMTVRKALTLIVAIWVCCGvCGVvFIVYSESKMVIVCLITMFFAMmLLMGTLYvHMFLFARLHVkRIAALPPADGVAPQQHSC  274 hMC3-R
DRYVTIFYALRYHSIMTVRKALTLIgvIWVCCGICGVMFIiYSESKMVIVCLITMFFAMVLLMGTLYIHMFLFARLHVQRIAvLPPAgvVAPQQHSC  237 mMC3-R
DRYVTIFYALRYHSIMTVRKALsLIVAIWVCCGICGVMFIVYSESKMVIVCLITMFFAMVLLMGTLYIHMFLFARLHVQRIAALPPADGVAPQQHSC  237 rMC3-R
                PKC

          ------ VI ------                ------ VII ------
          MKGAVTITILLGVFIFCWAPFFLHLVLIITCPTNPYCtCYTAHFNTYLVLIMCNSVIDPLIYAFRSLELRNTFrEILCGCNGMNLG  360 hMC3-R
          MKGAVTITILLGVFIFCWAPFFLHLVLIITCPTNPYCICYTAHFNTYLVLIMCNSVIDPLIYAFRSLELRNTFKEILCGCNsMNLG  323 mMC3-R
          MKGAVTITILLGVFIFCWAPFFLHLVLIITCPTNPYCICYTAHFNTYLVLIMCNSVIDPLIYAFRSLELRNTFKEILCGCNGMNVG  323 rMC3-R
                                                                                PKC   * *
```

Fig. 1. Amino acid alignment of the human, mouse, and rat melanocortin-3 receptors. Identical residues between two or more species are indicated by capital letters. Predicted transmembrane domains (I–VII) are indicated by solid bars. Potential protein kinase C (PKC) phosphorylation (solid underline), glycosylation (dashed underline), and palmitoylation (*) sites are indicated.

influence MC3-R activity, since, pharmacologically, the rodent and human MC3 receptors behave similarly *(see below)*. Since the extended amino terminus of the human MC3-R is predicted based upon an in-frame ATG codon in the genomic sequence of this intronless gene, the bonafide translation initiation codon may be the second ATG codon in the human gene. Furthermore, mutational data indicates that either translation initiation codon may be used in the human gene without affecting the binding activity of the resultant MC3-R in vitro *(14)*. Purification and characterization of human MC3-R protein may be necessary to clarify this discrepancy. Perhaps more importantly, in the absence of a selective MC3-R antibody, this extended amino terminus represents a region of the molecule which may be tagged with an epitope without compromising biologic activity.

4. Tissue-Specific Expression of MC3-R mRNA

4.1. Central Sites of Expression

Northern blot hybridization experiments demonstrated that the greatest expression of the MC3-R gene is in the brain, with two mRNA species of approximately 2.0 and 2.5 kb detected in rat hypothalamic poly(A)+ RNA, but not in other brain areas *(3)*. However, using the more sensitive technique of *in situ* hybridization, a thorough examination of MC3-R mRNA distribution in the rat brain demonstrated specifically labeled nuclei outside of the hypothalamus *(3)* (*see* Table 1), whereas analysis of MC3-R mRNA in the mouse brain also revealed additional expression sites in the thalamus, hippocampus, and cortex *(4)*.

Not surprisingly, MC3-R mRNA is found primarily in areas of the brain that receive direct innervation from proopiomelanocortin (POMC) immunoreactive neurons *(15)*. However, the arcuate nucleus, which contains all of the forebrain POMC-expressing neurons, displays moderate levels of MC3-R mRNA, while the nucleus of the solitary tract (NTS) containing the other central POMC expressing neurons *(16)* apparently does not express MC3-R mRNA. Figure 2 depicts the sites of expression of MC3-R mRNA in the rat brain, in addition to displaying hypothalamic and brainstem POMC neurons and their projections. MC3-R mRNA is also found in the anterior amygdala, hippocampus (CA1-3), and piriform cortex, which are regions not known to contain Nle^4, D-Phe^7-melanocyte-stimulating hormone (NDP-MSH) binding sites *(17)*. These sites potentially represent areas of the brain that send projections (and MC3-R protein) to POMC presumptive terminal fields originating from either the hypothalamus or the brainstem.

The expression of MC3-R mRNA in regions of the brain such as the anteroventral periventricular nucleus and posterior hypothalamic area suggest that MC3-Rs may play a role in cardiovascular and thermoregulatory control

Table 1
Distribution of MC3-R mRNA in the Rat Brain

Region	Signal	Specifically Labeled Nuclei
Hypothalamus	+++	Dorsomedial part of the ventromedial nucleus
	++	Arcuate nucleus
	++	Posterior hypothalamic area
	++	Anteroventral preoptic nucleus
	+(+)	Anterior hypothalamic nucleus
	+(+)	Lateral hypothalamic area
	+(+)	Medial preoptic nucleus
	+(+)	Lateral preoptic area
	+(+)	Ventral part of the premammillary nucleus
	+	Supramammillary nucleus
	+	Anteroventral periventricular nucleus
	+	Preoptic periventricular nucleus
	+	Posterior periventricular nucleus
	(+)	Dorsal part of the premammillary nucleus
Thalamus	++	Medial habenular nucleus
	+(+)	Paraventricular nucleus
	+	Central medial nucleus
	+	Rhomboid nucleus
Septum	++	Intermediate part of the lateral nucleus
	+	Dorsomedial nucleus of the bed nuclei of the stria terminalis
	+	Anterolateral nucleus of the bed nuclei of the stria terminalis
Hippocampus	+	CA1-3
Olfactory cortex	+	Piriform cortex
Amygdala	+	Anterior amygdaloid area
Other	++(+)	Ventral tegmental area
	++(+)	Central linear nucleus of raphe
	++	Interfasicular nuclei
	+	Periaqueductal gray
	(+)	Substantial inominata

The complete distribution of MC3-R mRNA was detected in various brain regions by *in situ* hybridization. Semiquantitative estimates of the signal are indicated: + (weak), ++ (moderate), and +++ (strong), with parentheses indicating intermediate levels.

From ref. *3*, with permission.

(18), while expression in the dorsomedial and ventromedial hypothalamic nuclei suggest a potential role for MC3 receptors in ingestive behaviors.

Since pharmacologic data indicates that the MC3-R may be the mediator of γ-MSH activity (see below), the distribution of γ-MSH in the central nervous

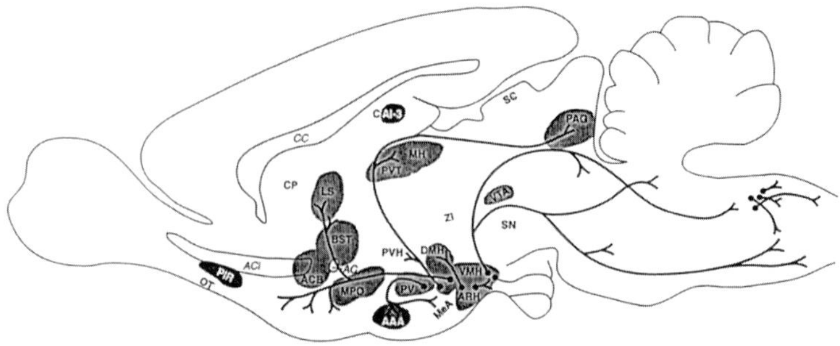

MC3-R mRNA

Fig. 2. Localization of melanocortin-3 receptor (MC3-R) messenger RNA (mRNA) in the rat brain. The major regions of MC3-R mRNA expression are shown by shaded areas projected onto a diagram of a rat brain sagittal section. Dots indicate the proopiomelanocortin (POMC) neurons in the hypothalamus and brainstem, and lines indicate the projections from these neurons. Dark shading indicates areas of MC3-R mRNA that do not correspond with regions of Nle^4, D-Phe^7-melanocyte-stimulating hormone (NDP-MSH) binding. AAA, anterior amygdala; AC, anterior commissure; ACB, nucleus accumbens; ACi, Anterior commissure, intrabulbar; ARH, arcuate nucleus hypothalamus; BST, bed nuclei of the stria terminalis; CA1-3, hippocampus; CC, corpus callosum; CP, caudate putamen; DMH, dorsomedial nucleus, hypothalamus; LS, lateral septal area; MeA, medial amygdala; MH, medial habenula; MPO, medial preoptic area; OT, olfactory tubercle; PAG, periaquaductal gray; PIR, piriform cortex; PV, periventricular zone; PVH, paraventricular nucleus, hypothalmus; PVT, paraventricular nucleus, thalamus; SC, superior colliculus; SN, substantia nigra; VMH, ventromedial nucleus, hypothalamus, VTA, ventral tegmental area; ZI, zona incerta. (From ref. *18*, with permission.)

system might be expected to coincide with MC3-R expression sites. In fact, like ACTH-immunoreactive areas, a high density of γ-MSH immunopositive fibers have been localized to regions in the rat brain such as the limbic system (septum, bed nucleus of the stria terminalis, medial amygdala, and thalamic periventricular nucleus), and the hypothalamus (preoptic, periventricular, paraventricular, and dorsomedial nuclei) *(19,20)*. However, γ-MSH immunopositive fibers in medullary cardiovascular control centers (ventrolateral medulla and commissural NTS) do not colocalize with MC3-R mRNA *(20)* nor with NDP-MSH binding sites *(17)*. Intense γ-MSH immunoreactivity has been found in blood vessels in the rat brain, but was not found to coincide with developmental or adult expression patterns for the MC3-R *(21)*.

Further research regarding the regulation of MC3-R gene expression has been hampered by the lack of an appropriate cell culture model since to date, there are no cell lines known to express endogenous MC3 receptors.

4.2. Peripheral Sites of Expression

In addition to the primarily central nervous system pattern of MC3-R mRNA expression, *in situ* hybridization of developing mouse embryo spinal cords has also demonstrated MC3-R mRNA expression in the peripheral nervous system *(W. Chen, unpublished observations)*. Northern analysis of poly(A)+ RNA has further established the presence of MC3-R transcripts of the appropriate size in human placenta, while MC3-R expression was also detected in several human gut tissues including the stomach, duodenum, and pancreas using a combination of RT-PCR and Southern blotting techniques *(4)*. In another study, PCR analysis of human tissues similarly detected MC3-R cDNA in the heart, while Southern blotting of amplified cDNA detected expression in the testis, ovary, mammary gland, skeletal muscle, and kidney *(22)*. This latter result is particularly intriguing since γ-MSH alters cardiovascular activity *(23–26)*, and pharmacologic data would also suggest that functional MC3 receptors reside in the kidney *(see below)*.

5. Pharmacologic Activity of MC3-R In Vitro

5.1. Melanocortin Agonists

As is the case for all melanocortin receptors, the MC3-R is functionally coupled through Gs to activate adenylyl cyclase and elevate intracellular cAMP production in response to stimulation by melanocortin peptides *(3–5)*. However, in contrast to the other melanocortin receptors, the MC3-R is also reported to be coupled to Gq, with modest activation of inositol 1,3,4-trisphosphate turnover and induction of intracellular calcium $[Ca^{2+}]_i$ in response to stimulation with α-MSH *(27)*. Although unique among the melanocortin receptors in its ability to respond to physiologic levels of γ-MSH, the MC3-R does not show apparent selectivity in its response to stimulation by the various melanocortin peptides α-, β-, γ-MSH nor ACTH *(3–5)*. Tables 2 and 3 summarize the pharmacologic activities of the human, mouse, and rat MC3 receptors obtained from heterologous expression systems.

5.2. Synthetic Agonists and Antagonists

Since many biologic activities have been ascribed to melanocortin peptides, recognition of the melanocortin receptor subtypes has led to the search for potent and specific agonists and antagonists in hopes of assigning function to each melanocortin receptor. The natural γ-MSH ligands are still the most selective agonists of the MC3-R, being approx 100× more potent at the MC3-R than the MC4-R. The introduction of bulky aromatic amino acids at position 7 of a synthetic cyclic α-MSH agonist (Ac-Nle^4-*c*[Asp^5,D-Phe^7, Lys^{10}]α-MSH[4–10]-NH_2) led to the discovery of potent antagonists of the

Table 2
Comparative Binding Activities of the Human, Mouse, and Rat MC3-R

Ligand	Human MC3-R (K_i)[a]	Mouse MC3-R (IC_{50})[b]	Rat MC3-R (K_i)[c]
α-MSH	20.7 ± 3.7	26	0.52 ± 0.44
β-MSH	13.4 ± 6.4	22	—
γ-MSH	17.7 ± 1.9	9	0.44 ± 0.62
ACTH-[1–39]	86.9 ± 23.9	12	—
NDP-MSH	0.22 ± 0.03	1.8	0.10 ± 0.18

Nanomolar values obtained from competition binding studies with ^{125}I-NDP-MSH.
[a]Data complied from ref. *57*.
[b]Data complied from ref. *5*.
[c]Data complied from ref. *3*.

Table 3
Functional Coupling of the Human, Mouse, and Rat MC3-R to Adenylyl Cyclase

Ligand	Human MC3-R[a]	Mouse MC3-R[b]	Rat MC3-R[c]
α-MSH	0.67 ± .36	1.15	3.8 ± 1.45
β-MSH	—	1.04	—
γ-MSH	—	0.56	3.8 ± 1.45
ACTH-[1–39]	—	3.05	3.8 ± 1.45
NDP-MSH	0.13 ± .03	0.58	1.6 ± 0.27

EC_{50} (n*M*) values obtained from accumulated intracellular cAMP levels.
[a]Data compiled from ref. *28*.
[b]Data compiled from ref. *5*.
[c]Data compiled from ref. *3*.

MC3-R and MC4-R *(28)*. Two of these compounds, SHU8914 (pI) and SHU9119 [D-Nal(2)] were identified as full agonists of the MC1-R and MC5-R, weak partial agonists of hMC3-R (EC_{50} 1134 ± 197 nM and 2813 ± 575n*M*, respectively), and subsequently characterized as potent antagonists of the hMC3-R (pA_2=8.3, each compound) as well as antagonists of the hMC4-R (pA_2=9.7 and 9.3, respectively). SHU9005, a linear iodo-substituted α-MSH analog, has also been characterized as a potent antagonist of the rat MC3-R (pA_2= 8.6) and the mouse MC4-R, but a full agonist of the human MC4-R, human MC1-R, and mouse MC5-R *(Kesterson and Cone, unpublished observations)*. Although unable to unequivocally discriminate between the rodent neural melanocortin receptor subtypes, these antagonists of MC3 and MC4 receptors have now been used in vivo to define melanocortin pathways which influence physiologic control of feeding *(29)*, cardiovascular activity *(26)*, thermoregulation *(30)*, and natriuresis (see below). Meanwhile, using a vari-

Table 4
Species-Dependent Binding Activities of Agouti Signaling Peptide

Ligand	Human MC4-R (K_i)[a]	Human MC3-R (K_i)[a]	Mouse MC4-R (IC_{50})[b]	Rat MC3-R (IC_{50})[b]
Human ASP	70 ± 18	140 ± 56	—	—
Murine ASP	54 ± 18	190 ± 74	3.9 ± 0.6	>100

Nanomolar values obtained from competition binding studies with ^{125}I-NDP-MSH.
[a]Data compiled from ref. *38*.
[b]Data compiled from ref. *29*.
[c]Data compiled from ref. *56*.

ety of synthetic melanocortin derivatives of ACTH[4–10] with partially selective antagonist activity at the neural melanocortin receptors, the phenomenon of α-MSH-induced excessive grooming behavior has also tentatively been ascribed to MC4-R and not MC3-R *(31)*.

5.3. Agouti and Agouti-Related Transcript Peptides

The genetic locus *agouti* encodes for the naturally occurring peptide antagonist of the MC1-R *(32)*, which when ectopically expressed in the C57BL/6J-A^y mouse generates a unique phenotype. Not only is there a complete pheomelanization of the coat, but additional characteristics include hyperglycemia, hyperinsulinemia in males, late-onset obesity, and increased linear growth (reviewed in ref. *33*). These latter attributes appear to be solely due to agouti signaling protein's (ASP) ability to antagonize the MC4-R, since the genetic deletion of the murine MC4-R demonstrated a phenotype that is virtually indistinguishable from the C57BL/6J-A^y mouse, albeit without the yellowing of the fur *(34)*.

Other research has suggested an alternative view of ASP action based on the observations that ASP is capable of inducing increased $[Ca^{2+}]^i$ in skeletal muscle cultures *(35)*, as well as in human embryonic kidney cells (HEK-293 cells) transfected with either the human MC1-R or human MC3-R *(36)*. Coupled with the report that the MC3-R is also functionally linked to Gq *(27)*, these data suggest that the MC3-R is a potential candidate for a receptor mediating the effects of ASP since inhibition binding assays also indicate that ASP may have some limited affinity for MC3-R, depending upon the species source of both ligand and receptor. However, as can be seen in Table 4, the relative affinity of ASP for MC3-R is always significantly lower than when compared to the affinity for MC4-R. It must be kept in mind that the measurements of the relative affinity of ASP for the various melanocortin receptors have been determined with baculovirus-expressed ASP protein *(29,32,37–39)*, which may not reflect the true characteristics of the mammalian protein, and for which there have been no adequate controls to determine nonspecific interactions.

Since murine ASP is normally only expressed during a short period in the hair growth cycle *(40)*, the normal functional significance, if any, of affinity for the MC3-R remains to be determined. However, since human ASP mRNA is expressed in testis, ovary, heart, and kidney *(41)* (as is human MC3-R mRNA), these tissues represent potentially relevant sites of expression and thereby regulation of MC3-R activity by ASP. Perhaps more interesting is the recent isolation of ART, a novel agouti-related transcript (also known as AGRP or agouti related protein), which in humans is expressed primarily in the adrenal gland, subthalamic nucleus, and hypothalamus, with a lower level of expression occurring in testis, lung, and kidney *(42)*. Remarkably, *in situ* histochemistry also demonstrated that the murine ART homolog is centrally expressed primarily in the arcuate nucleus of the hypothalamus, and is elevated in the murine models of obesity, which are deficient in leptin signaling *(42)*.

Recombinant human ART protein, or AGRP, was subsequently found to bind in vitro to the human MC3-R and the human MC4-R with high affinity (IC_{50} = 1.1 ± 0.5 n*M* and 0.5 ± 0.1 n*M*, respectively), and to a lessor extent the human MC5-R (IC_{50} > 40 n*M*) *(43)*. Functional activation curves indicate that human AGRP is a potent antagonist of the human MC3 and MC4 receptors, a limited antagonist of the MC5-R, but is not an antagonist of the human MC1 or MC2 receptors *(44)*. In the case of the MC4-R, dose-response curves representing stimulation of cAMP production by α-MSH in the presence of AGRP are not consistent with a competitive antagonism model, which suggests that other proteins may be involved in AGRP inhibition of MC4-R activity.

Remarkably, when human or murine AGRP is overexpressed in vivo using a β-actin promoter, transgenic AGRP mice are phenotypically similar to MC4-R-deficient and C57BL/6J-A^y animals *(44,45)*. When compared to nontransgenic littermates, AGRP overexpressing animals are hyperinsulinemic, hyperglycemic (males only), hyperphagic, and obese. Moreover, AGRP-overexpressing animals display pancreatic-islet hypertrophy similar to C57BL/6J-A^y mice, but do not show yellowing of the fur, thus indicating that AGRP does not influence MC1-R function in vivo. Unfortunately, no additional unique phenotype has been ascribed to AGRP transgenic mice, which might lead to an understanding of the physiologic role of MC3 receptors.

5.4. Chimeric Receptors

Since the melanocortin receptor subtypes each display a unique pharmacologic response to various endogenous and exogenous ligands, chimeric receptor studies have been initiated in order to determine key domains involved in functional coupling and in ligand recognition. To define receptor domains necessary for agouti interaction with melanocortin receptors, chimeric MC3/MC4 receptors were created because agouti binds with higher affinity to human MC4-R than rat MC3-R. As seen in Fig. 3, competition binding studies using ^{125}I-NDP-α-MSH and purified murine agouti protein indicate that one domain required for the high-affinity binding site of agouti resides within the carboxy terminus of the MC4 receptor *(Kesterson, Adan, and Cone, unpublished observations)*. By contrast, MC1-R in vitro mutagenesis studies have suggested that key determinants of α-MSH ligand binding reside within the amino-terminus of the melanocortin receptors (*see* Chapter 10).

The generation of chimeric MC3/MC1 receptors by substituting domains TM4, EL2, or TM5 of the human MC3-R with corresponding domains of the MC1-R did not substantially effect binding of α-MSH or NDP-α-MSH; that is, the chimeric receptors pharmacologically behaved as MC3 receptors *(46)*. Since the MC1-R has a 100-fold higher affinity for α-MSH, and the chimeric MC3/MC1 receptors did not acquire MC1-R-like activity, the authors concluded that TM4, EL2, and TM5 do not directly participate in ligand binding. However, this conclusion presumes conserved amino acids that were not altered by the chimeric receptors do not participate in ligand binding. Another assumption is that single residues involved in ligand binding behave independently, whereas these results do not address the likelihood that multiple residues (or domains) acting in concert actually confer the higher affinity state.

6. Physiology of γ-MSH

6.1. Cardiovascular Effects and Unidentified Receptors

The initial observation that γ^2-MSH possessed pressor and cardioaccelerator activities 10-fold more potent than ACTH[4–10] was made following intravenous administration of POMC peptides in conscious rats (*23;* reviewed in ref. *24*). Subsequent research demonstrated that these hemodynamic effects of γ^2-MSH were dependent upon the state of arousal or sympathetic tone of the animal, since when under deep anesthesia, γ^2-MSH produced a depressor effect and slight bradycardia *(47)*. Peripheral administration of γ^2-MSH, either intracisternal or intravenous, induced a greater response in both blood pressure and heart rate than intracerebroventricular (icv) administration *(48)*, thereby indicating that the central nervous system (CNS) may not be the principal target of γ^2-MSH action. Another interpretation of this data would suggest that the hindbrain is a potential site of action of γ^2-MSH, possibly through either afferent innervation of the nucleus tractus solitarius (NTS) from arterial baroreceptors, or possibly due to the lack of a blood–brain barrier in circumventricular regions such as the area postrema. However, when γ^2-MSH was injected directly into the NTS, an unexpected decrease in blood pressure

IDENTIFICATION OF MC4 RECEPTOR DOMAINS REQUIRED FOR AGOUTI FUNCTION

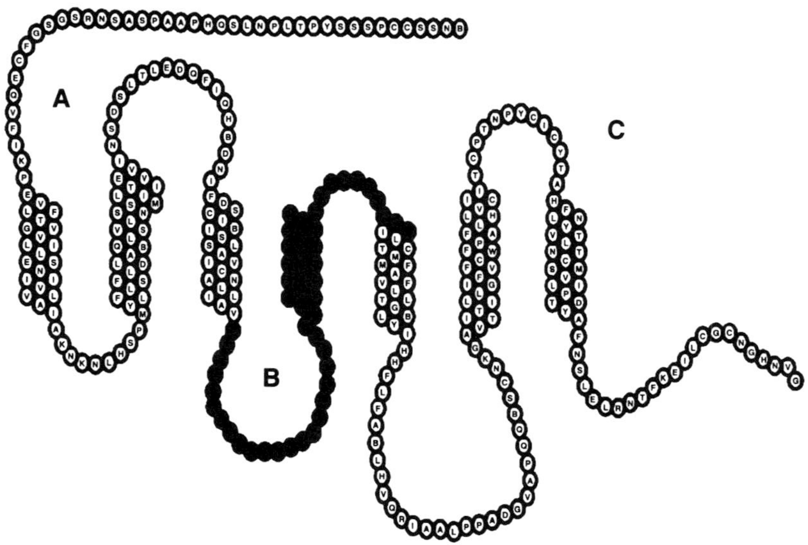

MC3 Receptor / MC4 Receptor Chimeras

Receptor	AGOUTI Antagonism
MC3R	no
MC4R	yes
4a3b4c	yes
3a4bc	yes
3ab4c	yes
4ab3c	no

Fig. 3. Agouti binding maps to the carboxy-terminus of the MCR-4. Shown is a schematic of human MC4-R/rat MC3-R chimeric receptor 4a3b4c which depicts domain A (amino-terminus through TM3), domain B (intracellular loop 2, TM4, and extracellular loop 2) and domain C (TM5 through carboxy-terminus). Competition binding studies using ^{125}I-NDP-α-MSH indicate that only chimeric receptors maintaining MC4-R domain C are inhibited with purified murine agouti protein (40n*M*).

and heart rate resulted *(47)*. Similarly, α-MSH was found to be more potent than γ^2-MSH in eliciting this response when microinjected into the medullary dorsovagal complex (DVC), an area that includes the NTS and the dorsal motor nucleus of the vagus *(26)*. In contrast to γ^2-MSH, α-MSH did not affect blood pressure nor heart rate when administered peripherally.

The opposing cardiovascular responses to γ-MSH, which are dependent on the route of administration, suggests that either γ-MSH is acting on a single melanocortin receptor that serves contrasting functions dependent upon its site of expression, or is acting on different melanocortin receptors (e.g., MC3-R versus MC4-R) that have opposing actions. The MC3-R is a possible candidate to mediate some of these cardiovascular responses to melanocortins and γ-MSH, since in vitro, MC3-Rs uniquely respond to physiologic levels of γ-MSH. Since the hypotensive and bradycardic effects induced by α-MSH microinjected into the DVC can be completely inhibited by pre-treatment with the antagonist SHU9119 *(26)*, either MC3 or MC4 receptors are mediating the central cardiovascular responses to melanocortins. However, since neither SHU9119 nor SHU9005 (both potent antagonists of the rat MC3-R) were able to inhibit the peripheral pressor and tachycardic effects of γ-MSH, there is likely an unidentified "melanocortin" receptor yet to be discovered.

6.2. Natriuretic Effects

Acute unilateral nephrectomy (AUN) induces an increase in both potassium and sodium excretion by the remaining kidney through an adaptive mechanism, which is dependent upon intact pituitary function *(49)*, as well as innervation of both kidneys prior to AUN *(50)*. An initial screen of POMC peptides as candidate mediators of natriuresis identified the N-terminal fragment of POMC, but not the endorphin encoding region of POMC based upon immunoreactive activity found in serum following AUN *(49)*. Direct evidence for the involvement of γ-MSH comes from studies in which the infusion of γ-MSH into the renal artery induces natriuresis in the ipsilateral, but not contralateral kidney *(51,52)*. Further research demonstrated that while all of the MSH peptides have some natriuretic activity, an antibody specific to γ-MSH was able to block the experimental induction of natriuresis by AUN, thereby suggesting a specific role for γ-MSH in this experimental system *(51)*.

In order to identify the melanocortin receptor which might be mediating the natriuretic effects of γ-MSH, Humphreys and colleagues *(53)* have induced natriuresis either by AUN, or by administration of γ-MSH agonists in the presence of the MC3-R and MC4-R selective antagonist SHU9119. Figure 4 shows that an increase in sodium excretion ($U_{Na}V$) induced by intravenous injection of NDP-γ-MSH is completely blocked by infusion of SHU9119 into the renal

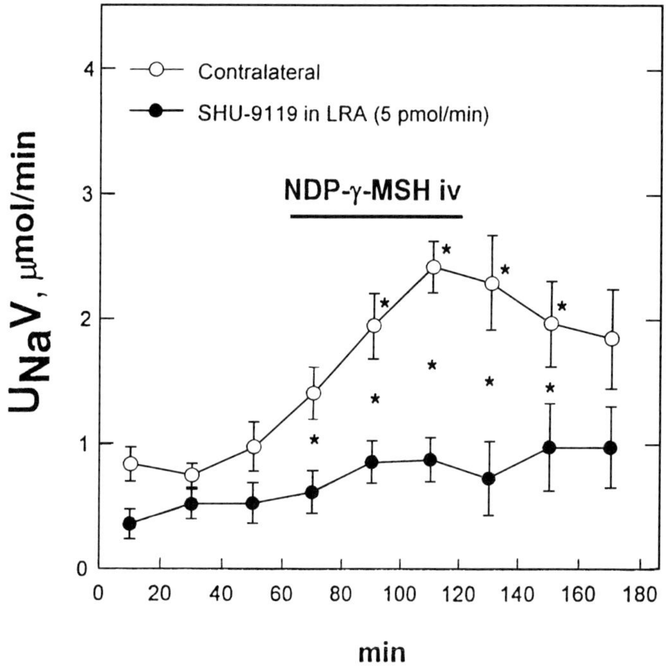

Fig. 4. Effect of intravenous NDP-γ-MSH (2 pmol/min) on $U_{Na}V$ during continuous infusion of SHU9119 into the left renal artery (5 pmol/min). NDP-γ-MSH was infused for the second hour of the 3-h experiment, indicated by the horizontal line, and led to a large increase in $U_{Na}V$ from the right kidney. SHU9119 completely prevented natriuresis from the left kidney, but had no effect on the contralateral kidney. (Courtesy of Mike Humphreys.)

artery of the ipsilateral kidney at 5 pmol/min; however, infusion of vehicle or SHU9119 at the lower dose of 1 pmol/min was ineffective (data not shown). Additionally, increased $U_{Na}V$ could also be blocked by SHU9005, a potent antagonist of the rat MC3-R (pA_2=8.6) and a full agonist of the human MC4-R, human MC1-R, and mouse MC5-R *(Kesterson and Cone, unpublished observations)*.

Altogether, these data suggest a model for AUN in which MC3 receptors respond to elevations of plasma γ-MSH by mediating a signal to the kidney to increase sodium excretion. Since denervation ablates the natriuretic response, MC3 receptors are possibly restricted to renal nerve termini, which would explain the inability to previously detect specific binding of γ-MSH in rat kidney *(54)*. Further research will, however, be necessary to determine if this reflex pathway for regulating sodium metabolism plays any role in the normal physiologic control of sodium balance.

7. Perspectives

Although we are still left without a thorough understanding of the physiologic role of the MC3-R, several lines of evidence would suggest that we may already have clues to the function of this challenging receptor. Since the initial cloning of the MC3-R as a candidate gene for non-insulin-dependent diabetes mellitus, there has been little research to follow up on this linkage. However, recent data suggest that this association may be worth reexamining. For instance, overexpression of AGRP (the naturally occurring MC4-R and MC3-R antagonist) in mice results in an obese and diabetic phenotype which is similar to that of MC4-R deficient animals, but does not include any additional reported phenotype. Furthermore, conservation as well as the functional importance of melanocortinergic signaling in humans has now been established, based upon the identification and characterization of a remarkable set of patients with defective POMC alleles *(55)*. Mutations in the human POMC gene lead to severe early-onset obesity, adrenal insufficiency, and red hair, all of which can be accounted for by our present understanding of the physiologic role for the MC4-R, MC2-R, and MC1-R, respectively. As is the case for the AGRP-overexpressing mice, the presumed loss of MC3-R activity does not induce any additional recognizable phenotypic changes. This implies that either the MC3-R could be involved in modulating feeding behavior pathways, or that the MC3-R may play a more subtle physiologic role that is presently masked in these animals (e.g., regulating sodium metabolism).

In summary, there has been a resurgence in research interest in melanocortins and their receptors, brought about by the cloning and definition of the biologic function of four of the five murine receptors. This interest is now leading to the development of specific agonists and antagonists of the melanocortin receptors by the pharmaceutical industry. In the near future, I hope that this will enable researchers to more precisely define a physiologic role for the MC3-R.

References

1. Mountjoy, K. G., Robbins, L. S., Mortrud, M. T., and Cone, R. D. (1992) The cloning of a family of genes that encode the melanocortin receptors. *Science* **257,** 1248–1251.
2. Chhajlani, V. and Wikberg, J. E. (1992) Molecular cloning and expression of the human melanocyte stimulating hormone receptor cDNA. *FEBS Lett.* **309,** 417–420.
3. Roselli–Rehfuss, L., Mountjoy, K. G., Robbins, L. S., Mortrud, M. T., Low, M. J., Tatro, J. B., Entwistle, M. L., Simerly, R. B., and Cone, R. D. (1993) Identification of a receptor for gamma melanotropin and other proopiomelanocortin peptides in the hypothalamus and limbic system. *Proc. Natl. Acad. Sci. U. S. A.* **90,** 8856–8860.

4. Gantz, I., Konda, Y., Tashiro, T., Shimoto, Y., Miwa, H., Munzert, G., Watson S. J., DelValle, J., and Yamada, T. (1993) Molecular cloning of a novel melanocortin receptor. *J. Biol. Chem.* **268,** 8246–8250.
5. Desarnaud, F., Labbe, O., Eggerickx, D., Vassart, G., and Parmentier, M. (1994) Molecular cloning, functional expression and pharmacological characterization of a mouse melanocortin receptor gene. *Biochem. J.* **299,** 367–373.
6. Bell, G. I., Xiang, K., Newman, M. V., Wu, S., Wright, L. G., Fajans, S. S., Spielman, R. S., and Cox, N. J. (1991) Gene for non–insulin dependent diabetes mellitus (maturity–onset diabetes of the young subtype) is linked to DNA polymorphism on human chromosome 20q. *Proc. Natl. Acad. Sci. U. S. A.* **88,** 1484–1488.
7. Yamada, Y., Xiang, K., Bell, G. I., Seino, S., and Nishi, M. (1992) Dinucleotide repeat polymorphism in a gene on chromosome 20 encoding a G–protein coupled receptor (D20S32e). *Nucleic Acids Res.* **19,** 2519.
8. Magenis, R. E., Smith, L., Nadeau, J. H., Johnson, K. R., Mountjoy, K. G., and Cone, R. D. (1994) Mapping of the ACTH, MSH, and neural (MC3 and MC4) melanocortin receptors in the mouse and human. *Mamm. Genome* **5,** 503–508.
9. Malafosse, A., Leboyer, M., Dulac, O., Navelet, Y., Plouin, P., Beck, C., LaklouH., Mouchnino, G., Grandscene, P., Vallee, L., Guilloud–Bataille, M., Samolyk, D., Baldy–Moulinier, M., Feingold, J., and Mallet, J. (1992) Confirmation of linkage of benign familial neonatal convulsions to D20S19 and D20S20. *Hum. Genet.* **89,** 54–58.
10. Biervert, C., Schroeder, B. C., Kubisch, C., et al. (1998) A potassium channel mutation in neonatal human epilepsy. *Science* **279,** 403–406.
11. Singh, N. A., Charlier, C., Stauffer, D., et al. (1998) A novel potassium channel gene, KCNQ2, is mutated in an inherited epilepsy of newborns. *Nature Genet.* **1,** 25–29.
12. O'Dowd, B. F., Hnatowich, M., Caron, M. G., Lefkowitz, R. J., and Bouvier, M. (1989) Palmitoylation of the human beta 2–adrenergic receptor. Mutation of Cys341 in the carboxyl tail leads to an uncoupled nonpalmitoylated form of the receptor. *J. Biol. Chem.* **264,** 7564–7569.
13. Sahm, U. G., Qarawi, M. A., Olivier, G. W., Ahmed, A. R., Branch, S. K., Moss, S. H., and Pouton, C. W. (1994) The melanocortin (MC3) receptor from rat hypothalamus: photoaffinity labelling and binding of alanine–substituted alpha–MSH analogues [published erratum appears in *FEBS Lett* 1994 Sep 5; 351(2):295]. *FEBS Lett.* **350,** 29–32.
14. Schioth, H. B., Muceniece, R., Wikberg, J. E., and Szardenings, M. (1996) Alternative translation initiation codon for the human melanocortin MC3 receptor does not affect the ligand binding. *Eur. J. Pharmacolo.* **314,** 381–384.
15. Jacobowitz, D. M. and O'Donohue, T. L. (1978) Alpha–Melanocyte stimulating hormone: immunohistochemical identification and mapping in neurons of rat brain. *Proc. Natl. Acad. Sci. U. S. A.* **75,** 6300–6304.
16. Bronstein, D. M., Schafer, M. K., Watson, S. J., and Akil, H. (1992) Evidence that beta–endorphin is synthesized in cells in the nucleus tractus solitarius: detection of POMC mRNA. *Brain Res.* **587,** 269–275.
17. Tatro, J. B. and Entwistle, M. L. (1994) Heterogeneity of brain melanocortin receptors suggested by differential ligand binding in situ. *Brain Res.* **635,** 148–158.
18. Low, M. J., Simerly, R. B., and Cone, R. D. (1994) Receptors for the melanocortin peptides in the central nervous system. *Curr. Opin. Endrocrinol. Diabetes* **1,** 79–88.

19. Kawai, Y., Inagaki, S., Shiosaka, S., Shibasaki, T., Ling, N., Tohyama, M., and Shiotani, Y. (1984) The distribution and projection of gamma–melanocyte stimulating hormone in the rat brain: an immunohistochemical analysis. *Brain Res.* **297,** 21–32.
20. Fodor, M., Sluiter, A., Frankhuijzen–Sierevogel, A., Wiegant, V. M., Hoogerhout, P., de Wildt, D. J., and Versteeg, D. H. (1996) Distribution of Lys–gamma 2–melanocyte–stimulating hormone– (Lys–gamma 2–MSH)–like immunoreactivity in neuronal elements in the brain and peripheral tissues of the rat. *Brain Res.* **731,** 182–189.
21. Xia, Y. and Wikberg, J. E. S. (1997) Postnatal expression of melanocortin–3 receptor in rat diencephalon and mesencephalon. *Neuropharmacology* **36,** 217–224.
22. Chhajlani, V. (1996) Distribution of cDNA for melanocortin receptor subtypes in human tissues. *Biochem. Mole. Biol. Int.* **38,** 73–80.
23. Klein, M. C., Hutchins, P. M., Lymangrover, J. R., and Gruber, K. A. (1985) Pressor and cardioaccelerator effects of gamma MSH and related peptides. *Life Sci.* **36,** 769–775.
24. Gruber, K. A. and Callahan, M. F. (1989) ACTH–(4–10) through gamma–MSH: evidence for a new class of central autonomic nervous system–regulating peptides. *Am. J. Physiol.* **257,** R681–R694.
25. Van Bergen, P., Janssen, P. M., Hoogerhout, P., de Wildt, D. J., and Versteeg, D. H. (1995) Cardiovascular effects of gamma–MSH/ACTH–like peptides: structure–activity relationship. *Eur. J. Pharmacol.* **294,** 795–803.
26. Li, S. J., Varga, K., Archer, P., Hruby, V. J., Sharma, S. D., Kesterson, R. A., Cone, R. D., and Kunos, G. (1996) Melanocortin antagonists define two distinct pathways of cardiovascular control by alpha–and gamma–melanocyte–stimulating hormones. *J. Neurosci.* **16,** 5182–5188.
27. Konda, Y., Gantz, I., DelValle, J., Shimoto, Y., Miwa, H., and Yamada, T. (1994) Interaction of dual intracellular signaling pathways activated by the melanocortin–3 receptor. *J. Biol. Chem.* **269,** 13162–13166.
28. Hruby, V. J., Lu, D., Sharma, S. D., Castrucci, A. L., Kesterson, R. A., al–Obeidi, F. A., and Cone, R. D. (1995) Cyclic lactam alpha–melanotropin analogues of Ac–Nle4–cyclo[Asp5, D–Phe7,Lys10] alpha–melanocyte–stimulating hormone–(4–10)–NH2 with bulky aromatic amino acids at position 7 show high antagonist potency and selectivity at specific melanocortin receptors. *J. Med. Chem.* **38,** 3454–3461.
29. Fan, W., Boston, B. A., Kesterson, R. A., Hruby, V. J., and Cone, R. D. (1997) Role of melanocortinergic neurons in feeding and the agouti obesity syndrome. *Nature* **385,** 165–168.
30. Huang, Q. H., Entwistle, M. L., Alvaro, J. D., Duman, R. S., Hruby, V. J., and Tatro, J. B. (1997) Antipyretic role of endogenous melanocortins mediated by central melanocortin receptors during endotoxin–induced fever. *J. Neurosci.* **17,** 3343–3351.
31. Adan, R. A., Oosterom, J., Ludvigsdottir, G., Brakkee, J. H., Burbach, J. P., and Gispen, W. H. (1994) Identification of antagonists for melanocortin MC3, MC4 and MC5 receptors. *Euro. J. Pharmacol.* **269,** 331–337.
32. Lu, D., Willard, D., Patel, I. R., Kadwell, S., Overton, L., Kost, T., Luther, M., Woychik, R. P., Wilkison, W. O., and et al. (1994) Agouti protein is an antagonist of the melanocyte–stimulating–hormone receptor. *Nature* **371,** 799–802.
33. Yen, T. T., Gill, A. M., Frigeri, L. G., Barsh, G. S., and Wolff, G. L. (1994) Obesity, diabetes, and neoplasia in yellow A(vy)/– mice: ectopic expression of the agouti gene. *FASEB J.* **8,** 479–488.

34. Huszar, D., Lynch, C. A., Fairchild–Huntress, V., Dunmore, J. H., Fang, Q., Gu, W., Kesterson, R. A., Boston, B. A., Cone, R. D., Smith, F. J., Burn, P., and Lee, F. (1997) Targeted disruption of the melanocortin–4 receptor results in obesity in mice. *Cell* **88,** 131–141.
35. Zemel, M. B., Kim, J. H., Woychik, R. P., Michaud, E. J., Kadwell, S. H., and Patel, I. R. (1995) Agouti regulation of intracellular calcium: role in the insulin resistance of viable yellow mice. *Proc. Natl. Acad. Sci. U. S. A.* **92,** 4733–4737.
36. Kim, J. H., Kiefer, L. L., Woychik, R. P., Wilkison, W. O., Truesdale, A., Ittoop, O., Nichols, J., and Zemel, M. B. (1997) Agouti regulation of intracellular calcium: role of melanocortin receptors. *Am. J. Physiol.* **272,** E379–E384.
37. Willard, D. H., Bodnar, W., Harris, C., Kiefer, L., Nichols, J. S., Blanchard, S., Moyer, M., Burkhart, W., Weiel, J., and et al. (1995) Agouti structure and function: characterization of a potent alpha–melanocyte stimulating hormone receptor antagonist. *Biochemistry* **34,** 12,341–12,346.
38. Kiefer, L. L., Ittoop, O. R., Bunce, K., Truesdale, A. T., Willard, D. H., Nichols, J. S., Mountjoy, K., Chen, W. J., and Wilkison, W. O. (1997) Mutations in the carboxyl terminus of the agouti protein decrease agouti inhibition of ligand binding to the melanocortin receptors. *Biochemistry* **36,** 2084–2090.
39. Ying–Kui, Y., Ollmann, M. M., Wilson, B. D., Dickinson, C., Yamada, T., Barsh, G. S., and Gantz, I. (1997) Effects of recombinant agouti-signaling protein on melanocortin action. *Mol. Endocrinol.* **11,** 274–280.
40. Bultman, S. J., Michaud, E. J., and Woychik, R. P. (1992) Molecular characterization of the mouse agouti locus. *Cell* **71,** 1195–1204.
41. Wilson, B. D., Ollmann, M. M., Kang, L., Stoffel, M., Bell, G. I., and Barsh, G. S. (1995) Structure and function of ASP, the human homolog of the mouse agouti gene. *Human Mol. Gene.* **4,** 223–230.
42. Shutter, J. R., Graham, M., Kinsey, A. C., Scully, S., Luthy, R., and Stark, K. L. (1997) Hypothalamic expression of ART, a novel gene related to agouti, is up–regulated in obese and diabetic mutant mice. *Genes Dev.* **11,** 593–602.
43. Fong, T. M., Mao, C., MacNeil, T., Kalyani, R., Smith, T., Weinberg, D., Tota, M. R., and Van der Ploeg, L. H. T. (1997) ART (protein product of agouti–related transcript) as an antagonist of MC–3 and MC–4 receptors. *Biochem. Biophys. Res. Commun.* **237,** 629–631.
44. Ollmann, M. M., Wilson, B. D., Yang, Y. K., Kerns, J. A., Chen, Y., Gantz, I., and Barsh, G. S. (1997) Antagonism of central melanocortin receptors in vitro and in vivo by agouti–related protein. *Science* **278,** 135–138
45. Graham, M., Shutter, J. R., Sarmiento, U., Sarosi, I., and Stark, K. L. (1997) Overexpression of Agrt leads to obesity in transgenic mice. *Nat. Genet.* **17,** 273,274.
46. Schioth, H.B., Muceniece, R., Szardenings, M., Prusis, P., and Wikberg, J.E. (1996) Evidence indicating that the TM4, EL2, and TM5 of the melanocortin 3 receptor do not participate in ligand binding. *Biochem. Biophys. Res. Commun.* **229,** 687–692.
47. De Wildt, D. J., van der Ven, J. C., van Bergen, P., de Lang, H., and Versteeg, D. H. G. (1994) A hypotensive and bradycardic action of γ2–melanocyte–stimulating hormone (γ2–MSH) microinjected into the nucleus tractus solitarii of the rat. *Arch. Pharmacol.* **349,** 50–56.
48. Versteeg, D. H., Krugers, H., Meichow, C., De Lang, H., and de Wildt, D. J. (1993) Effect of ACTH–(4–10) and γ^2–MSH on blood pressure after intracerebroventricular and intracisternal administration. *J. Cardiovasc. Pharmacol.* **21,** 907–911.

49. Lin, S. Y., Wiedemann, E., and Humphreys, M. H. (1985) Role of the pituitary in reflex natriuresis following acute unilateral nephrectomy. *Am. J. Physiol.* **249,** 390–395.
50. Ribstein, J. and Humphreys, M. H. (1984) Renal nerves and cation excretion after acute reduction in functioning renal mass in the rat. *Am. J. Physiol.* **246,** F260–F265.
51. Lin, S. Y., Chaves, C., Wiedemann, E., and Humphreys, M. H. (1987) A gamma–melanocyte stimulating hormone–like peptide causes reflex natriuresis after acute unilateral nephrectomy. *Hypertension* **10,** 619–627.
52. Lymangrover, J. R., Buckalew, V. M., Harris, J., Klein, M. C., and Gruber, K. A. (1985) Gamma–2MSH is natriuretic in the rat. *Endocrinology* **116,** 1227–1229.
53. Ni, X. P., Kesterson, R. A., Sharma, S. D., Hruby, V. J., Cone, R. D., Wiedemann, E., and Humphreys, M. H. (1997) Prevention of reflex natriuresis after acute unilateral nephrectomy by melanocortin receptor antagonists. (1998) *Am. J. Physiol.* **274,** R931–R938.
54. Pedersen, R. C. and Brownie, A. C. (1983) Lys–gamma 3–melanotropin binds with high affinity to the rat adrenal cortex. *Endocrinology* **112,** 1279–1287.
55. Krude, H., Biebermann, H., Luck, W., Horn, R., Brabant, G., and Gruters, A. (1998) Severe early–onset obesity, adrenal insufficiency and red hair pigmentation caused by POMC mutations in humans. *Nat. Genet.* **19,** 155–157.
56. Yang, Y. K., Ollmann, M. M. M., Wilson, B. D., et al. (1997) Effects of recombinant agouti-signaling protein on melanocortin action. *Molec. Endocrinol.* **11,** 274–280.
57. Schioth, H. B., Muceniece, R., Wikberg, J. E., and Chhajlani, V. (1995) Characterization of melanocortin receptor subtypes by radioligand binding analysis. *Eur. J. Pharmacol.* **288,** 311–317.

CHAPTER 14

The Melanocortin-4 Receptor

Roger D. Cone

1. Introduction

After cloning of the melanocyte MC1-R *(1,2)* and adrenocortical MC2-R *(2)*, interest in the possibility of unique neural homologs of these receptors grew from observations of central effects of melanocortins, such as effects on learning and memory (reviewed in ref. *3*) and temperature control *(4)*. Furthermore, the *in situ* ligand binding experiments of Tatro had demonstrated the presence of high-affinity binding sites for (^{125}I-NDP-MSH) in rat brain *(5)*, and these as well as the physiologic experiments suggested these sites were encoded by pharmacologically distinct melanocortin receptors.

Degenerate polymerase chain reaction (PCR) and homology screening approaches (Chapter 7) have now produced two neural melanocortin receptors, the MC3-R (Chapter 13) and MC4-R. MC5-R mRNA (Chapter 15) has been reported in total brain *(6)* and in cerebellum *(7)*; however, bona fide MC5-R binding sites in brain have yet to be verified, and in situ hybridization data is not yet available to confirm neuronal or glial expression of this receptor.

The effects of melanocortins on learning and memory that motivated much of the initial work on the neural receptors may not, in the final analysis, be mediated by melanocortin receptors. Structure–activity relationship studies led to a synthetic "melanocortin" peptide, ORG2766, that was very active in depressing extinction of learned behavior in avoidance assays *(8)*, however, this peptide has now been demonstrated to have virtually no affinity for the central melanocortin receptors *(9,10)*.

Rather, the unexpected finding of a role for the MC4-R in energy homeostasis has been largely responsible for the renewed interest in melanocortins in general, and the MC4-R in particular *(11,12)*. This aspect of MC4-R function shall be the

The Melanocortin Receptors Ed.: R. D. Cone

focus of this chapter, although it is important to keep in mind that the broad distribution of the MC4-R mRNA in the brain suggests complex roles for this receptor in neuroendocrine and autonomic control.

2. Structure of the Melanocortin-4 Receptor

The MC4-R has now been cloned from mouse, human, hamster, rat, and chicken (Fig. 1). The receptor is a 332 amino acid long 7-membrane spanning protein encoded by a single coding exon. Like all the melanocortin receptors, the MC4-R lacks the conserved cysteines in extracellular loops 1 and 2 thought to form a disulfide bond in most G protein-coupled receptors (GPCRs). The receptor amino acid sequence is highly conserved, with 93% amino acid identity between rat and human, and 87% identity between chicken and human. Residues in predicted transmembrane domains are virtually identical across all MC4 receptors cloned thus far. Evolutionarily, the MC4-R seems more related to the MC3-R (55% overall identity) and MC5-R (61%) than the peripheral MC1-R (47%) or MC2-R (46%).

There are few data regarding the function of residues or domains of the MC4-R in ligand binding or receptor signaling. In one study, a MC4-R Ile137Thr variant found in an obese patient was expressed and found to bind ^{125}I-NDP-MSH with lower affinity than the wild-type receptor (K^d 9n*M* vs 1.2n*M*) *(13)*. The mutant receptor had an EC_{50} for elevation of intracellular cAMP by α-MSH that was approximately 15 times that of the EC_{50} for the wild-type receptor (37n*M* vs 2.5 n*M*). The nonconservative Ile137Thr change occurs in the third membrane spanning domain of the MC4-R, a domain shown in the MC1-R to be involved in ligand binding and receptor activation *(14)*. Val103Ile and Thr112Met mutant receptors were also characterized in this study, but did not differ pharmacologically from the wild-type receptor after transfection into the heterologous 293 cell line. The potential role of the MC4-R in human obesity is discussed in Subheading 5.3. below.

A second study examined the role of the relatively nonconserved amino terminal domains of the MC1-R, MC3-R, MC4-R, and MC5-R, and demonstrated that the bulk of the MC4-R amino terminal extracellular domain (aa 1–28) could be truncated without affecting the expression, ligand binding affinity, or activation of the receptor in COS cells *(15)*. Truncation of residues 1–34 resulted in a construct that did not produce any detectable receptor.

Of course, there are unique properties of the MC4-R that deserve structure–activity analysis. Most interestingly, the MC4-R binds both melanocortin antagonists agouti *(16)*, and agouti-related protein *(17)*, while MC3-R only binds AGRP, and MC1-R only binds agouti (Fig. 2). The MC5-R does not appear to bind agouti *(6,18)*, and has only low affinity for AGRP

```
Human     MVNSTHRGMHTSLHLWNRSSYRLHSNASESLGKGYSDGGCYEQL
Rat       MNSTHHHGMYTSLHLWNRSSHGLHGNASESLGKGHSDGGCYEQL
Hamster   MNSTHHHGMHTSLHFWNRSTYGLHSNASEPLGKGYSEGGCYEQL
Chicken   MNFTQHRGTLQPLHFWNQSN-GLHRGASEPSAKGHSSGGCYEQL

         TMI                          TMII

FVSPEVFVTLGVISLLENILVIVAIAKNKNLHSPMYFFICSLAVADMLVSVSNGSETI
FVSPEVFVTLGVISLLENILVIVAIAKNKNLHSPMYFFICSLAVADMLVSVSNGSETI
FVSPEVFVTLGVISLLENILVIVAIAKNKNLHSPMYFFICSLAVADMLVSVSNGSETI
FVSPEVFVTLGIISLLENVLVIVAIAKNKNLHSPMYFFICSLAVADMLVSVSNGSETI

                                      TMIII

IITLLNSTDTDAQSFTVNIDNVIDSVICSSLLASICSLLSIAVDRYFTIFYALQYHNI
VITLLNSTDTDAQSFTVNIDNVIDSVICSSLLASICSLLSIAVDRYFTIFYALQYHNI
VITLLNSTDTDAQSFTVNIDNVIDSVICSSLLASICSLLSIAVDRYFTIFYALQYHNI
VITLLNNTDTDAQSFTINIDNVIDSVICSSLLASICSLLSIAVDRYFTIFYALQYHNI

             TMIV                        TMV

MTVKRVGISISCIWAACTVSGILFIIYSDSSAVIICLITMFFTMLALMASLYVHMFLM
MTVRRVGIIISCIWAACTVSGVLFIIYSDSSAVIICLITMFFTMLVLMASLYVHMFLM
MTVKRVGIIISCIWAVCTVSGVLFIIYSDSSAVIICLITVFFTMLALMASLYVHMFLM
MTVKRVGVIITCIWAACTVSGILFIIYSDSSVVIICLISMFFTMLILMASLYVHMFMM

                                            TMVI

ARLHIKRIAVLPGTGAIRQGANMKGAITLTILIGVFVVCWAPFFLHLIFYISCPQNPY
ARLHIKRIAVLPGTGTIRQGANMKGAITLTILIGVFVVCWAPFFLHLLFYISCPQNPY
ARLHIKRIAVLPGTGTIRQGANMKGAITLTILIGVFVVCWAPFFLHLIFYISCPQNPY
ARMHIKKIAVLPGTGPIRQGANMKGAITLTILIGVFVVCWAPFFLHLIFYISCPYNPY

              TMVII

CVCFMSHFNLYLILIMCNSIIDPLIYALRSQELRKTFKEIICCYPLGGLCDLSSRY
CVCFMSHFNLYLILIMCNAVIDPLIYALRSQELRKTFKEIICFYPLGGICELPGRY
CVCFMSHFNLYLILIMCNSIIDPLIYALRSQELRKTFKEIICCYPLGGLCDLSSRY
CVCFMSHFNFYLILIMCNSIIDPLIYAFRSQELRKTFKEIICCCNLRGLCDLPGKY
```

Fig. 1. Amino acid sequences of the MC4-R (from Genbank). Approximate location of the transmembrane domains is indicated.

(~10^{-7}*M*) *(17,19,20)*. The MC2-R does not appear to bind AGRP *(20)*, however, one report suggests that agouti may be a noncompetitive antagonist of the MC2-R *(21)*.

```
               Putative Signal Sequence

mAgouti   MDVTRLLLAT LVSFLCFFTV HSH
hAgouti   MDVTRLLLAT LLVFLCFFTA NSH
mAGRP     MLTAMLLSCV LLLALPPTLG VQMG
hAGRP     MLTAAVLSCA LLLALPATRG AQMG


mAgouti   LALEETL GDDRSLRSNS SMNSLDFSSV SIVALNKKSK KISRKEAEKR KRSSKKKASM  80
hAgouti   LPPEEKL RDDRSLRSNS SVNLLDVPSV SIVALNKKSK QIGRKAAEK- KRSSKKEASM  79
mAGRP      VAPLKG IRRPDQALFP EFPGLSLNG- LKKTTADRAE EVLLQKAEAL AEVLDPQNRE  79
hAGRP      LAPMEG IRRPDQALLP ELPGLGLRAP LKKTTAEQAE EDLLQEAQAL AEVLDLQDRE  80


mAgouti   KKVARPPPPS --PCVATRDS CKPPAPACCD PCASCQCRFF GSACTCRVLN PNC          131
hAgouti   KKVVRPRTPL SAPCVATRNS CKPPAPACCD PCASCQCRFF RSACSCRVLS LNC          132
mAGRP     -------SRS PRRCVRLHES CLGQQVPCCD PCATCYCRFF NAFCYCRKLG TATNLCSRT   132
hAGRP     -------PRS|SRRCVRLHES CLGQQVPCCD PCATCYCRFF NAFCYCRKLG TAMNPCSRT|  133

               hAGRP Antibody Immunoreactive Fragment (83 to 132-NH2)
```

Fig. 2. Amino acid sequences of the agouti and agouti-related proteins from mouse and human. Protein sequence is divided according to domains; first row: putative signal sequence, second row: charged amino terminal domains, third row: cystein-rich conotoxin domain. Bold type indicates disulfide-bonded cysteine residues. Boxed sequence indicates the pharmacologically active AGRP[83–132] fragment against which antibodies have been raised.

3. Expression of the Melanocortin-4 Receptor

3.1. Expression in the Adult Rodent

Expression of the MC4-R appears to be highly localized to the nervous system *(22,23)*. Even using polymerase chain reaction, MC4-R mRNA was undetectable in a wide range of peripheral human tissues, with the exception of a very faint signal in pituitary *(24)*.

Within the adult rat brain, *in situ* hybridization demonstrated widespread expression of the MC4-R mRNA, with some areas of moderate expression within every major division of the central nervous system, including the cortex, thalamus, hypothalamus, and brainstem *(9)*. Rat brain nuclei containing MC4-R mRNA are listed in Table 1. The distribution of MC4-R expression contrasts with that of MC3-R, which has a much more limited distribution in the brain, being expressed primarily within the hypothalamus, and limited regions of the thalamus and brainstem *(10)*.

The widespread distribution of MC4-R expression implies complex roles for the receptor in a wide variety of physiologic processes, as discussed previously *(23)*. For example, the receptor is expressed in a number of areas involved in the processing of visual and auditory information (superior colliculus, auditory regions of the isocortex), as well as somatomotor control (caudoputamen, nucleus accumbens, substantia nigra, red nucleus). Taken as a whole, however, the neuroanatomic distribution of the receptor defines circuitry involved in autonomic and neuroendocrine functions, with the highest densities of expression in the hypothalamus, septal region, and brainstem. For example, within the paraventricular nucleus of the hypothalamus (PVH), the receptor mRNA is found within both parvicellular and magnocellular neurons. Additionally, the receptor is found in the dorsal zone of the medial parvicellular PVH, which contains CRH neurons that project to the median eminence, and is also found in the lateral parvicellular part and ventral zone of the medial parvicellular part of the PVH, both of which have descending projections mediating autonomic responses. Interestingly, a preliminary confocal microscopic analysis of the distribution of MC4-R protein in the mouse and rat demonstrated a high density of MC4-R immunoreactivity in nerve fibers in the medial parvicellular PVH, suggesting that many MC4-R-expressing neurons may send receptor-containing fibers to this site.

More recently, expression of proopiomelanocortin (POMC) and MC4-R in the adult rat spinal cord has also been demonstrated *(25)*. MC4-R mRNA was detected by Rnase protection in samples from rat lumbar spinal cord, while MC3-R and MC5-R mRNAs were undetectable. Species differences may exist, however, since MC5-R mRNA and binding sites have been demonstrated in mouse spinal cord *(26)*. The existence of POMC expression and MC4-R/MC5-R expression in spinal cord suggests the existence of a functional melanocortin system within this region.

3.2. Developmental Expression in the Rodent

The POMC system is one of the earliest peptidergic systems to be expressed in the rat brain, with POMC immunoreactivity occurring after the first appearance of the arcuate nucleus neurons at E12.5 *(27,28)*. Early reports demonstrated the existence of MSH binding sites during rat development in the CNS as well as cranial and sympathetic ganglia using in situ binding of the broad melanocortin agonist ^{125}I-NDP-MSH *(29,30)*. More recently, the specific distribution of MC4-R and MC3-R mRNAs was characterized in the developing rat by *in situ* hybridization *(31,32)*. The developing sympathetic nervous system (E14–E20) showed high levels of MC4-R mRNA in regions such as the sympathetic trunk, superior cervical, and paravertebral ganglia *(32)*. mRNA was also seen in some highly innervated tissues such as the adrenal and kidney. Widespread expression was also seen in the developing spinal cord. The MC4-R was the sole neural melanocortin receptor expressed during fetal development, appearing in sensory trigeminal nuclei (E16), the dorsal motor nucleus of vagus (E16), cranial nerve ganglia (E16), inferior olive (E18) and cerebellum (E18), striatum (E16), and entorhinal

(text continued on p. 416)

Table 1
Distribution of MC4-R mRNA in the Rat CNS

CNS	Signal[a]
I. FOREBRAIN (FB)	
A. Isocortex (ISO)	
1. *Motor areas* (MO)	
a. primary motor area (MOp)	++
b. secondary motor areas (MOs)	++
2. *Agranular insular area* (Al)	
a. dorsal part (Ald)	++(+)
b. ventral part (Alv) (layers 2,3 & 5)	++(+)
c. posterior part (Alp)	++(+)
3. *Anterior cingulate area* (ACA)	
a. dorsal part (ACAd) (layer 6A)	++
b. ventral part (ACAv)	+(+)
4. *Auditory areas* (AUD)(Primary, dorsal, ventral)	+++
5. *Infralimbic area* (ILA)	++
6. *Orbital area* (ORB)	
a. ventral part (ORBv)	+++
b. ventrolateral part (ORBvl)	++
7. *Retrosplenial area* (RSP)	
a. dorsal part (RSPd) (deeper layers)	+(+)
8. *Ventral temporal association areas* (TEv)	+++
9. *Claustrum* (CLA)	+(+)
B. Olfactory Cortex (OLF)	
1. *Accessory olfactory bulb* (AOB)	
a. mitral layer (AOBmi)	+
2. *Anterior olfactory nucleus* (AON)	
a. dorsal part (AONd)	++
b. lateral part (AONl)	++
c. medial part (AONm)	+
d. posteroventral part (AONpv)	+
3. *Taenia tecta* (TT)	
a. dorsal part (TTd)	+++(+)
4. *Olfactory tubercle* (OT)	
a. pyramidal layer (OT2)	+++
5. *Piriform area* (PIR)	
a. pyramidal layer (PIR2)	+(+)
6. *Postpiriform transition area* (TR)	+
C. Hippocampal Formation (cortex) (HPF)	
1. *Retrohippocampal region* (RHP)	
a. entorhinal area (ENT)	
(1) lateral part (ENTl) (layer 2)	+++
b. parasubiculum (PAR)	++

Table 1 *(continued)*

CNS	Signal[a]
c. subiculum (SUB)	
(1) ventral part (SUBv)	++(+)
2. *Hippocampal region* (HIP)	
a. Ammon's horn (CA)	
(1) field CA1 (CA1)	+(+)
(2) field CA2 (CA2)	++
(3) field CA3 (CA3)	+++
D. AMYGDALA (AMY)	
1. *Bed nucleus of the accessory olfactory tract* (BA)	+++
2. *Medial nucleus of the amygdala* (MEA)	
a. anterodorsal part (MEAad)	+++
b. posterodorsal part (MEApd)	++
c. posteroventral part	++
3. *Cortical nucleus of the amygdala* (COA)	
a. anterior part (COAa)	++
b. posterior part (COAp)	
(1) medial zone (COApm)	+++
4. *Anterior amygdaloid area* (AAA)	++
5. *Central nucleus of the amygdala* (CEA)	
a. medial part (CEAm)	++
b. lateral part (CEAl)	+
c. capsular part (CEAc)	+++
6. *Basolateral nucleus of the amygdala* (BLA)	
a. posterior part (BLAp)	++
7. *Basomedial nucleus of the amygdala* (BMA)	
a. anterior part (BMAa)	++(+)
b. posterior part (BMAp)	(+)
8. *Posterior nucleus of the amygdala* (PA)	+
E. SEPTAL REGION (SEP)	
1. *Lateral septal nucleus*	
a. dorsal part (LSd)	++
b. intermediate part (LSi)	++++
c. ventral part (LSv)	++
2. *Medial septal complex* (MSC)	
a. medial septal nucleus (MS)	+++
b. nucleus of the diagonal band (NDB)	+++
3. *Bed nuclei of the stria terminalis* (BST)	
a. anterior division (BSTa)	
(1) anterodorsal area (BSTad)	++(+)
(2) anterolateral area (BSTal)	++(+)
(3) anteroventral area (BSTav)	++(+)

(continued)

Table 1 *(continued)*

CNS	Signal[a]
(4) rhomboid nucleus (BSTrh)	++(+)
(5) dorsomedial nucleus (BSTdm)	++(+)
(6) dorsolateral nucleus (BSTdl)	++(+)
(7) ventral nucleus (BSTv)	++(+)
(8) magnocellular nucleus (BSTmg)	++(+)
b. posterior division (BSTp)	
(1) principal nucleus (BSTpr)	+++
(2) interfascicular nucleus (BSTif)	++(+)
(3) transverse nucleus (BSTtr)	++(+)
4. *Septohippocampal nucleus* (SH)	+++(+)
5. *Subfornical organ* (SFO)	+++(+)
F. CORPUS STRIATUM (CSTR)	
1. *Striatum* (STR)	
a. caudoputamen (CP)	++
b. nucleus accumbens (ACB)	++
c. fundus of the striatum (FS)	+++
2. *Pallidum* (PAL)	
a. magnocellular preoptic nucleus (MA)	+
G. THALAMUS (TH)	
1. *Dorsal thalamus* (DOR)	
a. midline group of the dorsal thalamus (MID)	
(1) nucleus reuniens (RE)	+
b. lateral group of the dorsal thalamus (LAT)	
(1) suprageniculate nucleus (SGN)	+++
2. *Ventral thalamus* (VNT)	
a. zona incerta (ZI)	++
b. peripeduncular nucleus (PP)	++(+)
c. subparafascicular nucleus (SPF)	
(1) magnocellular part (SPFm)	++
H. HYPOTHALAMUS (HY)	
1. *Periventricular zone of the hypothalamus* (PVZ)	
a. suprachiasmatic preoptic nucleus (PSCH)	+++
b. median preoptic nucleus (MEPO)	++
c. anteroventral periventricular nucleus (AVPv)	++++
d. preoptic periventricular nucleus (PVpo)	+(+)
e. supraoptic nucleus (SO)	+++(+)
(1) accessory supraoptic group (ASO)	
(a) nucleus circularis (NC)	+++
f. paraventricular nucleus of the hypothalamus (PVH)	
(1) descending division (PVHd)	
(a) medial parvicellular part, ventral zone (PVHmpv)	++(+)

Table 1 *(continued)*

CNS	Signal[a]
(b) lateral parvicellular part (PVHlp)	++(+)
(2) magnocellular division (PVHm)	
(a) anterior magnocellular part (PVHam)	++
(b) posterior magnocellular part (PVHpm)	++
(3) parvicellular division (PHVp)	
(a) anterior parvicellular part (PHVap)	++
(b) medial parvicellular part, dorsal zone (PVHmpd)	+++
(c) periventricular part (PHVpv)	+
g. anterior periventricular nucleus of the hypothalamus (PVa)	++
h. arcuate nucleus of the hypothalamus (ARH)	+
i. posterior periventricular nucleus of the hypothalamus (PVp)	++
2. *Medial zone of the hypothalamus* (MEZ)	
a. medial preoptic area (MPO)	
(1) medial preoptic nucleus (MPN)	
(a) lateral part (MPNl)	++
(b) medial part (MPNm)	++++
(c) central part (MPNc)	+++
b. anterodorsal preoptic nucleus (ADP)	+
c. anteroventral preoptic nucleus (AVP)	+++(+)
d. posterodorsal preoptic nucleus (PD)	+
e. anterior hypothalamic area (AHA)	
(1) anterior hypothalamic nucleus (AHN)	
(a) anterior part (AHNa)	++(+)
(b) central part (AHNc)	++(+)
(c) posterior part (AHNp)	++(+)
f. tuberal area of the hypothalamus (TUA)	
(1) ventromedial nucleus of the hypothalamus (VMH)	
(a) dorsomedial part (VMHdm)	+
(b) ventrolateral part (VMHvl)	+++
(2) dorsomedial nucleus of the hypothalamus (DMH)	
(a) anterior part (DMHa)	+++
(b) posterior part (DMHp)	+
(c) ventral part (DMHv)	+
(3) ventral premammillary nucleus (PMv)	++
g. mammillary body (MBO)	
(1) tuberomammillary nucleus (TM)	
(a) dorsal part (TMd)	+++

(continued)

Table 1 *(continued)*

CNS	Signal[a]
(b) ventral part (TMv)	++
(2) medial mammillary nucleus (MM)	++
h. posterior hypothalamic nucleus (PH)	++(+)
3. *Lateral zone of the hypothalamus* (LZ)	
a. lateral preoptic area (LPO)	++(+)
b. lateral hypothalamic area (LHA)	+++
II. BRAINSTEM (BS)	
A. SENSORY	
1. *Visual*	
a. superior colliculus (SC)	
(1) optic layer (SCop)	++++
(2) intermediate gray layer (SCig)	++
(3) deep gray layer (SCdg)	++
b. pretectal region (PRT)	
(1) nucleus of the optic tract (NOT)	++++
(2) posterior pretectal nucleus (PPT)	+
(3) nucleus of the posterior commissure (NPC)	++(+)
(4) anterior pretectal nucleus (APN)	+
(5) medial pretectal area (MPT)	+
c. medial terminal nucleus of the accessory optic tract (MT)	++
2. *Somatosensory*	
a. spinal nucleus of the trigeminal (SPV)	
(1) caudal part (SPVC)	++(+)
3. *Auditory*	
a. nucleus of the lateral lemniscus (NLL)	+
b. inferior colliculus (IC)	
(1) external nucleus (ICe)	+
4. *Gustatory*	
a. nucleus of the solitary tract, rostral zone of medial part (NTSm)	+
5. *Visceral*	
a. nucleus of the solitary tract (NTS)	
(1) medial part, caudal zone (NTSm)	+
b. parabrachial nucleus (PB)	
(1) medial division (PBm)	
(a) medial part (PBmm)	++
(2) lateral division (PBl)	+
(a) central lateral part (PBlc)	+
(b) external lateral part (PBle)	++
B. MOTOR	
1. *Viscera*	
a. inferior salivatory nucleus (ISN)	+

Table 1 *(continued)*

CNS	Signal[a]
b. dorsal motor nucleus of the vagus nerve (DMX)	++++
c. nucleus ambiguus, ventral division (AMBv)	++
2. *Extrapyramidal*	
a. substantia nigra (SN)	
(1) compact part (SNc)	++
(2) reticular part (SNr)	++
b. ventral tegmental area (VTA)	+
C. Pre- and Postcerebellar Nuclei	
1. *Red nucleus* (RN)	+++
D. Reticular Core	
1. *Central gray of the brain* (CGB)	
a. periaqueductal gray (PAG)	++
b. interstitial nucleus of Cajal (INC)	+
c. dorsal tegmental nucleus (DTN)	+
2. *Raphé* (RA)	
a. superior central nucleus raphé (CS)	
(1) medial part (CSm)	+(+)
(2) lateral part (CSl)	+(+)
b. dorsal nucleus raphé (DR)	+
c. nucleus raphé magnus (RM)	+
d. nucleus raphé pallidus (RPA)	++
3. *Reticular formation* (RET)	
a. mesencephalic reticular nucleus (MRN)	++
(1) retrorubral area (RR)	++
b. pedunculopontine nucleus (PPN)	++
c. pontine reticular nucleus (PRN)	
(1) caudal part (RPNc)	+
d. gigantocellular reticular nucleus (GRN)	++
e. paragigantocellular reticular nucleus (PGRN)	
(1) lateral part (PGRNl)	++(+)
f. magnocellular reticular nucleus (MARN)	++(+)
g. supratrigeminal nucleus (SUT)	+(+)
h. parvicellular reticular nucleus (PARN)	+++
i. medullary reticular nucleus (MDRN)	
(1) dorsal part (MDRNd)	++
(2) ventral part (MDRNv)	++
III. SPINAL CORD (SP)	
A. Dorsal Horn of the Spinal Cord (DH)	
1. *Substantia gelatinosa of the spinal cord* (SGE)	++(+)

[a]Semiquantitative estimates of the signals are indicated: + (weak), ++ (moderate), +++ (strong), with parentheses indicating intermediate levels. Reprinted from *(9)*, with permission from the Endocrine Society.

cortex (E22) *(31)*. MC3-R mRNA appeared in previously characterized sites such as the ventromedial hypothalamic nucleus and arcuate nucleus during the postnatal period.

3.3. Regulation of MC4-R Expression

It is quite likely that α-MSH and β-endorphin may be co-released from POMC terminals at MC4-R expressing neurons, thus it is interesting to consider a potential role of the MC4-R in opioid action. Data in support of this come from studies demonstrating that melanocortins can antagonize opioid tolerance and dependence *(33,34)*. In this regard, Alvara and colleagues *(35)* examined levels of MC4-R mRNA in rats receiving chronic morphine administration. Morphine treatment under conditions known to induce tolerance and dependence was found to produce a 20–30% reduction in MC4-R mRNA in the striatum and periaquaductal gray after 5 days, and in the nucleus accumbens and olfactory tubercle after 1–3 days. Importantly, no change in MC4-R mRNA levels were seen in several other brain regions, suggesting a specific response in MC4-R expression in relevant brain regions in a model of opioid addiction. A 50% reduction in ^{125}I-NDP-MSH binding was demonstrated after opioid treatment in the ventrolateral striatum in this study. The proposed role of melanocortins in opiate addiction has recently been reviewed elsewhere *(36)*.

A promising approach being used to study MC4-R binding sites in rat brain sections utilizes ^{125}I-NDP-MSH binding in the presence of high concentrations of γ^2-MSH to block MC3-R binding sites *(37)*. In this study, a food restriction paradigm producing a 14% weight loss resulted in a 20–60% increase in putative MC4-R binding specifically in the ventromedial, arcuate, and dorsomedial hypothalamic nuclei, and in the median eminence.

4. Pharmacology of the Melanocortin 4 Receptor

4.1. Melanocortin Agonists

The human and mouse MC4-R couples to Gs and activation of adenylyl cyclase in heterologous cell lines. No data yet exist regarding the coupling of the MC4-R in the CNS or in neurons. The order of potency for activation of the receptor by native melanocortin peptides is desacetyl-α-MSH>/= ACTH1-39>/= α-MSH=β-MSH>>γ^2-MSH (Table 2). In contrast to the MC3-R, the synthetic melanocortin agonists NDP-MSH and MTII are approximately 50–100 times more potent at the MC4-R than the native α-MSH ligands. Another novel pharmacologic feature of the MC4-R is that it is selectively activated by α-MSH versus γ^2-MSH, while the MC3-R is nearly equipotently activated by α-MSH and γ^2-MSH. It is important to note, however, that while γ^2-MSH is

Table 2
Pharmacologic properties of MC4-R Agonists

Ligand	HMC4-R (EC_{50})	HMC3-R (EC_{50})	HMC4-R (K_i)	HMC3-R (K_i)
NDP-MSH	1×10^{-11}	0.13×10^{-9}	2.2×10^{-9}	0.22×10^{-9}
Desacetyl-α-MSH	5×10^{-10}		5.7×10^{-7}	3.7×10^{-9}
α-MSH	1.5×10^{-9}	0.67×10^{-9}	6.4×10^{-7}	2.1×10^{-8}
ACTH	6.8×10^{-10}			
γ2-MSH	$>10^{-7}$			

EC_{50} values are from refs. *9* and *39*, and K_i values are from ref. *41*.

100-fold less active than α-MSH, it is nonetheless a full agonist of the MC4-R, and thus may be a bona fide ligand of the MC4-R in vivo if expressed at high enough levels. Finally, it is interesting to note that the α-MSH ligands have significantly greater affinity and potency at the MC3-R than the MC4-R. These data tend to suggest that the MC3-R may serve as an auto-receptor.

4.2. Melanocortin Antagonists

4.2.1. Synthetic Antagonists

The discovery of the neural melanocortin receptors led to a search for specific melanocortin antagonists to probe the physiologic roles of these new receptors. The first antagonists reported for the neural receptors were linear peptides analogs of the ACTH[4–10] sequence *(38)*. Three peptides in particular, [I-Phe^{7}]ACTH[4–10], Pro^{8,10},Gly^{9}]ACTH[4–10], and [D-Arg^{8}]-ACTH[4–10] were found to antagonize the MC4-R. At high doses (15 μg), coinjection these analogs were able to inhibit excessive grooming behavior induced by intracerebroventricular injection of 1.5 μg of α-MSH. The general utility of these antagonists was questioned in this report, however, due to the low affinity of ACTH[4–10] for the MC4-R, and thus the low potency of the resulting antagonists.

The discovery of SHU9119, Ac-Nle^{4}-*c*[Asp^{5},D-Nal(2)^{7}, Lys^{10}]α-MSH[4–10]-NH_2, produced the first high-affinity melanocortin antagonist *(39)*, which has now been demonstrated to be useful in a number of physiologic assays across multiple species (see Subheading 5 below). The cyclic lactam heptapeptide template on which this analog was based, MTII (Ac-Nle^{4}-*c*[Asp^{5},(D-Phe^{7}, Lys^{10}]α-MSH[4–10]-NH_2), had previously been demonstrated to be a stable and potent melanocortin agonist *(40)*. In addition to insertion of D-Nal(2), insertion of D-iodophenylalanine at position 7 also led to a melanocortin antagonist, and these data suggest that increasing the bulk of the amino acid moiety at position 7, while retaining the aromatic character, is responsible for the conversion to an antagonist. SHU9119 is a high-affinity antagonist of the hMC3-R (pA2 = 8.3, IC_{50} = 3.3n*M*) and hMC4-R (pA2 = 9.3, IC_{50} = 1.8n*M*) *(39)*, and demonstrates similar properties at the rodent MC3-R and MC4-R *(11)*. Small partial agonist activity of the compound is seen at the MC3-R. This compound is a potent full agonist at the MC1-R and MC5-R *(39)*.

The discovery of antagonism via the D-Nal^7 insertion has led to the characterization of a new series of cyclic melanocortin analogs, some of which have some valuable new properties *(41–43)*. The properties of these MC4-R antagonists are compared in Table 3. HS014, for example, has a 17-fold selectivity for MC4-R over MC3-R, compared with the 10-fold selectivity seen with SHU9119, and also is a less potent agonist than SHU9119 of the MC1-R and MC5-R *(42)*. This compound appears to have a 10-fold lower affinity, however, for the MC4-R compared with SHU9119 *(41)*. HS024 has a comparable affinity for the MC4-R as SHU9119 and appears to have a 20-fold MC4-R/MC3-R selectivity. This compound is particularly interesting because as a consequence of an insertion of Arg at position 5 it now is a high-affinity antagonist of the MC1-R, MC3-R, MC4-R and MC5-R. This compound is as potent as SHU9119 in the stimulation of food intake, and may have lower toxicity. HS028 has an even greater selectivity for the MC4-R (80-fold), and has only a threefold lower affinity for MC4-R than SHU9119. This compound retains partial agonist properties at the MC1-R and MC5-R, is a potent antagonist of both MC3-R and MC4-R, and may have some slight agonist activity at the MC4-R. Chronic high-dose ICV administration of HS028 for 7 days produced sustained increases in 24-h food intake and weight gain in rats *(43)*. The selectivity for these compounds is based upon K^i values from competition binding studies using displacement of the synthetic ligand ^{125}I-NDP-MSH. Ultimately, it will be valuable to know the relative antagonist efficacy of these compounds generated by comparing pA2 values obtained from examining their ability to block activation of the receptor by the endogenous ligand desacetyl-α-MSH. While SHU9119 has an equivalent affinity for MC3-R and MC4-R, it is a 10-fold better antagonist of the MC4-R when examined functionally by this method *(39)*.

4.2.2. Agouti

The small proteins encoded by the agouti and agouti-related protein (AGRP) genes are endogenous antagonists of the MC1-R and MC3-R/MC4-R, respectively. *Agouti* was originally characterized as a gene locus involved in regulating pigmentation, and in a variety of mammalian species agouti, depending on the degree of dominance of the allele, acts to block the synthesis of eumelanin, or brown-black pigment normally under the positive control of melanocyte stimulating hormone and its receptor *(44)*. Thus, recessive agouti alleles lead to dark black coat colors while the most dominant alleles result in yellow or red coat colors. Remarkably, however, dominant alleles of agouti in the mouse (e.g., A^y, A^{VY}) also cause an obesity syndrome (Table 4) characterized by obesity (60–70 g at maturity) associated with hyperphagia *(45)*, mild hyperinsulinemia (2–5 times normal) *(46)*, and normal reproductive and adrenal axes *(47)*. A neuroendocrine change unique to the agouti obesity syndrome is increased somatic growth *(48)*; A^y animals, for example, are 10–15% longer, as measured by fibula or nose-to-anus lengths, than their wild-type counterparts *(12)*. Both the genetic and physiologic parameters of the agouti obesity syndrome are thus much more reminiscent of common forms of human obesity than the obesity syndrome seen in the leptin-deficient *ob/ob* mouse.

Agouti was cloned from the mouse in 1992 *(49,50)*, and determined to encode a novel 131 amino acid peptide with a putative signal peptide at the amino terminus, a basic rich region of approximately 50 amino acids, and a cysteine rich domain with homology to the cysteine repeat motif found in peptide toxins *(51)*, such as the conotoxins and agatoxins *(52)* (Fig. 2). Early parabiosis experiments demonstrated that agouti was not a hormone, since it was unable to transmit its effects from one mouse to another *(48)*. Furthermore, skin transplantation experiments suggested that agouti was produced by hair follicle cells and acted in *trans* on adjacent melanocytes to somehow regulate eumelanin synthesis *(53,54)*. These data, along with the structure of the gene, suggested that agouti is a paracrine factor that might act directly on the melanocyte to somehow block MSH action. Pharmacologic characterization of recombinant murine agouti produced in the baculovirus system demonstrated that agouti acts as a high-affinity competitive antagonist (K^i = 6×10^{-10}) of the murine MSH or MC1 receptor *(16)*. Remarkably, however, agouti was also found to be a specific high-affinity antagonist of the hypothalamic MC4-R *(16)*. The murine agouti peptide was found to have little to no antagonist activity at the MC3 or MC5 melanocortin receptor subtypes. Agouti was thus the first example of an endogenous high-affinity antagonist of a G protein-coupled receptor. Based on the fact that murine agouti is normally made only in the skin, it was proposed early on that the normal role of the peptide was the regulation of eumelanin synthesis, and that aberrant ectopic expression in the central nervous system (CNS) was responsible for the agouti obesity syndrome.

The agouti protein sequence is highly conserved in mammals (Fig. 2), and genetic evidence links the gene to the regulation of the eumelanin/pheomelanin switch in other animals in addition to the mouse, such as the fox *(55)*. Although the majority of work on agouti has been performed in the mouse, there may be some very interesting species differences in the pharmacologic properties, distribution of expression, and function of agouti. For example, while agouti alleles with dominant properties appear to exist in

Table 3A
Structure and Properties of Synthetic MC4-R Antagonists

SHU9119	Ac-Nle[4]-*c*[Asp[5],D-Nal(2)[7], Lys[10]]-α-MSH [4–10]-NH_2
HS964	*c*[Ac-Cys[4],D-Nal(2)[7], Cys[11]]-α-MSH [4–11]-NH_2
HS014	*c*[Ac-Cys[11], D-Nal(2)[14], Cys[18],Asp[22]]-β-MSH-[11–22]-NH_2
HS024	*c*[Ac-Cys[3],Nle[4],Arg[5],D-Nal(2)[7], Cys[11]]-α-MSH[3–11]-NH_2
HS028	*c*[Ac-Cys[11],diCl-D-Phe[14], Cys[18],Asp[22]]-β-MSH-[11–22]-NH_2

Position no.	1	2	3	4	5	6	7	8	9	10	11	12	13	14	15
Peptide															
α-MSH	Ser	Tyr	Ser	Met	Glu	His	Phe	Arg	Trp	Gly	Lys	Pro	Val		
SHU9119				Nle	Asp	His	D-Nal	Arg	Trp	Lys					
HS964				Cys	Glu	His	D-Nal	Arg	Trp	Gly	Cys				
HS014				Cys	Glu	His	D-Nal	Arg	Trp	Gly	Cys	Pro	Pro	Lys	Asp
HS024			Cys	Nle	Arg	His	D-Nal	Arg	Trp	Gly	Cys				
HS028				Cys	Glu	His	*d*Cl-D-Phe	Arg	Trp	Gly	Cys	Pro	Pro	Lys	Asp

Table 3B
Properties of Synthetic MC4-R Antagonists

Compound	MC1-R	MC3-R	MC4-R	MC5-R
SHU9119	0.71	1.2 (pA2 = 8.3)	0.36 (pA2 = 9.3)	1.12
HS964	1460	281	23.2	164
HS014	108	54	3.2	694
HS024	18.6	5.45	0.29	3.29
HS028	60	74	0.95	211

K_i values (n*M*) are from refs. *39* and *41*.

Table 4
Phenotype of the Agouti Obesity Syndrome

Strain	C57BL/6J	C57BL/6J-A^Y	C57BL/6J-*ob*/*ob*
Mature weight	30–35 g	45–55 g	60–80 g
Insulin	0.3–0.5 ng/mL	1–2 ng/mL	30–50 ng/mL
Leptin	4–5 ng/mL	10–15 ng/mL	0
Glucose	100–120 mg/dL	150–160 mg/dL	250–400 mg/dL
Glucocorticoids	20–50 ng/mL	25–50 ng/mL	300–400 ng/mL
Body length	Normal	+ 10–15%	– 5–10%
Reproductive axis	Normal	Normal	Infertile
Food intake	4–4.5 g/24 h	5.5–6 g/24 h	6–7 g/24 h
Metabolic rate	Normal	– 10%	– 25–40%

other species, such as the red fox, no other cases of obesity linked to yellow or red (pheomelanized) coat colors have been reported. This could be due to absence of expression of agouti in the brain in these animals. Alternatively, perhaps the agouti/MC4-R interaction does not produce competitive antagonism in all species. Pharmacologic differences in the protein may also exist; while the murine protein is a competitive antagonist, evidence exists for inverse agonist activity in the fox *(55)*. Biochemical data on the structure of the agouti protein are discussed in detail in Chapter 16.

The agouti coding sequence in the human is 85% identical to the mouse *(56)*, the protein has comparable pharmacologic properties to the murine protein *(21)*, and transgenic mice overexpressing the human protein exhibit the agouti obesity syndrome *(57)*. Nevertheless, the role of agouti in the human is a particularly interesting case, because there is no wild-type agouti pigmentation phenotype seen in humans. Furthermore, unlike the mouse, in which the wild-type agouti gene is expressed specifically in the skin in a tightly controlled developmental pattern *(49,50)*, full-length agouti mRNA has not yet been demonstrated in the human. A 403-bp fragment of the mRNA has been detected in adipose tissue and testes by polymerase chain reaction *(56)*.

The expression of human agouti mRNA in the adipocyte is interesting in regard to the etiology of the agouti obesity syndrome. Data now clearly show that aberrant antagonism of the hypothalamic MC4-R by ectopic expression of agouti is responsible for the syndrome. Nevertheless, some activities for agouti other than competitive antagonism of the hypothalamic MC4-R have been proposed, such as an MC4-R independent effect of the protein on intracellular calcium concentration *(57a–61)*. Targeted overexpression of agouti in the adipocyte in transgenic mice using the aP2 promoter yielded a strain of mice that were extremely sensitive to daily insulin injections (0.5–2.0 units per day per mouse, SC), demonstrating a 1.7-fold increase in the rate of weight gain over a 2- wk period of insulin administration *(60)*. These data suggest potential roles for the agouti protein outside of competitive antagonism of the MC1-R and MC4-R, neither of which appear to be expressed in adipocytes *(61a)*.

4.2.3. *Agouti-Related Protein*

The fact that the MC4-R binds the agouti peptide at high affinity when this peptide is normally only expressed in the skin can be put in context by the discovery of a brain homolog of agouti called agouti-related transcript (ART) *(62)* or agouti-related protein (AGRP) *(19)*. A fragment of the AGRP cDNA first appeared in the expressed sequence tag database, and the full-length sequence was subsequently found to encode a 132 amino acid peptide that is 80% identical to agouti in the cysteine motif domain, largely unrelated in the amino-terminal domain, and also contains a signal peptide (Fig. 2). The AGRP mRNA is expressed centrally almost exclusively in the arcuate nucleus of the hypothalamus, and is found in the adrenal cortex and medulla as well *(19,62)*. The expression of AGRP mRNA in the arcuate nucleus suggests strongly that AGRP could be released at many of the same sites to which POMC neurons project and release melanocortin agonists.

Detailed information on the neuroanatomic distribution of AGRP is now available *(63–67)*. AGRP mRNA in the rat is found almost exclusively in neuropeptide Y(NPY)-positive neurons, and >95% of NPY arcuate neurons contain AGRP *(63,64)*. A polyclonal antibody against the carboxy-terminal 83–132 fragment of AGRP has also been used to characterize the distribution of AGRP-immunoreactive fibers in the rat *(66)* mouse *(63,67)*, and rhesus monkey *(66)*. The major fiber tracts are well-conserved across species, with dense projections originating in the arcuate nucleus and proceeding along the third ventricle (Fig. 3). Dense fiber bundles are also visible in the paraventricular, dorsomedial, and posterior nuclei in the hypothalamus, and in the bed nucleus of the stria terminalis and lateral septal nucleus of the septal region, and in some brainstem

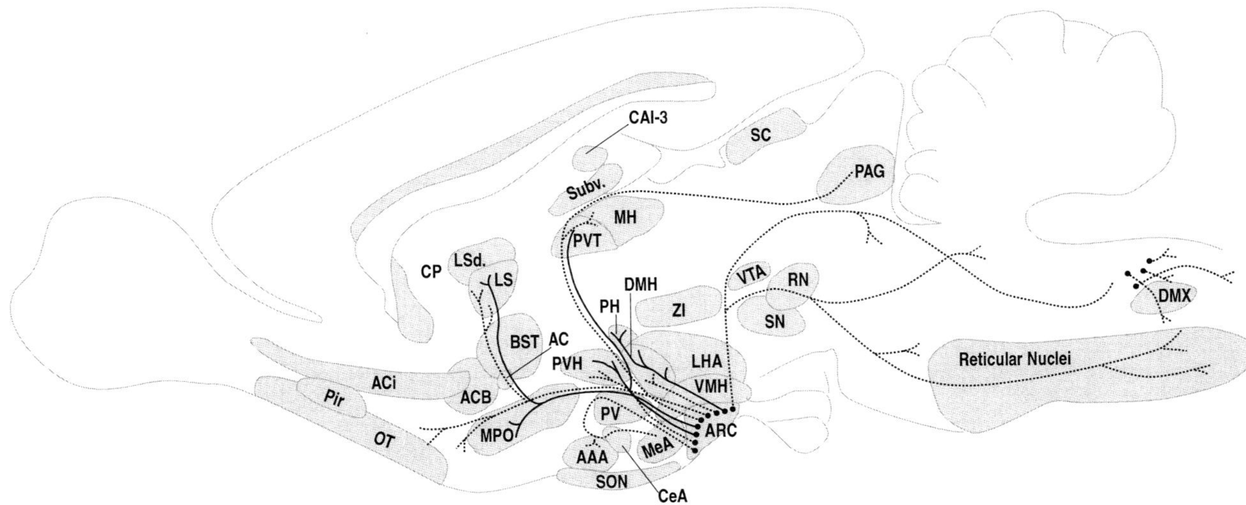

Fig. 3. Sagittal section of the rat brain schematically indicating the distribution of POMC-immunoreactive (dashed lines) and AGRP-immunoreactive (solid lines) neuronal fibers. Reprinted from *(66)* with permission from the Endocrine Society.

regions such as the parabrachial nucleus. AGRP-containing fibers are not visualized in a number of areas, such as the amygdala and thalamus, that express MC3-R and MC4-R mRNA and receive innervation from the proopiomelanocortin (POMC) neurons that serve as the source of melanocortin agonists. Thus, AGRP is most likely to be involved in modulating a conserved subset of the physiologic functions of central melanocortin peptides.

Based on the particular distribution of AGRP neurons those functions are likely to include the central control of energy homeostasis. Some interesting data on the regulation of AGRP in murine models of obesity is already available. AGRP immunoreactivity disappears in monosodium glutamate-treated mice *(63)*, as has been demonstrated previously for arcuate NPY immunoreactivity. AGRP-immunoreactive fiber density was demonstrated to decrease more than 40% in some brain regions (DMH, ARC, PAG, PBN) in the anorectic *anx/anx* mouse *(68,69)*, while ARC mRNA levels remained constant *(63)*. In the medial ARC a small percentage of AGRP neurons (10–25%) have been shown to express the long form of the leptin receptor *(67)*. AGRP mRNA is found to be elevated 5–10 times in the leptin-deficient versus wild-type C57Bl/6J mouse *(70)*, and is suppressed to normal levels following leptin treatment of these animals. Furthermore, AGRP is elevated 13-fold following a 48-h fast.

Pharmacologic characterization of recombinant AGRP protein, produced in the baculovirus system, has demonstrated that this peptide is a specific high-affinity competitive antagonist of the MC3-R (K^i = 3-4n*M*) and MC4-R (K^i = 2.5n*M*) *(17,19,20)*. One group has shown some limited antagonist activity of AGRP at the human MC5-R (IC_{50} ~ 300n*M*) *(20)*. While the protein is not thought to be processed in vivo, the carboxy-terminal 83–132 fragment has been demonstrated to retain the binding affinity of the full-length protein. The K^i value at the *Xenopus* MC1-R, determined by Schild regression analysis, was estimated to be 0.7n*M* *(72)*. Furthermore, AGRP[83–132], prepared synthetically and folded in vitro, was demonstrated to block ^{125}I-NDP-MSH binding to the hMC3-R (3.4n*M*) and hMC4-R (12.8n*M*) *(73)* in a manner comparable to that shown previously for the full length protein (IC_{50}s= 1.0n*M* and 3.2n*M*, respectively) *(17)*. Rossi et al. *(73)* also demonstrated a potent ability of this synthetic peptide to stimulate increased food intake for up to 24-h in the rat following administration into the third ventricle.

Biochemical characterization of two forms of bacterially produced AGRP, lacking amino terminal residues 1–5 or 1–65, has yielded a method for folding the bacterial protein so that it retains activity comparable to the full-length protein produced in baculovirus-infected insect cells *(74)*. After refolding, the protein appears as a fully oxidized monomer, suggesting the presence of five disulfide bonds. Stepwise reduction and alkylation of the protein has demonstrated the following disulfide-bonded cysteine pairs: C85–C109,

C90–C97, C74–C88, C67–C82, C81–C99, where amino acid 1 is the alanine at position 21 following the signal peptide.

The arrangement of disulfide bonds described above would result in a loop between C90 and C97 of AGRP, CRFFNAFC, that has been proposed to mimic the HFRW pharmacophore of the melanocortin agonists *(76)*. A series of cyclic peptides containing this motif were found to be full antagonists of ^{125}I-NDP-MSH binding to the hMC4-R, providing compelling evidence for this hypothesis. For example, *Ycyclo*[CRFFNAFC]Y had an IC_{50} for ^{125}I-NDP-MSH binding to the hMC4-R of 57n*M*, and a K_i by Schild analysis of approximately 800n*M*, compared with values of 1–10n*M* shown for the full-length protein and 83–132 peptide. The cyclic peptides had 30 to 100-fold lower affinities for the hMC3-R.

Two independent laboratories have created transgenic mice in which AGRP expression is driven by the β-actin promoter, and like the A^Y mice and the MC4-R-KO mice, these mice develop the characteristic features of the agouti obesity syndrome *(19,70)*. Thus, these findings infer the existence of a more complex neuronal system that involves a large number of downstream MC4-R and MC3-R containing neurons being coordinately regulated by both melanocortin agonists released by arcuate POMC neurons and the AGRP antagonist released by NPY/AGRP arcuate nucleus neurons. Additional work will be required to define the MC3-R and MC4-R sites at which melanocortins alone, AGRP alone, or both peptides are released.

5. Function of the Melanocortin 4 Receptor

5.1. Identification of a Role of the MC4-R in Energy Homeostasis

Three observations led to the melanocortin hypothesis, the idea that aberrant antagonism of the central MC4-R is the direct cause of the agouti obesity syndrome: (i) dominant agouti alleles leading to obesity result from agouti promoter rearrangements that direct expression of the peptide outside of the skin *(49,50,77)*; (ii) agouti is an antagonist of the MC4-R *(16)*; and (iii) the MC4-R is expressed in several brain regions known to be involved in the regulation of feeding and metabolism, including the paraventricular nucleus (PVN), arcuate nucleus (ARH), lateral hypothalamic area (LHA), and dorsomedial nucleus (DMH) *(9)*. Two experimental approaches, described below, were then used to generate support for this hypothesis.

5.1.1. Creation of the MC4-R Knockout Mouse

One experimental approach involved deletion of the MC4-R from the mouse genome by homologous recombination in ES cells *(12)*. Homozygous knockout animals were found to recapitulate all the unique hallmarks of the agouti obesity syndrome, mild hyperphagia and hyperinsulinemia, hyperglycemia limited to males, obesity, increased linear growth, and normal reproductive and adrenal axes. Remarkably, heterozygous loss of the MC4-R produced a phenotype intermediate to the wild type and knockout animals in every respect. Thus, the MC4-R is not an on–off switch, but rather, appears to act as a rheostat on feeding, metabolism, and growth.

5.1.2. Pharmacologic Studies

In another set of experiments, the cyclic heptapeptide antagonist of the central MC3-R and MC4-R receptors *(39)*, SHU9119, along with a related heptapeptide agonist *(78,79)*, MTII, were administered intracerebroventricularly to normal C57Bl/6J mice *(11)*. The agonist, MTII, was found to potently inhibit feeding in fasted mice while administration of the antagonist, SHU9119, just before lights out led to a 29% mean increase in 4-h food intake. The antagonist data argues that the endogenous POMC neurons exert a tonic inhibitory effect on feeding and energy storage via their release of desacetyl-α-MSH, the primary melanocortin cleavage product in the brain, at downstream sites containing MC4-R and possibly MC3-R. Support for the argument that POMC neurons regulate metabolism as well as feeding behavior is implied by the observation that A^{VY} animals pair-fed to limit caloric intake to that in sex- and age-matched lean animals still become obese *(80,81)*. Direct effects of the melanocortin pathway on insulin release and metabolic rate and sympathetic outflow are discussed in more detail below.

5.2. The Role of the MC4-R and the POMC System in the Normal Regulation of Energy Homeostasis

5.2.1. Inputs to the Melanocortin System

Since a role for the POMC neurons in energy homeostasis is a somewhat recent finding, there is only limited information available regarding relevant physiologic inputs to POMC neurons and MC4-R activation. One source of information may be found in studies of the regulation of central POMC mRNA and peptides. Neuroanatomic considerations and studies of POMC gene expression may also provide some clues regarding inputs to POMC neurons. For example, POMC neurons are adjacent to NPY neurons and the two sets of neurons may make synaptic connections. Intracerebroventricular administration of NPY inhibits the release of α-MSH *(82)* and reduces POMC mRNA levels *(83)*. Fasting reduces POMC mRNA *(84,85)*, and reduced levels of POMC in the *ob/ob* mouse can be overcome by administration of leptin *(86)*. Administration of leptin appears to induce POMC mRNA levels primarily in the rostral arcuate neurons *(87)*. POMC mRNA levels are reduced by glucocorticoids *(88)*, increased by testosterone *(89)*, and reduced during lactation *(90)*, commensurate with the concept that POMC is an endogenous regulator of feeding that responds appropriately to hormonal signals known to increase or decrease feeding.

Despite the fact that numerous observations regarding POMC regulation fit nicely with the hypothesis that POMC is an endogenous regulator of feeding, the degree of alteration of POMC gene expression in most paradigms is small. For example, in one study, after either food restriction or a 5-d fast POMC mRNA levels in the rat are were only reduced around 24% *(84,85)*. Of course, gene knockout experiments have demonstrated a gene dosage effect for the MC4-R implying a lack of "spare receptors " in this signaling system. Thus, just as animals with one intact MC4-R gene are more obese and have significantly higher serum insulin than animals with two intact gene copies, 20% changes in the levels of POMC mRNA, and presumably peptide, may thus have significant biologic consequences. While the downregulation of POMC gene expression by fasting or leptin deficiency is small, Hagan et al. *(91)* have recently demonstrated a 185% increase in arcuate POMC mRNA in rats overfed for a period of ten days to 105% of control body weight, suggesting a more regulated response of POMC to nutritional excess.

In contrast to the modest regulation of the POMC gene, the AGRP gene is very significantly regulated by metabolic state. Levels of AGRP mRNA in the arcuate appear to be upregulated 10-fold in the *ob/ob* mouse, relative to wild-type animals *(19,62)*, and are apparently upregulated 13-fold by a 2-d fast in wild-type mice *(71)*. Thus, like the pigmentation system *(92)*, POMC may serve to provide rather constitutive levels of agonist, while the regulation of this system may derive more from variable levels of the AGRP antagonist.

A rather interesting controversy has developed concerning the role of the melanocortin system and MC4-R in mediating the central actions of leptin. The response to reduction or absence of serum leptin mimics the complex adaptive neuroendocrine changes that occur during starvation: hyperphagia, hypercortisolism, infertility, and depressed metabolic rate *(93)*. It is unlikely that the the melanocortin system is singularly responsible for transmitting these signals in response to reduced leptin since the agouti obesity syndrome is less severe than the obesity seen in the *ob/ob* mice, and the reproductive and adrenal axes are basically normal in animals lacking the MC4-R *(12)*. Following the discovery of the role of the melanocortin system in the agouti obesity syndrome, however, it was proposed that while NPY is required for the full orexigenic effects of reduced leptin *(93a)*, the melanocortin pathway must be required for the anorexigenic effects of elevated leptin *(94,95)*. The natural extension of this hypothesis is that the A^Y mouse is obese because inhibition of MC4-R signaling is a genetic roadblock to leptin action. This hypothesis may be partially based upon the observation that mRNA for the long form of the leptin receptor is found in at least 40% of POMC neurons, implying that leptin may be an important hormonal input to these neurons *(96)*. Furthermore, central resistance to the anorexigenic effects of leptin has been reported in the A^Y mouse *(97)*, although resistance to leptin is commonly seen in obesity.

However, there are questions about this hypothesis based on the observation that NPY gene expression in the arcuate nucleus is not elevated by antagonism or deletion of the MC4-R *(98)*, implying that the NPY gene in the arcuate nucleus in these animals was sensing and being suppressed by leptin. To examine this question in more detail, A^Y and *ob*/*ob* mice were crossed to eventually yield the *ob*/*ob* A^Y mouse to look for additivity of the phenotypes *(99)*. To look at the direct central effects of each lesion, animals were adrenalectomized and replaced with normal levels of glucocorticoids in the drinking water, since the elevated glucocorticoids in the *ob*/*ob* mice have potent central and peripheral catabolic actions. Characterization of the double mutant demonstrated the effects of MC4-R inhibition and the absence of leptin to be additive on weight gain and serum insulin. Furthermore, absence of the leptin gene in the A^Y background restored full leptin sensitivity to mice, implying that obesity in this model is independent of the leptin pathway, and that the resistance to leptin results from classic desensitization to leptin action.

In apparent contradiction to this however, preadministration of the MC3-R and MC4-R antagonist SHU9119 was found to block the acute inhibition of feeding resulting from central administration of leptin in the rat *(100,101)*. We have also been able to repeat this observation in the mouse. While these data appear contradictory at first glance, there are many potential interpretations. For example, the actions of leptin on feeding are most likely multifactorial, and the different obeservations described above may be highlighting short-term versus long-term actions of leptin on feeding behavior. Alternatively, one of the results described above may be artifactual, for example, the antagonist result may be due to nonphysiologic blockade of melanocortin receptors due to high dosage, or alternatively, the genetic approach may produce artifactual results due to developmental defects in the hypothalamus resulting from one or both of the mutations present from birth in these strains. Further complicating these data, recent findings show that leptin-induced weight loss and inhibition of feeding is attenuated by the CRH antagonist α-helical CRH[9–41] *(102)* as well as the GLP-I antagonist exendin[9–39] *(103)*. Thus, multiple anorexigenic neuropeptides appear to act downstream of leptin. Recent observations on the behavior of the MC4-R-KO mouse tend to support the hypothesis that there are multiple redundant systems downstream of the anorexigenic actions of leptin *(104)*. This work demonstrated that the melanocortin agonist MTII had a greatly reduced ability to inhibit food intake in the MC4-R-KO, suggesting that the majority of the inhibitory effect of central MSH on feeding is mediated through MC4-R and not MC3-R signaling. Additionally, while obese MC4-R-KO mice were resistant to leptin, young (13 to 16-wk-old) MC4-R-KO remained sensitive to the ability of leptin to inhibit feeding and weight gain. Even the obese MC4-R-KO mice remained sensitive to the anorectic actions of urocortin and ciliary neurotrophic factor, supporting the concept of multiple redundant pathways inhibiting food intake. Determining the precise contribution of the melanocortin system to leptin action, and the specific roles of the MC3-R and MC4-R, remains an important goal.

5.2.2. Potential Downstream Effectors of the Central Melanocortin System

A very active area of investigation involves characterization of pathways by which the melanocortin system may exert its effects on feeding behavior, serum insulin levels *(105)*, metabolic rate *(105)*, and somatic growth. Early hypotheses were based upon the projections of POMC and AGRP neurons, and the sites of expression of the MC3-R and MC4-R receptors. In relationship to brain regions known to regulate feeding, it is interesting to note that POMC neurons send dense projections to the paraventricular nucleus (PVH), dorsomedial hypothalamic nucleus (DMH), ventromedial hypothalamic nucleus (VMH), lateral hypothalamic area (LHA), and dorsal motor nucleus of the vagus, the brainstem (DMX) *(106–111)*. As described above, AGRP/NPY fibers send projections to most of the same sites. The PVH is probably an important site of POMC and AGRP action at the MC4-R, since microinjection into the PVH of most known orexigenic agents, including galanin, NPY, norepinephrine, γ-aminobutyric acid (GABA), and opioids will stimulate feeding *(112)*. Recently, effects of melanocortins on feeding behavior following stereotaxic administration into the PVH have also been demonstrated *(105,113)*. From stereotaxic injection studies, the most potent site of NPY's orexigenic action has been localized to the PVH and the perifornical area (PFH) *(114,115)*. Likewise, the PVH receives catecholaminergic projections from the brainstem known to regulate feeding *(116)*, and noncatecholaminergic brainstem projections containing the anorexigenic peptide GLP-1 also extensively innervate the PVH and ARC *(117)*. Finally, the PVH appears to be the most potent site of the anorexigenic actions of CRH and possibly urocortin *(118)*. MC4-R mRNA is found in all three subdivisions (magnocellular, parvicellular, descending) of the PVN *(9)*.

NPY potently stimulates feeding, alters body temperature, and increases plasma insulin levels following administration within the PVH. Recent data demonstrate that intra-PVH administration of melanocortins also effects insulin release, tissue insulin sensitivity, and basal metabolic rate, as measured by indirect calorimetry *(105)*. Direct measurement of sympathetic nerve activity in the rat has also demonstrated that the melanocortin agonist MTII is capable of stimulating sympathetic nerve traffic to brown fat, renal, and lumbar nerves *(119)*. While leptin and MTII could both mediate thermogenic effects via increased sympathetic outflow to brown fat in the rodent, the melanocortin antagonist SHU9119 blocked MTII-stimulated outflow to brown fat, but not leptin-stimulated activity. SHU9119 did block the ability of leptin to increase sympathetic nerve activity to renal and lumbar beds, suggesting that the leptin and melanocortin sensitive pathways mediating sympathetic outflow are overlapping, yet not identical.

To test the hypothesis that the PVH integrates the orexigenic NPY signal and anorexigenic MSH signal, and also integrates opposing effects of NPY and MSH on metabolism, the effect of stereotaxic coinjection of the peptides within the PVH was examined. MTII (0.3 nmol) completely suppressed the ability of NPY (0.14 nmol) to stimulate food intake in this paradigm *(105)*.

Next, a potential cellular basis for integration of the NPY and α-MSH signals was investigated *(105)*. Whole-cell recordings were made from a subset of PVH parvocellular neurons which demonstrated an inhibitory $GABA_A$ synaptic response to electrical stimulation in the PVH, registered as an outward synaptic current (IPSC). Bath application of MTII (0.1–100n*M*) caused a concentration-dependent increase in the amplitude of the current (mean increase = 25 ± 4.47%; n=11, $p < .0002$). In 23/24 neurons responsive to MTII, application of 100n*M* NPY caused an inhibition of the IPSC (28.65 ± 2.57%; $p < .0001$). α-MSH, the endogenous melanocortin agonist, caused an increase in the IPSC of 31.8 ± 9.88% (30n*M*, n=5). The actions of α-MSH could be prevented by pretreatment with the endogenous melanocortin antagonist AGRP (10n*M*; $p < .001$). Taken together, these data suggest the possibility of a cellular basis for the integration of the NPY and MSH signals in the regulation of energy homeostasis (Fig. 4). While the PVH appears to be an important site for this integration, it is quite possible that a distributed group of brain centers may be important in this activity. For example, the dorsal motor nucleus of the vagus in the brainstem is one of the heaviest sites of MC4-R mRNA expression *(9)*. SHU9119 administration into the fourth ventricle, designed to specifically block brainstem MC4-R sites, stimulated 24-h food intake in the rat as efficaciously as did third ventricle administration *(120)*.

5.3. The MC4-R and the Genetics of Human Obesity

In contrast to the mouse, previous studies had not identified simple single gene obesity syndromes in the human until 1997. Nevertheless, obesity has a heritability of 50–80%, according to twin studies, and obesity is seen as part of some complex genetic syndromes, such as Bardet-Biedl and Prader-Willi syndromes. Following on findings made first in rodents, inherited obesity syndromes in humans have now been reported to be caused by mutations

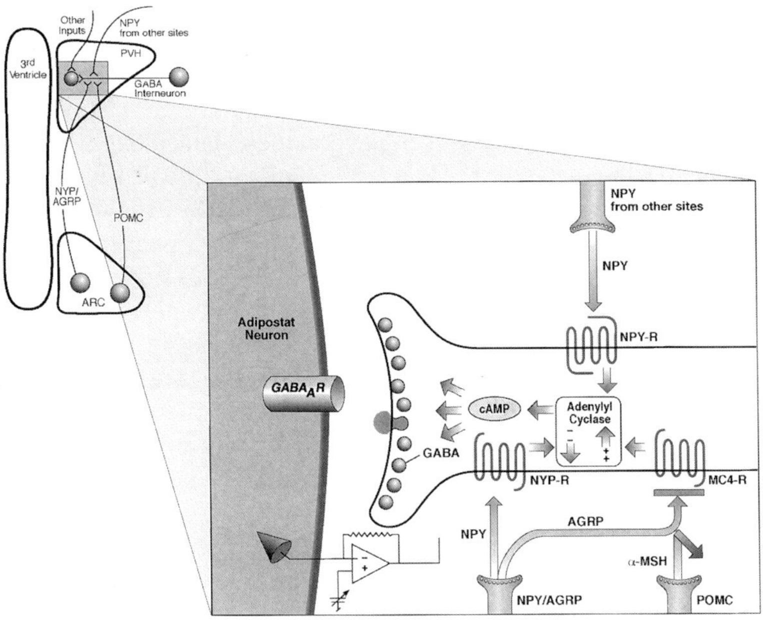

Fig. 4. Model for the integration of information from α-MSH, NPY, and AGRP at GABAergic neurons upstream of the adipostat. (Top Left) Arcuate POMC neurons, arcuate NPY/AGRP neurons, and NPY neurons from other sites such as the brainstem project to GABAergic interneurons in the medial parvocellular PVH. These neurons provide inhibitory input to the adipostat neurons characterized here. (Inset) Melanocortin receptors and NPY receptors in the GABA interneurons may regulate GABA release directly via their opposing action on adenylate cyclase.

in leptin *(121)*, the leptin receptor *(122)*, and prohormone convertase 1 *(123)*. While these cases are apparently extremely rare, they are very important in that they demonstrate conservation of function in humans.

More recently, mutations in the human POMC gene have been demonstrated to be associated with a novel syndrome encompassing adrenal insufficiency due to absence of ACTH, red hair due to an absence of MSH, and severe early-onset obesity, presumably due to an absence of hypothalamic POMC peptides *(124)*. This finding implies that the central POMC system, including the central MC3-R and MC4-R, serves a similar function in humans as shown experimentally in the mouse and rat.

One patient was found to be a composite heterozygote for a nonsense and an insertion mutation in POMC, both of which prevent the production of α-MSH and ACTH. A second patient was homozygous for a mutation that disrupted the translational start site of POMC. Both patients were found to have red hair, adrenal insufficiency, and severe obesity, confirming the role of ACTH in adrenal function, and demonstrating for the first time in humans that α-MSH/ACTH are required for eumelanin synthesis and development of a normal body mass index.

While this syndrome is likely to be rare, these data, along with the data from the mouse showing that haploinsufficiency for the MC4-R causes obesity *(12)*, argue that variant alelles of POMC or the MC4-R might act like dominant quantitative trait loci for obesity *(19,70)*. The linkage (LOD = 4.95) of serum leptin levels and human obesity to a chromosomal locus near POMC on chromosome 2 in a large Mexican-American population *(125)* provides additional support for the hypothesis that variant alleles of POMC could be contributing to more common forms of human obesity. This linkage has also been identified in an African-American *(126)* and French population *(127)*. The linkage of common obesity to POMC does not appear to be associated with coding sequence changes *(128)*, thus if the POMC gene is involved, then promoter or splicing mutations are implied.

More recently, mutations in the MC4-R itself (Table 5) have been found to be associated with obesity *(129,130)*. One group identified a heterozygous frameshift mutation in codon 211 of the receptor in a patient from a cohort of extremely obese children (mean body mass index (BMI) = 34 kg/mg^2 at <10 yr) *(130)*. The proband's father passed on the mutation, and exhibited a BMI of 41. A second report identified a heterozygous 4bp insertion at codon 244 in a 20-yr-old proband with a BMI of 30 *(129)*. This mutation cosegrated with severe obesity over three generations in this family (LOD =1.5). A stop codon at amino acid 35 has also been reported in two unrelated children with severe obesity (BMIs = 31, 46) *(128)*.

A large number of missense mutations in the receptor have also been identified. Originally, a Val103Ile change was reported to be present at a frequency of about 4% in both obese and nonobese British subjects *(131)*. Hinney and colleagues *(132)* have identified a large number of heterozygous missense mutations in obese but not nonobese children, and some of the nonconservative changes found might be expected to disrupt receptor function causing haploinsufficiency. Finally, a novel missense mutation (Ile137Thr) identified in an extremely obese adult (BMI=57) has been characterized pharmacologically and has a greatly reduced affinity (Kd=9nm, versus K^d= 1.2n*M* in the wild-type) for the synthetic ligand NDP-α-MSH *(133)*. A statistically significant association of this allele with obesity in other members of the proband's family could not be proven, however this could have been due to the small sample size.

At this time, the frequency of haploinsufficiency of the MC4-R (deletion, insertion, and nonsense mutants) in all obese subjects studies (*n*=452) is approximately 1%. If one were to (i.) restrict this analysis to morbid obesity

Table 5
Allelic Variants of the Human MC4-R

Receptor Variant	Comments	Author/Reference
Δ CTCT, nt 633	Severe obesity, 4 yr old (BMI = 28), and father (BMI = 41)	Yeo et al. *(130)*
GATT insertion, nt 732	Severe obesity, 3 generations, (BMI = 30–57)	Vaisse et al. *(129)*
Val103Ile	No association with obesity	Gotoda et al. *(131)* Hinney et al. *(132)* Gu et al. *(133)*
Stop codon, Tyr35	Single obese child/adolescent	Hinney et al. *(132)*
Ser30Phe	Single obese child/adolescent	Hinney et al. *(132)*
Asp37Val	Single obese child/adolescent	Hinney et al. *(132)*
Pro78Leu	Single obese child/adolescent	Hinney et al. *(132)*
Thr112Met	Single obese child/adolescent	Hinney et al. *(132)* Gu et al. *(133)*
Arg165Trp	Single obese child/adolescent	Hinney et al. *(132)*
Gly252Ser	Single obese child/adolescent	Hinney et al. *(132)*
Ile317Thr	Single obese child/adolescent	Hinney et al. *(132)*
Ile251Leu	Obese and normal children	Hinney et al. *(132)*
Silent change, C579T	Normal child/adolescent	Hinney et al. *(132)*
Ile137Thr	Obese adult (BMI = 57)	Gu et al. *(133)*

in children, (ii) include the likelihood that some of the missense mutations produce haploinsufficiency, and (iii) consider the fact that mutations in noncoding sequences of the receptor have not yet been examined it becomes possible to hypothesize that a very significant percentage of severe childhood obesity may be due to haploinsufficiency of the MC4-R.

5.4. The Role of Mahogany in MC4-R Action

The murine *mahogany* (*mg*) and *mahoganoid* (*md*) loci were identified several decades ago as recessive suppressors of A^Y-induced pigmentation that were able to shift melanogenesis from the pheomelanin (red/yellow pigment) pathway toward eumelanin (black/brown pigment) production *(134,135)*. These genes were mapped to chromosome 2 and 16, respectively. Two mutations have been identified at the murine *mahogany* locus, *mg*, which originated in the LDJ/Le background, and mg^{3J} in the C3HeB/FeJ background

(136). The *md* coat color mutation originated in the C3H/HeJ genetic background *(136)*. Recently, *mg* and *md* were found to suppress not only the yellow coat color but also the A^Y-induced increase in body weight *(136)*, suggesting that these genes are required for *agouti* action both in the skin as well as in the brain. Additional studies demonstrated that *mg/mg* suppresses virtually the entire agouti (A^Y) obesity phenotype, restoring insulin, glucose and leptin levels to normal, and reducing somatic growth to normal as well *(137)*. However, *mg/mg* did not suppress hyperphagia in the A^Y mouse. Furthermore, *mg/mg* was also shown to induce increased activity, increased basal metabolic rate, and hyperphagia in the normal C57BL/6J mouse *(137)*. Since agouti is not normally expressed in the brain, it is possible that the wild type *mahogany* gene is necessary for the function of the brain agouti homolog, AGRP. In the absence of the MC4-R antagonist AGRP, chronic MC4-R stimulation by α-MSH would result, producing a hyperactive hypermetabolic state. The hyperphagia could be secondary to the increase in energy expenditure, as the animal attempts to maintain energy homeostasis.

The recent cloning of the *mahogany* gene provides an interesting tool to test this model, and to generally attempt to determine the role of mahogany in the MC4-R – AGRP interaction *(138,139)*. The gene, isolated by two groups using positional cloning methods, encodes a transmembrane form of the attractin gene, a member of the CUB family of cell–adhesion proteins previously demonstrated to be involved in cell–cell interactions between T cells and monocytes *(140)*. The *mahogany* sequence predicts a 1336 amino acid protein containing a large extracellular domain encoding three epidermal growth factor (EGF) domains, two lamininlike EGF repeats, a CUB domain, two plexinlike repeats, one C-type lectin domain, and seven consecutive Kelch repeats. This is followed by a single membrane spanning domain, and a small putative intracellular signalling domain of 126 amino acids, containing no recognizable motifs. Curiously, while the phenotype of the *mg/mg* can be completely understood thus far via defective agouti/AGRP signaling, the gene is expressed in a wide variety of tissues, and the attractin splice form has already been suggested to play a melanocortin-independent role in immune function. One proposed mechanism for *mahogany* function has been that of a low-affinity coreceptor for agouti and AGRP binding to target cells *(138)*. An alternative proposal is that *mahogany* may be a protein broadly involved in a class of cell–cell interaction that is required for the action of agouti and AGRP in trans on melanocytes and MC3-R/MC4-R neurons, respectively (Fig. 5).

5.5. Additional Roles of the MC4-R

The neuroanatomic distribution of MC4-R mRNA characterized by *in situ* hybridization in the rat suggested a multitude of roles for the receptor

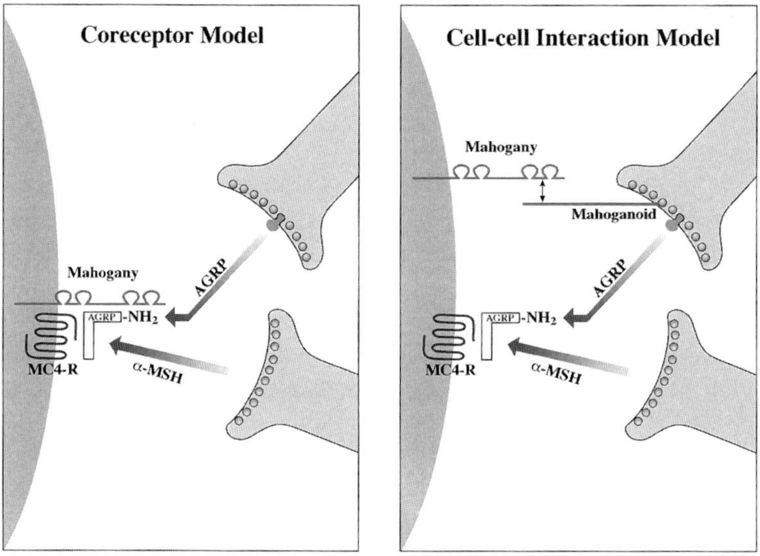

Fig. 5. Alternative models for the role of *mahogany* in AGRP function. Coreceptor model (left): *Mahogany* and the MC4-R form a receptor complex, with the MC4-R providing a high-affinity interaction with the cysteine-rich domain of AGRP or agouti, and *mahogany* providing a low affinity interaction with the amino-terminal domains of AGRP or agouti. Cell–cell interaction model (right): *Mahogany* is a cell–adhesion molecule involved in forming the necessary cell–cell interaction to allow AGRP or agouti to act at the MC4-R. *Mahoganoid*, a gene with similar properties to *mahogany*, is a candidate target for *mahogany*.

in autonomic and neuroendocrine function *(9)*. Indeed, though the majority of attention has been paid to the role of the receptor in energy homeostasis, additional roles for the MC4-R and/or MC3-R are becoming apparent, primarily through studies utilizing the neural melanocortin receptor antagonist, SHU9119.

5.5.1. Neuroimmunomodulatory Roles

Melanocortins have been known for some time to have antiinflammatory and antipyretic activities (reviewed in ref. *141*). It has long been thought that a major component of this activity is centrally mediated, although arguments for peripheral actions were made based upon the ability of peripherally administered α-MSH to block a fever induced by bacterial lipopolysaccharide (LPS). Recently, however, even systemic α-MSH was demonstrated to suppress LPS-mediated fever via central melanocortin receptors, since ICV administration of the MC3-R/MC4-R antagonist, SHU9119, blocked the antipyretic activity of peripherally administered α-MSH *(142)*. Centrally administered α-MSH has potent activity in a wide variety of antiinflammatory assays, and the data above suggest that the MC3-R and MC4-R may be mediating these effects, possibly through sympathetic outflow to the vasculature and effects on functions such as extravasation of immune cells, as well as through interaction with central pathways utilized by cytokines. With regard to central cytokine action, SHU9119 has been utilized to show that endogenous melanocortin pathways appear to be involved in suppression of LPS-induced fever, but appear to contribute to LPS-induced anorexia *(142)*. Thus, the MC4-R may play a complex role in mediating diverse aspects of the acute phase response. Interestingly, brain microvascular endothelial cells themselves have been demonstrated, by ligand binding and photoaffinity labeling with NDP-MSH derivatives, to express melanocortin receptors, although the subtype was not identified *(143)*.

5.5.2. Role of the MC4-R in Cardiovascular Homeostasis

SHU9119 has also been used to identify a potential role for the MC4-R in cardiovascular homeostasis *(144)*. The first suggestion of a role for the MC4-R in cardiovascular homeostasis came from the finding that the site of highest MC4-R mRNA was in the medulary dorsovagal complex *(9)*, the site of the first synapse of the baroreceptor reflex and a site of descending POMC fibers from the arcuate nucleus. MC3-R mRNA was not seen in this region *(10)*. Previous studies had demonstrated that activation of arcuate POMC neurons produced hypotension and bradycardia *(145–147)*, but that only a portion of this effect was mediated by release of β-endorphin. The MC4-R was implicated in this pathway when it was demonstrated that α-MSH also elicited hypotension and bradycardia when injected into the medulary dorsovagal complex, and that this activity was blocked by coinjection of SHU9119 *(144)*.

5.5.3. Role of the MC4-R in Grooming Behavior and the H–P–A Axis

Central administration of melanocortins has also been demonstrated to stimulate grooming behavior in rodents, and to activate the hypothalamic–pituitary–adrenal (H–P–A) axis, independently from their behavioral actions *(148)*. The melanocortin-3 specific agonist Lys-γ^2-MSH was unable to elicit either of these responses, and both grooming behavior and H–P–A axis activation by ACTH[1–24] were blocked by coadministration of SHU9119, implicating the MC4-R in both of these activities *(149)*. Additional roles for the melanocortin system in neuroendocrine function derive from the high levels of MSH binding seen in the median eminence *(5,37)*, the high level of expression of AGRP in fibers in the median eminence *(5,37,66)*, and the inhibition of the preovulatory LH and prolactin surge following α-MSH administraiton into the median eminence *(150)*.

6. Perspectives

Interest in the MC4-R has been greatly stimulated by the finding of a role for this receptor in the regulation of energy homeostasis. To point out some glaring voids in our knowledge, we do not yet know where the receptor protein is made, how it signals, the various ways in which it effects synaptic transmission, and the full breadth of its physiological functions. Multiple pharmaceutical companies are pursuing this receptor as a drug target for the treatment of obesity, so it is hoped that specific small molecule MC4-R agonists and antagonists will help advance the field.

What is clear from both the distribution of the receptor, and the limited data available from studies with the antagonist SHU9119, is that the MC4-R is likely to have multiple roles in the regulation of autonomic outflow, neuroendocrine function, and behavior. Whether this will preclude the use of drugs acting at the MC4-R for the treatment of obesity, or for other clinical applications is hard to predict at this time. While this review may truly point out how very little is yet known about the MC4-R, the glimpses that have been provided suggest a good deal of exciting work ahead for those in the field.

References

1. Chhajlani, V. and Wikberg, J. E. S. (1992) Molecular cloning and expression of the human melanocyte stimulating hormone receptor cDNA. *FEBS Lett.* **309,** 417–420.
2. Mountjoy, K. G., Robbins, L. S., Mortrud, M. T., and Cone, R. D. (1992) The cloning of a family of genes that encode the melanocortin receptors. *Science* **257,** 543–546.
3. DeWied, D. and Jolles, J. (1982) Neuropeptides derived from pro-opiocortin: behavioral, phsyiological, and neurochemical effects. *Physiol. Rev.* **62,** 977–1059.
4. Murphy, M. T., Richards, D. B., and Lipton, J. M. (1983) Antipyretic potency of centrally administered α-melanocyte stimulating hormone. *Science* **221,** 192–194.
5. Tatro, J. B. (1990) Melanotropn receptors in the brain are differentially distributed and recognize both corticotropin and α-melanocyte stimulating hormone. *Brain Res.* **536,** 124–132.
6. Gantz, I., Shimoto, Y., Konda, Y., Miwa, H., Dickinson, C. J., and Yamada, T. (1994) Molecular cloning, expression, and characterization of a fifth melanocortin receptor. *Biochem. Biophys. Res. Commun.* **200,** 1214–1220.
7. Labbe, O., Desarnaud, F., Eggerickx, D., Vassart, G., and Parmentier, M. (1994) Molecular cloning of a mouse melanocortin 5 receptor gene widely expressed in peripheral tissues. *Biochemistry* **33,** 4543–4549.
8. DeWied, D. and Wolterink, G. (1988) Structure-activity studies on the neuroactive and neurotropic effects related to ACTH. *Ann. N. Y. Acad. Sci.* **525,** 130–140.
9. Mountjoy, K. G., Mortrud, M. T., Low, M. J., Simerly, R. B., and Cone, R. D. (1994) Localization of the melanocortin-4 receptor (MCR-4) in neuroendocrine and autonomic control circuits in the brain. *Mol. Endocrinol.* **8,** 1298–1308.

10. Roselli-Rehfuss, L., Mountjoy, K. G., Robbins, L. S., Mortrud, M. T., Low, M. J., Tatro, J. B., Entwistle, M. L., Simerly, R., and Cone, R. D. (1993) Identification of a receptor for γ-MSH and other proopiomelanocortin peptides in the hypothalamus and limbic system. *Proc. Natl. Acad. Sci. U. S. A.* **90,** 8856–8860.
11. Fan, W., Boston, B. A., Kesterson, R. A., Hruby, V. J., and Cone, R. D. (1997) Role of melanocortinergic neurons in feeding and the agouti obesity syndrome. *Nature* **385,** 165–168.
12. Huszar, D., Lynch, C. A., Fairchild-Huntress, V., Dunmore, J. H., Fang, Q., Berkemeir, L. R., Gu, W., Kesterson, R. A., Boston, B. A., Cone, R. D., Smith, F. J., Campfield, L. A., Burn, P., and Lee, F. (1997) Targeted disruption of the melanocortin-4 receptor results in obesity in mice. *Cell* **88,** 131–141.
13. Gu, W., Tu, Z., Kleyn, P. W., at al. (1999). Identification and functional analysis of novel human melanocortin-4 receptor variants. *Diabetes* **48,** 635–639.
14. Lu, D., Vage, D. I., and Cone, R. D. (1998) A ligand-mimetic model for the constitutive activation of the melanocortin-1 receptor. *Mol. Endocrinol.* **12,** 592–604.
15. Schioth, H. B., Petersson, S., Muceniece, R., Szardenings, M., and Wikberg, J. E. S. (1997) Deletions of the N-terminal regions of the human melanocortin receptors *FEBS Lett.* **410,** 223–228.
16. Lu, D., Willard, D., Patel, I. R., Kadwell, S., Overton, L., Kost, T., Luther, M., Chen, W., Woychik, R. P. Wilkison, W. O., and Cone, R. D. (1994) Agouti protein is an antagonist of the melanocyte-stimulating hormone receptor *Nature* **371,** 799–802.
17. Fong, T. M., Mao, C., MacNeil, C., Kalyani, R., Smith, T., Weinberg, D., Tota, M. R., and Van der Ploeg, L. H. (1997) ART (protein product of the agouti-related transcript) as an antagonist of MC-3 and MC-4 receptors. *Biochem. Biophys. Res. Commun.* **237,** 629–631.
18. Mountjoy, K. G., Willlard, D. H., and Wilkison, W. O. (1999) Agouti antagonism of melanocortin-4 receptor: greater effect with desacetyl-α-MSH. *Endocrinology* **140,** 2167–2172.
19. Ollmann, M. M., Wilson, B. D., Yang, Y.-K., Kerns, J. A., Chen, Y., Gantz, I., and Barsh, G. S. (1997) Antagonism of central melanocortin receptors in vitro and in vivo by agouti-related protein. *Science* **278,** 135–137.
20. Yang, Y.-K., Thompson, D. A., Dickinson, C. J., Wilken, J., Barsh, G. S., Kent., S. B. H., and Gantz, I. (1999) Characterization of agouti-related protein binding to melanocortin receptors. *Mol. Endocrinol.* **13,** 148–155.
21. Yang, Y., Ollmann, M. M., Wilson, B. D., Dickinson, C., Yamada, T., Barsh, G. S., and Gantz, I. (1997) Effects of recombinant agouti-signaling protein on melanocortin action. *Mol. Endocrinol.* **11,** 274–280.
22. Gantz, I., Miwa, H., Konda, Y., Shimoto, Y., Tashiro, T., Watson, S. J., DelValle, J., and Yamada, T. (1993) Molecular cloning, expression, and gene localization of a fourth melanocortin receptor. *J. Biol. Chem.* **268,** 15,174–15,179.
23. Low, M. J., Simerly, R. B., and Cone, R. D. (1994) Receptors for the melanocortin peptides in the central nervous system. *Curr. Opin. Endocrinol. Diab.* **1,** 79–88.
24. Chhajlani, V. (1996) Distribution of cDNA for melanocortin receptor subtypes in human tissues *Biochem. Mol. Biol. Int.* **38,** 73–80.
25. Kraan, V. D., Tatro, J. B., Entwistle, M. L., Brakkee, J. H., Burbach, J. P. H., Adan, R. A. H., and Gispen, W. H. (1999) Expression of melanocortin and

proopiomelanocortin in the rat spinal cord in relation to the neurotrophic effects of melanocortins. *Mol. Brain Res.* **63,** 276–286.

26. Chen, W., Kelly, M. A., Opitz-Araya, X., Thomas, R. E. Low, M. J., and Cone, R. D. (1997) Exocrine gland dysfunction in MC5-R deficient mice: evidence for coordinated regulation of exocrine gland function by melanocortin peptides. *Cell* **91,** 789–798.
27. Elkabes, S., Loh, Y. P., Nieburgs, A., and Wray, S. (1989) Prenatal ontogenesis of pro-opiomelanocortin in the mouse central nervous system and pituitary: gland: an in situ hybridization and immunocytochemical study. *Brain Res. Dev. Brain Res.* **46,** 85–95.
28. Schwartzberg, D. G., and Nakane, P. K. (1982) Ontogenesis of adrenocorticotropin-related peptide determinants in the hypothalamus and pituitary gland of rat. *Endocrinology* **110,** 855–864.
29. Lichtensteiger, W., Hanimann, B., Schlumph, M., Siegrist, W., and Eberle, A. N. (1993) Pre-and postnatal ontogeny of [^{125}I] Nle^4, D-PHE^7-α-MSH binding sites in rat brain. *Ann. N. Y. Acad. Sci.* **680,** 652–654.
30. Lichtensteiger, W., Hanimann, B., Siegrist, W., and Eberle, A. N. (1996) Region- and stage-specific patterns of melanocortin receptor ontogeny in rat central nervous system, cranial nerve ganglia and sympathetic ganglia. *Dev. Brain. Res.* **91,** 93–110.
31. Kistler-Heer, V., Lauber, M. E., and Lichtensteiger, W. (1998) Different developmental patterns of melanocortin MC3 and MC4 receptor mRNA: predominance of MC4 in fetal rat nervous system. *J. Neuroendocrinol.* **10,** 133–146.
32. Mountjoy, K. G. and Wild, J. M. (1998) Melanocortin-4 receptor mRNA expression in the developing autonomic and central nervous systems. *Dev. Brain. Res.* **107,** 309–314.
33. Contreras, P. C., and Takemori, A. E. (1984) Antagonism of morphine-induced analgesia, tolerance, and dependence by α-melanocyte-stimulating hormone. *J. Pharmacol. Exp. Ther.* **229,** 21–26.
34. Szekely, J. I., Miglecz, E., Dunai-Kovacs, Z., Tarnawa, I., Ronai, A. Z., Graf, L., and Bajusz, S. (1979) Attenuation of morphine tolerance and dependence by α-melanocyte stimulating hormone (α-MSH). *Life Sci.* **24,** 1931–1938.
35. Alvaro, J. D., Tatro, J. B., Quillan, J. M., Fogliano, M., Eisenhard, M., Lerner, M. R., Nestler, E. J., and Duman, R. S. (1996) Morphine down-regulates melanocortin-4 receptor expression in brain regions that mediate opiate addiction. *Mol. Pharmacol.* **50,** 583–591.
36. Alvaro, J. D., Tatro, J. B., and Duman, R. S. (1997) Melanocortins and opiate addiction *Life Sci.* **61,** 1–9.
37. Harrold, J.A., Widdowson, P.S., and Williams, G. (1999) Altered energy balance causes selective changes in melanocortin-4 (MCR4-R), but not melanocortin-3 (MC3-R), receptors in specific hypothalamic regions. *Diabetes* **48,** 267–271.
38. Adan, R. A H., Oosterom, J., Ludvigsdotter, G., Brakkee, J. H., Burbach, J. P. H., and Gispen, W. H. (1994) Identification of antagonists for melanocortin MC3, MC4 and MC5 receptors. *Eur. J. Pharm.* **269,** 331–337.
39. Hruby, V. J., Lu, D., Sharma, S. D., Castrucci, A. L., Kesterson, R. A., Al-Obeidi, F. A., Hadley, M. E., and Cone, R. D. (1995) Cyclic lactam α-melanotropin analogues of Ac-Nle^4[Asp^4, D-Phe^7 Lys^{10}]α-MSH (4–10)-NH_2 with bulky aromatic amino acids at position 7 show high antagonist potency and selectivity at specific melanocortin receptors. *J. Med. Chem.* **38,** 3454–3461.

40. Al-Obeidi, F., Castrucci, A. M. d. L., HAdley, M. E., and Hruby, V. J. (1989) Potent and prolonged acting cycle lactam analogues of α-melanotropin:design based on molecular dynamics. *J. Med. Chem.* **32,** 2555–2561.
41. Kask, A., Mutulis, F., Muceniece, R., Pahkla, R., Mutule, I., Wikberg, J. E. S., Rago, L., and Schioth, H. B. (1998) Discovery of a novel superpotent and selective melanocortin-4 receptor antagonist (HS024):evaluation in vitro and in vivo. *Endocrinology* **139,** 5006–5014.
42. Schioth, H. B., Mutulis, F., Muceniece, R., Prusis, P., and Wilberg, J. E. S. (1998) Discovery of novel melanocortin-4 receptor selective MSH analogues. *Br. J. Pharamacol.* **124,** 75–82.
43. Skuladottir, G. V., Jonsson, L., Skarphedinsson, J. O., Mutulis, F., Muceniece, R., Raine, A., Mutule, I., Helgason, J., Prusis, P., Wikberg, J. E. S., and Schioth, H. B. (1999) Long term orexigenic ffect of a novel melanocortin 4 receptor selective antagonist. *Br. J. Pharm.* **126,** 27–34.
44. Lu, D., Chen, W., and Cone, R. D. (1998) Regulation of melanogenesis by the MSH receptor in *The Pigmentary System.* (Nordlund, J., Boissy, R., Hearing, V., King, R., and Ortonne, J.-P., eds.) Oxford University Press, New York.
45. Wolff, G. L., Roberts, D. W., and Galbraith, D. B. (1986) Prenatal determination of obesity, tumor susceptibility, and coat color pattern in viable yellow (A^{vy}/a) mice. *J. Heredity* **77**, 151–158,
46. Frigeri, L. G., Wolff, G. L., and Robel, G. (1983) Impairment of glucose tolerance in yellow (Avy/A) (Balb/c xVY) F-1 hybrid mice by hyperglycemic peptide(s) from human pituitary glands. *Endocrinology* **113,** 2097–2105.
47. Wolff, G. L. and Flack, J. D. (1971) Genetic regulation of plasma corticosterone concentration and its response to castration and allogeneic tumor growth in the mouse. *Nature* **232,** 181–182.
48. Wolff, G. L. (1963) Growth of inbred yellow (Ay) and non-yellow (aa) mice in parabiosis. *Genetics* **48,** 1041–1058.
49. Bultman, S. J., Michaud, E. J., and Woychik, R. P. (1992) Molecular characterization of the mouse agouti locus. *Cell* **71,** 1195–1204.
50. Miller, M. W., Duhl, D. M. J., Vrieling, H., Cordes, S. P., Ollmann, M. M., Winkes, B. M., and Barsh, G. S. (1993) Cloning of the mouse *agouti* gene predicts a novel secreted protein ubiquitously expressed in mice carrying the *lethal yellow* (A^y) mutation. *Genes Dev.* **7,** 454–467.
51. Manne, J., Argeson, A. C., and Siracusa, L. D. (1995) Mechanisms for the pleiotropic effects of the agouti gene. *Proc. NAtl. Acad. Sci. U. S. A.* **92,** 4721–4724.
52. Olivera, B. M., Miljanich, G. P., Ramachandran, J., and Adams, M. E. (1994) Calcium channel diversity and neurotransmitter release: the w-Conotoxins and w-Agatoxins. *Annu. Rev. Biochem.* **63,** 823–867.
53. Silvers, W. K. (1958) An experimental approach to action of genes at the agouti locus in the mouse. III. Transplants of newborn A^w-, A-, a^t- skin to A^y-, A^w-, A-, and *aa* hosts. *J. Exp. Zool.* **137,** 189–196.
54. Silvers, W. K. and Russel, E. S. (1955) An experimental approach to action of genes at the agouti locus of the mouse. *J. Exp. Zool.* **130,** 199–220.
55. Vage, D. I., Lu, D., Klungland, H., Lien, S., Adalsteinsson, S., and Cone, R. D. (1997) A non-epistatic interaction of *agouti* and *extension* in the fox, *Vulpes vulpes. Nat. Genet.* **15,** 311–315.

56. Kwon, H. Y., Bultman, S. J., Loffler, C., Chen, W.-J., Furdon, P. J., Powell, J. G., Usala, A.-L., Wilkison, W., Hansmann, I., and Woychik, R. P. (1994) Molecular structure and chromosomal mapping of the human homolog of the agouti gene. *Proc. Natl. Acad. Sci. U. S. A.* **91,** 9760–9764.
57. Klebig, M. L., Wilkinson, J. E., Geisler, J. G., and Woychik, R. P. (1995) Ectopic expression of the agouti gene in transgenic mice causes obesity, features of type II diabetes, and yellow fur. *Proc. Natl. Acad. Sci. U. S. A.* **92,** 4728–4732.

57a. Jones, B. H., Kim, J. H., Zemel, M. B., Woychik, R. P., Michaud, E. J., Wilkison, W. O., and Moustaid, N. (1996) Upregulation of adipocyte metabolism by agouti protein: possible paracrine actions in yellow mouse obesity. *Am. J. Physiol.* **270,** E192–E196.

58. Kim, J. H., Kiefer, L. L., Woychik, R. P., Wilkison, W. O., Truesdale, A., Ittoop, O., Willard, D., Nichols, J., and Zemel, M. B. (1997) Agouti regulation of intracellular calcium: role of melanocortin receptors. *Am J. Physiol.* **272,** E379–384.
59. Zemel, M. B. (1998) Nutritional and endocrine modulation of intracellular calcium: implications in obesity, insulin resistance and hypertension. *Mol. Cell. Biochem.* **188,** 129–136.
60. Mynatt, R. L., Miltenberger, R. J., Klebig, M. L., Zemel, M. B., Wilkison, J. E., Wilkison, W. O., and Woychik, R. P. (1997) Combined effects of insulin treatment and adipose tissue-specific agouti expression on the development of obesity. *Proc. Natl. Acad. Sci. U. S. A.* **94,** 919–922.
61. Zemel, M. B., Kim, J. H., Woychik, R. P., Michaud, E. J., Kadwell, S. H., Patel, I. R., and Wilkison, W. O. (1995) Agouti regulation of intracellular calcium: role in the insulin resistance of viable yellow mice. *Proc. Natl. Acad. Sci. U. S. A.* **92,** 4733–4737.

61a. Boston, B. A. and Cone, R. D. (1996) Characterization of melanocortin receptor subtype expression in murine adipose tissues and in the 3T3-L1 cell line. *Endocrinology* **137,** 2043–2050.

62. Shutter, J. R., Graham, M., Kinsey, A. C., Scully, S., Luthy, R., and Stark, K. L. (1997) Hypothalamic expression ART, a novel gene related to agouti, is upregulated in oese and diabetic mutant mice. *Genes Dev.* **11,** 593–602.
63. Broberger, C., De Lecea, L., Sutcliffe, J. G., and Hokfelt, T. (1998) Hypocretin/orexin- and melanin-concentrating hormone-expressing cells form distinct populations in the rodent lateral hypothalamus: relationship to the neuropeptide Y and agouti gene-related protein systems. *J. Comp. Neurol.* **402,** 460–474.
64. Chen, P., Li, C., Haskell-Luevano, C., Cone, R. D., and Smith, M. S. (1998) Altered expression of agouti-related protein and its colocalization with neuropeptide in the arcuate nucleus of the hypothalamus. *Endocrinology* **140,** 2645–2650.
65. Hahn, T. M., Breininger, J. F., Baskin, D. G., and Schwartz, M. W. (1999) Coexpression of Agrp and NPY in fasting-activated hypothalamic neurons. *Nat. Neurosci.* **1,** 271–272.
66. Haskell-Luevano, C., Chen, P., Li, C., Chang, K., Smith, M. S. Cameron, J. L., and Cone, R. D. (1999) Characterization of the neuroanatomical distribution of agouti-related protein (AGRP) immunoreactivity in the rhesus monkey and the rat. *Endocrinology* **140,** 1408–1415.
67. Wilson, B. D., Kaelin, C. B., Ollmann, M. M., Gantz, I., Watson, S. J., and Barsh, G. S. (1999) Physiological and anatomical circuitry between agouti-related protein and leptin signaling. *Endocrinology* **140,** 2387–2397.

68. Broberger, C., Johnsen, J., Schalling, M., and Hokfelt, T. (1997) Hypothalamic neurohistochemistry of the murine anorexia (anx/anx) mutation: altered processing of neuropeptide Y in the arcuate nucleus. *J. Comp. Neurol.* **13,** 124–135.
69. Maltais, L. J., Lane, P. W., and Beamer, W. G. (1984) Anorexia, a recessive mutation causing starvation in preweanling mice. *J. Hered.* **75,** 468–472.
70. Graham, M., Shuttre, J. R., Sarmieto, U., Sarosi, I., and Stark, K. L. (1997) Overexpression of Agrt leads to obesity in transgenic mice. *Nat. Genet.* **17,** 273–274.
71. Mizuno, T. M. and Mobbs, C. V. (1999) Hypothalamic agouti-related messenger ribonucleic acid is inhibited by leptin and stimulated by fasting. *Endocrinology* **140,** 814–817.
72. Quillan, J. M., Sadee, W., Wei, E. T., Jimenez, C., Ji, L., and Chang, J. K. (1998) A synthetic human agouti-related protein-(83-132)-NH2 fragment is a potent inhibitor of melanocortin receptor function. *FEBS Lett.* **428,** 59–62.
73. Rossi, M., Kim, M. S., Morgan, D. G., Small, C. J., Edwards, C. M., Sunter, D., Abusnana, S., Goldstone, A. P., Russel, S. H., Stanley, S. A., Smith, D. M., Yagaloff, K., Ghatei, M. A., and Bloom, S. R. (1998) A C-terminal fragment of agouti-related protein increases feeding and antagonizes the effect of alpha-melanocyte stimulating hormone in vivo. *Endocrinology* **139,** 4428–4431.
74. Rosenfeld, R. D., Zeni, L., Welcher, A. A., Narhi, L. O., Hale, C., Marasco, J., Delaney, J., Gleason, T., Philo, J. S., Katta, V., Hui, J., Baumgartner, J., Graham, M., Stark, K. L., and Karbon, W. (1998) Biochemical, biophysical, and pharmacological characterization bacterially expressed human agouti-related protein. *Biochemistry* **37,** 16,041–16,052.
75. Bures, E. J., Hui, J. O., Young, Y., Chow, D. T., Katta, V., Rohde, M. F., Zeni, L., Rosenfeld, R. D., Stark, K. L., and Haniu, M. (1998) Determination of disulfide structure in agouti-related protein (AGRP) by stepwise reduction and alkylation. *Biochemistry* **37,** 12,172–12,177.
76. Tota, M. R., Smith, T. S., Mao, C., MacNeil, T., Mosley, R. T., Van der Ploeg, L. H. T., and Fong, T. M. (1999) Molecular interaction of agouti protein and agouti-related protein with human melanocortin receptors. *Biochemistry* **38,** 897–904.
77. Michaud, E. J., Bultman, S. J., Stubbs, L. J., and Woychik, R. P. (1993) The embryonic lethality of homozygous lethal yellow mice (Ay/Ay) is associated with the disruption of a novel RNA-binding protein. *Genes Dev.* **7,** 1203–1213.
78. Al-Obeidi, F., Hadley, M. E., Pettitt, B.-M., and Hruby, V. J. (1989) Design of a new class of superpotent cyclic α-melanotropins based on quenched dynamic simulations. *J. Am. Chem. Soc.* **111,** 3413–3416.
79. Sugg, E. E., Castrucci, A. M., Hadley, M. E., van Binst, G., and Hruby, V. J., (1988) Cyclic lactam analogues of Ac-[Nle^4]alpha-MSH4-11-NH2. *Biochemistry* **27,** 8181–8188.
80. Frigeri, L. G., Wolff, G. L., and Teguh, C. (1988) Differential responses of yellow Avy/A agouti A/a (Balb/c x VY) F-1 hybrid mice to the same diets: glucose tolerance, weight gain, and adipocyte cellularity. *Int. J. Obesity***12,** 305–320.
81. Yen, T. T., McKee, M. M., and Stamm, N. B. (1984) Thermogenesis and weight control. *Int. J. Obesity* **8(Suppl. 1),** 65–78.
82. Jegou, S., Blasquez, C., Delbende, C., Bunel, D. T., and Vaudry, H. (1993) Regulation of alpha-melanocyte-stimulating hormone release from hypothalamic neurons. *Ann. N. Y. Acad. Sci.* **680,** 260–278.

83. Garcia de Yebenes, E. and Pelletier, G. (1994) Negative regulation of proopiomelanocortin gene expression by $GABA_A$ receptor activation in the rat arcuate nucleus. *Peptides* **15,** 615–618.
84. Bergendahl, M., Wiemann, J. N., Clifton, D. K., Huhtaniemi, I., and Steiner, R. A., (1992) Short-term starvation decreases POMC mRNA but does not alter GnRH mRNA in the brain of adult male rats. *Neuroendocrinology* **56,** 913–920.
85. Brady, L. S., Smith, M. A., Gold, P. W., and Herkenham, M. (1990) Altered expression of hypothalmic neuropeptide mRNAs in food-restricted and food-deprived rats. *Neuroendocrinology* **52,** 441–447.
86. Thornton, J. E., Cheung, C. C., Clifton, D. K., and Steiner, R. A., (1997) Regulation of hypothalamic proopiomelanocortin mRNA by leptin in ob/ob mice. *Endocrinology* **138,** 5063–5066.
87. Schwartz, M. W., Seeley, R. J., Woods, S. C., Weigle, D. S, Campfield, L. A., Burn, P., and Baskin, D. G. (1997) Leptin increases hypothalamic pro-opiomelanocortin mRNA expression in the rostral arcuate nucleus. *Diabetes* **46,** 2119–2123.
88. Beaulieu, S, Gagne, B., and Barden, N. (1988) Glucocorticoid regulation of proopiomelanocortin messenger ribonucleic acid content of rat hypothalamus. *Mol. Endocrinol.* **2,** 727–731.
89. Chowen-Breed, J., Fraser, H. M., Vician, L., Damassa, D. A., Clifton, D. K., and Steiner, R. A. (1989) Testosterone regulation of proopiomelanocortin messenger ribonucleic acid in the arcuate nucleus of the male rat. *Endocrinology* **124,** 1697–1702.
90. Smith, M. S. (1993) Lactation alters neuropeptide-Y and proopiomelanocortin gene expression in the arcuate nucleus of the rat. *Endocrinology* **133,** 1258–1265.
91. Hagan, M. M., Rushing, P. A., Schwartz, M. W., Yagaloff, K. A., Burn, P., Woods, S. C., and Seely, R. J. (1999) Role of the CNS melanocortin system in the response to overfeeing. *J. Neurosci.* **19,** 2362–2367.
92. Cone, R. D., Lu, D., Chen, W., Koppula, S., Vage, D. I., Klungland, H., Boston, B., Orth, D. N., Pouton, C., and Kesterson, R. A. (1996) The melanocortin receptors: agonists, antagonists, and the hormonal control of pigmentation. *Recent Prog. Horm. Res.* **51,** 287–318.
93. Ahima, R., Prabakaran, D., Mantzoros, C., Qu, D., Lowell, B., T, M.- F., and Flier, S. (1996) Role of leptin in the neuroendocrine response to fasting. *Nature* **382,** 250–252.
93a. Erickson, J., Hollopeter, G., and Palmiter, J. D. (1996) Attenuation of the obesity syndrone of ob/ob mice by the loss of neuropeptide Y. *Science* **274,** 1704–1707.
94. Friedman, J. M. (1997) The alphabet of weight control. *Nature* **385,** 119–120.
95. Gura, T. (1997) Obesity sheds its secrets. *Science* **275,** 751–753.
96. Cheung, C. C., Clifton, D. K., and Steiner, R. A. (1997) Proopiomelanocortin neurons are direct targets for leptin in the hypothalamus. *Endocrinology* **138,** 4489–4492.
97. Halaas, J. L., Boozer, C.,Blair-West, J., Fidahusein, N., Denton, D. A., and Friedman, J. M. (1997) Physiological response to the long-term peripheral and central leptin infusion in lean and obese mice. *Proc. Natl. Acad. Sci. U. S. A.* **94,** 8878–8883.
98. Kesterson, R. A., Huszar, D., Lynch, C. A., Simerly, R. B., and Cone, R. D. (1997) Induction of neuropeptide Y gene expression in the dorsal medial hypothalamic nucleus in two models of the agouti obesity syndrome. *Mol. Endocrinol.* **11,** 630-637.

99. Boston, B. A., Blaydon, K. M., Varnerin, J., and Cone, R. D. (1997) Independent and additive effects of central POMC and leptin pathways on murine obesity. *Science* **278,** 1641–1644.
100. Seeley, R. J., Yagaloff, K. A., Fischer, S. L., Burn, P., Thiele, T. E., van Dijk, G., Baskin, D. G., and Schwartz, M. W. (1997) Melanocortin receptors in leptin effects. *Nature* **390,** 349.
101. Thiele, T., van Dijk, G., Yagaloff, K. A., Fischer, S. L., Schwartz, M., Burn, P., and Seeley, R. J. (1998) Central infusion of melanocortin agonist MTII in rats: assessment of c-Fos expression and taste aversion. *Am. J. Physiol.* **274,** R248–R254.
102. Uehara, Y., Shimizu, H., Ohtani, K., Sato, N., and Mori, M. (1998) Hypothalamic corticotropin-releasing hormone is a mediator of the anorexigenic effect of leptin. *Diabetes* **47,** 890–893.
103. Goldstone, A. P., Mercer, J. G., Gunn, I., Moar, K. M., Edwards, C. M., Rossi, M., Howard, J. K., Rasheed, S., Turton, M. D., Small, C., Heath, M. M., O'Shea, D., Steere, J., Meeran, K., Ghatei, M. A., Hoggard, N., and Bloom, S. R. (1997) Leptin interacts with glucagon-like peptide-1 neurons to reduce food intake and body weight in rodents. *FEBS Lett.* **415,** 134–138.
104. Marsh, D. J., Hollopeter, G., Huszar, D., Laufer, R., Yagaloff, K. A., Fisher, S. L., Burn, P., and Palmiter, R. D. (1999) Response of melanocortin-4 receptor-deficient mice to anorectic and orexigenic peptides. *Nat. Genet.* **21,** 119–122.
105. Cowley, M. A., Prunchuk, N., Fan, W., Dinulescu, D. M., Colmers, W. F., and Cone, R. D. (1999) Integration of NPY, AGRP, and melanocortin signals in the hypothalamic paraventricular nucleus: evidence of a cellular basis for the adipostat. *Neuron* **24,** *155–163.*
106. Bronstein, D. M., Schafer, M. K., Watson, S. J., and AKil, H. (1992) Evidence that beta-endorphin is synthesized in cells in the nucleus tractus solitarius: detection of POMC mRNA. *Brain Res.* **587,** 269–275.
107. Jacobowitz, D. M. and O'Donohue, T. L. (1978) α-Melanocyte-stimulating hormone: immunohistochemical identification and mapping in neurons of rat brain. *Proc. Natl. Acad. Sci. U. S. A.* **75,** 6300–6304.
108. Joseph, S. A., Pilcher, W. H., and Bennet-Clarke, C. (1983) Immunocytochemical localization of ACTH parikarya in nucleus tractus solitarius : evidence for a second opiocortin neuronal system. *Neurosci. Lett.* **38,** 221–225.
109. Nilaver, G., Zimmerman, E. A., Defendi, R., Liotta, A., Krieger, D. T., and Brownstein, M. J. (1979) Adrenocorticotropin and beta-lipotropin in the hypothalamus: localization in the same arcuate neurons by sequential immunocytochemical procedures. *J. Cell Biol.* **81,** 50–58.
110. Palkovits, M., Mezey, E., and Eskay, R. L. (1987) Pro-opiomelanocortin-derived peptides. (ACTH/beta-endorphin/alpha-MSH) in brainstem baroreceptor areas of the rat. *Brain Res.* **436,** 323–328.
111. Watson, S. J., Akil, H., Richard, C. W., and Barchas, J. D. (1978) Evidence for two separate opiate peptide neuronal systems and the coexistence of beta-lipotropin, beta-endorphin, and ACTH immunoreactivities in the same hypothalamic neurons. *Nature* **275,** 226–228.
112. Woods, S. C., Seeley, R. J., Porte, D. Jr., and Schwartz, M. W. (1998) Signals that regulate food intake and energy homeostasis. *Science* **280,** 1378–1382.
113. Giraudo, S. Q., Billington, C. J., and Levine, A. S. (1998) Feeding effects of hypothalamic injection of melanocortin 4 receptor ligands. *Brain Res.* **809,** 302–306.

114. Stanley, B. G., Chin, A. S., and Leibowitz, S. F. (1985) Feeding and drinking elicited by central injection of neuropeptide Y: evidence for a hypothalamic site of action. *Brain Res. Bull.* **14,** 521–524.
115. Stanley, B. G., Magdalin, W., Seirafi, A., Thomas, W. J., and Leibowitz, S. F. (1993) The perifornical area: the major focus of (a) patchily distributed hypothalamic neuropeptide Y-sensitive feeding system(s). *Brain Res.* **604,** 304–317.
116. Ahlshog, J. E. and Hoebel, B. G. (1973) Overeating and obesity from damage to a noradrenergic system in the rat brain. *Science* **182,** 166–169.
117. Jin, S. L., Han, V. K., Simmons, J. G., Towle, A. C., Lauder, J. M., and Lund, P. K. (1988) Distribution of glucagon-like peptide I (GLP-I), glucagon, and glicentin in the rat brain, an immunocytochemical study. *J. Comp. Neurol.* **271,** 519–532.
118. Heinrichs, S. C., Menzaghi, F., Pich, E. M., Hauger, R. L., and Koob, G. F. (1993) Corticotropin-releasing factor in paraventricular nucleus modulates feeding induced by neuropeptide Y. *Brain Res.* **611,** 18–24.
119. Haynes, W. G., Morgan, D. A., Djalali, A., Sivitz, W. I., and Mark, A. L. (1999) Interactions between the melanocortin system and leptin in control of sympathetic nerve traffic. *Hypertension* **33,** 542–547.
120. Grill, H. J, Ginsberg, A. B., Seeley, R. J., and Kaplan, J. M. (1998) Brainstem application of melanocortin receptor ligands produces long-lasting effects on feeding and body weight. *J. Neurosci.* **18,** 10,128–10,135.
121. Montague, C. T., Farooqi, I. S., Whitehead, J. P., Soos, M. A., Rau, H., Wareham, N. J., Sewter, C. P., Digby, J. E., Mohammed, S. N., Hurst, J. A., Cheetham, C. H., Earley, A. R., Barnett, A. H., Prins, J. B., and O'Rahilly, S. (1997) Congenital leptin deficiency is associated with severe early-onset obesity in humans. *Nature* **387,** 903–908.
122. Clement, K., Vaisse, C., Lahlou, N., Cabrol, S., Pelloux, V., Cassuto, D., Gourmelen, M., Dina, C., Chambaz, J., Lacorte, J. M., Basdevant, A., Bougneres, P., Lebouc, Y., Froguel, P., and Guy-Grand, B. (1998) A mutation in the human leptin receptor gene causes obesity and pituitary dysfunction. *Nature* **392,** 398–401.
123. Jackson, R., Creemers, J., Ohagi, S., Raffin-Sanson, M. L., Sanders, L., Montague, C. T., Hutton, J. C., and O'Rahilly, S. (1997) Obesity and impaired prohormone processing associated with mutations in the human prohormone convertase 1 gene. *Nat. Genet.* **16,** 303–306.
124. Krude, H., Biebermann, H., Luck, W., Horn, R., Brabant, G., and Gruters, A. (1998) Severe early-onset obesity, adrenal insufficiency and red hair pigmentation caused by POMC mutations in humans. *Nat. Genet.* **19,** 155–157.
125. Comuzzie, A. G., Hixson, J. E., Almasy, L., Mitchell, B. D., Mahaney, M. C., Dyer, T. D., Stern, M. P., MacCluer, J. W., and Blangero, J. (1997) A major quantitative trait locus determining serum leptin levels and fat mass is located on human chromosome 2. *Nat. Genet.* **15,** 273–276.
126. Rotimi, C. N., Comuzzie, A. G., Lowe, W. L., Luke, A., Blangero, J., and Cooper, R. S. (1999) The quantitative trait locus on chromosome 2 for serum leptin levels is confirmed in African-Americans. *Diabetes* **48,** 643–644.
127. Hager, J., Dina, C., Francke, S., Dubois, S., Houari, M., Vatin, V., Vaillant, E., Lorentz, N., Basdevant, A., Clement, K., Guy-Grand, B., and Froguel, P. (1998) A genome-wide scan for human obesity genes reveals a major susceptibility locus on chromosome 10. *Nat. Genet.* **20,** 304–308.

128. Hinney, A., Becker, I., Heibult, O., Nottebom, K., Schmidt, A., Ziegler, A., Mayer, H., Siegfried, W., Blum, W. F., Remschmidt, H., and Hebebrand, J. (1998) Systematic mutation screening of the pro-opiomelanocortin gene: identification of several genetic variants including three different insertions, one nonsense and two missense point mutations in probands of different weight extremes. *J. Clin. Endocrinol. Metab.* **83,** 3737–3741.
129. Vaisse, C., Clement, K., Guy-Grand, B., and Froguel, P. (1998) A frameshift mutation in human MC4R is associated with a dominant form of obesity. *Nat. Genet.* **20,** 113–114.
130. Yeo, G. S. H., Farooqi, I. S., Aminian, S., Halsall, D. J., Stanhope, R. G., and O'Rahilly, S. (1998) A frameshift mutation in MC4R associated with dominantly inherited human obesity. *Nat. Genet.* **20,** 111–112.
131. Gotoda, T., Scott, J., and Aitman, T. J. (1997) Molecular screening of the human melanocortin-4 receptor gene: identification of a missense variant showing no association with obesity, plasma glucose, or insulin. *Diabetologia* **40,** 976–979.
132. Hinney, A., Schmidt, A., Nottebom, K., Heibult, O., Becker, I., Ziegler, A., Gerger, G., Sina, M., Gorg, T., Mayer, H., Siegfried, W., Fichter, M., Remschmidt, H., and Hebebrand, J. (1999) Several mutations in the melanocortin-4 receptor gene including a nonsense and a frameshift mutation associated with dominantly inherited obesity in humans. *J. Clin. Endocrinol. Metab.* **84,** 1483–1486.
133. Gu, W., Tu, Z., Kleyn, P. W., Kissebah, A., Duprat, L., Lee, J., Chin, W., Maruti, S., Deng, N., Fisher, S. L., Franco, L. S., Burn, P., Yagaloff, K. A., Nathan, J., Heymsfield, S., Albu, J., Pi-Sunyer, F. X., and Allison, D. B. (1999) Identification and functional analysis of novel human melanocortin-4 receptor variants. *Diabetes* **48,** 635–639.
134. Lane, P. W. (1960) New mouse mutants. *Mouse News Lett.* **22,** 35.
135. Lane, P. W. and Green, M. C. (1960) Mahogany, a recessive color mutation in linkage group V of the mouse. *J. Hered.* **51,** 228–230.
136. Miller, K. A., Gunn, T. M., Carrasquillo, M. M., Lamoreux, M. L., Galbraith, D. B., and Barsh, G. S. (1997) Genetic studies of the mouse mutations *mahogany* and *mahoganoid. Genetics.* **146,** 1407–1415.
137. Dinulescu, D. M., Fan, W., Boston, B. A., McCall, K., Lamoreux, M. L., Moore, K. J., Montagno, J., and Cone, R. D. (1998) Mahogany (mg) stimulates feeding and increases basal metabolic rate independent of its suppression of agouti. *Proc. Natl. Acad. Sci. U. S. A.* **95,** 12,707–12, 712.
138. Gunn, T. M., Miller, K. A., He, L., Hyman, R. W., Davis, R. W., Azarani, A., Schlossman, S. F., Duke-Cohan, J. S., and Barsh, G. S. (1999) The mouse mahogany locus encodes a transmembrane form of human attractin. *Nature* **398,** 1521–1526.
139. Nagle, D. L., McGrail, S. H., Vitale, J., Woolf, E. Z., Dussault, B. J. Jr., DiRocco, L., Holmgren, L., Montagno, J., Bork, P., Huszar, D., Fairchild-Huntress, V., Ge, P., Keilty, J., Ebeling, C., Baldini, L., Gilchrist, J., Burn, P., Carlson, G. A., and Moore, K. J. (1999) The mahogany protein is a receptor involved in suppression of obesity. *Nature* **398,** 148–152.
140. Duke-Cohan, J. S., Gu, J., McLaughlin, D. F., Xu, Y., Freeman, G. J., and Schlossman, S. F. (1998) Attractin (DPPT-L), a member of the CUB family of cell adhesions and guidance proteins, is secreted by activated human T lymphocytes and modulates immune cell interactions. *Proc. Natl. Acad. Sci. U. S. A.* **95,** 11,336–11,341.

141. Tatro, J. B. (1997) Receptor biology of the melanocortins, a family of neuroimmunomodulatory peptides. *Neuroimmunomodulation* **3,** 259–284.
142. Huang, Q. H., Hruby, V. J., and Tatro, J. B. (1999) Role of central melanocortins in endotoxin-induced anorexia. *Am. J. Physiol.* **276,** R864–R871.
143. De Angelis, E., Sahm, U. G., Ahmed, A. R. H., Olivier, G. W. J., Notarianni, L. J., Branch, S. K., Moss, S. H., and Pouton, C. W. (1995) Identification of a melanocortin receptor expressed by murine brain microvascular endothelial cells in culture. *Microvasc. Res.* **50,** 25–34.
144. Li, S.-J., Varga, K., Archer, P., Hruby, V. J., Sharma, S. D., Kesterson, R. A., Cone, R. D., and Kunos, G. (1996) Melanocortin antagonists define two distinct pathways of cardiovascular control by α-and γ-melanocyte-stimulating hormones. *J. Neurosci.* **16,** 5182–5188.
145. Mastrianni, J. A., Abbott, F. V., and Kunos, G. (1989) Activation of central μ-opioid receptors is involved in clonidine analgesia in rats. *Brain Res.* **479,** 283–289.
146. Mastrianni, J. A., Palkovits, M., and Kunos, G. (1989) Activation of brainstem endorphinergic neurons causes cardiovascular depression, and facilitates baroreflex bradycardia. *Neuroscience* **33,** 559–566.
147. Mosqueda-Garcia, R., Eskay, R., Zamir, N., Palkovits, M., and Kunos, G. (1986) Opioid-mediated cardiovascular effects of clonidine in spontaneously hypertensive rats : elimination by neonatal treatment with monosodium glutamate. *Endocrinology* **118,** 1814–1822.
148. Weigant, V. M., Jolles, J., Colbern, D. L., Zimmerman, E., and Gispen, W. H. (1979) Intracerebroventricular ACTH activates the pituitary adrenal system: dissociation from a behavioral response. *Life Sci.* **25,** 1791–1796.
149. Von Frijtag, J. C., Croiset, G., Gispen, W. H., Adan, R. A., and Wiegant, V. M. (1998) The role of central melanocortin receptors in the activation of the hypothalamus-pituitary-adrenal-axis and the induction of excessive grooming. *Br. J. Pharmacol.* **123,** 1503–1508.
150. Scimonelli, T. and Celis, M. E. (1990) A central action of alpha-melanocyte-stimulating hormone on serum levels of LH and prolactin in rats. *J. Endocrinol.* **124,** 127–132.

CHAPTER 15

The Melanocortin-5 Receptor

Wenbiao Chen

1. Introduction

Melanocortins are a subset of peptides derived from the proopiomelanocortin (POMC) gene product, including ACTH, α-, β-, and γ-MSH *(1,2)*. The name refers to the two principal activities of these peptides, that is, regulation of pigmentation in skin and hair and steroidogenesis in the adrenal cortex.

Over the past four decades, a number of other physiological and behavioral activities have been attributed to melanocortins (for an earlier review, *see* ref. *3*). While some are based on the ability of melanocortins to restore the abnormalities resulting from hypophysectomy, most of these functions of melanocortins were effects induced by administration of melanocortin and related peptides, usually in pharmacological doses. Unlike steroidogenesis and pigmentation, the other functions of melanocortins are not well characterized. In many cases, the site and the molecular mechanism of action are not clear.

Melanocortin peptides exert their functions by activating their receptors on the plasma membrane. Five melanocortin receptors have been cloned. They are all G protein-coupled receptors (GPCRs) that stimulate cAMP production upon activation. These receptors are named MC1-R to MC5-R, according to their order of being cloned *(4)*. MC1-R and MC2-R are the two classic melanocortin receptors that mediate the regulation of pigmentation by MSH and steroidogenesis by ACTH, respectively *(5,6)*. MC3-R is a receptor expressed in the brain and placenta. It is the only cloned receptor that has equal affinity for both α-and γ-MSH, and is therefore a likely candidate for mediating natriuretic and cardiovascular effects attributed to both peptides *(7–10)*. MC4-R is another neural melanocortin receptor recently found to regulate feeding and metabolism in mice *(11–14)*. More detailed reviews on the pharmacology and functions of MC1-R to MC4-R can be found elseswhere in this volume. This

The Melanocortin Receptors Ed.: R. D. Cone

chapter focuses on the molecular and pharmacological characteristics and physiological functions of MC5-R.

2. Molecular Cloning and Characterization of the MC5-R

Several groups independently cloned the MC5-R from mouse, rat, sheep, and human largely based on sequence homology *(15–20)*. In most cases, a fragment of MC5-R was obtained by polymerase chain reaction (PCR) using primers that span the third and sixth transmembrane domains designed on the basis of all or a subset of known GPCRs. The fragments then served as probes to obtain the entire coding sequence from cDNA or genomic libraries. Using degenerate primers based on all known GPCRs, Barrett et al. *(16)* obtained an ovine MC5-R fragment, and Fathi and coworkers *(20)* cloned a mouse MC5-R fragment. Gantz and colleagues *(17)* acquired a murine MC5-R fragment from genomic DNA using degenerate primers based on MC1-R to MC4-R. Chhajilani and collaborators *(15)* cloned a human MC5-R fragment using primers based on human MC1-R sequence. Griffon et al. *(18)*, however, obtained a rat MC5-R in the process of searching for sequences that hybridize to a rat dopamine D3 probe at low stringency. By contrast, Labbe and coworkers *(19)* cloned the entire coding sequence of murine MC5-R directly from a genomic library by using human MC3-R as a probe. We obtained murine MC5-R using a similar approach *(21,22)*.

There is an extensive sequence homology between the MC5-R and other melanocortin receptors. The identity between MC5-R and other melanocortin receptors ranges from 46% to 61%, with MC2-R and MC4-R at the lower and upper ends, respectively. The highest homology resides in the transmembrane segments and the intracellular regions.

The MC5 receptors from different species display even greater identity. For example, there is 96% identity and 99.5% similarity between the rat and mouse MC5-R. The overall identity between human and mouse MC5-R is 88% at the amino acid level (Fig. 1). Such a high degree of conservation suggests that MC5-R may play an important physiological function.

3. Pharmacological Properties of MC5-R

MC5-R is responsive to all melanocortins at physiological levels except γ-MSH. This property is similar to MC1-R and MC4-R. Although discrepancy exists among the data from different laboratories about the EC_{50} values of various melanocortins for stimulation of cAMP production via the MC5-R (*see* Table 1), all reports agree upon the order of potency for melanocortins: NDP-α-MSH > ACTH[1–24] ≥ α-MSH > ACTH[1–39] = β-MSH >> γ-MSH.

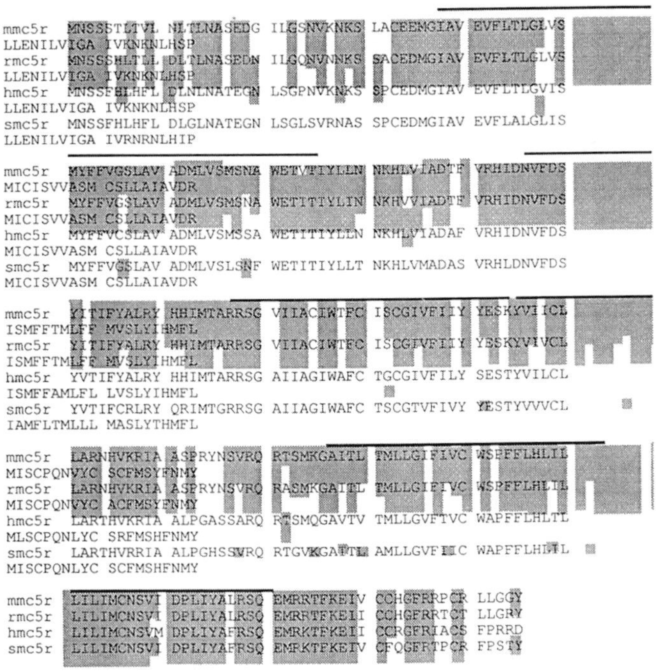

Fig. 1. Amino acid sequence alignment of all cloned MC5-Rs. Lightly shaded residues are conserved in most of MC5-Rs. Heavily shaded ones are common to most of the GPCRs. Bars span putative transmembrane domains according to Baldwin *(98)*.

The discrepancy in EC_{50} values may result from differences in cell lines that are used to express the receptor, and the species from which MC5-R is derived. Fathi et al. *(20)* found that α-MSH is 27 times more potent at the murine MC5-R than the human MC5-R. The high EC_{50} (~ 50n*M*) for γ-MSH makes MC5-R a unlikely candidate for mediating effects ascribed to γ-MSH. ACTH[4–10], a commonly used compound that is active in a number of behavioral assays, is not an effective ligand for MC5-R *(17,18)*, nor does it display high affinity for the receptor (IC_{50} is about 125μ*M*) *(19)*. However, as pharmacological doses have been used in most studies, ACTH[4–10] induced changes may still be mediated by the MC5-R.

One interesting pharmacologic feature of the MC5-R is the marked difference between the EC_{50} and IC_{50} values (Table 1) *(17,19)*. This differs markedly from MC1-R *(23,24)*, MC2-R *(25,26)*, MC3-R *(19)* and MC4-R *(27)*, whose EC_{50} and IC_{50} values are much closer. This may be due to more efficient coupling of the MC5-R to G proteins *(19)*.

Table 1
A Summary of Pharmacological Data of MC5-R Reported by Various Laboratories

	Expression		EC_{50}/IC_{50} Values (n*M*)						
Species	Cell Line	Transfection	NDP-α-MSH	α-MSH	β-MSH	γ-MSH	ACTH (1–24)	ACTH (1–39)	Reference
Human	cos-1	transient	1.8	51.6	209	816	30	na	Fathi et al. *(20)*
	cos-7		2.39	8,240	14,400	>100,000	na	17,000	Schioth et al. *(23)*
	cos-1	transient	na/na	1.9/na	na/na	na/na	na/na	na/na	Fathi et al. *(20)*
Mouse	CHO	stable	0.05/1.1	1.1/62.5	6.5/212	42.9/1270	na/na	6.0/236	Labbe et al. *(19)*
	L	stable	na/na	0.2/60*	1.4/na*	35/na*	na/na	5.5/300*	Gantz et al. *(17)*
Rat	CHO	stable	1.0	0.58	12	45	0.46	6.2	Griffon et al. *(18)*
Sheep	cos-7	transient	na	10	10	na	na	10	Barret et al. *(16)*

*Estimated based on the published graph; na, not available in the reference.

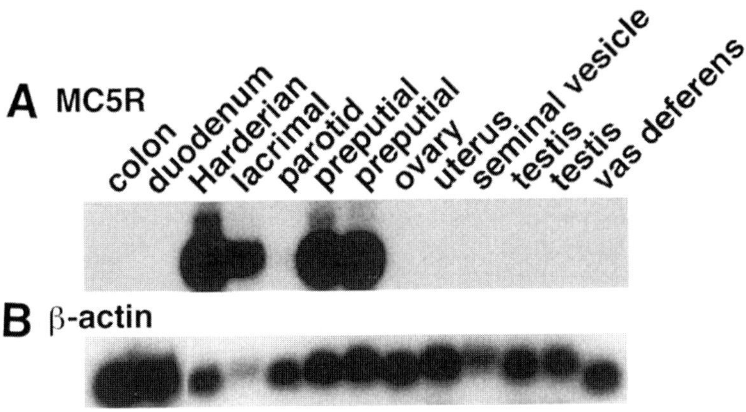

Fig. 2. MC5-R is highly abundant in preputial, harderian, and lacrimal glands. Reprinted, with permission, from *(22)*.

Another peculiarity of MC5-R is its response to NDP-α-MSH. Data reported by Fathi et al. *(20)* indicated that NDP-α-MSH is a partial agonist at the mouse MC5-R. The partial agonism of NDP-α-MSH was also observed by the author in the murine MC5-R expressed in HEK293 cells *(21)*, and in exocrine tissues *(22)*. However, Boston and Cone *(28)* demonstrated antagonism of NDP-α-MSH against α-MSH in differentiated 3T3-L1 cells. This action is possibly on the MC5-R, as it is the only α-MSH-responsive melanocortin receptor found in the cells. It is possible that the 3T3-L1 cells harbor a mutant MC5-R that displays an antagonistic response toward NDP-α-MSH. Alternatively, the MC5-R mRNA in the cells may not produce functional protein. Rather, the peculiar pharmacological property may be a feature of a novel melanocortin receptor. Whether the peculiar response to NDP-α-MSH of the MC5-R is only confined to the murine receptor, or is true to MC5 receptors from other species remains to be tested.

3. Tissue Distribution of the MC5-R mRNA

The MC5-R is found in a wide range of tissues. Northern analysis by different laboratories has detected MC5-R mRNA in skeletal muscle, adipose tissues, brain, lung, adrenal, and stomach *(17–20,28)*. We have found MC5-R mRNA in skin tissues at a level comparable to skeletal muscle. Additionally, we have found much higher levels of MC5-R mRNA in several exocrine tissues, such as Harderian, lacrimal, and preputial glands (Fig. 2). The level of MC5-R mRNA in preputial gland is at least 30 times higher than in the skin. Upon further analysis by in situ hybridization, we demonstrated that MC5-R is specifically expressed in sebaceous gland in skin *(22)*. Similar results were reported by van der Kraan et al. *(99)*. In addition, they detected MC5-R mRNA in the prostate gland and pancreas. The expression is restricted to secretory epithelia of those exocrine glands. *In situ* hybridization analysis also revealed MC5-R mRNA in the adrenal cortex of adult rats, in both zona glomerolusa and zona fasciculata and in the submaxilary gland of neonatal rat *(18)*. Using more sensitive methods, such as RNase protection assay (RPA) and RT-PCR, however, a large number of tissues were found to be MC5-R positive. These include cerebral cortex, pons, medulla, cerebellum, hypothalamus, hipocampus, midbrain, striatum, olfactory tubercle, olfactory bulb, pituitary, thyroid, tongue, thymus, spleen, bone marrow, kidney, and testis *(15,16, 18–20)*. The functional significance of these low levels of MC5-R mRNA awaits further examination. It is unclear whether MC5-R is distributed evenly in the tissues at low levels or is highly abundant in a small fraction of cells in the tissues. In this regard, it would be useful to perform *in situ* hybridization analysis in the MC5-R positive tissues. The result would provide clues about the function of the receptor in these tissues.

4. Physiological Functions of the MC5-R

The highly conserved amino acid sequence cross-species suggests that MC5-R may play some essential function in mammals. Although predicting the exact function for the MC5-R has been made difficult by virtue of the wide expression of this receptor, some speculations have been made, based on its pharmacologic properties and its site of expression. Among these speculative MC5-R functions are (i) the neuro/myotropic effects of melanocortin peptides on developing and regenerating neuromuscular systems *(17,20)*; (ii) melanocortin-induced gastric effect and aldosterone secretion *(18)*; and (iii) melanocortin elicited antiinflammation *(19)*. Additionally, systemic administration of melanocortins elicits a number of other physiological and behavioral changes independent of adrenal gland function. As very little melanocortin peptide crosses the blood–brain barrier *(29)*, these effects are conceivably due to activation of melanocortin receptors outside of blood–brain barrier. Although there are some domains of CNS that are outside of the barrier, many of these effects are presumably noncentrally mediated, possibly via MC5-R.

To define the physiological function of the MC5-R, we generated MC5-R deficient (MC5-R-KO) mice by gene targeting. In these mice, a 600 bp segment of the MC5-R gene was deleted. This fragment contains the coding sequence for the N-terminal one third of MC5-R plus 200 bp of immediate 5' sequence. The mutation has been bred congenically into two strains, 129 SV and C57Bl/6J. This deletion results in a nonfunctional receptor as indicated by the complete lack of ^{125}I-NDP-α-MSH binding by the membranes of skeletal muscle, Harderian, lacrimal and preputial glands from mice that are homozy-

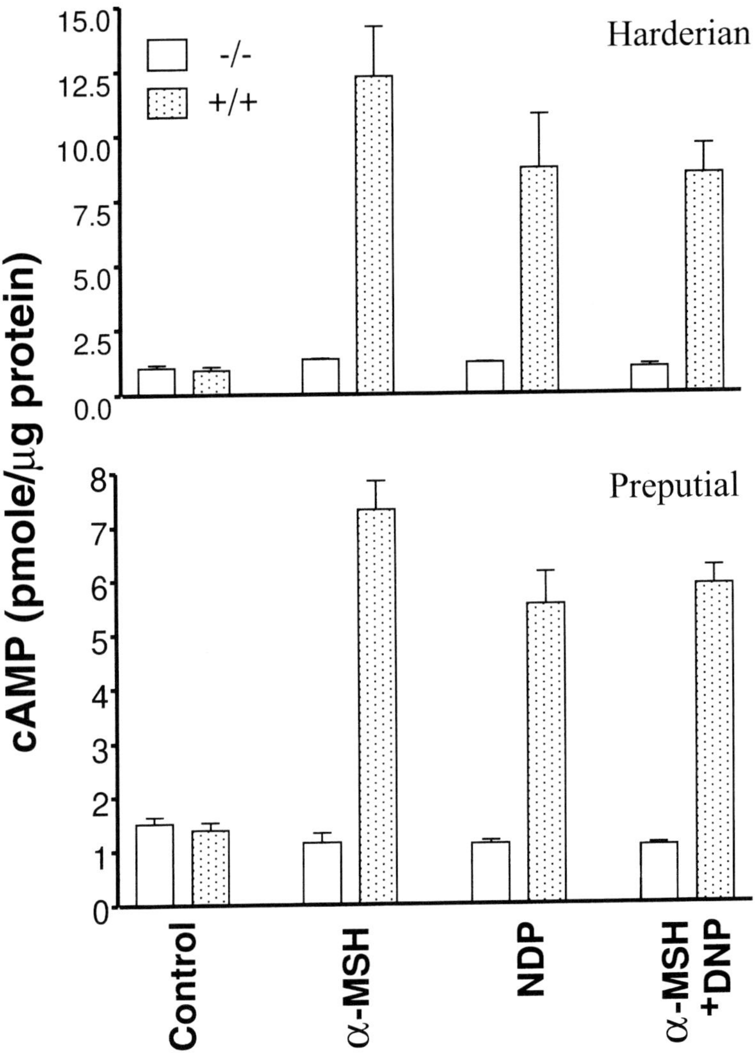

Fig. 3. There is no α-MSH and NDP-α-MSH stimulation of cAMP production in harderian (upper panel) and preputial glands (lower panel) from MC5-R-KO mice. Error bars stand for SEM. Reprinted, with permission, from *(22)*.

gous for the mutated MC5-R allele, while the same membranes from wild-type mice exhibited strong specific binding. In Harderian and preputial cells from MC5-R-deficient mice, melanocortin peptides failed to stimulated cAMP production, whereas up to 20-fold increase of cAMP was detected in these exocrine cells from wild-type mice (Fig. 3). Therefore, MC5-R is the only melanocortin receptor expressed in these exocrine glands. Indeed, none of the other four melanocortin receptors were detectable by northern analysis.

MC5-R-null mice reproduce and thrive normally. There was no obvious anatomical, or behavioral abnormalities. Therefore, MC5-R is not essential for development and daily life under laboratory conditions. However, the availability of MC5-R-deficient adult mice and the nonoverlapping expression of MC5-R with other melanocortin receptors made defining the physiological functions of the receptor possible.

4.1. MC5-R and α-MSH-Induced Antiinflammation

One suggested function of MC5-R is mediation of antiinflammatory effects of melanocortins, as its mRNA is found in spleen and bone marrow *(19)*. Outside the antiinflammatory activity of ACTH secondary to glucocorticoid production *(30)*, α-MSH also potently blocks inflammation. The antiinflammatory action of α-MSH is thought to be a result of inhibiting interleukin 1 beta (IL-1β) and tumor necrosis factor (TNF) production and/or activity. α-MSH inhibits several IL-1β elicited effects. For example, it blocked IL-1β elicited HPA axis activation *(31)*, hypothermia, and elevation of serum amyloid P and circulating neutrophils *(32)*. α-MSH also antagonized IL-1β induced acute inflammation and hypersensitivity *(33)*. Intraperitoneal injection of α-MSH reduced κ-carrageenan induced paw edema and arachidonic acid induced ear swelling in a dose responsive fashion in the mouse *(34)*. High dose of the endotoxin lipopolysaccharide (LPS) results in lethality due primarily to massive TNF production. α-MSH and its derivative HP228 reduced LPS-induced TNF synthesis and lethality in mice *(35,36)*, and LPS-induced nitric oxide synthesis stimulation *(37)*. However, there is some inconsistency in the literature as to whether the melanogenic pharmacophore is necessary for the antiinflammatory actions. Several reports from Lipton and colleagues *(38)* indicated that some antiinflammatory activity resides in the three residues at the C-terminus (α-MSH [11–13]). It is possible that α-MSH may act upon a receptor outside the melanocortin receptor family to block inflammation. Another uncertainty is the site of action. Several lines of evidence suggests that the antiinflammatory activity of α-MSH is largely centrally mediated *(39)*. Central administration of α-MSH effectively inhibited IL-1 induced ear inflammation and κ-carrageenan elicited hind paw edema. This action is mediated by descending β-adrenergic neurons. Both spinal transection and a β_2-adrenergic antagonist markedly reduced central α-MSH activity. Furthermore, the antiinflammatory action of peripherally administered α-MSH is also largely dependent on an intact spinal cord, suggesting it may also be centrally mediated. When the κ-carrageenan induced hind paw edema assay and arachidonic acid elicited ear swelling assay were performed, both wild-type and MC5-R-KO mice exhibited reduced inflammation after α-MSH injection *(W. Chen and R. Cone, unpublished observations)*. Therefore, MC5-R is not essential for the antiinflammatory activity of α-MSH. It is possible, however, that MC5-R is responsible for a small portion of the antiinflammatory activity. The two assays may not be sensitive enough to reveal subtle differences between control and mutant mice. Further studies, such as testing in spinal transected mice, may help clarify the role of MC5-R in antiinflammation.

4.2. MC5-R and Nerve Regeneration

Another plausible function of MC5-R is mediation of neuro/myocyte tropic activities of melanocortins. Both α-MSH and ACTH are mitogenic for satellite cells from skeletal muscle *(40,41)*. As satellite cells are implicated in muscle regeneration, melanocortin-stimulated proliferation may therefore be a compensatory mechanism for muscle damage. In fact, Hughes and colleagues *(42,43)* observed the increase of both the number of binding sites and the quantity of α-MSH/ACTH immunoactivity in skeletal muscle in mice with muscle pathologies. In several nerve regeneration models, α-MSH-like compounds promote regeneration after nerve crush *(44,45)*. In addition, when administered during development, melanocortins and related peptides accelerate maturation of the neuromuscular system *(46)*. The presence of MC5-R in skeletal muscles make it a likely candidate for mediating these neuromuscular effects of α-MSH and its derivatives *(20)*. However, it remains unclear whether these melanocortin effects have any physiologic relevance. In addition, these effects may involve other unknown receptors. In most of the experiments, ACTH/MSH[4–10] analogs were used. These analogues have essentially no activity on MC5-R, or the other four melanocortin receptors. The importance of MC5-R in nerve regeneration was evaluated in a sciatic nerve regeneration paradigm. Preliminary data indicates that there is no difference in the rate of nerve regeneration between control and mutant MC5-RKO mice judged by footprints as well as the number and size of nerve sprouting events *(W. Chen and R. Cone, unpublished observations)*. MC5-R may therefore play little physiologic role in peripheral nerve regeneration. No overt developmental abnormalities in neuromuscular system were observed in MC5-R-KO mice, indicating the receptor is not essential for the maturation of neuromuscular system. However, these data do not exclude MC5-R as mediator of the observed neurotropic activity of melanocortins. The rate of regeneration as well as maturation of the neuromuscular system after α-MSH administration may be different between wild-type and MC5-R-KO mice. The role of MC5-R in melanocortin-induced myocyte proliferation awaits to be established.

4.3. MC5-R and Control of Weight Homeostasis

It was proposed that MC5-R may be partially responsible for melanocortin-induced lipolysis in adipocytes. α-MSH and ACTH have potent lipolytic activity on adipocytes *(47)*. The pharmacologic profiles of this melanocortin effect varies among species. In rat adipocytes, α-MSH is much less potent than ACTH. By contrast, α-MSH is several times more potent than ACTH in rabbit fat cells. MC5-R, as well as MC2-R, are expressed, albeit at low levels, in mouse fat tissues. The difference between species may be a result of different levels of expression of MC5-R and MC2-R in adipocytes *(28)*. In species with low levels of MC2-R in fat tissues, MC5-R may play an important role in lipolysis. This hypothetical function of MC5-R was a potential explanation for obesity in A^y mice. Although ineffective in inhibition of MC5-R signaling in vitro *(48)*, agouti may block MC5-R activation in vivo. However, no difference in weight was found between wild-type A^y and MC5-R-deficient A^y mice, disproving the hypothetical role of MC5-R in the A^y obesity syndrome. In fact, data show that the MC4-R is the primary target of agouti in the induction of obesity *(13,14)*.

4.4. Defective Water Repulsion and Thermoregulation in MC5-R-Deficient Mice

One striking difference between wild-type and MC5-R- deficient mice emerged in an accident in which one cage was flooded due to drinking water leakage. Rescued mice fell into two groups: those with hair that dried quickly and those that dried more slowly. It turned out that all the slow driers are MC5-R-deficient, and the others wild-type. When placed in a cage without bedding after swimming, wild-type mice dried in about 25 min on average after a 3-min swim at 32°C. By contrast, it took MC5-R-KO mice more than 40 min to dry (Fig. 4). There were at least two possibilities to explain the difference. The mutant mice may either absorb more water after swimming, or evaporate the water slower due to a lower core temperature, or both. To determine the primary cause of the difference, the weight and core temperature of wild-type and mutant mice were monitored before and after 3 minutes swimming in 32°C water.

The longer drying time in the mutant mice is due to impaired water repulsion. MC5-R-KO mice absorbed almost twice as much water as the wild-type controls. The rate of evaporation, however, was comparable. Although the average core body temperature before swimming are the same, there is a significant difference in the core body temperature after swimming. The core body temperature decreased 2°C during the 3-min swim at 32°C in mutant mice, compared to 0.7°C in the controls. This is presumably a consequence of increased heat loss in the water by the mutant mice. In addition, core body temperature dropped another 0.5°C before recovering in MC5-R-KO mice,

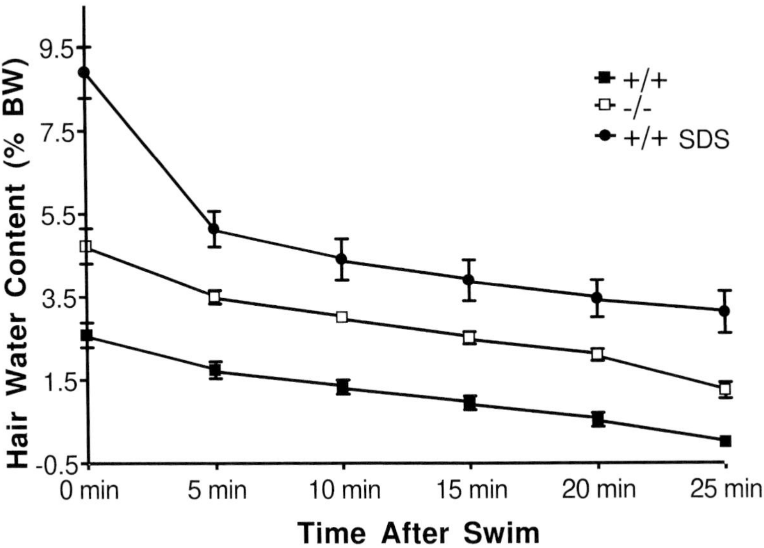

Fig. 4. Impaired water repulsion in MC5-R-KO mice. MC5-R-KO mice absorb more water during the swim than wild-type controls. Removal of hair lipids with 5% SDS wash increases water adsorption in wild-type mice. Error bars stand for SEM of 12 MC5-R-KO mice, 11 control mice, and 6 SDS washed wild-type mice. Data were obtained from mice in 129SV background. Reprinted, with permission, from *(22)*.

whereas no further decline was seen in wild-type mice. The core body temperature returned to baseline in 20 min in wild-type mice, at which time the body temperature in MC5-R-KO mice was still 1.5°C below normal. Furthermore, when challenged with cold air (5–6°C cold room), mutant and wild-type also exhibited remarkable difference in their core body temperature. Wild-type mice increased core temperature slightly at the beginning, and maintained above-normal temperature for at least 3 h. By contrast, MC5-R-KO mice underwent a mild hypothermia. The cold air-and water-induced hypothermia indicate that MC5-R deficiency in mice results in impaired insulation by the hair. As similar situations are frequently encountered by mice living in the wild, it is conceivable that MC5-R-KO mice may not survive well in the wild compared to their wild-type littermates. Therefore, MC5-R is important in thermal regulation in mice, and possibly other rodents.

The impaired insulation in MC5-R-KO mice is due to reduced hair lipid production. Removal of lipids by a 5% sodium dodecyl sulfate (SDS) wash increased water absorption after swimming and lowered the core temperature of wild-type mice placed at 5–6°C. In fact, when hair lipid content was measured, it was found that there was a 15–20% reduction of acetone extractable materials, both in male and female MC5-R-KO mutants. This deficiency is largely due to a 50% decrease in sterol esters, as revealed by thin layer chromatography (TLC) analysis of the hair lipids. Sterol esters constitute more than 26% of the total acetone extractable lipids in wild-type mice, but only about 13% in the mutants. No significant decrease of other classes of hair lipids was found in MC5-R mutant mice. As sterol esters are the most hydrophobic hair lipids, its deficiency may largely explain the impaired water repulsion in MC5-R-KO mice. There is some uncertainty, however, about whether deficiency in hair lipids is the only factor for the observed insulation impairment. When similarly washed with 5% SDS, MC5-R-KO mice still absorb more water after a swimming, and are more hypothermic in cold than wild-type counterparts. It should be pointed out that SDS wash can only remove about 50% of hair lipids. Therefore, the failure of SDS wash to remove the difference between wild-type and mutant mice does not disprove the notion that decreased sterol esters account for most, if not all, the difference in water repulsion and heat insulation in MC5-R-KO mice.

The observed phenotype in hair lipids suggested that MC5-R may be the mediator of the sebotropic activity of α-MSH found by Thody and coworkers *(49)* in the early 1970s. In a serious of experiments, Thody and coworkers elegantly demonstrated that α-MSH is a sebotropic hormone. They initially discovered that hypophysectomy decreases the secretion of hair lipids (sebum) in rats *(50)*. Removal of the neurointermediate lobe of the pituitary resulted in similar deficiency *(49)*. The reduction was fully restored by concomitant α-MSH and androgen administration *(51,52)*, possibly through the stimulation of lipogenesis *(53)*. Application of α-MSH alone only slightly improved sebum secretion. The same was true for testosterone. The physiological significance of this melanocortin activity was not clear. Nor was the molecular mechanism of this action. Although α-MSH-regulated lipids are thought to originate from the sebaceous glands, there was no direct evidence for α-MSH action on the gland. Recent demonstration of MC5-R expression in skin indicates that α-MSH may influence hair lipid synthesis/secretion by its direct action in skin. In fact, *in situ* hybridization studies revealed specific expression of MC5-R in sebaceous gland in the skin, but not in any other cell types. Taken together, these data suggest that α-MSH stimulate sebum production by activating the MC5-R in the sebaceous gland.

The high level of MC5-R expression in sebaceous gland prompted the investigation of other exocrine glands. It was found that MC5-R mRNA is abundant in several other exocrine glands. These include preputial gland, a specialized sebaceous gland, Harderian gland, and lacrimal gland. The following sections discuss the role of MC5-R in these glands.

4.5. MC5-R Regulates Protein Secretion by the Lacrimal Gland

The lacrimal gland is the major source for the protein-rich aqueous layer of tear film. This layer plays an essential role in lubricating the eye surface, as well as protecting the eyes from pathogenic insults by enzymatic and immunological mechanisms (reviewed in ref. *54*). Lacrimal secretion is under neuronal and hormonal control. It has been reported by Jahn et al *(55)*, Salomon and colleagues *(56)*, and Tatro's group *(57)* that α-MSH and ACTH stimulate protein secretion in rat lacrimal gland. Similar effects of α-MSH and ACTH was observed in mice lacrimal gland explants with an EC_{50} of 4n*M* for ACTH *(21)*. This stimulation of protein secretion by α-MSH and ACTH is absent in MC5-R-KO animals, indicating a role for MC5-R in melanocortin stimulated tear secretion. The physiological significance of melanocortin elicited tear secretion however, is unclear. As tear proteins consist of immunoglobulins as well as enzymes that destroy pathogens, and pathogen infection elevates serum α-MSH and ACTH, it is likely that melanocortin-mediated protein secretion in lacrimal gland is an integral part of stress response that helps animals overcome hostile insults. Future studies should compare the ability of the eyes of wild-type and MC5-R-KO mice to combat infection of pathogens. It will also be interesting to examine the corneal histology of MC5-R-KO mice to see if there is any abnormality due to lack of melanocortin induced lacrimal secretion under stress.

4.6. MC5-R is Required for Porphyrin Production in the Harderian Gland

The Harderian gland is a bilobular retroorbital structure that secretes primarily two products, lipids and porphyrins, onto the eyes, and into the nasal cavity *(58)*. Most vertebrates, with the exception of primates, have Harderian glands, although their functional role is not understood. One proposed function of the Harderian gland is lubrication of the eyes. There is almost a perfect correlation between the presence of a nictating membrane and the Harderian gland. In rodents, the lipid components not only are secreted onto the eye surface, but also distributed along the coat of the animal by grooming behaviors. The Harderian lipids may thus also play an important thermoregulatory role, as has been demonstrated in mongolian gerbils *(59,60)*. At least in some species, Harderian gland secretions also contains pheromones *(61)*. Harderian porphyrins may play some role in phototransduction as they emit fluorescence when excited by UV light. They may also protect eyes from UV irradiation by absorbing UV light *(62)*. Very little is known about the regulatory mechanisms that govern Harderian gland secretion. Since porphyrins are cosecreted in abundance with lipids and other Harderian components, and can be easily quantified by measuring absorbency at 402 nm (Fig. 5) or fluorescence at 602 nm when excited at 402 nm, they are excellent markers of Harderian secretion.

Absorbance Spectrum of Harderian Gland Extracts

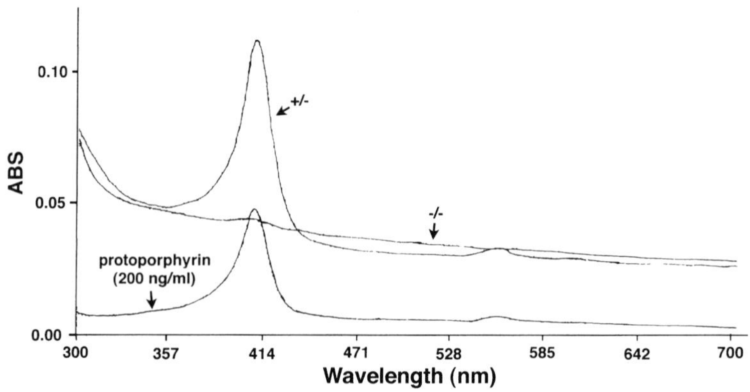

Fig. 5. MC5-R is involved in porphyrin synthesis and release in Harderian gland. Reprinted, with permission, from *(22)*.

MC5-R mRNA is highly abundant in Harderian gland and is required for both the synthesis and secretion of porphyrins. The amount of Harderian porphyrins differs from strain to strain, and is sexually dimorphic. Male Harderian gland contains less porphyrins than female *(58)*. In the two strains that harbor the MC5-R mutation, 129 SV mice have more Harderian porphyrins than C57Bl/6J. In 1 h of partial restraint stress Harderian porphyrin content is significantly increased in wild-type C57Bl/6J mice, but not in their MC5-R-KO littermates. There seems to be a secretion phase prior to the increased biosynthesis. In wild-type 129 SV females, 15 minutes of partial immobilization significantly decreases Harderian porphyrins. However, the porphyrins are restored to their normal levels due to increased synthesis. The importance of MC5-R for Harderian porphyrin synthesis is further demonstrated in 129 SV MC5-R-KO mice. Porphyrin levels are dramatically decreased in both male and female mutant mice (Fig. 5).

Previous studies have shown that Harderian porphyrin synthesis is under pituitary control *(58,63)*. Hypophysectomy blunts porphyrin synthesis in hamsters and mice. It appears that prolactin plays an important role for Harderian porphyrin synthesis. Supplementation of prolactin in hypophysectomized mice partially restored porphyrin content *(64)*. Now, melanocortins are the second class of pituitary hormones that regulate Harderian porphyrin synthesis. Another known regulator of Harderian porphyrin synthesis is androgen. Unlike the situation in sebaceous gland, androgen counteracts melanocortins and inhibits porphyrin biosynthesis. Androgen inhibition is thought to be the reason behind sexual dimorphism of Harderian porphyrin. The molecular mechanism for the MC5-R-mediated porphyrin synthesis and secretion is unknown. In the case of synthesis, there may be an analogy to the melanocyte, in which α-MSH regulates tyrosinase, the rate simulating enzyme in pigment synthesis. It is possible that MC5-R stimulation of cAMP regulates biosynthesis in the Harderian gland by controlling the activity of a key enzyme in the porphyrin biosynthesis pathway.

4.7. MC5-R and Pheromone Secretion

The preputial gland is a specialized sebaceous gland implicated in pheromone production in rodents *(65–67)*. Like the sebaceous gland, the preputial gland is mainly engaged in synthesis of fatty acids. Moreover, the lipid content of this gland can also be stimulated by α-MSH *(51)*. These lipids are secreted into the urethra and serve as the major chemosignal in urine. Several lines of evidence indicate that melanocortin peptides also play a role in regulation of preputial secretion. When injected into male mice, α-MSH elicits aggression of conspecific animals toward recipients. The behavioral changes differ according to the social status of the recipient. When the recipient is a socially dominant one, α-MSH lowers its aggressiveness towards subordinates *(68)*. In some cases, α-MSH reverses the social order. When α-MSH is injected into subordinate mice, the recipient suffers significantly more attacks from the cagemate. The aggression was induced by olfactory cues from urine *(68,69)*. Injection of α-MSH into female rats increase the release of sexual attractants *(70)*. It also stimulates or inhibits their sexual behaviors depending on their state of sexual acceptance *(71,72)*. The opposite effects of α-MSH may be a result of its differential interaction with sex hormones. By the same token, α-MSH also decreases active social interaction in rats *(73)*. However, it is not clear whether olfactory cues are involved in this case.

Melanocortin-regulated pheromone secretion is physiologically relevant. Stress increases serum levels of ACTH and α-MSH, both of which are MC5-R ligands. It has been know for some time that stressed animals can be recognized by conspecifics. Both rats and mice can distinguish odors of stressed animals from unstressed ones. Odors from stressed animals elicit aversive response in the conspecific. Thus, the odors are thought to contain some "alarm substances." For example, nonstressed rats were able to recognize and avoid odors from stressed rats *(74,75)*. The odors are presumably from the body surface and/or urine *(76)*. Odors from stressed rats also induced analgesia in nonstressed conspecifics, a common reaction to stress *(77)*. In another paradigm, rats exude an "alarm substance" into the water that alerted subsequent rats after a forced swim *(78)*. The secretion depends on the pituitary but not on the adrenal gland *(79,80)*. High-dose ACTH in hypophysectomized rats produced the same substance *(79)*. It is uncertain which receptor may mediate the secretion of the materials, as the site of release is not known. Similarly in the mouse, stress provided an olfactory cue that causes aversion in nonstressed conspecifics *(81,82)*. Experience of conspecific stress odors alters both cellular and humoral immune responses *(83)*. Furthermore, severe stress induces aggression of cohorts *(84)*, a behavior that is recapitulated by α-MSH injection.

MC5-R is the prime candidate for mediating the secretion of stress-induced alarm substances. The receptor is expressed at very high levels in preputial gland. In addition, it is the only melanocortin receptor in the gland. Neither α-MSH nor ACTH can stimulate cAMP accumulation in preputial gland from MC5-R-KO mice. Nevertheless, the hypothetical role of MC5-R in stress pheromone secretion remains to be confirmed.

It is conceivable that the ability of producing stress pheromones is very important in evolution. This is true for all nonhuman organisms among which vocal communication is not well developed, and communication is largely through chemosignals. In general, stress pheromones elicit aversion. This benefits the species in several ways. First, they would limit disease transmission. Second, they may help conspecifics avoid a stressful environment, or prepare conspecifics to cope with stress. In addition, they may provide cues for females to recognize socially dominant males, since in general, subordinates experience more stress. As a consequence, the overall quality of the offspring is improved. Therefore, MC5-R may link stress and pheromone release, and enable animals to "smell" stress.

4.8. Possible Functions of MC5-R in Spinal Cord

MC5-R mRNA and protein were relatively abundant in spinal cord by northern analysis and radioactive ligand binding studies. It may thus be involved in spinal cord function. Northern analysis indicates that the level of MC5-R mRNA is only slightly less abundant than skeletal muscle *(W. Chen and R. Cone, unpublished results)*. No other melanocortin receptor can be detected by Northern analysis. It appears that MC5-R is the major melanocortin receptor in the adult spinal cord. This is supported by ^{125}I-NDP-α-MSH binding studies, in which marked decrease of binding was found in spinal cord membranes from MC5-R-KO mice in comparison to wild-type mice. The minor component of NDP-α-MSH binding sites are likely MC4-R. MC4-R has been found by in situ hybridization in dorsal root ganglia (DRG) in rat and mouse, both fetus and adult *(12,85)*.

MC5-R may regulate both sensory and motor functions in spinal cord. It may mediate the described antagonistic effect of melanocortins against morphine/β-endorphin. Melanocortins, including ACTH, ACTH[1–24], and β-MSH, when administered intrathecally, or in medium of spinal cord explants, have been shown to block the effects of opiates in spinal cord *(86–88)*. A spinal melanocortin receptor(s) is likely the target of these peptides. It is therefore interesting to investigate the response of MC5-R-KO mice to stress and opiates. Another potential role for spinal MC5-R is regulation of motoneuron activity. Melanocortin peptides increase the amplitudes of endplate potentials and frequency of miniature endplate potentials in frog neuromuscular preparations. The increase is moderate (both are about 80%), lasts several hours after ACTH exposure, and results from activation of presynaptic melanocortin receptors. ACTH, ACTH[1–24] and α-MSH are equally effective, whereas ACTH[4–10] is ineffective *(89)*. It is conceivable that the moderate increase of synaptic efficacy by melanocortin peptides would significantly enhance muscle strength. The response would therefore help in coping with stress *(89)*. A similar increase was found in the mouse neuromuscular junction. In this case, melanocortins have been shown to decrease the number of failures following stimulation of the phrenic nerve *(90)*. This may be the reason behind the observed tropic activity of melanocortins on endplate of rat and rabbits *(91,92)*. *In situ* hybridization analysis of MC5-R mRNA in spinal cord should provide additional clues about the functions of the receptor in spinal cord.

Stress-elevated serum melanocortins may not be the only ligand source for melanocortin receptor in motoneurons. A series of studies by Hughes and coworkers *(93–96)* have demonstrated the expression of POMC mRNA and proteins in motoneurons. In addition, the expression is markedly increased after nerve transection, and in mice with muscular dystrophy, with motoneuron disease, with diabetes mellitus, as well as in mice treated with a motoneuron toxin. Therefore, under stress, melanocortins may increase the efficacy of the neuromuscular junction via a presynaptic receptor, possibly MC5-R, by both endocrine and autocrine mechanisms. It is now possible to test this hypothesis in MC5-R-KO mice.

5. Conclusions

MC5-R seems to be involved in motivating multiple systems to cope with stress. The receptor is specifically expressed at high levels in multiple exocrine glands, including sebaceous, preputial, lacrimal and Harderian gland. The receptor plays an important role in production and/or secretion of the major products in these glands. Loss of MC5-R function in mouse results in decreased synthesis of sterol esters in sebaceous gland and porphyrins in Harderian gland, respectively. MC5-R also mediates melanocortin-induced protein secretion in lacrimal gland. It is a likely mediator of melanocortin-induced potentiation of transmitter release from motoneurons. All these MC5-R mediated responses could be argued to provide advantages in stressful environments.

6. Future Directions

Future studies should aim for a comprehensive understanding of MC5-R function using MC5-R-KO mice as a model. At least three directions emerge in this chapter: (i). testing the hypothetical roles based on MC5-R expression, such as pheromone release in preputial gland and sensory and motor function in spinal cord; (ii). identifying additional MC5-R expressing cells by *in situ* hybridization to identify other potential functions of the receptor; (iii). understanding the molecular mechanisms underlying the observed phenotypes, such as identification of enzyme(s) that MC5-R acts on to stimulate biosynthesis of porphyrins and sterol esters in Harderian gland and sebaceous gland, respectively. One potential hurdle for uncovering all the MC5-R functions would be functional redundancy among the melanocortin receptors. For instance, both MC1-R and MC5-R have been found in the immune systems *(19,97)*. In addition, the exact sites of MC5-R in brain are not characterized. It is possible that MC5-R may overlap somewhat with MC3-R or MC4-R in the brain. Since MC1-R-and MC4-R-deficient mice are available, and MC3-R mutant mice should also be generated in the near future, any functional redundancy could be revealed by generating mice with double mutations.

References

1. Nakanishi, S., Inoue, A., Kita, T., Nakamura, M., Chang, A. C., Cohen, S. N., and Numa, S. (1979) Nucleotide sequence of cloned cDNA for bovine corticotropin–beta–lipotropin precursor. *Nature* **278,** 423–427.
2. Roberts, J. L., Seeburg, P. H., Shine, J., Herbert, E., Baxter, J. D., and Goodman, H. M. (1979) Corticotropin and beta–endorphin: construction and analysis of recombinant DNA complementary to mRNA for the common precursor. *Proc. Natl. Acad. Sci. U. S. A.* **76,** 2153–2157.
3. Eberle, A. N. (1988) *The Melanotropins* Karger, Basel.
4. Cone, R. D., Lu, D., Koppula, S., Vage, D. I., Klungland, H., Boston, B., Chen, W., Orth, D. N., Pouton, C., and Kesterson, R. A. (1996) The melanocortin receptors: agonists, antagonists, and the hormonal control of pigmentation. *Recent. Prog. Horm. Res.* **51,** 287–317.
5. Mountjoy, K. G., Robbins, L. S., Mortrud, M. T., and Cone, R. D. (1992) The cloning of a family of genes that encode the melanocortin receptors. *Science* **257,** 1248–1251.
6. Chhajlani, V. and Wikberg, J. E. (1992) Molecular cloning and expression of the human melanocyte stimulating hormone receptor cDNA. *FEBS Lett.* **309,** 417–420.
7. Roselli–Rehfuss, L., Mountjoy, K. G., Robbins, L. S., Mortrud, M. T., Low, M. J., Tatro, J. B., Entwistle, M. L., Simerly, R. B., and Cone, R. D. (1993) Identification of a receptor for gamma melanotropin and other proopiomelanocortin peptides in the hypothalamus and limbic system. *Proc. Natl. Acad. Sci. U. S. A.* **90,** 8856–8860.
8. Gantz, I., Konda, Y., Tashiro, T., Shimoto, Y., Miwa, H., Munzert, G., Watson, S. J., DelValle, J., and Yamada, T. (1993) Molecular cloning of a novel melanocortin receptor. *J. Biol. Chem.* **268,** 8246–8250.

9. De Wildt, D. J., Krugers, H., Kasbergen, C. M., De Lang, H., and Versteeg, D. H. (1993) The hemodynamic effects of gamma 2–melanocyte–stimulating hormone and related melanotropins depend on the arousal potential of the rat. *Eur. J. Pharmacol.* **233,** 157–164.
10. Valentin, J. P., Wiedemann, E., and Humphreys, M. H. (1993) Natriuretic properties of melanocyte–stimulating hormones. *J. Cardiovasc. Pharmacol.* **22,** S114–8.
11. Gantz, I., Miwa, H., Konda, Y., Shimoto, Y., Tashiro, T., Watson, S. J., DelValle, J., and Yamada, T. (1993b) Molecular cloning, expression, and gene localization of a fourth melanocortin receptor. *J. Biol. Chem.* **268,** 15174–15179.
12. Mountjoy, K. G., Mortrud, M. T., Low, M. J., Simerly, R. B., and Cone, R. D. (1994) Localization of the melanocortin–4 receptor (MC4–R) in neuroendocrine and autonomic control circuits in the brain. *Mol. Endocrinol.* **8,** 1298–1308.
13. Fan, W., Boston, B. A., Kesterson, R. A., Hruby, V. J., and Cone, R. D. (1997) Role of melanocortinergic neurons in feeding and the agouti obesity syndrome. *Nature* **385,** 165–168.
14. Huszar, D., Lynch, C. A., Fairchild–Huntress, V., Dunmore, J. H., Fang, Q., Berkemeier, L. R., Gu, W., Kesterson, R. A., Boston, B. A., Cone, R. D., Smith, F. J., Campfield, L. A., Burn, P., and Lee, F. (1997) Targeted disruption of the melanocortin–4 receptor results in obesity in mice. *Cell* **88,** 131–141.
15. Chhajlani, V., Muceniece, R., and Wikberg, J. E. (1993) Molecular cloning of a novel human melanocortin receptor. *Biochem. Biophys. Res. Commun.* **195,** 866–873.
16. Barrett, P., MacDonald, A., Helliwell, R., Davidson, G., and Morgan, P. (1994) Cloning and expression of a new member of the melanocyte–stimulating hormone receptor family. *J. Mol. Endocrinol.* **12,** 203–213.
17. Gantz, I., Shimoto, Y., Konda, Y., Miwa, H., Dickinson, C. J., and Yamada, T. (1994) Molecular cloning, expression, and characterization of a fifth melanocortin receptor. *Biochem. Biophys. Res. Commun.* **200,** 1214–1220.
18. Griffon, N., Mignon, V., Facchinetti, P., Diaz, J., Schwartz, J. C., and Sokoloff, P. (1994) Molecular cloning and characterization of the rat fifth melanocortin receptor. *Biochem. Biophys. Res. Comm.* **200,** 1007–1014.
19. Labbe, O., Desarnaud, F., Eggerickx, D., Vassart, G., and Parmentier, M. (1994) Molecular cloning of a mouse melanocortin 5 receptor gene widely expressed in peripheral tissues. *Biochemistry* **33,** 4543–4549.
20. Fathi, Z., Iben, L. G., and Parker, E. M. (1995) Cloning, expression, and tissue distribution of a fifth melanocortin receptor subtype. *Neurochem. Res.* **20,** 107–113.
21. Chen, W. (1997) Doctoral dissertation. Molecular characterization of melanocortin receptors: role of MC5R in exocrine gland function. Oregon Health Sciences University.
22. Chen, W., Kelly, M. A., Optiz–Araya, X., Low, M. J., and Cone, R. D. (1997) Exocrine gland dysfunction in MC5R-deficient mice: evidence for coordinated regulation of exocrine gland function by melanocortin peptides. *Cell* **91**, 789–798.
23. Schioth, H. B., Muceniece, R., Wikberg, J. E., and Chhajlani, V. (1995) Characterizations of melanocortin receptor subtypes by radioligand binding analysis. *Eur. J. Pharmacol.* **288,** 311–317.
24. Lu, D. (1997) Doctoral dissertation, Antagonists, antagonists, and constitute activation of the melanocortin-1 receptor (MCR1) : a dissertation. Oregon Health Sciences University.

25. Cammas, F. M., Kapas, S., Barker, S., and Clark, A. J. (1995) Cloning, characterization and expression of a functional mouse ACTH receptor. *Biochem. Biophys. Res. Commun.* **212,** 912–918.
26. Kapas, S., Cammas, F. M., Hinson, J. P., and Clark, A. J. (1996) Agonist and receptor binding properties of adrenocorticotropin peptides using the cloned mouse adrenocorticotropin receptor expressed in a stably transfected hela cell line. *Endocrinology* **137,** 3291–3294.
27. Adan, R. A., Cone, R. D., Burbach, J. P., and Gispen, W. H. (1994) Differential effects of melanocortin peptides on neural melanocortin receptors. *Mol. Pharmacol.* **46,** 1182–1190.
28. Boston, B. A. and Cone, R. D. (1996) Characterization of melanocortin receptor subtype expression in murine adipose tissues and in the 3T3–L1 cell line. *Endocrinology* **137,** 2043–2050.
29. Wilson, J. F., Anderson, S., Snook, G., and Llewellyn, K. D. (1984) Quantification of the permeability of the blood–CSF barrier to alpha–MSH in the rat. *Peptides* **5,** 681–685.
30. Batemen, A., Signh, A., Kral, T., and Solomon, S. (1989) The immune–hypothalamic–pituitary–adrenal axis. *Endocr. Rev.* **10,** 92–112.
31. Daynes, R. A., Robertson, B.A., Cho, B.H., Burmham, D.K., and Newton, R. (1987) α–Melanocyte–stimulating hormone exhibits target cell selectivity in its capacity to affect interleukin 1–inducible responses in vivo and in vitro. *J. Immunol.* **139,** 103–109.
32. Robertson, B. A., Gahring, L.C., and Daynes, R.A. (1986) Nuropeptide regulation of interleukin–1 activities: capacity of α–melanocyte stimulating hormone to inhibit interleukin–1–inducible respomses in vivo and in vitro exhibits target cell selectivity. *Inflammation* **10,** 371–385.
33. Hiltz, M. E., Catania, A., and Lipton, J. M. (1992) Alpha–MSH peptides inhibit acute inflammation induced in mice by rIL–1 beta, rIL–6, rTNF–alpha and endogenous pyrogen but not that caused by LTB4, PAF and rIL–8. *Cytokine* **4,** 320–328.
34. Lipton, J. M., Ceriani, G., Macaluso, A., McCoy, D., Carnes, K., Biltz, J., and Catania, A. (1994) Antiinflammatory effects of the neuropeptide alpha–MSH in acute, chronic, and systemic inflammation. *Ann. N. Y. Acad. Sci.* **741,** 137–148.
35. Girten, B., McPherson, S., Garcia, A., McDowell, R., and Tuttle, R. (1994) comparative effects of α–MSH and [Nle4,D–phe7]–α–MSH in LPS induced endotoxema. *FASEB J.* **8,** A643.
36. Girten, B., Omholt, P., Lee, M., McPherson, S., McDowell, R., and Tutle, R. (1995) Comparative effects of α–MSH, NDP and HP228 in pain, inflammation, and endotoxema. *FASEB J.* **9,** A955.
37. Abou–Mohamed, G., Papapetropoulos, A., Ulrich, D., Catravas, J. D., Tuttle, R. R., and Caldwell, R. W. (1995) HP–228, a novel synthetic peptide, inhibits the induction of nitric oxide synthase in vivo but not in vitro. *J. Pharmacol. Exp. Ther.* **275,** 584–591.
38. Hiltz, M. E. and Lipton, J. M. (1989) Antiinflammatory activity of a COOH–terminal fragment of the neuropeptide alpha–MSH. *FASEB J.* **3,** 2282–2284.
39. Macaluso, A., McCoy, D., Ceriani, G., Watanabe, T., Biltz, J., Catania, A., and Lipton, J. M. (1994) Antiinflammatory influences of alpha–MSH molecules: central neurogenic and peripheral actions. *J. Neurosci.* **14,** 2377–2382.

40. Cossu, G., Cusella–De–Angelis, M.G., Senni, M.I., De Angelis L., Vivarelli, E., Vella, S., Bouche, M., Boitani, C., and Molinaro, M. (1989) Adrenocorticotropin is a specific mitogen for ,a,alian myogenic cells. *Dev. Biol.* **131,** 331–336.
41. De Angelis, L., Cusella–De Angelis, M. G., Bouche, M., Vivarelli, E., Boitani, C., Molinaro, M., and Cossu, G. (1992) ACTH–like peptides in postimplantation mouse embryos: a possible role in myoblast proliferation and muscle histogenesis. *Dev. Biol.* **151,** 446–458.
42. Hughes, S., Smith, M. E., and Bailey, C. J. (1992) Beta–endorphin and corticotropin immunoreactivity and specific binding in the neuromuscular system of obese–diabetic mice. *Neuroscience* **48,** 463–468.
43. Smith, M. E. and Hughes, S. (1994) POMC neuropeptides and their receptors in the neuromuscular system of wobbler mice. *J. Neurol. Sci.* **124,** 56–58.
44. Bijlsma, W. A., Schotman, P., Jennekens, F.G.I., Gispen, W.H., and de Wied, D. (1983) The enhanced recovery of sensorimotor function in rats is related to the melanotropic moiety of ACTH/MSH neuropeptides. *Eur. J. Pharmacol.* **92,** 231–236.
45. Strand, F. L. and Kung, T.T. (1980) ACTH accelerates neuromuscular function following crushing of peripheral nerve. *Peptides* **1,** 135–138.
46. Strand, F. L., Lee, S. J., Lee, T. S., Zuccarelli, L. A., Antonawich, F. J., Kume, J., and Williams, K. A. (1993) Non–corticotropic ACTH peptides modulate nerve development and regeneration. *Rev. Neurosci.* **4,** 321–363.
47. Ramachandran, J., Farmer, S.W., Liles, S., Li, C.H. (1976) Comparison of the steriodogenic and melanotropic activities of corticotropin, α–melanotropin and analogs with their lipolytic activities in rat and rabbit adipocytes. *Biochim Biophys Acta* **428,** 339–346.
48. Lu, D., Willard, D., Patel, I. R., Kadwell, S., Overton, L., Kost, T., Luther, M., Chen, W., Woychik, R. P., and Wilkison, W. O. (1994) Agouti protein is an antagonist of the melanocyte–stimulating–hormone receptor. *Nature* **371,** 799–802.
49. Thody, A. J. and Shuster, S. (1973) Possible role of MSH in the mammal. *Nature* **245,** 207–209.
50. Thody, A. J. and Shuster, S. (1970) The pituitary and sebaceous gland activity. *J. Endocrinol.* **48,** 139–40.
51. Ebling, F. J., Ebling, E., Randall, V., and Skinner, J. (1975) The synergistic action of alpha–melanocyte–stimulating hormone and testosterone of the sebaceous, prostate, preputial, Harderian and lachrymal glands, seminal vesicles and brown adipose tissue in the hypophysectomized–castrated rat. *J. Endocrinol.* **66,** 407–412.
52. Thody, A. J., Cooper, M. F., Bowden, P. E., Meddis, D., and Shuster, S. (1976) Effect of alpha–melanocyte–stimulating hormone and testosterone on cutaneous and modified sebaceous glands in the rat. *J. Endocrinol.* **71,** 279–288.
53. Hay, J. B., Meddis, D., Thody, A. J., and Shuster, S. (1982) Mechanism of action of alpha–melanocyte–stimulating hormone in rat preputial glands: the role of androgen metabolism. *J. Endocrinol.* **94,** 289–294.
54. Dartt, D. A. (1994) Regulation of tear secretion. *Adv. Exp. Med. Biol.* **350,** 1–9.
55. Jahn, R., Padel, U., Porsch, P. H., and Soling, H. D. (1982) Adrenocorticotropic hormone and α–melanocyte–stimulating hormone induce secretion and protein phosphorylation in the rat lacrimal gland by activation of a cAMP–dependent pathway. *Eur. J. Biochem.* **126,** 623–629.
56. Leiba, H., Garty, N. B., Schmidt–Sole, J., Piterman, O., Azrad, A., and Salomon, Y. (1990) The melanocortin receptor in the rat lacrimal gland: a model system for the

study of MSH (melanocyte stimulating hormone) as a potential neurotransmitter. *Eur. J. Pharmacol.* **181,** 71–82.

57. Entwistle, M. L., Hann, L. E., Sullivan, D. A., and Tatro, J. B., (1990) Characterization of functional melanotropin receptors in lacrimal glands of the rat. *Peptides* **11,** 477–483.
58. Paynes, A. P. (1994) The harderian gland: a terecentennial review. *J. Anat.* **185,** 1–49.
59. Thiessen, D. D. and Kittrell, M. W. (1980) The harderian gland and thermoregulation in the gerbil (*Meriones unguiculatus*). *Physiol. Behav.* **24,** 417–424.
60. Harlow, H. J. (1984) The influence of harderian gland removal and fur lipid removal on heat loss and water flux to and from skin of muskrats, *Ondatra zibethicus. Phsyiol. Zool.* **57,** 349–356.
61. Thiessen, D. D. and Harriman, A. E. (1986) Harderian gland exudates in the male Meriones unguiculatus regulate female proceptive behavior, agression, and investigation. *J. Comp. Psychol.* **100,** 85–87.
62. Spike, R. C., Payne, A. P., and Moore, M. R. (1992) Porphyrins and their possible significance in Harderian glands, in *Harderian Glands: Porphyrin Metabolism, Behavioral and Endocrine Effects* (Webb, S. M., Hoffman, R. A., Puig–Domingo, M. L., and Reiter, R. J. eds.), Springer–Verlag, New York pp. 165 – 193.
63. Rodriguez, C., Menendez–Pelaez, A., Reiter, R. J., Buzzell, G. R., and Vaughan, M. K. (1992) Porphyrin metabolism in the harderian glands of golden hamster: in vivo regualtion by testicular hormones, light conditions, pineal gland, and pituitary hormone. *Proc. Soc. Exp. Med.* **200,** 25–29.
64. Buzzell, G. R., Hoffman, R. A., Vaughan, M. K., and Reiter, R. J. (1992) hypophesectomy prevents castration–induced increase in porphyrin concentrations in the harderian glands of male golden hamster: a possible role for prolactin. *J. Endocrinol.* **133,** 29–35.
65. Bronson, F. H., and Caroom, D. (1971) Preputial gland of the male mouse: attractant function. *J. Reprod. Fertil.* **25,** 279–282.
66. Orsulak, P. J. and Gawienowski, A. M. (1972) Olfactory preferences for the rat preputial gland. *Biol. Reprod.* **6,** 219–223.
67. Chipman, R. K. and Alberecht, E. D. (1974) The relationship of the male preputial gland to the acce;eration of oestrus in the laboratory mouse. *J. Reprod. Fertil.* **38,** 91–96.
68. Nowell, N. W., Thody, A. J., and Woodley, R. (1980) alpha–Melanocyte stimulating hormone and aggressive behavior in the male mouse. *Physiol. Behav.* **24,** 5–9.
69. Nowell, N. W., Thody, A. J., and Woodley, R. (1980) The source of an aggression–promoting olfactory cue, released by alpha–melanocyte stimulating hormone, in the male mouse. *Peptides* **1,** 69–72.
70. Thody, A. J., Donohoe, S. M., and Shuster, S. (1981) alpha–Melanocyte stimulating hormone and the release of sex attractant odors in the female rat. *Peptides* **2,** 125–129.
71. Thody, A. J., Wilson, C. A., and Everard, D. (1981) alpha–Melanocyte stimulating hormone stimulates sexual behaviour in the female rat. *Psychopharmacology* **74,** 153–156.
72. Thody, A. J. and Wilson, C. A. (1983) Melanocyte stimulating hormone and the inhibition of sexual behaviour in the female rat. *Physiol. Behav.* **31,** 67–72.
73. File, S. E. (1978) ACTH, but not corticosterone impairs habituation and reduces exploration. *Pharmacol. Biochem. Behav.* **9,** 161–166.

74. Mackay–Sim, A. and Li\aing, D.G. (1980) Discrimination of odors from stressed rats by non–stressed rats. *Physiol. Behav.* **24,** 699–704.
75. Valenta, J. G. and Rigby, M.K. (1968) Discrimination of the odor of stressed rats. *Science* **161,** 599–601.
76. Mackay–Sim, A. and Laing, D. G. (1981) The sources of odors from stressed rats. *Physiol. Behav.* **27,** 511–513.
77. Famselow, M. S. (1985) Odors released by stressed rats produce opioid analgesia in unstressed rats. *Behav. Neurosci.* **99,** 589–592.
78. Abel, E. L. and Bilitzke, P. J. (1990) A possible alarm substance in the forced swimming test. *Physiol. Behav.* **48,** 233–239.
79. Abel, E. L. (1994) The pituitary mediates production or release of an alarm chemosignal in rats. *Horm. Behav.* **28,** 139–145.
80. Abel, E. L. and Bilitzke, P. J. (1992) Adrenal activity does not mediate alarm substance reaction in the forced swim test. *Psychoneuroendocrinology* **17,** 255–259.
81. Carr, W. J., Martorano, R.D., and Krames, L. (1970) Responses of mice to odors associated with stress. *J. Comp. Physiol. Psychol.* **71,** 223–228.
82. Rottman, S. J. and Snowndon, C.T. (1972) Demonstration and analysis of an alarm pheromone in mice. *J. Comp. Physiol. Psychol.* **81,** 483–490.
83. Cocke, R., Moynihan, J.A., Cohen, N., Grota, L.J., and Ader, R. (1993) Exposure to conspecific alarm chemosignals alters immune responses in Balb/c mice. *Brain, Behav. Immunity* **7,** 36–46.
84. Mugford, R. A. and Nowell, N.W. (1971) Shock–induced release of the preputial gland secretions that elicit fighting in mice. *J. Endocrinol.* **51,** xvi–xvii.
85. Lichtensteiger, W., Hanimann, B., Siegrist, W., and Eberle, A.N. (1996) Region– and stage–specific patterns of melanocortin receptor ontogeny in rat central nervous system, cranial nerve ganglia and sympathetic ganglia. Brain Research. Developmental *Brain Res.* **91,** 93–110.
86. Zimmermann, E. and Krivoy, W. (1973) Antagonism betwwen morphine and the polypeptides ACTH, ACTH1–24 and beta–MSH in the nervous system. *Prog. Brain Res.* **39,** 383–394.
87. Smock, T. and Fields, H. L. (1981) ACTH1–24 blocks opiate – induced analgesia in the rat. *Brain Res.* **212,** 202–206.
88. Belcher, G., Smock, T., and Fields, H. L. (1982) Effects of intrathecal ACTH on opiate analgesia in the rat. *Brain Res.* **247,** 373–377.
89. Johnston, M. F., Kravitz, E. A., Meiri, H., and Rahamimoff, R. (1983) Adrenocorticotropic hormone causes long–lasting potentiation of transmitter release from frog motor nerve terminals. *Science* **220,** 1071–1072.
90. Davies, D. A. Smith, M. E., (1994) ACTH (4–10) increases quantal content at the mouse neuromuscular junction. *Brain Res.* **637,** 328–330.
91. Shapiro, M. S., Namba, T., Grob, D. (1968) The effect of corticotropin on the neuromuscular junction: morphologic studies in rabbits. *Neurology* **18,** 1018–1022.
92. Strand, F. L., Williams, K. A., Alves, S. E., Antonawich, F. J., Lee, T. S., Lee, S. J., Kume, J., and Zuccarelli, L. A. (1994) Melanocortins as factors in somatic neuromuscular growth and regeneration. *Pharmacol. Ther.* **62,** 1–27
93. Hughes, S. and Smith, M. E. (1988) Effect of nerve section on beta–endorphin and alpha–melanotropin immunoreactivity in motor nerves of normal and dystrophic mice. *Neurosci. Lett.* **92,** 1–7.

94. Hughes, S. and Smith, M. E. (1989) Proopiomelanocortin–derived peptides in mice with motoneurone disease. *Neurosci. Lett.* **103,** 169–173.
95. Hughes, S., and Smith, M. E. (1993) Upregulation of the pro–opiomelanocortin gene in spinal motoneurones in muscular dystrophy in mice. *Neurosci. Lett.* **163,** 205–207.
96. Hughes, S. and Smith, M. E. (1994) Upregulation of the pro–opiomelanocortin gene in motoneurons after nerve section in mice. *Brain Res. Mol. Brain Res.* **25,** 41–49.
97. Star, R. A., Rajora, N., Huang, J., Stock, R. C., Catania, A., and Lipton, J. M. (1995) Evidence of autocrine modulation of macrophage nitric oxide synthase by alpha–melanocyte–stimulating hormone. *Proc. Natl. Acad. Sci. U. S. A.* **92,** 8016–8020.
98. Baldwin, J. M. (1993) The probable arrangement of the helices in G protein–coupled receptors. *EMBO J.* **12,** 1693–1703.
99. van der Kraan, M., Adan, R. A., Entwistle, M. L., Gispen, W. H., Burback, J. P., and Tatro, J. B. (1998) Expression of melanocortin-5 receptor in secretory epithelia support a functional role in exocrine and endocrine glands. *Endocrinology* **139,** 2348–2355.

Part V

Receptor Regulation

CHAPTER 16

Regulation of the Melanocortin Receptors by Agouti

William O. Wilkison

1. Introduction

The melanocortin family of receptors has been implicated in the regulation of a number of physiologic systems. Despite the cloning and characterization of these receptors, little is known about their regulation. I will summarize in this chapter what is known about a novel regulator of melanocortin receptor activity, the agouti gene product. Not only does the action of agouti on these receptors explain or clarify the physiologic role of some of these receptors, agouti function and regulation also imparts new possibilities for these receptors having a role in processes such as energy homeostasis.

2. Agouti

2.1. *Cloning*

The existence of the agouti gene has been postulated as long ago as the late 1800s as a genetic determinant which imparted varied coat color on mice *(1,2)*. In the search for the agouti gene, Woychik and colleagues *(3–5)* were able to take advantage of a radiation-induced mutation in a limb deformity gene that also had an absence of yellow pigmentation. Both loci are localized to chromosome 2 but are fairly far apart on the chromosome. Suspecting that the loss of pigmentation may have been due to a chromosomal rearrangement that blocked expression of the putative agouti gene product, subsequent mapping by using limb deformity gene markers and other positional markers allowed these investigators to clone a segment of DNA corresponding to the agouti structural gene. Subsequent work as described below has confirmed the identity of this gene as responsible for the coat color variations as well as other phenotypes.

The Melanocortin Receptors Ed.: R. D. Cone

MDVTRLLLATLVSFLCFFTVHS HLALEETLGD

DRSLRSNSSMNSLDFSSVSIVALN KKSKKI

SRKEAEKRKRSSKKKASMKKVAR PPPPSPC

VATRDSCKPPAPACCDPCASCQCRFFGSAC

TCRVLNPNC

Fig. 1. Primary structure of the murine agouti protein. The predicted amino acid sequence of the murine agouti protein is shown. The underlined portion of the sequence indicates the signal sequence that is processed for secretion of the mature agouti protein. The N residue *(40)* in red is the *N*-glycosylation site. The region in pink from amino acid 57 to 76 designates the basic region. The cysteine residues (in yellow) indicate the carboxyl terminal residues (77 to 131) that are disulfide-bonded in the protein.

2.2. *Protein Structure*

The agouti gene encodes for a 131 amino acid polypeptide *(5,6)*. This protein (Fig. 1) is a secreted protein, as predicted by the consensus signal sequence, the verification of the amino terminal residue of purified recombinant protein as histidine *(7)* and the identification of agouti in the serum of animals overexpressing the protein (data not shown). In fact, the purification of recombinant mouse and human agouti using the agouti signal sequence from various expression systems (such as CV1 cells and insect cells infected with baculovirus) is performed using the media, not the cell cytosol or particulate fractions. The protein can be arbitrarily divided into three domains or motifs. The amino terminal portion (from amino acid 23 to 56) bears a glycosylation site which appears functional based on analysis of recombinant protein isolated from insect cells *(7)*. The middle or "basic" region (from amino acid 57 to 76) of the polypeptide contains a stretch of amino acids with a high percentage (16 out of 29) of lysines and arginines. This region facilitates the purification of the recombinant and natural protein by cation exchange chromatography.

The carboxyl-terminal portion of the protein has some intriguing properties and homologies. A proline-rich region that has some functional significance separates the basic domain from this region (*see* Subheading 3.4.). This portion (from amino acid 77 to 131) of the protein contains 10 cysteine residues. These residues appear to be disulfide-bonded, at least in the recombinant protein, and some sense of the disulfide-bonding pattern is known *(7)*. An intriguing homology has been identified with the disulfide bonding pattern and cysteine spacing of the agouti residues corresponding to that of the calcium channel blocking neurotoxins, including conotoxin, agatoxin and plectoxin

```
        23         33         43         53
murine  HLALEETLGD DRSLRSNSSM NSLDFSSVSI VALNKKSKKI
human   ..PP..K.R. .........V .L..VP.... ........Q.

        63         73         83
        SRKEAEKRKR SSKKKASMKK VARPPPPS--PC
        G..A...-.. ....E..... VV..RT.LSA..

        93         103        113        123
        VATRDSCKPP APACCDPCAS CQCRFFGSAC TCRVLNPNC
        ....N..... .......... ......R... S....SL..
```

Fig. 2. Comparison of murine and human agouti primary sequences. The predicted primary sequences for both the human and mouse mature agouti protein were aligned. The dashes between murine residues 91 and 92 indicate the lack of residues for allowance to line up the sequences. The dashes in the human sequence denotes the loss of one amino acid. Note the complete conservation of the cysteine residues and their spacing.

(7,8). The functional relevance of this homology is not clear, although agouti does seem to have some effects on calcium mobilization (*see* Subheading 4.2.).

2.3. Human Homology

The human agouti gene has been cloned and the primary sequence for the human protein deduced *(9,10)*. The gene reveals remarkable conservation (87%) of DNA sequence and equal conservation (89%) of the amino acid sequence (Fig. 2). The major difference between the human and murine proteins is a one amino acid deletion in the human gene. There is complete conservation of the cysteine spacing (and disulfide-bonding pattern) in the human protein. However, there are differences in the activity of these two proteins as discussed in the next section. The human gene is located at chromosome 20q13, which is also a region of the human chromosome associated with genetic inheritance of obesity and diabetes susceptibility *(9)*.

The human pattern of distribution of the agouti mRNA is very different than that of the mouse. Agouti is primarily localized in skin in the wild-type mouse and is temporally regulated *(5,6)*. It is only in various mutant strains that agouti is ubiquitously expressed, where the varied phenotypes of agouti overexpression are observed. In humans, agouti is expressed in the testes, foreskin, and adipose tissue, *(5,6)* as detected by reverse transcriptase polymerase chain reaction. Wilson et al. *(10)* see agouti mRNA expression by Northern blot analysis in heart and ovaries with lower levels of expression in the liver and kidney, while Kwon et al. *(9)* failed to see agouti mRNA in these organs.

3. Melanocortin Receptor Biochemistry

3.1. Agouti Is a Competitive Antagonist of MC1-R

As agouti is a gene involved in coat coloration and a primary regulatory molecule of coat color via the alpha melanocyte stimulating hormone (α-MSH) in rodents *(11)*, it seemed logical that agouti may function to interfere with the action of α-MSH. This was borne out by the experiments done by Lu et al. *(12)* where recombinant agouti was partially purified and shown to antagonize both the binding of a radiolabeled ligand to the MSH receptor (MC1-R) as well as block the accumulation of cAMP in response to a variety of melanocortin ligands. Importantly, agouti had no effect on thyroid stimulating hormone stimulated cAMP production, suggesting a direct effect of agouti on the melanocortin receptors. Later work, using agouti protein purified to homogeneity, confirmed the agouti antagonism of α–MSH action on MC1-R *(7)*.

Studies performed by Blanchard et al. *(41)* showed that agouti was a competitive inhibitor of ligand binding and activation. The K_i for agouti antagonism of binding (1.9n*M*) was shown to be equivalent to that of the IC_{50} for antagonism of cAMP production (0.9n*M*), arguing for a competitive inhibitor effect. Also, these effects were independent of the melanocortin ligand used, were not influenced by preincubation of either agonist or antagonist, and were consistent only with a competitive model of inhibition.

3.2. Agouti Antagonism of the Other Receptor Family Members

In addition to the MC1-R, there are four other known members of the melanocortin receptor family, designated MC2-R to MC5-R. The genes for all of these receptors have been cloned, and with the exception of the MC2-R, have been expressed in recombinant HEK293 cell lines *(13–15)*. Since antagonism of MC1-R by agouti was sufficient to explain the coat coloration effects of agouti mutations but not the other phenotypes observed, it was of interest to examine the effect of the agouti protein on the other MCRs.

Lu et al. *(12)* had shown the effect of recombinant murine agouti on α-MSH activation of the human MC4-R, the rat MC3-R, and the murine MC5-R. Murine agouti had little or no effect on the MC3-R or MC5-R. However, the protein was able to antagonize α-MSH activation of the MC4-R, shifting the EC_{50} for α-MSH from 4.9 n*M* to 33 n*M*.

Further analysis by Kiefer et al. *(16)* showed that murine agouti was a potent antagonist of [^{125}I]NDP-MSH binding to B16F10 cells (expressing predominantly MC1-R) and HEK293 cells expressing murine MC3-R, MC4-R, and MC5-R. Agouti was most potent in antagonizing the MC1-R (K_i= 2.6), while having a rank order of MC4-R (K_i=54), MC3-R (K_i=190), and MC5-R (K_i=1200).

3.3. Effects of Human Versus Mouse Agouti

Examination of recombinant human agouti on these various systems gives a slightly different picture. Work by Kiefer et al. *(16)* show that human agouti (K_i=23 n*M*) is a less potent ligand binding antagonist against the human MC1-R than the murine agouti (K_i=2.1). However, the order of potency against the human receptors expressed in HEK293 cells remains the same (MC1-R > MC4-R > MC3-R > MC5-R) and, with exception of the MC1-R, the K_i's are basically the same.

Yang et al. see a different pattern of inhibition. Measuring antagonism of α-MSH-induced cAMP accumulation against the human receptors expressed in OS3 cells, a rank order potency of MC4-R > MC1-R > MC5-R > MC3-R was observed. Displacement of [^{125}I]NDP-MSH from these cell lines by agouti also maintained this rank order of potency. These discrepancies may be due to use of different cell lines or the form of recombinant agouti protein. Based on the data, it is not possible to directly compare the K_i values generated between the two groups.

It is also clear that, except for MC1-R, human agouti protein may not exhibit competitive inhibition on these receptors. Agouti may antagonize the functional response of MC2-R and MC4-R to melanocortin ligands with a combination of competitive and noncompetitive kinetics *(17)*. This has also been observed by Mountjoy and Wong *(18)* in that human agouti displays noncompetitive kinetics with respect to activation of human MC4-R expressed in HEK293 cells *(18a)*. These issues will only be resolved with the development of labeled agouti protein to directly measure the binding affinity of that ligand as well as agouti produced in vivo and in other recombinant systems.

3.4. Localization of Agouti Active Residues

The α-MSH peptide is small (13 amino acids) and several synthetic analogs of α-MSH, including NDP-MSH, are even smaller (7 residues), yet the agouti protein comprises 131 amino acids. However, there are no regions of primary sequence on agouti showing homology to the melanocortin peptides. Therefore, it was of interest to determine which portion of the agouti polypeptide was necessary for the MCR antagonism activity observed.

Analysis of mutant agouti cDNAs was described by Perry et al. *(19)* using transgenic animals expressing agouti cDNA under control of the β-actin promoter. Using a mutant allele of agouti with a point mutation in the signal peptide region, they found that this cDNA failed to produce active protein, as evidenced by lack of yellow coat color and obesity. It was presumed that the mutation failed to produce secreted protein due to inefficient processing of the signal peptide.

The first report of localizing the active site of agouti antagonism of MCRs was Willard et al. *(7)*. The agouti polypeptide was digested by a variety of proteases and it was discovered that the digestion products of lysC failed to inactivate the agouti antagonism activity. Subsequent mass spectroscopy analysis revealed that this protease generated a cleavage product consisting of the carboxyl terminal domain (val83-cys131). The isolated C-terminal fragment retained α-MSH antagonism equipotent to the mature agouti polypeptide. Further analysis by Kiefer et al. *(16)* revealed this C-terminal fragment was equipotent with mature agouti in antagonism of NDP-MSH stimulation of cAMP accumulation in and antagonism of [^{125}I]NDP-MSH binding to B16F10 cells.

Further confirmation of the active domain of murine agouti being the carboxyl-terminal region was done using deletion constructs of agouti. Expression and purification of proteins bearing a deletion of the basic region (Δasn56-pro86) and deletion of the carboxyl terminal region (Δpro89-cys131) allowed characterization of these polypeptides on the MCRs. The carboxyl terminal deletion protein lost binding antagonism activity by two orders of magnitude and the basic region deletion only lost significant activity against the MC3-R *(16)*.

Although it was clear that the majority of the antagonism determinants were localized in the carboxyl-terminal region of agouti, there was still detectable antagonism activity in the carboxyl terminal deletion construct. It was significant that the proteolytic construct retained 3 amino acids additional to the basic construct deletion. Site-directed mutagenesis performed on the full-length agouti, specifically targeting the residues spanning val83 to Pro89, revealed that val83 was an important residue for antagonism by agouti to MC1-R. In addition, arg85, pro86, and pro89 were identified as important determinants for selectivity of antagonism. Their K_i's were essentially unchanged at MC1-R, while these mutant proteins have increased K_i's (6-to 10-fold) relative to the wild-type protein at MC3-R and MC4-R *(16)*.

However, the main determinant of activity for MCR antagonism is localized in the carboxyl terminal region of agouti. Systematic mutational analysis (alanine-scanning) was performed on the full-length murine agouti *(43)*. These 24 mutants were expressed and purified for assay against the murine MCRs. As shown in Fig. 3, only three residues appeared to be critical determinants of agouti antagonism: arg116 ($K_i > 650$ n*M*), phe118 ($K_i = 220$ n*M*), and asp108 ($K_i = 34$ n*M*). All other mutations gave insignificant changes in ligand binding antagonism activity against the murine MC1-R.

Further mutagenesis was done to analyze the effect of these residues. Substitution of arg116 with a histidine or lysine residue allowed some activity to be regained but not to the level of wild type (arg116lys; K_i=30), indicating a basic charge is essential for activity with arginine being optimal. Substitution of phe118 with a tryptophan residue also allowed some recovery of activity (phe118trp; K_i~2), indicating the necessity of a hydrophobic region for activity.

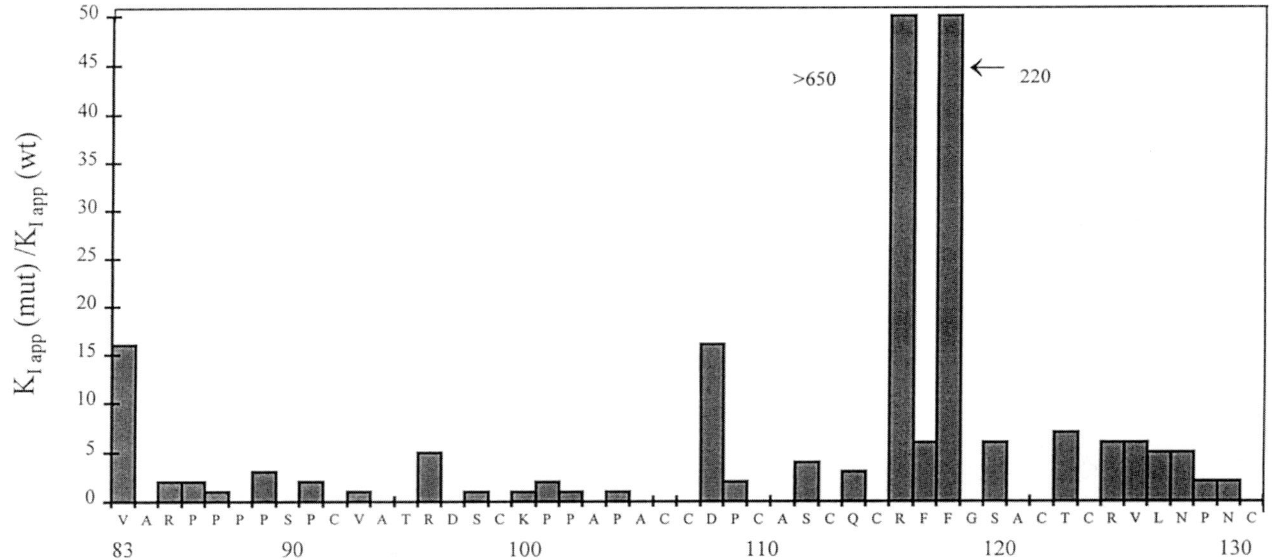

Fig. 3. Residues asp108, arg116, and phe118 are critical determinants of agouti antagonism activity on MC1-R. Agouti cDNA was subjected to alanine-scanning mutagenesis, expression in baculovirus/*Trichiplusia ni* cells, and partially purified and concentration determined as described *(16)*. The K_i for agouti mutant proteins was determined against [^{125}I]NDP-MSH binding on B16F10 cells essentially as described *(41)*. The ratio of the K_i for the mutant protein/the K_i of the wild-type protein [K_{Iapp} (mut)/K_{Iapp}(wt)] calculated and plotted against the residues of the mutated agouti proteins. Bars indicate residues actually mutated to alanine and assayed.

Mutation of other residues resulted in modifications of the selectivity of agouti antagonism against the various receptors. Mutagenesis of phe117 to tryptophan resulted in a protein that exhibited more potent antagonism against MC5-R and less potent antagonism against MC4-R. Likewise, mutation of thr123 to glycine resulted in a protein with similar potency against MC4-R and MC5-R while losing potency against MC3-R and MC1-R.

This data allowed us to generate a computer graphic image of the carboxyl terminal region of agouti *(43)*, using the ω-conotoxin structure *(20)* as a general template (Fig. 4). As shown, the three primary residues determining agouti activity are grouped on one side of the molecule (indicated in red). This suggests that one side of the molecule interacts with the MC1-R for antagonism and suggests that, in conjunction with the kinetic data, that these residues probably interact with the binding site of α-MSH to allow competitive inhibition of this peptide. As for the other receptors, further mutational/biochemical and structural analysis must be performed in order to determine the mechanism of agouti inhibition of these receptors.

4. Biologic Relevance

4.1. Pigmentation

As mentioned previously, the agouti gene had been identified as a locus involved in the regulation of coat color primarily in rodents. The cloning and characterization of the agouti gene, along with the discovery of the ability of this gene product to antagonize the α-MSH receptor, has allowed a rational and likely explanation for the effects of agouti on the regulation of pelage.

Many mutations have been identified that show unusual yellow coat color in mice. We now know that almost all of these phenotypes are the result of mutations in the agouti gene, primarily the promoter *(5,21)*. In A^y, for instance, an 18-kb deletion occurs in the region of the agouti promoter and a gene, *RALY*, which is located upstream of the agouti gene. The result of this deletion is control of expression of the normal agouti gene by the *RALY* gene promoter, leading to ubiquitous and unregulated expression *(5)*. The question was how overexpression of the normal gene product would lead to yellow coat color.

Fig. 4. *(opposite page)* Homology model of residues 92-125 of murine agouti. A computer homology model of the carboxyl terminal region of murine agouti was generated based on the ω-conotoxin structure *(42)*. Residues that result in large increases in K_{Iapp} [K_{Iapp} (mut)/K_{Iapp}(wt)~15] at MC1-R when mutated to alanine are shown in pink. Residues resulting in very large increase [(K_{Iapp} (mut)/K_{Iapp}(wt)>40] are shown in purple. Two faces of this model is presented. Note that the active residues are localized to one side of the agouti protein.

4A

4B

The normal mechanism of eumelanin or black pigment production is via the increase of intracellular cAMP levels by activation of the MC1-R by α-MSH. It is now accepted that the agouti gene product acts to block the formation of eumelanin by its ability to antagonize α–MSH-mediated increases in intracellular cAMP levels. The default mechanism of the melanocyte is to then produce phaeomelanin or yellow pigment.

This hypothesis is consistent with a large array of data. The normal expression pattern of agouti during mouse development is during day 4 to day 6 in skin, at which time pheomelanin is produced in animals with agouti or wild-type pelage *(5,6)*. This yellow band of pigment is responsible for the agouti phenotype. Expression of agouti in the ventral skin areas of mice with the black and tan (a^t) phenotype correlates with the appearance of yellow pigment only on the ventral side. Mutations in the agouti promoter region (such as *a*) that decrease normal agouti mRNA expression levels fail to allow the animal to express any pheomelanin, resulting in jet-black animals *(22)*. Simple overexpression of agouti in transgenic mice leads to animals with bright yellow coats and the level of mRNA expression roughly correlates with the amount of yellow pelage *(19,23)*. Experiments performed by Shimizu and others *(24,25)* have shown that injection of α–MSH and other melanocortin peptides into A^y or A^{vy} mice allow the development of black fur, indicating the melanocortin peptide can overcome the agouti effect. This is exactly what would be predicted using the competitive inhibitor model for agouti antagonism of MC1-R.

However, recent work suggests that the model may not be that simple. Two reports suggest that agouti not only antagonizes α-MSH mediated cAMP increases but may also have direct effects on either the MC1-R or the melanocytes themselves. Specifically, agouti may mediate the down regulation of the MC1-R, either via internalization or desensitization *(26)*. The effect of this down regulation would be to decrease the responsiveness of the melanocytes, or in this case the B16F10 cells, to α-MSH. Also, agouti may act as an inverse agonist, as it can block melanogenesis mediated not only by α-MSH but also forskolin, cholera toxin, or pertussis toxin. This effect appears to be inverse agonism, since agouti is unable to affect a variant cell line which lacks the MC1-R *(27)*. It is possible agouti may be mediating these effects via its ability to mobilize intracellular calcium in an MCR-dependent manner (*see* Subheading 4.2.).

Agouti's role in human pigmentation is unknown. Agouti mRNA is detectable in human foreskin, but the expression levels do not correlate with degree of pigmentation *(10)*. There are no known genetic disorders of pigmentation that are associated with the 20q13 region where agouti is located. Further studies will be required to determine the role, if any, of human agouti in pigmentation regulation.

4.2. Signal Transduction

In addition to the ability to block melanocortin-mediated cAMP accumulation, agouti has the ability to mobilize intracellular calcium. This possibility was originally explored by Zemel et al. *(28)* due to the intriguing structural homology of the agouti carboxyl terminal region to calcium channel blockers. Recombinant agouti was incubated with a muscle cell line and it was found that agouti stimulated intracellular calcium mobilization. This mobilization was dependent on the presence of extracellular calcium and appeared to involve primarily influx of external calcium as opposed to mobilization of intracellular stores. This correlated with the elevated intracellular calcium concentrations in the tissues of the A^y mice, which ectopically express the agouti polypeptide. Additional work indicated this agouti effect was dose-responsive and was also observed in a vascular smooth muscle cell line and 3T3-L1 adipocytes. The EC_{50} for the effect of murine agouti on L6 myocytes was about 62n*M*, a little higher than that seen for agouti effects on MC1-R, but comparable to that seen for MC4-R and MC3-R.

Further analysis of this mechanism indicated it was an MCR-dependent event *(29)*. Measurements of calcium mobilization in HEK293 cells stably transfected with human MC1-R, MC3-R, and MC5-R revealed that MC1-R and MC3-R receptors responded to human agouti (20n*M*) by elevating calcium 60–70n*M* over the baseline concentration. Interestingly, MC5-R failed to respond to agouti with respect to calcium mobilization, as well as the HEK293 nontransfected cell line. A dose response of murine agouti on the MC1-R stably transfected cell line gave an EC_{50}=18n*M*, somewhat higher than the predicted binding affinity of agouti. Although agouti appears to mobilize calcium in an MCR-dependent mechanism (at least in HEK293 cells), the mode of activation may not be through the same site as melanocortin binding.

4.3. Obesity and Type II Diabetes

Of major interest to the academic and industrial community is the phenotype observed in mice overexpressing the agouti protein. In addition to the yellow pelage, these mice exhibit obesity and insulin resistance *(30)*. These phenotypes have been shown to be a direct result of agouti overexpression as evidenced by transgenic animals expressing agouti ubiquitously under control of heterologous promoters *(19,23)*.

Due to the biochemical action of the agouti protein, the mechanism by which agouti mediates this phenotype is probably via the MCRs. It is clear that MC1-R is not responsible for these phenotypes, as crossing constitutively active MC1-R mutant mice (e^{so}) with the A^y phenotype results in black animals that get obese *(31)*. Also, MC1-R null mutant animals are yellow but fail to gain weight or become insulin-resistant *(32)*.

A more likely candidate based on the biochemical analysis of agouti antagonism is the MC4-R. This receptor is localized in the central nervous system in rodents *(33)* and has a broader distribution in man *(34)*. Agouti is a potent antagonist for melanocortin mediated cAMP elevation via this receptor. Recently, two publications have described strong evidence for the MC4-R being the receptor that mediates the obesity and insulin resistance of agouti overexpression. A knockout of MC4-R in mice results in animals that become obese and insulin-resistant *(35)*. Additionally, intracerebroventricular administration of a peptide antagonist of MC4-R leads to hyperphagia while administration of an MC4-R selective agonist in yellow mice relieves hyperphagia *(36)*. These data clearly implicate the MC4-R as a mediator of food intake and possibly glucose homeostasis, indicating an important role of melanocortins in energy regulation.

Additional work suggests a peripheral role of agouti in mediating insulin resistance. Mynatt et al. *(37)* have generated transgenic mice expressing agouti under control of the aP2 promoter, an adipocyte-specific promoter. Northern blots of these animals indicate agouti expression limited to brown and white fat depots. Normally, the animals do not exhibit any disease phenotype, but upon challenging the animals with insulin injections, the aP2 transgenics begin to gain weight at rates much higher than their nontransgenic littermates. Since agouti is probably not produced in the central nervous system in these animals, it is likely the weight gain and insulin resistance observed by insulin challenge is due to an effect of agouti on a peripheral tissue such as adipose or muscle *(37)*. Liver effects can be ruled out since agouti expression in the liver fails to initiate any phenotypic changes similar to that seen in the A^y mice, despite the production of active protein in this transgenic animal *(Wilkison and Mynatt, unpublished results)*.

The above conclusions are complemented by the report that agouti-mediated obesity can be reversed by calcium channel blockers *(38)*. Mice expressing agouti under control of the β-actin promoter gain weight and have increased fatty acid synthase activity in adipose tissue. Treatment of these animals with nifedipine for four weeks caused an 18% decrease in fat pad weight along with a 74% decrease in fatty acid synthase activity. These workers had previously shown that agouti can increase triglyceride accumulation and fatty acid synthase activity in adipocyte systems in vitro and that these effects can, in part, be attenuated with calcium channel blockers *(39)*. Thus, agouti appears to be mediating its effects on obesity and type II diabetes via melanocortin receptors. It is also possible that peripheral receptors play a contributory role to these phenotypes.

The situation in man is less clear. Agouti is expressed in adipose tissue, testes, and foreskin. The regulation of agouti expression is not known in man. While genetic mapping places the human agouti gene near a region in

```
magouti  VARPPPPS--PC VATRDSCKPP
hagouti  VVRPRTPLSAPC VATRNSCKPP
mART     PQNRESRSPRRC VRLHESCLGQ
hART     LQDREPRSSRRC VRLHESCLGQ

         APACCDPCAS CQCRFFGSAC TCRVLNPNC
         APACCDPCAS CGCRFFRSAC SCRVLSLNC
         QVPCCDPCAT CYCRFFNAFC YCRKLGTAT NLCSRT
         QVPCCDPCAT CYCRFFNAFC YCRKLGTAM NPCSRT
```

Fig. 5. Critical residues for MCR antagonism activity are conserved between agouti and ART. Primary predicted sequences for human and murine agouti and ART are shown. The sequences are aligned around the cysteine residue spacing. The cysteine residue spacing is conserved for all residues except the carboxyl terminal most cysteine (cys131 for agouti and cys129 for ART). Critical binding determinants are shown in bold blue.

chromosome 20 associated with obesity and diabetes, no association or linkage studies have correlated agouti polymorphisms with these disease states to date. The tissue distribution of the human MCRs differs greatly from the rodent. By RT-PCR, all five MCRs have mRNA in adipose tissue *(34)*. In addition, the physiologic roles of MC3-R, MC4-R, and MC5-R are far from defined. It is possible that agouti regulates MCRs expressed in human adipose tissue, but no definitive studies have presented data to address this possibility.

Finally, at least one new member of the agouti family, designated ART (agouti-related transcript) or AGRP (agouti-related protein) has been cloned *(40)*. This mRNA is present in the central nervous system in rodents and a human homolog of this gene has been identified. (*See* Fig. 5.) The human mRNA is also present in the central nervous system. AGRP is a potent antagonist of the neural MC3 and MC4 receptors (*see* Chapter 14).

Thus, we can summarize by saying that agouti most likely mediates its effects on hyperphagia and weight gain through the MCRs, with MC4-R having a clear role in feeding behavior and possibly another MCR with a more peripheral expression pattern, regulating other energy homeostasis mechanisms. The identification of new agouti family members leaves open the possibility of having different agouti-like molecules regulating different MCRs. A whole new arena of receptor regulation has been uncovered by the discovery and characterization of these endogenous receptor modulators.

References

1. Morse, H. C., III (1981) The laboratory mouse–a historical perspective, in *The Mouse in Biochemical Research* (Foster, H. L., Small, J. D., and Fox, J. G., eds.), Academic Press, New York, pp. 1–16.
2. Green, M. C. (1989) Catalog of mutant genes and polymorphic loci, in *Genetic Variants and Strains of the Laboratory Mouse* (Lyon, M. F. and Searle, A. G., eds.), Oxford University Press, Oxford, UK, pp. 17–20.
3. Woychik, R. P., Generoso, W. M., Russell, L. B., Cain, K. T., Cacheiro, N. L. A., Bultman, S. J., Selby, P. B., Dickinson, M. E., Hogan, B. L. M., and Rutledge, J. C. (1990) Molecular and genetic characterization of a radiation–induced structural rearrangement in mouse chromosome 2 causing mutations at the *limb deformity* and *agouti* loci. *Proc. Natl. Acad. Sci. U. S. A.* **87,** 2588–2592.
4. Bultman, S. J., Russell, L. B., Gutierrez–Espeleta, G. A., and Woychik, R. P. (1991) Molecular characterization of a region of DNA associated with mutations at the *agouti* locus in the mouse. *Proc. Natl. Acad. Sci. U. S. A.* **88,** 8062–8066.
5. Bultman, S. J., Michaud, E. J., and Woychik, R. P. (1992) Molecular characterization of the mouse *agouti* locus. *Cell* **71,** 1195–1204.
6. Miller, M. W., Duhl, D. M. J., Vrieling, H., Cordes, S. P., Ollmann, M. M., Winkes, B. M., and Barsh, G. S. (1993) Cloning of the mouse *agouti* gene predicts a secreted protein ubiquitously expressed in mice carrying the *lethal yellow* mutation. *Genes & Development* **7,** 454–467.
7. Willard, D. H., Bodnar, W., Harris, C., Kiefer, L., Nichols, J. S., Blanchard, S., Hoffman, C., Moyer, M., Burkhart, W., Weiel, J., Luther, M. A., Wilkison, W. O., and Rocque, W. J. (1995) Agouti structure and function: characterization of a potent α–melanocyte stimulating hormone receptor antagonist. *Biochemistry* **34,** 12,341–12,346.
8. Manne, J., Argeson, A. C., and Siracusa, L. (1995) Mechanism for the pleiotropic effects of the *agouti* gene. *Proc. Natl. Acad. Sci. U. S. A.* **92,** 4721–4724.
9. Kwon, H.–Y., Bultman, S. J., Loffler, C., Chen, W.–J., Furdon, P. J., Powell, J. G., Usala, A.–L., Wilkison, W. O., Hansmann, I., Woychik, R. P. (1994) Molecular structure and chromosomal mapping of the human homolog of the agouti gene. *Proc. Natl. Acad. Sci. U. S. A.* **91,** 9760–9764.
10. Wilson, B. D., Ollmann, M. M., Kang, L., Stoffel, M., Bell, G. I., and Barsh, G. S. (1995) Structure and function of *ASP*, the human homolog of the mouse *agouti* gene. *Hum. Mol. Genet.* **4,** 223–230.
11. Jackson, I. J. (1993) More to colour than meets the eye. *Curr. Biol.* **3,** 518–521.
12. Lu, D., Willard, D., Patel, I. R., Kadwell, S., Overton, L., Kost, T., Luther, M., Chen, W., Woychik, R. P., Wilkison, W. O., and Cone, R. D. (1994) Agouti protein is an antagonist of the melanocyte–stimulating–hormone receptor. *Nature* **371,** 799–802.
13. Mountjoy, K. G., Robbins, L. S., Mortrud, M. T., and Cone, R. D. (1992) The cloning of a family of genes that encode the melanocortin receptors. *Science* **257,** 1248–1251.
14. Roselli–Rehfuss, L., Mountjoy, K. G., Robbins, L. S., Mortrud, M. T., Low, M. J., Tatro, J. B., Entwistle, M. L., Simerly, R. B., and Cone, R. D. (1993) Identification of a receptor for γ melanotropin and other proopiomelanocortin peptides in the hypothalamus and limbic system. *Proc. Natl. Acad. Sci. U. S. A.* **90,** 8856–8860.

15. Labbe, O., Desarnaud, F., Eggerickx, D., Vassart, G., and Parmentier, M. (1994) Molecular cloning of a mouse melanocortin 5 receptor gene widely expressed in peripheral tissues. *Biochemistry* **33,** 4543–4549.
16. Kiefer, L. L., Ittoop, O. R. R., Bunce, K., Truesdale, A. T., Willard, D. H., Nichols, J. S., Blanchard, S. G., Mountjoy, K., Chen, W.–J., and Wilkison, W. O. (1997) Mutations in the carboxyl terminus of the agouti protein decrease agouti inhibition of ligand binding to the melanocortin receptors. *Biochemistry* **36,** 2084–2090.
17. Yang, Y.–K., Ollmann, M. M., Wilson, B. D., Dickinson, C., Yamada, T., Barsh, G. S., and Gantz, I. (1997) Effects of recombinant agouti–signaling protein on melanocortin action. *Mol. Endocrinol.* **11,** 274–280.
18. Mountjoy, K. G. and Wong, J. (1997) Obesity, diabetes and functions for proopiomelanocortin–derived peptides. *Mol.Cell. Endocrinol.* **128,** 171–177.
18a. Mountjoy, K. G., Willard, D. H., and Wilkison, W. O. (1999) Agouti antagonism of melanocortin-4 receptor. Greater effect with desacetyl-α-melanocyte stimulating hormone (MSH) than with α-MSH. *Endocrinology* **140(5),** 2167–2172.
19. Perry, W. L., Nakamura, T., Swing, D.A., Secrest, L., Eagleson, B., Hustad, C. M., Copeland, N. G., Jenkins, N. A. (1996) Coupled site-directed mutagenesis/transgenesis identifies important functional domains of the mouse agouti protein. *Genetics* **144(1),** 255–264.
20. Olivera, B. M., Miljanich, G. P., Ramachandran, J., and Adams, M. E. (1994) Calcium channel diversity and neurotransmitter release: the omega-conotoxins and omega-agatoxins. *Ann. Rev. Biochem.* **63,** 823–867.
21. Duhl, D. M. J., Vrieling, H., Miller, K. A., Wolff, G. L., and Barsh, G. S. (1994) Neomorphic *agouti* mutations in obese yellow mice. *Nat.Genet.* **8,** 59–65.
22. Siracusa, L. D. (1994) The *agouti* gene:turned on to yellow. *Trends Genet.* **10,** 423–428.
23. Klebig, M. L., Wilkinson, J. E., Geisler, J. G., and Woychik, R. P. (1995) Ectopic expression of the *agouti* gene in transgenic mice causes obesity, features of type II diabetes, and yellow fur. *Proc. Natl. Acad. Sci. U. S. A.* **92,** 4728–4732.
24. Tamate, H. B. and Takeuchi, T. (1981) Induction of the shift in melanin synthesis in lethal yellow (A^y/a) mice in vitro. *Dev. Genet.* **2,** 349–356.
25. Shimizu, H., Shargill, N.S., Bray, G.A., Yen, T.T., and Gessellchen, P.D. (1989) Effects of MSH on food intake, body weight, and coat color of the yellow obese mouse. *Life Sci.* **45,** 543–552.
26. Siegrist, W., Willard, D. H., Wilkison, W. O., and Eberle, A. N. (1996) Agouti protein inhibits growth of B16 melanoma cells *in vitro* by acting through melanocortin receptors. *Biochem. Biophys. Res. Commun.* **218,** 171–175.
27. Siegrist, W., Drozdz, R., Cotti, R., Willard, D. H., Wilkison, W. O., and Eberle, A. N. (1997) Interactions of α–melanotropin and agouti on B16 melanoma cells: evidence for inverse agonism of agouti. *J. Recept. Signal Transduct. Res.* **17,** 75–98.
28. Zemel, M. B., Kim, J. H., Woychik, R. P., Michaud, E. J., Kadwell, S. H., Patel, I. R., and Wilkison, W. O. (1995) Agouti regulation of intracellular calcium:role in the insulin resistance of viable yellow mice. *Proc. Natl. Acad. Sci. U. S. A.* **92,** 4728–4732.
29. Kim, J. H., Kiefer, L. L., Woychik, R. P., Wilkison, W. O., Truesdale, A., Ittoop, O., Willard, D., Nichols, J., and Zemel, M. B. (1997) Agouti regulation of intracellular calcium:role of melanocortin receptors. *Am. J. Physiol.* **272,** E379–E384.

30. Yen, T. T. (1988) The viable yellow obese–diabetic mouse. *Nutrition* **4,** 457–459.
31. Wolff, G. L., Galbraith, D. B., Domon, O. E., and Row, J. M. (1978) Phaeomelanin synthesis and obesity in mice. *J. Hered.* **69,** 295–298.
32. Hauschka, T. S., Jacobs, B. B., and Holdridge, B. A. (1968) Recessive yellow and its interaction with belted in the mouse. *J. Hered.* **59,** 339–341.
33. Gantz, I., Miwa, H., Konda, Y., Shimoto, Y., Tashiro, T., Watson, S. J., DelValle, J., and Yamada, T. (1993) Molecular cloning, expression, and gene localization of a fourth melanocortin receptor. *J. Biol. Chem.* **268,** 15,174–15,179.
34. Chagnon, Y. C., Persusse, L., Chagnon, M., Bricault, A.–M., Nadeau, A., Chen, W.–J., Wilkison, W. O., and Bouchard, C. (1997) Linkage and association studies between the melanocortin receptors 3, 4, and 5 genes and obesity–related phenotypes in the Quebec Family Study. *Hum. Mol. Genet.* **3,** 663–673.
35. Huszar, D., Lynch, C. A., Fairchild–Huntress, V., Dunmore, J. H., Fang, Q., Berkemeier, L. R., Gu, W., Kesterson, R. A., Boston, B. A., Cone, R. D., Smith, F. J., Campfield, L. A., Burn, P., and Lee, F. (1997) Targeted disruption of the melanocortin–4 receptor results in obesity in mice. *Cell* **88,** 131–141.
36. Fan, W., Boston, B. A., Kesterson, R. A., Hruby, V. J., and Cone, R. D. (1997) Role of melanocortinergic neurons in feeding and the *agouti* obesity syndrome. *Nature* **385,** 165–168.
37. Mynatt, R. L., Miltenberger, R. J., Kleibig, M. L., Zemel, M. B., Wilkinson, J. E., Wilkison, W. O., and Woychik, R. P. (1997) Combined effects of insulin treatment and adipose tissue–specific agouti expression on the development of obesity. *Proc. Natl. Acad. Sci. U. S. A.* **94,** 919–922.
38. Kim, J. H., Mynatt, R. L., Moore, J. W., Woychik, R. P., Moustaid, N., and Zemel, M. B. (1996) The effects of calcium channel blockade on *agouti*–induced obesity. *FASEB J.* **10,** 1646–1652.
39. Jones, B. H., Kim, J.–H., Zemel, M. B., Woychik, R. P., Michaud, E. J., Wilkison, W. O., and Moustaid, N. (1996)Upregulation of adipocyte metabolism by agouti protein:possible paracrine actions in obesity of the yellow mouse. *Am. J. Physiol.* **270,** E190–E192.
40. Shutter, J. R., Graham, M., Kinsey, A. C., Scully, S., Luthy, R., and Stark, K. L. (1997) Hypothalamic expression of ART, a novel gene related to agouti, is up–regulated in *obese* and *diabetic* mutant mice. *Genes Dev.* **11,** 593–602.
41. Blanchard, S. G., Harris, C. O., Ittoop, O. R. R., Nichols, J. S., Parks, D. J., Truesdale, A. T., and Wilkison, W. O. (1995) Agouti antagonism of melanocortin binding and action in the $B_{16}F_{10}$ murine melanoma cell line. *Biochemistry* **34,** 10,406–10,411.
42. Davis, J. H., Bradley, E. K., Miljanich, G. P., Nadasdi, L., Ramachandran, J., and Basus, V. J. (1993) Solution structure of ω–conotoxin GVIA using 2–D NMR spectroscopy and relaxation matrix analysis. *Biochemistry* **32,** 7396–7405.
43. Kiefer, L. L., Veal, J. M., Mountjoy, K. G., and Wilkison, W. O. (1998) Melanocortin receptor binding determinants in the agouti protein. *Biochemistry* **37(4),** 991–997.

Chapter 17

Melanocortins and Melanoma

Alex N. Eberle, Sylvie Froidevaux, and Walter Siegrist

1. Introduction

Cutaneous melanoma is the cancer with the steepest increase in incidence in the Caucasian population *(1)* and is currently the most common cancer among young adults *(2)*. Mortality rates are increasing correspondingly, and the disease still leads to death in one of every four to five patients. Ultraviolet light exposure has been identified as the main exogenous risk factor. A highly pigmented skin type protects from the deleterious effects of ultraviolet irradiation and is associated, consequently, with a lower risk. As the melanocortins are well-known stimulators of melanogenesis not only in melanocytes but also in melanoma cells, the question arises as to whether these peptides have a protective function or represent an additional risk factor for melanoma development. Experimental investigations in vivo were initiated by Lee et al. *(3)* who were the first to demonstrate that daily injections of α-melanocyte-stimulating hormone (α-MSH) into B16 tumor-bearing mice not only induced a marked increase in tyrosinase activity and melanogenesis of the tumors but also had a tendency to retard proliferation of the tumors. This growth retardation was shown to be negatively correlated with the metastatic potential of the cells *(4)*: B16-F1 cells (→low metastatic potential) were more affected by α-MSH than B16-F5 cells (→intermediate metastatic potential). The growth of B16-F10 cells (→high metastatic potential) was not affected by α-MSH, although the number of MSH receptors did not differ significantly between F1 and F10 cells *(5)*. This indicates that the response of melanoma cells to melanocortin peptides is complex and not simply a question of MSH receptor numbers expressed on the cell surface. This chapter reviews the literature of the past ten years by addressing the following topics: the functional effects of MSH peptides on melanogenesis and intracellular signaling

The Melanocortin Receptors Ed.: R. D. Cone

in melanoma cells as well as on cell proliferation and metastasis; the regulation of MSH (melanocortin 1 [MC1]) receptor expression on melanoma cells; the role of ectopically produced proopiomelanocortin (POMC) peptides, and finally the potential application of MSH peptides to tumor targeting and therapy. The literature published before 1988 was extensively covered in an earlier review *(5)*.

2. Effects of Melanocortins on Melanoma Cell Differentiation

2.1. Regulation of Melanogenesis by Melanocortins

Pigmentation (melanogenesis) represents an important differentiation factor for melanocytes and melanoma cells. Among the many factors regulating melanogenesis in pigment cells, the melanocortins and their natural antagonists play a pivotal role. The marked agonist properties of MSH peptides in inducing melanogenesis in rodent and human melanocytes have been well documented both in vitro and in vivo *(6)* and confirmed for various melanoma cells *(5)*. MSH-antagonist properties, that is, inhibition of hormone-induced melanogenesis, were reported for the naturally occurring agouti protein and melatonin. Whereas agouti protein has been clearly proven to be an MSH antagonist *(7)* or inverse agonist (*see* Subheading 2.3.), the role of melatonin as an inhibitor of MSH-induced melanin production is less well defined: earlier observations showed only slight or negligible effects of melatonin on melanogenesis in B16 mouse melanoma cells (cf. ref. *5*) but a more recent report by Valverde et al. *(8)* noted that 10^{-4} M melatonin completely blocked the melanogenic effect of 10^{-6} M α-MSH in B16 cells by inducing MC1 receptor downregulation and inhibition of de novo synthesis of tyrosinase. The latter was also seen at considerably lower melatonin concentrations. It is likely that different subclones of B16 cells differ in their responsiveness to melatonin, particularly when applied at pharmacologic concentrations.

The physiologic role and pharmacologic effects of MSH peptides on melanoma cells were investigated extensively with different subclones of B16 and Cloudman S91 mouse melanoma cells (cf. ref. *5*). From these studies, it was concluded that

1. Melanoma cells express only one type of functional MSH receptor (→MC1 receptor).
2. MC1 receptors are coupled to the G_s/adenylate cyclase-cyclic adenosine monophosphate/protein kinase A (cAMP/PKA) signaling pathway.
3. Hormone-induced melanogenesis is dependent on de novo synthesis of tyrosinase, the key enzyme for melanin synthesis exhibiting both tyrosine hydroxylase and dopa oxidase activity.

4. Melanotic melanoma cells synthesize melanin also in the absence of hormone, but usually at a much lower rate.
5. Amelanotic melanoma cells, such as B16-G4F cells *(9)* or AM-7AS Cloudman S91 cells *(10)*, have much lower tyrosinase activity and mRNA than normal (melanotic) cells.

In a detailed study on synergism and partial antagonism of modulators of the PKA signaling pathway, Siegrist et al. *(11)* demonstrated that for example cholera toxin and forskolin mimic MSH-induced melanogenesis with 10-fold lower and, respectively, 3000-fold higher EC_{50} than α-MSH (Table 1) but with similar kinetics. These findings were paralleled by the determinations of cAMP formed in the adenylate cyclase assay. On the other hand, pertussis toxin led to a much slower and less prominent increase in intracellular cAMP *(C. Bagutti, unpublished observations)*, inducing a delayed, smaller and more variable melanin response when tested under the same conditions (Table 1). Both toxins synergistically potentiated the MSH effect, but high concentrations of forskolin, which induces very high levels of cAMP, partly antagonized the melanogenic effect of α-MSH *(11)*.

Exposure of cultured Cloudman S91 cells to *ultraviolet B (UVB) irradiation* was found to stimulate the production of mRNAs for both MC1 receptor and POMC as well as the biosynthesis and release of MSH and ACTH peptides *(12)*. A similar production and release of melanocortin peptides was also reported for transformed PAM 212 mouse keratinocytes *(12)*. From these results it was concluded that the effects of UVB light on cutaneous melanogenesis are mediated through a series of coordinated events in which MC1 receptors and POMC-derived peptides play a central role.

Fuller and Meyskens *(13)* were the first to show an effect of MSH on cultured human melanocytes and melanoma cells by demonstrating activation of tyrosinase and eumelanogenesis. Similar results were reported later *(14,15)* but it was also shown that the effect of MSH on many melanoma cells is much weaker or even absent as compared to its effect in vivo *(16)* and probably depends on the culture conditions and origin of the cells. Frequently human melanoma cells do not respond to MSH by melanin production in vitro *(17)* although when injected into experimental animals many of these nonpigmented cell lines form pigmented tumors. Similarly, Syrian hamster W1-1-1 melanoma cells were not responsive to MSH with respect to tyrosinase activation and melanogenesis when kept in culture *(18)*. On the other hand, several human melanoma cell lines have been established which respond very well to α-MSH and cAMP elevating agents by increasing tyrosinase activity and melanogenesis,for example, the human HBL cell line *(19)*.

Detailed *structure–activity studies* with regard to peptide-induced tyrosinase activation or melanin formation are presented elsewhere in this

Table 1
Effects of α-MSH, Agouti Protein and Signaling Modulators on B16 Cell Growth, Melanin Production, and MC1-R Binding

	Growth Inhibition		Melanin Production		MC1-R Downregulation	
	Effect[a] (%)	EC_{50}[b] (n*M*)	Effect[c] (%)	EC_{50}[b] (n*M*)	Effect[d] (%)	EC_{50}[b] (n*M*)
α-MSH	23[e] (±5)	0.65 (±0.24)	100[e]	0.22 (±0.17)	80[e] (±3)	0.28 (±0.09)
Agouti protein	43[e] (±6)	13 (±2.5)	–393[c] (±11)	12 (±3.6)	70[e] (±5.5)	—
Forskolin	67[e] (±6)	10700 (±3400)	112[g] (±7)	650 (±410)	28[e] (±12)	63 (±34)
Cholera toxin	19[e] (±6)	0.044 (±0.065)	107[h] (±10)	0.015 (±0.018)	85[e] (±1)	0.018 (±0.001)
Pertussis toxin	22[e] (±9)	0.0020 (±0.0018)	28[i] (±24)	0.0033 (±0.0019)	0	—
TPA	34[f] (±5)	0.76 (±0.24)	0	—	39[e] (±6)	13 (±5)

Melanin production and cell growth were assessed after incubation for 72 h at 37°C. Modulation of MC1-R levels was measured after 24 h of incubation at 37°C. Data are expressed as the mean ± SD of three to nine separate experiments, each performed in triplicates or quadruplicates. MC1-R binding constants were 2.3 ± 0.2 nmol/L for α-MSH and 3.7 nmol/L for agouti protein using [^{125}I]-[Nle^4, D-Phe^7]-α-MSH as radioligand. Data are from ref. *11* and from unpublished results.

[a]Maximal effect on cell growth; percent reduction of absorbance (A_{650}) as compared to controls.

[b]Concentration inducing a half-maximal response.

[c]Maximal melanin production; 100% refers to the increase in absorbance induced by supramaximal concentrations of α-MSH (typically at 1.0 A_{405}); 0% corresponds to the absorbance of control cultures (typically at 0.3 A_{405}). For agouti protein, % inhibition of melanin production was determined after 7 days of culture by comparing the A_{405}-value with that of the ratio between α-MSH-stimulated and constitutive melanogenesis.

[d]Maximal effect on MC1-R downregulation; 0% corresponds to the MC1-R level in control cultures.

[e]$p < 0.001$ *vs* control.

[f]$p < 0.01$ *vs* control.

[g]$p < 0.001$ *vs* control; p < 0.01 *vs* α-MSH.

[h]$p < 0.001$ *vs* control; not significant *vs* α-MSH.

[i]$p < 0.01$ *vs* control; p < 0.001 *vs* α-MSH.

volume. Briefly, the most potent melanogenic peptide for mouse melanoma cells were those analogs of α-MSH containing a D-phenylalanine residue at position 7, for example, [Nle^4, D-Phe^7]-α-MSH, whereas in human melanoma cells [β-Ala^1]-ACTH[1–17]-N-$(CH_2)_4$-NH_2 was more potent *(20,21)*. In B16 cells, the melanogenic activity of melanocortin peptides usually paralleled their potency with respect to induction of metastases *(22)*. Of the naturally occurring melanocortins and some important fragments, the potency order was α-MSH > β-MSH > ACTH[1–24] > desacetyl-α-MSH > ACTH[1–39]> ACTH[4–10] *(21,22)*. Although similar data were found with a B16-F10 subclone *(23)*, the relative potency may, however, vary from species to species and cell line to cell line indicating slight differences in the recogntion/stability of the peptides in the different systems *(Eberle, unpublished results)*. γ_1-MSH and γ_2-MSH did not alter tyrosinase activity in hamster and mouse melanoma cell lines but γ_3-MSH at $10^{-5}M$ induced tyrosinase activity *(24)*. While the latter inhibited β-MSH-induced tyrosinase, γ_2-MSH potentiated β-MSH *(24)*. Similar potentiation and inhibition of α-MSH had also been observed for high concentrations of α-MSH fragments *(5)*.

2.2. *Different Signaling Pathways Controling Melanogenesis*

Although the G_s/adenylate cyclase-cAMP/PKA pathway is thought to be the main signaling route to activation of tyrosinase and melanogenesis, it is now known that other signaling molecules also play an important role in the different steps between MC1 receptor activation and melanin production. For example, early responses to MC1-R activation in Cloudman S91 and B16-F1 mouse melanoma cells includes the phosphorylation of a 34 kDa membrane protein which was found to peak after 10 min of hormonal stimulation *(25)*, thus slightly preceding maximal cAMP formation found after 20 min *(26)*. This is followed by the translocation of soluble PKC activity from the cytoplasm to the membrane fraction with its maximum after 60 min of hormonal stimulation *(27)* and de novo mRNA and protein synthesis in the following few hours *(28)*. Prolonged incubation of Cloudman S91 cells with α-MSH induced a large but transient increase in tyrosinase mRNA abundance as well as enzyme activity with a maximum at 60 h after MSH stimulation *(29)*. The effect of MSH on the initial gene transcription was independent of ongoing protein synthesis. In parallel, an increase in the level of the beta isoform of protein kinase C (PKC) was observed *(30)*. When Cloudman S91 cells were treated with phorbol dibutyrate, 95% of the PKC activity was lost within 48 h and the α-MSH-induced melanogenesis was completely blocked as was the induction of tyrosinase mRNA and protein *(30)*. This confirms an earlier study with B16 melanoma cells where the phorbol ester TPA (12-*O*-tetradecanoyl phorbol acetate) was found to lower basal tyrosinase activity and to partly inhibit the increase in tyrosinase activity (i.e., tyrosinase mRNA levels) in cells treated with either MSH peptides, dibutyryl-cAMP or IBMX *(30)*. A similar inhibitory effect of nanomolar concentrations of TPA was also found in the melanin assay but TPA had no effect on constitutive melanin production in B16-F1 cells *(11)*. The selective PKC inhibitor CGP 41251, a derivative of staurosporine, potentiated α-MSH-induced melanogensis in B16-F1 cells as demonstrated by an 8-fold reduction of EC_{50} for α-MSH *(32)*. In summary, induction of murine melanogenesis by α-MSH involves up-regulation of tyrosinase mRNA and protein mediated in part by the PKC-dependent pathway, associated with an up-regulation of the beta isoform of PKC, previously demonstrated to specifically activate tyrosinase in human melanocytes. It has also been reported that α-MSH and the [Nle^4, D-Phe^7]-α-MSH analog promote a larger induction of tyrosine hydroxylase activity than of dopa oxidase activity *(33,34)*, demonstrating the existence of two isoforms of the tyrosinase enzyme which are regulated differently by melanocortins. On the other hand, dopachrome tautomerase activity was decreased by α-MSH and cAMP-elevating agents in cultures of B16-F10 cells *(35)*, but there was no correlation between tyrosinase activation and tautomerase inhibition.

The involvement of a second signaling pathway regulating tyrosinase activity was also postulated by a study of the relationship between the metastatic potential of B16 cell lines and their melanin production *(36)*. Although B16-F1 cells (→lowest metastatic potential), F10 and F10C1 cells (→highest metastatic potential) produced equally pigmented tumors in vivo, the cells differed in their melanogenic response to cAMP-elevating agents in vitro. The least metastatic cells (F1) produced the least agonist-induced cAMP level but this was sufficient to induce the greatest tyrosinase activation and melanin production of the cell lines tested. Conversely, the more metastatic cells (F10C1) produced higher levels of cAMP but a lower tyrosinase activation and melanin production in response to MSH *(36)*. It was concluded that agonist-stimulated cAMP production is not the only mediator for melanogenesis in highly metastatic B16 melanoma cells.

A third important cosignaling molecule in melanocortin-induced effects is *calcium*, which has been shown to be indispensable for the action of α-MSH in both melanophores and melanoma cells *(37,38)*. MSH-receptor binding is dependent on extracellular Ca^{2+} and postreceptor activation of intracellular signaling pathways also requires Ca^{2+} *(39,40)*. Calmodulin (CaM) appears to play a role in MSH receptor function since synthetic CaM-binding peptides (representing the CaM-binding domain for CaM-dependent enzymes) inhibited MC1-R in B16-M2R cells *(41)*. These earlier findings were confirmed by a more recent study on the Ca^{2+} requirement for tyrosinase activation and melanin formation in B16 cells following stimulation with cAMP-elevating agents *(42)*. A minimum of 0.4–0.6 m*M* Ca^{2+} in the extracellular medium was required for a maximum tyrosinase response, whereas iontophoretic application of Ca^{2+} into the cells inhibited tyrosinase activity *(42)*. The Ca^{2+}-lowering agent TMB8 stimulated tyrosinase activity and significantly increased the sensitivity and maximum melanogenic response to α-MSH as well as the secretion of melanin into the medium. Similarly, the Ca^{2+} channel blocker verapamil markedly enhanced melanogenesis but did not alter the metastatic potential of the cells *(43)*. It seems therefore that calcium is required for several steps in melanogenesis.

Other intracellular responses to MC1-R stimulation include the transient induction of c-fos mRNA in Cloudman S-91 cells *(44)* and the activation of the mitogen-activated protein (MAP) kinase, p44mapk, by cAMP-dependent activation of MAP kinase kinase in B16 cells *(45)*. In these cells, cAMP-elevating agents induced a translocation of p44mapk to the nucleus and an activation of the transcription factor AP-1 which, in turn, may stimulate tyrosinase expression through interaction with specific DNA sequences present in the mouse tyrosinase promoter *(45)*.

trans-Retinoic acid was reported to inhibit MSH-stimulated melanogenesis in both Cloudman S91 mouse melanoma and Bomirski hamster melanoma cells by blocking the induction of tyrosinase and dopachrome tautomerase activity *(46)*, whereas hexamethylene bisacetamide, sodium butyrate, and dimethylsulfoxide only inhibited MSH-induced tyrosinase activity *(47)*. Retinoic acid and hexamethylene bisacetamide appeared to arrest melanosomal maturation. The radical scavanger pyrroloquinoline quinone (PQQ, a bacterial redox coenzyme) inhibited the expression of tyrosinase mRNA at a postreceptor level *(48)*. All these agents represent useful tools for the study of the different steps of melanogenesis and they all reduce the endogenous antioxidant activity in melanoma cells since the α-MSH-induced increase of tyrosinase activity in melanoma cells is regarded to lead to increased utilization of the superoxide O_2^{-1} ion and hence to provide melanoma cells and melanocytes with a unique endogenous anti-oxidant mechanism *(49)*.

2.3. *Regulation of Melanogenesis by Agouti Protein*

It is has been shown that in mammalian melanocytes of skin or hair, the ratio of eumelanin and phaeomelanin synthesis is regulated by α-MSH and agouti protein (AP): whereas MSH preferentially increases the synthesis of eumelanin by activating MC1-R, the expression of agouti protein correlates with the formation of pheomelanins (see chapter 16, this volume). Recombinant mouse agouti protein *(7,50)* and the human homolog, agouti signaling peptide (ASP) *(51)*, were shown to inhibit both MSH binding to MC1-R and MSH-induced cAMP formation and hence were thought to stimulate phaeomelanogenesis by blocking the activation of MC1-R. However, agouti protein reduced both eumelanin and phaeomelanin production in B16-F1 mouse melanoma cells

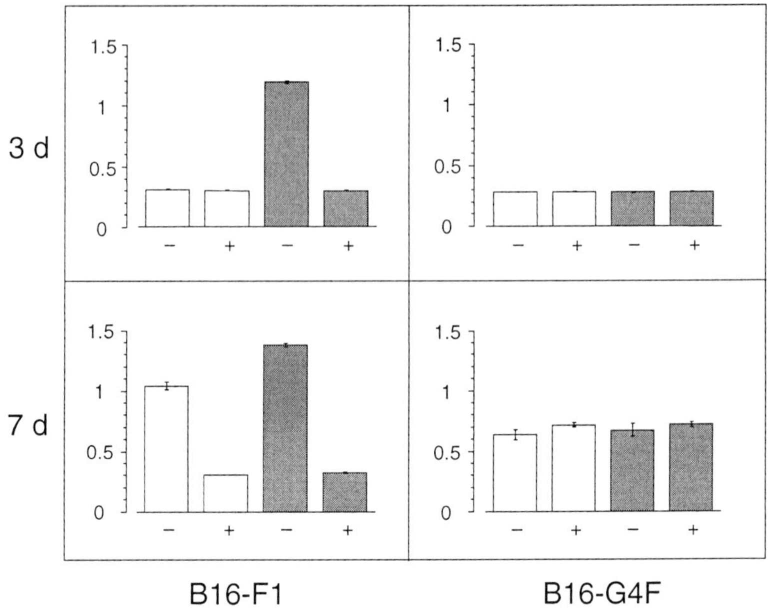

Fig. 1. Effects of agouti and α-MSH on melanin production by MC1 receptor positive (F1, left panels) and negative (G4F, right panels) B16 cells. Ordinates represent absorbance at 405 nm. Cultures were assessed in the absence (open bars) or presence (filled bars) of 1n*M* α-MSH, and in the absence (-) or presence (+) of 100n*M* agouti. Cells were incubated for 3 days (upper panels) and 7 days (lower panels). Data of a representative experiment are shown as means ± SD of triplicate values. (From ref. *53*, with permission.)

whether α-MSH was present or absent *(52)*. Such dose-dependent inhibition or arrest of constitutive (basal) melanogenesis in B16-F1 melanoma cells by agouti protein as well as the competitive inhibition of MSH-induced melanin production had already been reported earlier *(53,54)* and been shown to depend on the expression of MC1-R by the cells *(53)*. In B16-G4F cells that lack MC1-R, there was no effect on melanogenesis by agouti protein (Fig. 1) *(53)*. On the other hand, agouti protein unexpectedly inhibited forskolin-, cholera and pertussis toxin-induced melanogenesis in B16-F1 cells *(11)*. In particular, the inhibition of pertussis toxin was very effective. The dose-dependent reduction of constitutive melanogenesis of B16-F1 cells was paralleled by inhibition of adenylate cyclase and, accordingly, the dose-dependent inhibition of hormone-stimulated melanogenesis by agouti protein was explained by reduced cAMP production *(Eberle, unpublished results)*. Taken together, these results indicate that agouti protein is in fact an inverse agonist for

MC1-R which not only blocks MSH-receptor binding but also affects the postreceptor signaling pathway.

There are other mechanisms independent of agouti protein to block melanogenesis in melanoma cells. For example, transfection of the class I major histocompatibility complex (MHC) genes H-2Kb or H-2Kd into BL6 mouse melanoma cells, a subclone of B16-F10 cells lacking expression of class I H-2 genes, resulted in the loss of melanin production by complete downregulation of the entire melanogenic pathway, including inhibition of tyrosinase and MC1 receptor gene expression, cAMP responses and melanosomal biogenesis *(55)*. Other genes, such as H-2Dd, H-2Ld, or H-2IAk did not alter the pigmented phenotype.

2.4. Effect of Melanocortins on Dendrite Formation

Another important differentiation factor of melanocytes and melanoma cells is dendrite formation and extension, which comprise a characteristic morphology and functional activity of (normal) melanocytes in the skin, such as the ability to transfer melanosomes into neighboring keratinocytes. In vitro, the morphology of melanocytes and melanoma cells usually differs from that observed in vivo. MSH peptides *(5,56)*, dibutyryl-cAMP or IBMX *(57,58)*, and the PKC inhibitor CGP 41251 *(32)* induce morphologic changes of melanoma cells such as an increase dendrite formation and swelling of the cells. Hormonal stimulation of B16-F1 cells also leads to aggregation of the cells, concomitant with melanin formation and release of melanin *(5)*. Stimulation by dibutyryl-cAMP, IBMX or forskolin of B16-G4F cells, which lack MC1-R, did not induce cell aggregation *(Eberle, unpublished results)*. The effect of α-MSH on increased dendricity suggested a potential role for this peptide in melanocyte–matrix interactions and in pigment transfer through reorganization of the actin stress fiber cytoskeleton *(59)*. In B16-F10 melanoma cells, α-MSH also led to a significant increase in myosin-V expression, a protein thought to act as motor for melanosome translocation *(60)*. Different human melanoma cell lines incubated with 100n*M* [Nle^4, D-Phe^7]-α-MSH for a prolonged period underwent morphologic differentiation, that is, swelling of the cells and increased dendrite formation *(61)*.

3. Effects of Melanocortins on Melanoma Proliferation and Metastasis

3.1. Regulation of Mouse Melanoma Cell Proliferation

Both stimulatory and inhibitory effects of MSH on the growth of cultured rodent melanoma cells and of melanoma tumors in experimental animals have been reported *(5)* and there was an inverse correlation between differentiation (pigmentation) and proliferation in vitro amongst melanoma cell lines with various degrees of metastatic potential (B16-F10 and sublines from JB/MS melanoma) *(62)*. Some authors observed stimulated anchorage-independent growth of Cloudman S91 melanoma cells *(63,64)*, whereas others described growth inhibition by MSH in Cloudman S91 cells *(65)* and Bomirski hamster melanoma *(66)*. These divergent growth responses may be explained by a variable dominance between a primary growth-promoting effect and a secondary growth-inhibiting effect of MSH due to potentially cytotoxic intermediary products of increased melanin synthesis. The latter is particularly evident in cultured mouse B16-F1 melanoma cells where α-MSH showed an antiproliferative effect *(11,53)*. This growth-inhibiting effect of α-MSH was mimicked by cholera toxin (but not pertussis toxin) and by forskolin but antagonized by TPA *(11)*. TPA alone inhibited B16-F1 cell growth *(11)* and the PKC inhibitor GCP 41251 also antagonized cell proliferation *(32)*. The proliferation of B16-M2R melanoma cells was blocked by agents that stimulate cAMP production but enhanced by TPA *(67)*. Thus, whereas stimulation of PKA led to the same result in different cell lines, activation of PKC did not yield a uniform effect in different clones of B16 cells.

Agouti protein, which is devoid of any melanogenic activity and thought to be an inverse agonist for the MC1-R (*see* Subheading 2.3.) unexpectedly showed an antiproliferative effect on B16-F1 cells similar to that of α-MSH with a half-maximal effective concentration of 13n*M* and a maximal 43% growth inhibition at 100n*M (11,53)*. It seems therefore that, although MC1-R is indispensable for mediating both MSH- and agouti-regulated cell proliferation and melanogenesis, there is a functional branching of the signaling cascade, after the stimulation/inhibition of MC1-R, which is responsible for the differential regulation of the two effects.

The *prostaglandins* PGE_1 and PGE_2, which increase tyrosinase activity in Cloudman S91 and B16-F1 cells, were found to inhibit cell proliferation by blocking the progression of the cells from G2 phase of the cell cycle into M or G1 (68). 2',5'-dideoxyadenosine (DDA), an inhibitor of adenylate cyclase, which enhanced the melanogenic response of Cloudman S91 cells to PGE_1, PGE_2, α-MSH, IBMX or dibutyryl-cAMP, also augmented the effect of PGE_1 and PGE_2 on the cell cycle. Whereas DDA and α-MSH had no effect on the cell cycle, in combination they recruited more cells in the G2 phase than untreated controls *(68)*. D-α-Tocopheryl succinate induced growth inhibition of B16 cells and reduced basal and MSH-stimulated adenylate cyclase *(69)*. Retinoic acid also inhibited growth of B16 cells similar to prostaglandin A_2. The latter did not change basal or MSH-stimulated adenylate cyclase activity whereas retinoic acid affected cAMP levels.

3.2. Regulation of Human Melanoma Cell Proliferation

α-MSH is clearly a growth-stimulatory signal for *human melanocytes* in culture owing to its ability to induce cAMP production *(70,71)*. On the other hand, many human *melanoma* cells grow in vitro independently of MSH or any other cAMP-elevating agents *(72,73)*. Whether this also applies to growth of human melanocytes or human melanoma in vivo has not yet been investigated. On the other hand, Jiang et al. *(61)* reported that prolonged incubation in vitro of different human melanoma cell lines, melanotic or amelanotic, in the presence of 100n*M* [Nle^4, D-Phe^7]-α-MSH led to a decreased cell number and that this effect of [Nle^4, D-Phe^7]-α-MSH was independent of its melanogenic action which was limited to just some of these cell lines. However, in an intracutaneous murine model of melanoma cell tumor growth in vivo, [Nle^4, D-Phe^7]-α-MSH did not decrease the growth of the primary tumor *(74)*.

The PKC inhibitor CGP 41251 reduced the proliferation of human D10 and HBL melanoma cells as well as mouse B16-F1 cells *(32)*; no synergistic or antagonistic effect with MSH peptides was noted. A similar inhibition of cell proliferation was also observed for some retinoic acid analogs *(75)*: RARγ-selective retinoids exerted the most prominent growth effects, with up to 68% and 69% inhibition of human D10 and Cloudman S91 mouse melanoma cells, respectively. An RXR-selective compound had a much weaker effect. Growth inhibition by RARα- and RARβ-selective compounds was even below 10% in both cell types. A different selectivity profile of retinoids was found for receptor regulation (*see* Subheading 4.2.).

Whereas normal human melanocytes maintained in chemically defined media in vitro require IGF-I (or insulin), bFGF, TPA, and α-MSH for growth, nevus cells were shown to grow in the absence of bFGF and primary human melanoma cells only required one growth factor such as IGF-I (or insulin) for continuous proliferation *(76)*. On the other hand, metastatic human melanoma cells were able to proliferate, after a short adaptation period, in medium depleted of any growth factor and other proteins. Doubling times were somewhat longer (30–60%) as compared to those maintained in fetal calf serum (FCS)-containing medium *(76)*. This growth autonomy of human melanoma cells is apparently due to endogenous production of growth factors, for example, transforming growth factor-alpha (TGF-α), and the expression of the corresponding receptor, for example, epidermal growth factor (EGF)/TGF-α receptor *(76)*.

3.3. Role of Melanocortins and MC1 Receptors in Metastasis

It is believed that MSH possesses the capacity to regulate not only melanogenesis but also other factors critical to the metastatic growth of melanoma cells. There is one study in which B16-F10 melanoma cells and sublines generated from the JB/MS melanoma were pretreated with α-MSH in vitro and then injected into experimental animals *(62)*: no effect was found for the growth of these cells as subcutaneous primary tumor but the pretreatment decreased the number of pulmonary metastases in most of these cell lines. On the other hand, no consistent correlation between hormonal responsiveness and metastatic capacity was found with mouse K1735 cells *(77)*.

Other authors found a positive correlation between MSH-induced cAMP accumulation and the formation of pulmonary metastases after intravenous injection of different clones of B16 mouse melanoma *(78)*. A more differentiated phenotype induced by MSH-treatment of B16-F1 and B16-F10 cells was associated with a higher rate of experimental pulmonary metastasis *(79)*. Identical observations were made in our own laboratory *(Froidevaux, unpublished results)*. The stimulation of the experimental metastatic potential by α-MSH in several sublines of B16 melanoma could be prevented by prolonged exposure of the cells to TPA, suggesting an involvement of PKC in MSH action *(80)*. A structure-activity study of melanocortin peptides to which B16-F1 cells were exposed for 48 h preceding injection into mice showed that the potency order of the different peptides paralleled their melanogenic activity: [Nle^4, D-Phe^7]-α-MSH > α-MSH > β-MSH > ACTH[1–24] > desacetyl-α-MSH > ACTH[1–39] *(22)*.

In B16-F1 cells, α-MSH up-regulated and Ca^{2+} downregulated the expression of MTS1, a metastasis associated gene that codes for a Ca^{2+} binding protein of the S-100 family and that is related to cell proliferation, cancer metastasis and invasion *(41,81)*. Upregulation of 18A2/MTS1 led to changes in cytoskeletal dynamics of B16-F1 cells, as demonstrated by the patchy focal redistribution of CD44v6, an isoform of the transmembrane cell adhesion-mediating protein CD44 *(81)*. It is possible that through this induction of patching of CD44, α-MSH could provide discrete and strong adhesive foci promoting cell adhesion and invasive behavior.

MC1 receptor variants with known mutations in the second and seventh transmembrane domain were found to be more common in melanoma patients than in normal controls *(82)*. For example, the Asp84Glu variant was only present in melanoma cases and appears to be of particular significance. Therefore, variants of the MC1 receptor gene may be causally associated with the development of melanoma *(82)*. However, the molecular mechanism for a possible association of melanoma with MC1 receptor mutants is not yet known.

While B16-BL6 mouse melanoma cells constitutively produce melanin and express high levels of MC1 receptor mRNA regardless of the site of growth, metastatic K-1735 mouse melanoma cells, which are amelanotic in culture, did not form pigmented tumors in the subcutis of syngeneic mice but produced melanotic brain metastases when injected into the internal carotid artery *(83)*.

When transplanted back into the subcutis, isolated K-1735 cells from the brain tumors became amelanotic and unresponsive to α-MSH. Thus, the phenotype of these metastatic cells directly correlated with the level of MC1 expression which appears to be influenced by the specific organ environment *(83)*.

From the data presented above, it is concluded that mechanism of how MSH peptides and MC1 receptors are involved in the process of metastasis is not yet solved because some of the reports from different laboratories are conflicting. One of the reasons is the difficulty of studying the development of melanoma tumors which is a very slow process and dependent on many different factors. Nevertheless, there is no doubt that melanocortins and MC1-R are involved in the metastatic process, most likely also in human melanoma.

4. Expression and Regulation of MC1 Receptors in Melanoma

4.1. MC1 Receptor Expression on Melanoma Cells

The first determinations of MSH receptor expression on *mouse melanoma* plasma membranes were done by Siegrist et al. *(84)* who reported a single class of MSH binding sites on B16-F1 (B_{max} approx. 10,000 sites/cell; K_d approx. 1–2 nmol/L). Similar results were found for Cloudman S91 cells (Fig. 2). Biochemical analysis of these receptors by photocrosslinking revealed a band of approx. 45 kDa (85,86). Some mouse melanoma cells, for example, B16-M2R, appeared to have a receptor with a slightly different molecular weight *(87)*. On the other hand, lectin-resistant B16-Wa4 cells (with low content of sialic acid residues) recognized α-MSH with no difference as compared to B16-F1 cells *(88)* but the apparent size of the MC1 receptor was about 3 kDa smaller in W4 cells, as determined by photocrosslinking studies *(86)*. Stimulation of adenylate cyclase activity by α-MSH was the same in Wa4 and F1 cells whereas VIP-or PEG1-induced stimulation was reduced in Wa4 cells *(88)*.

The presence of high-affinity MSH receptors on *human melanoma* cell lines has been confirmed by binding studies with [^{125}I]-labeled α-MSH ligands *(17,19,89)*. Scatchard analysis revealed that most human melanoma cell lines contain between a few hundred and a few thousand receptors per cell, with dissociation constants in the nanomolar or subnanomolar range (Fig. 2) *(17)*. Similarly, 300–800 receptors were reported in nomal human melanoctyes *(70,90)*. Different degrees of MSH receptor expression were found also on surgical melanoma specimens from different patients investigated by autoradiography *(91)*. Biochemical analysis of human MC1-R on human melanoma cells by photocrosslinking revealed size of about 45 kDa which corresponded with that of mouse melanoma cells *(86)* but in some cell lines, such as HBL cells, a higher molecular weight was determined *(Eberle, unpublished results)*.

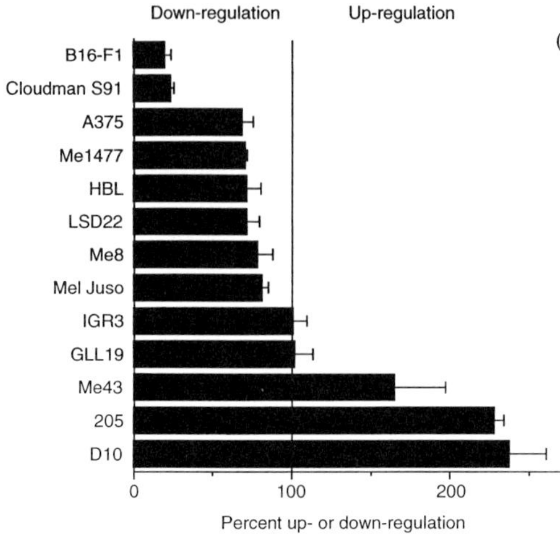

	MC1 receptors (sites per cell ± SD)	K_D (pmoles/l ± SD)
B16-F1	9570 ± 1750	1310 ± 310
Cloudman S91	4660 ± 120	820 ± 120
A375	608 ± 134	410 ± 123
Me1477	574 ± 304	420 ± 319
HBL	6616 ± 1012	210 ± 33
LSD22	1518 ± 729	1040 ± 614
Me8	1281 ± 500	2870 ± 1091
Mel Juso	391 ± 160	770 ± 370
IGR3	704 ± 232	850 ± 349
GLL19	467 ± 229	460 ± 331
Me43	478 ± 306	280 ± 216
205	1053 ± 147	220 ± 614
D10	1590 ± 510	195 ± 42

Fig. 2. Homologous regulation of MC1 receptors in mouse (B16-F1, Cloudman S91) and human melanoma cell lines (all others) and receptor numbers and dissociation constants for α-MSH. For receptor regulation studies, the cells were incubated with 30n*M* α-MSH for 2 days, detached and washed with ice-cold acidic buffer and receptor binding determined with [^{125}I]-[Nle^4, D-Phe^7]-α-MSH. Receptor numbers and binding constants were determined in saturation assays using [^{125}I]-α-MSH. (Adapted from refs. *17* and *21*.)

The postulated *cell cycle-dependence* of MSH receptor expression in Cloudman S91 mouse melanoma cell and of their hormonal responsiveness *(92,93)* could not be confirmed by others *(94)*. More subtle approaches to arrest cells in specific phases of the cell cycle, such as arrest of Cloudman S91 cells in the S2/M cell phase following UVB irradiation *(95)*, may give a clearer answer into a possible cell-cycle dependence of MC1-R expression. It has already been pointed out that simultaneous stimulation of MC1-R and inhibition of adenylate cylcase recruits melanoma cells preferentially in the G2 cell phase (*see* Subheading 3.1.).

The presence of *intracellular binding sites* for MSH in melanoma cells was demonstrated by Orlow et al. *(96)* in Cloudman S91 cells. Similar results were obtained with B16-F1 cells by Froidevaux et al. *(unpublished results)* who performed binding studies with [^{125}I]-[Nle^4, D-Phe^7]-α-MSH as radioligand and different membrane fractions prepared from B16-F1 melanoma tumors grown in experimental animals: MSH binding sites were present on plasma membranes and internal vesicles in similar quantities. However, the affinity for α-MSH was 3-to 6-fold higher in the plasma membrane fraction as compared to internal vesicles. The intracellular fraction of MSH receptors may originate from both internalized membrane receptor and newly synthesized receptor. Internalization of MC1-R was also reported after interaction of melanoma cells with α-MSH analogs such as [^{3}H]-Ac-[Nle^4, D-Phe^7]-α-MSH[4–11]-NH_2 *(97)*, [^{125}I]-α-MSH *(17)*, multivalent fluorescent MSH-macromolecular conjugates *(98)* or [^{125}I]-[Nle^4, D-Phe^7]-α-MSH *(20,99)*. No recycling of receptor could be detected in cells stimulated with [Nle^4, D-Phe^7]-α-MSH *(99)* and it should be noted that internalization of MC1-R was not found in all melanoma cells (*see* Subheading 4.2.). In cells where MC1-R was down-regulated, receptor internalization was rapid at 37°C: 60% internalization after a 2-h exposure of B16-F1 cells to 50n*M* α-MSH and 85–90% internalization after 10–20 h *(20)*.

4.2. MC1 Receptor Up-and Downregulation In Vitro

A detailed analysis of homologous and heterologous regulation of MC1-R in 2 mouse and 11 human melanoma cell lines was performed by Siegrist et al. *(21)*. α-MSH induced upregulation of its own receptors in three human cell lines and downregulation in six human and two mouse melanoma cell lines (Fig. 2). No regulation was observed in two human lines. Scatchard analysis revealed modulation of the number of receptors per cell without any change in affinity. The EC_{50}s for up-and downregulation were 1.6n*M* and 0.23n*M*, respectively. ACTH[1–17] and [Nle^4,D-Phe^7]-α-MSH were more potent, whereas ACTH[1–24], desacetyl-α-MSH, and [Nle^4]-α-MSH were less potent in receptor upregulation as compared to α-MSH. Downregulation, but not upregulation, could be fully mimicked by G_s protein activation and partially by elevation of intracellular cAMP with forskolin. Micromolar concentrations of forskolin, however, completely blocked the downregulation of MC1-R induced by α-MSH, cholera toxin or pertussis toxin *(11)*.

Other authors *(100)* also reported that [Nle^4, D-Phe^7]-α-MSH induced downregulation and rapid internalization of MC1 receptors into the lysosomal compartment of B16 melanoma cells where the ligand was degraded. Downregulation was found to persist as long as 96 h without replacement of the receptors. However, when MC1 receptors were removed by trypsin treatment, they were rapidly replaced *(100)*. Pharmacologic concentrations of melatonin were also reported to reduce the number of MC1 receptors on B16 melanoma cells by approx. 25% *(8)*.

PKC also seems to be involved in MC1 receptor regulation. TPA downregulated MC1-R in B16-F1 melanoma cells by about 40% *(11)* and in B16-M2R cells by about 85% *(67)*, whereas the PKC inhibitor CGP 41251 was found to upregulate MC1 receptors in mouse B16-F1 cells as well as human D10 cells *(32)*. On the other hand, MC1 receptors in human HBL cells were downregulated by CGP 41251 *(32)*. The effect of the PKC inhibitor was synergistic with α-MSH in the human cells but antagonistic in mouse cells.

Retinoic acid generally downregulated MC1 receptors in mouse and human melanoma cells but had no effect on HBL cells *(21,75)* but in one study

MC1-R upregulation was reported *(100)*. Retinoic acid receptor (RAR) subtype specificity was investigated with two cell lines, human D10 and Cloudman S91 cells *(75)*. In D10 cells, MC1-R downregulation was induced most effectively by an RARγ-selective retinoid (84%) but RARα-, RARβ-and RXR-selective agonists were much less potent. The pattern for MC1-R downregulation was completely different in Cloudman S91 cells. The RXR-selective compound was most active (85%), followed by the RARα-, RARγ-, and RARβ-selective agonists. Thus it seems that the different selectivity profiles for growth inhibition (see above) and MC1-R downregulation in Cloudman S91 cells are the result of independent regulatory mechanisms *(75)*. Independent regulation of MC1 receptor gene expression distinct from the regulation of the other melanocyte-specific genes was also postulated from studies with whole-cell hybrids and microcell hybrids between mouse fibroblasts and pigmented Syrian hamster melanoma cells *(101)*.

Heterologous receptor up-regulation was reported for (i) UVB-irradiation (10–20 mJ/cm^2), which led to a 2-to10-fold increase in β-MSH binding to Cloudman S91 cells, explaining the observed increase of melanin production after UVB-irradiation *(102)*; (ii) dialysis of fetal calf serum added to the culture medium of B16-F1 cells which led to increased expression of MC1 receptors *(103)*, and (iii) interferon-α, β, and γ, which upregulated MC1 receptors on murine melanoma cells by a factor of about 2.5 and, in combination with α-MSH, significantly increased melanin production as compared to cells treated with α-MSH alone *(104)*. In one report, interleukin-1 also upregulated MC1-R on Cloudman S91 cells *(105)*. A variant of human A375 melanoma cells, which is sensitive to the cytostatic effect of interleukin-1β (A375r-) and which does not express MC1 receptors, could be converted, by altering the culture conditions, to a cell variant resistant to interleukin-1β (A375r+) but expressing MC1 receptors *(106)*. However, MC1 receptor expression was dismissed as a factor involved in cytokine resistance. In summary, MSH receptors on melanoma cells are both positively and negatively regulated. Whereas PKA activation seems to be involved in receptor downregulation, the mechanism responsible for upregulation remains to be elucidated.

4.3. Role of Agouti in MC1 Receptor Regulation

Agouti protein was found to induce MC1-R downregulation in B16-F1 cells; the characteristics of this downregulation were virtually identical to those observed for the α-MSH-induced MC1-R down-regulation *(53)*. The concentration range at which agouti was effective was the same as that of α-MSH (3n*M*), which means that agouti affects receptor regulation at a 100-fold lower concentration than that required for inducing growth inhibition or for blocking α-MSH-induced melanogenesis. This is additional support for the finding (*see* Subheading 2.3.) that agouti protein is an inverse agonist for MC1-R and not just an antagonist.

4.4. MC1-R Negative Mouse Melanoma Expressing Human MC1-R

The human MC1-R was stably expressed in B16-G4F cells which are deficient of (mouse) MC1-R *(107)*. The K_d for [Nle^4, D-Phe^7]-α-MSH in four selected clones ranged from 0.187 to 0.705 nmol/L, thus corresponding to the K_d observed with the different human melanoma cell lines. Except for one clone, all transfectant cell lines produced melanin constitutively. The presence of α-MSH induced an additional dose-related but small increase in melanin production in these cells, which could be suppressed by the addition of specific α-MSH antibodies without altering the constitutive part of melanin production. Human and mouse agouti protein both reduced α-MSH-induced melanogenesis but did not alter constitutive melanogenesis. These results indicate that the human MC1-R expressed in these clones was constitutively activated and that its state of activation could be further increased by the hormone but not decreased by agouti. Thus, stable expression of the human melanoma MC1-R in a homologous mouse tissue may lead to constitutive activation of melanogenesis and hence provides a useful tool for the study of MC1-R function and coupling to the signal transduction cascade *(107)*.

4.5. In Vivo MC1 Receptor Regulation

Transplantation of melanoma cells into mice followed by injection of α-MSH revealed that the the kinetics and regulation of MC1 receptors in vivo is very similar to that found in vitro: MC1 receptors were downregulated on mouse B16-F1 cells but upregulated on human D10 cells *(Froidevaux and Eberle, unpublished results)*. Single injections of α-MSH produced a less prominent change in the receptor state than that observed in vitro, due to the short half-life of α-MSH.

5. Autocrine Melanocortin Production

5.1. Ectopic Production of POMC Peptides

There is some evidence for peripheral production of melanocortin peptides by skin keratinocytes, which suggests a paracrine mode of action in addition to the endocrine role of these peptides. POMC peptides were frequently detected by immunohistochemical staining in corporal skin affected by diseases, including basal cell carcinoma and melanoma, but not in normal skin, except for hair follicles of scalp skin *(108)*. Also, POMC products were consistently observed in keratinocytes and mononuclear cells at keloid lesions *(108)*. Schauer et al. *(109)* found constitutive expression of α-MSH- and ACTH-immunoreactivity that was upregulated after treatment of keratinocytes with either phorbol ester, UV light, or interleukin-1. POMC transcripts of variable length *(110)* and MSH-immunoreactivity *(111,112)* were found in several rodent and human melanoma cell lines, confirming a much earlier report on POMC-derived ACTH secreted by human melanoma cell lines *(113)*. However, the molecular identity of the ectopic MSH/ACTH-immunoreactivity was never clearly established but thought to be a protein with a higher molecular weight than the known melanocortin peptides. Some authors *(112,115)* reported that the MSH-immunoreactivity was associated with a less differentiated, invasive and metastatic phenotype, whereas others *(111)* found a correlation with a higher degree of pigmentation. Highly dendritic human melanocytes were shown to stain with a monoclonal antibody against human ACTH *(116)* and, when studied in short-term organ culture, positive melanocytes were seen after a pulse of UV light or Adriamycin treatment. In melanomas, isolated groups of melanocytes were also positive for ACTH. This indicates that POMC is processed differently in melanocytes and melanomas as compared to melanotrophic cells of the pars intermedia.

5.2. Occurrence of Melanocortin Peptides in Melanoma Cells and Tumors

Immunoreactive α-MSH was spontaneously released by human HBL melanoma cells which express a high number of MC1 receptors on their cell surface *(115)*. This release was significantly increased in the presence of the ACTH[4–10] fragment or β-MSH and blocked at low temperature. Human melanoma cells with a low number of MC1 receptors, such as IGR3, only released little immunoreactive α-MSH, but this was greatly enhanced after transfection of the MC1 gene into these cells *(115)*.

Immunohistochemical analysis of tumor sections of human cutaneous malignant melanoma of nodular type for occurrence of α-, β-, and γ_3-MSH demonstrated that the staining intensity was stronger the closer the cells were to the center of the tumor parenchyma and the larger or more poorly differentiated they were *(117)*. However, MSH expression was also seen in the peripheral part of the tumor and in perilesional tissues including epidermis, sweat glands, sebaceous glands, and hair follicles. Further studies are required to determine MSH-immunoreactivity also in sections of other types of melanoma tumors. Another study also noted considerable amounts of MSH immunoreactivity in human melanoma tumors ranging from 0.31 to 4.27 pmol/g of wet tissue *(118,119)* but suggested that this form of MSH is more hydrophobic and of higher molecular weight than α-MSH (*see* Subheading 5.1.). It is interesting to note that the plasma levels for α-MSH in melanoma patients was significantly higher (mean of 12.2 pmol/L for 37 patients as compared to 7.9 pmol/L for 38 control persons) even though the standard deviation was relatively high so that α-MSH may not serve as a typical tumor marker *(120)*.

In summary, it is possible that MSH peptides by autocrine and/or paracrine production (from melanoma cells or neighboring keratinocytes) are engaged in the regulation of differentiation (melanogenesis), proliferation, and metastasis of the tumor cells. However, further studies will be necessary to verify whether the production of POMC- (or melanocortin-) immunoreactivity could serve as an indicator for melanoma malignancy.

6. Melanocortin Peptides for Melanoma Tumor Targeting

Besides the understanding of the (patho-)physiologic role of melanocortin peptides in the control of differentiation and proliferation of melanoma cells, these peptides are also being studied as potential diagnostics and therapeutics for the detection and treatment of melanoma metastases. At present, there are still no efficient modalities for the treatment of recurrent melanoma and MSH peptides or mimetics are expected to become useful drugs to target melanoma. Disseminated microdeposits of melanoma cells are difficult to detect and are resistant to conventional cytotoxic therapy. A potent and specific targeting strategy would be of great value for both tumor localization and treatment, in particular by using MSH peptides labeled with diagnostic (e.g., ^{99m}Tc, ^{111}In, $^{67/68}$Ga, ^{64}Cu or ^{86}Y, ^{18}F) or therapeutic radionuclides (e.g., ^{90}Y, ^{67}Cu, ^{188}Re) or by employing peptide–toxin conjugates. For both of these approaches, the modification of the expression of MC1 receptors on melanoma in vitro and in vivo is of particular relevance.

6.1. Quantification of Melanocortin Receptors on Tumor Slices

Targeting studies require the analysis of expression and distribution of melanocortin receptors on melanoma tissue of experimental animals and melanoma patients. To this end, cryosections of solid melanoma tumors (mouse B16-F1, human D10, and HBL) grown on experimental animals were used for visualization of MC1 receptors by autoradiography with [^{125}I]-α-MSH and [^{125}I]-[Nle^4, D-Phe^7]-α-MSH tracers *(121,122)*. The presence of increasing concentrations of unlabeled α-MSH during incubation with tracer led to a dose-dependent displacement of the radioligand. Quantitative analysis of the autoradiograms produced dissociation constants which were comparable with those obtained with cell binding assays: K_d = 1.87 and 1.31 nmol/L for B16-F1 tumors and cells, respectively; 0.32 and 0.33 nmol/L for D10, and

2.24 and 1.36 nmol/L for HBL tumors and cells, respectively, and receptor densities paralleled those found on cultivated cells *(122)*. This indicates similar binding properties of α-MSH radioligands to both cultured melanoma cells and tissue sections of melanoma tumors. Similar binding characteristics were also observed with human melanoma tissue sections originating from biopsies of melanoma patients *(122)*.

Localization studies of hMC1 receptor on WM266-4 human melanoma cells carried out by applying an antipeptide antiserum specific for the cloned human MC1 receptor showed that most of the receptors were located on the plasma membrane but quantification was not possible *(123)*.

6.2. Melanocortin Peptides for Melanoma Diagnosis and Therapy

A bivalent α-MSH complex composed of two α-MSH molecules and the diethylenetriamine pentaacetic acid (DTPA) chelator for labeling with ^{111}In, [^{111}In]DTPA-bis-MSH, was synthesized, tested in vitro, and found to associate specifically with melanoma tissue in Cloudman S91 tumor-bearing mice *(124)*. A first clinical trial, in which this radiopeptide was administered to 15 patients with confirmed or suspected metastatic melanoma, showed that of lesions over 10 mm in diameter, 89% were detectable with whole-body γ-scanning *(125)*. Subsequently, shorter synthetic MSH peptide fragments with the general structure Nle-Asp-His-D-Phe-Arg-Trp-Lys(DIP)-NH_2 were studied in melanoma-bearing mice *(126)*. It was shown that a DTPA-mono-MSH derivative containing two diisopropyl groups (DIP) on the Lys^{11} side chain yielded a much lower non-specific accumulation of ^{111}In in the liver than the DTPA-bis-MSH peptide without DIP. Similar results were later reported with [^{111}In]DTPA-[Nle^4, D-Phe^7]-α-MSH containing one or two MSH molecules per DTPA residue *(127)*.

Other MSH peptides proposed for clinical studies include a derivative of [Nle^4, D-Phe^7]-α-MSH containing an iodobenzoic acid (IBA) residue on the Lys^{11} side chain, [Nle^4, D-Phe^7, Lys^{11}([$^{125/131}$I]IBA)]-α-MSH, which can be iodinated with either ^{125}I or ^{131}I and which shows a faster tissue clearance and a higher affinity than conventionally radioiodinated [Nle^4, D-Phe^7]-α-MSH *(128)*, and finally an α-MSH molecule extended by *N*-acetyl-Cys-Gly-Cys-Gly at its N-terminus for complexing rhenium *(129)*. At present, diagnostic MSH peptides containing novel chelating molecules for ^{111}In or ^{99m}Tc and therapeutic analogs for ^{90}Y or ^{87}Re resembling those reported for the somatostatin analog octreotide *(130)*, are being developed and tested.

6.3. Melanocortin–Toxin Conjugates for Tumor Therapy

Cytotoxic principles have been suggested for melanocortin receptor-targeted therapy. An MSH analog covalently linked to an antibody against the

T-cell receptor CD3 complex was used to direct cytotoxic T cells to melanoma target cells *(131)*. Whereas in vitro melanoma cell lysis was observed, the method has not yet found in vivo application. In another approach, an α-MSH-diphtheria toxin fusion protein was genetically engineered *(132)* but proved not to be resistant enough to proteolytic cleavage in vivo. A second construct containing the cytotoxic fragment A and the fragment B with a deletion between residues 387 and 485 proved to be resistant to proteolytic cleavage and was shown to exhibit cytotoxic activity against human and murine melanoma cells *(133)*. This effect was mediated by interaction with MC1 receptors *(134)*. Other workers coupled the alkylating anticancer drug melphalan to α-MSH fragments which exhibited significant antitumor activity when tested with L1210 leukemia or human amelanotic melanoma xenograft-bearing mice *(135)*. Depending on the site of introduction of the melphalan residue into the MSH fragments, the compounds were more or less specific for melanoma and acted either through an MC1-mediated mechanism or a receptor-independent mechanism *(136)*. Generally they were less cytotoxic to other cells than melphalan alone *(136,137)*. Similar observations were made with MSH fragments containing the difluoromethylornithine (DFMO) moiety *(138)*: although these latter complexes showed cytotoxic activity in vitro, their action did not seem to be mediated by MC1-R. The role of the MSH peptide was more that of an enhancer of the cytotoxic effect of the alkylating groups. It should be noted however that for these studies MSH fragments were chosen that had only limited biostability and relatively low MC1 receptor binding; more potent analogs may provide a much better tool for this approach.

Whereas some of these toxin–MSH conjugates or the chemically reactive MSH analogs may work well in vitro, their application in vivo is much more complex. The constructs must be stable enough to resist enzymatic degradation. Furthermore, they must be able to penetrate into the tumor tissue and should be hydrophilic in order to avoid accumulation in the liver. Novel MC1-specific ligands are required to fulfil these criteria. Attempts in targeted therapy for melanoma may also be based on gene therapy. A promising approach involves the application of melanocyte-specific expression cassettes using promoters for melanocyte-specific proteins *(139)*. This would allow targeted expression of gene constructs comprising, for example, immunity-stimulating cytokines or drug-activating enzymes. Melanocortin peptides may be useful as coregulators of melanoma in such an attempt.

Acknowledgments

This work was supported by the Swiss Cancer League, the Swiss National Science Foundation and the Roche Research Foundation.

References

1. Koh, H. K. (1991) Cutaneous melanoma. *N. Engl. J. Med.* **325,** 171–182.
2. Williams, M. L. and Pennella, R. (1994) Melanoma, melanocytic nevi, and other melanoma risk factors in children. *J. Pediatr.* **124,** 833–845.
3. Lee, T. H., Lee, M. S., and Lu, M. Y. (1972) Effects of α-MSH on melanogenesis, and tyrosinase of B16 melanoma. *Endocrinology* **91,** 1180–1188.
4. Niles, R. M. and Makarski, J. S. (1978) Control of melanogenesis in mouse melanoma cells of varying metastatic potential. *J. Natl. Cancer Inst.* **61,** 523–526.
5. Eberle, A. N. (1988) *The Melanotropins: Chemistry, Physiology, and Mechanisms of Action.* Karger, Basel.
6. Lerner, A. B. (1993) The discovery of the melanotropins. *Ann. N. Y. Acad. Sci.* **680,** 1–12.
7. Lu, D., Willard, D., Patel, I. R., Kadwell, S., Overton, L., Kost, T., Luther, M., Chen, W., Woychik, R. P., Wilkison, W. O., and Cone, R. D. (1994) Agouti protein is an antagonist of the melanocyte-stimulating-hormone receptor. *Nature* **371,** 799–802.
8. Valverde, P., Benedito, E., Solano, F., Oaknin, S., Lozano, J. A., and Garcia-Borron, J. C. (1995) Melatonin antagonizes α-melanocyte-stimulating hormone enhancement of melanogenesis in mouse melanoma cells by blocking the hormone-induced accumulation of the c locus tyrosinase. *Eur. J. Biochem.* **232,** 257–263.
9. Solca, F., Chluba-de Tapia, J., Iwata, K., and Eberle, A. N. (1993) B16-G4F mouse melanoma cells: an MSH receptor-deficient cell clone. *FEBS Lett.* **322,** 177–180.
10. Hoganson, G. E., Ledwitz-Rigby, F., Davidson, R. L., and Fuller, B. B. (1989) Regulation of tyrosinase mRNA levels in mouse melanoma cell clones by melanocyte-stimulating hormone, and cyclic AMP. *Somat. Cell. Mol. Genet.* **15,** 255–263.
11. Siegrist, W., Drozdz, R., Cotti, R., Willard, D. H., Wilkison, W. O., and Eberle, A. N. (1997) Interactions of α-melanotropin, and agouti on B16 melanoma cells: evidence for inverse agonism of agouti. *J. Recept. Signal Transd. Res.* **17,** 75–98.
12. Chakraborty, A. K., Slominski, A., Ermak, G., Hwang, J., and Pawelek, J. (1995) Ultraviolet B, and melanocyte-stimulating hormone (MSH) stimulate mRNA production for α-MSH receptors, and proopiomelanocortin-derived peptides in mouse melanoma cells, and transformed keratinocytes. *J. Invest. Dermatol.* **105,** 655–659.
13. Fuller, B. B., and Meyskens, F. L. Jr. (1981) Endocrine responsiveness in human melanocytes, and melanoma cells in culture. *J. Natl. Cancer Inst.* **66,** 799–802.
14. Hunt, G., Todd, C., Cresswell, J. E., and Thody, A. J. (1994) α-Melanocyte stimulating hormone, and its analogue Nle^4, D-Phe^7 α-MSH affect morphology, tyrosinase activity, and melanogenesis in cultured human melanocytes. *J. Cell Sci.* **107,** 205–211.
15. Hunt, G., Donathien, P. D., Lunec, J., Todd, C., Kyne, S., and Thody, A. J. (1994) Cultured human melanocytes respond to MSH peptides, and ACTH. *Pigment Cell Res.* **7,** 217–221.
16. Halaban, R. (1993) Growth regulation in normal, and malignant melanocytes. *Recent Results Cancer Res.* **128,** 133–150.
17. Siegrist, W., Solca, F., Stutz, S., Giuffrè, L., Carrel, S., Girard, J., and Eberle, A. N. (1989) Characterization of receptors for α-melanocyte-stimulating hormone on human melanoma cells. *Cancer Res.* **49,** 6352–6358.

18. Rauth, S., Hoganson, G. E., and Davidson, R. L. (1990) Bromodeoxyuridine- and cyclic AMP-mediated regulation of tyrosinase in Syrian hamster melanoma cells. *Somat. Cell. Mol. Genet.* **16,** 583–592.
19. Ghanem, G. E., Comunale, G., Libert, A., Vercammen-Grandjean, A., and Lejeune, F. J. (1988) Evidence for α-melanocyte-stimulating hormone (α-MSH) receptors on human malignant melanoma cells. *Int. J. Cancer* **41,** 248–255.
20. Siegrist, W. and Eberle, A. N. (1993) Homologous regulation of the MSH receptor in melanoma cells. *J. Recept. Res.* **13,** 263–281.
21. Siegrist, W., Stutz, S., and Eberle, A. N. (1994) Homologous, and heterologous regulation of α-melanocyte-stimulating hormone receptors in human, and mouse melanoma cell lines. *Cancer Res.* **54,** 2604–2610.
22. Lunec, J., Pieron, C., and Thody, A. J. (1992) MSH receptor expression, and the relationship to melanogenesis, and metastatic activity in B16 melanoma. *Melanoma Res.* **2,** 5–12.
23. Tatro, J. B., Entwistle, M. L., Lester, B. R., and Reichlin, S. (1990) Melanotropin receptors of murine melanoma characterized in cultured cells, and demonstrated in experimental tumors in situ. *Cancer Res.* **50,** 1237–1242.
24. Slominsik, A., Costantino, R., Wortsman, J., Paus, R., and Ling, N. (1992) Melanotropic activity of gamma MSH peptides in melanoma cells. *Life Sci.* **50,** 1103–1108.
25. De Graan, P. N. E., Brussaard, A. B., Gamboni, G., Girard, J., and Eberle, A. N. (1987) α-MSH-induced changes in protein phosphorylation of Cloudman S91 mouse melanoma cells. *Mol. Cell. Endocrinol.* **51,** 87–93.
26. Fuller, B. B., Lunsford, J. B., and Iman, D. S. (1987) α-Melanocyte-stimulating hormone regulation of tyrosinase in Cloudman S91 mouse melanoma cell culture. *J. Biol. Chem.* **262,** 4024–4033.
27. Buffey, J., Thody, A. J., Bleehen, S. S., and Mac Neil, S. (1992) α-Melanocyte-stimulating hormone stimulates protein kinase C activity in murine B16 melanoma. *J. Endocrinol.* **133,** 333–340.
28. Fuller, B. B., and Viskochil, D. H. (1979) The role of RNA, and protein synthesis in mediating the action of MSH on mouse melanoma cells. *Life Sci.* **24,** 2405–2416.
29. Rungta, D., Corn, T. D., and Fuller, B. B. (1996) Regulation of tyrosinase mRNA in mouse melanoma cells by α-melanocyte-stimulating hormone. *J. Invest. Dermatol.* **107,** 698–693.
30. Park, H. Y., Russakovsky, V., Ao, Y., Fernandez, E., and Gilchrest, B. A. (1996) α-Melanocyte stimulating hormone-induced pigmentation is blocked by depletion of protein kinase C. *Exp. Cell Res.* **25,** 70–79.
31. Fuller, B. B., Niekrasz, I., and Hoganson, G. E. (1990) Down-regulation of tyrosinase mRNA levels in melanoma cells by tumor promotors, and by insulin. *Mol. Cell. Endocrinol.* **72,** 81–87.
32. Siegrist, W., Sauter, P., and Eberle, A. N. (1995) A selective protein kinase C inhibitor (CGP 41251) positively, and negatively modulates melanoma cell MSH receptors. *J. Recept. Signal Transduct. Res.* **15,** 283–296.
33. Valverde, P., Garcia-Borron, J. C., Martinez-Liarte, J. H., Solano, F., and Lozano, J. A. (1992) Melanocyte stimulating hormone activation of tyrosinase in B16 mouse melanoma cells: evidence for a differential induction of two distinct isoenzymes. *FEBS Lett.* **304,** 114–118.

34. Valverde, P., Garcia-Borron, J. C., Jimenez-Cervantes, C., Solano, F., and Lozano, J. A. (1993) Tyrosinase isoenzymes in mammalian melanocytes. 2. Differential activation by α-melanocyte-stimulating hormone. *Eur. J. Biochem.* **217,** 541–548.
35. Martinez-Liarte, J. H., Solano, F., Garcia-Borron, J. C., Jara, J. R., and Lozano, J. A. (1992) α-MSH, and other melanogenic activators mediate opposite effects on tyrosinase, and dopachrome tautomerase in B16/F10 mouse melanoma cells. *J. Invest. Dermatol.* **99,** 435–439.
36. Hill, S. E., Rees, R. C., and Mac Neil, S. (1990) The regulation of cyclic AMP production, and the role of cyclic AMP in B16 melanoma cells of differing metastatic potential. *Clin. Exp. Metastasis* **8,** 475–489.
37. Salomon, Y. (1990) Melanocortin receptors: targets for control by extracellular calcium. *Mol. Cell. Endocrinol.* **70,** 139–145.
38. Eberle, A. N., Siegrist, W., Bagutti, C., Chluba-de Tapia, J., Solca, F., Wikberg, J. E. S., and Chhajlani, V. (1993) Receptors for melanocyte-stimulating hormone on melanoma cells. *Ann. N. Y. Acad. Sci.* **680,** 320–341.
39. De Graan, P. N. E., Eberle, A. N., and Van de Veerdonk, F. C. G. (1981) Photoaffinity labelling of MSH receptors reveals a dual role of calcium in melanophore stimulation. *FEBS Lett.* **129,** 113–116.
40. Gerst, J. E., Sole, J., and Salomon, Y. (1987) Dual regulation of β-melanotropin receptor function, and adenylate cyclase by calcium, and guanosine nucleotides in the M2R melanoma cell line. *Mol. Cell. Endocrinol.* **46,** 137–147.
41. Gerst, J. E. and Salomon, Y. (1988) A synthetic analog of the calmodulin-binding domain of myosin light chain kinase inhibits melanotropin receptor function, and activation of adenylate cyclase. *J. Biol. Chem.* **263,** 7073–7078.
42. Buffey, J. A., Edgecombe, M., and MacNeil, S. (1993) Calcium plays a complex role in the regulation of melanogenesis in murine B16 melanoma cells. *Pigment Cell Res.* **6,** 385–393.
43. Parker, C. and Sherbet, G. V. (1993) The Ca^{2+} channel blocker verapamil enhances melanogenesis without altering metastatic potential in the B16 murine melanoma. *Melanoma Res.* **3,** 347–350.
44. Gordon, L., Peacocke, M., and Gilchrest, B. A. (1992) Induction of c-fos but not c-myc in S-91 cells by melanization signals. *J. Dermatol. Sci.* **3,** 35–41.
45. Englaro, W., Rezzonico, R., Durand-Clement, M., Lallemand, D., Ortonne, J. P., and Ballotti, R. (1995) Mitogen-activated protein kinase pathway, and AP-1 are activated during cAMP-induced melanogenesis in B16 melanoma cells. *J. Biol. Chem.* **270,** 24,315–24,320.
46. Orlow, S. J., Chakraborty, A. K., and Pawelek, J. M. (1990) Retinoic acid is a potent inhibitor of inducible pigmentation in murine, and hamster melanoma cell lines. *J. Invest. Dermatol.* **94,** 461–464.
47. Orlow, S. J., Chakraborty, A. K., Boissy, R. E., and Pawelek, J. M. (1990). Inhibition of induced melanogenesis in Cloudman melanoma cells by four phenotypic modifiers. *Exp. Cell Res.* **191,** 209–218.
48. Kosano, H., Setogawa, T., Kobayashi, K., and Nishigori, H. (1995) Pyrroloquinoline quinone (PQQ) inhibits the expression of tyrosinase mRNA by α-melanocyte stimulating hormone in murine B16 melanoma cells. *Life Sci.* **56,** 1707–1713.
49. Valverde, P., Manning, P., McNeil, C. J., and Thody, A. J. (1996) Activation of tyrosinase reduces the cytotoxic effects of the superoxide anion in B16 mouse melanoma cells. *Pigment Cell Res.* **9,** 77–84.

44. Lu, D., Willard, D., Patel, I. R., Kadwell, S., Overton, L., Kost, T., Luther, M., Chen, W., Woychik, R. P., Wilkison, W. O., and Cone, R. D. (1994) Agouti protein is an antagonist of the melanocyte-stimulating hormone receptor. *Nature* **371,** 799–802.
45. Willard, D. H., Bodnar, W., Harris, C., Kiefer, L., Nichols, J. S., Blanchard, S., Hoffman, C., Moyer, M., Burkhart, W., Weiel, J., Luther, M. A., Wilkison, W. O., and Rocque, W. J. (1995) Agouti structure, and function: characterization of a potent α-melanocyte-stimulating hormone receptor antagonist. *Biochemistry* **34,** 12,341–12,346.
46. Wilson, B. D., Ollmann, M. M., Kang, L., Stoffel, M., Bell, G. I., and Barsh, G. S. (1995) Structure, and function of ASP, the human homolog of the mouse agouti gene. *Hum. Mol. Genet.* **4,** 223–230.
47. Graham, A., Wakamatsu, K., Hunt, G., Ito, S., and Thody, A. J. (1997) Agouti protein inhibits the production of eumelanin, and phaeomelanin in the presence, and absence of α-melanocyte-stimulating hormone. *Pigment Cell Res.* **10,** 298–303.
48. Siegrist, W., Willard, D. H., Wilkison, W. O., and Eberle, A. N. (1996) Agouti protein inhibits growth of B16 melanoma cells in vitro by acting through melanocortin receptors. *Biochem. Biophys. Res. Commun.* **5,** 171–175.
49. Hunt, G. and Thody, A. J. (1995) Agouti protein can act independently of melanocyte-stimulating hormone to inhibit melanogenesis. *J. Endocrinol.* **147,** R1–R4.
55. Prezioso, J. A., Hearing, V. J., Muller, J., Urabe, K., Wang, N., and Gorelik, E. (1995) Impairment of the melanogenic pathway in B16 melanoma cells transfected with class I H-2 genes. *Melanoma Res.* **5,** 15–25.
56. Preston, S. F., Volpi, M., and Pearson, C. M. (1987) Regulation of cell shape in the Cloudman melanoma cell line. *Proc. Natl. Acad. Sci. U. S. A.* **84,** 5247–5251.
57. Giuffrè, L., Schreyer, M., Mach, J. P., and Carrell, S. (1988) Cyclic AMP induces differentiation in vitro of human melanoma cells. *Cancer* **61,** 1132–1141.
58. Nakazawa, K., Damour, O., and Collombel, C. (1993) Modulation of normal human melanocyte dendricity by growth-promoting agents. *Pigment Cell Res.* **6,** 406–416.
59. Scott, G., Cassidy, L., and Abdel-Malek, Z. (1997) α-Melanocyte-stimulating hormone, and endothelin-1 have opposing effects on melanocyte adhesion, migration, and pp125FAK phosphorylation. *Exp. Cell Res.* **25,** 19–28.
60. Nascimento, A. A., Amaral, R. G., Bizario, J. C., Larson, R. E., and Espreafico, E. M. (1997). Subcellular localization of myosin-V in the B16 melanoma cells, a wild-type cell line for the dilute gene. *Mol. Biol. Cell* **8,** 1971–1988.
61. Jiang, J., Sharma, S. D., Nakamura, S., Lai, J. Y., Fink, J. L., Hruby,V. J., and Hadley, M. E. (1995) The melanotropic peptide, [Nle^4, D-Phe^7]-α-MSH, stimulates human melanoma tyrosinase activity, and inhibits cell proliferation. *Pigment Cell Res.* **8,** 314–323.
62. Kameyama, K., Vieira, W. D., Tsukamoto, K., Law, L. W., and Hearing, V. J. (1990) Differentiation, and the tumorigenic, and metastatic phenotype of murine melanoma cells. *Int. J. Cancer* **45,** 1151–1158.
63. Abdel Malek, Z. A., Hadley, M. E., Bregman, M. D., Meyskens, F. L., and Hruby, V. J. (1986) Actions of melanotropins on mouse melanoma cell growth in vitro. *J. Natl. Cancer Inst.* **76,** 857–863.
64. Bregman, M. D., Abdel Malek, Z. A., and Meyskens, F. L. (1985) Anchorage-independent growth of murine melanoma in serum-less media is dependent on insulin or melanocyte-stimulating hormone. *Exp. Cell Res.* **157,** 419–428.

65. Wick, M. M. (1981) Inhibition of clonogenic growth of melanoma cells by combination of melanocyte stimulating hormone, and theophylline. *J. Invest. Dermatol.* **77,** 253–255.
66. Slominski, A., Moellmann, G., and Kuklinska, E. (1989) MSH inhibits growth in a line of amelanotic hamster melanoma cells, and induces increases in cyclic AMP levels, and tyrosinase activity without inducing melanogenesis. *J. Cell Sci.* **92,** 551–559.
67. Gerst, J. E., Benezra, M., Schimmer, A., and Salomon, Y. (1989) Phorbol ester impairs melanotropin receptor function, and stimulates growth of cultured M2R melanoma cells. *Eur. J. Pharmacol.* **172,** 29–39.
68. Abdel Malek, Z. A., Swope, V. B., Trinkle, L. S., and Nordlund, J. J. (1989) Stimulation of Cloudman melanoma tyrosinase occurs predominantly in G2 phase of the cell cycle. *Exp. Cell Res.* **180,** 198–208.
69. Sahu, S. N., Edwards-Prasad, J., and Prasad, K. N. (1987) α-Tocopheryl succinate inhibits melanocyte-stimulating hormone (MSH)-sensitive adenylate cyclase activity in melanoma cells. *J. Cell Physiol.* **133,** 585–589.
70. De Luca, M., Siegrist, W., Bondanza, S., Mathor, M., Cancedda, R., and Eberle, A. N. (1993) α-Melanocyte stimulating hormone (α-MSH) stimulates normal human melanocyte growth by binding to high-affinity receptors. *J. Cell Sci.* **105,** 1079–1084.
71. Yaar, M. and Gilchrest, B. A. (1991) Human melanocyte growth, and differentiation: a decade of new data. *J. Invest. Dermatol.* **97,** 611–617.
72. Halaban, R., Tyrrell, L., Longley, J., Yarden, Y., and Rubin, J. (1993) Pigmentation, and proliferation of human melanocytes, and the effects of melanocyte-stimulating hormone, and ultraviolet B light. *Ann. N. Y. Acad. Sci.* **680,** 290–301.
73. Halaban, R. (1996) Growth factors, and melanomas. *Semin. Oncol.* **23,** 673–681.
74. Gehlsen, K. R., Hadley, M. E., Levine, N., Ray, C. G., and Hendrix, M. J. (1992) Effects of a melanotropic peptide on melanoma cell growth, metastasis, and invasion. *Pigment Cell Res.* **5,** 219–223.
75. Siegrist, W., Hintermann, E., Roggo, C., Apfel, C. M., Klaus, M., and Eberle, A. N. (1998) Melanoma cell growth inhibition, and melanocortin receptor downregulation induced by selective, and non-selective retinoids. *Melanoma Res.* **8,** 113–122.
76. Herlyn, M., Kath, R., Williams, N., Valyi-Nagi, I., and Rodeck, U. (1990) Growth-regulatory factors for normal, premalignant, and malignant human cells in vitro. *Adv. Cancer Res.* **54,** 213–234.
77. Sheppard, J. R., Lester, B., Doll, J., Buscarino, C., Gonzales, E., Greig, R., and Poste, G. (1986) Biochemical regulation of adenylate cyclase in murine melanoma clones with different metastatic properties. *Int. J. Cancer* **37,** 713–722.
78. Sheppard, J. R., Koestler, T. P., Corwin, S. P., Buscarino, C., Doll, J., Lester, B., Greig, R. G., and Poste, G. (1984) Experimental metastasis correlates with cyclic AMP accumulation in B16 melanoma clones. *Nature* **308,** 544–547.
79. Bennett, D. C., Dexter, T. J., Ormerod, E. J., and Hart, I. R. (1986) Increased experimental metastatic capacity of a murine melanoma following induction of differentiation. *Cancer Res.* **46,** 3239–3244.
80. Bennett, D. C., Holmes, A., Devlin, L., and Hart, I. R. (1994) Experimental metastasis, and differentiation of murine melanoma cells: actions, and interactions of factors affecting different intracellular signalling pathways. *Clin. Exp. Metastasis* **12,** 385–397.

81. Lakshmi, M. S., Parker, C., and Sherbet, G. V. (1997) Expression of the transmembrane glycoprotein CD44, and metastasis associated 18A2/MTS1 gene in B16 murine melanoma cells. *Anticancer Res.* **17,** 1451–3455.
82. Valverde, P., Healy, E., Sikkink, S., Haldane, F., Thody, A. J., Carothers, A., Jackson, I. J., and Rees, J. L. (1996) The Asp84Glu variant of the melanocortin-1 receptor (MC1R) is associated with melanoma. *Hum. Mol. Genet.* **5,** 1663–1666.
83. Radinsky, R., Beltran, P. J., Tsan, R., Zhang, R. Cone, R. D., and Fidler, I. J. (1995) Transcriptional induction of the melanocyte-stimulating hormone receptor in brain metastases of murine K-1735 melanoma. *Cancer Res.* **55,** 141–148.
84. Siegrist, W., Oestreicher, M., Stutz, S., Girard, J., and Eberle, A. N. (1988) Radioreceptor assay for α-MSH using mouse B16 melanoma cells. *J. Recept. Res.* **8,** 323–343.
85. Scimonelli, T. and Eberle, A. N. (1987) Photoaffinity labelling of melanoma cell MSH receptors. *FEBS Lett.* **226,** 134–138.
86. Solca, F., Siegrist, W., Drozdz, R., Girard, J., and Eberle, A. N. (1989) The receptor for α-melanotropin of mouse, and human melanoma cells: application of a potent α-melanotropin photoaffinity label. *J. Biol. Chem.* **264,** 14,277–14,281.
87. Solca, F. F. A., Salomon, Y., and Eberle, A. N. (1991) Heterogeneity of the MSH receptor among B16 murine melanoma subclones. *J. Recept. Res.* **11,** 379–390.
88. Damien, C., Robberecht, P., Hooghe, R., De Neef, P., and Christophe, J. (1989) Decreased adenylate cyclase activation by helodermin, and PGE1 in the lectin-resistant variant Wa4 of the mouse melanoma cell line B16. *Peptides* **10,** 1075–1079.
89. Siegrist, W., Bagutti, C., Solca, F., Girard, J., and Eberle, A. N. (1992) MSH receptors on mouse, and human melanoma cells: receptor identification, analysis, and quantification. *Prog. Histochem. Cytochem.* **26,** 110–118.
90. Donatien, P. D., Hunt, G., Pieron, C., Lunec, J., Taieb, A., and Thody, A. J. (1992) The expression of functional MSH receptors on cultured human melanocytes. *Arch. Dermatol. Res.* **284,** 424–426.
91. Tatro, J. B., Atkins, M., Mier, J. W., Hardarson, S., Wolfe, H., Smith, T., Entwistle, M. L., and Reichlin, S. (1990) Melanotropin receptors demonstrated in situ in human melanoma. *J. Clin. Invest.* **85,** 1825–1832.
92. Wong, G., Pawelek, J., Sansone, M., and Morowitz, J. (1974) Response of mouse melanoma cells to melanocyte stimulating hormone. *Nature* **248,** 351–354.
93. Varga, J. M., DiPasquale, A., Pawelek, J., McGuire, J. S., and Lerner, A. B. (1974) Regulation of melanocyte stimulating hormone action at the receptor level: discontinuous binding of hormone to synchronized mouse melanoma cells during the cell cycle. *Proc. Natl. Acad. Sci. U. S. A.* **71,** 1590–1593.
94. Fuller, B. B. and Brooks, B. A. (1980) Application of percent labeled mitoses (PLM) analysis to the investigation of melanoma cell responsiveness to MSH stimulation throughout the cell cycle. *Exp. Cell Res.* **126,** 183–190.
95. Bolognia, J. L., Sodi, S. A., Chakraborty, A. K., Fargnoli, M. C., and Pawelek, J. M. (1994) Effects of ultraviolet irradiation on the cell cycle. *Pigment Cell Res.* **7,** 320–325.
96. Orlow, S. J., Hotchkiss, S., and Pawelek, J. M. (1990) Internal binding sites for MSH: analyses in wild-type, and variant Cloudman melanoma cells. *J. Cell Physiol.* **142,** 129–136.
97. Panasci, L. C., McQuillan, A., and Kaufman, M. (1987) Biological activity, binding, and metabolic fate of Ac-[Nle^4, D-Phe^7]-α-MSH_{4-11}-NH_2 with the F1 variant of B16 melanoma cells. *J. Cell. Physiol.* **132,** 97–103.

98. Sharma, S. D., Granberry, M. E., Jiang, J., Leong, S. P., Hadley, M. E., and Hruby, V. J. (1994) Multivalent melanotropic peptide, and fluorescent macromolecular conjugates: new agents for characterization of melanotropin receptors. *Bioconjug. Chem.* **5,** 591–601.
99. Wong, W. and Minchin, R. F. (1996) Binding, and internalization of the melanocyte stimulating hormone receptor ligand [Nle^4, D-Phe^7]-α-MSH in B16 melanoma cells. *Int. J. Biochem. Cell. Biol.* **28,** 1223–1232.
100. Chakraborty, A. K., Orlow, S. J., and Pawelek, J. M. (1990) Stimulation of th receptor for melanocyte-stimulating hormone by retinoic acid. *FEBS Lett.* **276,** 205–208.
101. Powers, T. P. and Davidson, R. L. (1996) Coordinate extinction of melanocyte-specific gene expression in hybrid cells. *Somat. Cell. Mol. Genet.* **22,** 41–56.
102. Bolognia, J., Murray, M., and Pawelek, J. (1989) UVB-induced melanogenesis may be mediated through the MSH-receptor system. *J. Invest. Dermatol.* **92,** 651–656.
103. McQuillan, A., Hutchinson, M., and Panasci, L. C. (1989) Evidence for increased α-MSH receptor binding in the F1 variant of B16 melanoma cells grown in dialyzed fetal calf serum. *J. Cell. Physiol.* **141,** 281–283.
104. Kameyama, K., Tanaka, S., Ishida, Y., and Hearing, V. J. (1989) Interferons modulate the expression of hormone receptors on the surface of murine melanoma cells. *J. Clin. Invest.* **83,** 213–221.
105. Birchall, N., Orlow, S. J., Kupper, T., and Pawelek, J. (1991) Interactions between ultraviolet light, and interleukin-1 on MSH binding in both mouse melanoma, and human squamous carcinoma cells. *Biochem. Biophys. Res. Commun.* **175,** 839–845.
106. Baumann, J. B., Bagutti, C., Siegrist, W., Christen, E., Zumsteg, U., and Eberle, A. N. (1997) MSH receptors, and the response of human A375 melanoma cells to interleukin-1β. *J. Recept. Signal Transduct. Res.* **17,** 199–210.
107. Chluba-de Tapia, J., Bagutti, C., Cotti, R., and Eberle, A. N. (1996) Induction of constitutive melanogenesis in amelanotic mouse melanoma cells by transfection of the human melanocortin-1 receptor gene. *J. Cell Sci.* **109,** 2023–2030.
108. Slominski, A., Wortsman, J., Mazurkiewicz, J. E., Dietrich, J., Lawrence, K., Gorbani, A., and Paus, R. (1993) Detection of proopiomelanocortin-derived antigens in normal, and pathologic human skin. *J. Lab. Clin. Med.* **122,** 658–666.
109. Schauer, E., Trautinger, F., Kock, A., Schwarz, A., Bhardwaj, R., Simon, M., Ansel, J. C., Schwarz, T., and Luger, T. A. (1994) Proopiomelanocortin-derived peptides are synthesized, and released by human keratinocytes. *J. Clin. Invest.* **93,** 2258–2262.
110. Slominski, A. (1991) POMC gene expression in mouse, and hamster melanoma cells. *FEBS Lett.* **291,** 165–168.
111. Ghanem, G., Loir, B., Hadley, M., Abdel Malek, Z., Libert, A., Del Marmol, V., Lejeune, F., Lozano, J., and Garcia-Borron, J. C. (1992) Partial characterization of IR-α-MSH peptides found in melanoma tumors. *Peptides* **13,** 989–994.
112. Lunec. J., Pieron, C., Sherbet, G. V., and Thody, A. J. (1990) α-Melanocyte-stimulating hormone immunoreactivity in melanoma cells. *Pathobiology* **58,** 193–197.
113. Orth, D. N. (1973) Establishment of human malignant melanoma clonal cell lines that secrete ectopic adrenocorticotropin. *Nature New Biol.* **242,** 26–28.
114. Zubair, A., Lakshmi, M. S., and Sherbet, G. V. (1992) Expression of α-melanocyte stimulating hormone, and the invasive ability of the B16 murine melanoma. *Anticancer Res.* **12,** 399–402.

115. Loir, B., Bouchard, B., Morandini, R., Del Marmol, V., Deraemaecker, R., Garcia-Borron, J. C., and Ghanem, G. (1997) Immunoreactive α-melanotropin as an autocrine effector in human melanoma cells. *Eur. J. Biochem.* **244,** 923–930.
116. Iyengar, B. (1995) Corticotropin expression by human melanocytes in the skin. *Pigment Cell Res.* **8,** 142–146.
117. Liu, P. Y. and Johansson, O. (1995) Immunohistochemical evidence of α-, β-, and γ-melanocyte stimulating hormone in cutaneous malignant melanoma of nodular type. *J. Dermatol. Sci.* **10,** 203–212.
118. Ghanem, G. E., Verstegen, J., De Rijcke, S., Hanson, P., Van Onderbergen, A., Libert, A., Del Marmol, V., Arnould, R., Vercammen-Granjean, A., and Lejeune, F. (1989) Studies on factors influencing human plasma α-MSH. *Anticancer Res.* **9,** 1691–1696.
119. Ghanem, G. E., Verstegen, J., Libert, A., Arnould, R., and Lejeune, F. (1989) α-Melanocyte-stimulating hormone immunoreactivity in human melanoma metastses extracts. *Pigment Cell Res.* **2,** 519–523.
120. Schwartze, G. and Fiedler, H. (1994) The diagnostic significance of α-MSH in malignant melanoma of man. *Hautarzt* **45,** 468–470.
121. Siegrist, W., Girard, J., and Eberle, A. N. (1991) Quantification of MSH receptors on mouse melanoma tissue by receptor autoradiography. *J. Receptor Res.* **11,** 323–331.
122. Bagutti, C., Oestreicher, M., Siegrist, W., Oberholzer, M., and Eberle, A. N. (1995) α-MSH receptor autoradiography on mouse, and human melanoma tissue sections, and biopsies. *J. Recept. Signal Transduct. Res.* **15,** 427–442.
123. Muceniece, R. and Wikberg, J. E. (1996) Immunological localization of melanocortin 1 receptors on the cell surface of WM266-4 human melanoma cells. *Cancer Lett.* **98,** 157–162.
124. Bard, D. R., Knight, C. G., and Page-Thomas, D. P. (1990) A chelating derivative of α-melanocyte stimulating hormone as a potential imaging agent for malignant melanoma. *Br. J. Cancer* **62,** 919–922.
125. Wraight, E. P., Bard, D. R., Maughan, T. S., Knight, C. G., and Page-Thomas, D. P. (1992) The use of a chelating derivative of alpha melanocyte stimulating hormone for the clinical imaging of malignant melanoma. *Br. J. Radiol.* **65,** 112–118.
126. Bagutti, C., Stolz, B., Albert, R., Bruns, C., Pless, J., and Eberle, A. N. (1994) [^{111}In]-DTPA-labeled analogues of α-melanocyte-stimulating hormone for melanoma targeting: receptor binding in vitro, and in vivo. *Int. J. Cancer* **58,** 749–755.
127. Bard, D. R. (1995) An improved imaging agent for malignant melanoma, based on [Nle^4, D-Phe^7]-α-melanocyte stimulating hormone. *Nucl. Med. Commun.* **16,** 860–866.
128. Garg, P. K., Alston, K. L., Welsh, P. C., and Zalutsky, M. R. (1996) Enhanced binding, and inertness to dehalogenation of α-melanotropic peptides labeled using *N*-succinimidyl 3-iodobenzoate. *Bioconjug. Chem.* **7,** 233–239.
129. Giblin, M. F., Jurisson, S. S., and Quinn, T. P. (1997) Synthesis, and characterization of rhenium-complexed α-melanotropin analogs. *Bioconjug. Chem.* **8,** 347–353.
130. Otte, A., Jermann, E., Behe, M., Goetze, M., Bucher, H. C., Roser, H. W., Heppeler, A., Müller-Brand, J., and Mäcke, H. R. (1997) DOTATOC: a powerful new tool for receptor-mediated radionuclide therapy. *Eur. J. Nucl. Med.* **24,** 792–795.
131. Liu, M. A., Nussbaum, S. R., and Eisen, H. N. (1988) Hormone conjugated with antibody to CD3 mediates cytotoxic T cell lysis of human melanoma cells. *Science* **239,** 395–398.

132. Murphy, J. R., Bishai, W., Borowski, M., Miyanohara, A., Boyd, J., and Nagle, S. (1986) Genetic construction, expression, and melanoma-selective cytotoxicity of a diphtheria toxin-related α-melanocyte-stimulating hormone fusion protein. *Proc. Natl. Acad. Sci. U. S. A.* **83,** 8258–8262.
133. Wen, Z. L., Tao, X., Lakkis, F., Kiyokawa, T., and Murphy, J. R. (1991) Diphtheria toxin-related α-melanocyte-stimulating hormone fusion toxin: internal in-frame deletion from Thr^{387} to His^{485} results in the formation of a highly potent fusion toxin which is resistant to proteolytic degradation. *J. Biol. Chem.* **266,** 12,289–12,293.
134. Tatro, J. B., Wen, Z., Entwistle, M. L., Atkins, M. B., Smith, T. J., Reichlin, S., and Murphy, J. R. (1992) Interaction of an α-melanocyte-stimulating hormone-diphtheria toxin fusion protein with melanotropin receptors in human melanoma metastases. *Cancer Res.* **52,** 2545–2548.
135. Süli-Vargha, H., Jeney, A., Kopper, L., Olah, J., Lapis, K., Botyanszki, J., Csukas, I., Gyovari, B., and Medzihradszky, K. (1990) Investigations on the antitumor effect, and mutagenicity of α-MSH fragments containing melphalan. *Cancer Lett.* **54,** 157–162.
136. Morandini, R., Süli-Vargha, H., Libert, A., Loir, B., Botyanszki, J., Medzihradszky, K., and Ghanem, G. (1994) Receptor-mediated cytotoxicity of α-MSH fragments containing melphalan in a human melanoma cell line. *Int. J. Cancer* **56,** 129–133.
137. Ghanem, G. E., Libert, A., Arnould, R. Vercammen, A., and Lejeune, F. (1991) Human melanoma targeting with α-MSH-melphalan conjugate. *Melanoma Res.* **1,** 105–114.
138. Süli-Vargha, H., Morandini, R., Bodi, J., Nagy, L., Medzihradszky-Schweiger, H., and Ghanem, G. (1997) In vitro cytotoxic effect of difluoromethylornithine increased nonspecifically by peptide coupling. *J. Pharm. Sci.* **86,** 997–100.
139. Vile, R. G. and Hart, I. R. (1993) In vitro, and in vivo targeting of gene expression to melanoma cells. *Cancer Res.* **53,** 962–967.

CHAPTER 18

Regulation of the Mouse and Human Melanocortin-1 Receptor

Zalfa Abdel-Malek

1. Introduction

Decades before the molecular cloning of the melanocortin 1 receptor (MC1-R) gene, genetic studies on the coat color of mice concluded that the *extension* (*e*) locus codes for a receptor for melanocyte stimulating hormone (MSH) *(1,2)*. Activation of this receptor is known to regulate the switch from pheomelanin to eumelanin synthesis in mouse follicular melanocytes *(1–4)*. In addition, mutations at the *e* locus were found to be associated with either a reduction or an increase in eumelanin formation *(1,5,6)*. Since the 1970s numerous studies have focused on elucidating the mechanism of action of α-or β-MSH on the vertebrate pigmentary systems. In most cases, these studies relied on bioassays of lizard or frog skins, or utilized established mouse melanoma cell lines as an in vitro model to explore the role of MSH in mammalian pigmentation *(7–12)*. Comparative analysis of the MSH receptors expressed on pigment cells of different vertebrate species was based primarily on structure–function studies. In these, the relative potencies of physiologic melanotropic hormones or synthetic analogs of α-MSH were compared *(9,13–17)*. Most of what we currently know about the signaling pathway of α-MSH came from studies on the pigmentary effects of α-or β-MSH, particularly on mouse normal melanocytes or melanoma cell lines *(2,12,18–21)*.

In the mouse, a physiologic role for α-MSH in stimulating melanocyte differentiation and inducing eumelanin formation has long been acknowledged *(1–4)*. In newborn mice, α-MSH stimulates the differentiation of melanoblasts into melanocytes by inducing tyrosinase activity, formation and translocation of melanosomes, and increased dendritogenesis *(22a–c)*. However, a role for

The Melanocortin Receptors Ed.: R. D. Cone

α-MSH in regulating human cutaneous pigmentation remained controversial. The possibility that α-MSH is a physiologic regulator of human pigmentation was downplayed by the argument that humans virtually lack an intermediate pituitary lobe. In the early 1960s, it was demonstrated that injection of human volunteers with high concentrations of α-or β-MSH or ACTH resulted in skin darkening *(23,24)*. More recently, these findings were corroborated by the report that injection of human subjects with a potent synthetic analog of α-MSH, [Nle^4, D-Phe^7]-α-MSH (NDP-MSH), increased skin pigmentation, particularly in sun exposed sites *(25)*. Despite these reports, evidence for a direct response of human melanocytes to melanotropins remained lacking until recently *(26–30)*. The cloning of the human MC1-R and the finding that it is not only expressed by established mouse melanoma tumor cells but by normal human melanocytes as well rekindled the interest in exploring the responsiveness of these cells to melanotropins *(28,31–33)*. It also led to investigating the possible physiologic role of these hormones as regulators of human pigmentation.

2. Studies on Mouse Melanocytes and Melanoma Cells

2.1. Regulation of the MSH Receptor by the cAMP Pathway

For decades, established mouse melanoma cell lines have been extensively utilized to investigate the regulation of mammalian pigmentation in vitro *(10–13,16,20,21,34–39)*. The feasibility of maintaining these cells in culture and their profound response to α-or β-MSH allowed for elucidation of the mechanism of action of melanotropins and the regulation of the mouse MSH receptor. Early studies on the mouse Cloudman S91 melanoma cells revealed that treatment with α-MSH significantly and transiently increased intracellular cyclic adenosine monophosphate (cAMP) levels *(21,37)*. Treatment with cAMP inducers, such as cholera toxin, or with phosphodiesterase inhibitors, such as the methylxanthines, mimicked the melanogenic effect of α-MSH *(38,39)*. These along with studies on frog and lizard skins, as well as on mouse melanoblasts unequivocally proved that the melanogenic effect of MSH is mediated by activation of the cAMP pathway *(1,7,8,18,19,39)*. Activation of this pathway was later found to regulate the expression of the MSH receptor. Evidence for regulation of the mouse MSH receptor by cAMP inducers was first provided by the demonstration that cholera toxin or dibutyryl cAMP (db cAMP) increased the number of MSH receptors on Cloudman melanoma cells *(38)*. Recently, this was further corroborated by the finding that pretreatment of these cells with α-MSH upregulated the expression of the mRNA transcript of the MC1-R *(40)*.

Contrary to the above findings with Cloudman S91 melanoma cells, it was reported that B16 melanoma cells treated with α-MSH, cholera toxin, or forskolin demonstrated a reduction in MC1-R expression, which was evident as a decrease in receptor number per cell without reduction in receptor affinity *(41)*. Downregulation of the MC1-R was also observed following treatment with the phorbol ester 12-*O*-tetradecanoylphorbol-13-acetate (TPA) or the physiologic MSH antagonist the agouti signaling protein *(42)*. Pertussis toxin, which had a delayed stimulatory effect on melanogenesis through inhibition of G_i and the resulting accumulation of cAMP, had no effect on MC1-R expression. Compared to forskolin which directly activates adenylate cyclase, cholera toxin which ribosylates G_s was more effective in down regulating MC1-R. Based on this, it was suggested that down regulation of the receptor was not primarily mediated by stimulation of adenylate cyclase but by coupling of G_s to Ca^{+2} channels and phospholipase C_β, both of which were induced by cholera toxin.

2.2. Cell Cycle Dependent Expression of the MSH Receptor

The receptor for MSH was reported to be predominantly expressed during G2 phase of the cell cycle. Cloudman melanoma cells were found to express the highest number of receptors and to bind more MSH molecules during G2 phase *(43)*. This led to the conclusion that the MSH receptor exhibits positive cooperativity which might be induced by autocrine factors that are mainly synthesized during G2 phase *(44)*. Multiple irradiations of Cloudman melanoma cells with ultraviolet (UV) light resulted in their arrest in G2 as well as in increased binding capacity of α-MSH *(45,46)*. Furthermore, it was demonstrated that UV light, MSH, and dbcAMP increased the level of the MC1-R mRNA in Cloudman melanoma cells *(40)*. These results led to the proposal that the MSH receptor functions as a transducer of the effects of UV light on cutaneous melanocytes *(47)*.

2.3. Postinflammatory Mediators and MSH Receptor Regulation

We have reported that prostaglandin E1 stimulates melanogenesis in Cloudman melanoma cells and causes their arrest in G2 phase of the cell cycle *(48,49)*. Since this phase is thought to be the most permissible for MSH receptor expression, we propose that the inflammatory mediator prostaglandin E-1 increases the responsiveness of melanocytes to α-MSH by enhancing its binding to its receptor. This might be one mechanism for postinflammatory hyperpigmentation.

Another study which aimed at elucidating a mechanism for postinflammatory hyperpigmentation found that the inflammatory mediators interferon (IFN)-α, -β, and -γ increase the number of α-MSH binding sites in mouse JB/MS cell line *(50)*. None of the above three types of interferon alone had an effect on melanogenesis yet each one interacted synergistically with α-MSH to increase melanin production.

2.4. *Internal Binding Sites for MSH*

It has been shown that in Cloudman melanoma cells, the receptor for MSH is internalized after ligand binding and is localized to the Golgi apparatus *(51)*. More recently, it was proposed that internal binding sites for MSH exist and that they differ from the plasma membrane sites in their sedimentation on a sucrose gradient *(52)*. The availability of internal binding sites correlated directly with the response of Cloudman melanoma cells to MSH. Both the internal binding sites and the membrane receptors for MSH were found to have identical molecular weights (50–53 kDa) and common antigenic determinants *(53)*. Irradiation of these melanoma cells with UV light decreased the binding of MSH to internal binding sites and concomitantly increased its binding to membrane bound sites *(53)*.

3. The Human Melanocortin-1 Receptor

3.1. *Expression of the MC1-R on Normal Human Melanocytes*

Despite the demonstrations that injection of human volunteers with α-MSH, β-MSH, and adrenocorticotropin hormone (ACTH) resulted in skin darkening, it remained to be determined whether or not this was due to a direct effect of these melanocortins on human melanocytes *(23–25)*. Until the early 1980s an optimal in vitro model for human pigmentation was lacking. Finally, in 1982 the first medium capable of supporting the proliferation and long-term maintenance of normal human melanocytes in vitro was described *(54)*. This medium relied on the use of tumor promoting phorbol esters in conjunction with cholera toxin to enhance the proliferation of melanocytes in culture. Due to the irreversible effects of cholera toxin, its presence complicated, rather than facilitated, the studies on the response of human melanocytes to melanotropins. Using this medium, melanocytes failed to respond to α-MSH, which led some to conclude that human melanocytes lack the expression of MSH receptors and that melanocortins have no role in human pigmentation *(55–57)*. Thus, it is not surprising that only in the past 5 years, after modification of the melanocyte culture conditions, could investigators begin to characterize the human MSH receptor and explore how it is regulated *(26,28,33)*.

With the cloning of the MC1-R, it became evident that it is expressed by cultured normal human melanocytes *(28,31)*. In addition, stimulatory effects of α-MSH, ACTH, and to a lesser extent β-MSH, on melanocyte proliferation and melanogenesis have been documented *(26–30)*. These studies put to rest a long-standing controversy about the responsiveness of human melanocytes to melanotropins.

It has been suggested that the MC1-R is also expressed by human keratinocytes *(58)*. This was based on studies using epidermoid carcinoma cell lines, which were found to bind α-MSH, particularly after UV irradiation, and to respond to α-MSH with a dose-dependent increase in proliferation. The presence of MC1-R on these cells was demonstrated by reverse transcriptase-polymerase chain reaction (RT-PCR). Our findings, however, point to the lack of expression of functional MC1-R in normal human keratinocytes *(A. Tada, I. Suzuki, V. Swope, S. Boyce, and Z. Abdel-Malek, unpublished data)*. We found that primary cultures of human keratinocytes failed to bind α-MSH, ACTH, or β-MSH, or to respond to any of these melanotropins with an increase in intracellular cAMP. Moreover, we could not detect MC1-R mRNA in keratinocytes by Northern blot analysis even when the amount of total RNA used exceeded several folds the amount routinely used for the detection of MC1-R mRNA in melanocytes *(28)*.

In addition to melanocytes, human microvascular endothelial cells were reported to express MC1-R, as determined by RT-PCR *(59)*. Expression of the receptor was found to be upregulated by pretreatment with interleukin (IL)-1β or α-MSH, and α-MSH was shown to stimulate the production of IL-8 by these cells.

3.2. Regulation of MC1-R Expressed on Normal Human Melanocytes by α-MSH and ACTH, Other cAMP Inducers, UV Light, and Epidermal Paracrine Factors

We have demonstrated that treatment of normal human melanocytes in vitro with α-MSH or ACTH increased the mRNA level of the human MC1-R. This effect was evident within 4–6 h, continued to increase up to 9 h, and returned to steady state level within 24 h of treatment *(28)*. Whether or not the observed increase in MC1-R mRNA translates into expression of a higher number of MC1-R is to be determined. These results offer an explanation for the ability of melanocytes to respond to continued treatment with α-MSH or ACTH, and suggest positive autoregulation of the human MC1-R by its ligands.

In normal human melanocytes, we have found that activation of the cAMP pathway is prerequisite for UV induced melanogenesis *(60)*. Among the physiologic factors that stimulate melanogenesis, α-MSH and ACTH, and less so β-MSH, stimulate cAMP formation in human melanocytes, and as stated above, increase the expression of the MC1-R mRNA *(28)*. Ultraviolet light is known to stimulate the synthesis of α-MSH and ACTH by epidermal keratinocytes and melanocytes *(61–63)*. These results put together suggest that exposure of human skin to UV light results in upregulation of the MC1-R expression, at least partially by increasing melanotropin synthesis in the epidermis.

It has been reported that exposure of cultured human melanocytes to UV light or the inflammatory mediators tumor necrosis factor-α or γ-interferon increased MSH binding to its receptor *(64)*. A similar effect was observed upon increasing cAMP levels by treatment with cholera toxin or db cAMP. The phorbol ester TPA, which is commonly used as a mitogen for human melanocytes, reduced the binding of α-MSH to its receptor. In our laboratory, we found that a single irradiation with different doses of UVB light reduced the level of MC1-R mRNA *(M. C. Scott, I. Suzuki, and Z. Abdel-Malek, unpublished data)*. It is not known whether this reduction is due to decreased mRNA stability or reduced transcriptional rate of the MC1-R gene. We observed that UVB irradiated human melanocytes responded equally to α-MSH as their unirradiated counterparts with increased cAMP formation and melanogenesis *(60)*. This suggests the presence of spare MC1-R whose binding affinity is not diminished by UV treatment, and indicates that responsiveness to α-MSH does not absolutely require the transcription of the MC1-R gene.

An interesting finding is that the MC1-R mRNA was upregulated by endothelin-1, which is synthesized by human keratinocytes, particularly after UV exposure or IL-1 treatment *(65,66)*. Both α-MSH and endothelin-1 interact synergistically to enhance human melanocyte proliferation and modulate melanogenesis *(66,67)*. Based on this, we propose the following model for the regulation of the human MC1-R in the epidermis (Fig. 1). Exposure to UV rays from the sun stimulates the synthesis of IL-1 by human keratinocytes. Interleukin-1 in turn enhances the synthesis of endothelin-1 by keratinocytes, and α-MSH and ACTH by keratinocytes and melanocytes *(63,65)*. The direct effects of IL-1 on MC1-R expression are not known. However, endothelin-1, α-MSH and ACTH increase the expression of MC1-R mRNA and possibly enhance the responsiveness of melanocytes to melanotropins.

3.3. Regulation of the MC1-R Expressed on Human Melanoma Tumor Cells

Human melanoma cells are known to synthesize immunoreactive α-MSH *(68,69)*. The autoproduction of α-MSH is thought to contribute to the metastatic potential of melanoma tumors *(68)*. Interest in using α-MSH analogs for melanoma diagnosis and surveillance and in conjugating melanotropin analogs to chemotherapeutic drugs for targeted melanoma therapy (*see also* Chapter 17) made it important to characterize the human MSH receptor on these tumor cells *(70)*. It is known that human melanoma tumors express different numbers of MSH binding sites *(41)*. In one study, the possible regulation of MSH receptor expression by α-MSH was investigated using 11 different human melanoma cell lines *(41)*. Three of these cell lines responded to α-MSH with upregulation of the number of MSH receptors, 6 lines demonstrated a decrease in the number of binding sites, while two lines showed no change in receptor number following α-MSH treatment. The change in receptor number, when observed, was not accompanied by alteration in binding

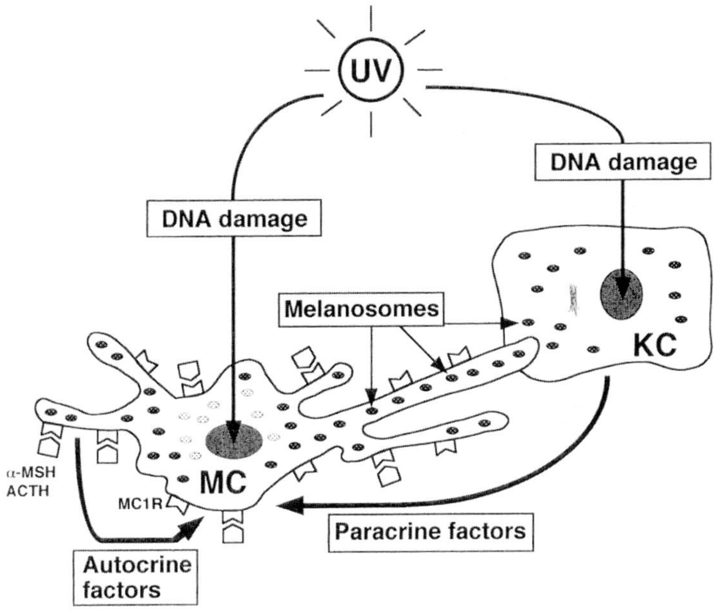

Fig. 1. Regulation of human MC1-R expression by UV light and paracrine/autocrine factors. Ultraviolet light has direct as well as indirect effects on epidermal melanocytes and keratinocytes. The direct effects are exemplified by DNA damage. The indirect effects include modulation of synthesis of soluble paracrine and autocrine factors. These factors include IL-1α, a primary cytokine which stimulates the synthesis of endothelin-1 as well as α-MSH and ACTH. All of these factors seem to upregulate MC1-R expression, resulting in increased responsiveness to melanotropins and stimulation of melanogenesis.

affinity. Moreover, the increase in receptor number was independent of protein synthesis and seemed to be due to recruitment of spare receptors *(71)*. Recently, certain variants of the MC1-R gene were found to be expressed in 20 out of 43 melanoma patients, compared to only 8 out of 44 controls *(72)*. It was postulated that these variants are associated with poor melanogenic response to UV light and possibly decreased responsiveness to α-MSH. The possible association of the MC1-R variants with increased risk of skin cancer, including melanoma, suggests that this gene might be a tumor susceptibility gene *(72,73)*.

3.4. Comparison of the Autoregulation of the Human MC1-R and Other G Protein Coupled Receptors

We have reported that treatment of human melanocytes with α-MSH or ACTH resulted in a transient increase in MC1-R mRNA within 6–8 hours

(28). A similar increase in MC2-R mRNA in adrenocortical cells was evident within 4–7 h of ACTH treatment *(74)*. In human adrenocortical cells, this effect was sustained for 24 hours, while in the mouse counterpart it lasted for 3 days. As with the MC1-R, the effect of ACTH on MC2-R mRNA was mediated by activation of the cAMP pathway, since forskolin treatment mimicked the effect of ACTH. In bovine fasciculata adrenal cells, pretreatment with ACTH resulted in increased ACTH binding, an effect that was mimicked by 8-bromo-cAMP *(75)*. It would be interesting to determine if the other melanocortin receptors, namely MC3-R, MC4-R, and MC5-R, are also autoregulated by melanotropins or up regulated by stimulation of the cAMP pathway. Since elevation of the MC2-R mRNA is expected to increase the number of MC2-R, we predict that this will be true for MC1-R as well *(76)*. In human melanocytes, the increase in MC1-R mRNA does not increase the responsiveness to α-MSH or ACTH, but seems to sustain the responsiveness to these hormones.

The human MC1-R differs from other G protein coupled receptors, such as the β_2 adrenergic receptors, that are known to undergo desensitization following agonist treatment *(77)*. Mechanisms by which desensitization occurs involve phosphorylation of the receptors on serine and threonine residues which impairs the coupling of the receptor to G proteins, and sequestration, which leads to receptor down regulation *(78)*. The rate of β_2 receptor gene expression was found to be stimulated by a brief treatment for 30 min with epinephrine or db cAMP, and to be reduced following treatment for 24 h with either agent. Prolonged treatment also resulted in a gradual reduction in β_2 adrenergic receptor number and in a decrease in agonist induced adenylate cyclase activity. Desensitization has also been shown to occur in the α_2 adrenergic receptor subtypes α2c10 and α2c2 upon coupling of these receptors to G_s *(78)*. As stated above, treatment of normal human melanocytes with either α-MSH or ACTH for 6–9 h increased the mRNA level of the MC1-R, and prolonged treatment did not result in loss of responsiveness of melanocytes to these hormones, suggesting lack of desensitization. Prolonged treatment of human melanocytes with α-MSH was associated with a sustained high level of intracellular cAMP and continued stimulation of proliferation and melanogenesis *(27,28)*.

4. Comparison of the Properties of the Mouse and Human MC1-R

The mouse and human MC1-R share only 76% homology *(79)*. When comparing the potencies of the different melanocortins in the functional coupling of these two receptors, the following differences became evident. The human MC1-R has a higher affinity than the mouse MC1-R for α-MSH and ACTH. The human MC1-R recognizes both melanocortins with an equal affinity, while the mouse MC1-R has a higher affinity for α-MSH than ACTH *(28,80)*. In human melanocytes, both hormones have the same EC_{50} values in cAMP radioimmunoassay and equivalent melanogenic and proliferative effects *(28)*. In addition, activation of the human MC1-R by α-MSH binding results in prolonged (longer than 24 h) stimulation of cAMP formation, while binding of α-MSH to the mouse MC1-R has a transient (about 2 h) effect on cAMP synthesis *(21,28,37)*. The human MC1-R seems to constitutively activate the cAMP signaling pathway *(81)*. Accordingly, it was proposed that the human MC1-R evolved to be "supersensitive" to the melanocortin peptides *(80)*. Interestingly, however, while NDP-MSH was 100-fold more potent than α-MSH in inducing melanogenesis in mouse Cloudman melanoma cells, it was only about 10-fold more potent than α-MSH in its ability to bind the human MC1-R, stimulate cAMP formation, and induce proliferation and melanogenesis in human melanocytes *(13,28)*.

A potential difference between the mouse and human MC1-R is that relatively few receptors are expressed per normal human melanocyte (about 700 binding sites/cell), and are required for full mitogenic and melanogenic stimulation *(33)*. Studies on mouse B16 melanoma cells showed the expression of about 10-fold higher number of receptors per cell, about 7000 binding sites *(33)*. So far, no studies have been carried out on normal mouse melanocytes to determine the number of receptors expressed per melanocyte. It is possible that in human melanocytes, activation of only a few receptors is sufficient for the biologic effects of α-MSH or ACTH, since the human MC1-R has a high affinity for these two ligands *(28)*.

Recently, we demonstrated that the human MC1-R is similar to its mouse counterpart in that its binding to α-MSH is blocked by the agouti signaling protein *(82,83)*. These results indicate that the functional relationship between the agouti and MC1-R gene products is similar in mice and humans.

5. Concluding Remarks

The past five years have witnessed several major advancements in the field of melanotropin research. These include the cloning and characterization of the melanocortin receptors *(31,32,84–87)*, the demonstration that human melanocytes respond to α-MSH and ACTH with increased melanogenesis and proliferation *(26–30)*, and the finding that these melanotropins are synthesized by epidermal keratinocytes and melanocytes, particularly in response to inflammation or UV irradiation *(61–63)*. Other developments in elucidating the regulation of human pigmentation include the identification of human MC1-R gene variants that are associated with skin type I or II phenotypes and possibly with melanoma formation *(72,73)*. Future studies will determine the consequence of these variants on the affinity of the MC1-R for α-MSH, the ability of the bound MC1-R to activate adenylate cyclase, and the regulation of receptor expression by ligand, other paracrine factors and UV light. These future studies will delineate the possible role of the human MC1-R gene as a susceptibility gene for skin cancers, including melanoma.

While significant advances have been made toward the understanding of the regulation of eumelanin synthesis, the regulation of pheomelanogenesis remains for the most part elusive. The demonstration that α-MSH induces eumelanin synthesis in normal human melanocytes indicates that the *extension* locus serves the same function in both mouse and human melanocytes *(88)*. The cloning of the human *agouti* gene and the purification of its product, the agouti signaling protein, has made it possible to investigate the potential role of this factor in the regulation of human pigmentation *(89,90)*. The agouti signaling protein acts as an inhibitor of α-MSH binding to the mouse as well as the human MC1-R *(82,83)*. In human melanocytes, agouti signaling protein also inhibits the α-MSH induced stimulation of cAMP formation, melanogenesis, and proliferation *(83)*. Future studies will be aimed at investigating the potential role of agouti signaling protein in inducing the switch to pheomelanin synthesis in human melanocytes. The significance of delineating the regulation of the eumelanin–pheomelanin switch lies in the importance of these two forms of melanin in photoprotection against sun-induced DNA damage and skin carcinogenesis.

Acknowledgments

I thank Itaru Suzuki, Sungbin Im, and Akihiro Tada for their contributions to the data presented in this manuscript regarding the human MC1-R. This work was supported in part by a grant (5 R01 ES06882) from the National Institute of Environmental Health Sciences (awarded to Zalfa Abdel-Malek, Ph.D.).

References

1. Tamate, H. B. and Takeuchi, T. (1984) Action of the *e* locus of mice in the response of phaeomelanic hair follicles to α–melanocyte–stimulating hormone in vitro. *Science* **224,** 1241–1242.
2. Takeuchi, T., Kobunai, T., and Yamamoto, H. (1989) Genetic control of signal transduction in mouse melanocytes. *J. Invest. Dermatol.* **92,** 239S–242S.
3. Geschwind, I. I., Huseby, R. A., and Nishioka, R. (1972) The effect of melanocyte–stimulating hormone on coat color in the mouse. *Rec. Prog. Hormone Res.* **28,** 91–130.
4. Burchill, S. A., Thody, A. J., and Ito, S. (1986) Melanocyte–stimulating hormone, tyrosinase activity and the regulation of eumelanogenesis and pheomelanogenesis in hair follicular melanocytes of the mouse. *J. Endocrinol.* **109,** 15–21.
5. Silvers, W. K. (1979) *The Coat Colors of Mice. A Model for Mammalian Gene Action and Interaction.* Springer–Verlag, New York.

6. Robbins, L. S., Nadeau, J. H., Johnson, K. R., Kelly, M. A., Roselli–Rehfuss, L., Baack, E., Mountjoy, K. G., and Cone, R. D. (1993) Pigmentation phenotypes of variant extension locus alleles result from point mutations that alter MSH receptor function. *Cell* **72,** 827–834.
7. Sawyer, T. K., Yang, Y. C. S., Bregman, M. D., Hruby, V. J., Heward, C. B., Fuller, B. B., and Hadley, M. E. (1979) Structure–function studies of melanophore stimulating hormones (α–MSH and β–MSH) and their analogs on melanoma plasma membrane adenylate cyclase: comparison with frog skin melanophores, in *Peptides: Structure and Biological Function* (Gross, E. and Meienhofer, J., eds.), Pierce Chemical Company, Rockford. pp. 1017–1020.
8. Hadley, M. E., Heward, C. B., Hruby, V. J., Sawyer, T. K., and Yang, Y. C.–S. (1981) Hormone receptors of vertebrate pigment cells, in *Pigment Cell 1981. Phenotypic Expression in Pigment Cells. Proc XIth International Pigment Cell Conference, Sendai, Japan, 1980* (Seiji, M., ed.), University of Tokyo Press, Tokyo, Japan, pp. 323–330.
9. Sawyer, T. K., Hruby, V. J., Wilkes, B. C., Draelos, M. T., Hadley, M. E., and Bergsneider, M. (1982) Comparative biological activities of highly potent active–site analogues of α–melanotropin. *J. Med. Chem.* **25,** 1022–1027.
10. Pawelek, J., Wong, G., Sansone, M., and Morowitz, J. (1973) Molecular biology of pigment cells: molecular controls in mammalian pigmentation. *Yale J. Biol. Med.* **46,** 430–443.
11. Fuller, B. B. and Viskochil, D. H. (1979) The role of RNA and protein synthesis in mediating the action of MSH on mouse melanoma cells. *Life Sci.* **24,** 2405–2416.
12. Pawelek, J. M. (1985) Studies on the Cloudman melanoma cell line as a model for the action of MSH. *Yale J. Biol. Med.* **58,** 571–578.
13. Marwan, M. M., Abdel–Malek, Z. A., Kreutzfeld, K. L., Castrucci, A. M., Hadley, M. E., Wilkes, B. C., and Hruby, V. J. (1985) Stimulation of S91 melanoma tyrosinase activity by superpotent α–melanotropins. *Mol. Cell Endocrinol.* **41,** 171–177.
14. Hruby, V. J., Wilkes, B. C., Hadley, M. D., Al–Obeidi, F., Sawyer, T. K., Staples, D. J., Vaux, A. E., Dym, O., Castrucci, A. M. L., Hintz, M. F., Priehm, J. P., and Rao, K. M. (1987) α–Melanotropin: the minimal active sequence in the frog skin bioassay. *J. Med. Chem.* **30,** 2126–2130.
15. Castrucci, A. M. L., Hadley, M. E., Sawyer, T. K., Wilkes, B. C., Al–Obeidi, F., Staples, D. J., De Vaux, A. E., Dym, O., Hintz, M. F., Riehm, J. P., Rao, K. R., and Hruby, V. J. (1989) α–Melanotropin: the minimal active sequence in the lizard skin bioassay. *Gen. Comp. Endocrinol.* **73,** 157–163.
16. Hadley, M. E., Abdel–Malek, Z. A., Kreutzfeld, K. L., Marwan, M. M., and Hruby, V. J. (1985) [Nle^4, D–Phe^7]–α–MSH: A superpotent melanotropin that "irreversibly" activates melanoma tyrosinase. *Endocr. Res.* **11,** 157–170.
17. Wilkes, B. C., Sawyer, T. K., Hruby, V. J., and Hadley, M. E. (1983) Differentiation of the structural features of melanotropins important for biological potency and prolonged activity *in vitro*. *Int. J. Pept. Protein Res.* **22,** 313–324.
18. Hirobe, T. and Takeuchi, T. (1977) Induction of melanogenesis in the epidermal melanoblasts of newborn mouse by MSH. *J. Embryol. Exp. Morphol.* **37,** 79–90.
19. Hirobe, T. and Takeuchi, T. (1977) Induction of melanogenesis in vitro in the epidermal melanoblasts of newborn mouse skin by MSH. *In Vitro* **13,** 311–315.

20. Kreiner, P. W., Gold, C. J., Keirns, J. J., Brock, W. A., and Bitensky, M. W. (1973) Hormonal control of melanocytes: MSH–sensitive adenyl cyclase in the Cloudman melanoma. *Yale J. Biol. Med.* **46,** 583–591.
21. Fuller, B. B., Lunsford, J. B., and Iman, D. S. (1987) Alpha–melanocyte–stimulating hormone regulation of tyrosinase in Cloudman S91 mouse melanoma cell cultures. *J. Biol. Chem.* **262,** 4024–4033.
22a. Hirobe, T. and Takeuchi,T. (1977) Induction of melanogenesis in the epidermal melanoblasts of newborn mouse skin by MSH. *J. Embryol. Exp. Morphol.* **37,** 79–90.
22b. Hirobe, T. and Takeuchi, T. (1978) Changes of organelles associated with the differentiation of epidermal melanocytes in the mouse. *J. Embryol. Exp. Morphol.* **43,** 107–121.
22c. Hirobe, T. (1992) Control of melanocyte proliferation and differentiation in the mouse epidermis. *Pigment Cell Res.* **5,** 1–11.
23. Lerner, A. B. and McGuire, J. S. (1961) Effect of alpha– and beta–melanocyte stimulating hormones on the skin colour of man. *Nature* **189,** 176–179.
24. Lerner, A. B. and McGuire, J. S. (1964) Melanocyte–stimulating hormone and adrenocorticotrophic hormone: their relation to pigmentation. *N. Engl. J. Med.* **270,** 539–546.
25. Levine, N., Sheftel, S. N., Eytan, T., Dorr, R. T., Hadley, M. E., Weinrach, J. C., Ertl, G. A., Toth, K., and Hruby, V. J. (1991) Induction of skin tanning by the subcutaneous administration of a potent synthetic melanotropin. *JAMA* **266,** 2730–2736.
26. De Luca, M., Siegrist, W., Bondanza, S., Mathor, M., Cancedda, R., and Eberle, A. N. (1993) αMelanocyte stimulating hormone (αMSH) stimulates normal human melanocyte growth by binding to high–affinity receptors. *J. Cell Sci.* **105,** 1079–1084.
27. Abdel–Malek, Z., Swope, V. B., Suzuki, I., Akcali, C., Harriger, M. D., Boyce, S. T., Urabe, K., and Hearing, V. J. (1995) Mitogenic and melanogenic stimulation of normal human melanocytes by melanotropic peptides. *Proc. Natl. Acad. Sci. U. S. A.* **92,** 1789–1793.
28. Suzuki, I., Cone, R., Im, S., Nordlund, J., and Abdel–Malek, Z. (1996) Binding capacity and activation of the MC1 receptors by melanotropic hormones correlate directly with their mitogenic and melanogenic effects on human melanocytes. *Endocrinology* **137,** 1627–1633.
29. Hunt, G., Todd, C., Cresswell, J. E., and Thody, A. J. (1994) α–Melanocyte stimulating hormone and its analogue Nle^4DPhe^7α–MSH affect morphology, tyrosinase activity and melanogenesis in cultured human melanocytes. *J. Cell Sci.* **107,** 205–211.
30. Hunt, G., Todd, C., Kyne, S., and Thody, A. J. (1994) ACTH stimulates melanogenesis in cultured human melanocytes. *J. Endocrinol.* **140,** R1–R3.
31. Mountjoy, K. G., Robbins, L. S., Mortrud, M. T., and Cone, R. D. (1992) The cloning of a family of genes that encode the melanocortin receptors. *Science* **257,** 1248–1251.
32. Chhajlani, V. and Wikberg, J. E. S. (1992) Molecular cloning and expression of the human melanocyte stimulating hormone receptor cDNA. *FEBS Lett.* **309,** 417–420.
33. Donatien, P. D., Hunt, G., Pieron, C., Lunec, J., Taïeb, A., and Thody, A. J. (1992) The expression of functional MSH receptors on cultured human melanocytes. *Arch. Dermatol. Res.* **284,** 424–426.
34. Wong, G. and Pawelek, J. (1973) Control of phenotypic expression of cultured melanoma cells by melanocyte stimulating hormones. *Nature New Biol.* **241,** 213–215.

35. Niles, R. M. and Makarski, J. S. (1978) Control of melanogenesis in mouse melanoma cells of varying metastatic potential. *J. Natl. Cancer Inst.* **61,** 523–526.
36. Aroca, P., Urabe, K., Kobayashi, T., Tsukamoto, K., and Hearing, V. J. (1993) Melanin biosynthesis patterns following hormonal stimulation. *J. Biol. Chem.* **268,** 25,650–25,655.
37. Wong, G., Pawelek, J., Sansone, M., and Morowitz, J. (1974) Response of mouse melanoma cells to melanocyte stimulating hormone. *Nature* **248,** 351–354.
38. DiPasquale, A., McGuire, J., and Varga, J. M. (1977) The number of receptors for β–melanocyte stimulating hormone in Cloudman melanoma cells is increased by dibutyryl adenosine 3':5'–cyclic monophosphate or cholera toxin. *Proc. Natl. Acad. Sci. U. S. A.* **74,** 601–605.
39. O'Keefe, E. and Cuatrecasas, P. (1974) Cholera toxin mimics melanocyte stimulating hormone in inducing differentiation in melanoma cells. *Proc. Natl. Acad. Sci. U. S. A.* **71,** 2500–2504.
40. Chakraborty, A., Slominski, A., Erinak, G., Hwang, J., and Pawelek, J. (1995) Ultraviolet B and melanocyte stimulating hormone (MSH) stimulate mRNA production for α–MSH receptors and proopiomelanocortin–derived peptides in mouse melanoma cells and transformed keratinocytes. *J. Invest. Dermatol.* **105,** 655–659.
41. Siegrist, W., Stutz, S., and Eberle, A. N. (1994) Homologous and heterologous regulation of α–melanocyte–stimulating hormone receptors in human and mouse melanoma cell lines. *Cancer Res.* **54,** 2604–2610.
42. Siegrist, W., Drozdz, R., Cotti, R., Willard, D. H., Wilkison, W. O., and Eberle, A. N. (1997) Interactions of α–melanotropin and agouti on B16 melanoma cells: evidence for inverse agonism of agouti. *J. Recept. Signal Trans. Res.* **17,** 75–98.
43. Varga, J. M., DiPasquale, A., Pawelek, J., McGuire, J. S., and Lerner, A. B. (1974) Regulation of melanocyte stimulating hormone action at the receptor level: discontinuous binding of hormone to synchronized mouse melanoma cells during the cell cycle. *Proc. Natl. Acad. Sci. U. S. A.* **71,** 1590–1593.
44. McLane, J. A. and Pawelek, J. M. (1988) Receptors for β melanocyte stimulating hormone exhibit positive cooperativity in synchronized melanoma cells. *Biochemistry* **27,** 3743–3747.
45. Bolognia, J., Murray, M., and Pawelek, J. (1989) UVB–induced melanogenesis may be mediated through the MSH–receptor system. *J. Invest. Dermatol.* **92,** 651–656.
46. Chakraborty, A. K. and Pawelek, J. M. (1992) Up–regulation of MSH receptors by MSH in Cloudman melanoma cells. *Biochem. Biophys. Res. Commun.* **188,** 1325–1331.
47. Pawelek, J. M., Chakraborty, A. K., Osber, M. P., Orlow, S. J., Min, K. K., Rosenzweig, K. E., and Bolognia, J. L. (1992) Molecular cascades in UV–induced melanogenesis: a central role for melanotropins? *Pigment Cell Res.* **5,** 348–356.
48. Abdel–Malek, Z., Swope, V. B., Amornsiripanitch, N., and Nordlund, J. J. (1987) In vitro modulation of proliferation and melanization of S91 melanoma cells by prostaglandins. *Cancer Res.* **47,** 3141–3146.
49. Abdel–Malek, Z. A., Ross, R., Pike, J. W., Trinkle, L., Swope, V., and Nordlund, J. J. (1988) Hormonal effects of vitamin D_3 on epidermal melanocytes. *J. Cell. Physiol.* **136,** 273–280.
50. Kameyama, K., Tanaka, S., Ishida, Y., and Hearing, V. J. (1989) Interferons modulate the expression of hormone receptors on the surface of murine melanoma cells. *J. Clin. Invest.* **83,** 213–221.

51. Varga, J. M., Moellmann, G. E., Fritsch, P., Godawska, E., and Lerner, A. B. (1976) Association of cell surface receptors for melanotropin with the Golgi region in mouse melanoma cells. *Proc. Natl. Acad. Sci. U. S. A.* **73,** 559–562.
52. Orlow, S. J., Hotchkiss, S., and Pawelek, J. M. (1990) Internal binding sites for MSH: analyses in wild–type and variant Cloudman melanoma cells. *J. Cell. Physiol.* **142,** 129–136.
53. Chakraborty, A. K., Orlow, S. J., Bolognia, J. L., and Pawelek, J. M. (1991) Structural/functional relationships between internal and external MSH receptors: modulation of expression in Cloudman melanoma cells by UVB radiation. *J. Cell. Physiol.* **147,** 1–6.
54. Eisinger, M. and Marko, O. (1982) Selective proliferation of normal human melanocytes in vitro in the presence of phorbol ester and cholera toxin. *Proc. Natl. Acad. Sci. U. S. A.* **79,** 2018–2022.
55. Halaban, R., Pomerantz, S. H., Marshall, S., Lambert, D. T., and Lerner, A. B. (1983) Regulation of tyrosinase in human melanocytes grown in culture. *J. Cell Biol.* **97,** 480–488.
56. Ranson, M., Posen, S., and Mason, R. S. (1988) Human melanocytes as a target tissue for hormones: in vitro studies with 1α–25,dihydroxyvitamin D3, α–melanocyte stimulating hormone, and β–estradiol. *J. Invest. Dermatol.* **91,** 593–598.
57. Friedman, P. S., Wren, F., Buffey, J., and McNeil, S. (1990) α–MSH causes a small rise in cAMP but has no effect on basal or ultraviolet–stimulated melanogenesis in human melanocytes. *Br. J. Dermatol.* **123,** 145–151.
58. Bhardwaj, R. S., Becher, E., Mahnke, K., Hartmeyer, M., Scholzen, T., Schwarz, T., and Luger, T. A. (1996) Evidence of the expression of a functional melanocortin receptor 1 by human keratinocytes. [Abstract]. *J. Invest. Dermatol.* **106,** 817.
59. Hartmeyer, M., Scholzen, T., Becher, E., Bhardwaj, R. S., Fastrich, M., Schwarz, T., and Luger, T. A. (1996) Human microvascular enothelial cells (HMEC–1) express the melanocortin receptor type 1 and produce increased levels of IL–8 upon stimulation with α–MSH. [Abstract]. *J. Invest. Dermatol.* **106,** 809.
60. Im, S., Moro, O., Medrano, E. E., Cornelius, J., Babcock, G., Nordlund, J., and Abdel–Malek, Z. (1998) Activation of the cAMP pathway by α–melanotropin mediates the response of human melanocytes to UVB radiation. *Cancer Res.* **58,** 47–54.
61. Schauer, E., Trautinger, F., Kock, A., Schwarz, A., Bhardwaj, R., Simon, M., Ansel, J. C., Schwarz, T., and Luger, T. A. (1994) Proopiomelanocortin–derived peptides are synthesized and released by human keratinocytes. *J. Clin. Invest.* **93,** 2258–2262.
62. Kippenberger, S., Bernd, A., Loitsch, S., Ramirez–Bosca, A., Bereiter–Hahn, J., and Holzmann, H. (1995) α–MSH is expressed in cultured human melanocytes and keratinocytes. *Eur. J. Dermatol.* **5,** 395–397.
63. Chakraborty, A. K., Funasaka, Y., Slominski, A., Ermak, G., Hwang, J., Pawelek, J. M., and Ichihashi, M. (1996) Production and release of proopiomelanocortin (POMC) derived peptides by human melanocytes and keratinocytes in culture: regulation by ultraviolet B. *Biochim. Biophys. Acta.* **1313,** 130–138.
64. Thody, A. J., Hunt, G., Donatien, P. D., and Todd, C. (1993) Human melanocytes express functional melanocyte–stimulating hormone receptors. *Ann. N. Y. Acad. Sci.* **680,** 381–390.

65. Imokawa, G., Yada, Y., and Miyagishi, M. (1992) Endothelins secreted from human keratinocytes are intrinsic mitogens for human melanocytes. *J. Biol. Chem.* **267,** 24,675–24,680.
66. Tada, A., Suzuki, I., Im, S., Davis, M. B., Nordlund, J. J., and Abdel–Malek, Z. M. (1998) Endothelin-1 is a paracrine growth factor that modulates melanogenesis of human melanocytes and participates in their response to ultraviolet radiation. *Cell Growth Diff.* **9,** 575–584.
67. Swope, V. B., Medrano, E. E., Smalara, D., and Abdel–Malek, Z. (1995) Long–term proliferation of human melanocytes is supported by the physiologic mitogens a–melanotropin, endothelin–1, and basic fibroblast growth factor. *Exp. Cell Res.* **217,** 453–459.
68. Lunec, J., Pieron, C., Sherbet, G. V., and Thody, A. J. (1990) Alpha–melanocyte–stimulating hormone immunoreactivity in melanoma cells. *Pathobiology* **58,** 193–197.
69. Ghanem, G., Loir, B., Hadley, M., Abdel–Malek, Z., Libert, A., Del Marmol, V., Lejeune, F., Lozano, J., and García–Borrón, J.–C. (1992) Partial characterization of IR–α–MSH peptides found in melanoma tumors. *Peptides* **13,** 989–994.
70. Hadley, M. E. and Dawson, B. V. (1988) Biomedical applications of synthetic melanotropins. *Pigment Cell Res.* **Suppl 1,** 69–78.
71. Siegrist, W. and Eberle, A. N. (1993) Homologous regulation of the MSH receptor in melanoma cells. *J. Recept. Res.* **13,** 263–281.
72. Valverde, P., Healy, E., Sikkink, S., Haldane, F., Thody, A. J., Carothers, A., Jackson, I. J., and Rees, J. L. (1996) The Asp84Glu variant of the melanocortin 1 receptor (*MC1R*) is associated with melanoma. *Hum. Mol. Genet.* **5,** 1663–1666.
73. Valverde, P., Healy, E., Jackson, I., Rees, J. L., and Thody, A. J. (1995) Variants of the melanocyte–stimulating hormone receptor gene are associated with red hair and fair skin in humans. *Nat. Genet.* **11,** 328–330.
74. Mountjoy, K. G., Bird, I. M., Rainey, W. E., and Cone, R. D. (1994) ACTH induces up–regulation of ACTH receptor mRNA in mouse and human adrenocortical cell lines. *Mol. Cell Endocrinol.* **99,** R17–R20.
75. Penhoat, A., Jaillard, C., and Saez, J. M. (1989) Corticotropin positively regulates its own receptors and cAMP response in cultured bovine adrenal cells. *Proc. Natl. Acad. Sci. U. S. A.* **86,** 4978–4981.
76. Rainey, W. E., Viard, I., and Saez, J. M. (1989) Transforming growth factor β treatment decreases ACTH receptors on ovine adrenocortical cells. *J. Biol. Chem.* **264,** 21,474–21,477.
77. Collins, S., Bouvier, M., Bolanowski, M. A., Caron, M. G., and Lefkowitz, R. J. (1989) cAMP stimulates transcription of the β_2–adrenergic receptor gene in response to short–term agonist exposure. *Proc. Natl. Acad. Sci. U. S. A.* **86,** 4853–4857.
78. Eason, M. G. and Liggett, S. B. (1992) Subtype–selective desensitization of α_2–adrenergic receptors. *J. Biol. Chem.* **267,** 25473–25479.
79. Cone, R. D., Mountjoy, K. G., Robbins, L. S., Nadeau, J. H., Johnson, K. R., Roselli–Rehfuss, L., and Mortrud, M. T. (1993) Cloning and functional characterization of a family of receptors for the melanotropic peptides. *Ann. N. Y. Acad. Sci.* **680,** 342–363.
80. Mountjoy, K. G. (1994) The human melanocyte stimulating hormone receptor has evolved to become "super–sensitive" to melanocortin peptides. *Mol. Cell Endocrinol.* **102,** R7–R11.

81. Chluba–de Tapia, J., Bagutti, C., Cotti, R., and Eberle, A. N. (1996) Induction of constitutive melanogenesis in amelanotic mouse melanoma cells by transfection of the human melanocortin–1 receptor gene. *J. Cell Sci.* **109,** 2023–2030.
82. Lu, D., Willard, D., Patel, I. R., Kadwell, S., Overton, L., Kost, T., Luther, M., Chen, W., Woychik, R. P., Wilkison, W. O., and Cone, R. D. (1994) Agouti protein is an antagonist of the melanocyte–stimulating–hormone receptor. *Nature* **371,** 799–802.
83. Suzuki, I., Tada, A., Ollmann, M. M., Barsh, G. S., Im, S., Lamoreux, M. L., Hearing, V. J., Nordlund, J., and Abdel–Malek, Z. A. (1997) Agouti signaling protein inhibits melanogenesis and the response of human melanocytes to α–melanotropin. *J. Invest. Dermatol.* **108,** 838–842.
84. Gantz, I., Konda, Y., Tashiro, T., Shimoto, Y., Miwa, H., Munzert, G., Watson, S. J., DelValle, J., and Yamada, T. (1993) Molecular cloning of a novel melanocortin receptor. *J. Biol. Chem.* **268,** 8246–8250.
85. Gantz, I., Miwa, H., Konda, Y., Shimoto, Y., Tashiro, T., Watson, S. J., DelValle, J., and Yamada, T. (1993) Molecular cloning, expression, and gene localization of a fourth melanocortin receptor. *J. Biol. Chem.* **268,** 15,174–15,179.
86. Roselli–Rehfuss, L., Mountjoy, K. G., Robbins, L. S., Mortrud, M. T., Low, M. J., Tatro, J. B., Entwistle, M. L., Simerly, R. B., and Cone, R. D. (1993) Identification of a receptor for gamma–melanotropin and other proopiomelanocortin peptides in the hypothalamus and limbic system. *Proc. Natl. Acad. Sci. U. S. A.* **90,** 8856–8860.
87. Labbé, O., Desarnaud, F., Eggerickx, D., Vassart, G., and Parmentier, M. (1994) Molecular cloning of a mouse melanocortin 5 receptor gene widely expressed in peripheral tissues. *Biochemistry* **33,** 4543–4549.
88. Hunt, G., Kyne, S., Wakamatsu, K., Ito, S., and Thody, A. J. (1995) Nle^{4}DPhe^7 α–Melanocyte–stimulating hormone increases the eumelanin: phaeomelanin ratio in cultured human melanocytes. *J. Invest. Dermatol.* **104,** 83–85.
89. Kwon, H. Y., Bultman, S. J., Löffler, C., Chen, W.–J., Furdon, P. J., Powell, J. G., Usala, A.–L., Wilkison, W., Hansmann, I., and Woychik, R. P. (1994) Molecular structure and chromosomal mapping of the human homolog of the agouti gene. *Proc. Natl. Acad. Sci. U. S. A.* **91,** 9760–9764.
90. Wilson, B. D., Ollmann, M. M., Kang, L., Stoffel, M., Bell, G. I., and Barsh, G. S. (1995) Structure and function of *ASP*, the human homolog of the mouse *agouti* gene. *Hum. Mol. Genet.* **4,** 223–230.

Part VI

Future Vistas

Chapter 19

Future Vistas

Roger D. Cone

1. Introduction

The cloning of the genes for the melanocortin receptors (Chapter 7) and of the agouti proteins (Chapter 14), as well as the identification of melanocortin receptor subtype specific agonists and antagonists (Chapter 8) have produced some powerful new tools for the study of the melanocortin physiology. These advances, as well as advances in understanding the physiological roles of the melanocortins, appear to be responsible for a dramatic ten-fold increase in publications on the topic over the last five years (Fig. 1), at least as searched under the keyword "melanocortin." The finding of a role for the MC4-R in energy homeostasis has, for the first time, attracted a very significant effort in the area from the pharmaceutical industry as well. One industry-watcher has told me that there are now about 40 biotechnology and pharmaceutical companies doing some sort of research on the MC4-R for the treatment of obesity.

While there is a resurgence of interest in the melanocortins, there nevertheless are many fascinating questions in melanocortin biology that remain unanswered. The list below is organized according to molecule, and represents not an exhaustive effort but simply a handful of mysteries that have piqued my interest.

2. The Melanocortin-1 Receptor

2.1. Pigmentation in Nonhuman Mammals

There are a wide variety of variant alleles of the MC1-R, characterized in the genetic literature as *extension* alleles *(1)*. Most of these alleles are straightforward null alleles of the MC1-R that promote pheomelanized coats, or dominant alleles that produce dark brown or black coats via mutations that

The Melanocortin Receptors Ed.: R. D. Cone
© Humana Press Inc., Totowa, NJ

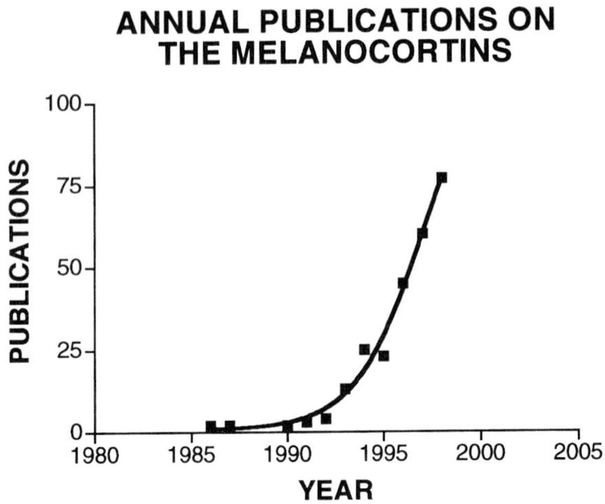

Fig. 1. Increase in annual number of publications on the melanocortins.

constitutively activate the MC1-R *(2–5)*. Different constitutively activating mutations are found in each of the four species characterized, mouse, cow, sheep, and fox, thus cloning of additional dominant MC1-R mutations is likely to continue to yield new information relevant to MC1-R structure and function.

Perhaps more novel yet are alleles at the *extension* locus which produce alternating patches of brown-black and yellow coat (e.g., tortoiseshell, e^p, in the guinea pig) or interspersed hairs containing only pheomelanin or eumelanin (e.g., brindled, e^{br}, in the dog). Variegated pigment patterns are often associated with heterozygosity of X-linked pigment genes, such as the x-linked *orange* locus in the cat, with the variable inactivation of the gene resulting from X chromosome inactivation *(6)*. Variegated coat colors resulting not from a X-linked gene, but rather from autosomal *extension* alleles occur in the rabbit, dog, cattle, pig, and guinea pig, and the mechanism(s) involved here are likely to be quite interesting.

2.2. Human Pigmentation

Chapter 11 discusses the human MC1-R, and makes the point that heterozygosity for variant alleles of the MC1-R is much more frequent in individuals with fair skin and red hair. Although red hair color is commonly inherited as an autosomal recessive trait, thus far the MC1-R does not appear to be the only gene causing the common inheritance of this trait. Since homozygous or compound heterozygous inheritance of nonfunctional MC1-R alleles does not fully explain common inheritance of red hair, why the increased frequency of variant alleles in those with red hair and fair skin?

3. The Melanocortin-2 Receptor

A curious aspect of the MC2-R that remains largely unpublished, but has been noted by several laboratories, is that the human MC2-R remains near impossible to express in heterologous cells. The receptor can be expressed in adrenocortical cells and in melanocytes, and it appears that the mouse MC2-R can be expressed in CHO cells *(7)*. It is possible that the human receptor requires an accessory protein for stable expression. Curiously, while many patients with ACTH resistance have been demonstrated to have mutations in the MC2-R, there exists a class of patients that have normal MC2-R sequences; in these patients the disease also appears to map to a different chromosome (*see* Chapter 12 for additional discussion). It is tempting to speculate that defects in an accessory protein necessary for human MC2-R expression may be responsible for a class of ACTH-resistant patients.

4. The Melanocortin-3 Receptor

It is fairly safe to say that we really do not yet know anything about the physiologic function of the MC3-R. It appears that γ-MSH stimulates natriuresis via MC3 receptors in or near the kidney *(8)*, but a physiologic role for γ-MSH in normal natriuresis remains unproven. The MC3-R is the only known melanocortin receptor that responds well to γ-MSH *(9,10)*, suggesting that the receptor may react specifically in response to release of γ-MSH, however, there are no specific data to support this latter hypothesis, and it should be remembered that the MC3-R binds α-MSH as well as it does γ-MSH.

Two observations support a role for the MC3-R in energy homeostasis. First, the receptor is expressed at the highest levels in the ventromedial hypothalamic nucleus and arcuate nucleus *(10)*, two regions known to be involved in the regulation of energy homeostasis. Second, the expression of agouti related protein is significantly upregulated by fasting *(11)*, and by the absence of leptin *(12,13)*, and this protein is a high-affinity antagonist of both the MC4-R and MC3-R *(14)*. Since the AGRP transgenic mouse should have blockade of the MC3-R and MC4-R *(12,15)*, one might expect an added phenotype or more severe obesity phenotype in this mouse in comparison to the MC4-R-KO mouse *(16)*. This is not the case; however, the animals have not yet been carefully compared side-by-side, and remain in different background strains, which would complicate such an analysis. Feeding in MC4-R-KO mice does not appear to be potently inhibited by the melanocortin agonist MTII *(17)*, however this does not mean that the MC3-R does not have an important role in energy homeostasis, since the melanocortin system appears to be involved in energy expenditure as well as energy intake. It will be very interesting to probe the role of the MC3-R in more detail in reference to regulation of metabolic rate and regulation of glucose homeostasis, for example.

5. The Melanocortin-4 Receptor

5.1. MC4-R in Energy Homeostasis

Disruption of MC4-R signaling causes obesity. Nevertheless, the normal physiologic roles of the MC4-R in energy homeostasis are not yet understood. What are the normal physiologic inputs to feeding and energy expenditure that utilize MC4-R pathways? What outputs to feeding behavior, energy expenditure, and energy partitioning depend on these pathways? Unlike most other genes involved in energy homeostasis, and unlike most other genetic disorders involving G protein-coupled receptors, obesity occurs when only a single copy of the MC4-R is lesioned. Why is MC4-R signaling sensitive to gene dosage?

5.2. Is the MC4-R Involved in Common Human Obesity?

Heterozygosity for MC4-R mutations appears to be associated with childhood obesity *(18,19)*. Do these mutations create nonfunctional receptors? Does heterozygosity for the MC4-R predispose these individuals to obesity? Why doesn't a single normal copy of the gene suffice to confer normal MC4-R signaling? How frequently are lesions in the MC4-R involved in childhood obesity?

5.3. Mechanisms of MC4-R Signaling

The MC4-R couples to G_s and activation of adenylyl cyclase in heterologous cells; however, nothing is known regarding the distribution of expression or mechanism of action of the MC4-R protein in neurons in vivo (Chapter 14). Where is the MC4-R protein expressed on neurons? Is it expressed at synapses or on cell bodies? Does it couple to G_s and/or other signaling pathways? Is it involved in presynaptic or postsynaptic modes of regulation of neurotransmission? Is the receptor itself desensitized or regulated in any important ways? How is it that both POMC and AGRP fibers access MC4-R sites? Clearly these some of the types of questions that will need to be addressed to better understand this receptor and its role in energy homeostasis.

5.4. Other Roles of the MC4-R

The MC4-R is currently attracting a tremendous amount of interest due to the fact that disruption of MC4-R function causes an obesity syndrome. It should be kept in mind, however, that this receptor is very widely expressed, albeit at low levels, being found in around 150 different brain nuclei. Potential roles for this receptor in cardiovascular homeostasis *(20)*, thermoregulation *(21)*, and grooming behavior *(22)* have already been demonstrated, thus the receptor may be more generally involved in the control of autonomic outflow to a number of varied physiologic systems.

6. The Melanocortin-5 Receptor

The MC5-R is widely expressed in exocrine glands *(23,24)*, but only limited data is available on exactly what exocrine gland products are dependent upon the MC5-R for regulated expression. The MC5-R is expressed at high levels in the lacrimal gland. Is it involved in the release of electrolytes and proteins present in tears? The receptor is also expressed at high levels in the prostate, so it will be very interesting to determine if it plays a role in normal prostate function, and in abnormal prostate growth in benign prostatic hyperplasia, and prostate cancer, the most common type of cancer in men. In rodents, the receptor is found at high levels in sebaceous glands, Harderian glands, and preputial glands, suggesting the receptor may play a role in the regulated release of pheromones. If this is the case, the receptor could provide a link between the stress axis and behavior. If the MC5-R is also involved in exocrine gland function in humans it could be a useful therapeutic target for disorders such as dry eye syndrome, acne, and blepheritis.

Finally, functional MC5 receptor appears to be expressed in spinal cord and muscle in the mouse *(23)*, and receptor mRNA has been demonstrated in brain *(25,26)*. One may speculate that this receptor may be involved in the neuroregenerative activities demonstrated for melanocortins (Chapter 4), but no data yet exists to support this conjecture. Additionally, one study was unable to detect MC5-R mRNA in rat spinal cord, yet clearly demonstrated the presence of MC4-R mRNA *(27)*, thus much remains to be resolved regarding expression and function of the MC5-R both in and outside of exocrine glands.

7. *Agouti*, Agouti-Related Protein, and *Mahogany*

7.1. Agouti

Much remains to be learned about agouti and agouti-related protein. Agouti acts as a paracrine factor and does not appear to circulate well. Thus, the agouti signal is probably tightly controlled not only at the synthesis stage, but also at stages relevant to release and termination of the agouti signal. Some hair shafts demonstrate a very distinct onset and termination of pheomelanin synthesis. How is this achieved? Is the agouti signal terminated by rapid and specific degradation of the peptide, or by internalization?

There is also little known regarding the binding site for the agouti protein on the MC1-R and MC4-R. While the cysteine-rich domain alone is capable of high-affinity binding to receptors, there may be important biological roles for the basic-rich amino terminal domain as well. Several groups have postulated that agouti may act at sites other than the MC1-R to activate pheomelanogenesis, and this also deserves added investigation.

7.2. Agouti-Related Protein

The same structural and pharmacologic issues described above exist for the hypothalamic agouti-related protein. Additionally, there are many neuroanatomic and physiologic issues outstanding for AGRP. How and where is AGRP released in the brain? How are AGRP and α-MSH released so as to act at the same MC4-R sites? Does AGRP act only to block α-MSH stimulation of the MC3-R and MC4-R, or can it also act as an inverse agonist of basal MC3-R and MC4-R activity? Does AGRP act at proteins other than the MC4-R and MC3-R? How is the AGRP gene regulated? Lastly, is AGRP secreted by the adrenal, and what is its role in the periphery?

7.3. Mahogany and the Agouti Suppressors

A number of genes have been identified over the years as suppressors of the dominant action of the agouti gene, including *mahogany (28)*, *mahoganoid (29)*, and *umbrous (30)*. The cloning of one of these, *mahogany*, raises many more questions than it answers (Chapter 14). *Mahogany* is required for the function of agouti in the skin, the function of agouti when aberrantly expressed in the brain, and probably the function of AGRP where it is normally made in the brain. Yet, *mahogany* does not appear to be expressed specifically at sites of melanocortin receptor expression. Furthermore, the extracellular portion of *mahogany*, known as attractin, is clearly involved in immune function, and expression of *mahogany* at high levels in the hippocampus implies a role for the protein in learning and memory. Perhaps *mahogany* acts as a receptor cofactor for the melanocortin receptors as well as many other receptors, and in the case of the former is a low-affinity agouti or AGRP binding factor. Alternatively, perhaps *mahogany* is involved in a variety of cell–cell interactions, as has been demonstrated for attractin in T cell–macrophage interactions, and this specific cell–cell interaction is required for the appropriate connections necessary for AGRP to be released at MC4-R-containing sites. Given the complexity of the *mahogany* protein, it is likely to be involved in numerous protein–protein interactions and numerous modes of function.

References

1. Searle, A. G. (1968) *Comparative Genetics of Coat Colors in Mammals* Logos Press, London.
2. Klungland, H., Vage, D. I., Gomez-Raya, L., Adelsteinsson, S., and Lien, S. (1995) The role of melanocyte-stimulating hormone (MSH) receptor in bovine coat color determination. *Mamm. Genome* **6,** 636–639.
3. Robbins, L. S., Nadeau, J. H., Johnson, K. R., Kelly, M. A., Roselli-Rehfuss, L., Baack, E., Mountjoy, K. G., and Cone, R. D. (1993) Pigmentation phenotypes of variant extension locus alleles result from point mutations that alter MSH receptor function. *Cell* **72,** 827–834.

4. Vage, D., Klungland, H., Lu, D., and Cone, R. (1999) Molecular and pharmacological characterization of dominantblack coat color in sheep. *Mamm. Genome* **10,** 39–43.
5. Vage, D. I., Lu, D., Klungland, H., Lien, S., Adalsteinsson, S., and Cone, R.D. (1997) A non-epistatic interaction of *agouti* and *extension* in the fox, *Vulpes vulpes. Nat. Genet.* **15,** 311–315.
6. Lyon, M. F. (1961) Gene action in the X-chromosome of the mouse *(Mus musculus L.). Nature* **190,** 372–373.
7. Kapas, S., Cammas, F. M., Hinson, J. P., and Clark, A. J. L. (1996) Agonist and receptor binding properties of adrenocorticotropin peptides using the cloned mouse adrenocorticotropin receptor expressed in a stably transfected HeLa cell line. *Endocrinology* **137,** 3291–3294.
8. Ni, X.-P., Kesterson, R. A., Sharma, S. D., Hruby, V. J., Cone, R. D., Wiedemann, E., and Humphreys, M. H. (1998) Prevention of reflex natriuresis after acute unilateral nephrectomy by melanocortin receptor antagonists. *Am. J. Physiol.* **274,** R931–R938.
9. Gantz, I., Konda, Y., Tashiro, T., Shimoto, Y., Miwa, H., Munzert, G., Watson, S. J., DelValle, J., and Yamada, T. (1993) Molecular cloning of a novel melanocortin receptor. *J. Biol. Chem.* **268,** 8246–8250.
10. Roselli-Rehfuss, L., Mountjoy, K. G., Robbins, L. S., Mortrud, M. T., Low, M. J., Tatro, J. B., Entwistle, M. L., Simerly, R., and Cone, R. D. (1993) Identification of a receptor for γ-MSH and other proopiomelanocortin peptides in the hypothalamus and limbic system. *Proc. Natl. Acad. Sci. U. S. A.* **90,** 8856–8860.
11. Mizuno, T. M. and Mobbs, C. V. (1999) Hypothalamic agouti-related messenger ribonucleic acid is inhibited by leptin and stimulated by fasting. *Endocrinology* **140,** 814–817.
12. Ollmann, M. M., Wilson, B. D., Yang, Y.-K., Kerns, J. A., Chen, Y., Gantz, I., and Barsh, G. S. (1997) Antagonism of central melanocortin receptors in vitro and in vivo by agouti-related protein. *Science* **278,** 135–137.
13. Shutter, J. R., Graham, M., Kinsey, A. C., Scully, S., Luthy, R., and Stark, K. L. (1997) Hypothalamic expression of ART, a novel gene related to agouti, is upregulated in *obese* and *diabetic* mutant mice. *Genes Dev.* **11,** 593–602.
14. Fong, T. M., Mao, C., MacNeil, C., Kalyani, R., Smith, T., Weinberg, D., Tota, M. R., and Van der Ploeg, L. H. (1997) ART (protein product of agouti-related transcript) as an antagonist of MC-3 and MC-4 receptors. *Biochem. Biophys. Res. Commun.* **237,** 629–631.
15. Graham, M., Shuttre, J. R., Sarmiento, U., Sarosi, I., and Stark, K. L. (1997) Overexpression of Agrt leads to obesity in transgenic mice. *Nat. Genet.* **17,** 273–274.
16. Huszar, D., Lynch, C. A., Fairchild-Huntress, V., Dunmore, J. H., Fang, Q., Berkemeier, L. R., Gu, W., Kesterson, R. A., Boston, B. A., Cone, R. D., Smith, F. J., Campfield, L. A., Burn, P., and Lee, F. (1997) Targeted disruption of the melanocortin-4 receptor results in obesity in mice. *Cell* **88,** 131–141.
17. Marsh, D. J., Hollopeter, G., Huszar, D., Laufer, R., Yagaloff, K. A., Fisher, S. L., Burn, P., and Palmiter, R. D. (1999) Response of melanocortin-4 receptor-deficient mice to anorectic and orexigenic peptides. *Nat. Genet.* **21,** 119–122.
18. Vaisse, C., Clement, K., Guy-Grand, B., and Froguel, P. (1998) A frameshift mutation in human MC4R is associated with a dominant form of obesity. *Nat. Genet.* **20,** 113–114.

19. Yeo, G. S. H., Farooqi, I. S., Aminian, S., Halsall, D. J., Stanhope, R. G., and O'Rahilly, S. (1998) A frameshift mutation in MC4R associated with dominantly inherited human obesity. *Nat. Genet.* **20,** 111–112.
20. Li, S.-J., Varga, K., Archer, P., Hruby, V. J., Sharma, S. D., Kesterson, R. A., Cone, R. D., and Kunos, G. (1996) Melanocortin antagonists define two distinct pathways of cardiovascular control by α- and γ-melanopcyte-stimulating hormones. *J. Neurosci.* **16,** 5182–5188.
21. Huang, Q.-H., Entwistle, M. L., Alvaro, J. D., Duman, R. S., Hruby, V. J., and Tatro, J. B. (1997) Antipyretic role of endogenous melanocortins mediated by central melanocortin receptors during endotoxin-induced fever. *J. Neurosci.* **17,** 3343–3351.
22. Von Frijtag, J. C., Croiset, G., Gispen, W. H., Adan, R. A., and Wiegant, V. M. (1998) The role of central melanocortin receptors in the activation of the hypothalamus-pituitary-adrenal-axis and the induction of excessive grooming. *Br. J. Pharmacol.* **123,** 1503–1508.
23. Chen, W., Kelly, M. A., Opitz-Araya, X., Thomas, R. E., Low, M. J., and Cone, R. D. (1997) Exocrine gland dysfunction in MC5-R deficient mice: evidence for coordinated regulation of exocrine gland function by melanocortin peptides. *Cell* **91,** 789–798.
24. van der Kraan, M., Adan, R. A. H., Entwistle, M. L., Gispen, W. H., Burbach, J. P. H., and Tatro, J. B. (1998) Expression of melanocortin-5 receptor in secretory epithelia supports a functional role in exocrine and endocrine glands. *Endocrinology* **139,** 2348–2355.
25. Fathi, Z., Iben, L. G., and Parker, E. M. (1995) Cloning, expression, and tissue distribution of a fifth melanocortin receptor subtype. *Neurochem. Res.* **20,** 107–113.
26. Labbe, O., Desarnaud, F., Eggerickx, D., Vassart, G., and Parmentier, M. (1994) Molecular cloning of a mouse melanocortin 5 receptor gene widely expressed in peripheral tissues. *Biochemistry* **33,** 4543–4549.
27. Kraan, v. d., Tatro, J. B., Entwistle, M. L., Brakkee, J. H., Burbach, J. P. H., Adan, R. A. H., and Gispen, W. H. (1999) Expression of melanocortin receptors and proopiomelanocortin in the rat spinal cord in relation to the neurotrophic effects of melanocortins. *Mol. Brain Res.* **63,** 276–286.
28. Lane, P. W. (1960) New mouse mutants. *Mouse News Lett.* **22,** 35.
29. Lane, P. W. and Green, M. C. (1960) Mahogany, a recessive color mutation in linkage group V of the mouse. *J. Hered.* **51,** 228–230.
30. Mather, K. and North, S. B. (1940) Umbrous: a case of dominance modification in mice. *J. Genet.* **40,** 229–241.

Index

N

O

P

R

S

T

U

Y

Z